THE AMINOGLYCOSIDES

KIDNEY DISEASE

Editors

J. Stewart Cameron
Professor in Renal Medicine
Clinical Science Laboratories
Guy's Hospital
London, England

Richard J. Glassock
Professor of Medicine
UCLA School of Medicine and
Chief, Division of Nephrology
and Hypertension
Harbor-UCLA Medical Center
Torrance, California

Andrew Whelton
Associate Professor of Medicine
The Johns Hopkins University School of Medicine
Baltimore, Maryland

Charles van Ypersele de Strihou
Professor of Medicine and
Chief, Division of Nephrology
University of Louvain
Clinic St-Luc
Brussels, Belgium

Other Volumes in Preparation

THE AMINOGLYCOSIDES

Microbiology, Clinical Use, and Toxicology

Edited by

Andrew Whelton, M.D.
School of Medicine
The Johns Hopkins University
Baltimore, Maryland

Harold C. Neu, M.D.
College of Physicians and Surgeons
Columbia University
New York, New York

MARCEL DEKKER, INC.   New York and Basel

Library of Congress Cataloging in Publication Data

Main entry under title:

The Aminoglycosides.

(Kidney disease; 2)
Includes index.
1. Aminoglycosides—Physiological effect. 2. Amino-
glycosides—Toxicology. 3. Micro-organisms—Effect of
antibiotics on. I. Whelton, Andrew [date].
II. Neu, Harold Conrad. III. Series.
QP925.A55A443   615'.329   81-15202
ISBN 0-8247-1364-8   AACR2

Marcel Dekker, Inc.
270 Madison Avenue, New York, New York 10016

Current printing (last digit):
10 9 8 7 6 5 4 3

Printed in the United States of America

ABOUT THE SERIES

This book is the second in a series of texts on nephrology, intended to cover all aspects of that discipline. Nephrology has grown at a great rate over the past decade, and the available knowledge has outstripped the texts available. The discipline has, however, attained such maturity that a description of any one of its parts becomes more than ephemeral—or worse, outdated before it even appears. We aim to fill some of the many gaps that exist, not only for the specialist in nephrology, but also for the general physician interested in nephrology. In this description, we include the study and management of hypertension, since these two are interwoven at so many points. We welcome comments on the series as it evolves, or suggestions for topics that might be considered for future volumes.

J. Stewart Cameron
London, England

Richard J. Glassock
Los Angeles, California

Andrew Whelton
Baltimore, Maryland

Charles van Ypersele de Strihou
Brussels, Belgium

PREFACE

More than three decades have elapsed since the first clinical introduction of the amino-glycoside class of antibiotics. Their continued popularity over the years has been ensured largely as the result of the sequential development of new aminoglycosides with wider spectra of antibacterial activity and greater clinical efficacy. It is of interest to note that in the past 12 months, the National Disease and Therapeutic Index, IMS America Ltd., reported that 4.3 million courses of aminoglycoside antibiotics were prescribed in the United States alone.*

All clinicians who have used these clinically invaluable drugs will recall that there are important dose-related toxic side effects that can be encountered with all the aminogly-cosides. In recent years the latter side effects, together with the identification of increased clinical efficacy, have been the focus of intensive basic and clinical investigation. These lines of investigation have included molecular manipulation of the nucleus of aminoglyco-sides; pharmacokinetic study of these compounds in serum and various body tissues; and comparative efficacy and toxicity studies in experimental animals and in human subjects, to mention but a few of the study techniques employed.

These new and important developments related to the aminoglycoside class of anti-biotics have been reported in a widely divergent series of medical publications. Since no single source reference on the topic existed, we felt that the time to prepare such a com-prehensive work was long overdue. Without the continued support and enthusiasm of the contributing authors it would not have been possible to generate such a text. We are therefore most grateful to all the authors who have given of their valuable time and invalu-able expertise in writing the multidisciplinary chapters that form the backbone of this reference book.

This book is dedicated to the clinical scientists of the past, in recognition of their foresight, which has led to the excellent drugs of today, and is likewise dedicated to the future investigators who will lead us forward to safer and more effective antibiotic therapy. It also goes without saying that many other individuals have made significant contributions to the development and preparation of the text. Although all these individuals cannot be mentioned here, we particularly wish to thank Robin J. Siegert and Robert L. Stout for their tireless help throughout the production of this book.

Andrew Whelton, M.D.   Harold C. Neu, M.D.
Baltimore, Maryland   New York, New York

*Data copyright property of National Disease and Therapeutic Index, IMS America Ltd., Ambler, Penn-sylvania, 1981. Reproduced here with permission.

CONTRIBUTORS

Gerald B. Appel, M.D. Department of Medicine, Columbia University College of Physicians and Surgeons, Columbia Presbyterian Medical Center, New York, New York

Michael Barza, M.D. Infectious Disease Division, Department of Medicine, Tufts University School of Medicine, Tufts-New England Medical Center Hospital, Boston, Massachusetts

J. P. Bendirdjian, M.D. Department of Nephrology, Université de Rouen, Hôpital de Bois-guillaume, Bois Guillaume, France

C. L. Bendush, M.D. Clinical Investigation Division, Eli Lilly and Company, Indianapolis, Indiana

William M. Bennett, M.D. Division of Nephrology, Department of Medicine, University of Oregon Health Sciences Center, Portland, Oregon

Gerald P. Bodey, Sr., M.D. Chemotherapy Branch, Department of Medicine, and Section of Infectious Diseases, Department of Developmental Therapeutics, University of Texas System Cancer Center, M.D. Anderson Hospital and Tumor Institute, Houston, Texas

Robert E. Brummett, Ph.D. Department of Otolaryngology, University of Oregon Health Sciences Center School of Medicine, Portland, Oregon

J. P. Fillastre, M.D. Groupe de Physiopathologie Tissulaire, Université de Rouen, Hôpital de Boisguillaume, Bois Guillaume, France

B. Foucher, M.D. Université de Rouen, Hôpital de Boisguillaume, Bois Guillaume, France

Kaye E. Fox, Ph.D. Department of Otolaryngology, University of Oregon Health Sciences Center School of Medicine, Portland, Oregon

Haruhide Kawabe, Ph.D. Department of Oral Bacteriology, Nippon Dental University, Niigata City, Japan

Jean Klastersky, M.D. Department of Medicine, Institut Jules Bordet, Brussels, Belgium

Michael W. Lauermann, M.D.* Department of Infectious Disease, Tufts-New England Medical Center Hospital, Boston, Massachusetts

Felix Leitner, M.D. Department of Microbiological Research, Bristol Laboratories, Syracuse, New York

*Dr. Lauermann is now with Long Beach Memorial Hospital and St. Mary's Hospital of Long Beach, Long Beach, California.

Paul S. Lietman, M.D., Ph.D. Departments of Medicine, Pharmacology and Experimental Therapeutics, and Pediatrics, The Johns Hopkins University School of Medicine, Baltimore, Maryland

Friedrich C. Luft, M.D. Department of Medicine, Indiana University School of Medicine, Indianapolis, Indiana

George H. McCracken, Jr., M.D. Department of Pediatrics, University of Texas Health Sciences Center at Dallas, Dallas, Texas

George H. Miller, Ph.D. Department of Chemotherapy, Schering Corporation, Bloomfield, New Jersey

Susumu Mitsuhashi, Ph.D. Department of Microbiology, Laboratory of Bacterial Resistance, Gunma University School of Medicine, Maebashi City, Japan

Robert C. Moellering, Jr., M.D. Department of Medicine, Harvard Medical School, and Massachusetts General Hospital, Boston, Massachusetts

Thomas R. Moench, M.D. Division of Internal Medicine, Department of Medicine, The Johns Hopkins Hospital, Baltimore, Maryland

A. Werner Mondorf, M.D. University of Frankfurt Center of Internal Medicine, Frankfurt, Federal Republic of Germany

J. P. Morin, Ph.D. Groupe de Physiopathologie Tissulaire, Université de Rouen, Hôpital de Boisguillaume, Bois Guillaume, France

Tattanahalli L. Nagabhushan, Ph.D. Department of Antiinfectives Chemistry, Schering Corporation, Bloomfield, New Jersey

Harold C. Neu, M.D. Departments of Medicine and Pharmacology, College of Physicians and Surgeons, Columbia University, New York, New York

Steen Olsen, M.D. Department of Pathology, University Institute of Pathology, Kommunehospitalet, Aarhus, Denmark

Richard A. Parker, M.D. Division of Nephrology, Department of Internal Medicine, University of Oregon Health Sciences Center, Portland, Oregon

George A. Porter, M.D. Department of Medicine, University of Oregon Health Sciences Center, Portland, Oregon

Kenneth E. Price, Ph.D. Department of Research, Bristol Laboratories, Syracuse, New York

Jerome J. Schentag, Pharm. D. Department of Pharmaceutics and Pharmacy, State University of New York at Buffalo, and Millard Fillmore Hospital, Buffalo, New York

Jane D. Siegel, M.D. Department of Pediatrics, University of Texas Health Sciences Center at Dallas, Dallas, Texas

Frederic J. Silverblatt, M.D. Infectious Diseases Section, Department of Medicine, San Fernando Valley Medical Program, University of California, Los Angeles, California

Craig R. Smith, M.D. Department of Medicine, The Johns Hopkins University School of Medicine, Baltimore, Maryland

Kim Solez, M.D. Department of Pathology and Medicine, The Johns Hopkins University School of Medicine, Baltimore, Maryland

W. Leigh Thompson, M.D., Ph.D. Department of Medicine and Pharmacology, Case Western Reserve University, Cleveland, Ohio

Jean-Pierre Thys, M.D. Infectious Diseases Unit, Saint-Pierre University Hospital, Brussels, Belgium

Marvin J. Weinstein, Ph.D. Department of Microbiology, Schering Corporation, Bloomfield, New Jersey

Andrew Whelton, M.D. Department of Medicine, The Johns Hopkins University School of Medicine, Baltimore, Maryland

CONTENTS

THE AMINOGLYCOSIDES

CHEMISTRY AND MICROBIOLOGY

1

STRUCTURE-ACTIVITY RELATIONSHIPS
IN AMINOGLYCOSIDE-AMINOCYCLITOL
ANTIBIOTICS

TATTANAHALLI L. NAGABHUSHAN, GEORGE H.
MILLER, and MARVIN J. WEINSTEIN
Schering Corporation, Bloomfield, New Jersey

In recent years several excellent reviews of the chemistry and biological properties of aminoglycoside-aminocyclitol antibiotics have been published [3,10,32,36]. In this chapter we have reviewed the rationale for the structural modifications of aminoglycosides and have relied mainly on work carried out in our laboratories. No attempt has been made to catalog the structures and biological activities of all aminoglycoside-aminocyclitol antibiotics published to date.

The aminoglycoside antibiotics are the oldest known class of broad-spectrum antibacterials with activity against gram-positive and gram-negative organisms as well as Mycobacteria. They are, however, not active against anaerobic bacteria or fungi. There are currently a number of aminoglycoside antibiotics in clinical practice and except for streptomycin and dihydrostreptomycin, which are derivatives of streptamine, they are all glycosides of 2-deoxystreptamine [3].

Stereochemically, the streptamines are interesting molecules. They have two pairs of enantiotopic centers about a σ plane which become diastereotopic in a chiral (optically active) environment. The numbering of the ring atoms is shown in Figure 1. Numbering begins with the carbon bearing the nitrogen atom possessing the R configuration and proceeds through the carbon atom bearing the nitrogen atom, which has the S configuration

3

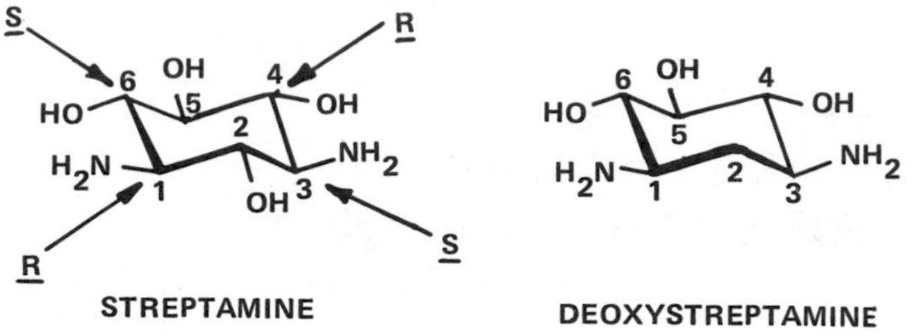

Figure 1   Numbering of streptamine and deoxystreptamine carbon atoms.

[15]. Although several pseudodisaccharides exhibit antimicrobial properties, a basic structural unit required for therapeutically useful potency and spectrum appears to lie in the pseudotrisaccharide structures, composed of 4,6-di-O-glycosyl (the kanamycins, gentamicin), 4,5-di-O-glycosyl (ribostamycin), and 4-O-glycosyl (the streptomycins) derivatives of diaminocyclitols. Since the chirality at carbon-4 of the streptamines is R and that at carbon-6 is S, and R takes precedence over S, the sugar attached to the 4-position receives primed numbers and the sugar attached to the 6-position receives doubly primed numbers. Similarly, in the 4,5-di-O-glycosyl derivatives, the sugar attached to the 5-position receives doubly primed numbers since the one attached to a center with a lower number takes precedence over the other attached to a center with a higher number [9,15].

I.   Structure and Conformation of Kanamycin A

Modern aminoglycoside chemotherapy began with the discovery of the kanamycins in 1957 by Umezawa and co-workers and subsequent introduction of kanamycin A in the clinic in the early sixties [42]. The kanamycins are produced by *Streptomyces kanamyceticus* and differ from the classical aminoglycoside antibiotics of the streptomycin or neomycin class in structure, biological activity, and toxicity. Kanamycin A is a deoxystreptamine containing antibiotic in which the latter is linked in the α configuration to 6-amino-6-deoxy-D-glucose at the 4-position and to 3-amino-3-deoxy-D-glucose at the 6-position. The structure of kanamycin A is shown in Figure 2 in a conformational drawing which depicts the shape of the molecule as delineated from single-crystal x-ray structure analysis [12], proton and carbon-13 magnetic resonance spectroscopy [15,18,19], and calculations of nonbonded steric interactions with the use of Buckingham atom-atom potentials [18–20]. As seen more clearly in the space-filling model (Figure 3), the molecule adopts a crescent-shaped conformation, with the oxygen atom O_5' pointing toward the nitrogen atom N_3 and the oxygen atom O_5'' pointing toward the oxygen atom O_5. In this conformation, the four amino groups of kanamycin A are all on the convex side of the molecule, a spatial disposition that is probably required for biological activity [18,19]. This conformational preference in kanamycin A is presumably influenced not only by favorable steric factors but also by stabilizing electronic forces operative in the exo-anomeric effect exhibited by both glycosidic linkages [15]. Other deoxystreptamine antibiotics which are closely related to kanamycin A in structure are expected to exhibit similar conformational properties. Several nuclear magnetic resonance spectroscopy studies indicate that this may indeed be the case [18,19].

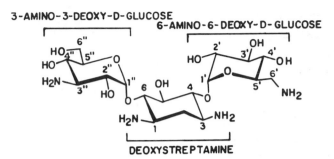

Figure 2   Structure of kanamycin A.

II. Structure-Activity Relationship of the Kanamycins Against Sensitive Organisms

In Figure 4 the structures of some of the natural and semisynthetic variants of kanamycin A are shown. Kanamycin A, like other aminoglycoside-aminocyclitol antibiotics, is bactericical and is active against a number of gram-positive and gram-negative organisms, including *Mycobacterium tuberculosis*. Although the incidence of resistance to this drug has risen considerably since its introduction in the clinic, it is still active against a variety of pathogens and finds continued use in the clinic. The minimum inhibitory concentrations (MIC) of kanamycin A and clinically important derivatives against a number of sensitive organisms are given in Table 1.

Structural changes in kanamycin A can produce diverse effects on its microbiological

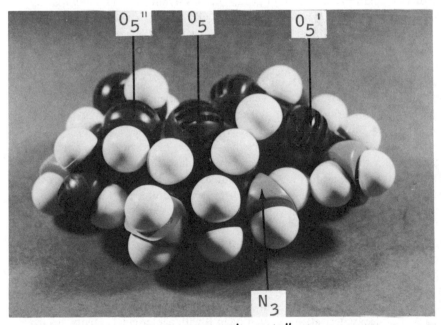

Figure 3   Molecular model showing the $O_5' \rightarrow N_3/O_5'' \rightarrow O_5$ crescent conformation of kanamycin A.

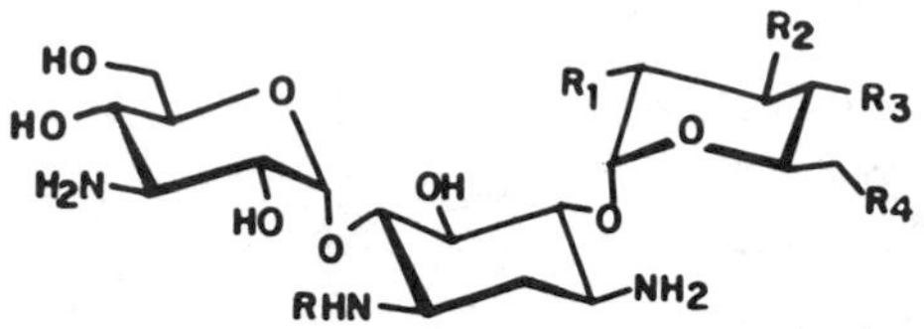

ANTIBIOTIC	SOURCE	R	R_1	R_2	R_3	R_4
KANAMYCIN A	n	H	OH	OH	OH	NH_2
KANAMYCIN B	n	H	NH_2	OH	OH	NH_2
KANAMYCIN C	n	H	NH_2	OH	OH	OH
TOBRAMYCIN	n	H	NH_2	H	OH	NH_2
DIBEKACIN	ss	H	NH_2	H	H	NH_2
AMIKACIN	ss	HABA	OH	OH	OH	NH_2
BUTAKACIN	ss	HAB	OH	OH	OH	NH_2

Figure 4   Structures of important kanamycin A derivatives. n, natural; ss, semisynthetic; HABA, S-4-amino-2-hydroxybutyryl; HAB, S-4-amino-2-hydroxybutyl.

activities, such as increasing the potency, decreasing the potency, broadening the spectrum of activity without significantly changing the potency against sensitive organisms, and broadening the spectrum of activity while enhancing the potency against sensitive organisms. For example, as reported in the literature and as seen partially in Table 1, replacement of the equatorial $2'$-hydroxyl group of kanamycin A by an amino group (kanamycin B) increases the in vitro potency significantly against sensitive strains [43]. However, the presence of an amino group at the $2'$-position as in kanamycin B without an amino group at the $6'$-position is not sufficient to achieve high levels of activity, as evidenced by the much lower potency of kanamycin C against all sensitive strains. Kanamycin C is a derivative of kanamycin B in which the $6'$-amino group has been replaced by a hydroxyl group, and it is the least active of the three kanamycins.

The recently introduced drug amikacin is a semisynthetic derivative of kanamycin A which has an S-4-amino-2-hydroxybutyryl (HABA) side chain at the N_1 position [11]. This change in kanamycin A, while it enhances the spectrum of activity, does not alter to any appreciable extent the potency against sensitive strains other than *Pseudomonas,* as seen in Table 1. Its activity against many resistant strains is, however, changed, as will be discussed later. Similar microbiological properties are exhibited by butakacin, the 1-N-S-(4-amino-2-hydroxybutyl)-kanamycin A currently under study in the United Kingdom [39]. In contrast to the modest changes in microbiological potency of kanamycin A effected through the above-mentioned changes, increases in potency against sensitive strains as well as activity against resistant organisms occurs when the $3'$-hydroxyl group of kanamycin B is replaced by a hydrogen atom (tobramycin) [40]. As seen in Table 1, tobramycin, produced by *Streptomyces tenebrarius* and introduced for therapeutic use only a few years ago, is more potent than the kanamycins, amikacin or butakacin. Dibekacin is a $3',4'$-dideoxy derivative of kanamycin B with bactericidal activity against kanamycin-sensitive organisms similar to those of tobramycin [44]. The lack of $3',4'$-

Table 1 Comparison of the in Vitro Activities of Aminoglycosides Against Sensitive Bacteria[a]

	Number of strains	Kanamycin A	Amikacin	Butakacin	Kanamycin B	Tobramycin	Dibekacin
Bacillus subtilis	1	0.25	0.125	<0.06	<0.06	<0.06	0.125
Staphylococcus aureus	1	2	0.5	0.5	0.5	0.25	0.25
Enterobacter	1	4	1	2	1	1	2
Escherichia coli	4	1	0.5	0.5	0.5	0.25	0.5
Klebsiella pneumoniae	2	1	1	1	1	0.5	1
Proteus mirabilis	2	1	1	2	1	0.5	1
Proteus indole positive	2	1	2	1	4	4	8
Pseudomonas	5	16	1	0.5	16	0.25	0.5
Salmonella	1	4	4	4	2	4	4
Serratia	2	0.5	1	2	4	2	4

[a]Median MIC values (μg/ml, 24 hr), Mueller-Hinton agar.

hydroxyls makes dibekacin the first semisynthetic structural counterpart of the genta-micins (C_{1a}) to arrive on the market.

III. Mechanisms of Enzymatic Inactivation of Aminoglycoside-Aminocyclitol Antibiotics

Before proceeding with an analysis of the structure-activity relationships, it may be advantageous to review the biochemical mechanisms of bacterial resistance to amino-glycoside-aminocyclitol antibiotics. The increasing number of semisynthetic aminoglyco-sides reported in recent years with activities against resistant gram-negative strains has been due to an understanding, at the molecular level, of the mechanisms of enzymatic modification of aminoglycoside antibiotics by resistant organisms. The idea that trans-ferable antibiotic resistance in gram-negative organisms is due to an extranuclear genetic material called an R factor was developed independently in 1959 and 1960 by two groups of Japanese workers led by Ochiai and Akiba, respectively [33,2]. Following this early work in 1965, Okamoto and Suzuki discovered chloramphenicol acetyltransferase in an *Escherichia coli* that carried an R factor [34]. Although these authors suggested that the same organism produced enzymes that catalyzed the reaction of kanamycin A with acetyl coenzyme A and the reaction of streptomycin with adenosine-5'-triphosphate, it was not until 1967 that the first mechanisms of clinical resistance to aminoglycoside-aminocyclitol antibiotics were elucidated by Umezawa and co-workers [45,46]. They showed, by isolation and structure determination, that *E. coli* K12 R5 N-acetylated kanamycin A at the 6'-N position [45]. The R factor in this strain (R5) had been trans-ferred from a naturally isolated drug resistant strain. The resulting product, 6'-N-acetyl-kanamycin A, was devoid of antibacterial properties. Umezawa and co-workers demon-strated further that another enzyme produced by *E. coli* K12 ML1629 phosphorylated kanamycin A at the 3'-position [46]. This strain was obtained by transfer of an R factor from a naturally isolated, drug-resistant strain of *E. coli* K12 ML1410. The structure of the inactivated kanamycin A-3'-O-phosphate was established by proton magnetic reso-nance spectroscopy. Since these initial discoveries, several other important modes of enzymatic inactivation of aminoglycoside-aminocyclitol antibiotics have been discovered. Knowledge of the resistance mechanism involved has been useful in the design of novel semisynthetic aminoglycoside-aminocyclitol antibiotics for the treatment of infections caused by resistant organisms [3,8]. The currently recognized mechanisms of bacterial enzymatic inactivation of aminoglycoside-aminocyclitol antibiotics are summarized in Figure 5 using kanamycin B as an ideal substrate. They include acetylating enzymes (AAC), phosphorylating enzymes (APH), and nucleotidylating enzymes (ANT). A more detailed account of resistance mechanisms can be found in a recent review by Davies [8].

IV. Structure-Activity Relationship of the Kanamycins Against Resistant Organisms

As seen in Figure 5, kanamycin B is a substrate for N-acetylating enzymes at the 2'-, 6'-, and 3-positions, for phosphorylating enzymes at the 3'- and 2''-positions, and for adenylylating enzymes at the 4'- and 2''-positions. It is to be noted that the existence of intriguing differences in substrate specificities have permitted finer classification of particular enzymes. An illustration of this point appears in Table 2, where the suscepti-

Table 2 Comparison of the in Vitro Activities of Aminoglycosides Against Resistant Strains[a]

Resistance[b] mechanism	Number of strains	Karamycin A	Amikacin	Butakacin	Kanamycin B	Tobramycin	Dibekacin
APH(3')- I	8	>64	1	2	>32	4	8
APH(3')-II	3	>64	4	8	>32	2	8
APH(3')-III	1	>64	2	4	>32	2	4
ANT(2'')	5	>64	2	0.5	>32	32	>32
ANT(4')	2	>64	32	32	>32	>32	8
AAC(3)-I	4	>64[c]	2	2	>32[c]	1	2
AAC(3)-Ia	2	64	2	2	32	1	2
AAC(3)-II	4	>64	2	4	>32	32	>32
AAC(2') + APH(3')	2	64	1	2	>32	16	>32
AAC(6')-I	4	>64	32	>32	>32	32	>32
AAC(6')-II	2	>64	16	>32	>32	>32	>32
APH(2'') + AAC(6')	1	>64	8	>32	>32	16	32
Permeability	2	>64	>32	>32	>32	16	32

[a]Median MIC values (μg/ml, 24 hr), Mueller-Hinton agar.
[b]The resistance mechanism AAC(3)-Ia is a subgroup of AAC(3)-I that is resistant to netilmicin.
[c]One E. coli (JR 88) known to contain AAC(3)-I enzyme is sensitive to kanamycin A and kanamycin B; the three Pseudomonas strains were resistant to both.

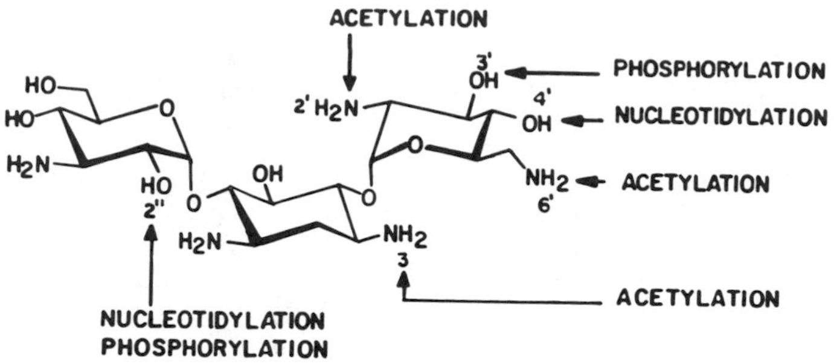

Figure 5   Sites of modification of kanamycin B by inactivating enzymes.

bility patterns of some kanamycin antibiotics to typical resistant strains containing various inactivating enzymes are shown. As mentioned earlier, kanamycin A, like its B analog, is a substrate for all of the inactivating enzymes except AAC(2'). This is understandable since it does not have an amino group at the 2'-position. Kanamycin C, on the other hand, is not active against 2'-N acetylators but is more active than kanamycin A against the 6'-N-acetylating organisms, since it lacks an amino group at that position [35]. Similarly, tobramycin, which has no hydroxyl group at the 3'-position, is active against strains carrying all types of 3'-phosphorylating enzyme. Dibekacin, which does not have hydroxyl groups at both the 3'- and 4'-positions, is also active against strains containing the ANT(4') enzymes.

The kanamycin family of antibiotics seems to be less susceptible to 3-N-acetylating enzymes than does the gentamicin family. This may be seen in Table 2, where both tobramycin and dibekacin are seen to be active against strains containing AAC(3)-I and AAC(3)-Ia enzymes, which confer resistance to gentamicin. Since the strains used in this experiment were predominately *Pseudomonas,* kanamycin does not appear to be active, but it is known to be a poor substrate for these enzymes in vitro [36].

Strains that contain AAC(2'), AAC(6')-I, AAC(6')-II, AAC(3)-II, ANT(2''), or APH(2'') enzymes are resistant to both tobramycin and dibekacin.

The elucidation in 1971 of the structures of the butirosins, the 1-N-S-HABA derivatives of ribostamycin and xylostacin [51], together with the earlier observation that in contrast to ribostamycin they were active against *P. aeruginosa* carrying 3'-phosphorylating enzymes [APH(3')-I], led to the design and semisynthesis of amikacin by Kawaguchi and co-workers in 1972 [11]. The microbiological results were probably more exciting than expected since amikacin proved to be active against all kanamycin-resistant organisms with the exception of those that produced AAC(6')-I and the more recently identified APH(3')-IV and ANT(4') enzymes. Amikacin has the broadest spectrum of activity against resistant organisms of any aminoglycoside-aminocyclitol antibiotic on the market today. Particularly rewarding was the finding that the drug was active against organisms containing ANT(2'') enzymes, as there was no relationship directly deducible between the activity of the butirosins against these organisms and amikacin. In the butirosins there is no sugar attached to the 6-position and thus no site of inactivation for ANT(2'') enzymes. It appears, therefore, that the presence of the side chain at the 1-position in the proximity of the 2''-hydroxyl group must sterically hinder the approach of the nucleotidylating agent.

Nagabhushan and co-workers have shown, from carbon-13 magnetic resonance spectroscopic studies, that the pseudotrisaccharide unit of amikacin, except for small changes in the torsion angle about the glycosidic linkage at the 6-position, basically maintains the $O_5' \to N_3/O_5'' \to O_5$ crescent conformation in which the 2''-hydroxyl group and the 1-amino group are close to each other (Figure 6) [18,19]. Furthermore, they showed from measurements of vicinal ^{13}C-^{1}H coupling between the carbonyl carbon nucleus and H_1 in a number of 1-N-acyl derivatives of deoxystreptamine and kanamycin A that the carbonyl group, like that in many peptides and amides, maintains a near syn-periplanar relationship with the C_1-H_1 bond [18]. This normal conformational preference of the amide linkage, together with the preferred rotamer about the C_1'''-C_2''' bond should, to a large extent, describe the conformation of the side chain in amikacin. The preferred rotamer about the C_1'''-C_2''' bond is probably influenced by the hydrogen bonding between the hydroxyl group of the side chain and the donor or acceptor group in the 3-amino-3-deoxy-D-glucose moiety. The barrier to rotation about the C_2'''-C_3''' and C_3'''-C_4''' bonds must be small to permit the ready availability of the required rotamer for binding at the receptor site.

Whereas prevention of nucleotidylation at the 2''-position by the 1-N substituent is readily rationalized on steric grounds as mentioned above, it is not easy to understand

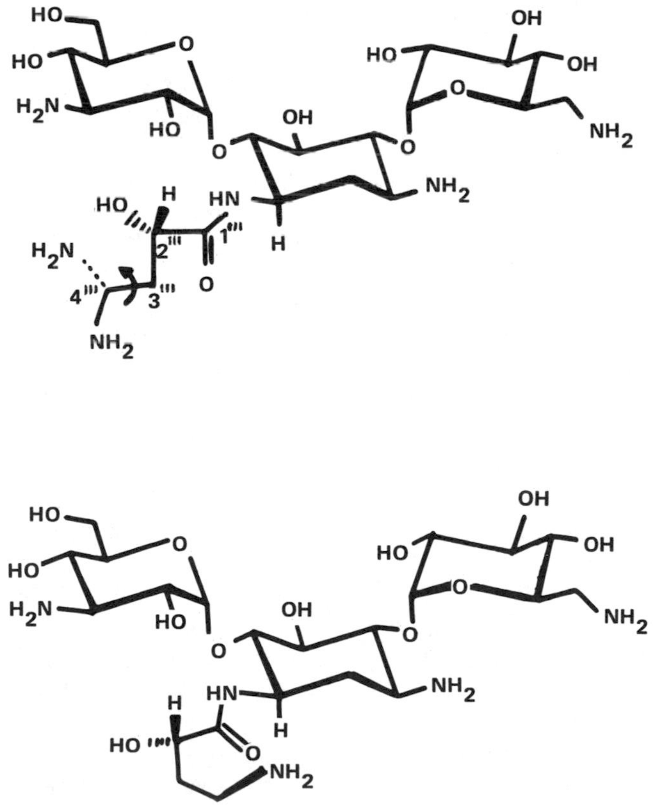

Figure 6   Possible conformers of amikacin in which the 4'''-amino group mimics the 3''-amino group (top) or the 1-amino group (bottom).

why introduction of the HABA side chain at the 1-position prevents phosphorylation at the 3'-position, a center that is 9 atoms (from the concave side) or 11 atoms (from the convex side) away. It appears not to be due to steric shielding of the group by the side chain. A possible explanation (hypothesis) might be that in one of the conformers the 4-amino group of the side chain and 3-amino group of deoxystreptamine are nearly the same distance apart as the 1- and the 3-amino groups of kanamycin A (Figure 6). The receptor sites of the modifying enzymes may recognize this new pair of amino groups rather than the 1- and 3-amino groups of deoxystreptamine. However, when bound at the receptor site in this way, the 3'-hydroxyl group would not be expected to be in the same place in space as, for instance, in the kanamycin A-enzyme complex. Thus, phosphorylation could not occur.

Another hypothesis is that the 4'''-amino group actually mimics the 3''-amino group from an alternative conformation. Presumably, this would also lead to an enzyme-substrate complex in which the 3'-hydroxyl group is displaced from its normal position in space and phosphorylation would not be expected to take place (Figure 6).

Butakacin, which lacks a carbonyl group in the side chain, can still maintain a conformation similar to the HABA of amikacin through similar hydrogen-bonding properties of the α-hydroxyl group. The biological properties of butakacin are very similar to those of amikacin (Tables 1 and 2).

V.   Structure-Activity Relationship of the Gentamicins

The biggest breakthrough in the aminoglycoside chemotherapy of severe gram-negative infections came in 1963 with the discovery of the gentamicins by Weinstein and co-workers [47]. These antibiotics, which were introduced in the clinic in the late sixties, were unique in that they were the first aminoglycosides to be isolated from a source other than *Streptomyces,* viz. *Micromonospora* (*purpurea* and *echinospora*). Although the gentamicins C_1, C_2, and C_{1a} are the most important members of the gentamicin family, striking structural similarities exist between the compounds of the kanamycin family and some of the gentamicins which warrant their discussion first.

The principal difference between the kanamycins and the gentamicins lies in the structure of the amino sugar attached to the 6-position of the deoxystreptamine unit. Whereas in the kanamycins this sugar is 3-amino-3-deoxy-α-D-glucose (kanosamine), in the gentamicins it is 3-methylamino-3-deoxy-4-C-methyl-β-L-arabinose (garosamine). The structures of some of the important gentamicins are shown in Figure 7, and the biological activities are presented in Tables 3 and 4.

Comparison of the structure of gentamicin B with that of kanamycin A (Figures 4 and 7) [48] suggests that these two antibiotics should show similar microbiological activities, except perhaps against organisms carrying ANT(2'') enzymes. Against these strains gentamicin B could be expected to be more active, since intuition suggests that the presence of a methyl group at the 3''-position should offer steric resistance to the approach of the adenylylating agent. However, this does not happen, and both antibiotics are readily inactivated by ANT(2'') enzymes. In general, both compounds have very similar antimicrobial properties. As expected from their structures, antibiotics JI 20A [41] and X_2 [4] are similar in bactericidal activity to kanamycin B and kanamycin C, respectively.

Gentamicin C_{1a} is the 3', 4'-dideoxy derivative of antibiotic JI 20A and, together with

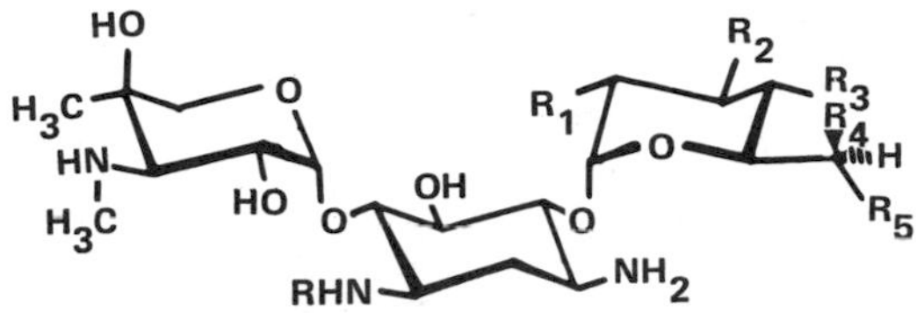

ANTIBIOTIC	SOURCE	R	R_1	R_2	R_3	R_4	R_5
GENTAMICIN B	n	H	OH	OH	OH	H	NH_2
JI 20 A	n	H	NH_2	OH	OH	H	NH_2
X_2	n	H	NH_2	OH	OH	H	OH
C_{1a}	n	H	NH_2	H	H	H	NH_2
C_1	n	H	NH_2	H	H	CH_3	$NHCH_3$
C_2	n	H	NH_2	H	H	CH_3	NH_2
C_{2b}	n, ss	H	NH_2	H	H	H	$NHCH_3$
SCH 20287	ss	HABA	OH	OH	OH	H	NH_2
SCH 21420	ss	HAPA	OH	OH	OH	H	NH_2
SCH 23722	ss	HAPA	OH	OH	OH	H	NHEt
SCH 24443	ss	HAPA	OH	OH	OH	CH_3	NH_2
SCH 21211	ss	H	H	OH	OH	H	NH_2
SCH 21768	ss	H	H	H	OH	H	NH_2
SCH 23200	ss	H	H	H	H	H	NH_2
SCH 23456	ss	H	HO	H	H	H	NH_2

Figure 7   Structures of gentamicin B and derivatives.

gentamicins C_1 and C_2, one of the first dideoxy derivatives to be marketed. As seen in Tables 3 and 4, the compound is highly potent against sensitive strains and bacteria containing all types of 3'-phosphorylating enzymes. In common with other aminoglycoside antibiotics of this type, gentamicin C_{1a} is not active against organisms containing APH(2''), ANT(2''), AAC(3), AAC(2'), or AAC(6') enzymes (Table 4). The recognition that the gentamicins were highly active against *P. aeruginosa* paved the way for the later realization that the lack of activity of the kanamycins against these pathogens might be due to the presence in these strains of 3'-phosphorylating enzymes. The synthesis of dibekacin was undoubtedly prompted by this logic as well as by the fact that the gentamicin C's were highly successful in the clinic.

Although the in vitro microbiological activities of gentamicins C_1 and C_2 are similar to those of C_{1a} (Tables 3 and 4), the presence of a C-methyl group and an N-methyl group in gentamicin C_1 at the 6'-position makes it active against some 6'-N-acetylating strains. Gentamicin C_2, which lacks an N-methyl group at the 6'-position, is, relative to gentamicin C_1, weakly active against these organisms. On the other hand, the absence of a C-methyl group at the 6'-position is not reflected in the activity of gentamicin C_{2b} [5] against AAC(6')-I. It appears, therefore, that an N-methyl group at the 6'-position is all that is needed to provide this activity. It is of interest to note that changing the configuration of gentamicin C_2 at the 6'-position to S (gentamicin C_{2a}) does not alter the spectrum or potency [4].

Table 3 Comparison of the in Vitro Activities of Aminoglycosides Against Sensitive Bacteria[a]

	Number of strains	Antibiotics			Gentamicins				Sch 21420
		Gentamicin B	JI 20A	X_2	C_1a	C_1	C_2	C_2b	
Bacillus subtilis	1	0.125	0.125	1	<0.06	<0.06	<0.06	<0.06	<0.06
Staphylococcus aureus	1	0.5	0.25	4	0.125	0.25	0.125	0.125	0.5
Enterobacter	2	1	0.5	4	0.25	0.5	0.5	0.5	0.5
Escherichia coli	4	1	0.5	4	0.5	0.5	1	0.25	0.5
Klebsiella pneumoniae	2	1	0.5	4	0.5	0.5	1	0.5	1
Proteus mirabilis	2	2	1	8	1	2	2	0.5	2
Proteus indole positive	2	1	1	16	2	4	2	4	2
Pseudomonas	5	16	16	>64	2	4	2	2	2
Salmonella	1	0.5	0.25	4	0.5	1	1	0.5	0.5
Serratia	2	4	2	8	4	1	1	1	4

[a]Median MIC values (μg/ml, 24 hr), Mueller-Hinton agar.

14

Table 4 Comparison of the in Vitro Activities of Aminoglycosides Against Resistant Strains[a]

Resistance[b] mechanism	Number of strains	Antibiotics				Gentamicins			Sch 21420
		Gentamicin B	JI 20A	X_2	C_1a	C_1	C_2	C_2b	
APH(3')-I	8	>64	64	>64	2	4	1	4	1
APH(3')-II	3	64	64	>64	2	4	2	2	4
APH(3')-III	1	>64	>64	>64	2	2	2	2	2
ANT(2'')	5	>64	64	>64	64	>64	>64	64	2
ANT(4')	2	>64	>64	>64	2	4	2	2	32
AAC(3)-I	4	64[c]	64	>64	64	>64	>64	>64	2
AAC(3)-Ia	2	>64	>64	>64	>64	>64	>64	>64	8
AAC(3)-II	4	>64	>64	>64	>64	>64	>64	>64	2
AAC(2')+APH(3')	2	>64	>64	>64	>64	>64	>64	>64	4
AAC(6')-I	4	64	32	>64	32	2	8	4	16
AAC(6')-II	2	>64	>64	>64	>64	>64	>64	>64	16
APH(2'')+AAC(6')	1	>64	>64	>64	>64	>64	64	>64	4
Permeability	2	>64	>64	>64	>64	>64	>64	>64	>64

[a]Median MIC values (μg/ml, 24 hr), Mueller-Hinton agar.
[b]The resistance mechanism AAC(3)-Ia is a subgroup of AAC(3)-I that is resistant to netilmicin.
[c]One E. coli (JR 88) known to contain AAC(3)-I enzyme is sensitive to gentamicin B; the three Pseudomonas strains are resistant.

As expected from the structural similarities between gentamicin B and kanamycin A, conversion of gentamicin B to the 1-N-S-HABA derivative (*Sch 20287*) resulted in a compound with a microbiological profile similar to that of amikacin [21]. More surprisingly, however, in contrast to the significantly higher potency displayed by the S-isomer over the R-isomer in the case of kanamycin A, both the S and R isomers are of equal potency in the case of gentamicin B [22]. Furthermore, the same relationship is observed in the one-carbon-lower homolog, 1-N-(S-3-amino-2-hydroxypropionyl)gentamicin B (1-N-S-HAPA-B; *Sch 21420*) (Tables 3 and 4), although both the diastereo isomers are more potent than *Sch 20287* [21,22]. The potencies of the R and the S isomers of the propionyl analogs of kanamycin A are nearly the same [13]. The biological properties of 1-N-(R,S-3-amino-2-hydroxypropyl)gentamicin B (*Sch 27658*), the butakacin analog of gentamicin B, showed no surprises. It was nearly identical to *Sch 21420* in spectrum and potency [22].

In an attempt to expand the spectrum of *Sch 21420* to obtain activity against AAC(6′)-I organisms, the 6′-N-ethyl derivative was prepared. The results of testing are disappointing in that this modification drastically reduced the drug's potency [23]. Although the same trend is seen in the methylation of amikacin at the 6′-N position (Chapter 2), the reduction in potency is not as dramatic as with ethylation of *Sch 21420*, presumably because in this series the smaller steric demand of the methyl group is preferred. It is interesting to note that the introduction of the HAPA side chain at the 1-position of gentamicin B prevents 6′-N acetylation by types I and II organisms (Table 4). Presumably, the rationale outlined above for the activity of amikacin against 3′-phosphorylators is also operative for some (6′-II)-acetylating strains but not for others (6′-I).

No significant loss in either potency or spectrum occurs when the S-HAPA chain is attached to the 1-position of gentamicin B_1, the 6′-R-C-methyl derivative of gentamicin B. The 1-N-S-HAPA-gentamicin B_1 possesses slightly better activity than *Sch 21420* does against 6′-N-acetylating strains [24].

As mentioned earlier in connection with tobramycin, deoxygenation of a hydroxyl group involved in enzymatic modification in the 4-O-glycosyl moiety leads to more potent derivatives. As shown by Wright and Lee, improvement in potency is also achieved when the 2′-hydroxyl group of gentamicin B, which is not involved in any type of enzymatic modification, is deoxygenated [52]. Thus, 2′-deoxygentamicin B is more potent than is gentamicin B. However, removal of both 2′- and 3′-oxygen functions of gentamicin B results in a dramatic increment in potency while increasing the activity against 3′-phosphorylating bacteria. The 2′,3′-dideoxy gentamicin B is similar to sisomicin in its biological activity. Wright and Lee have also demonstrated that replacement of the 2′-amino group of gentamicin C_{1a} by a hydrogen atom results in a compound, 2′,3′,4′-trideoxygentamicin B, which is similar to gentamicin C_{1a} in biological activity except that it is active against strains containing AAC(2′) enzymes. Maintenance of a hydroxyl group at the 2′-position of gentamicin B while replacing the 3′- and 4′-hydroxyl groups by hydrogen atoms also leads to a more potent compound with a broader spectrum, as shown by McCombie [16].

The consideration that an α-D-furanosyl moiety at the 4-position of deoxystreptamine might by and large retain the conformational preference of an α-D-pyranosyl function at the same place led McCombie to design and synthesize the 4-O-5′-amino-2′,3′,5′-trideoxy-α-D-furanosyl derivative of garamine (Figure 8) [16]. This compound has a potency and a spectrum of activity similar to 2′, 3′-4′-trideoxygentamicin B. The 1-N-S-HAPA derivatives of 2′,3′-dideoxygentamicin B, 3′,4′-dideoxygentamicin B, and the aforementioned furanosyl analog had biological properties similar to those of *Sch 21420* [16,53].

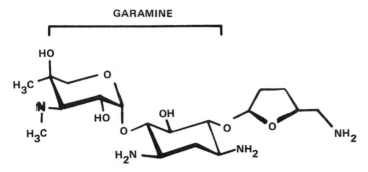

Figure 8   Presumed conformation of 4-O-5'-amino-(2',3',5'-trideoxy-α-D-ribofuranosyl)-garamine.

VI.   Structure-Activity Relationship of the Sisomicins

The discovery of a second family of clinically important and structurally unique amino-glycoside-aminocyclitol antibiotics produced by *Micromonospora* was announced by Weinstein and co-workers in 1970 [49]. The structures of the three most important members of this class are shown in Figure 9, and their biological activities are given in Tables 5 and 6. Sisomicin, the major product of fermentation of *Micromonospora inyoen-sis* and the latest aminoglycoside antibiotic to be marketed, has a spectrum of activity similar to that of gentamicin C_{1a} but is more potent (Tables 5 and 6). It is interesting to note that the presence of a double bond between C_4' and C_5' must change the shape of the ring from a normal chair conformation to that of a half-chair (Figure 10), yet biologi-cal activity is not lost. The configuration at C_5' in gentamicin C_{1a} is critical to its micro-biological activity, since the C_5' epimer of gentamicin C_{1a}, the product of hydrogenation of sisomicin, is totally inactive [38]. Although the reason for this is not clear, it is proba-ble that the loss of activity is not really due to inversion of configuration, but to the conformational change that occurs on epimerization at C_5' (Figure 11), leading to an equatorial glycoside of the β-L-configuration [14].

Antibiotic G-52 is the 6'-N-methyl derivative of sisomicin and as expected from

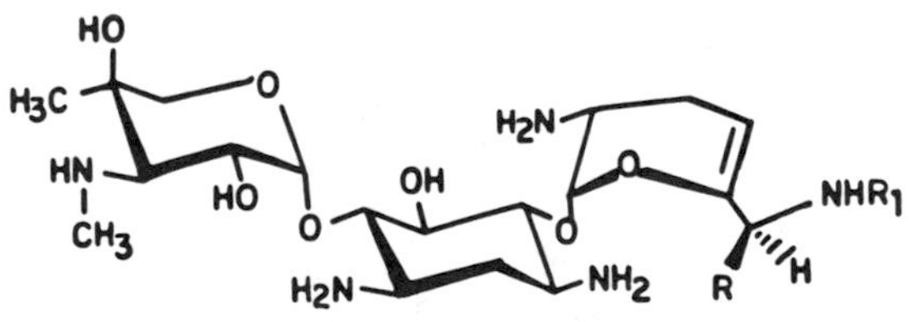

ANTIBIOTIC	R	R₁
SISOMICIN	H	H
G-52	H	CH₃
VERDAMICIN	CH₃	H

Figure 9   Structures of sisomicin, antibiotic G-52, and verdamicin produced by *M. inyo-ensis.*

Table 5 Comparison of the in Vitro Activities of Aminoglycosides Against Sensitive Bacteria[a,b]

	Number of strains	Sisomicin	Verdamicin	Antibiotic G-52	Netilmicin	Sch 21562	Sch 27083	Sch 22591	Sch 27082	Sch 27598
Bacillus subtilis	1	<0.06	<0.06	<0.06	<0.06	<0.06	<0.06	<0.06	<0.06	0.125
Staphylococcus aureus	1	0.125	0.125	0.125	<0.06	0.125	0.125	<0.06	0.25	0.5
Enterobacter	2	0.25	0.5	0.5	0.25	0.5	0.5	0.25	0.5	0.25
Escherichia coli	4	0.5	0.5	0.5	0.25	0.5	0.5	0.5	0.5	0.5
Klebsiella pneumoniae	2	0.5	0.5	0.5	0.25	0.5	0.5	0.5	0.5	0.25
Proteus mirabilis	2	0.5	0.5	2	0.5	2	2	1	1	1
Proteus indole positive	2	1	1	4	2	4	2	0.5	1	0.5
Pseudomonas	5	1	2	2	1	2	1	0.5	1	0.5
Salmonella	1	0.5	0.5	0.5	0.5	0.5	0.5	0.5	0.5	0.5
Serratia	2	1	1	0.5	2	0.5	1	1	1	2

[a]Median MIC values (μg/ml, 24 hr), Mueller-Hinton agar.
[b]*Sch 21562*, 6'-N-ethyl netilmicin; *Sch 27083*, 1-N-S-HAPA-sisomicin; *Sch 22591*, 5-epi-sisomicin; *Sch 27082*, 1-N-S-HAPA-5-epi-sisomicin; *Sch 27598*, 1-N-S-HAPA-5-epi-gentamicin B.

Table 6 Comparison of the in Vitro Activities of Aminoglycosides Against Resistant Strains[a]

Resistance[b] mechanism	Number of strains	Sisomicin	Verdamicin	Antibiotic G-52	Netilmicin	Sch 21562	Sch 27083	Sch 22591	Sch 27082	Sch 27598
APH(3')-I	8	2	2	4	2	4	1	1	2	1
APH(3')-II	3	1	4	2	1	4	4	2	2	2
APH(3')-III	1	1	2	2	1	2	2	2	2	2
ANT(2'')	5	32	64	>32	1	2	1	2	0.5	0.5
ANT(4')	2	2	4	2	2	4	4	2	4	16
AAC(3)-I	4	>64	>64	>32	4	32	2	2	4	1
AAC(3)-Ia	2	>64	>64	>32	>64	>64	4	4	4	2
AAC(3)-II	4	>64	>64	>32	64	>64	4	8	2	2
AAC(2')+APH(3')	2	64	32	>32	32	>64	4	2	1	1
AAC(6')-I	4	32	8	4	64	4	16	2	8	8
AAC(6')-II	2	>64	>64	>32	>64	>64	>64	>64	16	8
APH(2'')+AAC(6')	1	>64	>64	>32	16	>64	16	16	16	0.5
Permeability	2	64	>64	>32	>64	>64	>64	64	64	64

[a]Median MIC values (μg/ml, 24 hr), Mueller-Hinton agar.
[b]The resistance mechanism AAC(3)-Ia is a subgroup of AAC(3)-I that is resistant to netilmicin.

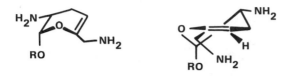

Figure 10   Approximate conformation of the cyclic vinyl ether moiety in sisomicin.

experience with the gentamicin series, its biological activity (Tables 5 and 6) is similar
to that of sisomicin except against organisms carrying 6'-N-acetylating enzymes [6].
Similarly, verdamicin, the 6'-C-methyl derivative of sisomicin, is active against some
6'-N-acetylating strains but appears less potent against these strains than does antibiotic
G-52.

The development by Wright of selective N-alkylation under protonated conditions
[54] and of the transition-metal complexing technique by Nagabhushan and co-workers
[25–27] has allowed extensive chemical modification of several aminoglycoside antibiotics.
Selective alkylation and acylation of the nitrogen functions of sisomicin has led to a large
number of new semisynthetic antibiotics, shown in Figure 12. The biological activities
of a few of these derivatives are given in Tables 5 and 6.

Let us first examine the effect of alkylation at the 6'-N-position of sisomicin. Increasing the length of the linear alkyl chain to four carbon atoms does not alter the potency
or spectrum of antibiotic G-52 in a significant manner. Although only one example is
available, introduction of a primary amino group at the end of the alkyl chain, as in
6'-N-(4-aminobutyl)sisomicin, does enhance the activity against organisms containing
AAC(6')-II enzymes. Branched-chain alkyls and aralkyls at the 6'-N position decrease
the potency [28].

De-N-methylation of sisomicin at the 3"-position has no effect on either the potency
or the spectrum of activity [29]. Although it was thought that introduction of alkyl
groups bigger than the methyl at the 3"-position would prevent enzymatic adenylylation
at the vicinal 2"-position, no change in the activity against organisms containing the
ANT(2") enzyme is observed when the methyl group of sisomicin is replaced by ethyl.
This has been more conclusively demonstrated in the gentamicin C_2 series, which has
longer chain lengths [29].

A dramatic effect on the spectrum of activity of sisomicin is obtained by introduction
of alkyl groups at the 1-position [55]. Thus, the 1-N-ethyl derivative of sisomicin, netilmicin, which is now in phase III clinical trials, is active not only against strains containing ANT(2") enzymes but also against those containing AAC(3)-I enzymes. Here, too,
extension of the chain length does not significantly alter either the potency or the spectrum
of activity relative to netilmicin. Similar modifications in the verdamicin structure produce
predictable activity profiles. Surprisingly, removal of the 3"-methyl group in netilmicin
results in loss of activity against ANT(2")-resistant strains [30]. Increasing the steric
bulk of netilmicin at the 3"-position by exchanging the methyl group for an ethyl group
results in a loss of potency, but the spectrum of activity is not altered [30]. Netilmicin
can be modified in a predictable manner to expand its spectrum of activity against
resistant strains carrying AAC(2') enzymes or those containing AAC(6') enzymes. Thus,
2'-N-ethyl netilmicin, although somewhat less potent than netilmicin, is active against
AAC(2')-containing strains, while 6'-N-ethyl-netilmicin picks up activity against AAC(6')
organisms in a similar way [56].

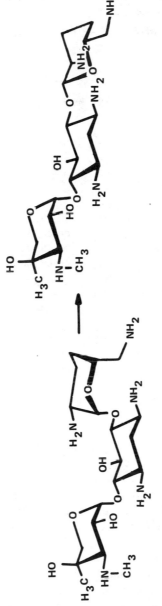

Figure 11 Conformation of 4-O-2,6-diamino-(2,3,4,6-tetradeoxy-β-L-erythro-hexopyranosyl)garamine.

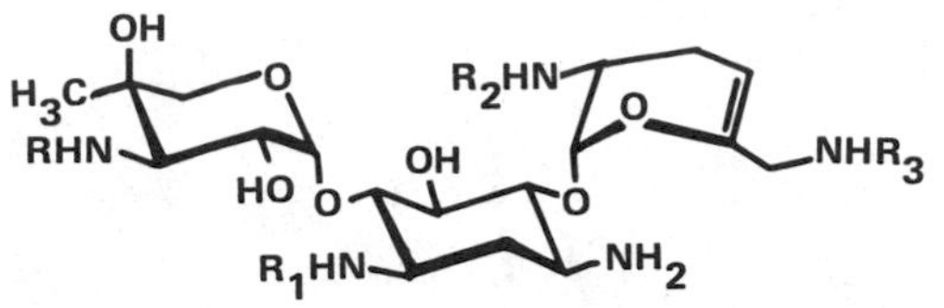

ANTIBIOTIC	R	R_1	R_2	R_3
SISOMICIN	CH_3	H	H	H
G52	CH_3	H	H	CH_3
6'-N-ETHYLSISOMICIN	CH_3	H	H	CH_3CH_2
6'-N-PROPYLSISOMICIN	CH_3	H	H	$CH_3CH_2CH_2$
6'-N-BUTYLSISOMICIN	CH_3	H	H	$CH_3CH_2CH_2CH_2$
6'-N-(4-AMINOBUTYL)SISOMICIN	CH_3	H	H	$CH_2CH_2CH_2CH_2NH_2$
66 40G	H	H	H	H
3"-N-ETHYL-66 40G	CH_3CH_2	H	H	H
NETILMICIN	CH_3	CH_3CH_2	H	H
1-N-PROPYLSISOMICIN	CH_3	$CH_3CH_2CH_2$	H	H
1-N-BUTYLSISOMICIN	CH_3	$CH_3CH_2CH_2CH_2$	H	H
1-N-(4-AMINOBUTYL)SISOMICIN	CH_3	$CH_2CH_2CH_2CH_2NH_2$	H	H
3"-De-N-METHYLNETILMICIN	H	CH_3CH_2	H	H
3"-De-N-METHYL-3"-N-ETHYLNETILMICIN	CH_3CH_2	CH_3CH_2	H	H
2'-N-ETHYLNETILMICIN	CH_3	CH_3CH_2	CH_3CH_2	H
6'-N-ETHYLNETILMICIN	CH_3	CH_3CH_2	H	CH_3CH_2
1-N-ACETYLSISOMICIN	CH_3	CH_3CO	H	H
1-N-PROPIONYLSISOMICIN	CH_3	CH_3CH_2CO	H	H
1-N-(S-3-AMINO-2-HYDROXYPROPIONYL)-SISOMICIN	CH_3	HAPA	H	H

Figure 12　Semisynthetic derivatives of sisomicin.

In contrast to both kanamycin A and gentamicin B, sisomicin is unique in that the introduction of simple acyl groups (e.g., acetyl) at the 1-position produces highly potent derivatives (Figure 12) with activities similar to those of netilmicin [57]. Contrary to earlier observations, Rane and Daniels [37] have recently shown that 1-N-S-HAPA-sisomicin is similar to *Sch 21420* in its spectrum of activity and is slightly more potent. Also unique in the sisomicin series are the broad-spectral and highly potent activity of the 1-N-peptidyl derivatives, which are comparable to netilmicin in their activities against resistant strains (Figure 12) [1].

So far, we have been discussing modifications at nitrogen functions or removal of hydroxyl groups to achieve either potency enhancement or broadening of the spectrum of activity. A novel way of achieving these objectives, demonstrated by Daniels and co-workers [7], involved inversion of configuration of the hydroxyl group at the 5-position of deoxystreptamine (Figure 13). Thus, as shown in Tables 5 and 6, 5-episisomicin is not only more potent than sisomicin but is also active against bacteria resistant to siso-micin via the ANT(2″), AAC(2′), and AAC(3) modes of inactivations. Although an explanation to account for such a dramatic effect upon inversion of configuration of one of the hydroxyl groups is not readily available, removal of an equatorial substituent from the carbon β to the aglycon carbon is expected to offer greater freedom for rotation

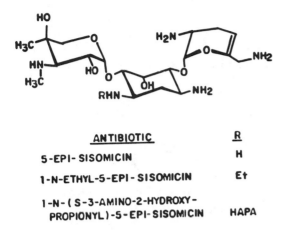

ANTIBIOTIC	R
5-EPI- SISOMICIN	H
1-N-ETHYL-5-EPI-SISOMICIN	Et
1-N-(S-3-AMINO-2-HYDROXY-PROPIONYL)-5-EPI-SISOMICIN	HAPA

Figure 13   Structures of 5-epi-sisomicin and derivatives.

about the aglycon carbon and the glycosidic oxygen bond without significantly affecting the exo-anomeric angle. Net rotation of the deoxystreptamine ring about the two aglycon carbon-oxygen bonds changes the positions of the 2″-hydroxyl group and the 2′-amino group relative to the amino groups of deoxystreptamine, thus perturbing the stereochemistry that seems critical for enzymic modifications at these centers. The axial hydroxyl group at C_5 could play the crucial role of stabilizing the new conformations through intramolecular hydrogen bonding, so that, at any given time, an abundance of the highly active rotameric form is available.

The activities of the alkylation and acylation products of 5-epi-sisomicin showed no surprises. The 1-N-S-HAPA derivative of 5-epi-sisomicin recently synthesized by Rane and Daniels [37] has even a broader spectrum of activity than that of the corresponding sisomicin analog. In fact, in our hands, it is active against all enzymatic mechanisms of resistance known to date (Table 5 and 6).

In contrast to the dramatic effect produced by inversion of configuration at C_5 in sisomicin, practically no change in either potency or spectrum is achieved upon inversion of the C_5 configuration in gentamicin B. However, conversion of 5-epi-gentamicin B to the corresponding 1-N-S-HAPA derivative results in a *Sch 21420*-like compound with significantly enhanced potency (Tables 5 and 6) [31].

VII.   Conclusions

Structural modifications of aminoglycosides-aminocyclitol antibiotics carried out to date were intended to enhance the spectrum of activity against resistant strains. Progress in the seventies was more rapid than in the earlier years because of the development of new chemical methods and an understanding of the mechanisms of resistance at a molecular level. Basically, four approaches were undertaken to enhance the spectrum of activity of an aminoglycoside antibiotic. These were: removal of the functional group involved in enzymatic inactivation [e.g., tobramycin, dibekacin, 2′-desaminogentamicin C_{1a} (*Sch 23200*)], steric shielding of a group involved in enzymatic alteration by modification of a proximal functionality (netilmicin), incorporation of the foregoing principle also to include inhibition of enzymatic modification at positions remote from the site of modification (amikacin, *Sch 21420,* butakacin), and changing the glycosidic conformation of the molecule to prevent enzymatic inactivation at several sites (5-epi-sisomicin). A combination of some of these approaches has already lead to compounds active against organisms capable of modifying amikacin (*Sch 27082*).

Another area that will receive attention in the future involves the mechanism of permeability-resistant strains and the design of structures to overcome this mode of resistance.

An equally important area will be an understanding of the mechanisms of nephrotoxicity and ototoxicity at a molecular level and development of a rational approach to the syntheses of less toxic aminoglycoside antibiotics. To date, the only semisynthetic aminoglycoside that has been evaluated for reduced nephrotoxicity and ototoxicity is netilmicin. No structural rationale has yet been advanced for the reduced toxicity observed in animal models.

Acknowledgments

The authors wish to thank P.J.L. Daniels, J. J. Wright, S. McCombie, D. F. Rane, A. Mallams, and F. Menzel for their contribution. We acknowledge the clerical support of M. M. Caruso and P. Blunt.

References

1. Afonso, A.: Personal communication.
2. Akiba, T., Koyama, K., Ishii, Y., Kimura, S., and Kukushima, T.: *Nippon Iji Shimpo (Tokyo),* No. 1886:45, 1960. (In Japanese.)
3. Daniels, P.J.L.: Antibiotics (aminoglycosides). In *Kirk–Othmer: Encyclopedia of Chemical Technology* (3rd ed.), Vol. 2. New York: Wiley, 1978, pp. 819–852.
4. Daniels, P.J.L.: Personal communication.
5. Daniels, P.J.L., Luce, C., Nagabhushan, T. L., Jaret, R. S., Schumacher, D., Reimann, H., and Ilavsky, J.: The gentamicin antibiotics: 6. Gentamicin C_{2b}, and aminoglycoside antibiotic produced by *Micromonospora purpurea* mutant JI-33. *J. Antibiot. (Tokyo)* 28:35–41, 1975.
6. Daniels, P.J.L., Jaret, R. S., Nagabhushan, T. L., and Turner, W. N.: The structure of antibiotic G-52, a new aminocyclitol-aminoglycoside antibiotic produced by *Micromonospora zionensis. J. Antibiot. (Tokyo)* 29:488–491, 1976.
7. Daniels, P.J.L., and Rane, D. F.: Synthetic and mutasynthetic antibiotics related to sisomicin. In *Microbiology,* edited by Schessinger, D. Washington, D.C., ASM Publications, 1979, pp. 314–317.
8. Davies, J., and Smith, D. I.: Plasmid-determined resistance to antimicrobial agents. *Annu. Rev. Microbiol.* 32:469–518, 1978.
9. Hitchens, M., and Rinehart, K. L., Jr.: Chemistry of the neomycins: XII. The absolute configuration of deoxystreptamine in the neomycins, paromomycins and kanamycins. *J. Am. Chem. Soc.* 85:1547–1548, 1963.
10. Kawaguchi, H.: Recent progress in aminoglycoside antibiotics. *J. Antibiot. (Tokyo)* 30 (Suppl.):190–200, 1977.
11. Kawaguchi, H., Naito, T., Nakagawa, S., and Fujisawa, K.: BB-K8, a new semisynthetic aminoglycoside. *J. Antibiot. (Tokyo)* 25:695–708, 1972.
12. Koyama, G., Iitaka, Y., Maeda, K., and Umezawa, H.: The crystal structure of kanamycin. *Tetrahedron Lett.,* 1875–1879, 1968.

13. Kondo, S., Iinuma, K., Hamada, M., Maeda, K., and Umezawa, H.: Synthesis of iso-seryl derivatives of kanamycins and their antibacterial activities. *J. Antibiot. (Tokyo)* 27:90–93, 1974.

14. Kugleman, M., Mallams, A. K., Vernay, H. F., Crowe, D. F., and Tanabe, M.: Semi-synthetic aminoglycoside antibacterials: Part 1. Preparation of selectively protected garamine derivatives. *J. Chem. Soc. [Perkin 1]*, 1088–1097, 1976.

15. Lemieux, R. U., Nagabhushan, T. L., Clemetson, K. J., and Tucker, L.C.N.: The synthesis of kanamycin analogs: I. α-D-Glucopyranosyl derivatives of deoxystrepta-mine. *Can. J. Chem.* 51:53–66, 1973.

16. McCombie, S.: Personal communication.

17. Miller, G. H., Chiu, P.J.S., and Waitz, J. A.: Biological activity of *Sch 21420*, the 1-N-S-α-hydroxy-β-aminopropionyl-derivative of gentamicin B. *J. Antibiot. (Tokyo)* 31:688–696, 1978.

18. Nagabhushan, T. L., Cooper, A. B., Brambilla, R., and Morton, J. B.: Paper No. 215, 58th Chemical Conference and Exhibition of the Chemical Institute of Canada, Toronto, 1974.

19. Nagabhushan, T. L., Cooper, A. B., Daniels, P.J.L., Morton, J. B., and Brambilla, R.: Paper No. Carb 5, 2nd Joint Conference of the Chemical Institute of Canada and the American Chemical Society, Montreal, 1977.

20. Nagabhushan, T. L., and Jeffrey, G. A.: Unpublished results.

21. Nagabhushan, T. L., Cooper, A. B., Tsai, H., Daniels, P.J.L., and Miller, G. H.: The syntheses and biological properties of 1-N-(S-4-amino-2-hydroxybutyryl)-gentamicin B and 1-N-(S-3-amino-2-hydroxypropionyl)-gentamicin B. *J. Antibiot. (Tokyo)* 31:681–687, 1978.

22. Nagabhushan, T. L., Cooper, A. B., and Miller, G. H.: Unpublished results.

23. Nagabhushan, T. L., Turner, W. N., and Miller, G. H.: Unpublished results.

24. Nagabhushan, T. L., Tsai, H., and Miller, G. H.: Unpublished results.

25. Nagabhushan, T. L., Cooper, A. B., Turner, W. N., Tsai, H., McCombie, S., Mallams, A. K., Rane, D., Wright, J. J., Reichert, P., Boxler, D. L., and Weinstein, J.: Inter-action of vicinal and non-vicinal aminohydroxy group pairs in aminoglycoside-aminocyclitol antibiotics with transition metal cations. Selective N-protection. *J. Am. Chem. Soc.* 100:5253–5254, 1978.

26. Nagabhushan, T. L., Cooper, A. B., Turner, W. N., and Tsai, H.: Interaction of vicinal and non-vicinal amino and hydroxy group pairs in aminoglycoside antibiotics with divalent-transition metal cations. Selective N-blocking and conversion to bio-logically active derivatives, Part I. Paper No. B. 36, 9th International Carbohydrate Symposium, London, April 1978.

27. Nagabhushan, T. L, Cooper, A. B., Turner, W. N., Tsai, H., McCombie, S., Mallams, A. K., Rane, D., Wright, J. J., Reichert, P., Weinstein, J., and Boxler, D. L.: Inter-action of vicinal and non-vicinal amino and hydroxy group pairs in aminoglycoside antibiotics with divalent-transition metal cations. Selective N-blocking and conver-sion to biologically active derivatives, Part II. Paper No. B. 37, 9th International Carbohydrate Symposium, London, April 1978.

28. Nagabhushan, T. L., Turner, W. N., Daniels, P.J.L., and Miller, G. H.: Unpublished results.

29. Nagabhushan, T. L., Wright, J. J., Cooper, A. B., Turner, W. N., and Miller, G. H.: Chemical modification of some gentamicins and sisomicin at the 3″-position. *J. Anti-biot. (Tokyo)* 31:43–54, 1978.

30. Nagabhushan, T. L., and Cooper, A. B.: Unpublished results.

31. Nagabhushan, T. L., Cooper, A. B., and Miller, G. H.: Unpublished results.

32. Nara, T.: Aminoglycoside antibiotics. In *Annual Reports on Fermentation Processes,* Vol. 1, edited by Perlman, D. New York, Academic Press, 1977, pp. 299–366.

33. Ochiai, K., Totani, T., and Toshiki, Y.: *Nippon Iji Shimpo (Tokyo)*, No. 1837:25, 1959. (In Japanese.)

34. Okamoto, S., and Suzuki, Y.: Chloramphenicol-, dihydrostreptamycin-, and kana-mycin-inactivating enzymes from multiple drug-resistant *Escherichia coli* carrying episome 'R'. *Nature,* 208:1301–1303, 1965.

35. Okanishi, M., Kondo, S., Suzuki, Y., Okamoto, S., and Umezawa, H.: Studies on inactivation of kanamycin and resistance of *E. coli. J. Antibiot. (Tokyo)* 20:132–135, 1967.

36. Price, K. E., Godfrey, J. C., and Kawaguchi, H.: Effect of structural modifications on the biological properties of aminoglycoside antibiotics containing 2-deoxystrepta-mine. *Structure-Activity Relationships Among the Semisynthetic Antibiotics,* edited by Perlman, D. New York, Academic Press, 1977, pp. 239–395.

37. Rane, D. F., and Daniels, P.J.L.: Personal communication.

38. Reimann, H., Jaret, R. S., and Cooper, D. J.: Sisomicin. Stereochemistry and attach-ment of the unsaturated sugar moiety. *Chem. Commun.,* 924–925, 1971.

39. Richardson, K., Jevons, S., Moore, J. W., Ross, B. C., and Wright, J. R.: Synthesis and antibacterial activities of 1-N-(S-ω-amino-2-hydroxyalkyl)-kanamycin A deriva-tives. *J. Antibiot. (Tokyo)* 30:843–846, 1977.

40. Stark, W. M., Hoehm, M. M., and Knox, N. G.: Nebramycin, a new broad-spectrum antibiotic complex: I. Direction and biosynthesis. *Antimicrob. Agents Chemother.,* 314–323, 1967.

41. U.S. Patent 3,903,072 (Sept. 2, 1975), Ilavsky, J., Bayan, A. P., Charney, W., and Reimann, H. (to Schering Corporation).

42. Umezawa, H., Ueda, M., Maeda, K., Yagishita, K., Kondo, S., Okami, Y., Utahara, R., Osato, Y., Nitta, K., and Takeuchi, T.: Production and isolation of a new anti-biotic, kanamycin. *J. Antibiot. (Tokyo) Ser. A* 10:181–189, 1957.

43. Umezawa, H.: Progress of fundamental studies on kanamycins. *Asian Med. J.* 11:291–301, 1968.

44. Umezawa, H., Umezawa, S., Tsuchiya, T., and Okazaki, Y.: 3',4'-Dideoxy-kanamycin B active against kanamycin-resistant *Escherichia coli* and *Pseudomonas aeruginosa. J. Antibiot. (Tokyo)* 24:485–487, 1971.

45. Umezawa, H., Okanishi, M., Utahara, R., Maeda, K., and Kondo, S.: Isolation and structure of kanamycin inactivated by cell free system of kanamycin-resistant *E. coli. J. Antibiot. (Tokyo)* 20:136–141, 1967.

46. Umezawa, H., Okanishi, M., Kondo, S., Hamana, K., Utahara, R., Maeda, K., and Mitsuhashi, S.: Phosphorylative inactivation of aminoglycoside antibiotics by *Escheri-chia coli* carrying R factor. *Science* 157:1559–1561, 1967.

47. Weinstein, M. J., Leudemann, G. M., Oden, E. M., and Wagman, G. H.: Gentamicin, a new broad-spectrum antibiotic complex. *Antimicrob. Agents Chemother.,* 1–7, 1963.

48. Weinstein, J., Cooper, D. J., and Daniels, P.J.L.: Abstr. 9, 12th Interscience Confer-ence on Antimicrobial Agents and Chemotherapy, Atlantic City, N.J., October 1972.

49. Weinstein, M. J., Marquez, J. A., Testa, R. T., Wagman, G. H., Oden, E. M., and Waitz, J. A.: Antibiotic 66-40, a new *Micromonospora*-produced aminoglycoside antibiotic. *J. Antibiot. (Tokyo)* 23:551–554, 1970.

50. Weinstein, M. J., Wagman, G. H., Marquez, J. A., Testa, R. T., and Waitz, J. A.: Verdamicin, a new broad spectrum aminoglycoside antibiotic. *Antimicrob. Agents Chemother.,* 7, 246–249, 1975.

51. Woo, P.W.K., Dion, W., and Bartz, Q. R.: Butirosins A and B, aminoglycoside anti-biotics: III. Structures. *Tetrahedron Lett.,* 2625–2628, 1971.

52. Wright, J. J., and Lee, P.: The synthesis and conformational properties of novel 2'-unsubstituted aminoglycoside antibiotics of the gentamicin-sisomicin class. Paper No. B. 38, 9th International Symposium on Carbohydrate Chemistry, London, April 1978.

53. Wright, J. J.: Personal communication.
54. Wright, J. J.: Synthesis of 1-N-ethylsisomicin: a broad spectrum semisynthetic amino-glycoside antibiotic. *Chem. Commun.*, 206–208, 1976.
55. Wright, J. J.: Paper No. 91, Abstracts of the 15th Interscience Conference on Anti-microbial Agents and Chemotherapy, Washington, D.C., September 1975.
56. Wright, J. J., and Jaret, R. S.: Unpublished results.
57. Wright, J. J., Cooper, A., Daniels, P.J.L., Nagabhushan, T. L., Rane, D., Turner, W. N., and Weinstein, J.: Selective N-acylation of gentamicin antibiotics. Synthesis of 1-N-acyl derivatives. *J. Antibiot. (Tokyo)* 29:714–719, 1976.

2

AMINOGLYCOSIDES UNDER DEVELOPMENT

FELIX LEITNER and KENNETH E. PRICE
Bristol Laboratories, Syracuse, New York

Since the advent of streptomycin, numerous other aminoglycoside antibiotics have been added to the roster of chemotherapeutic agents, and the improvements in antibacterial activity and spectrum of successive drugs are impressive indeed. But little, if any, progress was achieved in alleviating the major shortcoming that characterizes this class of antibiotics: nephro- and ototoxicity. To date, any increment in activity has been essentially proportional to an increase in toxicity, and therefore no significant improvement in therapeutic index has been obtained. Because of toxicity, aminoglycosides, at least in the

United States, are reserved for severe and life-threatening infections, and their use requires close monitoring of drug concentrations in blood.

Another problem, common to most antibiotics, is the emergence of resistant bacterial strains through selective pressure caused by drug use. Resistance to aminoglycosides is due primarily to elaboration of drug-modifying enzymes. Recognition of the mode of resistance, coupled with the fortuitous discovery of the protection that a 1-N-[(S)-2-hydroxy-4-aminobutyryl] side chain confers against enzymatic inactivation, prompted the design of new aminoglycosides active against many of the otherwise-resistant strains. The new aminoglycosides were prepared from established ones either by removing or alkylating functional groups that are targets of modifying enzymes or by acylating, and more recently alkylating, the 1-amino group with an appropriate radical. Two such antibiotics, dibekacin and amikacin, have been used successfully to treat infections caused by strains resistant to other aminoglycosides.

Present-day development of new aminoglycosides centers around the problems of toxicity and enzymatic inactivation. Attempts to solve the second problem are guided by past experience, but information useful in directing efforts to reduce toxicity is scanty.

This discussion of aminoglycosides under development is divided into three sections according to the means by which antibiotics were obtained: screening of fermentation broths, mutational biosynthesis, and chemical modification. Dibekacin, sisomicin, and netilmicin, although not available in the United States, are in clinical use abroad. For this reason they are not included in this review.

I.   Naturally Occurring Aminoglycoside Antibiotics

Despite the notable success that the design of new aminoglycoside antibiotics by chemical manipulation of current ones has enjoyed of late, the search for naturally occurring aminoglycosides has continued unabated. Screening has recently led to the discovery of aminoglycosides with novel structures as well as of previously unknown structural relatives of major drugs. Since the latter differ little in biological properties from their older relatives, it is the discovery of compounds of a novel type that renders screening rewarding. Fortimicin A and seldomycin 5 are examples of novel aminoglycosides, whereas sagamicin, verdamicin, and G-52, related to gentamicin and sisomicin, are of a now-traditional type.

A.   Fortimicin A

Fortimicin A and B are the major components of an antibiotic complex produced by soil isolates of *Micromonospora olivoasterospora* obtained from several locations in Japan and the United States [38,73]. The two antibiotics are pseudodisaccharides containing fortamine, a novel diaminocyclitol [13] (Figure 1). They differ chemically in that a glycyl radical is present only in fortimicin A. In addition, the free bases of the two aminoglycosides are distinct in the conformation of the aminocyclitol, but a conformational inversion in fortimicin B sulfate obviates this difference. Fortimicin A is unstable under alkaline conditions and is converted to the B component by loss of the glycyl radical [44]. Since fortimicin B is only weakly active, fortimicin A alone will be considered.

Against organisms that have no aminoglycoside-modifying enzymes, fortimicin A resembles kanamycin in spectrum but is generally two- to threefold less active under

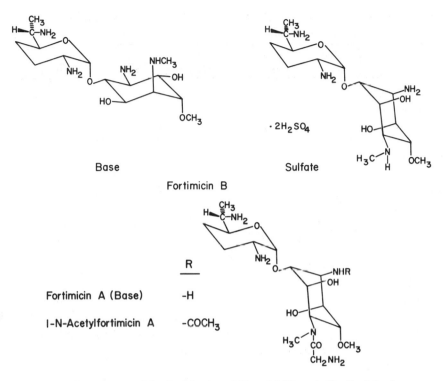

Fortimicin B

	R
Fortimicin A (Base)	-H
I-N-Acetylfortimicin A	-COCH₃

Figure 1 Structures of fortimicin A and B and 1-N-acetylfortimicin A.

standard assay conditions (Table 1). Like kanamycin, fortimicin is weakly active against *Pseudomonas aeruginosa* [16,22b]. Since fortimicin A lacks the functional groups affected by the aminoglycoside phosphoryltransferase APH(3′) and nucleotidyltransferases ANT(2″) and ANT(4′), it is active against organisms elaborating these enzymes* (Table 2). Fortimicin A is also active against bacteria containing the aminoglycoside acetyltransferase AAC(2′) and against strains of Enterobacteriaceae that form AAC(6′). By contrast, AAC(6′)-producing staphylococci are resistant. The broadly specific variants of 3-N-acetyltransferase, AAC(3)-II and AAC(3)-III, do not confer resistance to fortimicin, whereas AAC(3)-I does [38,40a]. The latter was thought to be of narrow specificity, accepting only gentamicin and the closely related antibiotics sagamicin and sisomicin as substrates. Now, however, the specificity of this enzyme appears to be considerably broader than was previously recognized, for not only fortimicin, but also seldomycin 5, is modified by the enzyme. The product of the enzymatic reaction with fortimicin has been identified as 1-N-acetylfortimicin A [40a,56] (Figure 1).

In the treatment of mice infected with *Staphylococcus aureus* strain Smith or with various Enterobacteriaceae, fortimicin A is comparable in efficacy to kanamycin A and amikacin [38,40a].

According to the data of Nara et al. [38], the acute toxicity of fortimicin after intravenous (IV) administration to mice is similar to that of kanamycin (Table 3). Kimura et al. [26a], however, found fortimicin considerably more toxic. The difference in the way of re-

*The nomenclature for the aminoglycoside-modifying enzymes is that used by Price et al. [48].

Table 1 Antibiotic Spectrum of Fortimicin A Against Organisms Lacking Known Aminoglycoside-Modifying Enzymes

Organism	Number of strains	MIC (μg/ml)[a]	
		Fortimicin A	Kanamycin A
Escherichia coli	20	6.5	2.6
Klebsiella pneumoniae	12	6	3
Enterobacter aerogenes	11	7.8	3
Enterobacter cloacae	10	9.2	4.6
Serratia marcescens	16	2.8	3
Proteus mirabilis	8	4	1.4
Proteus vulgaris	8	4.8	1.2
Proteus morganii	13	6.5	2
Proteus rettgeri	10	2.6	0.7
Providencia stuartii	20	2.2	0.8
Salmonella sp.	9	5.4	4
Pseudomonas aeruginosa	16	72	>78
Staphylococcus aureus	10	1.9	1.1
Staphylococcus epidermidis	5	1	0.8

[a]Mean values. The assays were performed in Mueller-Hinton medium. The inoculum was obtained by diluting an overnight broth culture of the organism 500-fold for *Pseudomonas aeruginosa* and 50-fold for all other organisms.

cording their results (as fortimicin sulfate in one instance, as base in the other) does not account for the discrepancy. The nephrotoxic potential of fortimicin A in the dog and rat was assessed as about one-seventh that of gentamicin [26a]. Fortimicin was also much less nephrotoxic than gentamicin in the cat [73a]. Furthermore, rats recovered more rapidly from renal damage after exposure to fortimicin than after treatment with amikacin [52a]. The vestibular toxicity of fortimicin in the cat [73a] and its cochlear toxicity in the guinea pig [52a] were also relatively low.

Several minor components of the fortimicin complex were recently identified [12a, 21a,57d,57f]. All are related structurally to the two major components and, except for fortimicin D (6′-demethylfortimicin A), are less active than fortimicin A. Numerous analogs of fortimicin A were prepared by 4-N-acylation or alkylation of fortimicin B [55a,57e]. Two of these, 4-N-[(S)-2-hydroxy-4-aminobutyl]- and 4-N[(S)-2-hydroxy-4-(methylamino)butyl]fortimicin B, are reportedly as active as fortimicin A [55a]. 2-Deoxyfortimicin A, obtained by a chemical modification of fortimicin A, is slightly more active than the parent compound, but, like the latter, lacks activity against organisms that elaborate AAC(3)-I [57b]. Recently, several naturally occurring antibiotics related to the fortimicins were described: sporaricin A and B produced by a soil isolate, *Saccharopolyspora hirsuta* subsp. *kobensis* nov. subsp. [11a,11c,21d,26b]; istamycin A and B formed by *Streptomyces tenjimariensis* nov. sp. [44a]; sannamycin A and B accumulated by *Streptomyces sannanensis* nov. sp. [11b,67a]; dactimicin (SF-2052) found in the fermentation broth of *Dactylosporangium matsuzakiense* nov. sp. [21b]; and N-formyl-fortimicin A detected, next to fortimicin A, in the fermentation broth of a new isolate

Table 2   Antibiotic Activity of Fortimicin A Against Organisms Producing
Aminoglycoside-Modifying Enzymes

Aminoglycoside-modifying enzyme	Organism	Number of strains	MIC (μg/ml)[a]	
			Fortimicin A	Kanamycin A
APH(3')-I	*Escherichia coli*	3	2	>125
	Klebsiella pneumoniae	2	3.4	>125
	Enterobacter cloacae	2	5.7	>125
	Serratia marcescens	2	5.7	>125
	Proteus rettgeri	1	2	32
APH(3')-II	*Escherichia coli*	2	5.7	>125
	Enterobacter cloacae	1	8	>125
APH(3')-III	*Staphylococcus aureus*	1	2	>125
	Staphylococcus epidermidis	1	0.7	>125
ANT(2″)	*Escherichia coli*	3	2	> 79
	Klebsiella pneumoniae	3	5	>125
	Enterobacter cloacae	1	8	>125
	Serratia marcescens	3	8	>100
ANT(4')	*Staphylococcus aureus*	2	4	53
	Staphylococcus epidermidis	2	1.4	53
AAC(2')	*Proteus rettgeri*	2	6	3.2
	Providencia stuartii	5	6.5	1.7
AAC(6')	*Escherichia coli*	1	2	8
	Enterobacter aerogenes	1	4	125
	Serratia marcescens	3	5	16
	Proteus rettgeri	1	4	125
AAC(6') + APH(3')	*Staphylococcus epidermidis*	3	>125	>125
AAC(3)-I	*Escherichia coli*	1	> 63	4
	Klebsiella pneumoniae	3	> 63	3.2
AAC(3)-II	*Escherichia coli*	1	4	>125
	Klebsiella pneumoniae	1	2	>125
	Serratia marcescens	1	8	>125
AAC(3)-III	*Klebsiella pneumoniae*	2	16	> 53
	Serratia marcescens	2	10	>125

[a]Mean values where applicable. Assay conditions as in Table 1.

of *M. olivoasterospora* [21c]. Sporaricin B and sannamycin B lack a glycyl radical and,
like fortimicin B, are only weakly active [11a,11b]. N-formylfortimicin A [4-N-(formami-
doacetyl)fortimicin B] is considerably less active than fortimicin A [21c]. The other
antibiotics, although differing somewhat in antibacterial potency, resemble fortimicin A
in spectrum and activity.

Table 3   Acute Toxicity in Mice of Various Naturally Occurring Aminoglycosides After Intravenous, Intraperitoneal, Subcutaneous, or Intramuscular Administration

Antibiotic	LD_{50} (mg/kg)[a]				Reference
	IV	IP	SC	IM	
Fortimicin A (sulfate)	380		400		38
Fortimicin A (base)	97			491	26a
Seldomycin 5 (sulfate)	365				39
Kanamycin A (base)	350				39
Sagamicin	93				43
Gentamicin C	75	440	480		70
Verdamicin	48	200	300		70
G-52	50	200	400		28
Sisomicin	34	221	288		70

[a] LD_{50} is the dose (mg/kg), administered by the indicated route, resulting in the death of half the animals treated.

B.   Seldomycin 5

Four new aminoglycosides, seldomycin factors 1, 2, 3, and 5 (originally coded XK-88-1, -2, -3, and -5) were discovered in the fermentation broth of a soil isolate, *Streptomyces hofunensis* nov. sp. (Nara et al. [39]). Factor 5 [32] (Figure 2), the most active of the seldomycins, was selected for further investigation.

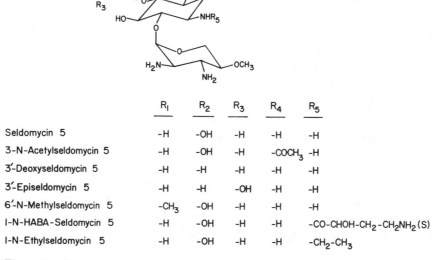

	R_1	R_2	R_3	R_4	R_5
Seldomycin 5	-H	-OH	-H	-H	-H
3-N-Acetylseldomycin 5	-H	-OH	-H	-COCH₃	-H
3′-Deoxyseldomycin 5	-H	-H	-H	-H	-H
3′-Episeldomycin 5	-H	-H	-OH	-H	-H
6′-N-Methylseldomycin 5	-CH₃	-OH	-H	-H	-H
I-N-HABA-Seldomycin 5	-H	-OH	-H	-H	-CO-CHOH-CH₂-CH₂NH₂ (S)
I-N-Ethylseldomycin 5	-H	-OH	-H	-H	-CH₂-CH₃

Figure 2   Structures of seldomycin 5 and derivatives.

Very little published information is available on the biological properties of seldomycin 5. Its intrinsic activity resembles that of kanamycin against *S. aureus* and the Enterobacteriaceae, and, like kanamycin, it lacks useful activity against *P. aeruginosa* [39]. The antibiotic is inactive against organisms producing AAC(3)-I, AAC(2′), and APH(3′)-I, but it is active against those that harbor APH(3′)-II and ANT(2″) [4, 29,31,57]. Its activity against AAC(6′)-containing organisms is erratic. The product resulting from the action of AAC(3)-I on seldomycin 5 has been identified as 3-N-acetylseldomycin 5 [56] (Figure 2). The acute toxicity of seldomycin 5, when given IV to mice, is similar to that of kanamycin A (Table 3).

Seldomycin has been altered chemically with the main objective of reducing vulnerability to modifying enzymes. Such conventional derivatives as 3′-deoxy-, 3′-epi-, 6′-N-methyl-, 1-N-[(S)-2-hydroxy-4-aminobutyryl]-, and 1-N-ethylseldomycin 5 were prepared (Figure 2). Deoxyseldomycin, except for slightly better activity against *P. aeruginosa*, does not differ substantially from the parent compound in intrinsic activity, but episeldomycin is a less active antibiotic [4,29,31]. Predictably, either modification enhances activity against APH(3′)-I producing organisms. However, epimerization also increases acute toxicity in mice by over threefold [29]. Methylseldomycin is on the average half as active as the parent compound, but the modification confers activity against AAC(6′)-producing organisms [76]. Biological data are not available on the 1-N-modified seldomycins [30,72].

C. Sagamicin (Gentamicin C_{2b})

The three major components of the gentamicin C complex differ in the degree of methylation at the 6′-position of the purpurosamine unit: gentamicin C_1 contains a 6′-C,6′-N-dimethylpurpurosamine; gentamicin C_2, a 6′-C-methylpurpurosamine; and gentamicin C_{1a}, an unmethylated purpurosamine (Figure 3). A mutant of the gentamicin C-accumulating strain *Micromonospora purpurea*, obtained in a search for improved producers, forms primarily gentamicin C_{1a} and no detectable amounts of gentamicin C_1 and C_2. A minor but substantial product of this mutant was identified as 6′-N-methylgentamicin C_{1a} and named gentamicin C_{2b} [7] (Figure 3). Minute concentrations of this compound are also formed by the parent strain.

Along with gentamicin C_{1a}, 6′-N-methylgentamicin C_{1a} was independently identified as a major component of an antibiotic complex elaborated by strains of *Micromonospora sagamiensis* var. *nonreducans* nov. sp. and termed sagamicin (XK-62-2; KW-1062) [12,37, 43,60]. The identity of sagamicin with gentamicin C_{2b} has been confirmed [7].

The association of gentamicin C_{2b} and C_{1a} is not a fortuitous one, for C_{1a} is the immediate precursor of C_{2b}, the end product of one of the two branches of the biosynthetic pathway for gentamicin postulated by Testa and Tilley [61]; the other branch leads through gentamicin C_2 to gentamicin C_1.

Sagamicin closely resembles gentamicin C in antibiotic activity and spectrum [7,42,43]. However, owing to the protection conferred by the 6′-N-methyl group against acetylation by AAC(6′), sagamicin is active against organisms that elaborate this enzyme and are therefore resistant to gentamicin C_{1a} [43]. Consequently, sagamicin is also more effective than the gentamicin C complex against such organisms [7,43].

In vivo, sagamicin and gentamicin are nearly equal in efficacy as determined by treatment of experimentally infected mice [36,54]. Data on the comparative pharmaco-

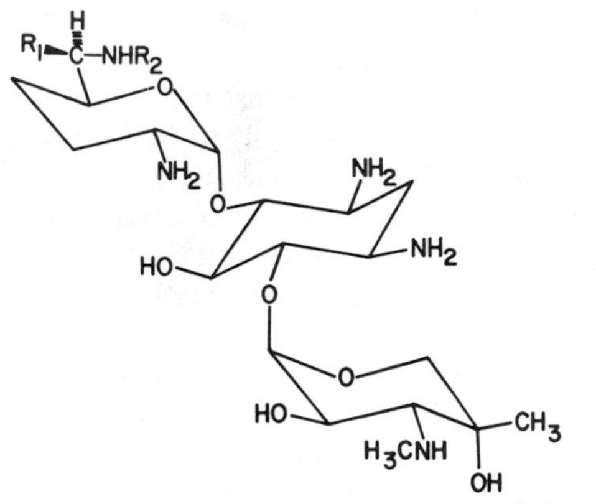

	R_1	R_2
Sagamicin (Gentamicin C_{2b})	-H	-CH$_3$
Gentamicin C_{1a}	-H	-H
Gentamicin C_2	-CH$_3$	-H
Gentamicin C_1	-CH$_3$	-CH$_3$

Figure 3   Structures of sagamicin, gentamicin C_{1a}, C_2, and C_1.

kinetic behavior of the two drugs when given subcutaneously (SC) to mice are conflicting. Whereas Goto et al. [19] report lower peak levels and a more gradual decline of drug concentration in serum with sagamicin, Sato and Marumo [55] detected virtually no difference between the two drugs. After intramuscular (IM) administration to rats, dogs, and humans, plasma concentrations of sagamicin and gentamicin were nearly identical [55,58].

The acute toxicity of sagamicin, measured after IV administration to the mouse, was slightly lower than that of gentamicin (Table 3). Sagamicin was less nephrotoxic than gentamicin to the rat and rabbit [42,58,60]. Its vestibular and cochlear toxicity, determined in the guinea pig and cat, was also lower [1,20].

Sagamicin is in clinical trial in Japan and an extensive literature is available in Japanese, with abstracts in English on the subject [*Chemotherapy (Tokyo)*, September 1977]. Ohkoshi et al. [41] reported (in English) on the efficacy and safety of the drug in a patient population of over 700. Most of the patients suffered from complicated infections with underlying diseases and 65% were over 50 years old. The overall efficacy was rated as 69%. The incidence of adverse reactions was 3% and of abnormal laboratory findings 7%. Hearing impairment and dizziness occurred in 6 of the 775 patients evaluable for safety.

D. Verdamicin (6'-C-Methylsisomicin) and G-52 (6'-N-Methylsisomicin)

In contrast to gentamicin C, sisomicin is a single chemical entity and may be viewed as the 4',5'-dehydro analog of gentamicin C_{1a}. Subsequent to the discovery of sisomicin, the

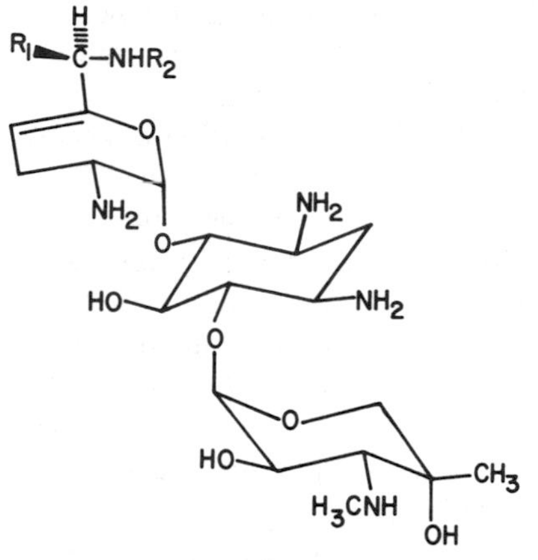

	R_I	R_2
Verdamicin	$-CH_3$	$-H$
G-52	$-H$	$-CH_3$
Sisomicin	$-H$	$-H$

Figure 4 Structures of verdamicin, G-52, and sisomicin.

corresponding analogs of gentamicin C_2 and sagamicin were identified as metabolites of two new species of *Micromonospora* (Figure 4). The former was named verdamicin, after the gray-green color of the producing organism, *M. grisea* [70]; the latter, formed by a strain of *M. zionensis*, received the code name G-52 [6,28]. Sisomicin is coproduced with either one of the two new aminoglycosides. G-52 was also obtained by chemical manipulation of 66-40C, a dimeric aminoglycoside found as a minor component in the submerged fermentation of the sisomicin-producing strain of *M. inoyensis* [11].

Verdamicin and G-52 resemble sisomicin in antibiotic spectrum and activity, except that G-52 is markedly and verdamicin moderately more active against AAC(6')-producing organisms [11,23,28,45,70,74]. Nevertheless, verdamicin is vulnerable to 6'-N-acetylation, and this property has been used for its radioenzymatic determination with AAC(6') from *Escherichia coli* W 677/R5 [3]. Verdamicin also closely resembles sisomicin in therapeutic efficacy. A large number of strains belonging to a variety of bacterial species was used for comparing the two antibiotics in the treatment of experimentally infected mice [70]. On a much more modest scale, G-52 was compared with gentamicin and the two compounds differed little in efficacy [28]. Verdamicin, G-52, and sisomicin have similar acute toxic liabilities for mice (Table 3), but the vestibular toxicity of verdamicin in cats is lower than that of gentamicin [70].

II. Mutasynthetic Aminoglycoside Antibiotics

Mutational biosynthesis is a recently developed procedure for obtaining new antibiotics from mutants of an organism that produces a known antibiotic. Mutants selected for this purpose are unable to form a biosynthetic intermediate of a particular type, one identical with a subunit of the antibiotic, and they will produce no antibiotic unless supplied with the subunit. It was found that such mutants (idiotrophs) are able to form new (mutasynthetic) antibiotics from exogenous analogs (mutasynthons) of the native subunit. Idiotrophs known to date are rather selective and accept only a small number of analogs as substitutes.

Much of the work on mutational biosynthesis was done with organisms that produce aminoglycosides. Mutasynthetic antibiotics related to neomycin, paromomycin, kanamycin, ribostamycin, butirosin, sisomicin, and gentamicin were obtained from 2-deoxystreptamine-requiring mutants of the appropriate organism, and an antibiotic related to streptomycin was formed by a streptidine-requiring mutant of *Streptomyces griseus* [51]. Mutants blocked beyond 2-deoxystreptamine, and thus requiring a larger fragment (e.g., neamine) for antibiotic production, were also isolated [59]. In some instances, differences between the mutasynthetic and natural antibiotic go beyond the expected substitution of mutasynthon for native subunit. For example, an idiotroph of *Streptomyces kanamyceticus,* when presented with 2-epistreptamine, formed 6'-deamino-6'-hydroxy-2-epi-hydroxykanamycin A instead of the expected 2-epi-hydroxykanamycin A [51].

Compared to the chemical modification of an antibiotic, mutasynthesis often leads to the desired product through a smaller number of steps and is therefore the preferred procedure, despite the fact that acceptance of an analog by the idiotroph and the structure of the resulting product are not predictable. However, since the overall yield by mutational biosynthesis is frequently low, it is not unusual for an antibiotic discovered by mutasynthesis to be subsequently prepared in a more convenient way by chemical means [8].

Among the antibiotics obtained by mutational biosynthesis, several related to sisomicin and gentamicin are of interest within the context of this review.

A. Mutasynthetic Antibiotics Related to Sisomicin

A 2-deoxystreptamine-requiring mutant of the sisomicin-producing organism *M. inyoensis* yielded a number of new sisomicin-related antibiotics from analogs of the missing subunit.

Addition of streptamine to the culture afforded Mu-1, the expected sisomicin analog, as the major component, along with Mu-1a and Mu-1b as minor components of an antibiotic complex [8] (Figure 5). In vitro, Mu-1 is less active than sisomicin, except against ANT(2'')-forming organisms, which are susceptible to this compound but not to sisomicin [62]. In a preliminary study, Mu-1 was found less nephro- and ototoxic than sisomicin in cats [8].

When grown in a medium supplemented with 2,5-dideoxystreptamine, the mutant formed two antibiotics: a minor product, Mu-2, which was the expected 5-deoxysisomicin (Figure 5), and an entirely unexpected major product, Mu-2a, identified as 5-deoxygentamicin A [8,62]. Mu-2 is generally less active in vitro than sisomicin, but the compound is effective against AAC(3)-producing strains [8]. Mu-2a is only weakly active. Lack of the 5-hydroxyl function does not result in lower toxicity. (These early mutasynthetic analogs of sisomicin were originally named mutamicins [62], but to avoid confusion with Mutamycin, the proprietary name of mitomycin C, they are now referred to as Mu [8]).

The most important mutasynthetic analog of sisomicin, 5-episisomicin (Mu-6), was

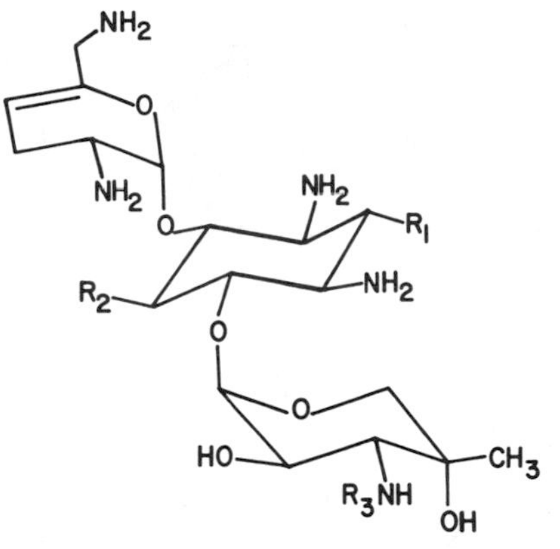

	R_1	R_2	R_3
2-Hydroxysisomicin (Mu-I)	-OH	-OH	-CH$_3$
Mu - Ia	-OH	-OH	-COCH$_3$
Mu - Ib	-OH	-OH	-H
5-Deoxysisomicin (Mu-2)	-H	-H	-CH$_3$
Sisomicin	-H	-OH	-CH$_3$

Figure 5   Structures of mutasynthetic antibiotics related to sisomicin.

obtained as essentially a single component from 2-deoxy-5-epistreptamine. This compound is under extensive investigation as a potentially marketable drug. Since 5-epi-sisomicin is currently made from sisomicin by chemical modification, its discussion is deferred to Section III.

B.   Mutasynthetic Antibiotics Related to Gentamicin

An ideotroph of a gentamicin-producing strain of *M. purpurea* was obtained by two sequential mutations, selecting first for 2-deoxystreptamine dependence, then for improved antibiotic production [52]. The idiotroph afforded the usual mixture of gentamicin C_{1a}, C_1, and C_2 from exogenous 2-deoxystreptamine. When the culture was supplemented with streptamine, a mixture of 2-hydroxygentamicin C_1 and C_2 with only a trace of C_{1a} was obtained (Figure 6). From 2,5-dideoxystreptamine, on the other hand, the mutant formed a complex consisting once again of three components: 5-deoxygentamicin C_1, C_2, and C_{1a} (Figure 6).

    In the main, the biological properties of these two mutasynthetic complexes parallel those of the corresponding sisomicin analogs. The 2-hydroxygentamicin complex, as well as its individual components, are slightly less active in vitro than are the respective gentamicins against Enterobacteriaceae, except for ANT(2")-forming organisms, which are more susceptible to the unnatural antibiotics [3a,10,52]. Against gentamicin-sensitive strains of *Pseudomonas*, 2-hydroxygentamicin C is fourfold less active than gentamicin C [3a]. The therapeutic efficacy of 2-hydroxygentamicin was assessed against experimenal

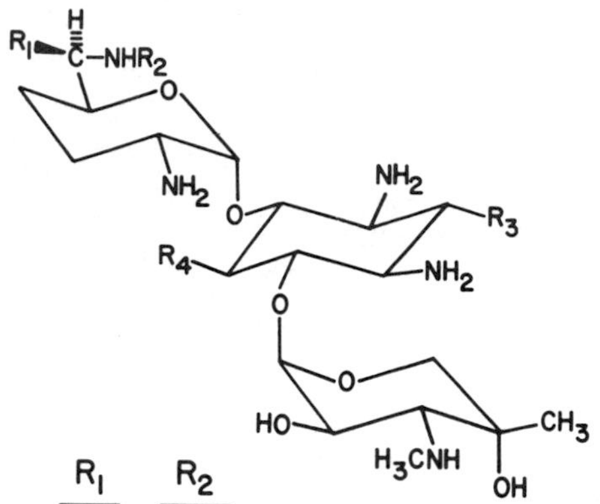

	R_1	R_2
C_{1a}	-H	-H
C_1	$-CH_3$	$-CH_3$
C_2	$-CH_3$	-H

	R_3	R_4
2-Hydroxygentamicin ($C_1 + C_2$)	-OH	-OH
5-Deoxygentamicin ($C_{1a} + C_1 + C_2$)	-H	-H
Gentamicin ($C_{1a} + C_1 + C_2$)	-H	-OH

Figure 6 Structures of mutasynthetic antibiotics related to gentamicin.

infections of mice. Overall, the compound was somewhat less effective than gentamicin [3a]. The acute toxicity of 2-hydroxygentamicin, determined after IV administration to mice, was only half that of gentamicin [52]. More relevant for the clinical potential of 2-hydroxygentamicin is the fact that its cochlear toxicity in the guinea pig, its vestibular and nephrotoxicity in the cat, and its nephrotoxicity in the rat are an estimated four to six times lower than those for gentamicin [3a,9].

5-Deoxygentamicin C resembles gentamicin C in activity. It was, however, more active against a few gentamicin-resistant strains: an AAC(3)-producing *E. coli,* and AAC(2′)-forming *Providencia,* and an AAC(6′)-forming *P. aeruginosa* [52]. In contrast to 2-hydroxygentamicin, 5-deoxygentamicin given IV to mice is more toxic than gentamicin by two- to threefold [52].

III. Aminoglycoside Antibiotics by Chemical Modification

Antibiotics described in this section were prepared either by chemical modification of a fermentation-derived antibiotic or by subsequent derivatization of a commercially available semisynthetic aminoglycoside. The main objective of these modifications was to obtain derivatives insensitive to aminoglycoside-modifying enzymes and therefore active against organisms resistant to many of the commercially available aminoglycosides. Secondary objectives were greater intrinsic activity and reduced oto- and nephrotoxicity for humans.

A. 5-Episisomicin (*Sch 22591*)

5-Episisomicin (Figure 7) was discovered by mutational biosynthesis (see Section II) but is now obtained from sisomicin by chemical modification [8]. According to Waitz et al. [66], the intrinsic activity of episisomicin is comparable to that of gentamicin against staphylococci and most of the Enterobacteriaceae, but the compound is two- to fourfold more active than gentamicin against *Providencia, Proteus rettgeri,* and *P. aeruginosa.* These results were confirmed by Fu and Neu [14] and Sanders et al. [53].

We have compared the intrinsic activity of episisomicin with that of a number of other aminoglycosides, some commercially available, others experimental. Data are summarized in Table 4. Overall, episisomicin resembles gentamicin and tobramycin in potency, but it is more active than either compound against *Serratia marcescens, Providencia stuartii,* and *P. rettgeri* and more active than gentamicin against *P. aeruginosa* and *S. aureus.*

Investigations by Waitz et al. [66], Fu and Neu [14], Kabins and Nathan [24], and Watanakunakorn [68] show that episisomicin is active against organisms that produce the following types of aminoglycoside-modifying enzymes: APH(3')-I, APH(3')-II, ANT(2''), AAC(2'), and AAC(3)-I. The activity of the compound is, however, significantly reduced against most organisms that produce AAC(6'), AAC(3)-II, and—according to our own data—AAC(3)-III (Table 5). The latter differs from AAC(3)-II in that it accepts only 4,6-diglycosylated-2-deoxystreptamines as substrates, whereas the type II enzyme recognizes also 4,5-diglycosylated deoxystreptamines, such as neomycin and paromomycin. Thus, epimerization of the 5-hydroxyl group of sisomicin yields significant advantages since the activity of the new molecule is affected by only three of the seven enzymes that modify sisomicin. Nevertheless, 5-episisomicin is not fully inert to modification by ANT(2''), AAC(2'), and AAC(3)-I. The activity of the drug against strains harboring these enzymes is due to the slow rate of enzymic modification at low antibiotic concen-

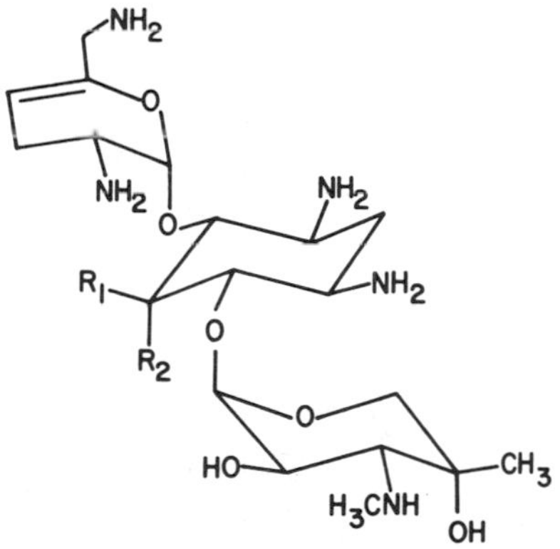

	R_1	R_2
5-Episisomicin	-H	-OH
Sisomicin	-OH	-H

Figure 7   Structures of 5-episisomicin and sisomicin.

Table 4 Antibiotic Spectrum of Various Aminoglycosides Against Organisms Lacking Known Aminoglycoside-Modifying Enzymes

Organism	Number of strains	MIC (μg/ml)[a]									
		Kana-mycin A	Tobra-mycin	Genta-micin C	Siso-micin	Netil-micin	5-Episiso-micin	1-N-HABA-dibekacin	Buti-kacin	Ami-kacin	1-N-HAPA-gentamicin B
Escherichia coli	10	4	0.7	0.6	0.4	0.5	0.4	1	3	2	0.8
Klebsiella pneumoniae	10	2.6	0.7	0.8	0.4	0.5	0.4	1.5	3.3	1.7	1.1
Enterobacter aerogenes	10	2.6	1.2	0.7	0.7	0.5	0.6	1.5	2.8	1.7	1.3
Enterobacter cloacae	10	3.5	1.6	1.4	0.9	0.4	0.5	2.3	3.5	2.5	2.3
Serratia marcescens	10	2.5	2.8	1.5	1.9	3	0.4	4.6	4.3	2	2.5
Proteus mirabilis	8	1.4	0.4	0.5	0.4	0.8	0.6	1.7	1.8	1.2	1.8
Proteus vulgaris	6	1.1	0.4	0.4	0.2	0.4	0.4	0.9	1.4	1	0.9
Proteus morganii	10	2	0.3	0.4	0.2	0.7	0.5	1.7	2.6	0.9	1.5
Proteus rettgeri	7	0.6	0.7	0.8	0.6	1	0.3	0.7	0.9	0.9	1.1
Providencia stuartii	10	1.1	3.5	3.7	2.6	5.7	0.2	1.6	1.2	1.2	1.5
Salmonella sp.	9	4	0.8	0.7	0.4	0.6	0.4	3.2	2.7	1.6	1.4
Pseudomonas aeruginosa	10	>32	0.4	2.3	0.9	7	0.9	3.5	4.3	3	4.9
Staphylococcus aureus	9	1.2	0.2	0.3	0.2	0.3	0.09	0.3	0.8	1.1	1.2
Staphylococcus epidermidis	4	0.7	0.06	0.08	0.05	0.06	0.03	0.08	0.3	0.5	0.4

[a]Mean values. Assay conditions as in Table 1.

trations as reflected in relatively high K_m values [65a]. Like aminoglycosides in general, episisomicin is ineffective against strains whose ability to accumulate the antibiotic is impaired.

Waitz et al. [66] have shown that the potent antibacterial activity of episisomicin is also demonstrable in vivo. When the compound was administered SC to mice infected experimentally with *P. aeruginosa*, the average PD_{50} (dose protecting 50% of the infected mice) obtained for episisomicin with 12 different strains was twofold lower than tobramycin's, fourfold lower than sisomicin's, and sixfold lower than gentamicin's. In similar studies conducted by Goering et al. [18], with 12 strains of *Pseudomonas,* the mean PD_{50} for episisomicin was comparable to those of tobramycin and sisomicin but significantly better than that of gentamicin.

The acute toxicity of episisomicin administered IV, SC, or intraperitoneally (IP) to mice is about the same as that of sisomicin and almost twice that of gentamicin [66]. The vestibular toxicity in cats and nephrotoxicity in rats of episisomicin and gentamicin are nearly identical [66]. The compound's absorption rate, peak serum concentration, and biological half-life in serum after IM administration to a dog appeared to be similar to those obtained with other aminoglycosides [66].

B.   1-N-[(S)-2-Hydroxy-3-aminopropionyl]gentamicin B *(Sch 21420)*

Gentamicin B is coproduced as minor component with the gentamicin C complex in fermentations of *M. purpurea* [67]. The synthesis of 1-N-[(S)-2-hydroxy-3-aminopropionyl]-gentamicin B (HAPA-gentamicin B) from gentamicin B was described by Nagabhushan et al. [34]. The structures of the two compounds are shown in Figure 8.

HAPA-gentamicin B was no doubt designed with amikacin as model. It is therefore not surprising that the biological properties of these structurally related agents are very similar. In several studies the activities of the two compounds in vitro against aminoglycoside-susceptible strains of staphylococci, Enterobacteriaceae, and *P. aeruginosa* were virtually the same. Only small differences were noted: amikacin was up to twofold more active than HAPA-gentamicin B against *P. aeruginosa* [33,40,75], and conversely, some strains of Enterobacteriaceae were more susceptible to HAPA-gentamicin B than to amikacin [15,22a,53,75]. We found comparable potencies (MIC differences $\leqslant$ twofold) for the two compounds against all species tested (Table 4).

Miller et al. [33] determined the activity of HAPA-gentamicin B against aminoglycoside-resistant bacteria. Like amikacin, this antibiotic was active against strains that produce APH(3')-I, APH(3')-II, ANT(2''), AAC(3)-I, AAC(3)-II, and AAC(2'), whereas the activity of both antibiotics was reduced against some organisms that produce AAC(6'). Other investigators, including ourselves, confirmed these results [25,40,69] (Table 5). We found in addition that HAPA-gentamicin B, like amikacin, appears to be a substrate for the adenylylating enzyme of staphylococci, ANT(4'). However, MIC values of amikacin for some AAC(6') producers are moderately higher than those of HAPA-gentamicin B.

Kabins and Nathan [25] and Jones et al. [22a] found, as had previously been reported for amikacin by Price et al. [46], that resistance to HAPA-gentamicin B is nearly always limited to organisms with a reduced ability to transport active antibiotic into the cell.

In mice experimentally infected with aminoglycoside-sensitive organisms, Miller et al. [33] demonstrated that gentamicin and tobramycin were generally more efficacious than HAPA-gentamicin B and amikacin. This superiority of gentamicin was confirmed in a

Table 5   Antibiotic Activity of Various Aminoglycosides Against Organisms Producing Aminoglycoside-Modifying Enzymes

Aminoglycoside-modifying enzyme	Organism	Number of strains	MIC (μg/ml)[a]									
			Kana-mycin A	Tobra-mycin	Genta-micin C	Siso-micin	Netil-micin	5-Epi-siso-micin	1-N-HABA-dibe-kacin	Buti-kacin	Ami-kacin	1-N-HAPA-genta-micin B
APH(3')-I	Coliforms[b]	6	>32	1.3	0.71	0.4	0.6	0.3	0.9	2.2	1.8	1
	Serratia marcescens	2	>32	2	2	0.7	2	0.3	1.4	4	2.8	1.4
	Proteus rettgeri	1	>32	1	4	1	4	0.3	1	1	1	0.5
APH(3')-II	Coliforms	3	>32	0.8	0.6	0.3	0.3	0.5	1.3	3.2	2	0.8
	Pseudomonas aeruginosa	1	>32	2	4	2	16	2	8	8	4	8
APH(3')-III	Staphylococcus aureus	1	>32	0.5	0.25	0.13	0.25	0.25	0.5	4	1	4
	Staphylococcus epidermidis	1	>32	0.13	0.06	0.032	0.063	0.016	0.13	0.5	0.5	0.5
AAC(2')	Proteus rettgeri	2	4	22.5	32	>23	>23	0.5	11.3	4	2.8	5.7
	Providencia stuartii	5	1.5	16	18.4	14	32	1.2	4.6	2.3	2.3	1.5
AAC(3)-I	Coliforms	4	3.4	0.7	32	16	0.8	0.5	1	2.3	2.3	0.8
	Pseudomonas aeruginosa	1	>32	2	>32	>32	16	1	4	8	4	8
AAC(3)-II	Coliforms	2	>32	>32	>32	>32	0.4	1.4	1	2	1	0.5
	Serratia marcescens	1	>32	>32	>32	>32	4	4	4	4	4	4
	Pseudomonas aeruginosa	3	>32	>32	>32	>32	12.7	>32	8	10	10	10
AAC(3)-III	Klebsiella pneumoniae	2	>32	>32	>32	>32	>32	5.7	2	1.4	4	2
	Serratia marcescens	2	>32	>32	>32	>32	11.3	>32	4	4	4	2.8
	Pseudomonas aeruginosa	1	>32	>32	>32	>32	>32	8	4	4	4	4

Enzyme	Species	n										
AAC(6')	*Escherichia coli*	1	8	4	0.5	0.3	4	0.3	2	32	2	2
	Enterobacter aerogenes	1	>32	32	2	32	>32	2	32	>32	32	8
	Serratia marcescens	3	16	>32	>32	>32	>32	8	>32	>32	8	2
	Serratia sp.	3	32	>32	2	32	>32	8	>32	>32	20	2.5
	Proteus rettgeri	1	>32	>32	>32	>32	>32	>32	16	>32	16	8
ANT(2'')	Coliforms	7	>32	23.8	26	10.8	0.5	0.7	0.8	1.8	1.6	0.7
	Serratia marcescens	3	>32	>32	>32	25.4	4	1.6	5	8	4	1.6
	Pseudomonas aeruginosa	5	>32	>32	>32	>32	9.2	1.7	5.3	3.5	4.6	6.1
ANT(4')	*Staphylococcus aureus*	2	>32	>32	0.35	0.3	0.5	0.4	1	32	4	8
	Staphylococcus epidermidis	2	>32	>32	0.13	0.09	0.13	0.06	0.4	32	5.7	16
ANT(4') + APH(3')	*Staphylococcus aureus*	1	>32	>32	0.5	0.25	0.5	0.5	2	>32	16	32
APH(3') + AAC(6')	*Staphylococcus epidermidis*	3	>32	20.2	>32	>32	4	5	0.9	>32	6.4	4.8
AAC(6') + APH(2'')	*Staphylococcus aureus*	5	>32	>32	>32	>32	4.6	6.1	1.2	>32	6.4	3
	Staphylococcus epidermidis	1	>32	32	>32	>32	8	8	2	>32	4	2
AAC(6') + APH(2'') + ANT(2'')	*Staphylococcus aureus*	3	>32	>32	32	>32	1.6	2	0.5	>32	5	3.2

[a] Mean values where applicable. Assay conditions as in Table 1.
[b] *E. coli, K. pneumoniae, and E. cloacae.*

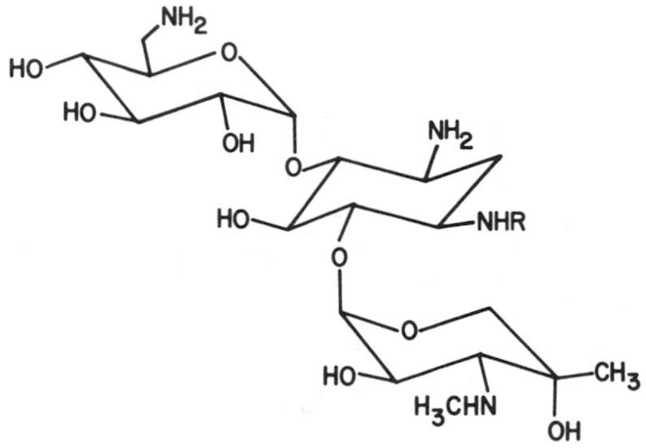

	R
I-N-HAPA-Gentamicin B	$-CO-CHOH-CH_2NH_2$ (S)
Gentamicin B	-H

Figure 8   Structures of 1-N-HAPA-gentamicin B and gentamicin B.

more recent study [17]. In the case of infections caused by resistant strains, however, amikacin and HAPA-gentamicin B were markedly more active than gentamicin and tobramycin [33]. HAPA-gentamicin B and amikacin were comparably effective in the treatment of infections caused by 24 of the 31 organisms used, but amikacin was appreciably more active (from three- to >sixtyfold) against infections induced by the other seven. Although six of the seven strains produced $ANT(2'')$, it cannot be concluded that the enzyme was responsible for the difference, since the two compounds were similarly effective in treating infections caused by 10 other organisms known to produce this enzyme.

The acute toxicity in mice of HAPA-gentamicin B was determined by Miller et al. [33]. When given IV the LD_{50} of HAPA-gentamicin B was 330 mg/kg compared to 75 mg/kg for gentamicin. Amikacin was not used in this study but has been shown by Price et al. [47] to have an IV LD_{50} in mice of 300 mg/kg versus 79 mg/kg for gentamicin. Thus, the acute IV LD_{50} values of HAPA-gentamicin B and amikacin are very similar. Furthermore, Miller et al. reported that the IP and SC LD_{50} values of HAPA-gentamicin B were at least 10 times higher than those of gentamicin. Serum- and urine-level determinations of HAPA-gentamicin B in mice and rats, after SC administration, and dogs, after IV administration, showed that this aminoglycoside is similar to all others in pharmacokinetic behavior [33].

The relative renal toxicity of HAPA-gentamicin B, gentamicin, and amikacin was also investigated by Miller and his associates [33]. The drugs were administered to rats IM once or twice daily for 7–14 days at doses well above the established or projected daily human dosage. Using glucosuria and changes in the glomerular filtration rate as indices of nephrotoxicity, they estimated that the potential of HAPA-gentamicin B for producing kidney damage is about one-fourth that of amikacin and one-twelfth that of gentamicin. Rankin et al. [49a], in a comparison of the functional and structural effects of HAPA-gentamicin B and amikacin on the kidneys of rats, confirmed the lower nephrotoxic

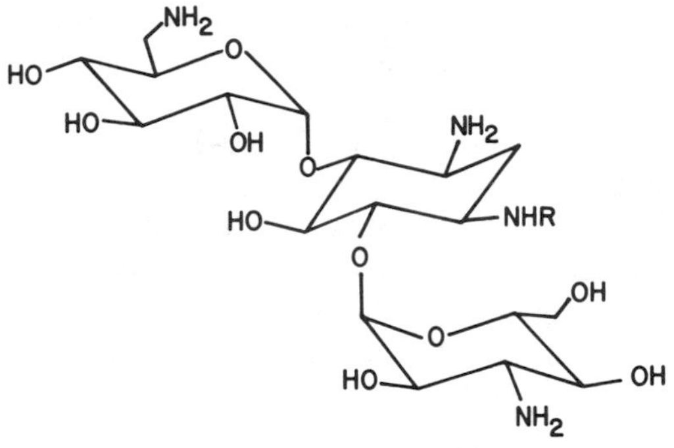

	R
Butikacin	$-CH_2-CHOH-CH_2-CH_2NH_2$ (S)
Amikacin	$-CO-CHOH-CH_2-CH_2NH_2$ (S)

Figure 9 Structures of butikacin and amikacin.

liability of HAPA-gentamicin B. Should these results be predictive of nephrotoxic potential for humans, HAPA-gentamicin B would have a greater therapeutic index than any of the presently available aminoglycosides.

C. Butikacin (UK 18892)

Richardson and associates [50] prepared 1-N-[(S)-2-hydroxy-4-aminobutyl]kanamycin A by diborane reduction of amikacin. The structure of this new compound, named butikacin, is shown in Figure 9.

Richardson et al. [50], Wise and Andrews [71], and Jevons et al. [22] reported that the activity of butikacin against aminoglycoside-susceptible strains of *S. aureus, P. aeruginosa,* and various Enterobacteriaceae is very similar to that of amikacin. Nevertheless, small differences were noted: according to Wise and Andrews, butikacin was nearly twice as active as amikacin against *P. aeruginosa* whereas, according to Jevons et al., amikacin was more active against a small number of staphylococcal strains. Our results confirm this similarity of butikacin and amikacin against susceptible strains, but we found that whenever differences do occur, they nearly always favor amikacin (Table 4).

Butikacin was active against the large majority of nearly 300 aminoglycoside-resistant organisms [22] and against selected strains known to produce AAC(3), ANT(2″), APH(3′)-I, or APH(3′)-II [50]. Further studies by others have shown that butikacin is not modified by AAC(3), AAC(2′), ANT(2″), and APH(3′)-I, but it is altered by AAC(6′) and ANT(4′) [2,71]. These are precisely the same two enzymes that modify amikacin. We have confirmed that of the known aminoglycoside-modifying enzymes, only AAC(6′) and ANT(4′) adversely affect the antibiotic activity of butikacin (Table 5). We noticed in addition that the reduction of activity caused by these enzymes is markedly more pronounced with butikacin than with amikacin.

Butikacin and amikacin were about equally effective in the treatment of mice experimentally infected with a variety of aminoglycoside-sensitive organisms [50]. Butikacin also showed excellent therapeutic efficacy in infections produced by bacterial strains highly resistant to kanamycin or gentamicin [2].

The acute toxicity of butikacin after IV administration to mice was 20–30% higher than that of amikacin (H. Kawaguchi, personal communication).

In a single-blind pharmacokinetic study of butikacin and amikacin with adult male volunteers there was no significant difference between the two compounds in any of the pharmacokinetic parameters examined [26]. No significant untoward reactions occurred with either drug, although butikacin reportedly caused less pain at the site of injection. When given IM to healthy adult males, at a daily dose of 10 mg/kg for 3 days, butikacin increased the urinary excretion of alanine aminopeptidase to a lesser extent than did amikacin [57a]. An elevated excretion of this enzyme indicates a lesion to the proximal tubules.

D. Propikacin (*UK 31214*)

The program of 1-N-alkyl modifications of the kanamycins, undertaken by Richardson and his associates [49c], yielded, besides butikacin, another derivative of interest, 1-N-(1,3-dihydroxy-2-propyl)kanamycin B, or propikacin (Figure 10).

Against most organisms that lack aminoglycoside-modifying enzymes, propikacin is only half as active as kanamycin B; however, against *P. aeruginosa,* it is four to eight times more active [49c]. Propikacin is effective against bacterial strains known to produce AAC(3), ANT(2″), APH(3′)-I, and APH(3′)-II, but not against those that elaborate AAC(6′) and AAC(2′) [49b,49c,71a]. Overall, propikacin resembles amikacin in anti-

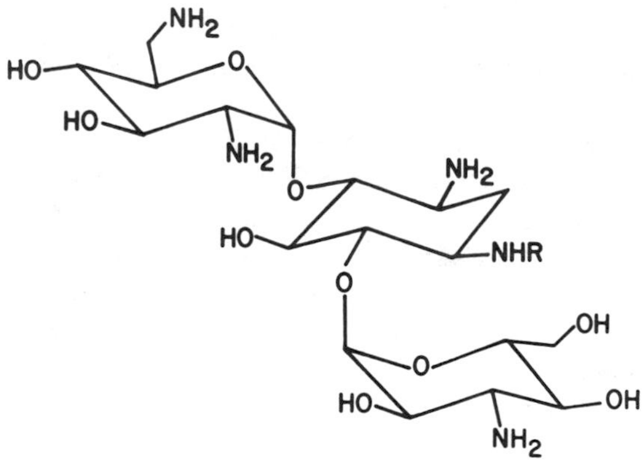

	R
Propikacin	$-CH(CH_2OH)_2$
Kanamycin B	$-H$

Figure 10   Structures of propikacin and kanamycin B.

bacterial activity and spectrum, except that amikacin is two to eight times more active against *S. marcescens, P. stuartii,* and *Proteus mirabilis,* and, in addition, is effective against AAC(2')-producing organisms [2b,71a].

The therapeutic efficacy of propikacin in experimentally infected mice is consistent with its activity in vitro. According to Brammer et al. [2b], kanamycin B is generally more effective than propikacin against experimental infections caused by organisms susceptible to kanamycin, but infections due to *P. aeruginosa* or to organisms that harbor AAC(3), ANT(2''), and APH(3') are refractory to kanamycin B and susceptible to treatment with propikacin.

In the dog and in healthy adult males, the pharmacokinetic properties of propikacin are comparable to those of amikacin [2b,70a]. In the newborn guinea pig, propikacin is less audiotoxic than amikacin or gentamicin [2a]. The nephrotoxicity of propikacin in healthy adults is similar to that of butikacin [57a].

E. 6'-N-Methylamikacin (BB-K 28) and 4'-Deoxy-
    6'-N-methylamikacin (BB-K-311)

Amikacin is inactivated by two aminoglycoside-modifying enzymes: AAC(6'), detected in strains of a variety of bacterial species, and ANT(4'), found to date only in staphylococci [49]. In an attempt to protect the molecule against enzymatic inactivation, one or both of the relevant functional groups of amikacin were modified to yield two derivatives: 6'-N-methylamikacin and 4'-deoxy-6'-N-methylamikacin, the first expected to be inert to AAC(6'), the second to both enzymes.

Methylamikacin (Figure 11), made independently in two laboratories [48,64], is in-

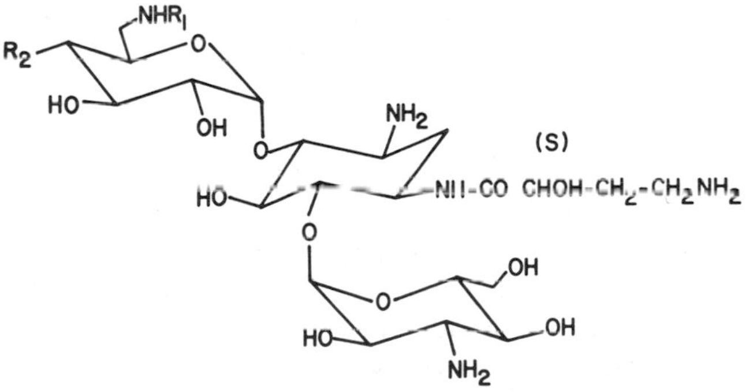

	R_1	R_2
6'-N-Methylamikacin	-CH$_3$	-OH
4'-Deoxy-6'-N-Methylamikacin	-CH$_3$	-H
Amikacin	-H	-OH

Figure 11   Structures of 6'-N-methylamikacin, 4'-deoxy-6'-N-methylamikacin, and amikacin.

Table 6   Antibiotic Spectrum of Methylamikacin and Amikacin in Mueller-Hinton Broth

Organism	Number of strains	Resistance mechanism	MIC[a] (μg/ml) Methyl-amikacin	Amikacin
Staphylococcus aureus	1	_[b]	1.0	0.5
Escherichia coli	2	—	2.4	1.0
Enterobacter cloacae	1	—	2.0	1.0
Klebsiella pneumoniae	1	—	0.5	0.5
Proteus rettgeri	1	—	0.13	0.25
Providencia stuartii	1	—	0.5	0.25
Serratia marcescens	1	—	2.0	0.5
Pseudomonas aeruginosa	2	—	1.0	0.5
Escherichia coli	1	APH(3′)	8.0	2.0
Enterobacter cloacae	2	APH(3′)	4.0	1.0
Proteus rettgeri	1	AAC(2′)	8.0	4.0
Escherichia coli	1	AAC(3)-I	8.0	4.0
Pseudomonas aeruginosa	2	AAC(3)-I + APH(3′)	1.0	0.5
Pseudomonas aeruginosa	1	AAC(3)-II	4.0	2.0
Escherichia coli	1	AAC(3)-III	4.0	1.0
Escherichia coli	1	AAC(6′)	4.0	32.0
Pseudomonas aeruginosa	1	AAC(6′)	0.5	4.0
Escherichia coli	1	ANT(2″)	4.0	2.0
Staphylococcus aureus	1	ANT(4′)	8.0	1.0
Staphylococcus aureus	1	ANT(4′)	>125.0	16.0
Staphylococcus aureus	1	ANT(4′) + APH(3′)	125.0	16.0
Pseudomonas aeruginosa	1	Transport mutant	63.0	16.0

[a]Mean values where applicable.

[b]—, no resistance mechanism present

deed active against AAC(6′)-producing strains of susceptible bacterial species [64] (Table 6), and radioenzymatic determinations confirmed that it is not affected by 6′-N-acetyltransferases [27a]. The compound is, as expected, vulnerable to ANT(4′) and therefore inactive against strains that form this enzyme. Methylation of the 6′-amine does, however, affect activity adversely, for intrinsically, methylamikacin is only one-half as active as amikacin against staphylococci and Enterobacteriaceae and one-third as active against *P. aeruginosa* (Table 6).

Deoxymethylamikacin (Figure 11) was synthesized by Naito et al. [35]. The antibiotic is markedly more active than amikacin against strains that harbor AAC(6′) and ANT(4′), although against amikacin-sensitive organisms, whether they are enzyme producers or not, the derivative is only one-fourth as active as the parent compound [35] (Table 7). By radioenzymatic assay it was confirmed that deoxymethylamikacin is not a substrate for *any* of the known aminoglycoside-modifying enzymes [27a].

No data on the biological properties of methylamikacin and deoxymethylamikacin in vivo are available.

Table 7   Antibiotic Spectrum of Deoxymethylamikacin and Amikacin in Mueller-Hinton
Broth

Organism	Number of strains	Resistance mechanism	MIC[a] (μg/ml) Deoxymethyl-amikacin	Amikacin
Staphylococcus aureus	1	$-^b$	1.0	0.5
Escherichia coli	2	–	8.0	1.0
Enterobacter cloacae	1	–	4.0	1.0
Klebsiella pneumoniae	1	–	1.0	0.5
Proteus rettgeri	1	–	1.0	0.5
Providencia stuartii	1	–	2.0	0.5
Serratia marcescens	1	–	4.0	1.0
Pseudomonas aeruginosa	2	–	4.0	0.5
Escherichia coli	1	APH(3′)	8.0	2.0
Enterobacter cloacae	2	APH(3′)	4.0	1.0
Proteus rettgeri	1	AAC(2′)	8.0	2.0
Escherichia coli	1	AAC(3)-I	8.0	2.0
Pseudomonas aeruginosa	1	AAC(3)-I + APH(3′)	16.0	2.0
Pseudomonas aeruginosa	1	AAC(3)-II	16.0	2.0
Escherichia coli	1	AAC(3)-III	8.0	1.0
Escherichia coli	1	AAC(6′)	8.0	32.0
Pseudomonas aeruginosa	1	AAC(6′)	16.0	8.0
Escherichia coli	1	ANT(2″)	4.0	2.0
Escherichia coli	1	ANT(2″) + APH(3′)	8.0	32.0
Staphylococcus aureus	1	ANT(4′)	8.0	2.0
Staphylococcus aureus	1	ANT(4′) + APH(3′)	8.0	16.0
Pseudomonas aeruginosa	1	Transport mutant	>63.0	16.0

[a]Mean values where applicable.
[b]–, no resistance mechanism present.

F. 1-N-[(S)-2-Hydroxy-4-aminobutyryl]dibekacin (HABA-dibekacin)

Dibekacin is a chemically modified aminoglycoside derivative prepared by reduction of
the 3′-and 4′-hydroxyl groups of kanamycin B [65]. This modification yielded a compound that has significant activity against P. aeruginosa strains and many other kanamycin-
resistant organisms. Because it is currently being marketed in Japan and several other
countries outside the United States, its properties will not be reviewed in this chapter on
developmental aminoglycosides. However, dibekacin can appropriately be considered a
"parent" compound since it has been subjected to additional derivatization in an effort
to further increase its resistance to attack by inactivating enzymes.

One such derivative, HABA-dibekacin, was prepared by acylation of dibekacin at the
1-N position with 2-hydroxy-4-aminobutyric acid using the active ester method [27].

Structures of dibekacin and its 1-N-acyl derivative, HABA-dibekacin, are shown in
Figure 12.

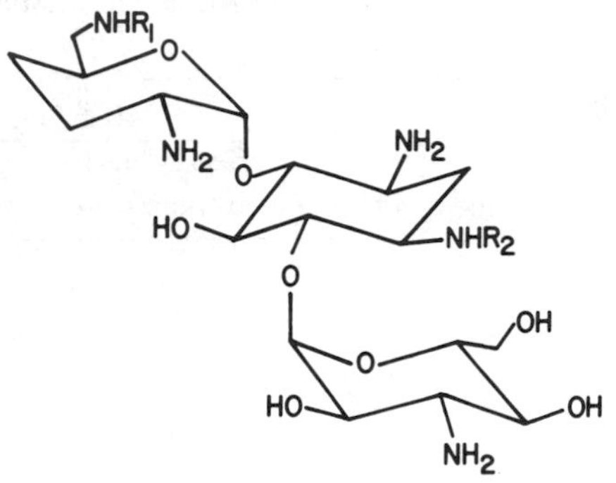

	R_1	R_2
l-N-HABA-Dibekacin	-H	$-CO-CHOH-CH_2-CH_2NH_2$ (S)
l-N-HABA-6'-N-Methyldibekacin	$-CH_3$	$-CO-CHOH-CH_2-CH_2NH_2$ (S)
Dibekacin	-H	-H

Figure 12   Structures of 1-N-HABA-dibekacin, 1-N-HABA-6'-N-methyldibekacin, and dibekacin.

Kondo et al. [27] demonstrated that HABA-dibekacin has a high level of activity against a representative spectrum of kanamycin-sensitive and -resistant Enterobacteriaceae and many strains of *P. aeruginosa*. Its level of intrinsic potency relative to that of a number of other aminoglycosides has been investigated by the present authors (Table 4).

The activity of HABA-dibekacin is very similar to that of amikacin, with two minor exceptions. It appears to be slightly less active than amikacin against *S. marcescens* but is somewhat more potent against staphylococci.

HABA-dibekacin was found to be active against strains that produce APH(3')-I, APH(3')-II, and ANT(2''), but ineffective against an AAC(6')-producing strain of *P. aeruginosa* [27]. Additional information regarding HABA-dibekacin's response to these enzymes, as well as to a number of others, has been obtained by the authors (Table 5).

Results show that like amikacin, whose activity is affected by two enzymes [AAC(6') and ANT(4')], HABA-dibekacin's activity is also affected by only two [AAC(2') and AAC(6')]. As is the case with butikacin, MIC values of HABA-dibekacin for organisms producing AAC(6') appear to be somewhat higher than those of amikacin.

No data relating to the in vivo properties of HABA-dibekacin are available at this time.

G.   1-N-[(S)-2-Hydroxy-4-aminobutyryl]-6'-N-methyldibekacin (HABA-methyldibekacin)

A derivative of dibekacin modified by acylation at the 1-N position and also by mono-methylation of the 6'-amine has been prepared at the Institute of Microbial Chemistry in Tokyo by Umezawa and co-workers [63]. As was the case with methylamikacin and

Table 8   Antibiotic Spectrum of HABA-methyldibekacin and Dibekacin in Mueller-Hinton Broth

Organism	Number of strains	Resistance mechanism	MIC[a] (μg/ml) HABA-methyl-dibekacin	Dibekacin
Staphylococcus aureus	1	—[b]	0.7	0.5
Escherichia coli	2	—	2.8	1.0
Enterobacter cloacae	1	—	2.0	1.0
Klebsiella pneumoniae	1	—	0.5	0.13
Proteus mirabilis	1	—	8.0	2.0
Proteus rettgeri	1	— —	1.0	0.25
Providencia stuartii	1	—	2.0	2.0
Serratia marcescens	1	—	2.0	4.0
Pseudomonas aeruginosa	2	—	1.0	0.25
Escherichia coli	2	APH(3′)	4.0	2.0
Enterobacter cloacae	2	APH(3′)	1.4	0.71
Escherichia coli	1	ANT(2″)	2.0	125.0
Escherichia coli	1	ANT(2″) + APH(3′)	4.0	125.0
Providencia stuartii	1	AAC(2′)	16.0	32.0
Proteus rettgeri	1	AAC(2′)	63.0	125.0
Escherichia coli	1	AAC(6′)	4.0	63.0
Pseudomonas aeruginosa	1	AAC(6′)	0.5	32.0
Escherichia coli	1	AAC(3)-I	4.0	2.0
Enterobacter cloacae	1	AAC(3)-I	2.0	0.5
Pseudomonas aeruginosa	1	AAC(3)-II	2.0	>125.0
Escherichia coli	1	AAC(3)-III	2.0	32.0
Pseudomonas aeruginosa	1	Transport mutant	>63.0	16.0

[a]Mean values where applicable.
[b]—, no resistance mechanism present.

deoxymethylamikacin, the 6⁺-amine of HABA-dibekacin was alkylated to block the attack of AAC(6′). The structures of dibekacin and its 1-N-acyl, 6′-N-methyl derivative (HABA-methyldibekacin) are shown in Figure 12. HABA-methyldibekacin demonstrated good activity against a broad spectrum of aminoglycoside-sensitive and -resistant enterobacteriaceae, *P. aeruginosa,* and various gram-positive organisms [63]. The effect of this dual modification on the in vitro antibacterial activity of dibekacin as determined by the present authors is shown in Table 8.

Data obtained with wild-type organisms that lack a resistance mechanism show that dibekacin's intrinsic potency is two- to fourfold greater than that of HABA-methyldibekacin. However, the extent of HABA-methyldibekacin's resistance to inactivation by enzymes is markedly greater than that of dibekacin. It would appear that HABA-methyldibekacin is refractory to the action of all the enzymes produced by the organisms tested, except for AAC(2′), an enzyme whose production is almost totally restricted to strains of *P. rettgeri* and *P. stuartii.* The activity of dibekacin, on the other hand, was affected by ANT(2″), AAC(2′), AAC(6′), AAC(3)-II, and AAC(3)-III. The effectiveness of both compounds was reduced in tests against a *P. aeruginosa* strain that has an impaired ability to transport and accumulate aminoglycosides intracellularly.

No published data on HABA-methyldibekacin's in vivo properties are available at the present time.

H.   5″-Amino-5″-deoxybutirosin A (AD-BTN)

This semisynthetic derivative of butirosin A was prepared by Culbertson et al. [5] of Parke-Davis and Co. Butirosins A and B are coproduced by antibiotic-producing strains of *Bacillus circulans* [21]. Structurally, these two isomers differ from each other in that butirosin A contains a 5-O-xylosyl and butirosin B, a 5-O-ribosyl moiety. Structures of the parent compound, butirosin A, and its aminodeoxy derivative, AD-BTN, are shown in Figure 13.

Culbertson and co-workers [5] showed in preliminary in vitro susceptibility tests that AD-BTN's intrinsic potency was about twice that of the parent compound. This initial finding was confirmed in part by Heifetz and co-workers [21]. As can be seen in Table 9, which is reproduced from their paper, the antibacterial activity (expressed as the geometric mean MIC) of AD-BTN was generally comparable to that of butirosin against most species of Enterobacteriaceae, but was significantly improved against strains of *P. aeruginosa* and *S. marcescens*.

Although the spectrum of antibacterial activity of AD-BTN was similar to that of gentamicin, its overall level of intrinsic activity was only one-half to one-fourth as great.

AD-BTN proved to be surprisingly effective against 82 gentamicin-resistant strains of *P. aeruginosa* [21]. At a concentration of 12.5 μg/ml, 80% of the strains were inhibited by AD-BTN as compared to 43% by butirosin and only 12% by gentamicin. Unfortunately, the authors provided no information as to the source of the strains or the nature of their mechanism(s) of resistance to gentamicin. Nevertheless, it is likely that virtually all of them produced one or more enzymes that can inactivate gentamicin

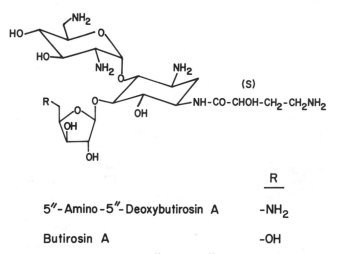

	R
5″-Amino-5″-Deoxybutirosin A	-NH₂
Butirosin A	-OH

Figure 13   Structures of 5″-amino-5″-deoxybutirosin A and butirosin A.

can inactivate gentamicin but which have no effect on the other two aminoglycosides. Butirosin (and probably AD-BTN as well) is known to be refractory to APH(3')-I, ANT(2''), AAC(3)-I, AAC(3)-II, and AAC(3)-III, but susceptible to APH(3')-II, AAC(2'), and AAC(6'). All the enzymes listed above, with the exception of APH(3')-I and APH(3')-II, can inactivate gentamicin [48].

In experimental infections of mice treated by the SC route, AD-BTN was found to be about one-third as effective as gentamicin and comparable in activity to butirosin when aminoglycoside-susceptible challenge strains of Enterobacteriaceae were used. Against infections caused by most strains of gentamicin- and butirosin-resistant Enterobacteriaceae, PD_{50} values were generally comparable for AD-BTN and gentamicin, but markedly higher for butirosin. AD-BTN's efficacy in infections caused by $P.$ $aeruginosa$ was about one-half that of gentamicin when the challenge organism was susceptible to both antibiotics, but generally superior if the infecting organism was resistant to gentamicin. In virtually all cases, AD-BTN's activity against $P.$ $aeruginosa$ was superior to that of butirosin [21].

The acute toxicities in mice of AD-BTN, butirosin, and gentamicin were determined by Heifetz et al. [21]. The results indicate that the introduction of an amino group into the 5''-position of butirosin causes a marked increase in toxicity. Acute IV LD_{50} values found for butirosin, AD-BTN, and gentamicin were 460, 126, and 50 mg/kg, respectively. Acute SC LD_{50} values were six- to eightfold higher than IV LD_{50} values for all three of the antibiotics. The enhanced toxicity of AD-BTN relative to butirosin is in agreement

Table 9 Comparative Antibacterial Spectra of Aminodeoxybutirosin, Butirosin, and Gentamicin in Vitro

Test organism	Number of susceptible[a] isolates	Geometric mean MIC[b] (μg/ml)		
		Aminodeoxy-butirosin	Butirosin	Gentamicin
Enterobacter cloacae	5	1.6	1.6	0.5
Escherichia coli	9	2.0	2.0	0.7
Klebsiella pneumoniae	10	0.9	1.1	0.8
Proteus mirabilis	2	6.2	12.5	1.6
Proteus, indole positive[c]	5	3.2	2.4	0.6
Pseudomonas aeruginosa	45	3.3	13.9	2.3
Salmonella spp.[d]	5	1.4	2.1	0.3
Serratia marcescens	6	3.5	8.9	1.1
Shigella spp.[e]	8	3.7	2.9	0.9
Staphylococcus aureus	2	0.6	1.1	0.1

[a]Selected for presentation because of inhibition by ≤25 μg of butirosin and ≤6.3 μg of gentamicin per milliliter.
[b]MIC, determined by agar dilution in MHA; compounds were handled in terms of base (assay) content.
[c]One isolate of $P.$ $morganii$ and two each of $P.$ $rettgeri$ and $P.$ $vulgaris$.
[d]Two isolates of $S.$ $enteritidis$ and one each of $S.$ $minnesota$, $S.$ $paratyphi$ A, and $S.$ $typhimurium$
[e]Two isolates of $S.$ $dysenteriae$, four of $S.$ $flexneri$, and three of $S.$ $sonnei$.

Source: Ref. 21. Used by permission.

with findings reported by Price et al. [47], who observed that all compounds in which an amine has been substituted for 6″-hydroxyl, in a C-6-O-(3-aminoglucose)-containing aminoglycoside or at the 5″-hydroxyl position in AD-BTN, have significantly increased acute toxicity. Although studies to determine AD-BTN's specific capability to cause oto- and nephrotoxicity have not yet been conducted, it is likely that these toxic liabilities of the 5″-amino derivative will be appreciably greater than those of its parent compound, butirosin.

IV. Conclusions

The antibiotics discussed in this chapter have been selected for a variety of reasons: novel chemical structure, advantages over a related major aminoglycoside, indications of intent to market, and potential for therapeutic advance.

Fortimicin A and, to a lesser extent, seldomycin 5 are aminoglycosides of a novel type. Although they do not appear to offer any advantage over drugs used currently, they are suitable for chemical manipulation.

Sagamicin is hardly distinguishable from gentamicin C except for lower oto- and nephrotoxicity in laboratory animals. Whether this will translate into an advantage of clinical significance remains to be seen. At any rate, the compound is in an advanced stage of development, as it has been in clinical trials in Japan for some time.

The 6′-C- and 6′-N-modified aminoglycosides—verdamicin, G-52, methylamikacin, deoxymethylamikacin, and HABA-methyldibekacin—have a modest advantage over the corresponding unmethylated compounds because of their broader protection against modifying enzymes, but often at the expense of intrinsic activity. Verdamicin was found to have a relatively low vestibular toxicity in cats. The mutasynthetic antibiotics 2-hydroxysisomicin and 2-hydroxygentamicin also display lower chronic toxicity—vestibular, cochlear, as well as renal—in animals compared to the parent compounds. However, there are no clear indications that any of these antibiotics are destined for clinical use.

Overall, butikacin, and propikacin appear to have no advantage over amikacin, and the advantage of AD-BTN over butirosin in terms of better activity and broader antibacterial spectrum may well be lost because of greater toxicity.

The discovery of 5-episisomicin appeared at first to provide a new tool for enhancing the activity of aminoglycosides against enzyme-producing resistant strains: the epimerization of the 5-hydroxyl group, which in itself is no target of known modifying enzymes. However, the effect of this alteration on activity and toxicity is not predictable. 5-Epikanamycin A and 5-epigentamicin B, for example, do not differ from their respective equatorial analogs in antibacterial activity and spectrum, whereas 1-N-HAPA-5-epigentamicin B, while displaying the same spectrum as HAPA-gentamicin B, is two to four times more active [5a,33a]. Conversely, 5-epikanamycin B is only one-fourth as active as kanamycin B [57c]. Furthermore, since 5-episisomicin resembles sisomicin in nephrotoxic potential, the considerably enhanced nephrotoxic liability of 1-N-HAPA-5-epigentamicin B and 1-N-HAPA-5-episisomicin in comparison with the corresponding equatorial analogs was unexpected [19a]. Thus, epimerization of the 5-hydroxyl group is not a broadly useful procedure for improving the properties of an aminoglycoside.

HABA-Dibekacin and HAPA-gentamicin B follow the by now classic design of amikacin with identical results: a considerably broadened protection against modifying enzymes. Gentamicin B was selected for chemical alteration because of its relatively low chronic toxicity in animals, an advantage that was maintained after 1-N acylation [33].

Hence, some of these compounds hold promise of a therapeutic advance: others appear to offer only an opportunity to gain a share of a lucrative market; still others are not likely to reach clinical trials but can serve as substrates for structural modifications that may lead to drugs with improved properties.

Acknowledgments

We are indebted to Theresa A. Pursiano for antibiotic activity determinations and to C. L. Heifetz and the American Society for Microbiology for permission to reproduce a table. We acknowledge the skillful clerical support of Maxine C. Postle and Carmen A. Cizek.

References

1. Akiyoshi, M., Yano, S., Shoji, T., Tajima, T., Imazawa, T., Hara, T., and Shimizu, M.: Animal test for evaluation of ototoxicity and safety of KW-1062. *Chemotherapy (Tokyo)* 25:1892–1914, 1977.
2. Andrews, R.J., Brammer, K.W., Cheeseman, H.E., and Jevons, S.: *UK 18892:* resistance to modification by aminoglycoside-inactivating enzymes. *Antimicrob. Agents Chemother.* 14:846–850, 1978.
2a. Baird, J.R.C., Carter, A.J., Cowley, D.M., and Henderson, J.T.: *UK 31214,* a new semisynthetic aminoglycoside: pharmacological profile. In *Current Chemotherapy and Infectious Disease. Proceedings of the 11th International Congress of Chemotherapy and the 19th Interscience Conference on Antimicrobial Agents and Chemotherapy,* Vol. 1, edited by Nelson, J.D., and Grassi, C. Washington, D.C., American Society for Microbiology, 1980, pp. 387–388.
2b. Brammer, K.W., Jevons, S., and Richardson, K.: *UK 31214,* a new semisynthetic aminoglycoside: activity against bacterial clinical isolates and pharmacokinetic properties. In *Current Chemotherapy and Infectious Disease. Proceedings of the 11th ICC and the 19th ICAAC,* Vol. 1, edited by Nelson, J.D., and Grassi, C. Washington, D.C., American Society for Microbiology, 1980, pp. 389–390.
3. Broughall, J.M., and Reeves, D.S.: Properties of the gentamicin acetyltransferase enzyme and application to the assay of aminoglycoside antibiotics. *Antimicrob. Agents Chemother.* 8:222–223, 1975.
3a. Came, P.E., O'Connor, J.R., Dobson, R.A., Wagner, R.B., and Fabian, R.J.: Antibacterial activities, nephrotoxicity, and ototoxicity of a new aminoglycoside, *Win 42122-2. Antimicrob. Agents Chemother.* 16:813–822, 1979.
4. Carney, R.E., McAlpine, J.B., Jackson, M., Stanaszek, R.S., Washburn, W.H., Cirovic, M., and Mueller, S.L.: Modification of seldomycin factor 5 at C-3'. *J. Antibiot. (Tokyo)* 31:441–450, 1978.
5. Culbertson, T.P., Watson, D.R., and Haskell, T.H.: 5"-Amino-5"-deoxybutirosin, a new semi-synthetic aminoglycoside antibiotic. *J. Antibiot. (Tokyo)* 26:790–793, 1973.
5a. Daniels, P.J.L., Cooper, A.B., McCombie, S.W., Nagabhushan, T.L., Rane, D.F., and Wright, J.J.: Some recent advances in the chemistry of antibiotics of the gentamicin series. *Jpn. J. Antibiot.* 32:S195–S204, 1979.
6. Daniels, P.J.L., Jaret, R.S., Nagabhushan, T.L., and Turner, W.N.: The structure of antibiotic G-52, a new aminocyclitol-aminoglycoside antibiotic produced by *Micromonospora zionensis. J. Antibiot. (Tokyo)* 29:488–491, 1976.

7. Daniels, P.J.L., Luce, C., Nagabhushan, T.L., Jaret, R.S., Schumacher, D., Reimann, H., and Ilavsky, J.: The gentamicin antibiotics: 6. Gentamicin C_{2b}, an aminoglycoside antibiotic produced by *Micromonospora purpurea mutant* JI-33. *J. Antibiot. (Tokyo)* 28:35–41, 1975.

8. Daniels, P.J.L., and Rane, D.F.: Synthetic and mutasynthetic antibiotics related to sisomicin. *Microbiology* (Wash., D.C.), 314–317, 1979.

9. Daum, S.J.: New gentamicin-type antibiotics produced by mutasynthesis. *Microbiology* (Wash., D.C.), 312–313, 1979.

10. Daum, S.J., Rosi, D., and Goss, W.A.: Mutational biosynthesis by idiotrophs of *Micromonospora purpurea:* II. Conversion of non-amino containing cyclitols to aminoglycoside antibiotics. *J. Antibiot. (Tokyo)* 30:98–105, 1977.

11. Davies, D.H., Mallams, A.K., Counelis, M., Loebenberg, D., Moss, E.L., Jr., and Waitz, J.A.: Semisynthetic aminoglycoside antibacterials: 6. Synthesis of sisomicin, antibiotic G-52, and novel 6'-substituted analogues of sisomicin from aminoglycoside 66-40C. *J. Med. Chem.* 21:189–193, 1978.

11a. Deushi, T., Iwasaki, A., Kamiya, K., Kunieda, T., Mizoguchi, T., Nakayama, M., Itoh, H., Mori, T., and Oda, T.: A new broad-spectrum aminoglycoside antibiotic complex, sporaricin: I. Fermentation, isolation and characterization. *J. Antibiot. (Tokyo)* 32:173–179, 1979.

11b. Deushi, T., Iwasaki, A., Kamiya, K., Mizoguchi, T., Nakayama, M., Itoh, H., and Mori, T.: New aminoglycoside antibiotics, sannamycin. *J. Antibiot. (Tokyo)* 32:1061–1065, 1979.

11c. Deushi, T., Nakayama, M., Watanabe, I., Mori, T., Naganawa, H., and Umezawa, H.: A new broad-spectrum aminoglycoside antibiotic complex, sporaricin: III. The structures of sporaricins A and B. *J. Antibiot. (Tokyo)* 32:187–192, 1979.

12. Egan, R.S., DeVault, R.L., Mueller, S.L., Levenberg, M.I., Sinclair, A.C., and Stanaszek, R.S.: A new antibiotic XK-62-2: III. The structure of XK-62-2, a new gentamicin C complex antibiotic. *J. Antibiot. (Tokyo)* 28:29–34, 1975.

12a. Egan, R.S., Stanaszek, R.S., Cirovic, M., Mueller, S.L., Martin, J.R., Collum, P., and Post, G.G.: The isolation and structure of fortimicin E. *J. Antibiot. (Tokyo),* in preparation.

13. Egan, R.S., Stanaszek, R.S., Cirovic, M., Mueller, S.L., Tadanier, J., Martin, J.R., Collum, P., Goldstein, A.W., De Vault, R.L., Sinclair, A.C., Fager, E.E., and Mitscher, L.A.: Fortimicins A and B, new aminoglycoside antibiotics: III. Structural identification. *J. Antibiot. (Tokyo)* 30:552–563, 1977.

14. Fu, K.P., and Neu, H.C.: Activity of 5-episisomicin compared with that of other aminoglycosides. *Antimicrob. Agents Chemother.* 14:194–200, 1978.

15. Gerlach, E., Barry, A., Fuchs, P., Jones, R., Gavan, T., Thornsberry, C., and Sommers, H.: Collaborative in vitro evaluation of *Sch 21420* compared with 5 other aminoglycosides. Abstr. 37, 18th Interscience Conference on Antimicrobial Agents and Chemotherapy, Atlanta, Ga., 1978.

16. Girolami, R.L., and Stamm, J.M.: Fortimicins A and B, new aminoglycoside antibiotics: IV. In vitro study of fortimicin A compared with other aminoglycosides. *J. Antibiot. (Tokyo)* 30:564–570, 1977.

17. Goering, R.V., Sanders, C.C., and Sanders, W.E., Jr.: In vivo analysis of structure-activity relationships among four aminoglycosides: gentamicin, netilmicin, 1-N HAPA gentamicin B, and amikacin. *Curr. Ther. Res.* 26:329–341, 1979.

18. Goering, R.V., Sanders, C.C., and Sanders, W.E., Jr.: Comparison of 5-episisomicin (*Sch 22591*), gentamicin, sisomicin, and tobramycin in treatment of experimental *Pseudomonas* infections in mice. *Antimicrob. Agents Chemother.* 14:824–828, 1978.

19. Goto, S., Ogawa, M., and Kaneko, Y.: Comparison of antibacterial activity of

KW-1062 with four aminoglycoside antibiotics in vitro and in vivo. *Chemotherapy (Tokyo)* 25:1808–1818, 1977.

19a. Hare, R.S., Schafer, T.W., Chiu, P.J., Sabatelli, F.J., Moss, E.L., Jr., and Miller, G.H.: Evaluation of new 1-N-substituted aminoglycosides against strains with known resistance mechanisms. In *Current Chemotherapy and Infectious Disease. Proceedings of the 11th ICC and the 19th ICAAC,* Vol. 1, edited by Nelson, J.D., and Grassi, C. Washington, D.C., American Society for Microbiology, 403–405, 1980.

20. Hashimoto, T., Ichikawa, S., and Kojimo, T.: The general pharmacological actions of KW-1062, a new aminoglycoside antibiotic. *Jpn. J. Antibiot.* 30:362–385, 1977.

21. Heifetz, C.L., Frey, C.A., Pearson, I.A., and Sesnie, J.C.: 5″-Amino-5″-deoxybutirosin, a semisynthetic analogue of butirosin A: antibacterial activity in vitro and in mice. *Antimicrob. Agents Chemother.* 10:543–548, 1976.

21a. Iida, T., Sato, M., Matsubara, I., Mori, Y., and Shirahata, K.: The structures of fortimicins C, D, and KE. *J. Antibiot. (Tokyo)* 32:1273–1279, 1979.

21b. Inouye, S., Ohba, K., Shomura, T., Kojima, M., Tsuruoka, T., Yoshida, J., Kato, N., Ito, M., Amano, S., Omoto, S., Ezaki, N., Ito, T., Niida, T., and Watanabe, K.: A novel aminoglycoside antibiotic, substance SF-2052. *J. Antibiot. (Tokyo)* 32:1354–1356, 1979.

21c. Inouye, S., Shomura, T., Ohba, K., Watanabe, H., Omoto, S., Tsuruoka, T., Kojima, M., and Niida, T.: Isolation and identification of N-formylfortimicin A. *J. Antibiot. (Tokyo)* 33:510–513, 1980.

21d. Iwasaki, A., Itoh, H., and Mori, T.: A new broad-spectrum aminoglycoside antibiotic complex, sporaricin: II. Taxonomic studies on the sporaricin producing strain *Saccharopolyspora hirsuta* subsp. *kobensis* nov. subsp. *J. Antibiot. (Tokyo)* 32:180–186, 1979.

22. Jevons, S., Cheeseman, H.E., and Brammer, K.W.: In vitro studies with *UK 18892,* a new aminoglycoside antibiotic. *Antimicrob. Agents Chemother.* 14:277–280, 1978. 1979.

22a. Jones, R.N., Barry, A.L., Fuchs, P.C., Gavan, T.L., Gerlach, E.H., Sommers, H., and Thornsberry, C.: 1-N-(S-3-Amino-2-hydroxypropionyl)gentamicin B (*Sch 21420*): a collaborative in vitro susceptibility comparison with amikacin and gentamicin against 12,984 clinical bacterial isolates. *Curr. Microbiol.* 1:359–364, 1978.

22b. Jones, R.N., Barry, A.L., Fuchs, P.C., Gavan, T.L., Sommers, H.M., and Gerlach, E.H.: Fortimicin A: collaborative in vitro susceptibility comparison with amikacin and gentamicin against 11,840 clinical bacterial isolates. *Antimicrob. Agents Chemother.* 16:823–828, 1979.

23. Jordan, G.W., and Hoeprich, P.D.: Susceptibility of three groups of *Staphylococcus aureus* to newer antimicrobial agents. *Antimicrob. Agents Chemother.* 11:7–12, 1977.

24. Kabins, S.A., and Nathan, C.: In vitro activity of 5-episisomicin in bacteria resistant to other aminoglycoside antibiotics. *Antimicrob. Agents Chemother.* 14:391–397, 1978.

25. Kabins, S.A., and Nathan, C.: In vitro activity in *Sch 21420,* derivative of gentamicin B, compared to that of amikacin. *Antimicrob. Agents Chemother.* 14:786–787, 1978.

26. Kendall, M.J., Wise, R., Andrews, J.M., and Bedford, K.A.: A pharmacological study of *UK 18892* and amikacin. *J. Antimicrob. Chemother.* 4:459–463, 1978.

26a. Kimura, E.T., Tekeli, S., Lewkowski, J.P., Majors, K.R., and Kesterson, J.W.: Acute and subacute toxicity studies with fortimicin A sulfate (*Abbott 44747*), a new aminoglycoside antibiotic. *Toxicol. Appl. Pharmacol.* 53:399–409, 1980.

26b. Kobayashi, F., Saino, Y., Koshi, T., and Hattori, Y.: In vitro and in vivo antimicrobial activities of sporaricin A, a new aminoglycoside. *Antimicrob. Agents Chemother.* 17:337–343, 1980.

27. Kondo, S., Iinuma, K., Yamamoto, H., Maeda, K., and Umezawa, H.: Syntheses of 1-N[(S)-4-amino-2-hydroxybutyryl]-kanamycin B and -3′,4′-dideoxykanamycin B active against kanamycin-resistant bacteria. *J. Antibiot. (Tokyo)* 26:412–415, 1973.

27a. Kresel, P.A., Pursiano, T.A., Price, K.E., Misiek, M., and Leitner, F.: BB-K311: a derivative of amikacin resistant to inactivation by all known aminoglycoside-modifying enzymes. In *Current Chemotherapy and Infectious Disease. Proceedings of the 11th ICC and the 19th ICAAC,* Vol. 1, edited by Nelson, J.D., and Grassi, C. Washington, D.C., American Society for Microbiology, 1980, pp. 413–415..

28. Marquez, J.A., Wagman, G.H., Testa, R.T., Waitz, J.A., and Weinstein, M.J.: A new broad spectrum aminoglycoside antibiotic, G-52, produced by *Micromonospora zionensis. J. Antibiot.* 29:483–487, 1976.

29. Matsushima, H., Kitaura, K., and Mori, Y.: Chemical transformation of seldomycin 5 into 3′-episeldomycin 5 and its antibacterial activity. *Bull. Chem. Soc. (Tokyo).* 50:3039–3042, 1977.

30. Matsushima, H., and Mori, Y.: 1-N(L-(-)-Alpha-hydroxy-gamma-aminobutyryl)-XK-88-5. Japanese Patent 78 90244, 1978.

31. Matsushima, H., Mori, Y., and Kitaura, K.: Synthesis of 3′-deoxyseldomycin 5. *J. Antibiot. (Tokyo)* 30:890-892, 1977.

32. McAlpine, J.B., Sinclair, A.C., Egan, R.S., DeVault, R.L., Stanaszek, R.S., Cirovic, M., Mueller, S.L., Goodley, P.C., Mauritz, R.J., Wideburg, N.E., Mitscher, L.A., Shirahata, K., Matsushima, H., Sato, S., and Iida, T.: A new aminoglycoside antibiotic complex—the seldomycins: IV. The structure of seldomycin factor 5. *J. Antibiot. (Tokyo)* 30:39–49, 1977.

33. Miller, G.H., Chiu, P.J.S., and Waitz, J.A.: Biological activity of *Sch 21420,* the 1-N-S-α-hydroxy-β-aminopropionyl derivative of gentamicin B. *J. Antibiot. (Tokyo)* 31:688–696, 1978.

33a. Nagabhushan, T.L., Cooper, A.B., and Miller, G.H.: Syntheses and biological properties of 5-epi-gentamicin B, 1-N-(S-3-amino-2-hydroxypropionyl)-5-epi-gentamicin B, and their kanamycin A analogs. In *Current Chemotherapy and Infectious Disease. Proceedings of the 11th ICC and the 19th ICAAC,* Vol. 1, edited by Nelson, J.D., and Grassi, C. Washington, D.C., American Society for Microbiology, 1980, pp. 411–413.

34. Nagabhushan, T.L., Cooper, A.B., Tsai, H., Daniels, P.J.L., and Miller, G.H.: The syntheses and biological properties of 1-N-(S-4-amino-2-hydroxybutyryl)-gentamicin B and 1-N-(S-3-amino-2-hydroxypropionyl)-gentamicin B. *J. Antibiot. (Tokyo)* 31:681–687, 1978.

35. Naito, T., Nakagawa, S., Toda, S., Fujisawa, K., and Kawaguchi, H.: Aminoglycoside antibiotics: XIII. Synthesis and activity of 4′-deoxy-6′-N-methylamikacin and related compounds. *J. Antibiot. (Tokyo)* 32:659–664, 1979.

36. Nakazawa, S., Sato, K., Yoshida, I., Saoshita, H., Nose, E., and Fukui, M.: Bacteriological evaluation of KW-1062, a new aminoglycoside antibiotic. *Chemotherapy (Tokyo)* 25:1828–1838, 1977.

37. Nara, T., Kawamoto, I., Okachi, R., Takasawa, S., Yamamoto, M., Sato, S., Sato, T., and Morikawa, A.: A new antibiotic XK-62-2 (sagamicin): II. Taxonomy of the producing organism, fermentative production and characterization of sagamicin. *J. Antibiot. (Tokyo)* 28:21–28, 1975.

38. Nara, T., Yamamoto, M., Kawamoto, I., Takayama, K., Okachi, R., Takasawa, S., Sato, T., and Sato, S.: Fortimicins A and B, new aminoglycoside antibiotics: I. Producing organism, fermentation and biological properties of fortimicins. *J. Antibiot. (Tokyo)* 30:533–540, 1977.

39. Nara, T., Yamamoto, M., Takasawa, S., Sato, S., Sato, T., Kawamoto, I., Okachi, R., Takahashi, I., and Morikawa, A.: A new aminoglycoside antibiotic complex—the seldomycins: I. Taxonomy, fermentation and antibacterial properties. *J. Antibiot. (Tokyo)* 30:17–24, 1977.

40. Neu, H.C., and Fu, K.P.: 1-N HAPA gentamicin B, a new aminoglycoside active against gentamicin resistant isolates—activity compared to other aminoglycosides. *J. Antibiot. (Tokyo)* 31:385–393, 1978.

40a. Ohashi, Y., Kawabe, H., Sato, K., Nakamura, N., Kurashige, S., and Mitsuhashi, S.: In vitro and in vivo antibacterial activity of KW-1070, a new aminoglycoside antibiotic. *Antimicrob. Agents Chemother.* 17:138–143, 1980.

41. Ohkoshi, M., Kawamura, N., Mashimo, K., Ishigami, J., and Miki, F.: Sagamicin: clinical study in 709 patients with bacterial diseases in Japan. In *Current Chemotherapy. Proceedings of the 10th International Congress of Chemotherapy*, Vol. 2, edited by Siegenthaler, W., and Lüthy, R. Washington, D.C., American Society for Microbiology, 1978, pp. 1038–1040.

42. Ohkoshi, M., Mashimo, K., Ishigami, J., and Miki, F.: Summary of basic and clinical studies on KW-1062 conducted in Japan. *Chemotherapy (Tokyo)* 25:1781–1800, 1977.

43. Okachi, R., Kawamoto, I., Takasawa, S., Yamamoto, M., Sato, S., Sato, T., and Nara, T.: A new antibiotic XK-62-2 (sagamicin). I. Isolation, physicochemical and antibacterial properties. *J. Antibiot (Tokyo)* 27:793–800, 1974.

44. Okachi, R., Takasawa, S., Sato, T., Sato, S., Yamamoto, M., Kawamoto, I., and Nara, T.: Fortimicins A and B, new aminoglycoside antibiotics: II. Isolation, physico-chemical and chromatographic properties. *J. Antibiot. (Tokyo)* 30:541–551, 1977.

44a. Okami, Y., Hotta, K., Yoshida, M., Ikeda, D., Kondo, S., and Umezawa, H.: New aminoglycoside antibiotics, istamycins A and B. *J. Antibiot. (Tokyo)* 32:964–966, 1979.

45. Phillips, I., Eykyn, S., King, B.A., Jenkins, C., Warren, C.A., and Shannon, K.P.: The in vitro antibacterial activity of nine aminoglycosides and spectinomycin on clinical isolates of common gram-negative bacteria. *J. Antimicrob. Chemother.* 3:403–410, 1977.

46. Price, K.E., DeFuria, M.D., and Pursiano, T.A.: Amikacin, an aminoglycoside with marked activity against antibiotic-resistant clinical isolates. *J. Infect. Dis.* 134 (Suppl.):S249–S261, 1976.

47. Price, K.E., Godfrey, J.C., and Kawaguchi, H.: Effect of structural modifications on the biological properties of aminoglycoside antibiotics containing 2-deoxystreptamine. In *Structure-Activity Relationships among the Semisynthetic Antibiotics*, edited by Perlman, D. New York, Academic Press, 1977, pp. 239–355.

48. Price, K.E., Godfrey, J.C., and Kawaguchi, H.: Effect of structural modifications on the biological properties of aminoglycoside antibiotics containing 2-deoxystreptamine (Supplement). In *Structure-Activity Relationships among the Semisynthetic Antibiotics*, edited by Perlman, D. New York, Academic Press, 1977, pp. 357–395.

49. Price, K.E., Casson, K., DeRegis, R.G., Kresel, P.A., Pursiano, T.A., and Leitner, F.: Amikacin: antimicrobial properties and resistance mechanisms affecting its activity. *Am. J. Med.* (Suppl. on U.S. Amikacin Symposium, Los Angeles, Calif., 1976), 14–24, 1977.

49a. Rankin, L.I., Luft, F.C., Yum, M.N., Sloan, R.S., Dinwiddie, C.B., Jr., and Isaacs, L.L.: Comparative nephrotoxicity of *Sch 21420* and amikacin in rats. In *Current Chemotherapy and Infectious Disease. Proceedings of the 11th ICC and the 19th ICAAC*, Vol. 1, edited by Nelson, J.D., and Grassi, C. Washington, D.C., American Society for Microbiology, 1980, pp. 589–590.

49b. Richardson, K., Brammer, K.W., and Jevons, S.: *UK 31214,* a new semisynthetic aminoglycoside: structure, synthesis, and in vitro evaluation. In *Current Chemotherapy and Infectious Disease. Proceedings of the 11th ICC and the 19th ICAAC,* Vol. 1, edited by Nelson, J.D., and Grassi, C. Washington, D.C., American Society for Microbiology, 1980, pp. 390–392.

49c. Richardson, K., Brammer, K.W., Jevons, S., Plews, R.M., and Wright, J.R.: Synthesis and antibacterial activity of 1-N-(1,3-dihydroxy-2-propyl) kanamycin B (*UK 31214*). *J. Antibiot. (Tokyo)* 32:973–977, 1979.

50. Richardson, K., Jevons, S., Moore, J.W., Ross, B.C., and Wright, J.R.: Synthesis and antibacterial activities of 1-N-[(S)-ω-amino-2-hydroxyalkyl]kanamycin A derivatives. *J. Antibiot. (Tokyo)* 30:843–846, 1977.

51. Rinehart, K.L., Jr.: Mutasynthesis of new antibiotics. *Pure Appl. Chem.* 49:1361–1384, 1977.

52. Rosi, D., Goss, W.A., and Daum, S.J.: Mutational biosynthesis by idiotrophs of *Micromonospora purpurea.* I. Conversion of aminocyclitols to new aminoglycoside antibiotics. *J. Antibiot. (Tokyo)* 30:88–97, 1977.

52a. Saito, A., Ueda, Y., and Akiyoshi, M.: Experimental studies on the ototoxicity and nephrotoxicity of fortimicin A. In *Current Chemotherapy and Infectious Disease. Proceedings of the 11th ICC and the 19th ICAAC,* Vol. 1, edited by Nelson, J.D., and Grassi, C. Washington, D.C., American Society for Microbiology, 1980, pp. 401–403.

53. Sanders, C.C., Sanders, W.E., Jr., and Goering, R.V.: In vitro studies with *Sch 21420* and *Sch 22591:* activity in comparison with six other aminoglycosides and synergy with penicillin against enterococci. *Antimicrob. Agents Chemother.* 14:178–184, 1978.

54. Sato, K., Araki, Y., Fukui, M., and Marumo, H.: Antibacterial activity of KW-1062, a new aminoglycoside antibiotic. *Chemotherapy (Tokyo)* 25:1844–1850, 1977.

55. Sato, K., and Marumo, H.: Absorption, distribution, excretion and metabolism of KW-1062. *Chemotherapy (Tokyo)* 25:1870–1874, 1977.

55a. Sato, M., and Mori, Y.: Chemical modification of fortimicins: preparation of 4-N-substituted fortimicin B. *J. Antibiot. (Tokyo)* 32:371–378, 1979.

56. Sato, S., Iida, T., Okachi, R., Shirahata, K., and Nara, T.: Enzymatic acetylation of fortimicin A and seldomycin factor 5 by aminoglycoside 3-acetyltransferase: I. [AAC(3)-I] of *E. coli* KY-8348. *J. Antibiot. (Tokyo)* 30:1025–1027, 1977.

57. Sato, S., Takasawa, S., Sato, T., Yamamoto, M., Okachi, R., Kawamoto, I., Iida, T., Morikawa, A., and Nara, T.: A new aminoglycoside antibiotic complex—the seldomycins: II. Isolation, physicochemical and chromatographic properties. *J. Antibiot. (Tokyo)* 30:25–30, 1977.

57a. Stefanescu, T., Mondorf, A.W., Franzke, T., and Scherberich, J.E.: Comparison of nephrotoxicity of two aminoglycosides, derivatives of kanamycin A and B, and amikacin. In *Current Chemotherapy and Infectious Disease. Proceedings of the 11th ICC and the 19th ICAAC,* Vol. 1, edited by Nelson, J.D., and Grassi, C. Washington, D.C., American Society for Microbiology, 1980, pp. 624–626.

57b. Suami, T.: Modifications of aminocyclitol antibiotics. *Jpn. J. Antibiot.* 32:S91–S102, 1979.

57c. Suami, T., and Nakamura, K.: Modification of aminocyclitol antibiotics: 7. Preparation of 5-epikanamycin B. *Bull. Chem. Soc. Jpn.* 52:955–956, 1979.

57d. Sugimoto, M., Ishii, S., Okachi, R., and Nara, T.: Fortimicins C, D and KE, new aminoglycoside antibiotics. *J. Antibiot. (Tokyo)* 32:868–873, 1979.

57e. Tadanier, J., Martin, J.R., Kurath, P., Goldstein, A.W., and Johnson, P.: 4-N-acylfortimicins B and the preparation of fortimicin A from fortimicin B. *Carbohydr. Res.* 79:91–102, 1980.

57f. Takahashi, K., Iida, T., Takazawa, S., Shimura, G., and Shirahata, K.: The structures of fortimicins. Abstr. Carb. 22, ACS/CSJ Chemical Congress, Honolulu, Hawaii, 1979.

58. Takeda, H., Kawashima, S., Wada, Z., Niwayama, M., Iwanaga, M., Kabasawa, T., Kinoshita, Y., Yamasaku, F., and Suzuki, Y.: Fundamental and clinical studies on KW-1062. *Chemotherapy (Tokyo)* 25:1943–1951, 1977.

59. Takeda, K., Okuno, S., Ohashi, Y., and Furumai, T.: Mutational biosynthesis of butirosin analogs: I. Conversion of neamine analogs into butirosin analogs by mutants of *Bacillus circulans. J. Antibiot. (Tokyo)* 31:1023–1030, 1978.

60. Tanaka, M., Deguchi, T., Ishii, A., Nakamura, N., Sato, K., Shimizu, M., and Hara, T.: Fundamental studies on sagamicin (KW-1062). In *Current Chemotherapy. Proceedings of the 10th International Congress of Chemotherapy*, Vol. 2, edited by Siegenthaler, W., and Lüthy, R. Washington, D.C., American Society for Microbiology, 1978, pp. 989–991.

61. Testa, R.T., and Tilley, B.C.: Biotransformation, a new approach to aminoglycoside biosynthesis: II. Gentamicin. *J. Antibiot. (Tokyo)* 29:140–146, 1976.

62. Testa, R.T., Wagman, G.H., Daniels, P.J.L., and Weinstein, M.J.: Mutamicins; biosynthetically created new sisomicin analogues. *J. Antibiot. (Tokyo)* 27:917–921, 1974.

63. Umezawa, H., Iinuma, K., Kondo, S., Hamada, M., and Maeda, K.: Synthesis of 1-N-acyl derivatives of 3',4'-dideoxy-6'-N-methylkanamycin B and their antibacterial activities. *J. Antibiot. (Tokyo)* 28:340–343, 1975.

64. Umezawa, H., Iinuma, K., Kondo, S., and Maeda, K.: Synthesis and antibacterial activity of 6'-N-alkyl derivatives of 1-N-[(S)-4-amino-2-hydroxybutyryl]-kanamycin. *J. Antibiot. (Tokyo)* 28:483–485, 1975.

65. Umezawa, H., Umezawa, S., Tsuchiya, T., and Okazaki, Y.: 3',4'-Dideoxy-kanamycin B active against kanamycin-resistant *Escherichia coli* and *Pseudomonas aeruginosa. J. Antibiot. (Tokyo)* 24:485–487, 1971.

65a. Vastola, A.P., Altschaefl, J., and Harford, S.: 5-Epi-sisomicin and 5-epi-gentamicin B: substrates for aminoglycoside-modifying enzymes that retain activity against aminoglycoside-resistant bacteria. *Antimicrob. Agents Chemother.* 17:798–802, 1980.

66. Waitz, J.A., Miller, G.H., Moss, E., Jr., and Chiu, P.J.S.: Chemotherapeutic evaluation of 5-episisomicin (*Sch 22591*), a new semisynthetic aminoglycoside. *Antimicrob. Agents Chemother.* 13:41–48, 1978.

67. Waitz, J.A., Moss, E.L., Jr., Oden, E.M., Wagman, G.H., and Weinstein, M.J.: Biological activity of *Sch 14342*, an aminoglycoside antibiotic coproduced in the gentamicin fermentation. *Antimicrob. Agents Chemother.* 2:464–469, 1972.

67a. Watanabe, I., Deushi, T., Yamaguchi, T., Kamiya, K., Nakayama, M., and Mori, T.: The structural elucidation of aminoglycoside antibiotics, sannamycins A and B. *J. Antibiot. (Tokyo)* 32:1066–1068, 1979.

68. Watanakunakorn, C.: Comparative in vitro activity of 5-epi-sisomicin and five other aminoglycosides. *J. Antimicrob. Chemother.* 4:474–477, 1978.

69. Watanakunakorn, C.: Comparative in vitro activity of a semisynthetic derivative of gentamicin B (*Sch 21420*) and five other aminoglycosides. *J. Antibiot. (Tokyo)* 31:1063–1064, 1978.

70. Weinstein, M.J., Wagman, G.H., Marquez, J.A., Testa, R.T., and Waitz, J.A.: Verdamicin, a new broad spectrum aminoglycoside antibiotic. *Antimicrob. Agents Chemother.* 7:246–249, 1975.

70a. Wilson, P., and Cookson, B.D.: Pharmacokinetics and toxicological screening of intramuscular and intravenous *UK 31214* in human volunteers. In *Current Chemotherapy and Infectious Disease. Proceedings of the 11th ICC and the 19th ICAAC,*

Vol. 1, edited by Nelson, J.D., and Grassi, C. Washington, D.C., American Society for Microbiology, 1980, pp. 392–393.

71. Wise, R., and Andrews, J.M.: *UK 18892,* a new aminoglycoside: an in vitro study. *Antimicrob. Agents Chemother.* 14:228–233, 1978.

71a. Wise, R., Andrews, J.M., and Bedford, K.A.: *UK 31214,* a new aminoglycoside and derivative of kanamycin B. *Antimicrob. Agents Chemother.* 17:298–301, 1980.

72. Wright, J.J., Daniels, P.J.L., Mallams, A.K., and Nagabhushan, T.L.: 1-N-Alkyl-aminoglycoside-XK-88 derivatives and methods for their manufacture. U.S. Patent 4,002,608, 1977.

73. Yamamoto, M., Okachi, R., Kawamoto, I., and Nara, T.: Fortimicin A production by *Micromonospora olivoasterospora* in a chemically defined medium. *J. Antibiot. (Tokyo)* 30:1064–1072, 1977.

73a. Yang, C.L., Buratto, B., Lehrer, S.B., Heyman, I.A., and Emerson, J.L.: Vestibular toxicity study with fortimicin A sulfate (*Abbott 44747*) in cats. In *Current Chemotherapy and Infectious Disease. Proceedings of the 11th ICC and the 19th ICAAC,* Vol. 1, edited by Nelson, J.D., and Grassi, C. Washington, D.C., American Society for Microbiology, 1980, pp. 395–397.

74. Yu, P.K.W., and Washington, J.A., II: Comparison of in vitro antibacterial activities of gentamicin and verdamicin. *Antimicrob. Agents Chemother.* 6:526–528, 1974.

75. Yu, P.K.W., and Washington, J.A., II: In vitro evaluation of a semisynthetic derivative of gentamicin B (*Sch 21420*). *Antimicrob. Agents Chemother.* 13:891–892, 1978.

76. (Authors not given.) 6'-N-Methyl-XK-88-5 en zijn bereiding. Dutch Patent 7708324, 1978.

3

CLINICAL MICROBIOLOGY AND THE IN VITRO ACTIVITY OF AMINOGLYCOSIDES

ROBERT C. MOELLERING, JR. Harvard Medical
School and Massachusetts General Hospital,
Boston, Massachusetts

There are significant differences in the spectra of activity of the various aminoglycosidic aminocyclitols. The effectiveness of an aminoglycosidic aminocyclitol against a given strain of bacterium is dependent upon the intrinsic activity of the drug and the resistance mechanisms possessed by the organism. Thus, to understand the antimicrobial spectra of the aminoglycosidic aminocyclitols, it is necessary to review their mechanism of action and the mechanism by which bacteria are resistant or develop resistance to them. This chapter considers these aspects of aminoglycoside activity when used alone and when employed in combination with other antimicrobials. This is followed by a review of a number of the considerations that are important in susceptibility testing and by an examination of the differences in in vitro effectiveness of the various compounds alone and in combination with other agents. The discussion in this chapter is limited to aminoglycosides that are presently in common clinical use in the United States and other parts of the world, omitting from major consideration the investigational compounds treated in Chapter 2.

I. Mechanism of Action

The aminoglycosidic aminocyclitols are bactericidal agents [171]. The largest portion of the efforts devoted to discovering the mechanism of the bactericidal action of the aminoglycosidic aminocyclitol antibiotics has been expended on streptomycin, which has been studied extensively since the early 1950s. The initial work concentrated upon a number of the physiological effects caused by streptomycin. Thus, it was shown that when growing bacteria were exposed to streptomycin, there was an almost immediate uptake of the drug, followed by an efflux of potassium from the bacterial cell [37]. A transient stimulation of RNA synthesis was followed by inhibition of protein synthesis and loss of viability [37]. Following additional exposure, increased streptomycin was bound to the cells, and there was impairment of respiration, inhibition of RNA and DNA synthesis, breakdown of RNA, and excretion of nucleotides from the cells [37,76]. Although early work suggested that streptomycin produced direct damage to the cell membrane as an explanation for some of the aforementioned phenomena [2], it soon became clear that the primary target of the aminoglycosidic aminocyclitols was the bacterial ribosome and that all of the other observed phenomena were in some way secondary to interaction with the ribosome [176,202]. It has been clearly shown that the 30S subunit of the ribosome is the site of aminoglycoside attachment [27]. Moreover, it has been shown that a single protein (designated P_{10} on the basis of its mobility on polyacrylamide gel electrophoresis) determines the streptomycin sensitivity of the 30S subunit [144]. Ribosomes from streptomycin-resistant mutants, which lack this protein or which have an altered P_{10} protein, fail to bind radiolabeled dihydrostreptomycin and are resistant to the in vitro effects of streptomycin and dihydrostreptomycin [144]. Once attached to the ribosome, streptomycin produces a number of important effects. Luzzatto et al. have suggested that streptomycin irrevocably blocks protein synthesis at the stage of initiation [104]. They base this on studies which showed that in bacteria exposed to streptomycin, there is a disappearance of polysomes with the accumulation of 70S monosomes (70S ribosomes attached to mRNA) which are thought to represent blocked initiation complexes. Modolell and Davis as well as Wallace and Davis have made additional observations concerning the effect of streptomycin on ribosomal protein synthesis [118,119,191]. These investigators have shown that streptomycin can block protein synthesis by inhibiting polypeptide chain elongation. In their model, streptomycin binds to the 30S ribosomal subunit and distorts the A site, resulting in impairment of its effective binding of both aminoacyl-tRNA and peptidyl-tRNA. It appears that aminoglycosides may thus inhibit the movement of peptidyl-tRNA associated with translocation by interfering with its exit from the ribosomal acceptor site [18]. This, in turn, may inhibit polypeptide chain elongation [18]. Streptomycin-bound ribosomes are not permanently frozen on mRNA as initiation complexes but can slowly dissociate from mRNA as 70S ribosomes. Indeed, the distortion of their A site should make this more likely to occur. Streptomycin-bound 70S ribosomes have an impaired ability to dissociate into 50S and 30S subunits. Aminoglycoside-bound 30S particles can form initiation complexes, but the end result is a cycle of initiation, blocked polypeptide chain extension, gradual release of ribosomes, and reinitiation.

Another consequence of the interaction of streptomycin with the ribosome is increased frequency of misreading of the genetic code due to incorrect codon-anticodon interaction [31]. This results in the formation of "nonsense" proteins; it can also manifest itself in vivo as "phenotypic suppression," which occurs when the aminoglycoside masks the phenotypic expression of certain mutations. The aminoglycosides vary in their ability to cause misreading [31], but this does not appear to correlate with their ability to kill bacteria. That property is probably related to the fact that binding of the aminoglycosides to ribosomes occurs with such great affinity that it is essentially irreversible

[192]. Thus, the bacterial cell is unable to make the proteins required for its vital processes after it has been exposed to lethal concentrations of the aminoglycosidic aminocyclitols.

The studies described above dealt primarily with streptomycin. It appears, however, that most of the concepts derived from these experiments may be applied to the other aminoglycosidic aminocyclitols (including the 2-deoxystreptamine-containing compounds). There are some differences with these agents, however. It is clear that ribosomal mutations involving the P_{10} protein do not result in resistance to the 2-deoxystreptamine-containing aminoglycosides. This is related, at least in part, to the fact that agents such as gentamicin, kanamycin, and tobramycin bind to multiple sites on the ribosome, including sites on both subunits [18]. As a consequence, these drugs can affect other steps in the translation process. They are also capable of producing more extensive misreading than is streptomycin. The end result, however, is the same after exposure of susceptible bacteria under appropriate conditions to any of these agents: irreversible binding of the aminoglycosidic aminocyclitol to the ribosome and cell death.

One important aspect of the bactericidal effect of aminoglycosides has recently been addressed by McDonald et al. [110]. Unlike many other antimicrobials that produce prolonged suppression of growth in susceptible organisms exposed to them, gentamicin, at least, produces little postexposure inhibition of growth among *Staphylococcus aureus*. Thus, those cells that are not immediately killed on brief exposure to gentamicin rapidly begin to grow when removed from inhibitory concentrations of the drug. The clinical significance and the mechanism of these differences in antimicrobial effectiveness and activity of various drugs remains to be determined.

Certain structural requirements for antimicrobial activity of the aminoglycosidic aminocyclitols have been determined by Davies and co-workers through degradation or chemical modification of these drugs. For streptomycin, optimal activity requires a free 3'-OH on the N-methyl-L-glucosamine residue and a 1-amidino group on the streptidine residue [29]. For the neomycins, cleavage of the molecule shows that the neamine residue is the site of activity. The number and location of amino groups on the sugars attached to the 2-deoxystreptamine ring have a profound effect on the biological activity of the aminoglycosidic aminocyclitols containing 2-deoxystreptamine. The presence of at least one unblocked amino group is needed for biological activity [11]. The addition of other substituents to the deoxyamino sugar combination can further increase the activity of these compounds. It appears that the amino groups are necessary in order for the aminoglycosides to bind to bacterial ribosomes.

To reach the ribosomes, the aminoglycosidic aminocyclitols must traverse the outer envelope and the cytoplasmic membrane of the bacterial cell. The process by which this occurs is the subject of Chapter 4; therefore, it is not reviewed in detail here. It is of relevance to note, however, that it appears that there are two energy-dependent phases of streptomycin uptake in *Escherichia coli* [70]. The second phase of uptake appears to be induced by interaction of the aminoglycosides with the ribosome, which results in a general induction of the polyamine transport system [69]. This autoinduction does not occur when *E. coli* that are ribosomally resistant to streptomycin are exposed to the drug [33], nor does it occur in bacteria that produce aminoglycoside-modifying enzymes because the modified aminoglycosides are not capable of reacting with the ribosome to produce autoinduction [70]. The actual uptake of aminoglycosides is an energy-dependent process that requires oxygen [14]. It does not occur under anaerobic conditions. As a result, facultative organisms are resistant to aminoglycosides under anaerobic conditions [187].

Streptococci, including the enterococci, are generally resistant to physiologic concentrations of the aminoglycosidic aminocyclitols. However, combinations of penicillins with these agents results in synergistic in vitro (and in vivo) killing [72]. The mechanism of this synergistic interaction is related to the fact that the cell wall of streptococci appears to act as a barrier to intracellular uptake of aminoglycosides. Agents that impair cell wall synthesis, including cycloserine, bacitracin, vancomycin, and the various penicillins and cephalosporins produce in vitro synergism when combined with aminoglycosides against enterococci and other streptococci [130,199]. Studies with radiolabeled streptomycin [127] and gentamicin [126] clearly demonstrate that exposure of enterococci to agents that interfere with cell wall synthesis results in enhanced intracellular uptake of aminoglycoside. The augmented bactericidal activity that ensues is almost certainly related to this interaction. Except for a study of penicillin-streptomycin synergism against *E. coli* [151], similar studies are not available for combinations of various penicillins and cephalosporins against staphylococci or gram-negative bacilli. Nonetheless, it is most likely that combinations of cell wall-active agents and aminoglycosides produce synergism against these organisms by a similar mechanism.

II. Mechanisms of Resistance

An understanding of the various mechanisms by which bacteria are or become resistant to the aminoglycosidic aminocyclitols is essential if one is to understand the basis for their differing spectra of activity. There are three types of resistance that account for the ability of certain strains of bacteria to grow in the presence of usually lethal concentrations of the aminoglycosidic aminocyclitols. These include (1) alteration of the target site (ribosomal resistance), (2) alteration in permeability, and (3) enzymatic modification [135]. The latter two mechanisms are discussed in detail in Chapters 4 and 5; thus, only a relatively brief exposition will be given here.

Ribosomal resistance to streptomycin occurs by single-step mutation at rates varying from 10^{-8} to 10^{-10} among gram-negative bacilli and enterococci exposed to this aminoglycoside in vitro [30,211]. Such mutants exhibit high levels of resistance to streptomycin [minimum inhibitory concentration (MIC) > 100 $\mu g/ml$] and are not cross-resistant to the 2-deoxystreptamine-containing aminoglycosides. As noted earlier, recent studies have demonstrated that this type of resistance is due to an alteration in a specific protein (P_{10}) of the 30S subunit of the ribosome [144]. This mutation results in decreased ribosomal binding of streptomycin with accompanying phenotypic resistance [80]. This type of alteration may account for resistance to penicillin-streptomycin synergism against some strains of enterococci [211], but it has rarely been demonstrated among clinical isolates of bacteria. Thus, at the present time it remains largely a laboratory curiosity. The same statement may be made for mutational resistance to kanamycin [30]. Although laboratory mutants ribosomally resistant to kanamycin have been derived [107], such organisms have not been found among clinical isolates [154].

Exposure of bacteria to gentamicin in vitro does not result in the development of single-step mutants with high-level resistance. Instead, one sees the development of stepwise resistance as the organisms are exposed to progressively higher concentrations of gentamicin [201]. These mutants have slower growth rates, lack virulence, have specific nutritional requirements, revert rapidly to sensitivity, and are cross-resistant to other aminoglycosides [71,201]. The molecular basis for this resistance is unknown, but it is most likely that it is due to changes in permeability. It seems much less likely that it is due to ribosomal resistance. Indeed, ribosomal resistance to gentamicin appears to be very infrequent. A recent report describes mutants of *E. coli* which were resistant to gentamicin and in which there was an altered ribosomal protein (L6). The significance of this observation is not clear, however, because these organisms also exhibited altered intracellular transport of gentamicin [16]. There is a single report of the clinical isolation of gentamicin-resistant strains of *Pseudomonas aeruginosa* (which emerged in a burn unit using topical gentamicin) that exhibited ribosomal resistance to this aminoglycoside [178]. However, the significance of this observation is somewhat doubtful because these organisms were subsequently shown to produce gentamicin-modifying enzymes [154]. Thus, experience to date suggests that ribosomal resistance to gentamicin (and other 2-deoxy-streptamine-containing aminoglycosides) is rarely an important cause of aminoglycoside resistance in a clinical setting.

Decreased intracellular uptake of aminoglycosidic aminocyclitols (see Chapter 4) appears to be more frequent than ribosomal alteration as a cause of resistance to aminoglycosides among clinical isolates of bacteria. Indeed, as has already been discussed, intracellular transport of these drugs is markedly impaired in the absence of oxygen [14]. Thus, all anaerobes are resistant to aminoglycosides (even though they contain ribosomes that are sensitive to aminoglycosides in vitro [13a]) and most do not possess the mechanisms to produce aminoglycoside-modifying enzymes. The latter statement may require further modification in the future, since very recent studies have demonstrated the presence of plasmids coding for the production of aminoglycoside phosphotransferase [APH(3')-I] in *Bacteroides ochraceus* [1a]. Facultative organisms also show increased resistance when grown under anaerobic conditions [187].

Streptococci are resistant to clinically achievable concentrations of aminoglycosides. In the case of the enterococci, it has been clearly shown that this resistance is due to decreased intracellular transport. It appears that the cell wall acts as a barrier to the entry of the aminoglycoside since any cell wall-active agent is capable of enhancing aminoglycoside uptake (and producing synergistic killing) in enterococci [126,127]. Streptococci other than enterococci have not been specifically studied, but it appears likely that their intrinsic resistance to aminoglycosides is due to a similar mechanism.

There have recently been a series of reports of the emergence of small-colony variants of certain organisms exposed to gentamicin in vitro or in vivo [99,112,116,137]. Whereas some of these appear to be related to the avirulent mutants described by Weinstein and co-workers [201], others have been shown to be virulent in experimental animals [137] and in humans [126]. Small-colony gentamicin-resistant variants of *S. aureus* seem to arise with surprising frequency when gentamicin is used as the sole agent to treat staphylococcal infections [112,116]. These organisms often differ from their parent strains in a number of characteristics, including colonial morphology, physiological properties, and antimicrobial susceptibility [117]. Miller et al. have shown that the mechanism of resistance in these organisms is related to decreased intracellular uptake of gentamicin [116]. A similar phenomenon has recently been seen in a strain of enterococcus isolated from a patient with enterococcal endocarditis. This organism, which grew as a small colony variant, was resistant to penicillin-gentamicin, but not penicillin-tobramycin synergism, and exhibited defective intracellular uptake of gentamicin, but not tobramycin, in the presence of penicillin [126].

In addition to the foregoing strains which are resistant to aminoglycosides and are phenotypically abnormal, it appears that phenotypically normal cells may also exhibit aminoglycoside resistance on the basis of abnormal intracellular transport [13,152]. In many of these organisms there is a broad spectrum of aminoglycoside resistance, which includes all of the clinically useful agents [13,152]. The frequency with which such organisms are found in nature is presently unknown, but it is very likely that they account for a large number of the organisms that are resistant to amikacin [152]. The biochemical basis for the defect in aminoglycoside transport in these strains has not yet been worked out in detail, but it may be due to mutation(s) that affect aerobic energy generation and energization of the bacterial membrane [14].

Present data concerning the mechanisms of aminoglycoside resistance among clinical isolates of bacteria suggest that the previously described mechanisms of resistance are found relatively infrequently among clinical isolates of bacteria. Instead, the most common type of resistance is that which is due to plasmid-mediated enzymatic modification of the aminoglycosidic aminocyclitols (see Chapter 6). There are three major classes of enzymes that catalyze modification of aminoglycosidic aminocyclitols either by acetylation of an amino group or by phosphorylation or adenylylation of hydroxyl groups on the molecule. There are a number of subtypes of these acetyltransferases, phosphotransferases, and adenylyltransferases, which vary in their substrate specificities. These are discussed in detail in Chapter 6. The enzymes are not secreted extracellularly and do not modify or inactivate extracellular antimicrobial agents. Instead, they are found in the periplasmic space of gram-negative bacilli [10] and in a still undefined intracellular location in gram-positive cocci [93]. The exact mechanism by which enzymatic modification of aminoglycosides renders them inactive has not been fully determined. It is possible that one (or more) of the following mechanisms is responsible for lack of activity:

1. Competition with unmodified aminoglycoside for intracellular transport
2. Ineffective binding to ribosomes
3. Inability to interfere to ribosomal function [28]

Recent studies by Bryan and co-workers suggest that resistance to dihydrostreptomycin in *E. coli* carrying an R factor mediating the adenylylation of streptomycin is due to inability of the adenylylated dihydrostreptomycin to react with ribosomes to inhibit protein synthesis. There is a net decrease in accumulation of dihydrostreptomycin in these cells as well, because the second energy-dependent phase of aminoglycoside uptake is determined by interaction of the aminoglycosides with ribosomes [33]. This is a plausible operational model and is consistent with other studies which show conclusively that modified aminoglycosides such as phosphorylated kanamycin [93,94], adenylylated streptomycin, and phosphorylated amikacin [94] are unable to inhibit ribosomal protein synthesis in an in vitro system.

The aminoglycosidic aminocyclitols vary in their ability to resist enzymatic inactivation. Of the compounds presently available for clinical use, amikacin is the most resistant to enzymatic modification [152]. Indeed, it is resistant to the adenylyltransferases [ANT(2″)], many of the acetyltransferases [AAC(3), AAC(2′)], and phosphotransferase [APH(3′)], which render certain strains of gram-negative bacilli resistant to gentamicin, tobramycin, and/or kanamycin [152]. Amikacin is subject to modification by a 6′-acetyltransferase [AAC(6′)] found in gram-negative bacilli [121,152]. In addition, some strains of staphylococci produce a 4′-adenylyltransferase [ANT(4′)] which modifies amikacin and produces low-level amikacin resistance [167]. Finally, some enterococci produce a phosphotransferase [APH(3′)-III] which is active against amikacin [93].

Resistance to antibiotic synergism in enterococci also appears to be related to the presence of plasmid-mediated aminoglycoside-modifying enzymes in many instances [92,93,203]. *Streptococcus faecalis* and *Streptococcus faecium* resistant to penicillin-streptomycin synergism have been shown to contain an adenylyltransferase with activity against streptomycin, while strains resistant to penicillin-kanamycin and penicillin-amikacin synergism produce a plasmid-mediated phosphotransferase [APH(3')] with an unusual degree of activity against amikacin [93,203]. Recently, it has been documented that strains of *S. faecium* synthesize a 6'-acetyltransferase [AAC(6')] with activity against tobramycin, kanamycin, netilmicin, and sisomicin [203]. This seems to account for the fact that such strains are consistently resistant to synergistic killing when exposed to combinations of penicillin with tobramycin, kanamycin, netilmicin, or sisomicin [123]. It seems likely that a similar mechanism may account for the resistance to synergism by carbenicillin-gentamicin combinations seen in strains of *P. aeruginosa* with high-level resistance to gentamicin [90] and may also account for other examples of this phenomenon among gram-negative bacilli, but studies to verify this have not been reported.

III.   In Vitro Activity of the Aminoglycosidic Aminocyclitols

From the foregoing discussion, it can be seen that there are subtle differences in the mechanisms of action and more dramatic differences in the mechanisms of resistance to the various aminoglycosidic aminocyclitols. In addition, these compounds differ in their "intrinsic activity." As a result, each has a different spectrum of activity. The in vitro activity of the aminoglycosides is quite dependent upon the methodology used to determine susceptibility. This aspect will be examined in some detail prior to a consideration of the individual spectra of activity of the various drugs. Finally, it must be emphasized that geographic differences in resistance patterns of clinical isolates of bacteria may also impinge upon the spectrum of activity of these agents [135].

A.   Susceptibility Testing

A number of different methods may be used to determine the in vitro activity of the aminoglycosidic aminocyclitols. Of these, the most commonly employed are the single-disk diffusion method [9,138] and the agar dilution and broth dilution methods [3,42]. Each of these methods has certain advantages and each presents specific problems when it is applied to the testing of aminoglycosidic aminocyclitols. Disk diffusion testing is still the most commonly employed method for routine determination of the aminoglycoside susceptibility of clinical isolates of bacteria. When strict attention to detail is followed, this method generally presents few problems. The in vitro effectiveness of the aminoglycosides is quite pH-dependent; indeed, these agents exhibit 80–100 times as much activity at pH 8.0 as at 5.0 [7,64]. Thus, the pH of the Mueller-Hinton agar plates used in disk diffusion testing should be between 7.3 and 7.5 [180]. The divalent cation content of the medium and, to a lesser extent, the presence of peptones and carbohydrates may also affect the activity of the aminoglycosides in vitro. The antagonistic effect of divalent cations is especially prominent, particularly when testing *P. aeruginosa* [34,44]. Calcium and magnesium have been most carefully studied in this regard, but the concentration of zinc, copper, and probably other divalent cations also affects the in vitro susceptibility of *P. aeruginosa* to aminoglycosides [193]. Indeed, recent studies suggest that divalent cation concentrations affect the susceptibilities of *P. aeruginosa* more than those of other organisms. The aminoglycoside susceptibility of *E. coli* is not affected to any significant degree by the concentration of divalent cations in the test medium [44,155] and non-fermenters other than *P. aeruginosa* are affected to a much lesser degree than *P. aeruginosa* [44]. The mechanism of the divalent cation effect appears related to the fact that high concentrations of these ions (especially calcium and magnesium) impair the binding of gentamicin (and presumably other aminoglycosides) to the cells of *P. aeruginosa* and in so doing, they decrease intracellular transport [155]. This effect is not seen with *E. coli* [155]. Because of the potential effect of alterations in media composition, it is recommended that control strains of *E. coli* and *P. aeruginosa* routinely be included with each batch of susceptibility determinations [138]. The recommended limits for control strains of *E. coli* and *P. aeruginosa* are given in Table 1.

The inoculum should be carefully controlled because very high inocula may result in "false resistance." However, there is much less inoculum effect with the aminoglycosides than with other antimicrobial agents such as the β-lactam antibiotics [28]. This may be related to the fact that mutation to resistance is rare and especially to the fact that the aminoglycoside-modifying enzymes in resistant strains are not secreted extracellularly and thus do not inactivate extracellular antibiotic [28].

Recent studies have raised significant questions as to the accuracy of the disk diffusion test in predicting susceptibility or resistance of the aminoglycosides. It has been noted that the diameters of the zones of inhibition for many clinical isolates of gram-negative bacteria cluster about the break point (12–13 mm) between susceptible and resistant for gentamicin and tobramycin [180]. In the latest standards approved by the National Committee for Clinical Laboratory Standards, this has been solved by including an intermediate zone of 13–14 mm for both gentamicin and tobramycin [138]. These same standards also recommend use of a 30 μg amikacin disk instead of the presently recom-

Table 1   Control Limits for Aminoglycosidic Aminocyclitols Against *E. coli* and
P. aeruginosa

Aminoglycoside	Disk content (μg)	Zone diameter (mm) for individual test	Zone diameter (mm) for mean of five values
E. coli (ATCC 25922)			
Amikacin	30	19-26	20.0-25.0
Gentamicin	10	19-26	20.2-24.8
Kanamycin	30	17-25	18.3-23.7
Neomycin	30	17-23	18.0-22.0
Streptomycin	10	12-20	13.3-18.7
Tobramycin	10	18-26	
P. aeruginosa (ATCC 27853)			
Amikacin	30	18-26	
Gentamicin	10	16-21	
Tobramycin	10	19-25	

Source: Data derived from [138].

mended 10 μg disk [138]. This recommendation was made because use of a 10 μg disk with standard criteria results in a high prevalence (20–50%) of strains which appear resistant by disk testing but are actually susceptible to amikacin when MIC is determined [129]. A recent multicenter collaborative study has confirmed that the use of 30 μg disks produces a much more accurate prediction of amikacin susceptibility [194].

Agar dilution and broth dilution tests are useful for quantitative susceptibility testing of the aminoglycosides. The aminoglycosidic aminocyclitols are very stable compounds (see Chapter 1). As a result, culture media containing these agents can be stored for weeks to months without significant loss of activity [160]. Agar-containing plates should be refrigerated at 4°C [160] and broth dilution tubes should be stored frozen at 0–20°C if not used immediately after preparation. Although aminoglycosidic aminocyclitols are very stable compounds, they bind with great affinity to cellulose and other materials used in filters [188]. This can lead to significant loss of activity if antibiotic standard solutions are purified by filtration.

Many of the problems relating to pH and cation content of the media that were discussed in relationship to disk diffusion testing also apply to quantitative susceptibility testing. Thus, the pH of the medium should be controlled to pH 7.3–7.5 in order to produce comparable results [180]. Mueller-Hinton agar should be used whenever possible for agar-dilution testing because the cation content of this medium approximates the recommended concentrations of 25 mg/L for magnesium and 50 mg/L for calcium [180]. Mueller-Hinton broth contains very little calcium or magnesium and must be supplemented by adding calcium chloride and magnesium chloride to obtain the desired concentrations of calcium and magnesium. This is especially important when performing broth dilution susceptibility testing for *P. aeruginosa*. For determining antimicrobial susceptibility or for examining synergism with β-lactams and aminoglycosides against enterococci and other streptococci, dextrose phosphate broth may be used in place of Mueller-Hinton broth [130]. There is evidence, however, that the use of different media may produce alterations in the susceptibility of enterococci to aminoglycosides [40]. The effect of media on synergy determination has not been extensively studied.

Exposure of aminoglycosidic aminocyclitols to high concentrations of penicillins such as carbenicillin or ticarcillin leads to inactivation of aminoglycosides in vitro [111]. Although this has been shown to occur in vivo, as well, especially in patients with impaired renal function, its exact clinical significance remains to be determined [157]. Because of this, however, mixtures of penicillins and aminoglycosides for in vitro synergy testing should not be made prior to the actual time of testing. The various aminoglycosidic aminocyclitols differ in their susceptibility to inactivation by carbenicillin and ticarcillin. Amikacin shows considerably greater resistance to inactivation under a variety of in vitro conditions than does tobramycin, sisomicin, or gentamicin [68].

The aminoglycosides are stable when mixed with a number of acrylic bone cements. When so incorporated, they are released slowly and in low concentrations into the surrounding media or tissues for many days [41,66].

B.   Spectrum of Activity

General Principles

The spectrum of activity of the aminoglycosidic aminocyclitols is the result of an interplay between the factors discussed earlier in the chapter. Because they are not transported into the cell and thus do not have access to bacterial ribosomes under anaerobic conditions, these agents have no useful activity against anaerobic organisms and are relatively inactive against facultative organisms under strictly anaerobic conditions [187]. It has been demonstrated that these agents may produce lethal enterocolitis in hamsters, presumably because they kill aerobic organisms but fail to suppress toxin-producing strains of *Clostridium difficile* [8].

Among gram-positive cocci, streptococci and *Listeria monocytogenes* are also generally resistant to the aminoglycosidic aminocyclitols, presumably because of poor intracellular penetration of these agents [125,127]. Staphylococci—both *S. aureus* and *S. epidermidis*—are susceptible to most of the aminoglycosides, especially the 2-deoxystreptamine-containing compounds. The clinical usefulness of these agents is limited by the emergence of resistant small-colony variants during therapy [112,116,117,137]. The 2-deoxystreptamine-containing agents are generally active against aerobic gram-positive bacilli such as diphtheroids [134] and against gram-negative cocci, including gonococci [46,47] and meningococci [62]. Most strains of *Hemophilus species* are susceptible in vitro to the newer aminoglycosides [40,62,109,181]. These agents are also active in vitro against mycoplasma [95,189]. They have little utility against *Pseudomonas pseudomallei* [52]. *Legionella pneumophila* have been shown to susceptible in vitro to streptomycin, kanamycin, gentamicin, and tobramycin [168,179] and in vivo to streptomycin and gentamicin [98], but the clinical significance of these observations remains to be determined.

Streptomycin is very effective against *Mycobacterium tuberculosis,* but none of the other aminoglycosides is as active as streptomycin [166,186]. Kanamycin is used as a second-line drug in the therapy of tuberculosis and has activity against some strains of atypical mycobacteria [200]. Earlier studies suggested that gentamicin may be quite active in vitro against *M. tuberculosis* [189], but not in vivo [166]. However, more recent studies show that gentamicin, tobramycin, and sisomicin have very little activity against *M. tuberculosis* or "atypical mycobacteria" [57]. Amikacin has recently been shown to have in vitro activity against *M. fortuitum* and *M. chelonei,* but the clinical significance of these observations is not known [24].

Nocardia species are generally resistant to the older aminoglycosides, including gentamicin and tobramycin [6]. Amikacin has been shown to have good in vitro activity against *N. asteroides* [25].

The major useful spectrum of activity of aminoglycosidic aminocyclitols is directed against aerobic and facultative gram-negative bacilli. Susceptibility patterns for the organisms most frequently isolated and tested in the clinical microbiology laboratory of the Massachusetts General Hospital are given in Table 2. These data are derived from disk diffusion testing. In general, streptomycin was the least active of the five agents tested, but even this drug inhibited the majority of most species of Enterobacteriaceae. Kanamycin was the next least active, primarily because the majority of strains of pseudomonads were resistant. Except for that, kanamycin compared surprisingly well with the other aminoglycosidic aminocyclitols against the clinical isolates in this series. The activity patterns of gentamicin and tobramycin were very similar. The major differences included the slightly greater activity of gentamicin against *Serratia species* and the enhanced effectiveness of tobramycin against *P. aeruginosa* and *Acinetobacter calcoaceticus* subsp. *anitratus.* Since the data for amikacin in Table 2 are derived from disk diffusion testing, the figures may be considered to represent the minimum effectiveness of the drug. As noted earlier, the use of a 10 μg amikacin disk with presently approved criteria results in an overestimation of resistance to amikacin [129]. However, in spite of that fact, the

Table 2   Susceptibility of Clinical Isolates of Bacteria to Five Aminoglycosidic
Aminocyclitols, Massachusetts General Hospital, January 1978–December, 1978

Organism	Number of unique isolates tested[a]	% Susceptible to:				
		Streptomycin	Kanamycin	Gentamicin	Tobramycin	Amikacin
Alcaligenes sp.	24	11	37	45	45	41
Aeromonas hydrophilia	17	80	88	100	100	94
Citrobacter freundii	190	78	79	88	86	98
Citrobacter diversus	55	91	96	100	100	100
Enterobacter agglomerans	39	79	79	87	84	100
Enterobacter aerogenes	368	93	96	97	98	98
Enterobacter cloacae	604	83	88	91	91	98
Enterobacter hafnia	29	100	96	100	100	100
E. coli	5764	64	85	98	98	98
Acinetobacter calcoaceticus var. *anitratus* (*Herellea* sp.)	694	37	83	54	77	55
Klebsiella pneumoniae	2010	81	83	87	87	99
Acinetobacter calcoaceticus var. *lwoffi* (*Mima* sp.)	22	77	90	90	90	90
Proteus mirabilis	1212	87	92	97	98	97
Proteus morgagnii	331	79	92	98	98	97
Proteus rettgeri	91	53	91	84	85	98
Proteus vulgaris	81	92	88	98	97	98
Providencia stuartii	39	31	74	64	64	94
Pseudomonas aeruginosa	1587	5	4	81	96	81
Pseudomonas maltophilia	133	45	31	54	60	47
Salmonella sp.	59	52	98	100	98	98
Serratia marcescens	389	63	80	86	81	93
Serratia liquefaciens	29	71	89	96	92	92
Shigella sp.	28	33	96	100	100	100

[a]Duplicate isolates from the same site and same patient have been eliminated.

enhanced activity of amikacin against gentamicin- and tobramycin-resistant strains of
Enterobacteriaceae can be deduced from these figures. Many of the strains of *P. aeruginosa* that appear resistant to amikacin by disk diffusion testing would be susceptible
if subjected to appropriate quantitative testing [129].

Table 3 presents the mean minimum inhibitory concentrations (MIC) of five "older"
aminoglycosides for various species of gram-negative bacilli. The comparative quantitative susceptibilities of 14 commonly isolated species of organisms to six "newer" parenteral aminoglycosides are given in Table 4. Although the figures in this table are useful
for comparative purposes, other investigators using different techniques and different
populations of organisms have suggested that the overall activity of these agents against
certain species of gram-negative organisms may be somewhat less than shown in Table 4
[32,53,79]. Our own studies and those of others suggest that enterococci are also somewhat more resistant to all these aminoglycosides than is shown in Table 4 [19,91,173].

Table 3   Susceptibility of Various Bacteria to Five Aminoglycosidic Aminocyclitols

Organism	Number of strains	Mean MIC (μg/ml)				
		Strepto-mycin	Neomy-cin	Kana-mycin	Paromo-mycin	Genta-micin
Staphylococcus aureus	29	2	0.5	1	1	0.125
Streptococcus faecalis	32	64	64	32	64	8
Escherichia coli	22	8	8	4	8	1
Klebsiella sp.	20	4	2	2	2	1
Aerobacter sp.	10	4	2	2	2	0.5
Proteus mirabilis	6	8	8	4	8	2
Proteus vulgaris	6	4	4	4	4	1
Proteus morganii	10	8	8	4	4	1
Proteus rettgeri	7	4	8	2	4	1
Pseudomonas aeruginosa	31	32	32	128	512	4
Salmonella sp.	14	16	2	2	2	1
Shigella sp.	17	8	8	4	8	2

Source: Data derived from [154]. Strains with unusual degrees of resistance were omitted from calculations of mean MIC.

In Vitro Effectiveness and Development of Resistance to the Major Parenteral Aminoglycosides

Streptomycin. Streptomycin was the first aminoglycoside to be employed for clinical purposes. Its utility, however, was rapidly eroded by the development of resistance [113]. Initially, it was particularly effective in vitro (MIC < 10 μg/ml) against *Bacillus anthracis, Klebsiella pneumoniae, Pasteurella (Yersinia) tularensis, H. influenzae, N. Gonorrhoeae, Pasteurella (Yersinia) pestis, Brucella* sp., *Salmonella* sp., *Shigella* sp. and *M. Tuberculosis* [207]. It is the most active of all the presently available aminoglycosides against *M. tuberculosis* and is still considered a major drug for the therapy of tuberculosis [5]. It is never employed alone for this purpose, because of the rapid development of resistance which occurs when it is used as a single agent for the treatment of tuberculosis [19].

Because of its excellent in vitro activity, streptomycin remains a drug of choice for the therapy of infections due to *Y. tularensis* and *Y. pestis* and is useful for infections caused by *Brucella* sp. [5]. When combined with penicillin, streptomycin produces synergistic killing of streptococci, including enterococci [130]. In recent years, however, there have emerged many strains of enterococci with high-level resistance to streptomycin (MIC > 2000 μg/ml) [19]. These are not killed synergistically in vitro by combinations of penicillin plus streptomycin [19], and in many locations account for as many as 50% of strains of enterococci encountered among clinical isolates [19,74]. To date, virtually all strains of *Streptococcus viridans,* including penicillin-resistant strains, have remained susceptible to the synergistic interaction of penicillin plus streptomycin in vitro [146].

For most other susceptible organisms, including the Enterobacteriaceae and pseudomonads, the development of resistance has been of such magnitude [113] that streptomycin is no longer utilized to treat infections due to these organisms. Perhaps related to the fact that the clinical use of streptomycin has markedly declined in recent years, data

Table 4   Susceptibility of Various Bacteria to Six Parenteral Aminoclycosidic Aminocyclitols

Organism	Number of strains	Gentamicin			Sisomicin			Netilmicin			Tobramycin			Amikacin			Kanamycin		
		50	75	100	50	75	100	50	75	100	50	75	100	50	75	100	50	75	100
Staphylococcus aureus	25	≤0.2	≤0.2	0.8	≤0.2	0.4	0.8	0.4	0.4	0.8	≤0.2	≤0.2	6.2	0.8	1.6	3.1	0.8	1.6	3.1
Enterococci	20	3.1	3.1	6.2	3.1	3.1	6.2	3.1	3.1	6.2	6.2	6.2	12.5	1.6	25	50	12.5	12.5	>50
Neisseria gonorrhoeae	20	6.2	6.2	12.5	3.1	3.1	12.5	3.1	6.2	12.5	6.2	6.2	12.5	25	25	50	12.5	12.5	25
Escherichia coli	40	0.8	1.6	6.2	1.6	1.6	6.2	0.8	1.6	6.2	1.6	3.1	12.5	3.1	6.2	12.5	6.2	12.5	>50
Klebsiella sp.	20	0.4	0.4	1.6	0.4	0.8	3.1	≤0.2	0.4	0.8	0.4	0.4	1.6	0.8	1.6	6.2	1.6	3.1	>50
Enterobacter sp.	40	0.4	0.8	1.6	0.4	0.4	12.5	0.4	0.4	6.2	0.8	1.6	6.2	1.6	1.6	6.2	1.6	3.1	>50
Citrobacter sp.	15	0.8	1.6	6.2	0.8	1.6	1.6	0.8	0.8	3.1	1.6	1.6	1.6	3.1	3.1	6.2	6.2	12.5	25
Serratia sp.	15	0.8	0.8	1.6	0.8	1.6	3.2	1.6	1.6	3.2	3.1	3.1	6.2	3.1	3.1	6.2	3.1	3.1	6.2
Proteus mirabilis	20	1.6	1.6	3.1	1.6	1.6	6.2	1.6	1.6	3.1	1.6	1.6	3.1	6.2	6.2	12.5	3.1	6.2	25
Proteus morganii	29	0.8	0.8	0.8	0.8	0.8	3.1	0.4	0.8	1.6	0.4	0.8	1.6	1.6	3.1	6.2	3.1	3.1	6.2
Proteus vulgaris	6	0.8	1.6	1.6	1.6	1.6	3.1	0.8	0.8	0.8	1.6	1.6	3.1	3.1	3.1	6.2	3.1	6.2	6.2
Proteus rettgeri	6	1.6	3.2	6.2	3.1	6.2	25	1.6	12.5	12.5	1.6	3.1	6.2	0.8	1.6	1.6	0.8	3.1	3.1
Providencia stuartii	8	1.6	3.1	6.2	1.6	1.6	6.2	3.1	3.1	12.5	1.6	3.1	12.5	0.8	3.1	3.1	0.8	1.6	3.1
Pseudomonas aeruginosa	40	0.4	0.8	3.1	≤0.2	0.4	1.6	0.4	0.8	6.2	≤0.2	≤0.2	1.6	0.4	0.8	12.5	–	–	–

MIC (μg/ml) for 50, 75, or 100% of strains

Source: Data derived from [165].

77

such as those in Table 2 suggest that the streptomycin-susceptible proportion among certain species of gram-negative bacilli (especially the Enterobacteriaceae) may be increasing. Indeed, in 1978 streptomycin inhibited more than 50% of almost all of the major species of Enterobacteriaceae isolated at the Massachusetts General Hospital. Virtually all strains of *P. aeruginosa* have remained resistant.

Kanamycin. As can be seen in Tables 2, 3, and 4, kanamycin is considerably more active than streptomycin against aerobic and facultative gram-negative bacilli. This is true whether one considers percentage of strains inhibited (Table 2) or mean MIC of susceptible strains (Tables 3 and 4). The reason for the higher prevalence of resistance to streptomycin than to kanamycin is not entirely clear. It has been suggested that this is related to the fact that single-step mutation to high-level resistance occurs with greater frequency when organisms are exposed to streptomycin than when exposed to kanamycin in vitro. However, it is not known if this occurs in vivo. Large-scale studies of the mechanisms of resistance to streptomycin among clinical isolates of bacteria would be required to answer this question; unfortunately, data from such studies presently do not exist.

The major defect in the spectrum of activity of kanamycin is its general lack of effectiveness against *P. aeruginosa* of plasmid-mediated enzymes such as APH(3'), AAC(3), and AAC(6') with activity against kanamycin [67,75] and to intrinsic resistance to the drug [28]. Widespread clinical use of kanamycin has led to the emergence of resistance among Enterobacteriaceae [26,51,58]. That this is often due to the selective pressure of heavy utilization of the drug is demonstrated by the fact that the percentage of susceptible strains may increase once the drug is no longer widely used in a given environment [51].

Gentamicin. Gentamicin is active against most strains of Enterobacteriaceae (Tables 2, 3, and 4) and unlike streptomycin and kanamycin, it is also quite effective against *P. aeruginosa*. Gentamicin is resistant to inactivation by several of the plasmid-mediated enzymes that attack kanamycin, including APH(3') and AAC(3)-III. This is responsible, in part, for its greater spectrum of activity. Virtually all strains of Enterobacteriaceae and *P. aeruginosa* proved susceptible to gentamicin when it was initially released for clinical use [189]. As a result, gentamicin largely replaced streptomycin and kanamycin for treatment of infections due to Enterobacteriaceae, pseudomonads, and other aerobic and facultative gram-negative bacilli in the United States and elsewhere. When combined with penicillin, gentamicin produces synergistic killing of virtually all strains of enterococci, including *S. faecalis* and *S. faecium* [19,123].

Emergence of resistance to gentamicin has generally not occurred as rapidly as that to streptomycin nor is it as widespread, but there has been a clear-cut increase in the number of gentamicin-resistant gram-negative organisms seen in many widely scattered institutions in the United States and Europe since the initial clinical use of the drug [12,50, 77,121,128,129,172]. Given the proper circumstances, these resistant organisms can be selected out rapidly in a local environment, even when gentamicin is used only for "prophylaxis" [158]. For the most part, gentamicin-resistant organisms have been confined to the larger "tertiary care" hospitals and have not yet been found in large numbers among "community isolates" of usually susceptible gram-negative bacteria [124]. As noted earlier, the emergence of resistant small colony variants during therapy has limited the usefulness of gentamicin as a single agent for the treatment of staphylococcal infections [99, 112, 116, 137]. More recently, staphylococci with plasmid-mediated resistance have also been seen in various centers [16a,35a].

Tobramycin. The spectrum of activity of tobramycin is generally similar to that of gentamicin (Tables 2 and 4). The major differences in activity relate to the fact that tobramycin is approximately two to four times as effective as gentamicin (on a weight basis) against *P. aeruginosa,* whereas gentamicin is more active against *Serratia* sp. [121, 140,165]. *Pseudomonas* species other than *P. aeruginosa* and *P. stutzeri* are often resistant to both gentamicin and tobramycin [121,184]. Against *A. calcoaceticus* subsp. *anitratus,* tobramycin is more active than gentamicin [121]. Both gentamicin and tobramycin show similar activity against *S. aureus, Hemophilus* sp., and *Neisseria* sp. [140]. Streptococci, including enterococci, are resistant to clinically achievable concentrations of tobramycin and gentamicin, but in combination with penicillin, tobramycin is as effective as gentamicin in producing synergistic killing of *S. faecalis,* but not *S. faecium* [123].

The emergence of tobramycin resistance seems to parallel that of gentamicin resistance in the clinical setting [121,128,129]. This is not surprising in view of the fact that except for their susceptibility to AAC(6'), AAC(3)-I, and ANT(4'), gentamicin and tobramycin do not differ in resistance to the important plasmid-mediated inactivating enzymes [167]. Most strains of *P. aeruginosa* with low-level resistance to gentamicin [121,140], and some gentamicin-resistant strains of *A. calcoaceticus* [121] are susceptible to tobramycin. With these exceptions, most other gentamicin-resistant organisms are cross-resistant to tobramycin [15,121,128,129,140].

Sisomicin. Sisomicin is closely related to gentamicin C_{1a}, differing from the latter only by the presence of a 3'-4' double bond. As a result of its close chemical similarity to gentamicin, its spectrum of activity is also quite similar to that of gentamicin [36,159]. The only major difference in spectrum is that sisomicin is approximately twice as active against *P. aeruginosa* (geometric mean MIC values for 28 susceptible strains of *P. aeruginosa:* gentamicin 0.76 $\mu g/ml$, sisomicin 0.49 $\mu g/ml$ [204]). It has been suggested that the enhanced susceptibility of *P. aeruginosa* to sisomicin is related to enhanced intracellular uptake of this drug compared with gentamicin [96]. Sisomicin is similar to or slightly less active than tobramycin in its efficacy against *P. aeruginosa* [20]. There are presently no data concerning the emergence of sisomicin-resistant organisms during clinical use, but in view of the similarity of this drug to gentamicin, it is likely that it will behave like gentamicin in this regard. Most tobramycin-resistant strains of *P. aeruginosa* are cross-resistant to sisomicin (as well as to streptomycin, kanamycin, gentamicin, and netilmicin) [210].

Amikacin. Like netilmicin, 3'-deoxykanamycin A and 3',4'-dideoxykanamycin B, amikacin is a semisynthetic aminoglycosidic aminocyclitol that was developed to overcome the problem of bacterial resistance to the older 2-deoxystreptamine-containing compounds [81]. It is the first semisynthetic aminoglycosidic aminocyclitol to attain widespread clinical use. Attachment of the amino acid L-α-hydroxybutyric acid to the C_1 group of the aminocyclitol ring of kanamycin A results in the formation of a compound (amikacin) that is resistant to all of the commonly occurring aminoglycoside-modifying enzymes except AAC(6') [81]. The intrinsic activity of amikacin is equal to that of kanamycin against susceptible organisms [81] (Table 4), but it also maintains activity against the majority of *P. aeruginosa* and Enterobacteriaceae that are resistant to streptomycin, kanamycin, tobramycin, sisomicin, and gentamicin [1,48,81,153,154,156, 165,171]. Because of this, amikacin has a very broad spectrum of activity against aerobic and facultative gram-negative bacilli (Tables 2 and 4). Among some nonfermenters, including *Pseudomonas* sp., *Acinetobacter* sp., *Flavobacterium* sp., and others, there does appear to be some cross-resistance between amikacin and other 2-deoxystreptamine-containing aminoglycosides such as tobramycin and gentamicin [129,152,153]. The mechanism of resistance in these organisms is most commonly related to a generalized defect in aminoglycoside permeability [152,153], but some strains of *P. aeruginosa* and *Acinetobacter* sp. have been shown to produce AAC(6′) with activity against amikacin [75,136,153]. Amikacin resistance due to the production of other modifying enzymes such as ANT(4′) found in some strains of staphylococci [167] and APH(3′) found among enterococci [93] has not yet been demonstrated to be a significant cause of amikacin resistance among gram-negative bacilli. Preliminary studies suggest that the clinical use of amikacin may be less likely to result in the emergence of aminoglycoside-resistant organisms than similar use of gentamicin or other 2-deoxystreptamine-containing agents [35], but there are presently insufficient data to allow the derivation of firm conclusions concerning this point.

Netilmicin. Netilmicin is the semisynthetic C_1-N-ethyl derivative of sisomicin. The addition of an ethyl group to the molecule confers on netilmicin resistance to modification by ANT(2″) [79,114,150]. Netilmicin also shows variable resistance to AAC(3)-I and AAC(3)-III, but not to AAC(3)-II, AAC(2′), or AAC(6′) [114]. Thus, it is more active than gentamicin or tobramycin against strains of *E. coli, K. pneumoniae,* and *Serratia marcescens,* as well as certain strains of *P. aeruginosa* that produce ANT(2″) [79,114,150]. In addition, it is also more active than gentamicin against gentamicin-resistant strains of *S. aureus* that produce both APH(2″)-I and AAC(6′) [150]. Against susceptible organisms it is generally equivalent in activity to gentamicin and sisomicin except that it is somewhat less active against *P. aeruginosa* (Table 4) [32,53,79,114, 150,165,173,195]. There is not enough experience with clinical use of netilmicin to allow meaningful assessment of its potential for the selection of resistant organisms. However, there is nothing thus far to suggest that it would behave differently from gentamicin in this regard.

IV. In Vitro Activity of Aminoglycosidic Aminocyclitols in Combination with Other Antimicrobial Agents

Although the aminoglycosidic aminocyclitols have been studied in combination with virtually all known antimicrobial agents, the most useful interaction occurs when these compounds are combined with the β-lactams or other cell wall-active antibiotics. The results of in vitro testing of antimicrobial combinations are extremely "method dependent" [122]. For testing combinations of aminoglycosides with β-lactams, both of which are bactericidal, the modified checkerboard titration or killing curve method seems to be optimal [122]. When these methods are utilized, such combinations are usually synergistic, additive, or indifferent. Antagonism is rarely, if ever, the result of such combination. The few reports of antagonism between *β-lactams* and aminoglycosides may be either the result of the use of other test methods, such as the cellophane transfer technique [103] or agar dilution [87] or the result of in vitro inactivation of the drugs [68,111,157]. When a given organism is highly resistant to one or both of the agents to which it is exposed in combination, the result is most likely to be indifference [19,90,139,190]. Combinations of penicillins and aminoglycosides have been most widely employed against enterococci [19] and *P. aeruginosa* [90], while combinations of cephalosporins and aminoglycosides have been studied most extensively against *K. pneumoniae* [86].

A list of the combinations of β-lactam and other cell wall-active agents with aminoglycosides that have been shown to be synergistic against various bacterial species is given in Table 5. This table does not represent an exhaustive list of all the published studies in this area, but is intended simply to give examples of the major types of interaction that have been studied. This table must be interpreted with some caution since the simple demonstration of synergy does not prove that such an interaction will be clinically useful. In some cases, the concentrations of drugs required to produce in vitro synergy are higher than those which can be attained clinically.

Combinations of aminoglycosidic aminocyclitols with bacteriostatic agents such as chloramphenicol or tetracycline generally result in the loss of bactericidal activity of the aminoglycoside, an example of in vitro antagonism [78,163]. Chloramphenicol-gentamicin combinations have been shown to exhibit in vitro antagonism against 35–45% of strains of *K. pneumoniae* [23] and 80% of *E. coli* [85]. Recent studies demonstrate that certain anticancer drugs (daunorubicin and cytosine arabinoside) may antagonize the bactericidal effect of gentamicin and amikacin in vitro against *K. pneumoniae* and *P. aeruginosa,* but not *E. coli* [133].

Combining aminoglycosides with clindamycin against gram-negative bacilli in vitro usually results in indifference, although some authors have suggested that such combinations may produce synergism [45,97] or antagonism [212], depending upon the definitions and test conditions employed. The use of clindamycin-gentamicin combinations against *S. viridans* has also produced conflicting results, with one study suggesting a synergistic interaction [38], whereas another showed antagonism against some strains [174]. Combinations of trimethoprim with amikacin have been shown to be synergistic in vitro against *E. coli, K. pneumoniae,* and *S. marcescens* [148]. Combinations of trimethoprim or trimethoprim-sulfamethoxazole with gentamicin have also demonstrated in vitro synergism against *K. pneumoniae* and *E. coli* [145].

V. Conclusions

The aminoglycosidic aminocyclitols have a wide spectrum of activity against staphylococci, diphtheroids, and aerobic or facultative gram-negative bacilli and diplococci. They also exhibit variable activity against mycobacteria and mycoplasma. The development of resistance has limited the clinical utility of streptomycin, kanamycin, and to a lesser extent gentamicin and tobramycin. There are presently not enough data to know whether widespread clinical use of amikacin will also result in the emergence of similar resistance among clinical isolates of bacteria.

By combining the aminoglycosides with other antimicrobial agents (particularly those which inhibit bacterial cell wall synthesis), the spectrum of activity of these drugs can be expanded to include certain organisms, such as the enterococci, which are resistant to aminoglycosides alone. Moreover, the utilization of synergistic combinations may enhance the bactericidal activity of these agents against susceptible organisms.

As noted in Chapter 2, more detailed knowledge of the mechanism of action of and the mechanisms of resistance to these compounds has allowed and will continue to make possible the rational development of new agents with enhanced spectra, especially against resistant organisms that produce inactivating enzymes. Such data should also allow the more rational enhancement of the spectrum of these agents through combinations with other antimicrobials to produce synergism while avoiding interactions that might be antagonistic.

Table 5 Synergistic Combinations of Aminoglycosides with β-Lactam (and Other Cell Wall-Active) Antibiotics

Organism	Synergistic combination[a]	References
Gram-positive cocci/bacilli		
Staphylococcus aureus (penicillin-susceptible)	Penicillin + GM, TM	162, 177
Staphylococcus aureus (penicillin-resistant)	Cefamandole + TM	185
	Cefazolin + GM, TM, KM, SM	196
	Cephalothin + GM, TM, KM, SM	196
	Nafcillin + GM, TM, KM, SM, SI, Net	117, 161, 183, 196, 197
	Oxacillin + GM, TM, SI, Net	196, 197
Staphylococcus aureus (methicillin-resistant)	Cephalothin + KM, GM	17, 82
	Oxacillin + GM	82
Staphylococcus epidermidis (methicillin-resistant)	Vancomycin + GM	102
Group B streptococci	Penicillin + GM	143, 170
Streptococcus viridans	Penicillin + SM, GM	146, 174, 205
Enterococci	Ampicillin + AM, GM, SM	74, 130, 209
	Bacitracin + SM	130
	Carbenicillin + GM	100
	Cephalothin + SM, KM, GM	199
	Cycloserine + SM	130
	Ly 127935 + GM	182
	Methicillin + GM	61, 198
	Nafcillin + GM, SI, Net	61, 198
	Oxacillin + GM	61, 105, 198
	Penicillin + SM, KM, GM, TM, AM, SI, Net, But	19, 63, 72, 91, 123, 131, 132, 164, 165, 173, 209
	Vancomycin + SM, GM	130
Listeria monocytogenes	Ampicillin + SM, GM	125
	Penicillin + SM, GM	43, 120
Diphtheroids	Penicillin + GM	134
Gram-negative bacilli		
Acinetobacter sp.	Carbenicillin + GM, TM, AM	60
Citrobacter sp.	Azlocillin + GM, AM, Net	142
	Mezloccilin + GM, AM, Net	142
Enterobacter sp.	Ampicillin + GM	83
	Azlocillin + GM, AM, Net	142
	Cefamandole + GM, AM	55
	Cephalothin + GM	83
	Mezlocillin + GM, AM, Net	142, 169
Escherichia coli	Ampicillin + GM	83
	Azlocillin + GM, AM, Net	142
	Carbenicillin + GM	87

Table 5 (continued)

Organism	Synergistic combination[a]	References
	Cefazolin + AM	59
	Cephalothin + GM, TM, AM, SI	59, 83, 108
	Mezlocillin + GM, AM, SI, Net	142, 169
Klebsiella pneumoniae	Azlocillin + GM, AM, Net	142
	Carbenicillin + GM	87
	Cefamandole + GM, TM, AM	55, 59, 185
	Cefazolin + GM, AM, SI	59, 86
	Cephaloridine + GM, KM, SM	103
	Cephalothin + KM, GM, TM, AM, SI	23, 59, 83, 108
	Mezlocillin + GM, AM, Net, SI	142, 169
Proteus mirabilis	Carbenicillin + GM	209
	Cephalothin + GM	83, 101
Indole-positive *Proteus* sp.	Azlocillin + GM, AM, Net	142
	Carbenicillin + TM	108
	Cephalothin + TM	108
	Mezlocillin + GM, AM, Net	142
Proteus sp.	Ampicillin + GM	83
	Carbenicillin + GM, TM, AM, SI	59, 84, 87
	Cefamandole + GM, TM, AM, SI	59
	Cefuroxime + GM, TM, AM, SI	59
	Cephalothin + GM, TM	59, 84
	Mezlocillin + SI	169
Providencia sp.	Carbenicillin + TM	108
	Cefazolin + GM, TM	73
	Cephalothin + GM, TM	73, 108
Pseudomonas aeruginosa	Azlocillin + GM, AM, Net	142
	BL-P 1654 + GM	190
	Carbenicillin + GM, TM, AM, SI, Net, 1-N-HAPA-gentamicin B	4, 21, 39, 54, 84, 87-90, 106, 141, 147, 149, 175, 208
	Ly 127935 + GM	182
	Mezlocillin + GM, TM, AM, SI, Net	142, 169
	Piperacillin + GM, TM, AM	21
	Ticarcillin + GM, TM, AM, SI	21, 22, 65, 147, 190, 206
Serratia marcescens	Azlocillin + GM, AM, Net	56, 142
	Carbenicillin + GM, TM, KM, AM, Net, 1-N-HAPA-gentamicin B	49, 56, 108, 115, 141
	Cefazolin + GM, TM, AM, SI	59
	Cephalothin + GM, TM, AM, SI	59
	Mezlocillin + GM, AM, SI, Net	56, 142, 169
	Ticarcillin + GM, AM, Net	56

[a] Abbreviations: AM, amikacin; But, butirosin; GM, gentamicin; KM, kanamycin; Net, netilmicin; SI, sisomicin; SM, streptomycin; TM, tobramycin.

References

1. Acar, J.F., Witchitz, J.L., Goldstein, F., Talbot, J.N., and LeGoffie, F.: Susceptibility of aminoglycoside-resistant gram-negative bacilli to amikacin: delineation of individual resistance patterns. *J. Infect. Dis.* 134(Suppl.):S280–S285, 1976.

1a. Altschafel, J.E., Davies, J., and Martinez, R.: Transposable neomycin-kanamycin resistance determinant from an anaerobe. Abstr. 491, Abstracts of the 11th International Congress of Chemotherapy and 19th Interscience Conference on Antimicrobial Agents and Chemotherapy, 1979.

2. Anand, N., and Davis, B.D.: Damage by streptomycin to the cell membrane of *Escherichia coli. Nature* 185:22–23, 1960.

3. Anderson, T.G.: Testing of susceptibility to antimicrobial agents and assay of antimicrobial agents in body fluids. In *Manual of Clinical Microbiology,* edited by Blair, J.E., Lennette, E.H., and Truant, J.P. Washington, D.C., American Society for Microbiology, 1970, pp. 299–310.

4. Andriole, V.T.: Synergy of carbenicillin and gentamicin in experimental infection with *Pseudomonas. J. Infect. Dis.* 124(Suppl.):S46–S55, 1971.

5. Anonymous. *Handbook of Antimicrobial Therapy. The Medical Letter on Drugs and Therapeutics,* New York, 1978.

6. Bach, M.C., Sabath, L.D., and Finland, M.: Susceptibility of *Nocardia asteroides* to 45 antimicrobial agents in vitro. *Antimicrob. Agents Chemother.* 3:1–8, 1973.

7. Barber, M., and Waterworth, P.M.: Activity of gentamicin against pseudomonas and hospital staphylococci. *Br. Med. J.* 1:203–205, 1966.

8. Bartlett, J.G., Chang, T.W., Moon, N., and Onderdonk, A.B.: Antibiotic-induced lethal enterocolitis in hamsters: studies with eleven agents and evidence to support the pathogenic role of toxin-producing *Clostridia. Am. J. Vet. Res.* 39:1525–1530, 1978.

9. Bauer, A.W., Kirby, W.M.M., Sherris, J.C., and Turck, M.: Antibiotic susceptibility testing by a standardized single disk method. *Am. J. Clin. Pathol.* 45:493–496, 1966.

10. Benveniste, R., and Davies, J.: Mechanisms of antibiotic resistance in bacteria. *Annu. Rev. Biochem.* 42:471–506, 1973.

11. Benveniste, R., and Davies, J.: Structure-activity relationships among the aminoglycoside antibiotics: role of hydroxyl and amino groups. *Antimicrob. Agents Chemother.* 4:402–409, 1973.

12. Block, C.S.: Gentamicin-resistant gram-negative bacilli in hospital patients. *S. Afr. Med. J.* 53:391–399, 1978.

13. Bryan, L.E., Haraphongse, R., and VanDenElzen, H.M.: Gentamicin resistance in clinical isolates of *Pseudomonas aeruginosa* associated with diminished gentamicin accumulation and no detectable enzymatic modification. *J. Antibiot. (Tokyo)* 29:743–753, 1976.

13a. Bryan, L.E., Koward, S.K., and VanDenElzen, H.M.: Mechanisms of aminoglycoside antibiotic resistance in anaerobic bacteria: *Clostridium perfringins* and *Bacteroides fragilis. Antimicrob. Agents Chemother.* 15:7–13, 1979.

14. Bryan, L.E., and VanDenElzen, H.M.: Effects of membrane-energy mutations and cations on streptomycin and gentamicin accumulation by bacteria: a model for entry of streptomycin and gentamicin in susceptible and resistant bacteria. *Antimicrob. Agents Chemother.* 12:163–177, 1977.

15. Brusch, J.L., Barza, M., Bergeron, M.G., and Weinstein, L.: Cross-resistance of *Pseudomonas* to gentamicin and tobramycin. *Antimicrob. Agents Chemother.* 1:280–281, 1972.

16. Buckel, P., Buchberger, A., Böck, A., and Wittmann, H.G.: Alteration of ribosomal

protein L6 in mutants of *Escherichia coli* resistant to gentamicin. *Mol. Genet.* 158:47–54, 1977.

16a. Buckwold, F.J., Albritton, W.L., Ronald, A.R., Lertzman, J., and Henriksen, R.: Investigations of the occurrence of gentamicin-resistant *Staphylococcus aureus*. *Antimicrob. Agents Chemother.* 15:152–156, 1979.

17. Bulger, R.J.: In vitro activity of cephalothin/kanamycin and methicillin/kanamycin combinations against methicillin-resistant *Staphylococcus aureus*. *Lancet* 1:17–19, 1967.

18. Cabanas, M.J., Vazquez, D., and Modolell, J.: Dual interference of hygromycin B with ribosomal translocation and with aminoacyl-tRNA recognition. *Eur. J. Biochem.* 87:21–27, 1978.

19. Calderwood, S.A., Wennersten, C., Moellering, R.C., Jr., Kunz, L.J., and Krogstad, D.J.: Resistance to six aminoglycosidic aminocyclitol antibiotics among enterococci: prevalence, evolution, and relationship to synergism with penicillin. *Antimicrob. Agents Chemother.* 12:401–405, 1977.

20. Campello, C., and Crevatin, E.: In vitro activity of gentamicin, sisomicin, and tobramycin against *Pseudomonas aeruginosa* strains from hospital sources. *Ann. Sclavo* 19:437–445, 1977.

21. Chanbusarakum, P., and Murray, P.R.: Analysis of the interactions between piperacillin, ticarcillin, or carbenicillin and aminoglycoside antibiotics. *Antimicrob. Agents Chemother.* 14:505–506, 1978.

22. Comber, K.R., Basker, M.J., Osborne, C.D., and Sutherland, R.: Synergy between ticarcillin and tobramycin against *Pseudomonas aeruginosa* and Enterobacteriaceae in vitro and in vivo. *Antimicrob. Agents Chemother.* 11:956–964, 1977.

23. D'Alessandri, R.M., McNeely, D.J., and Kluge, R.M.: Antibiotic synergy and antagonism against clinical isolates of *Klebsiella species*. *Antimicrob. Agents Chemother.* 10:889–892, 1976.

24. Dalovisio, J.R., and Pankey, G.A.: In vitro susceptibility of *Mycobacterium fortuitum* and *Mycobacterium chelonei* to amikacin. *J. Infect. Dis.* 137:318–321, 1978.

25. Dalovisio, J.R., and Pankey, G.A.: In vitro susceptibility of *Nocardia asteroides* to amikacin. *Antimicrob. Agents Chemother.* 13:128–129, 1978.

26. Damato, J.J., Eitzman, D.V., and Baer, H.: Persistence and dissemination in the community of R-factors of nosocomial origin. *J. Infect. Dis.* 129:205–209, 1974.

27. Davies, J.E.: Studies on the ribosomes of streptomycin-sensitive and resistant strains of *Escherichia coli*. *Proc. Natl. Acad. Sci. USA* 51:659–661, 1964.

28. Davies, J.: General mechanisms of antimicrobial resistance. *Rev. Infect. Dis.* 1:23–27, 1979.

29. Davies, J., Benveniste, R., Kvitek, K., Ozanne, B., and Yamada, T.: Aminoglycosides: biological effects of molecular manipulation. *J. Infect. Dis.* 119:351–354, 1969.

30. Davies, J., and Courvalin, P.: Mechanisms of resistance to aminoglycosides. *Am. J. Med.* 62:868–872, 1977.

31. Davies, J., Gorini, L., and Davis, B.D.: Misreading of RNA codewords induced by aminoglycoside antibiotics. *Mol. Pharmacol.* 1:93–106, 1965.

32. Dhawan, V., Marso, E., Martin, W.J., and Young, L.S.: In vitro studies with netilmicin compared with amikacin, gentamicin, and tobramycin. *Antimicrob. Agents Chemother.* 11:64–73, 1977.

33. Dickie, P., Bryan, L.E., and Pickard, M.A.: Effect of enzymatic adenylylation on dihydrostreptomycin accumulation in *Escherichia coli* carrying an R-factor: model explaining aminoglycoside resistance by inactivating mechanisms. *Antimicrob. Agents Chemother.* 14:569–580, 1978.

34. Donovick, R., Bayan, A.P., Conales, P., and Pansy, F.: The influence of certain substances on the activity of streptomycin: III. Differential effects of various

electrolytes on the action of streptomycin. *J. Bacteriol.* 56:125–137, 1948.

35. Douglas, R.G.: Personal communication, June 1979.

35a. Dowding, J.E.: Mechanisms of gentamicin resistance in *Staphylococcus aureus.*
Antimicrob. Agents Chemother. 11:47–50, 1977.

36. Drasar, F.A., Farrell, W., Maskell, J., and Williams, J.D.: Tobramycin, amikacin,
sisomicin, and gentamicin resistant gram-negative rods. *Br. Med. J.* 2:1284–1286,
1976.

37. Dubin, D.T., Hancock, R., and Davis, B.D.: The sequence of some effects of streptomycin in *Escherichia coli. Biochim. Biophys. Acta* 74:476–489, 1963.

38. Duperval, R., Bill, N.J., Geraci, J.E., and Washington, J.A., II: Bactericidal activity
of combinations of penicillin or clindamycin with gentamicin or streptomycin
against species of viridans streptococci. *Antimicrob. Agents Chemother.* 8:673–
676, 1975.

39. Eickhoff, T.C.: In vitro effects of carbenicillin combined with gentamicin or
polymyxin B against *Pseudomonas aeruginosa. Appl. Microbiol.* 18:469–473,
1969.

40. Eickhoff, T.C., and Ehret, J.M.: In vitro activity of netilmicin compared with
gentamicin, tobramycin, amikacin, and kanamycin. *Antimicrob. Agents Chemother.*
11:791–796, 1977.

41. Elson, R.A., Jephcott, A.E., McGechie, D.B., and Verettas, D.: Antibiotic-loaded
acrylic cement. *J. Bone Joint Surg.* [*Br.*] 59:200–205, 1977.

42. Ericsson, H.M., and Sherris, J.C.: Antibiotic sensitivity testing. Report of an international collaborative study. *Acta Pathol. Microbiol. Scand.* 271(Suppl.):1–90,
1971.

43. Espaze, E.P., Roubeix, Y.G., LeBerre, J.Y., and Courtieu, A.I.: In vitro susceptibility of *Listeria monocytogenes* to some antibiotics and their combinations. *Zentralbl. Bakteriol.* 240:76–85, 1978.

44. Fass, R.J., and Barnishan, J.: Effect of divalent cation concentrations on the
antibiotic susceptibilities of nonfermentors other than *Ps. aeruginosa. Antimicrob.
Agents Chemother.* 16:434–438, 1979.

45. Fass, R.J., Rotilie, C.A., and Prior, R.B.: Interaction of clindamycin and gentamicin in vitro. *Antimicrob. Agents Chemother.* 6:582–587, 1974.

46. Felarca, A.B., Laqui, E.M., and Ibarra, L.M.: Gentamicin in gonococcal urethritis
of Filipino males. *J. Infect. Dis.* 124(Suppl.):S287–S292, 1971.

47. Finland, M., Garner, C., Wilcox, C., and Sabath, L.D.: Susceptibility of *Neisseria
gonorrhoeae* to 66 antibacterial agents in vitro. *J. Am. Vener. Dis. Assoc.* 2:33–40,
1976.

48. Finland, M., Garner, C., Wilcox, C., and Sabath, L.D.: Susceptibility of recently
isolated bacteria to amikacin in vitro: comparisons with four other aminoglycoside
antibiotics. *J. Infect. Dis.* 134(Suppl.):S297–S306, 1976.

49. Flournoy, D.J.: Netilmicin synergy with carbenicillin or cefamandole against *Serratia. J. Antibiot. (Tokyo)* 31:131–134, 1978.

50. Forbes, I., Gray, A., Hurse, A., and Pavillard, R.: The emergence of gentamicin
resistant *Klebsiella* in a large general hospital. *Med. J. Aust.* 1:14–16, 1977.

51. Franco, J.A., Eitzman, D.V., and Baer, H.: Antibiotic usage and microbial resistance in an intensive care nursery. *Am. J. Dis. Child.* 126:318–321, 1973.

52. Franklin, M.: Effect of gentamicin on *Pseudomonas pseudomallei. J. Infect. Dis.*
124(Suppl.):S30–S32, 1971.

53. Fu, K.P., and Neu, H.C.: In vitro study of netilmicin compared with other aminoglycosides. *Antimicrob. Agents Chemother.* 10:526–534, 1976.

54. Fu, K.P., and Neu, H.C.: In vitro synergistic effect of netilmicin, a new aminoglycoside antibiotic. *Antimicrob. Agents Chemother.* 10:511–518, 1976.

55. Fu, K.P., and Neu, H.C.: A comparative study of the activity of cefamandole and other cephalosporins and analysis of the beta-lactamase stability and synergy of cefamandole with aminoglycosides. *J. Infect. Dis.* 137 (Suppl.):S38–S48, 1978.

56. Fu, K.P., and Neu, H.C.: The comparative synergistic activity of amikacin, gentamicin, netilmicin, and azlocillin, mezlocillin, carbenicillin and ticarcillin against *Serratia marcescens*. *J. Antibiot. (Tokyo)* 31:135–140, 1978.

57. Gangadharam, P.R., and Candler, E.R.: In vitro anti-mycobacterial activity of some new amino-glycoside antibiotics. *Tubercle* 58:35–38, 1977.

58. Gardner, P., and Smith, D.H.: Studies on the epidemiology of resistant (R) factors: I. Analysis of *Klebsiella* isolates in a general hospital. II. A prospective study of R-factor transfer in the host. *Ann. Intern. Med.* 71:1–9, 1969.

59. Gaya, H., Brown, E.M., Friedman, P., and Cox, S.E.M.: Comparative in vitro activity of cefuroxime and its interactions with aminoglycoside antibiotics. *Proc. R. Soc. Med.* 70(Suppl. 9):51–55, 1977.

60. Glew, R.H., Moellering, R.C., Jr., and Buettner, K.R.: In vitro synergism between carbenicillin and aminoglycosidic aminocyclitols against *Acinetobacter calcoaceticus* var. *anitratus*. *Antimicrob. Agents Chemother.* 11:1036–1041, 1977.

61. Glew, R.H., Moellering, R.C., Jr., and Wennersten, C.B.G.: Comparative synergistic activity of nafcillin, oxacillin, and methicillin in combination with gentamicin against enterococci. *Antimicrob. Agents Chemother.* 7:828–832, 1975.

62. Greenstone, G., Hammerberg, S., and Marks, M.I.: In vitro activity of netilmicin (*Sch 20569*) against bacterial isolates from ill children. *Chemotherapy* 24:29–33, 1978.

63. Gutschik, E., Jepsen, O.B., and Mortensen, I.: Effect of combinations of penicillin and aminoglycosides on *Streptococcus faecalis:* a comparative study of seven aminoglycoside antibiotics. *J. Infect. Dis.* 135:832–836, 1977.

64. Harris, H.W., Murray, R., Paine, T.F., Kilham, L., and Finland, M.: Streptomycin treatment of urinary tract infections, with special reference to the use of alkali. *Am. J. Med.* 2:229–250, 1947.

65. Heineman, H.S., and Lofton, W.M.: Unpredictable response of *Pseudomonas aeruginosa* to synergistic antibiotic combinations in vitro. *Antimicrob. Agents Chemother.* 13:827–831, 1978.

66. Hill, J., Klenerman, L., Trustey, S., and Blowers, R.: Diffusion of antibiotics from acrylic bone-cement in vitro. *J. Bone Joint Surg.* [*Br.*] 59:197–199, 1977.

67. Holmes, R.K., Minshew, B.H., Gould, K., and Sanford, J.: Resistance of *Pseudomonas aeruginosa* to gentamicin and related aminoglycoside antibiotics. *Antimicrob. Agents Chemother.* 6:253–262, 1974.

68. Holt, H.A., Broughall, J.M., McCarthy, M., and Reeves, D.S.: Interactions between aminoglycoside antibiotics and carbenicillin or ticarcillin. *Infection* 4:107–109, 1976.

69. Höltje, J.-V.: Streptomycin uptake via an inducible polyamine transport system in *Escherichia coli*. *Eur. J. Biochem.* 86:345–351, 1978.

70. Höltje, J.-V.: Induction of streptomycin uptake in resistant strains of *Escherichia coli*. *Antimicrob. Agents Chemother.* 15:177–181, 1979.

71. Houang, E.T., and Greenwood, D.: Aminoglycoside cross-resistance patterns of gentamicin-resistant bacteria. *J. Clin. Pathol.* 30:738–744, 1977.

72. Hunter, T.H.: Use of streptomycin in treatment of bacterial endocarditis. *Am. J. Med.* 2:436–442, 1947.

73. Hyams, P.J., Simberkoff, M.S., and Rahal, J.J., Jr.: Synergy between cephalosporin and aminoglycoside antibiotics against *Providencia* and *Proteus*. *Antimicrob. Agents Chemother.* 5:571–577, 1974.

74. Iannini, P.B., Ehret, J., and Eickhoff, T.C.: Effects of ampicillin-amikacin and

ampicillin-rifampin on enterococci. *Antimicrob. Agents Chemother.* 9:448–451, 1976.

75. Jacoby, G.A.: Properties of an R-plasmid in *Pseudomonas aeruginosa* producing amikacin (BB-K8), butirosin, kanamycin, tobramycin, and sisomicin resistance. *Antimicrob. Agents Chemother.* 6:807–810, 1974.

76. Jacoby, G.A., and Gorini, L.: The effect of streptomycin and other aminoglycoside antibiotics on protein synthesis. In *Antibiotics,* Vol. 1, *Mechanism of Action,* edited by Gottlieb, D., and Shaw, P.D. Berlin, Springer-Verlag, 1967, pp. 726–747.

77. Jaurequi, L., Cushing, R.D., and Lerner, A.M.: Gentamicin/amikacin-resistant gram-negative bacilli at Detroit General Hospital, 1975–1976. *Am. J. Med.* 62:882–888, 1977.

78. Jawetz, E., Gunnison, J.B., and Speck, R.S.: Studies on antibiotic synergism and antagonism: the interference of aureomycin, chloramphenicol, and terramycin with the action of streptomycin. *Am. J. Med. Sci.* 222:404–412, 1951.

79. Kabins, S.A., Nathan, C., and Cohen, S.: In vitro comparison of netilmicin, a semi-synthetic derivative of sisomicin and four other aminoglycoside antibiotics. *Antimicrob. Agents Chemother.* 10:139–145, 1976.

80. Kaji, H., and Tanaka, Y.: Binding of dihydrostreptomycin to ribosomal subunits. *J. Mol. Biol.* 32:221–230, 1968.

81. Kawaguchi, H.: Discovery, chemistry and activity of amikacin. *J. Infect. Dis.* 134 (Suppl.):S242–S248, 1976.

82. Klastersky, J.: Antibiotic susceptibility of oxacillin-resistant staphylococci. *Antimicrob. Agents Chemother.* 1:441–446, 1972.

83. Klastersky, J., Cappel, R., Swings, G., and Vandenborre, L.: Bacteriological and clinical activity of the ampicillin/gentamicin and cephalothin/gentamicin combinations. *Am. J. Med. Sci.* 262:283–290, 1971.

84. Klastersky, J., Henri, A., and Vandenborre, L.: Antimicrobial activity of tobramycin and gentamicin used in combination with cephalothin and carbenicillin. *Am. J. Med. Sci.* 266:13–21, 1973.

85. Klastersky, J., and Husson, M.: Bactericidal activity of the combinations of gentamicin with clindamycin or chloramphenicol against species of *Escherichia coli* and *Bacteroides fragilis. Antimicrob. Agents Chemother.* 12:135–138, 1977.

86. Klastersky, J., Meunier-Carpentier, F., Prevost, J.M., and Staquet, M.: Synergism between amikacin and cefazolin against *Klebsiella:* in vitro studies and effect on the bactericidal activity of serum. *J. Infect. Dis.* 134:271–276, 1976.

87. Klastersky, J., Swings, G., and Daneau, D.: Antimicrobial activity of the carbenicillin/gentamicin combination against gram-negative bacilli. *Am. J. Med. Sci.* 260: 373–379, 1970.

88. Klein, R.A., and Meyer, R.D.: Synergy studies with *Pseudomonas aeruginosa* resistant to gentamicin and/or carbenicillin. *Chemotherapy* 24:283–289, 1978.

89. Kluge, R.M., Standiford, H.C., Tatem, B., Young, V.M., Greene, W.H., Schimpff, S.C., Calia, F.M., and Hornick, R.B.: Comparative activity of tobramycin, amikacin, and gentamicin alone and with carbenicillin against *Pseudomonas aeruginosa. Antimicrob. Agents Chemother.* 6:442–446, 1974.

90. Kluge, R.M., Standiford, H.C., Tatem, B., Young, V.M., Schimpff, S.C., Greene, W.H., Calia, F.M., and Hornick, R.B.: The carbenicillin-gentamicin combination against *Pseudomonas aeruginosa:* correlation of effect with gentamicin sensitivity. *Ann. Intern. Med.* 81:584–587, 1974.

91. Korzeniowski, O.M., Wennersten, C., Moellering, R.C., Jr., and Sande, M.A.: Penicillin-netilmicin synergism against *Streptococcus faecalis. Antimicrob. Agents Chemother.* 13:430–434, 1978.

92. Krogstad, D.J., Korfhagen, T.R., Moellering, R.C., Jr., Wennersten, C., and Swartz,

M.N.: Plasmid-mediated resistance to antibiotic synergism in enterocci. *J. Clin. Invest.* 61:1645–1653, 1978.

93. Krogstad, D.J., Korfhagen, T.R., Moellering, R.C., Jr., Wennersten, C., Swartz, M.N., Perzynski, S., and Davies, J.: Aminoglycoside-inactivating enzymes in clinical isolates of *Streptococcus faecalis*. *J. Clin. Invest.* 62:480–486, 1978.

94. Krogstad, D.J., Shen, V., and Randall, A.T.: Enzymatically mediated aminoglycoside resistance in *Streptococcus faecalis*. Abstr. 494, Abstracts of the 11th International Congress of Chemotherapy and 19th Interscience Conference on Antimicrobial Agents and Chemotherapy, 1979.

95. Laborde, M.: Study of antibiotic sensitivity in 34 strains of large colony mycoplasma. *Pathol. Biol. (Paris)* 25:541–546, 1977.

96. Lee, B.K., Condon, R.G., Munayyer, H., and Weinstein, M.J.: Uptake of (methyl-14C)-sisomicin and (methyl-14C)-gentamicin into bacterial cells. *J. Antibiot. (Tokyo)* 31:141–146, 1978.

97. Leng, B., Meyers, B.R., Hirschman, S.Z., and Keusch, G.T.: Susceptibilities of gram-negative bacteria to combinations of antimicrobial agents in vitro. *Antimicrob. Agents Chemother.* 8:164–171, 1975.

98. Lewis, V.J., Thacker, W.L., Shepard, C.C., and McDade, J.E.: In vivo susceptibility of the Legionnaire's disease bacterium to ten antimicrobial agents. *Antimicrob. Agents Chemother.* 13:419–422, 1978.

99. Li, K., Farmer, J.J., III, and Coppola, A.: A novel type of resistant bacteria induced by gentamicin. *Trans. N.Y. Acad. Sci.* 36:396–415, 1974.

100. Libke, R.D., Regamey, C., Clarke, J.T., and Kirby, W.M.M.: Synergism of carbenicillin and gentamicin against enterococci. *Antimicrob. Agents Chemother.* 4:564–568, 1973.

101. Lode, H., Dalhoff, A., and Naumann, P.: Die Wirkung von Cepahlothin, Gentamycin, und ihrer Kombination gegen Bacterium *Proteus mirabilis*. *Int. J. Clin. Pharmacol.* 6:346–353, 1972.

102. Lowy, F.D., Walsh, J.A., Mayers, M.M., Klein, R.S., and Steigbigel, N.H.: Antibiotic activity in vitro against methicillin-resistant *Staphylococcus epidermidis* and therapy of an experimental infection. *Antimicrob. Agents Chemother.* 16:314–321, 1979.

103. Luboshitzky, R., Sacks, T., and Michel, J.: Bactericidal effect of combinations of antibiotics on *Klebsiella-Enterobacter-Serratia*. *Chemotherapy* 19:354–366, 1973.

104. Luzzatto, L., Apirion, D., and Schlessinger, D.: Mechanism of action of streptomycin in *E. coli*: interruption of the ribosome cycle at the initiation of protein synthesis. *Proc. Natl. Acad. Sci. USA* 60:873–880, 1968.

105. Marier, R.L., Joyce, N., and Andriole, V.T.: Synergism of oxacillin and gentamicin against enterococci. *Antimicrob. Agents Chemother.* 8:571–573, 1975.

106. Marks, M.I., Hammerberg, S., Greenstone, G., and Silver, B.: Activity of newer aminoglycosides and carbenicillins, alone and in combination, against gentamicin-resistant *Pseudomonas aeruginosa*. *Antimicrob. Agents Chemother.* 10:399–401, 1976.

107. Masukawa, H.: Localization of sensitivity to kanamycin and streptomycin in 30S ribosomal proteins of *E. coli*. *J. Antibiot. (Tokyo)* 22:612–623, 1969.

108. Mayama, M., Nagata, H., and Aoi, I.: In vitro antibacterial activity of tobramycin used in combination with cephalothin or carbenicillin. *Jpn. J. Antibiot.* 31:153–165, 1978.

109. Mayo, J.B., and McCarthy, L.R.: Antimicrobial susceptibility of *Haemophilus parainfluenzae*. *Antimicrob. Agents Chemother.* 11:844–847, 1977.

110. McDonald, P.J., Craig, W.A., and Kunin, C.M.: Persistent effect of antibiotics on *Staphylococcus aureus* after exposure for limited periods of time. *J. Infect. Dis.* 135:217–223, 1977.

111. McLaughlin, J.E., and Reeves, D.S.: Clinical and laboratory evidence for inactivation of gentamicin. *Lancet* 1:261–264, 1971.

112. Michel, J., Stessman, J., and Sacks, T.: Phenotypic variations in gentamicin resistant isolates of *Staphylococcus aureus*. *Chemotherapy* 24:314–320, 1978.

113. Miller, C.P., and Bohnhoff, M.: Development of streptomycin-resistant and streptomycin-dependent bacteria. In *Streptomycin, Nature and Practical Applications*, edited by Waksman, S.A. Baltimore, Md., Williams & Wilkins, 1949, Chap. 10.

114. Miller, G.H., Arcieri, G., Weinstein, M.J., and Waitz, J.A.: Biological activity of netilmicin, a broad-spectrum semisynthetic aminoglycoside antibiotic. *Antimicrob. Agents Chemother.* 10:827–836, 1976.

115. Miller, M.A., Mohammad, Y., Griffin, P.S., Bartlett, M., and Crane, J.K.: In vitro activity of cefamandole, cefoxitin, cefuroxime, and carbenicillin alone and in combination with aminoglycosides against *Serratia marcescens*. *Microbiol. Immunol.* 23:955–964, 1979.

116. Miller, M.H., Edberg, S.C., Wexler, M.A., and Steigbigel, N.H.: Mechanism of gentamicin-resistance in small-colony mutants of *S. aureus*. Abstr. 286, Abstracts of the 18th Interscience Conference on Antimicrobial Agents and Chemotherapy, 1978.

117. Miller, M.H., Wexler, M.A., and Steigbigel, N.H.: Single and combination antibiotic therapy of *Staphylococcus aureus* experimental endocarditis: emergence of gentamicin-resistant mutants. *Antimicrob. Agents Chemother.* 14:336–343, 1978.

118. Modolell, J., and Davis, B.D.: Mechanism of inhibition of ribosomes by streptomycin. *Nature* 224:345–348, 1969.

119. Modolell, J., and Davis, B.D.: Breakdown by streptomycin of initiation complexes formed on ribosomes of *Eschericia coli*. *Proc. Natl. Acad. Sci. USA* 65:1148–1155, 1970.

120. Mohan, K., Gordon, R.C., Beaman, T.C., Belding, R.C., Luecke, D., Edmiston, C., and Gerhardt, P.: Synergism of penicillin and gentamicin against *Listeria monocytogenes* in ex vivo hemodialysis culture. *J. Infect. Dis.* 135:51–54, 1977.

121. Moellering, R.C., Jr.: Microbiological considerations in the use of tobramycin and related aminoglycosidic aminocyclitol antibiotics. *Med. J. Aust.* 2(Suppl.1)4–8, 1977.

122. Moellering, R.C., Jr.: Antimicrobial synergism: an elusive concept. *J. Infect. Dis.* 140:640–643, 1979.

123. Moellering, R.C., Jr., Korzeniowski, O.M., Sande, M.A., and Wennersten, C.B.: Species-specific resistance to antimicrobial synergism in *Streptococcus faecium* and *Streptococcus faecalis*. *J. Infect. Dis.* 140:203–208, 1979.

124. Moellering, R.C., Jr., Kunz, L.J., Poitras, J.W., Cameron, M., and Mercier, B.A.: Microbiological basis for the rational use of prophylactic antibiotics. *South. Med. J.* 70(Suppl. 1):8–14, 1977.

125. Moellering, R.C., Jr., Medoff, G., Leech, I., Wennersten, C., and Kunz, L.J.: Antibiotic synergism against *Listeria monocytogenes*. *Antimicrob. Agents Chemother.* 1:30–34, 1972.

126. Moellering, R.C., Jr., Murray, B.E., Schoenbaum, S.C., Adler, J., and Wennersten, C.B.: A novel mechanism of resistance to penicillin-gentamicin synergism in *S. faecalis*. *J. Infect. Dis.* 141:81–86, 1980.

127. Moellering, R.C., Jr., and Weinberg, A.N.: Studies on antibiotic synergism against enterococci. II. Effect of various antibiotics on the uptake of ^{14}C-labeled streptomycin by enterococci. *J. Clin. Invest.* 50:2580–2584, 1971.

128. Moellering, R.C., Jr., Wennersten, C., and Kunz, L.J.: Emergence of gentamicin-resistant bacteria. Experience with tobramycin therapy of infections due to gentamicin-resistant organisms. *J. Infect. Dis.* 134(Suppl.):S40–S49, 1976.

129. Moellering, R.C., Jr., Wennersten, C., Kunz, L.J., and Poitras, J.W.: Resistance to

gentamicin, tobramycin, and amikacin among clinical isolates of bacteria. *Am. J. Med.* 62:873–881, 1977.

130. Moellering, R.C., Jr., Wennersten, C., and Weinberg, A.N.: Studies on antibiotic synergism against enterococci. I. Bacteriologic studies. *J. Lab. Clin. Med.* 77:821–828, 1971.

131. Moellering, R.C., Jr., Wennersten, C.B.G., and Weinberg, A.N.: Synergy of penicillin and gentamicin against enterococci. *J. Infect. Dis.* 124(Suppl.):S207–S209, 1971.

132. Moellering, R.C., Jr., Wennersten, C.B.G., and Weinstein, A.J.: Penicillin-tobramycin synergism against enterococci: a comparison with penicillin and gentamicin. *Antimicrob. Agents Chemother.* 3:526–529, 1973.

133. Moody, M., Morris, M.J., Young, V.M., Moye, L.A., III, Schimpff, S.C., and Wiernik, P.H.: Effect of two cancer chemotherapeutic agents on the antibacterial activity of three antimicrobial agents. *Antimicrob. Agents Chemother.* 14:737–742, 1978.

134. Murray, B.E., Karchmer, A.W., Moellering, R.C., Jr., and Kunz, L.J.: Diphtheroid prosthetic valve endocarditis: clinical presentation and therapy based on susceptibility to antimicrobial agents alone or in combination. Abstr. 275, Abstracts of the 18th Interscience Conference on Antimicrobial Agents and Chemotherapy, 1978.

135. Murray, B.E., and Moellering, R.C., Jr.: Patterns and mechanisms of antibiotic resistance. *Med. Clin. North Am.* 62:899–923, 1978.

136. Murray, B.E., and Moellering, R.C., Jr.: Aminoglycoside-modifying enzymes among clinical isolates of *Acinetobacter calcoaceticus* subsp. *anitratus* (*Herellea vaginicola*): explanation for high-level aminoglycoside resistance. *Antimicrob. Agents Chemother.* 15:190–199, 1979.

137. Musher, D.M., Baughn, R.E., Templeton, G.B., and Minuth, J.N.: Emergence of variant form of *Staphylococcus aureus* after exposure to gentamicin and infectivity of the variants in experimental animals. *J. Infect. Dis.* 136:360–369, 1977.

138. National Committee for Clinical Laboratory Standards. Performance standards for antimicrobial disc susceptibility tests. Approved Standard: ASM-2, Villanova, Pa., 1979.

139. Neu, H.C.: Effect of beta-lactamase location in *Escherichia coli* on penicillin synergy. *Appl. Microbiol.* 17:783–786, 1969.

140. Neu, H.C.: Tobramycin: an overview. *J. Infect. Dis.* 134(Suppl.):S3–S19, 1976.

141. Neu, H.C., and Fu, K.P.: 1-N HAPA gentamicin B, a new aminoglycoside active against gentamicin resistant isolates—activity compared to other aminoglycosides. *J. Antibiot. (Tokyo)* 31:385–393, 1978.

142. Neu, H.C., and Fu, K.P.: Synergy of azolocillin and mezlocillin compared with aminoglycoside antibiotics and cephalosporins. *Antimicrob. Agents Chemother.* 13:813–819, 1978.

143. Overturf, G.D., Horowitz, M., Wilkins, J., Leedom, J., and Steinberg, E.: Bactericidal studies of penicillin-gentamicin combinations against group B streptococci. *J. Antibiot. (Tokyo)* 30:513–518, 1977.

144. Ozaki, M., Mizuchima, S., and Nomura, M.: Identification and functional characterization of the protein controlled by the streptomycin-resistant locus in *E. coli.* *Nature* 222:333–339, 1969.

145. Paisley, J.W., and Washington, J.A., II: Synergistic activity of gentamicin with trimethoprim or sulfamethoxazole-trimethoprim against *Escherichia coli* and *Klebsiella pneumoniae.* *Antimicrob. Agents Chemother.* 14:656–658, 1978.

146. Parrillo, J.E., Borst, G.C., Mazur, M.H., Iannini, P., Klempner, M.S., Moellering, R.C., Jr., and Anderson, S.E.: Endocarditis due to resistant viridans streptococci during oral penicillin chemoprophylaxis. *N. Engl. J. Med.* 300:296–300, 1979.

147. Parry, M.F., and Neu, H.C.: A comparative study of ticarcillin plus tobramycin versus carbenicillin plus gentamicin for the treatment of serious infections due to

gram-negative bacilli. *Am. J. Med.* 64:961–966, 1978.

148. Parsley, T.L., Provonchee, R.B., Glicksman, C., and Zinner, S.H.: Synergistic activity of trimethoprim and amikacin against gram-negative bacilli. *Antimibrob. Agents Chemother.* 12:349–352, 1977.

149. Phair, J.P., Watanakunakorn, C., and Bannister, T.: In vitro susceptibility of *Pseudomonas aeruginosa* to carbenicillin and the combination of carbenicillin and gentamicin. *Appl. Microbiol.* 18:303–306, 1969.

150. Phillips, I., Smith, A., and Shannon, K.: Antibacterial activity of netilmicin, a new aminoglycoside antibiotic, compared with that of gentamicin. *Antimicrob. Agents Chemother.* 11:402–406, 1977.

151. Plotz, P.H., and Davis, B.D.: Synergism between streptomycin and penicillin: a proposed mechanism. *Science* 135:1067–1068, 1962.

152. Price, K.E., Casson, K., DeRegis, R.G., Kresel, P.A., Pursiano, T.A., and Leitner, F.: Amikacin: antimicrobial properties and resistance mechanisms affecting its activity. *Am. J. Med.* (Suppl. on U.S. Amikacin Symposium, Los Angeles, Calif., 1976) 14–24, 1977.

153. Price, K.E., DeFuria, M.D., and Pursiano, T.A.: Amikacin, an aminoglycoside with marked activity against antibiotic-resistant clinical isolates. *J. Infect. Dis.* 134(Suppl.): S249–S261, 1976.

154. Price, K.E., Godfrey, J.C., and Kawaguchi, H.: Effect of structural modifications on the biological properties of aminoglycoside antibiotics containing 2-deoxystreptamine. In *Structure-Activity Relationships Among the Semisynthetic Antibiotics,* edited by Perlman, D. New York, Academic Press, 1977, pp. 239–355.

155. Ramiriz-Ronda, C.H., Holmes, R.K., and Sanford, J.P.: Effects of divalent cations on binding of aminoglycoside antibiotics to human serum proteins and to bacteria. *Antimicrob. Agents Chemother.* 7:239–245, 1975.

156. Reynolds, A.V., Hamilton-Miller, J.M.T., and Brumfitt, W.: In vitro activity of amikacin and ten other aminoglycoside antibiotics against gentamicin-resistant bacterial strains. *J. Infect. Dis.* 134(Suppl.):S286–S290, 1976.

157. Riff, L.J., and Jackson, G.G.: Laboratory and clinical conditions for gentamicin inactivation by carbenicillin. *Arch. Intern. Med.* 130:887–891, 1972.

158. Roberts, N.J., Jr., and Douglas, R.G.: Gentamicin use and *Pseudomonas* and *Serratia* resistance: effect of a surgical prophylaxis regimen. *Antimicrob. Agents Chemother.* 13:214–220, 1978.

159. Rossi, A., Branca, G., Manzara, S., Morace, G., Polonelli, L., Spagnolo, N., and Ttonci, M.: Antibacterial activity of sisomicin. *J. Int. Med. Res.* 6:94–101, 1978.

160. Ryan, K.J., Needham, G.M., Dunsmoor, C.L., and Sherris, J.C.: Stability of antibiotics and chemotherapeutics in agar plates. *Appl. Microbiol.* 20:447–451, 1970.

161. Sande, M.A., and Courtney, K.B.: Nafcillin-gentamicin synergism in experimental staphylococcal endocarditis. *J. Lab. Clin. Med.* 88:118–124, 1976.

162. Sande, M.A., and Johnson, M.L.: Antimicrobial therapy of experimental endocarditis caused by *Staphylococcus aureus. J. Infect. Dis.* 131:367–375, 1975.

163. Sande, M.A., and Overton, J.W.: In vivo antagonism between gentamicin and chloramphenicol in neutropenic mice. *J. Infect. Dis.* 128:247–250, 1973.

164. Sanders, C.C.: Synergy of penicillin-netilmicin combinations against enterococci including strains highly resistant to streptomycin or kanamycin. *Antimicrob. Agents Chemother.* 12:195–200, 1977.

165. Sanders, C.C., Sanders, W.E., Jr., and Goering, R.V.: In vitro studies with *Sch 21420* and *Sch 22591*: activity in comparison with six other aminoglycosides and synergy with penicillin against enterococci. *Antimicrob. Agents Chemother.* 14: 178–184, 1978.

166. Sanders, W.E., Pejovic, I., Cacciatore, R., Valdez, H., and Dunbar, F.P.: Activity

of gentamicin against *Mycobacteria* in vitro and against *Mycobacterium tuberculosis* in mice. *J. Infect. Dis.* 124(Suppl.):S33–S36, 1971.

167. Santanam, P., and Kayser, F.H.: Tobramycin adenylyltransferase: a new aminoglycoside-inactivating enzyme from *Staphylococcus epidermidis*. *J. Infect. Dis.* 134(Suppl.):S33–S39, 1976.

168. Saravolatz, L.D., Pohlod, D.J., and Quinn, E.L.: In vitro susceptibility of *Legionella pneumophila*, serogroups I-IV. *J. Infect. Dis.* 140:251–252, 1979.

169. Schassan, H.H., Koperski, K., and Scherf, H.: Mezlocillin, a new acyl ureidopenicillin. Antimicrobial activity and combination effects with four aminoglycoside antibiotics. *Chemotherapy* 24:134–142, 1978.

170. Schauf, V., Deveikis, A., Riff, L., and Serota, A.: Antibiotic-killing kinetics of group B streptococci. *J. Pediatr.* 89:194–198, 1976.

171. Shah, R.M., Heetderks, G., and Stille, W.: Bactericidal activity of amikacin and gentamicin. *Chemotherapy* 23:260–266, 1977.

172. Siebert, W.T., Moreland, N.J., and Williams, T.W., Jr.: Resistance to gentamicin: a growing concern. *South. Med. J.* 70:289–292, 1977.

173. Smith, J.A., Morgan, J.R., and Mogyoros, M.: In vitro activity of netilmicin. *Antimicrob. Agents Chemother.* 11:362–364, 1977.

174. Snyder, R.J., Wilkowske, C.J., and Washington, J.A., II: Bactericidal activity of combinations of gentamicin with penicillin or clindamycin against *Streptococcus mutans*. *Antimicrob. Agents Chemother.* 7:333–335, 1975.

175. Sonne, M., and Jawetz, E.: Combined action of carbenicillin and gentamicin on *Pseudomonas aeruginosa* in vitro. *Appl. Microbiol.* 17:893–896, 1969.

176. Spotts, C.R., and Stanier, R.Y.: Mechanism of streptomycin action on bacteria: a unitary hypothesis. *Nature* 192:633–637, 1961.

177. Steigbigel, R.T., Greenman, R.L., and Remington, J.S.: Antibiotic combinations in the treatment of experimental *Staphylococcus aureus* infection. *J. Infect. Dis.* 131:245–251, 1975.

178. Tanaka, N.: Biochemical studies in gentamicin resistance. *J. Antibiot. (Tokyo)* 23:469–471, 1970.

179. Thornsberry, C., Baker, C.N., and Kirven, L.A.: In vitro activity of antimicrobial agents on Legionnaire's disease bacterium. *Antimicrob. Agents Chemother.* 13: 78–80, 1978.

180. Thornsberry, C., Gavan, T.L., Gerlach, E.H., and Sherris, J.C.: New developments in antimicrobial agent susceptibility testing. *Cumitech 6*, Coord. editor, Sherris, J.C. Washington, D.C., American Society for Microbiology, 1977, pp. 1–13.

181. Thornsberry, C., and Kirvin, L.A.: Antimicrobial susceptibility of *Haemophilus influenzae*. *Antimicrob. Agents Chemother.* 6:620–624, 1974.

182. Trager, G.M., White, G.W., Zimelis, V.M., and Panwalker, A.P.: LY-127935: a novel beta-lactam antibiotic with unusual antibacterial activity. *Antimicrob. Agents Chemother.* 16:297–300, 1979.

183. Tuazon, C.V., Lin, M.Y.C., Hill, R.M., and Sheagren, J.N.: In vitro nafcillin-gentamicin synergism against pathogenic strains of *Staphylococcus aureus. Curr. Ther. Res.* 23:760–766, 1978.

184. Uwaydah, M., and Taqi-Eddin, A.-R.: Susceptibility of nonfermentative gramnegative bacilli to tobramycin. *J. Infect. Dis.* 134(Suppl.):S28–S32, 1976.

185. Valenti, A.J., Siniscalchi, A.J., and Andriole, V.T.: In vitro synergy of cefamandole-tobramycin combinations. *Yale J. Biol. Med.* 50:177–182, 1977.

186. Van Scoy, R.E.: Antituberculous agents: isoniazid, rifampin, streptomycin, ethambutol. *Mayo Clin. Proc.* 52:694–700, 1977.

187. Verklin, R.M., Jr., and Mandell, G.L.: Alteration of effectiveness of antibiotics by anaerobiosis. *J. Lab. Clin. Med.* 89:65–71, 1977.

188. Wagman, G.H., Bailey, J.V., and Weinstein, M.J.: Binding of aminoglycoside antibiotics to filtration materials. *Antimicrob. Agents Chemother.* 7:316–319, 1975.

189. Waitz, J.A., and Weinstein, M.J.: Recent microbiologic studies with gentamicin. *J. Infect. Dis.* 119:355–360, 1969.

190. Wald, E.R., Standiford, H., Tatem, B.A., Calia, F.M., and Hornick, R.B.: BL-P1654 ticarcillin, and carbenicillin: in vitro comparison alone and in combination with gentamicin against *Pseudomonas aeruginosa. Antimicrob. Agents Chemother.* 7:336–340, 1975.

191. Wallace, B.J., and Davis, B.D.: Cyclic blockade of initiation sites by streptomycin-damaged ribosomes in *Escherichia coli*: an explanation for dominance of sensitivity. *J. Mol. Biol.* 75:377–390, 1973.

192. Wallace, B.J., Tai, P.C., and Davis, B.D.: Streptomycin and related antibiotics. In *Antibiotics: Mechanism of Action,* Vol. 5, Pt. 1, edited by Hahn, F.E., Berlin, Springer-Verlag, 1979, pp. 272–303.

193. Washington, J.A., II, Snyder, R.J., Kohner, P.C., Wiltse, C.G., Ilstrup, D.M., and McCall, J.T.: Effect of cation content of agar on activity of gentamicin, tobramycin, and amikacin against *Pseudomonas aeruginosa. J. Infect. Dis.* 137:103–111, 1978.

194. Washington, J.A., Yu, P.K.W., Gavan, T.L., Schoenknecht, F.D., and Thornsberry, C.: Interpretation of the disk diffusion susceptibility test for amikacin: report of a collaborative study. *Antimicrob. Agents Chemother.* 15:400–407, 1979.

195. Watanakunakorn, C.: Comparative in vitro activity of *Sch 20656*, netilmicin, gentamicin, and tobramycin. *Antimicrob. Agents Chemother.* 10:382–383, 1976.

196. Watanakunakorn, C., and Glotzbecker, C.: Enhancement of the effects of antistaphylococcal antibiotics by aminoglycosides. *Antimicrob. Agents Chemother.* 6:802–806, 1974.

197. Watanakunakorn, C., and Glotzbecker, C.: Enhancement of antistaphylococcal activity of nafcillin and oxacillin by sisomicin and netilmicin. *Antimicrob. Agents Chemother.* 12:346–348, 1977.

198. Watanakunakorn, C., and Glotzbecker, C.: In vitro activity of sisomicin and netilmicin alone and in combination with nafcillin, oxacillin, and methicillin against enterococci. *Chemotherapy* 24:236–239, 1978.

199. Weinstein, A.J., and Moellering, R.C., Jr.: Studies of cephalothin: aminoglycoside synergism against enterococci. *Antimicrob. Agents Chemother.* 7:522–529, 1975.

200. Weinstein, L.: Miscellaneous antimicrobial, antifungal and antiviral agents. In *The Pharmacological Basis of Therapeutics* (4th ed.), edited by Goodman, L.S., and Gilman, A. New York, Macmillan, 1970, pp. 1269–1310.

201. Weinstein, M.J.: The microbiology of gentamicin resistance. *Acta Pathol. Microbiol. Scand. Sect. B* 81(Suppl. 241):99–106, 1973.

202. Weisblum, B., and Davies, J.: Antibiotic inhibitors of the bacterial ribosome. *Bacteriol. Rev.* 32:493–528, 1968.

203. Wennersten, C.B., and Moellering, R.C., Jr.: Mechanisms of resistance to penicillin-aminoglycoside synergism in *S. faecium. Curr. Chemother.*, in press.

204. Wise, R., and Reeves, D.S.: Activity of sisomicin against gentamicin-sensitive organisms (letter). *Br. Med. J.* 1:288, 1977.

205. Wolfe, J.C., and Johnson, W.D., Jr.: Penicillin-sensitive streptococcal endocarditis. In vitro and clinical observations on penicillin-streptomycin therapy. *Ann. Intern. Med.* 81:178–181, 1974.

206. Yoshikawa, T.T., and Shibata, S.A.: In vitro antibacterial activity of amikacin and ticarcillin, alone and in combination against *Pseudomonas aeruginosa. Antimicrob. Agents Chemother.* 13:944–950, 1978.

207. Youmans, G.P., and Fisher, M.W.: Action of streptomycin on microorganisms

in vitro. In *Streptomycin, Nature and Practical Applications,* edited by Waksman, S.A. Baltimore, Md., Williams & Wilkins, 1949, Chap. 7.

208. Young, L.S.: Gentamicin: clinical use with carbenicillin and in vitro studies with recent isolates of *Pseudomonas aeruginosa. J. Infect. Dis.* 124(Suppl.):S202–S206, 1971.

209. Yourassowsky, E., Vanderlinden, M.P., and Schoutens, E.: A rapid, simple method for demonstrating synergy of amikacin and penicillin against various microorganisms. *J. Infect. Dis.* 134(Suppl.):S275–S279, 1976.

210. Yu, P.K., and Washington, J.A., II: Antimicrobial susceptibility of gentamicin-resistant *Pseudomonas aeruginosa. Mayo Clin. Proc.* 52:802–805, 1977.

211. Zimmermann, R.A., Moellering, R.C., Jr., and Weinberg, A.N.: Mechanism of resistance to antibiotic synergism in enterococci. *J. Bacteriol.* 105:873–879, 1971.

212. Zinner, S.H., Provonchee, R.B., Elias, K.S., and Peter, G.: Effect of clindamycin on the in vitro activity of amikacin and gentamicin against gram-negative bacilli. *Antimicrob. Agents Chemother.* 9:661–664, 1976.

4

AMINOGLYCOSIDE ANTIBIOTIC RESISTANCE IN BACTERIA

SUSUMU MITSUHASHI Gunma
University School of Medicine,
Maebashi City, Japan

HARUHIDE KAWABE Nippon Dental
University, Niigata City, Japan

The demonstration of the existence of the transferable (R) [77] and nontransferable (r) [80] drug-resistance plasmids, and of their wide distribution in gram-negative and gram-positive bacteria, has proven to be of great importance in both medicine and genetics. Following the discovery of drug-resistant strains of bacteria, many investigators in the medical and pharmaceutical sciences have sought to determine the biochemical mechanisms involved in such resistance.

The fact that many research workers did not notice the differences between in vitro-developed resistance and naturally occurring resistance in bacteria has hindered progress in the investigation of mechanisms of bacterial resistance that are of clinical importance. Therefore, studies on the biochemical mechanisms of drug resistance in naturally occurring resistant strains advanced very slowly, whereas biochemical studies on the mode of action of antibacterial agents made rapid progress.

In 1961, Miyamura [82] reported resistant dysentery bacilli isolated from patients in whom chloramphenicol (CM) was inactivated. Furthermore, Okamoto and Suzuki [90] observed that *Escherichia coli* carrying an R factor could inactivate streptomycin (SM), kanamycin (KM), and CM using adenosine triphosphate (ATP) or acetyl-coenzyme A (CoA) as a cofactor. The inactivation of KM by *E. coli* carrying R factor was later found to be due to the phosphorylation of the 3'-hydroxyl group of KMs. As a result of these findings, studies on the biochemical mechanisms of aminoglycoside resistance were initiated, and many types of drug inactivation were reported. The resistances described were found to be due to enzymatic inactivation of aminoglycoside antibiotics by phosphorylation, adenylylation, or acetylation of the hydroxyl or amino group of the drugs, using ATP or acetyl-CoA as coenzyme [5,22,23,25,73,79,118].

Owing to these findings, rapid progress has been made in studies of the biochemical methanisms involved. In vitro-developed resistance has been shown to differ from naturally occurring resistance in bacteria, and the mechanisms of plasmid-mediated resistance have been found to be the same in bacteria isolated in every country. Using the results of these studies we were able to develop a method of tailoring known chemotherapeutic agents to be effective against bacteria carrying plasmids. By modifying known chemotherapeutic agents, such as β-lactam antibiotics, aminoglycoside antibiotics, and nalidixic acid, we obtained many drugs that became effective against drug-resistant bacteria. Studies on the biochemical mechanisms of plasmid-mediated resistance have also provided a method for studying the origins of the drug-resistance determinants on plasmids and their evolutionary processes in microorganisms.

I. Epidemiology of Aminoglycoside Resistance in Bacteria

According to results based on surveys of *Staphylococcus aureus* strains isolated from 1961 to 1978, the isolation frequency of strains resistant to sulfanilamide (SA) was highest, followed by those resistant to penicillin G (PC), tetracycline (TC), and SM, in that order. A survey of 4290 *S. aureus* strains isolated between 1966 and 1972 indicated that strains carrying quadruple and triple resistance to the four drugs were isolated most frequently (43%), followed by those carrying double (42%) and single (15%) resistance [32,72,77,79a,80,81]. Correlation of the distribution frequency of SM and KM resistance in *S. aureus* strains with resistance patterns to TC, SM, PC, and SA is shown in Table 1. The isolation frequency of strains resistant to SM and KM showed the highest correlation with quadruple resistance, followed by triple, double, and single resistance, in that order. These results indicate that *S. aureus* strains easily develop resistance to aminoglycoside antibiotics in addition to already acquired multiple resistance when they are introduced in practical use.

Our surveys of resistance to TC, CM, SM, and SA in clinical isolates of gram-negative

Table 1   Streptomycin and Kanamycin Resistance in *S. aureus* Strains Collelated with Resistance Patterns to TC, SM, PC, and SA[a]

Resistance patterns[b]	Isolation frequency (%) of strains resistant to:	
	SM	KM
Quadruple	63.1	58.2
Triple	25.2	22.5
Double	11.6	15.6
Single	0	3.7

[a] Results are based on surveys from 4290 strains isolated between 1966 and 1972.
[b] Resistance to tetracycline (TC), streptomycin (SM), penicillin G (PC), and sulfanilamide (SA).

Table 2   Streptomycin Resistance Collelated with Resistance Patterns to TC, CM, SM, and SA[a]

Pattern of drug resistance[b]	Isolation frequency (%) of SM-resistant strains				
	E. coli	Shigella	Salmonella	Proteus	Klebsiella
Quadruple	61.6	91.1	13.8	62.2	67.2
Triple	16.4	6.2	57.8	27.1	22.1
Double	20.7	2.7	25.9	9.3	8.8
Single	1.3	0.1	2.4	1.3	1.8

[a]The results are based on surveys of 14,530 strains.
[b]Resistance to TC, CM, SM, and SA.

bacteria have disclosed the following: (1) a large proportion of the strains are quadruply resistant; (2) (CM.SM.SA) or (TC.SM.SA) resistance is prevalent among triply resistant strains; (3) most doubly resistant strains are (SM.SA)- or (TC.SA)-resistant; and (4) SA resistance is prevalent among singly resistant strains. Moreover, the proportion of singly CM- or SM-resistant strains is rather low (Table 2). Correlation of the distribution frequency of KM-resistant strains with resistance patterns to TC, CM, SM, and SA is shown in Table 3. The KM-resistant strains appeared in multiply resistant strains accompanied by quadruple and triple resistance to TC, CM, SM and SA; and singly and doubly KM-resistant strains were rather few in number [77,78,79a,80,81].

Demonstration frequencies of R plasmids encoding resistance to SM, KM, and gentamicin (GM) shown in Table 4 indicate that most SM, KM, and GM resistance in gram-negative bacteria is mediated by R plasmids. The remaining strains resistant to the drugs were found to be mostly due to the presence of nonconjugative (r) resistance plasmids [i.e., r(SM), r(KM), r(SM.SA), r(SM.SA.KM)] [72,78]. Similarly, most SM- and KM-resistant S. aureus strains were found to be due to the presence of r plasmids [i.e., r(SM), r(KM), and r(GM)] [32,72]. These results coincide with the findings that the biochemical mechanisms of aminoglycoside resistance in clinical isolates are due primarily to inactivation of the drugs and are different from those of chromosome-mediated resistance to the drugs.

Table 3   Isolation Frequency of Strains Resistant to KM in Relation to Resistance Patterns to TC, CM, SM, and SA[a]

Resistance patterns[b]	Isolation frequency (%) of KM-resistant strains				
	E. coli	Shigella	Salmonella	Proteus	Klebsiella
Quadruple	64.2	75.9	15.5	75.4	66.7
Triple	22.6	8.6	59.9	21.3	20.5
Double	7.5	10.3	20.9	3.3	6.4
Single	5.7	1.7	3.7	0	1.3
Sensitive	0	3.4	0	0	5.1

[a]The results are based on surveys of 14,530 strains of clinical isolates.
[b]Resistance to TC, CM, SM, and SA

Table 4 Demonstration Frequency of R Plasmids Carrying SM, KM, or GM Resistance from Resistant Strains[a]

Drug	Isolation frequency (%) of R plasmids encoding resistance to:							
	E. coli	Shigella	Salmonella	Proteus	K. pneumoniae	E. cloacae	S. marcescens	P. aeruginosa[b]
SM	61.0	85.3	57.3	53.1	60.0	52.0	70.6	23.0
KM	69.0	78.5	62.5	54.5	58.3	58.3	79.1	14.0
GM	89.0				66.0	78.6		26.0

[a]The results are based on surveys of 19,984 strains of clinical isolates.
[b]Highly resistant strains.

Streptomycin-phosphorylating enzyme was found from nonconjugative r(SM.SA) and conjugative R(SM.SA) and R(TC.SM.SA) plasmids. Streptomycin-adenylylating enzyme was found from r(SM/SP.SA), R(SM/SP.SA), R(TC.SM/SP.SA), R(CM.SM/SP.SA), and R(CM.TC.SM/SP.SA) plasmids encoding resistance to both SM and spectinomycin (SP). The mechanisms of SA resistance (SA_1) mediated by R(TC.CM.SM.SA), R(CM.SM.SA), and R(TC.SM.SA) are due primarily to the diminished uptake of the drug. By contrast, the mechanism of SA resistance (SA_2) mediated by r(SA), R(SA), r(SM.SA), R(SM.SA), and R(TC.SM.SA) is due mostly to the formation of SA-resistant dihydropteroate synthetase (DHPS) [78,80,86]. From these results, we proposed two pathways of the formation of multiple-resistance plasmids: (1) $r(SA_1)$ → $r(SM/SP.SA_1)$ → $R(TC.SM/SP.SA_1)$ or $R(CM.SM/SP.SA_1)$ → $R(TC.CM.SM/SP.SA_1)$, and (2) $r(SA_2)$ → $r(SM.SA_2)$ → $R(SM.SA_2)$ → $R(TC.SM.SA_2)$.

II. Aminoglycoside-Inactivating Enzymes

The mechanisms of aminoglycoside resistance in bacteria are known to be due primarily to the enzymatic inactivation of the drugs using nucleoside triphosphate or acetyl-CoA as coenzyme (i.e., nucleotidylation or acetylation). Aminoglycoside antibiotics are inactivated by phosphorylation or adenylylation of hydroxyl groups of the drugs in the presence of ATP. They are also inactivated by acetylation of their amino groups in the presence of acetyl-CoA. The inactivating enzymes are summarized in Table 5.

A. Aminoglycoside Antibiotic Phosphotransferase (APH)

The enzyme that catalyzes the phosphorylation of the 3'-hydroxyl group of aminogluco-side antibiotics was found in *E. coli* carrying an R factor [15,54,55,92,93,106,114,117, 125,127,128], *Pseudomonas aeruginosa* [27,28,49,52,55,62,109], *Staphylococcus aureus* [26,53], and *Providencia stuartii* [64].

Three types of the KM phosphotransferase [i.e., APH(3')] were confirmed in R-bearing *E. coli* strains resistant to KM. APH(3')-I phosphorylates the 3'-hydroxyl group of KM, neomycin (NM), and lividomycin (LV), but has no effect on butirosins (BTs). APH(3')-II phosphorylates the 3'-hydroxyl group of BTs, but is ineffective with LV. APH(3')-III phosphorylates the 3'-hydroxyl group of both BTs and the 5''-hydroxyl group of LV, in addition to KM and NM.

Ozanne et al. [93] reported that *E. coli* R+ strains resistant to SM were capable of phosphorylating the 3''-hydroxyl group of the drug. It was also found that crude extracts from *P. aeruginosa* strains phosphorylate DH-SM, the 6-hydroxyl group of SM being presumed to be phosphorylated [46].

KM Phosphotransferase I; APH(3')-I–KM, NM, RM, Paromomycin (PM), and LV Resistance

APH(3')-I phosphorylates the 3'-hydroxyl group of KMs and the 5''-hydroxyl group of LV, but does not affect the 3'-hydroxyl group of BTs. KM-A, -B, -C, ribostamycin (RM), gentamicin (GM)-A, NM, PM, neamine, and paromamine contain the 3'-hydroxyl group and are the substrates for this enzyme. $GM-C_1$, $-C_{1a}$, $-C_2$, sisomicin (SS), tobramycin (TM),

Table 5   Aminoglycoside-Modifying Enzymes in Plasmid-Containing Strains[a]

Aminoglycoside-modifying enzymes	Abbreviation	Substrates
1. Phosphorylation		
Aminoglycoside	APH(3')-I	KM, NM, PM, RM, LV(5''-OH)
phosphotransferase	APH(3')-II	KM, NM, PM, RM, BT
	APH(3')-III	KM, RM, BT, LV(5''-OH)
	APH(5'')	RM
	APH(2'')	GM, SS, KM, DKB, TM
	APH(3'')	SM
	APH(6)	SM
2. Acetylation		
Aminoglycoside	AAC(6')-1	KM, KM-B
acetyltransferase	AAC(6')-2	KM, KM-B, GM-C_{1a}, GM-C_2
		KM, KM-B, GM-C_{1a}, GM-C_2, DKB
	AAC(6')-3	KM, KM-B, GM-C_{1a}, GM-C_2, DKB,
	AAC(6')-4	AK
	AAC(3)-I	GM, SS
	AAC(3)-II	GM, SS, KM-B, TM
	AAC(3)-III	GM, SS, KM-A, -B, NM, PM, TM
	AAC(3)-IV	GM, KM-A, NM, TM, AK, RM, NTL, BT, APM
	AAC(2')	GM, KM-B, NM, RM, BT, DKB, LV(5''-OH)
3. Adenylyltransferase		
Aminoglycoside	AAD(2'')	GM, KM, DKB
adenylyltransferase	AAD(4')	KM, TM, NM, PM, BT, AK
	AAD(4',4'')	KM, NM, TM, GM-A, -B, BT, RM
	AAD(3'')	SM, SP
	AAD(6)	SM

[a] Abbreviations: KM, kanamycin; NM, neomycin; PM, paromomycin; RM, ribostamycin; LV, lividomycin; BT, butirosin; SM, streptomycin; GM, gentamicin; DKB, dibekacin; TM, tobramycin; AK, amikacin; SP, spectinomycin; SS, sisomicin; NTL, netilmicin; APM, apramycin; PH, phosphorylation; AC, acetylation; AD adenylylation.

and 3',4'-dideoxy-KM-B (DKB) [111], 3'-deoxy-KM-A [110], 3'-deoxy-KM-B [102], and 3'4'-dideoxyneamine [112] are not phosphorylated by APH(3')-I.

The LV-phosphorylating enzyme was demonstrated in LV-resistant *P. aeruginosa* strains [52] and in LV-resistant *E. coli* strains carrying the R factor [128], and the inactivated product of LV was a monophosphorylated LV, in which the D-ribose moiety of LV was phosphorylated [128]. Umezawa et al. [114] and Kondo et al. [55] indicated that the phosphorylated site of the inactivated LV was the 5''-hydroxyl group of the D-ribose moiety and confirmed the inactivated product by chemical synthesis of LV 5''-phosphate [130].

An epidemiological survey of KM-resistant strains isolated from clinical specimens shows KM-LV cross resistance in gram-negative bacteria and staphlococci. It was proved that this type of resistance is due to phosphorylation of the 3'-hydroxyl group of KM and the 5''-hydroxyl group of LV [114]. KM-A inhibits the phosphorylation of LV and LV inhibits the phosphorylation of KM by APH(3')-I.

RM contains two hydroxyl groups which are phosphorylated by phosphotransferase, the 3'-hydroxyl group of the 2, 6-diamino-2, 6-dioxy-D-glucose moiety, and the 5''-hydroxyl group of the ribose moiety. Phosphotransferase I phosphorylates the 3'-hydroxyl group of RM but does not affect the 5''-hydroxyl group. In a molecular model of RM, these hydroxyl groups are located close to each other. But the inactivated product of RM is RM 3'-phosphate [114] and the enzyme phosphorylates the 5''-hydroxyl group of 3',4'-dideoxy-RM, which lacks the 3'-hydroxyl group, the strains capable of producing APH(3')-I being resistant to 3',4'-dideoxy-RM [118]. (See Figure 1.)

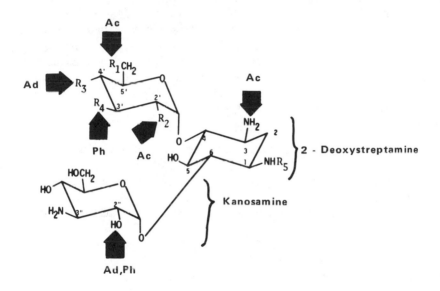

	R_1	R_2	R_3	R_4	R_5
Kanamycin A	NH_2	OH	OH	OH	H
B	NH_2	NH_2	OH	OH	H
C	OH	NH_2	OH	OH	H
Amikacin	NH_2	OH	OH	OH	$CO-CH(OH)-CH_2CH_2NH_2$
Dibekacin	NH_2	NH_2	H	H	H
Tobramycin	NH_2	NH_2	OH	H	H

Figure 1 Chemical structure of kanamycins. Arrows indicate the sites of phosphorylation (Ph), acetylation (Ac), and adenylylation (Ad).

KM Phosphotransferase II; APH(3')-II—KM, NM, RM, PM,
and BT Resistance

It was reported that another type of KM-NM phosphotransferase exists in *E. coli* carrying an R factor, called APH(3')-II. This enzyme phosphorylates the 3'-hydroxyl group of KMs, paromamine, neamine, NM, RM, and BTs, but does not affect the 5"-hydroxyl group of LV [125]. *E. coli* carrying the R factor, which encodes the production of APH(3')-II, is susceptible to LV, which is not phosphorylated by this enzyme. TM inhibits the phosphorylation of NM-B catalyzed by APH(3')-I, but has no effect on the reaction catalyzed by APH(3')-II [15]. Umezawa et al. [117] found that phosphotransferase II does not strongly phosphorylate the 3'-hydroxyl group of 4'-deoxy-KM-A compared with that of KM-A, suggesting that the 4'-hydroxyl group is involved in binding this enzyme to the drug. But APH(3')-I phosphorylates the 3'-hydroxyl group of 4'-dideoxy-KM-A as well as KM-A to the same degree. Umezawa et al. [116] found that if the enzymes are immobilized, the inactivated products are easily purified. APH(3')-II enzyme was purified by Matsuhashi et al. [67] and Goldman and Northrop [30]. The molecular weight of purified APH(3')-II enzyme was found to be 25,000 [67]. Matsuhashi et al. [65] compared APH(3')-I enzyme with APH(3')-II enzyme, indicating that APH(3')-I is different from APH(3')-II in chromatographic behavior, molecular weight, pH optimum, and Ki values. Furthermore, they reported no immunological cross reaction between anti-APH(3')-II and APH(3')-I or APH(3')-III [66].

KM Phosphotransferase III; APH(3')-III—KM, RM, BT,
and LV Resistance

Marengo et al. [64] reported that crude enzyme extracts of *P. stuartii*, resistant to KM, LV, and BT-B, phosphorylate KM, LM, and BT-B using ATP as a coenzyme. It is not certain whether the phosphorylation of these drugs is due to a single enzyme or to the presence of two enzymes. APH(3')-III phosphorylates the 3'-hydroxyl group of KM-A, RM, and BT-A and the 5"-hydroxyl group of LV. This enzyme was strongly inhibited by KM-A and RM at concentrations of 6 μM or more. The enzyme was labile and its molecular weight was estimated to be 25,500 using gel filtration [119].

Recently, Courvalin and Davies [18] reported a new type of APH(3') enzyme from *S. aureus*. This enzyme phosphorylates AK but does not confer detectable AK resistance to the strain, suggesting that the low AK resistance is due to a high K_m value for AK. This new type of APH(3') enzyme does not cross-react with an antiserum prepared against APH(3')-II enzyme. (See Figure 2.)

RM 5"-Phosphotransferase; APH(5")

RM was found to be converted to two kinds of inactivated RM [45]. RM 5"-phosphate accounted for approximately 99% of the inactivated RMs, and the remaining portion corresponded to RM 3"-phosphate. The hydroxyl group at the C5 position of the ribose moiety was preferentially phosphorylated by the inactivating enzyme in *P. aeruginosa*, even though the hydroxyl group at the C3 position of the 2,6-diaminoglucose moiety was available.

GM, KM-Phosphotransferase; APH(2")—GM, SS, KM, TM, DKB Resistance

GM-, KM-phosphorylating enzyme was reported in GM-, KM-resistant strains of *S. aureus* [29,43,61,99,100]. This enzyme inactivates GMs, SS, and KM, TM, and DKB

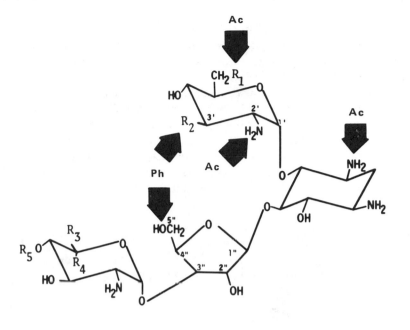

	R_1	R_2	R_3	R_4	R_5
Neomycin B	NH_2	OH	H	CH_2NH_2	H
Neomycin C	NH_2	OH	CH_2NH_2	H	H
Paromomycin I	OH	OH	H	CH_2NH_2	H
Paromomycin II	OH	OH	CH_2NH_2	H	H
Lividomycin A	OH	H	H	CH_2NH_2	mannose
Lividomycin B	OH	H	H	CH_2NH_2	H

Figure 2 Chemical structure of neomycin, paromomycin, and lividomycin.

by phosphorylation of the 2″-hydroxyl group of the drugs. The phosphorylated products by *S. aureus* were 2″-phosphoryl-SS [61] and 2″-phosphoryl-GM C_1 [43].

KM Phosphotransferase from Various Organisms

P. aeruginosa strains are mostly resistant to aminoglycoside antibiotics, such as KMs, NM, PM, and RM, and have enzymes that inactivate these antibiotics by phosphorylation of the 3′-hydroxyl group of the drugs. KM phosphotransferase was also found in *P. aeruginosa* [27,28,49,52,109], *S. aureus* [26,53], and *Providencia* [64]. Kida et al. [45] reported that a crude extract from *P. aeruginosa* preferentially phosphorylates the 5″-hydroxyl group rather than the 3″-hydroxyl group of RM. KM phosphotransferase has been found in KM-resistant staphylococci, and cross resistance between KMs and LV is frequently observed, suggesting the presence of APH(3′)-I. Kobayashi et al. [53] reported that the inactivated product of LV by an enzyme prepared from a KM-resistant

S. aureus strain was identical with that produced by *E. coli* carrying an R factor (i.e.,
5″-phosphoryl-LV). Butirosin-producing *Bacillus circulans* is known to contain an aminoglycoside phosphotransferase that is similar to the APH(3′) enzyme demonstrated from
clinical isolates resistant to the drug. Courvalin et al. [19] reported that the gene governing
the production of APH(3′) in BT-producing *B. circulans* can be introduced into *E. coli*,
where it is expressed as a determinant of resistance to aminoglycoside antibiotics. The
APH(3′) from *B. circulans* was purified in a homogeneous state and characterized. Anti-
APH(3′)-II or anti-APH(3′)-III serum did not cross-react with APH(3′) enzymes from
B. circulans or from NM-producing *Streptomyces fradiae* [25]. Studies of nucleic acid
homologies between genes coding for APH(3′) enzymes found no detectable homology
between the genes from antibiotic-producing strains and R plasmids in gram-negative
organisms [20]. (See Figure 3.)

SM Phosphotransferase

SM is also inactivated by phosphorylation of the drug by SM-resistant strains in the
presence of ATP (i.e., SM 3″-phosphotransferase and SM 6-phosphotransferase). SM-

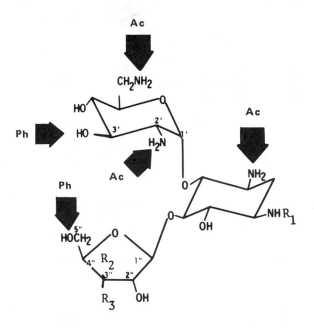

	R₁	R₂	R₃
Ribostamycin	H	H	OH
Butirosin A	-CO-CH(OH)-CH₂-CH₂NH₂	OH	H
Butirosin B	-CO-CH(OH)-CH₂-CH₂NH₂	H	OH

Figure 3   Chemical structure of ribostamycin and butirosin A.

phosphorylating enzymes have been reported in SM-resistant strains of *E. coli* [92,93, 116] and of *P. aeruginosa* [8,27,35,48,51].

SM 3″-Phosphotransferase; APH(3″)–SM Resistance. Ozanne et al. [93] reported that *E. coli* strains carrying an R factor which are SM-resistant and spectiomycin (SP)-sensitive have been found to phosphorylate the 3″-hydroxyl group of SM. Although it is known that the phosphorylating enzyme as well as the adenylylating enzyme reacts with the same hydroxyl group on N-methyl-L-glucosamine, the chemical structure of the inactivated product of SP has not been investigated fully. SM-resistant but SP-sensitive strains produce the SM-phosphorylating enzyme, and the strains resistant to both SM and SP produce the adenylylating enzyme [1]. It was reported that DH-SM was phosphorylated by *P. aeruginosa* [48], and the phosphorylated product was determined to be the 3″-hydroxyl group of DH-SM [35,84].

SM 6-Phosphotransferase; APH(6)–SM Resistance. This enzyme inactivates SM by phosphorylation of the hydroxyl group at the C_6 position of the N-methyl-L-glucosamine moiety of SM. Kida et al. [46] reported that crude extracts from *P. aeruginosa* phosphorylated DH-SM and the phosphorylated product showed the same Rf value as that of DH-SM 6-phosphate, which was formed by an extract prepared from a SM-producing strain of *Streptomyces,* as reported by Walker and Skorvaga [120].

SM 6-kinase is known to be present in extracts from SM-producing strains of *Streptomyces* [68], but SM 3″-kinase activity occurs in a nonstreptomycin-producing strain of *Streptomyces* [120]. 3″-Kinase from *Streptomyces* will be studied in the future to ascertain whether or not there is an evolutionary relationship between these enzymes [i.e., APH(6) from SM-resistant *P. aeruginosa* and APH(3″) from SM-resistant *Pseudomonas* and enteric bacteria carrying R plasmids] [4]. (See Figure 4.)

B.   Aminoglycoside Antibiotic Acetyltransferase (AAC)

Okamoto and Suzuki [90] reported that a cell-free extract from an *E. coli* strain carrying an R factor inactivated KM-A in the presence of acetyl-CoA. Umezawa et al. [107] purified the inactivated product of KM-A and determined that it was 6′-N-acetyl-KM-A.

Mitsuhashi [74] classified the aminoglycoside-aminocyclitol 6′-N-acetyltransferase into four groups [i.e., AAC(6′)-1, -2, -3, and -4] on the basis of substrate profiles and susceptibility to the drugs. The strains capable of producing AAC(6′)-1, -2, -3, and -4 are resistant, respectively, to (KMs-NM), (KMs, NM, GM-C_{1a}, -C_2), (KMs, NM, GM-C_{1a}, -C_2, DKB), and (KMs, NM, GM-C_{1a}, -C_2, DKB, amikacin (AK) [44]), resulting from the acetylation of these drugs using acetyl-CoA as a coenzyme.

GM 3-acetyltransferase in *P. aeruginosa* was reported by Mitsuhashi et al. [70], and was also demonstrated from R-bearing *Klebsiella* strains resistant to GM [47]. GM was inactivated by acetylation of the 2′-hydroxyl group of GM [17,83,129]. The aminoglycoside acetylatransferases (AACs) can be classed into three groups: AAC(6′), AAC(3), and AAC(2′).

Aminoglycoside 6′-N-Acetyltransferase; AAC(6′)–KM, GM, SS, DKB,
and AK Resistance

This enzyme inactivates KMs, GM-C_{1a}, -C_2, SS, DKB, and AK by acetylation of the 6′-amino group of the drugs. KM was found to be inactivated by *E. coli* carrying an R

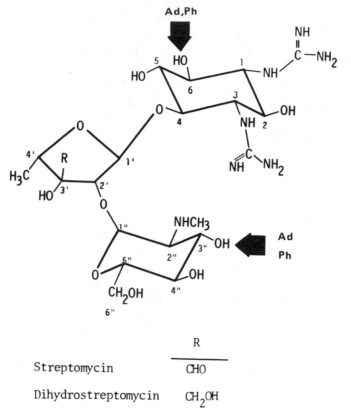

	R
Streptomycin	CHO
Dihydrostreptomycin	CH_2OH

Figure 4 Chemical structure of streptomycin.

factor [90,91], and the inactivated product of KM-A was determined to be 6′-N-acetyl-KM-A [107]. Similarly, Benveniste and Davies [2] purified the products acetylated by *E. coli* carrying an R factor (i.e., N-acetyl-KM-A, N-acetyl-KM-B, N-acetyl-NM-B, and N-acetyl-GM-C_{1a}). Moreover, KM 6′-N-acetyltransferase inactivates SS [87,88], KM-A [107], RM [131], DKB [37,124], and AK [41]. Based on these findings, Mitsuhashi [74] classified 6′-N-acetyltransferases encoded by R plasmids into four groups [i.e., AAC(6′) -1, -2, -3, and -4] on the basis of substrate profiles and susceptibility of R^+ strains to drugs. The enzyme AAC(6′)-1 acetylates KM-A, -B, and NM, but does not affect other aminoglycoside antibiotics, such as GMs, DKB, and AK. The enzyme AAC(6′)-2 from *Moraxella* strains [59] inactivates KM-A, -B, NM, and GM-C_{1a}, but does not affect DKB and AK. From our stock cultures we were also able to isolate two types of *P. aeruginosa* strains resistant to DKB—(DKB^r.AK^r.6′-methyl-DKB^s) and (DKB^r.AK^s.6′-methyl-DKB^r) (r, resistant; s, sensitive) [39–41,75,113] —which can inactivate the drugs by acetylation. Two acetylating enzymes were extracted and purified by affinity chromatography. Enzymological studies of the inactivation reaction and chemical studies of the inactivated products indicated that these drugs were inactivated by acetylation of the 6′-amino group. One enzyme, aminoglycoside 6′-N-acetyltransferase 3 [39], acetylates KM-A, -B, GM-C_{1a}, -C_2, DKB, and 6′-N-methyl-KM-B, but does not affect AK. Another enzyme, aminoglycoside 6′-N-acetyltransferase 4 [41], acetylates KM-A, -B, GM-C_{1a}, -C_2, DKB, and AK.

Based on epidemiological studies of aminoglycoside antibiotic resistance in *P. aeruginosa*, we were able to demonstrate (DKB^r.AK^s) or (DKB^r.AK^r) strains, whose resistance was encoded by R plasmids. AK was produced by acylation of the 1-amino group of KM-A with γ-amino-α-hydroxybutyric acid, but has the hydroxyl groups at the C3' and C2'' positions. The 3'-hydroxyl and 2''-hydroxyl groups of KM-A are known to be phosphorylated or adenylylated by enzymes from R-bearing strains, but strains capable of inactivating KM-A by 3'-phosphotransferase or 2''-adenylyltransferase were found to be susceptible to AK and could not inactivate the drug [95]. Furthermore, AK has an amino group at the C6' and C3 positions, but the strains possessing R plasmid encoding the formation of 6-N-acetyltransferases of AAC(6')-1, -2, and 3 are susceptible to AK. The strains carrying R plasmid encoding the formation of AAC(6')-4 are resistant to AK and inactivate the drug. These results indicate that the acylation of the 1-amino group of KM-A can protect the acetylation of the 6-amino group from some 6'-N-acetyltransferases, such as AAC(6')-1, -2, and -3.

Mitsuhashi et al. [76] reported on the biochemical mechanism of KM resistance in *Mycobacterium tuberculosis* isolated from tuberculosis patients. The chemical structure of the inactivated KM by *M. tuberculosis* was determined to be 6'-N-acetyl-KM by thin-layer chromatography using an authentic sample of 6'-N-acetyl-KM.

Aminoglycoside 3-Acetyltransferase

GM 3-Acetyltransferase I; AAC(3)-I—GM and SS Resistance. This enzyme inactivates GMs and SS by acetylation of the amino group at the C3 position of the 2-deoxystreptamine moiety. Mitsuhashi et al. [70,71] reported that the crude extracts from *P. aeruginosa* strains inactivate GM-C_1, -C_{1a}, and -C_2 in the presence of acetyl-CoA. The acetylated position of the inactivated product of GM-C_1 was concluded to be the amino group at the C_3 position of the 2-deoxystreptamine moiety of the drug [50]. A similar acetylating enzyme, AAC(3)-I, was also demonstrated in *E. coli* [115]. Brzezinska et al. [14] partially purified this enzyme and reported that the enzyme inactivates GMs and SS as excellent substrates, whereas closely related antibiotics, such as TM, KM-A, -B, and -C, are either poor substrates or are not acetylated. Williams and Northrop [121] purified and characterized AAC(3)-I by affinity and ion-exchange chromatography. Witchitsz [122] isolated *Enterobacter, Serratia, Klebsiella*, and *E. coli* strains possessing R-mediated resistance to GMs. They are resistant to GMs by acetylation of the amino group at the C3 position of the drugs [115]. The enzyme AAC(3)-I acetylates the 3-amino group of GM-C_1, -C_{1a}, -C_2, and SS, but does not affect KMs, TM, DKB, NM, and LV.

Aminoglycoside 3-Acetyltransferase II; AAC(3)-II—GM, SS, TM, and KM Resistance. Le Goffic et al. [58] described a GM acetyltransferase prepared from a *Klebsiella* strain possessing R-mediated resistance to GM, TM, and KM. This enzyme, AAC(3)-II, acetylates the amino group at the C3position of GM, SS, TM, and KM. It is different from AAC(3)-I in substrate profiles, the resulting resistance patterns, and isoelectric point—the isoelectric points of AAC(3)-I and -II being 7.4 and 6.4, respectively.

Aminoglycoside 3-Acetyltransferase III; AAC(3)-III—GM, SS, TM, KM, and PM Resistance. Biddlecome et al. [6] isolated AAC(3)-III enzyme from *P. aeruginosa*. AAC(3)-III has a much wider substrate range than does AAC(3)-II, and inactivates GM, SS, netilmicin (NTL), KM, TM, NM, and PM. Antiserum prepared against AAC(3)-I did not cross-react with AAC(3)-II, AAC(3)-III, or AAC(3)-IV [25].

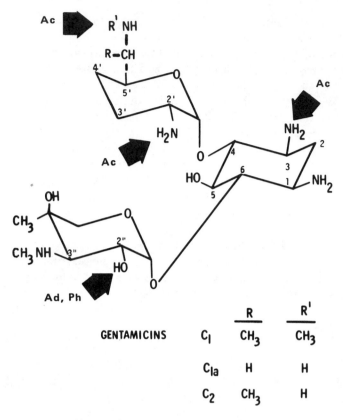

Figure 5   Chemical structure of gentamicins.

Aminoglycoside 3-Acetyltransferase IV; AAC(3)-IV—GM, KM, TM, Apramycin Resistance. Davies and O'Connor [24] reported a plasmid-determined AAC(3)-IV from strains of *E. coli, Klebsiella* sp., and *Arizonae* sp. resistant to apramycin (AP). AAC(3)-IV has the broadest substrate range among AAC(3) enzymes and inactivates GM, SS, NTL, KM, TM, NM, PM, and AP.

AAC(3)-I inactivates fortimicin (FT) by acetylation, but AAC(3)-II could not inactivate the drug [89]. It was also found that fortimicin is a substrate for AAC(3)-I, but not for any of other AAC(3) enzymes [i.e., AAC(3)-II, AAC(3)-III, and AAC(3)-IV] (see Figures 7 and 8) [25].

Aminoglycoside 2′-Acetyltransferase; AAC(2′)—LV, RM, BT, GM, KM-B, -C, DKB, and NM Resistance

This enzyme inactivates LV, RM, BT, GM, KM-B, -C, DKB, and NM by acetylation of the 2′-amino group of the drugs. *Providencia* strains resistant to many aminoglycoside antibiotics, except KM-A and AK, were isolated and their resistance was found to be due to acetylation of the 2′-amino group [17,83,129]. AAC(2′) catalyzes the acetylation of the 2′-amino group of GMs, KM-B, -C, TM, DKB, NM, BT, RM, and SS, but cannot inactivate KM-A and AK, which lack the 2′-amino group.

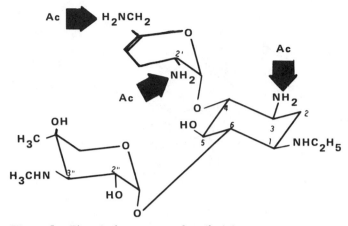

Figure 6 Chemical structure of netilmicin.

C. Aminoglycoside Antibiotic Adenylyltransferase (AAD)

SM Adenylyltransferase; AAD(3″)–SM, SP Resistance

This enzyme inactivates SM and SP by adenylylation of the 3″-hydroxyl group of SM and the 9-hydroxyl group of SP. Umezawa et al. [108] reported R-mediated SM resistance in *E. coli* resulting from adenylylation of the drug. The inactivated product of the drug was determined to be 3″-adenylyl SM [103,126].

A similar enzyme reported by Harwood and Smith [31] adenylylates SP, DH-SM, actinamine, SM, and bluensomycin [98]. Benveniste et al. [1] reported the enzymatic adenylylation of SM and SP by an *E. coli* R⁺ strain. They isolated a mutant sensitive to both SM and SP from an *E. coli* strain resistant to both drugs and showed that this mutant cannot produce an enzyme capable of adenylylating both drugs. But the revertant mutants resistant to SM and SP regained the activity of adenylylation of both drugs. Therefore, they concluded that the inactivation of the two drugs is catalyzed by the same enzyme.

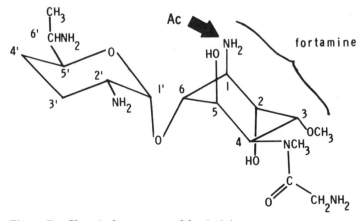

Figure 7 Chemical structure of fortimicin.

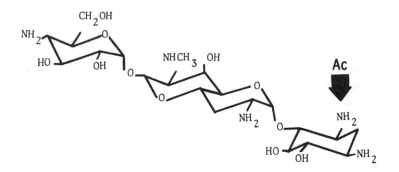

Figure 8   Chemical structure of apramycin.

An SM-sensitive and SP-resistant *S. aureus* strain was found to inactivate SP by adeny-lylation, although the adenylylated position of SP has not been established [38]. SP adenylyltransferase (AAD) [9] is produced by *S. aureus,* and inactivates SP but not SM [25,29].

SM 6-Adenylyltransferase; AAD(6)–SM Resistance

This enzyme inactivates SM by adenylylation of the 6-hydroxyl group of the drug. Kawabe and Mitsuhashi [36] reported intermediate SM resistance in *S. aureus* strains, which inactivate the drug by adenylylation. We have further disclosed by epidemiological surveys that there are three types of resistance to SM and SP in staphylococci: $(SM^s.SP^r)$, $(SM^r.SP^s)$, and $(SM^s.SP^r)$. The $(SM^s.SP^r)$ and (SM^rSP^s) mutants of *S. aureus* could be obtained from a strain $(SM^r.SP^r)$ by transduction or by elimination of resistance. *S. aureus* $(SM^r.SP^r)$ was found to inactivate both SM and SP by adenylylation. Genetic analysis disclosed that the genes governing resistance to SM and SP are located separately on different nonconjugative (r) plasmids. We purified an inactivated product of SM by *S. aureus* $(SM^r.SP^s)$ and showed it to be different from 3″-adenylyl SM using electro-phoresis [38]. Adenylylstreptidine was obtained from the methanolysis product of adenylyl SM and was determined to be 6-adenylylstreptidine by elemental analysis and periodate consumption. The adenylylated product of SM was, therefore, determined to be 6-adenylyl SM [101]. These facts indicate that there is a type of SM resistance in staphylococci resulting from the production of 6-adenylyl SM which is different from R-mediated (SM.SP) resistance resulting from adenylylation of the 3″-hydroxyl group of SM.

GM-KM 2″-Adenylyltransferase; AAD(2″)–GMs, KMs, TM, and DKB Resistance

This enzyme inactivates GMs, KMs, TM, and DKB by adenylylation of the 2″-hydroxyl group of the drugs. It was reported that *Klebsiella pneumoniae* carrying an R plasmid in-activated KM, GM-C_1, -C_{1a}, -C_2, SS [3,47], and DKB [85,123] by adenylylation of the drugs. The adenylylated DKB was concluded to be a monoadenylylated DKB by ultra-violet absorption at 260 nm, chemical analyses, and by the formation of both DKB and adenylic acid by hydrolyzation of the inactivated product by snake venom phosphodies-terase. Thus, the adenylyl DKB was found by NMR analysis to be DKB-2″-adenylate;

that is, the 2″-hydroxyl group of the 3-amino-3-deoxy-D-glucose moiety is adenylylated [85]. A synthesized 2″-deoxy-GM-C$_2$ was not inactivated by *K. pneumoniae* carrying an R plasmid encoding the production of AAD(2″) and inhibited the adenylylation of GMs by this enzyme [21], indicating that the 2″-hydroxyl group is adenylylated. A similar adenylylating enzyme was also demonstrated in *P. aeruginosa* [9,34].

TM 4′-Adenylyltransferase; AAD(4′)—TM, KM, NM, BTs, PM,
and AK Resistance

Santanam and Kayser [96] reported that *Staphylococcus epidermidis* strains resistant to aminoglycoside antibiotics have been shown to contain an enzyme which inactivates KMs, NM, BTs, PM, GM-A, AK, and TM by adenylylation. Le Goffic et al. [60] reported AAD(4′) which catalyzed the nucleotidylation of the 4′-hydroxyl group of the drugs. They purified the enzyme by affinity chromatography and characterized AAD(4′) enzyme. This enzyme also transfers the guanylyl and uridylyl groups, respectively, from guanosine triphosphate and uridine triphosphate to the 4′-hydroxyl group of the drugs. Recently, Santanam and Kayser [97] reported AAD(4′,4″) enzyme from *S. epidermidis*. This enzyme catalyzes the transfer of nucleoside monophosphate moiety from the nucleoside triphosphates to either the 4′-hydroxyl group or the 4″-hydroxyl group. They reported that the enzyme, modifying the two sites, appeared as a single and homogenious entity in affinity chromatography, in isoelectric focusing, and in gel filtration.

III. Aminoglycoside Resistance Mechanisms Without Inactivation

Three major mechanisms of resistance to aminoglycoside antibiotics are known: (1) the production of antibiotic-inactivating enzymes [5,23,25,79a,118]; (2) the change in one of the 30S ribosomal proteins, which is the site of action of the antibiotics [7,16,63, 94]; and (3) the decrease in membrane permeability of the drugs.

Tanaka [104] described GM resistance in a *P. aeruginosa* strain that did not inactivate the drug. Biochemical studies of this resistance disclosed that in vitro protein synthesis, using the ribosome of this strain, was not inhibited by GM, whereas ribosomes prepared from other *Pseudomonas* strains were sensitive to the drug, the resistance mechanism being due to a decrease in the ribosome's sensitivity to the drug.

The levels of SM resistance in *P. aeruginosa* were classified into three groups: susceptible, intermediate, and high-resistance [11,33,105]. There are two types of high SM resistance: (1) R-mediated resistance due to inactivation of the drug, and (2) a decrease in the ribosome sensitivity to SM. Intermediate SM resistance cannot be accounted for by either inactivation of the drug or decrease in the SM sensitivity of ribosome. R plasmids encoding intermediate SM resistance were not demonstrated from these strains, whose SM resistance was explainable by their reduced permeability to SM and hence by the diminished uptake of SM [105]. Kawabe et al. [42] reported that plasmid-mediated SM resistance in gram-negative enteric bacteria is due primarily to inactivation of the drug by phosphorylation or adenylylation. In *P. aeruginosa* strains, 20% of the R-mediated SM resistance is not due to inactivation of the drug. Biochemical studies of this type of SM resistance have disclosed the following: (1) a decrease in the accumulation of SM into the strains possessing this type of R plasmid, (2) enhancement of antibacterial activity of SM toward these strains by EDTA treatment, (3) increase in the whole-cell accumulation of SM into these strains after treatment with EDTA, and (4) no differences of [^{3}H] dihydro-SM binding to ribosomes prepared from R$^+$ and R$^-$ cells. We can conclude, therefore, that the mechanism of SM resistance mediated by this type of R plasmid is due to a decrease in the penetration of SM into bacterial cells [42].

Kono and O'Hara [57] reported that the mechanism of KM resistance in *P. aeruginosa* governed by an R plasmid was not due to the production of inactivating enzyme of the drug. This R plasmid did not produce any KM-inactivating enzymes, and its host ribosome was found to be sensitive to the drug.

They also reported that mechanisms of high SM resistance, showing a minimum inhibitory concentration (MIC) value of more than 51,200 μg/ml, were due to a decrease in the membrane permeability of the drug mediated by R plasmid and to the decrease in the ribosome sensitivity to the drug mediated by the chromosome [56].

Mills and Holloway [69] reported on mutants of *P. aeruginosa* which are tolerant (*tol*) to aeruginocin and also hypersensitive to aminoglycoside antibiotics. These mutants do not show any changes in susceptibility to a wide range of other toxic agents, including antibiotics and surfactants. This *tol* locus was mapped at 10 min from the FP2 origin. The hypersensitivity to aminoglycoside antibiotics appears to be due to increased permeability of the mutant for these antibiotics, since the mutant accumulates streptomycin at a faster rate than does the parent. These results indicate that the diminished uptake of aminoglycoside antibiotics is one of the mechanisms of resistance to the drugs, especially in *P. aeruginosa*.

Bryan and Van Den Elzen [10,12] reported that both GM and SM accumulation in drug-susceptible strains of *E. coli* and *P. aeruginosa* are reduced or prevented by a variety of inhibitors of electron transport, by agents that uncouple oxidative phosphorylation and electron transport, by a sulfhydryl poison, and by inhibitors of protein synthesis. These data demonstrate that the drug accumulation capable of causing cell death is caused by an active transport system. There are three phases of accumulation of SM or GM: energy-independent (initial), energy-dependent phase I (second phase), and energy-dependent phase II (third phase). The second phase of uptake kinetics represents transport of the aminoglycoside across the cell membrane by a process that uses energy obtained from electron transport and oxidative phosphorylation but not from membrane Mg-adenosine-5'-triphosphate (ATPase)-specified ATP hydrolysis [13].

The active transport of aminoglycoside antibiotics probably involves a system that transports structually related molecules into bacterial cells.

IV. Conclusions

According to studies of resistance mechanisms mediated by plasmids, we have noted that biochemical mechanisms of in vitro-developed resistant mutants are generally different from those of plasmid-mediated resistance, especially resistance to aminoglycoside antibiotics. It should be noted further that aminoglycoside-inactivating enzymes need coenzymes such as ATP and acetyl-CoA for inactivation of the drugs. We can postulate, therefore, that the origin of the aminoglycoside-resistant determinants on plasmids may be derived from the genes responsible for acetyltransferase, phosphotransferase, and nucleotidyltransferase, which are functional per se in normal bacterial cells.

We classified the aminoglycoside-inactivating enzymes mediated by plasmids [73,75, 79a]. Therefore, we can trace epidemiologically the resistance plasmids at hospitals, farms, and fish-culturing ponds by aminoglycoside-inactivating enzymes in addition to both drug-resistance patterns and incompatibility group. The plasmids encoding AAC(3) are rather few in Japan, although they have been demonstrated at higher frequencies in the United States and Europe. By contrast, the plasmids encoding AAC(6′) are predominant in Japan, probably because of the wide use of kanamycin, indicating the geographic epidemiology of resistance plasmids.

Plasmid epidemiology have also disclosed an increase in the resistance patterns of plasmid by expanding substrate profiles, for instance, from AAC(6′)-1 to AAC(6′)-2, from AAC(6′)-2 to AAC(6′)-3, and finally to AAC(6′)-4 after successive introduction of drugs such as KM, GM, DKB, and AK. After the introduction of DKB into practical use, R plasmids encoding AAC(6′), AAC(2′), APH(2″), and AAD(2″) have been isolated at higher frequencies in Japan, changing the patterns of resistance mediated by plasmids. Thus, the studies of the biochemical mechanisms of plasmid resistance have offered a way to know the evolutionary changes of R plasmids caused by the wide use of chemotherapeutic agents.

References

1. Benveniste, R., Yamada, T., and Davies, J.: Enzymatic adenylylation of streptomycin and spectinomycin by R-factor-resistant *Escherichia coli. Infect. Immun.* 1:109–119, 1970.
2. Benveniste, R., and Davies, J.: Enzymatic acetylation of aminoglycoside antibiotics by *Escherichia coli* carrying an R-factor. *Biochemistry.* 10:1787–1796, 1971.
3. Benveniste, R., and Davies, J.: R-factor mediated gentamicin resistance: a new enzyme which modifies aminoglycoside antibiotics. *FEBS Lett.* 14:293–296, 1971.
4. Benveniste, R., and Davies, J.: Aminoglycoside antibiotic-inactivating enzymes in *Actinomycetes* similar to those present in clinical isolates of antibiotic-resistant bacteria. *Proc. Natl. Acad. Sci. USA* 70:2276–2280, 1973.
5. Benveniste, R., and Davies, J.: Mechanisms of antibiotic resistance in bacteria. *Annu. Rev. Biochem.* 42:471–505, 1973.
6. Biddlecome, S., Haas, M., Davies, J., Miller, G.H., Rane, D.F., and Daniels, P.J.L.: Enzymatic modification of aminoglycoside antibiotics: a new 3-N-acetylating enzyme from a *Pseudomonas aeruginosa* isolate. *Antimicrob. Agents Chemother.* 9:951–955, 1976.
7. Bollen, A., Davies, J., Ozaki, M., and Mizushima, M.: Ribosomal protein conferring sensitivity to the antibiotic spectinomycin in *Escherichia coli. Science* 165:85–87, 1969.
8. Bryan, L.E., Van Den Elsen, H.M., and Tseng, J.T.: Transferable drug resistance in *Pseudomonas aeruginosa. Antimicrob. Agents Chemother.* 1:22–29, 1972.
9. Bryan, L.E., Shahrabadi, M.S., and Van Den Elzen, H.M.: Gentamicin resistance in *Pseudomonas aeruginosa:* R-factor-mediated resistance. *Antimicrob. Agents Chemother.* 6:191–199, 1974.
10. Bryan, L.E., and Van Den Elzen, H.M.: Gentamicin accumulation by sensitive strains of *Escherichia coli* and *Pseudomonas aeruginosa. J. Antibiot. (Tokyo)* 28:696–703, 1975.
11. Bryan, L.E., Van Den Elzen, H.M., and Shahrabadi, M.H.: The relationship of aminoglycoside permeability to streptomycin and gentamicin susceptibility of *Pseudomonas aeruginosa.* In *Microbial Drug Resistance,* edited by Mitsuhashi, S., and Hashimoto, H. Tokyo, University of Tokyo Press, 1975, p. 475.
12. Bryan, L.E., and Van Den Elzen, H.M.: Streptomycin accumulation in susceptible and resistant strains of *Escherichia coli* and *Pseudomonas aeruginosa. Antimicrob. Agents Chemother.* 9:928–938, 1976.

13. Bryan, L.E., and Van Den Elzen, H.M.: Effects of membrane-energy mutations and cations on streptomycin and gentamicin accumulation by bacteria: a model for entry of streptomycin and gentamicin in susceptible and resistant bacteria. *Antimicrob. Agents Chemother.* 12:163–177, 1977.

14. Brzezinska, M., Benveniste, R., Davies, J., Daniels, P.J.L., and Weinstein, J.: Gentamicin resistance in strains of *Pseudomonas aeruginosa* mediated by enzymatic N-acetylation of the deoxystreptamine moiety. *Biochemistry* 11:761–765, 1972.

15. Brzezinska, M., and Davies, J.: Two enzymes which phosphorylate neomycin and kanamycin in *Escherichia coli* strains carrying R factors. *Antimicrob. Agents Chemother.* 3:266–269, 1973.

16. Buckel, P., Buchberger, A., Böck, A., and Wittmann, H.G.: Alteration of ribosomal protein 16 in mutants of *Escherichia coli* resistant to gentamicin. *Mol. Gen. Genet.* 158:47–54, 1977.

17. Chevereau, M., Daniels, P.J.L., Davies, J., and LeGoffic, F.: Aminoglycoside resistance in bacteria mediated by gentamicin acetyltransferase II, an enzyme modifying the 2′-amino group of aminoglycoside antibiotics. *Biochemistry* 13:598–603, 1974.

18. Courvalin, P., and Davies, J.: Plasmid-mediated aminoglycoside phosphotransferase of broad substrate range that phosphorylates amikacin. *Antimicrob. Agents Chemother.* 11:619–624, 1977.

19. Courvalin, P., Weisblum, B., and Davies, J.: Aminoglycoside-modifying enzyme of an antibiotic-producing bacterium acts as a determinant of antibiotic resistance in *Escherichia coli*. *Proc. Natl. Acad. Sci. USA* 74:999–1003, 1977.

20. Courvalin, P., Fiandt, M., and Davies, J.: DNA relationships between genes coding for aminoglycoside-modifying enzymes from antibiotic-producing bacteria and R plasmids. In *Microbiology*, edited by Schlessinger, D. Washington, D.C., American Society for Microbiol., 1978, p. 19.

21. Daniels, P.J.L., Weinstein, J., Tkach, R.W., and Morton, J.: Gentamicin derivatives modified at the 2″-position. The preparation of 2″-epigentamicin C_1 and 2″-deoxygentamicin C_2. *J. Antibiot. (Tokyo)* 27:150–154, 1974.

22. Davies, J.: Bacterial resistance to aminoglycoside antibiotics. *J. Infect. Dis.* 124 (Suppl.)S7–S10, 1971.

23. Davies, J., Brzezinska, M., and Beneveniste, R.: R factors: biochemical mechanisms of resistance to aminoglycoside antibiotics. *Ann. N.Y. Acad. Sci.* 182:226–233, 1971.

24. Davies, J., and O'Connor, S.: Enzymatic modification of aminoglycoside antibiotics: 3-N-acetyltransferase with broad specificity that determines resistance to the novel aminoglycoside apramycin. *Antimicrob. Agents Chemother.* 14:69–72, 1978.

25. Davies, J., and Smith, D.I.: Plasmid-determined resistance to antimicrobial agents. *Annu. Rev. Microbiol.* 32:469–518, 1978.

26. Doi, O., Miyamoto, M., Tanaka, N., and Umezawa, H.: Inactivation and phosphorylation of kanamycin by drug-resistant *Staphylococcus aureus*. *Appl. Microbiol.* 16:1282–1284, 1968.

27. Doi, O., Ogura, M., Tanaka, N., and Umezawa, H.: Inactivation of kanamycin, neomycin, and streptomycin by enzymes obtained in cells of *Pseudomonas aeruginosa*. *Appl. Microbiol.* 16:1276–1281, 1968.

28. Doi, O., Kondo, S., Tanaka, N., and Umezawa, H.: Purification and properties of kanamycin-phosphorylating enzyme from *Pseudomonas aeruginosa*. *J. Antibiot. (Tokyo)* 22:273–282, 1969.

29. Dowding, J.E.: Mechanisms of gentamicin resistance in *Staphylococcus aureus*. *Antimicrob. Agents Chemother.* 11:47–50, 1977.

30. Goldman, P.R., and Northrop, D.B.: Purification and spectrophotometric assay of neomycin phosphotransferase II. *Biochem. Biophys. Res. Commun.* 69:230–236, 1976.

31. Harwood, J., and Smith, D.H.: Resistance factor-mediated streptomycin resistance. *J. Bacteriol.* 97:1262–1271, 1969.
32. Inoue, M., and Mitsuhashi, S.: The discovery of nonconjugative resistance (r) plasmids. In *Bacterial Drug Resistance*, edited by Mitsuhashi, S. Tokyo, Kodansha, 1979, p. 19.
33. Iyobe, S., Hasuda, K., Fuse, A., and Mitsuhashi, S.: Demonstration of R factors from *Pseudomonas aeruginosa. Antimicrob. Agents Chemoter.* 5:547–552, 1974.
34. Kabins, S., Nathan, C., and Cohen, S.: Gentamicin-adenylyltransferase activity as a cause of gentamicin resistance in clinical isolates of *Pseudomonas aeruginosa. Antimicrob. Agents Chemother.* 5:565–570, 1974.
35. Kawabe, H., Kobayashi, F., Yamaguchi, M., Utahara, R., and Mitsuhashi, S.: 3″-Phosphoryldihydrostreptomycin produced by the inactivating enzyme of *Pseudomonas aeruginosa. J. Antibiot. (Tokyo)* 24:651–652, 1971.
36. Kawabe, H., and Mitsuhashi, S.: Inactivation of dihydrostreptomycin by *Staphylococcus aureus. Jpn. J. Microbiol.* 15:545–548, 1971.
37. Kawabe, H., and Mitsuhashi, S.: Acetylation of dideoxykanamycin B by *Pseudomonas aeruginosa. Jpn. J. Microbiol.* 16:436–437, 1972.
38. Kawabe, H., Inoue, M., and Mitsuhashi, S.: Inactivation of dihydrostreptomycin and spectinomycin by *Staphylococcus aureus. Antimicrob. Agents Chemother.* 5:553–557, 1974.
39. Kawabe, H., Kondo, S., Umezawa, H., and Mitsuhashi, S.: R factor-mediated aminoglycoside antibiotic resistance in *Pseudomonas aeruginosa:* a new aminoglycoside 6′-N-acetyltransferase. *Antimicrob. Agents Chemother.* 7:494–499, 1975.
40. Kawabe, H., and Mitsuhashi, S.: 6′-N-acetyltransferase of aminoglycoside antibiotics in *Pseudomonas aeruginosa* carrying R factors. In *Microbial Drug Resistance*, edited by Mitsuhashi, S., and Hashimoto, H. Tokyo, University of Tokyo Press, 1975, p. 449.
41. Kawabe, H., Naito, T., and Mitsuhashi, S.: Acetylation of amikacin, a new semisynthetic antibiotic, by *Pseudomonas aeruginosa* carrying an R factor. *Antimicrob. Agents Chemother.* 7:50–54, 1975.
42. Kawabe, H., Umezawa, H., Iyobe, S., and Mitsuhashi, S.: Mechanisms of R-mediated streptomycin resistance in *Pseudomonas aeruginosa. Plasmid—Medical and Theoretical Aspects*, edited by Mitsuhashi, S., et al. Prague, Avicenum (Czechoslovak Medical Press), 1977, p. 365.
43. Kawabe, H., Naganawa, H., Kondo, S., Umezawa, H., and Mitsuhashi, S.: New plasmid-mediated phosphorylation of gentamicin C in *Staphylococcus aureus. Microbiol. Immunol.* 22:515–521, 1978.
44. Kawaguchi, H., Naito, T., Nakagawa, S., and Fujisawa, K.: BB-K8, a new semisynthetic aminoglycoside antibiotic. *J. Antibiot. (Tokyo)* 25:695–708, 1972.
45. Kida, M., Igarashi, S., Okutani, T., Asako, T., Hiraga, K., and Mitsuhashi, S.: Selective phosphorylation of the 5″-hydroxy group of ribostamycin by a new enzyme from *Pseudomonas aeruginosa. Antimicrob. Agents Chemother.* 5:92–94, 1974.
46. Kida, M., Asako, T., Yoneda, M., and Mitsuhashi, S.: Phosphorylation of dihydrostreptomycin by *Pseudomonase aeruginosa*. In *Microbial Drug Resistance*, edited by Mitsuhashi, S., and Hashimoto, H. Tokyo, University of Tokyo Press, 1975, p. 441.
47. Kobayashi, F., Yamaguchi, M., Eda, J., Higashi, F., and Mitsuhashi, S.: Enzymatic inactivation of gentamicin C components by cell-free extract from *Klebsiella pneumoniae. J. Antibiot. (Tokyo)* 24:719–721, 1971.
48. Kobayashi, F., Yamaguchi, M., and Mitsuhashi, S.: Inactivation of dihydrostreptomycin by *Pseudomonas aeruginosa. Jpn. J. Microbiol.* 15:381–382, 1971.
49. Kobayashi, F., Yamaguchi, M., and Mitsuhashi, S.: Phosphorylated inactivation of

aminoglycosidic antibiotics by *Pseudomonas aeruginosa. Jpn. J. Microbiol.* 15:
265–272, 1971.

50. Kobayashi, F., Yamaguchi, M., Eda, J., Hiramatsu, M., and Mitsuhashi, S.: Gentamicin
C-acetylating enzyme from resistant strains of *Pseudomonas aeruginosa* and the
structure of the acetylated product. *Gunma Rep. Med. Sci.* 5:291–301, 1972.

51. Kobayashi, F., Yamaguchi, M., Sato, J., and Mitsuhashi, S.: Purification and properties of dihydrostreptomycin-phosphorylating enzyme from *Pseudomonas aeruginosa.*
Jpn. J. Microbiol. 16:15–19, 1972.

52. Kobayashi, F., Yamaguchi, M., and Mitsuhashi, S.: Activity of lividomycin against
Pseudomonas aeruginosa: its inactivation by resistant strains. *Antimicrob. Agents
Chemother.* 1:17–21, 1972.

53. Kobayashi, F., Koshi, T., Eda, J., Yoshimura, Y., and Mitsuhashi, S.: Lividomycin
resistance in staphylococci by enzymatic phosphorylation. *Antimicrob. Agents
Chemother.* 4:1–5, 1973.

54. Kondo, S., Okanishi, M., Utahara, R., Maeda, K., and Umezawa, H.: Isolation of
kanamycin and paromamine inactivated by *E. coli* carrying R factor. *J. Antibiot.
(Tokyo)* 21:22–29, 1968.

55. Kondo, S., Yamamoto, H., Naganawa, H., Umezawa, H., and Mitsuhashi, S.: Isolation
and characterization of lividomycin A inactivated by *Pseudomonas aeruginosa*
and *Escherichia coli* carrying R factor. *J. Antibiot. (Tokyo)* 25:483–484, 1972.

56. Kono, M., and O'Hara, K.: Mechanisms of streptomycin (SM)-resistance of highly
SM-resistant *Pseudomonas aeruginosa* strains. *J. Antibiot. (Tokyo)* 29:169–175,
1976.

57. Kono, M., and O'Hara, K.: Kanamycin-resistance mechanism of *Pseudomonas aeruginosa* governed by an R-plasmid independently of inactivating enzymes. *J. Antibiot. (Tokyo)* 30:688–690, 1977.

58. Le Goffic, F., Martel, A., and Witchitz, J.: 3-N enzymatic acetylation of gentamicin,
tobramycin, and kanamycin by *Escherichia coli* carrying an R factor. *Antimicrob.
Agents Chemother.* 6:680–684, 1974.

59. Le Goffic, F.: Resistance of moraxella to tobramycin, kanamycin and BB-K8. In
Drug-Inactivating Enzymes and Other Problems of Resistant Bacteria, edited by
Mitsuhashi, S., et al. Prague, Avicenum (Czechoslovak Medical Press), 1975, p. 165.

60. Le Goffic, F., Martel, A., Capmau, M.L., Beca, B., Goebel, P., Chardan, H., Soussy,
C.J., Duval, J., and Bouanchaud, D.H.: New plasmid-mediated nucleotidylation of
aminoglycoside antibiotics in *Staphylococcus aureus. Antimicrob. Agents Chemother.*
10:258–264, 1976.

61. Le Goffic, F., Martel, A., Moreau, N., Capmau, M.L., Soussy, C.J., and Duval, J.:
2''-O-phosphorylation of gentamicin components by a *Staphylococcus aureus* strain
carrying a plasmid. *Antimicrob. Agents Chemother.* 12:26–30, 1977.

62. Maeda, K., Kondo, S., Okanishi, M., Utahara, R., and Umezawa, H.: Isolation of
paromamine inactivated by *Pseudomonas aeruginosa. J. Antibiot. (Tokyo)* 21:458–
459, 1968.

63. Masukawa, H.: Localization of sensitivity to kanamycin and streptomycin in 30S
ribosomal proteins of *Escherichia coli. J. Antibiot. (Tokyo)* 22:612–623, 1969.

64. Marengo, P.B., Chenowth, M.E., Overturf, G.D., and Wilkins, J.: Phosphorylation of
kanamycin, lividomycin A, and Butirosin B by *Providencia stuartii. Antimicrob.
Agents Chemother.* 6:821–824, 1974.

65. Matsuhashi, Y., Yagisawa, M., Kondo, S., Takeuchi, T., and Umezawa, H.: Aminoglycoside 3'-phosphotransferases I and II in *Pseudomonas aeruginosa. J. Antibiot.
(Tokyo)* 28:442–447, 1975.

66. Matsuhashi, Y., Sawa, T., Takeuchi, T., and Umezawa, H.: Immunological studies of
aminoglycoside 3'-phosphotransferases. *J. Antibiot. (Tokyo)* 29:1127–1128, 1976.

67. Matsuhashi, Y., Sawa, T., Takeuchi, T., and Umezawa, H.: Purification of amino-glycoside 3'-phosphotransferase II. *J. Antibiot. (Tokyo)* 24:204–206, 1976.
68. Miller, A.L., and Walker, J.B.: Enzymatic phosphorylation of streptomycin by extracts of streptomycin-producing strains of *Streptomyces. J. Bacteriol.* 99:401–405, 1969.
69. Mills, B.J., and Holloway, B.W.: Mutants of *Pseudomonas aeruginosa* that show specific hypersensitivity to aminoglycosides. *Antimicrob. Agents Chemother.* 10:411–416, 1976.
70. Mitsuhashi, S., Kobayashi, F., and Yamaguchi, M.: Enzymatic inactivation of gentamicin C components by cell-free extract from *Pseudomonas aeruginosa. J. Antibiot. (Tokyo)* 24:400–401, 1971.
71. Mitsuhashi, S., Koyayashi, F., Yamaguchi, M., O'Hara, K., and Kono, M.: Enzymatic inactivation of aminoglycoside antibiotics by resistant strains of bacteria. In *Bacterial Plasmids and Antibiotic Resistance,* edited by Krcmèry, V., et al. Prague, Avicenum (Czechoslovak Medical Press), 1972, p. 337.
72. Mitsuhashi, S., Inoue, M., Kawabe, H., Oshima, H., and Okubo, T.: Genetic and biochemical studies of drug resistance in staphylococci. In *Staphylococci and Staphylococcal Infections,* edited by Jeljaszewicz, J. Basel, Karger, 1973, p. 144.
73. Mitsuhashi, S., and Krcmèry, V.: Enzymatic inactivation of aminoglycoside antibiotics by bacteria bearing R plasmids. *Zentralbl. Bakteriol.,* 528, 1975.
74. Mitsuhashi, S.: Proposal for a rational nomenclature for phenotype, genotype and aminoglycoside-aminocyclitol modifying enzymes. In *Drug-Inactivating Enzymes and Other Problems of Resistant Bacteria,* edited by Mitsuhashi, S. et al. Prague, Avicenum (Czechoslovak. Medical Press), 1975, p. 115.
75. Mitsuhashi, S., and Kawabe, H.: Comparative studies of antibacterial activity of various aminoglycoside antibiotics. In *Drug-Inactivating Enzymes and Other Problems of Resistant Bacteria,* edited by Mitsuhashi, S. et al. Prague, Avicenum (Czechoslovak Medical Press), 1975, p. 157.
76. Mitsuhashi, S., Tanaka, T., Kawabe, H., and Umezawa, H.: Biochemical mechanism of kanamycin resistance in *Mycobacterium tuberculosis. Microbiol. Immunol.* 21: 325–327, 1977.
77. Mitsuhashi, S.: Epidemiology of bacterial drug resistance. In *R Factor—Drug Resistance Plasmid,* edited by Mitsuhashi, S. Tokyo, University of Tokyo Press, 1977, p. 3.
78. Mitsuhashi, S., Kawabe, H., Nagate, T., and Inoue, K.: Evolutional process of formation of multiple resistance plasmids. In *Topics in Infectious Diseases,* Vol 2, edited by Drews, J., and Hogenauer, G. New York, Springer-Verlag, 1977, p. 165.
79. Mitsuhashi, S., Yamagishi, S., Sawai, T., and Kawabe, H.: Biochemical mechanisms of plasmid-mediated resistance. In *R Factor-Drug Resistance Plasmid,* edited by Mitsuhashi, S. Tokyo, University of Tokyo Press, 1977, p. 195.
79a. Mitsuhashi, S.: Epidemiology and genetics of aminoglycoside antibiotic resistance in pathogenic bacteria. In *Drug Action and Drug Resistance in Bacteria, II: Aminoglycoside Antibiotics.* Tokyo, University of Tokyo Press/Baltimore, Md., University Park Press, 1975, p. 179.
80. Mitsuhashi, S.: Drug resistance of bacteria isolated from clinical specimens. In *Bacterial Drug Resistance,* edited by Mitsuhashi, S. Tokyo, Kodansha, 1979, p. 37. demic Press, 1979, p. 37.
81. Mitsuhashi, S.: Drug resistance pattern in bacteria. *Ibid.,* p. 83.
82. Miyamura, S.: Dysentery bacilli and its relation to the resistance. *Nihon Saikingaku Zasshi* 16:115–119, 1961. (In Japanese.)
83. Morel, C., Freymuth, F., and Villemon-Lemosquet, M.: Nouveau mécanisme enzymatique de résistance aux antibiotiques aminoglucosidiques. *Ann. Biol. Clin.* 31:353–357, 1973.

84. Naganawa, H., Kondo, S., Maeda, K., and Umezawa, H.: Structure determination of enzymatically phosphorylated products of aminoglycosidic antibiotics by proton magnetic resonance. *J. Antibiot. (Tokyo)* 24:823–829, 1971.
85. Naganawa, H., Yagisawa, M., Kondo, S., Takeuchi, T., and Umezawa, H.: The structure determination of an enzymatic inactivation product of 3',4'-dideoxykanamycin B. *J. Antibiot. (Tokyo)* 24:913–914, 1971.
86. Nagate, T., Inoue, M., Inoue, K., and Mitsuhashi, S.: Plasmid-mediated sulfanilamide resistance. *Microbiol. Immunol.* 22:367–375, 1978.
87. O'Hara, K., Kono, M., and Mitsuhashi, S.: Structure of enzymatically acetylated sisomicin by *Pseudomonas aeruginosa. J. Antibiot. (Tokyo)* 27:349–351, 1974.
88. O'Hara, K., Kono, M., and Mitsuhashi, S.: Enzymatic inactivation of a new aminoglycoside antibiotics, sisomicin, by resistant strains of *Pseudomonas aeruginosa. Antimicrob. Agents Chemother.* 5:558–561, 1974.
89. Ohashi, Y., Kawabe, H., Sato, K., Nakamura, N., and Mitsuhashi, S.: Fortimicin, a new aminoglycoside antibiotic. In *Microbial Drug Resistance,* Vol. 2, edited by Mitsuhashi, S. Tokyo, University of Tokyo Press, 1979, p. 243.
90. Okamoto, S., and Suzuki, Y.: Chloramphenicol-, dihydrostreptomycin- and kanamycin-inactivating enzymes from multiple drug-resistant *Escherichia coli* carrying episome "R." *Nature* 108:1301–1303, 1965.
91. Okanishi, M., Kondo, S., Suzuki, Y., Okamoto, S., and Umezawa, H.: Studies on inactivation of kanamycin and resistances of *E. coli. J. Antibiot. (Tokyo)* 20:132–135, 1967.
92. Okanishi, M., Kondo, S., Utahara, R., and Umezawa, H.: Phosphorylation and inactivation of aminoglycosidic antibiotics by *E. coli* carrying R factor. *J. Antibiot. (Tokyo)* 21:13–21, 1968.
93. Ozanne, B., Benveniste, R., Tipper, D., and Davies, J.: Aminoglycoside antibiotics: inactivation by phosphorylation in *Escherichia coli* carrying R factors. *J. Bacteriol.* 100:1144–1146, 1969.
94. Ozaki, M., Mizushima, S., and Nomura, M.: Identification and functional characterization of the protein controlled by streptomycin-resistant locus in *E. coli. Nature* 222:333–339, 1969.
95. Price, K.E., Chisholm, D.R., Misiek, M., Leitner, F., and Tsai, Y.H.: Microbiological evaluation of BB-K8, a new semisynthetic aminoglycoside. *J. Antibiot. (Tokyo)* 25:709–731, 1972.
96. Santanam, P., and Kayser, F.H.: Tobramycin adenylyltransferase: a new aminoglycoside-inactivating enzyme from *Staphylococcus epidermidis. J. Infect. Dis.* 134: S33–S39, 1976.
97. Santanam, P., and Kayser, F.H.: Purification and characterization of an aminoglycoside inactivating enzyme from *Staphylococcus epidermidis* FK109 that nucleotidylates the 4'- and 4"-hydroxyl group of the aminoglycoside antibiotics. *J. Antibiot. (Tokyo)* 31:343–351, 1978.
98. Smith, D.H., Janjigian, J.A., Prescott, N., and Anderson, P.W.: Resistance factor-mediated spectinomycin resistance. *Infect. Immun.* 1:120–127, 1970.
99. Soussy, C.J., Bouchaud, D.H., Fouace, D.J., Dublanchet, A., and Duval, J.: A gentamicin resistance plasmid in *Staphylococcus aureus. Ann. Microbiol. (Paris)* 126B: 91–92, 1975.
100. Speller, D.C.E., Raghunath, D., Stephens, M., Viant, A.C., Reeves, D.C., Wilkinson, P.J., Broughall, J.M., and Holt, H.A.: Epidemic infection by a gentamicin-resistant *Staphylococcus aureus* in three hospitals. *Lancet* 1:464–466, 1976.
101. Suzuki, I., Takahashi, N., Shirato, S., Kawabe, H., and Mitsuhashi, S.: Adenylylation of streptomycin by *Staphylococcus aureus:* a new streptomycin adenylyltransferase. In *Microbial Drug Resistance,* edited by Mitsuhashi, S., and Hashimoto, H.

Tokyo, University of Tokyo Press, 1975, p. 463.

102. Takagi, Y., Miyake, T., Tuchiya, T., Umezawa, S., and Umezawa, H.: Synthesis of 3'-deoxykanamycin B. *J. Antibiot. (Tokyo)* 26:403–406, 1973.

103. Takasawa, S., Utahara, R., Okanishi, M., Maeda, K., and Umezawa, H.: Studies on adenylylstreptomycin, a product of streptomycin inactivated by *E. coli* carrying the R factor. *J. Antibiot. (Tokyo)* 21:477–484, 1968.

104. Tanaka, N.: Biochemical studies on gentamicin resistance. *J. Antibiot. (Tokyo)* 23:469–471, 1970.

105. Tseng, J.-T., Bryan, L.E., and Van Den Elzen, H.M.: Mechanisms and spectrum of streptomycin resistance in a natural population of *Pseudomonas aeruginosa. Antimicrob. Agents Chemother.* 21:136–141, 1972.

106. Umezawa, H., Okanishi, M., Kondo, S., Hamana, K., Utahara, R., Maeda, K., and Mitsuhashi, S.: Phosphorylative inactivation of aminoglycosidic antibiotics by *Escherichia coli* carrying R factor. *Science* 157:1559–1561, 1967.

107. Umezawa, H., Okanishi, M., Utahara, R., Maeda, K., and Kondo, S.: Isolation and structure of kanamycin inactivated by a cell-free system of kanamycin-resistant *E. coli. J. Antibiot. (Tokyo)* 20:136–141, 1967.

108. Umezawa, H., Takasawa, S., Okanishi, M., and Utahara, R.: Adenylylstreptomycin, a product of streptomycin inactivated by *E. coli* carrying R factor. *J. Antibiot. (Tokyo)* 21:81–82, 1968.

109. Umezawa, H., Doi, O., Ogura, M., Kondo, S., and Tanaka, N.: Phosphorylation and inactivation of kanamycin by *Pseudomonas aeruginosa. J. Antibiot. (Tokyo)* 21:154–155, 1968.

110. Umezawa, S., Tsuchiya, T., Muto, R., Nishimura, Y., and Umezawa, H.: Synthesis of 3'-deoxykanamycin effective against kanamycin resistant *Escherichia coli* and *Pseudomonas aeruginosa. J. Antibiot. (Tokyo)* 24:274–275, 1971.

111. Umezawa, H., Umezawa, S., Tsuchiya, T., and Okazaki, Y.: 3',4'-Dideoxykanamycin B active against kanamycin-resistant *Escherichia coli* and *Pseudomonas aeruginosa. J. Antibiot. (Tokyo)* 24:485–487, 1971.

112. Umezawa, S., Tuchiya, T., Jikihara, T., and Umezawa, H.: Synthesis of 3',4'-dideoxyneamine active against kanamycin resistant *E. coli* and *P. aeruginosa. J. Antibiot. (Tokyo)* 24:711–712, 1971.

113. Umezawa, H., Nishimura, Y., Tsuchiya, T., and Umezawa, S.: Syntheses of 6'-N-methylkanamycin and 3',4'-dideoxy-6'-N-acetylating enzymes. *J. Antibiot. (Tokyo)* 25:743–745, 1972.

114. Umezawa, H., Yamamoto, H., Yagisawa, M., Kondo, S., Takeuchi, T., and Chabbert, Y.: Kanamycin phosphotransferase I: mechanism of cross resistance between kanamycin and lividomycin. *J. Antibiot. (Tokyo)* 26:407–411, 1973.

115. Umezawa, H., Yagisawa, M., Matsuhashi, Y., Naganawa, H., Yamamoto, H., Kondo, S., Takeuchi, T., and Chabbert, Y.: Gentamicin acetyltransferase in *Escherichia coli* carrying R-factor. *J. Antibiot. (Tokyo)* 26:612–614, 1973.

116. Umezawa, H., Matsuhashi, Y., Yagisawa, M., Yamamoto, H., Kondo, S., and Takeuchi, T.: Immobilization of phosphotransferases obtained from resistant bacteria. *J. Antibiot. (Tokyo)* 27:358–360, 1974.

117. Umezawa, S., Nishimura, Y., Hata, Y., Tsuchiya, T., Yagisawa, M., and Umezawa, H.: Synthesis of 4'-deoxykanamycin and its resistance to kanamycin phosphotransferase II. *J. Antibiot. (Tokyo)* 9:722–725, 1974.

118. Umezawa, H.: Biochemical mechanism of resistance to aminoglycosidic antibiotics. In *Drug Action and Drug Resistance in Bacteria, II. Aminoglycoside Antibiotics.* Tokyo, University of Tokyo Press/Baltimore, Md., University Park Press, 1975, p. 211.

119. Umezawa, Y., Yagisawa, M., Sawa, T., Takeuchi, T., Umezawa, H., Matsumoto, H.,

and Tazaki, T.: Aminoglycoside 3'-phosphotransferase III, a new phosphotransferase resistance mechanism. *J. Antibiot. (Tokyo)* 28:846–853, 1975.

120. Walker, J.B., and Skorvaga, M.: Phosphorylation of streptomycin and dihydro-streptomycin by *Streptomyces. J. Biol. Chem.* 248:2435–2440, 1973.

121. Williams, J.W., and Northrop, D.B.: Purification and properties of gentamicin acetyl-transferase I. *Biochemistry* 15:125–131, 1976.

122. Witchitz, J.L.: Plasmid-mediated gentamicin resistance not associated with kana-mycin resistance in *Enterobacteriaceae. J. Antibiot. (Tokyo)* 25:622–624, 1972.

123. Yagisawa, M., Naganawa, H., Kondo, S., Hamada, M., Takeuchi, T., and Umezawa, H.: Adenylyldideoxykanamycin B, a product of the inactivation of dideoxykanamy-cin B by *Escherichia coli* carrying R factor. *J. Antibiot. (Tokyo)* 24:911–912, 1971.

124. Yagisawa, M., Naganawa, H., Kondo, S., Takeuchi, T., and Umezawa, H.: 6'-N-acetylation of 3',4'-dideoxykanamycin B by an enzyme in a resistant strain of *Pseudomonas aeruginosa. J. Antibiot. (Tokyo)* 25:495–496, 1972.

125. Yagisawa, M., Yamamoto, H., Naganawa, H., Kondo, S., Takeuchi, T., and Umezawa, H.: A new enzyme in *Escherichia coli* carrying R factor phosphorylating 3'-hydroxyl of butirosin A, kanamycin, neamine and ribostamycin. *J. Antibiot. (Tokyo)* 25: 748–750, 1972.

126. Yamada, T., Tipper, D., and Davies, J.: Enzymatic inactivation of streptomycin by R factor-resistant *Escherichia coli. Nature* 219:288–291, 1968.

127. Yamaguchi, M., Kobayashi, F., and Mitsuhashi, S.: Antibacterial activity of livido-mycin toward R factor-resistant strains of *Escherichia coli. Antimicrob. Agents Chemother.* 1:139–142, 1972.

128. Yamaguchi, M., Koshi, T., Kobayashi, F., and Mitsuhashi, S.: Phosphorylation of lividomycin by *Escherichia coli* carrying an R factor. *Antimicrob. Agents Chemo-ther.* 2:142–146, 1972.

129. Yamaguchi, M., Mitsuhashi, S., Kobayashi, F., and Zenda, H.: A 2'-N-acetylating enzyme of aminoglycosides. *J. Antibiot. (Tokyo)* 27:507–515, 1974.

130. Yamamoto, H., Kondo, S., Maeda, K., and Umezawa, H.: Synthesis of lividomycin A 5''-phosphate, and enzymatically inactivated lividomycin A. *J. Antibiot. (Tokyo)* 25:485–486, 1972.

131. Yamamoto, H., Yagisawa, M., Naganawa, H., Kondo, S., Takeuchi, T., and Umezawa, H.: Kanamycin 6'-acetate and ribostamycin 6'-acetate, enzymatically inactivated products by *Pseudomonas aeruginosa. J. Antibiot. (Tokyo)* 25:746–747, 1972.

PHARMACOLOGY AND ASSAYS

5

PHARMACOLOGY OF AMINOGLYCOSIDES

HAROLD C. NEU   College of Physicians and Surgeons
Columbia University, New York, New York

It is only within the past few years that infectious disease experts have come to appreciate the importance of understanding pharmacokinetics in the administration of antibiotics. The fairly narrow toxic/therapeutic ratio of the aminoglycosides makes it essential that those who utilize these compounds in the treatment of serious infection do so in a manner that will provide adequate tissue and serum levels without provoking the toxic reactions that are discussed in detail in the other areas of this book. This chapter reviews briefly the basic concepts necessary to understand the pharmacology of antimicrobial agents and discusses in detail what is known currently about the pharmacology the aminoglycosides. The literature cited is not exhaustive, but rather has been selected to provide a balanced view of the differing proposals presented about the pharmacology of these agents.

I. Pharmacokinetic Principles

The mathematical expressions that are currently utilized in pharmacokinetics view the body as a series of interconnected compartments in which drugs are distributed. The compartments of the body are separated by barriers that impede the immediate diffusion of an agent from one compartment to another. In general, we view the bloodstream as a central compartment surrounded by peripheral (tissue) compartments. After administration of an antibiotic by the intravenous (IV) or intramuscular (IM) route in the case of the aminoglycosides, the compound is simultaneously distributed and eliminated. The elimination of a compound can be either by metabolism and/or by excretory means.

Most published pharmacokinetic analyses of new compounds are performed in healthy young males between the ages of 18 and 35 years. When antimicrobial agents are administered to patients, particularly agents such as the aminoglycosides, the population receiving the compounds are not such a homogeneous population, but differ with respect to weight, age, extracellular space, and secretory abilities. Thus, it is essential that a physician understand the effects of changes in body size, size of the central compartment, and excretory ability upon the pharmacokinetic parameters of a compound.

In general, it is wise to view the body as a two-compartment model. The serum concentration, C, of a compound can after IV administration be described by the sum of two exponentials: $C = A \cdot e^{-\alpha t} + B \cdot e^{-\beta t}$. Some time after administration a pseudo-steady state of equilibrium is reached between the two compartments and the serum concentration-time curve can be described by a single exponential: $C = B \cdot e^{-\beta t}$. The β is determined by utilizing multiple samples from the terminal logarithmic phase, which is linear. This avoids biased values for the parameters B, A, α, and β. Fortunately, the assay methods available to determine serum concentrations of the aminoglycosides are sufficiently sensitive so that the time-course curve can be followed for a sufficient time to determine adequately the overall elimination rate constant, β. Clearly aminoglycosides undergo removal from the body by a slowly declining γ phase; however, serum clearances represent good estimates, since K-Vd is very similar to the serum clearance calculated from the two-compartment analysis. Indeed, it is often necessary in a consideration of the kinetics of a drug given by IM injection to utilize a one-compartment open model, but the error associated with its use is usually only 15% at most. Unfortunately, most of the published studies of aminoglycoside pharmacokinetics have used a one-compartment open model to interpret the removal of drug in relation to renal function, and this may not be valid.

In some drugs it is necessary to make calculations for differing species of the drug molecules, such as with some of the cephalosporins, or for the degree of ionization of a compound. Fortunately, with the aminoglycosides only one species exists as far the pK of the compounds. This does not say that one should not consider the acid-base state of an area of the body, since the activity of any of the aminoglycosides is markedly reduced in an acid environment. In general, aminoglycosides undergo few drug-drug interactions within the body, but in the presence of high concentrations of β-lactams, particularly certain penicillins, the activity and pharmacokinetics of the aminoglycosides can be altered.

In classic kinetic analysis of the interaction of a compound and the part of the body for which it is designed, the body compartment contains the bioreceptors. In infections the bioreceptor is the microbial agent on which the antibacterial agent is supposed to act. Since bacterial infections can occur in many different parts of the body and be disseminated via the blood and lymphatic systems, it is difficult to determine the precise concentration needed to eliminate an organism in each compartment or the concentration achieved in each of the compartments. Furthermore, bacteria can be engulfed by phagocytic cells that do not kill the microorganism, but which may if some of the antimicrobial agent enters the cell on its own or is brought in with the microorganism. All of these theoretical but important facets of the action of antimicrobial agents are as yet unknown for the aminoglycosides.

In a sense a focus of infection is a separate pharmacokinetic compartment which can be viewed in the usual manner as transport via a first-order process in which $dT''/dt = k''Vo_{c_p}$, where c_p is the concentration of drug not bound to protein, T'' is the amount of drug transported into the infected site, t is time, and k'' is a rate constant for the transport. However, with infection we often are dealing with nonvascularized foci such as abscesses and exudates, and in that situation the compounds will follow Fick's law of diffusion, in which $dn/dt = Dq \cdot de/dx$, where dn/dt is the number of molecules that pass through area q per unit time when the concentration c and distance x are known, and the diffusion constant D has the dimensions cm^2/sec.

With the aminoglycosides this simplistic approach may not work, since other macromolecules, such as nucleic acid material or nucleohistones from disrupted cells, slow the diffusion by interacting with the diffusing molecules. Pore size and electrostatic forces of association and repulsion also get to alter transport. Thus, even with water there is slower transport. Dettli [25] calculated the time it would take streptomycin, using a diffusion constant of $D = 2 \times 10^{-6}$ cm^2/sec, to pass through water. He estimated that $D \sim 7\sqrt{M}$ $(cm^2/24\ hr)$, where M is the molecular weight. Thus, it would take streptomycin 1 day to reach the center of a 5 mm abscess or necrotic focus and 3.5 months to reach the center of an abscess, necrotic area, or a caseous tuberculous focus. Although aminoglycosides are not destroyed in abscesses as are many of the β-lactams, the degree of their inhibition by cellular materials is unknown, since the transport out of such areas may be different for the different species of compound—bound and free.

The length of time that a microorganism must be exposed to an antibacterial agent is not known. For the β-lactams the organism must be in a state of growth so that the penicillin or cephalosporin binds to the β-lactam receptors, which are involved with cell wall synthesis. Although aminoglycosides can kill resting organisms, there is a surface transport system that moves the compounds across the outer membrane through the periplasmic space and across the cytoplasmic membrane to reach the ribosomal receptor proteins. We know that this process occurs poorly if at all under anaerobic conditions. Thus, even though drug may be delivered to an area, it will not be active because it will not enter the receptor in the infecting organism. Thus more factors than mere distribution and excretion must always be considered when viewing the pharmacokinetic parameters of antibacterial agents.

II. General Properties of Aminoglycosides

Aminoglycosides are all water-soluble, stable over a wide pH range, and relatively heat resistant. Aminoglycosides have been used orally, parenterally, and have been incorporated into various creams and ointments, and absorbed to dextran beads, methylacrylate [92,102], and utilized as a replacement for joints. In each situation they are released to exert their antibacterial action. Aminoglycosides can be inactivated in vitro by penicillins such as carbenicillin or ticarcillin, but such inactivation is less likely to occur in serum [32,72,77,96,121], except in the presence of renal failure [123]. Amikacin is the least inactivated by penicillins.

A. Absorption

Aminoglycosides generally are administered by the intravenous or intramuscular route. Absorption via the gastrointestinal tract is erratic but can be of a degree to cause toxicity such as has been noted after the use of neomycin [12,58,60,91] to treat hepatic coma and in children treated with gentamicin for necrotizing enterocolitis. Furthermore, all the agents can be absorbed via the skin to produce appreciable serum levels when there is denudation of the ectoderm such as occurs in severe burns or in certain infections. This fact should be considered whenever the drugs are used topically. Absorption after aero-solization in nontracheotimized individuals is essentially nil, whereas some absorption does occur when the drugs are instilled into the trachea—less, however, than when they are injected into other body compartments [115]. Aminoglycosides are readily absorbed from the peritoneal cavity, and appreciable serum and tissue levels were produced when kanamycin or neomycin was used to irrigate the peritoneum after abdominal surgery [23,44,74]. Bladder instillation of neomycin [16] or other aminoglycosides does not result in any major absorption. The systemic concentration achieved when aminoglyco-

Table 1   Pharmacologic Properties of Gentamicin, Tobramycin, Netilmicin, and Sisomicin

Variable	Route	Dose (mg/kg body weight)	Value
Peak serum level (μg/ml)	Intramuscular	1	4
	Intramuscular	2	6–8
	Intravenous (2 min)	1	12–20
	Intravenous (30-min infusion)	1.5	4–6
Serum level, 6 hr (μg/ml)	Intramuscular		0.5–1
Creatinine clearance, ≥100 ml/1.73 m^2	Intravenous (rapid)		0.1
Serum half-life (hr)			
Initial phase			2
Second phase			27–200
Volume of distribution (% body weight)			20–30
Protein binding			0
Clearance on hemodialysis, 6 hr (%)			50
Peritoneal dialysis, removed (mg/2 L)			1
Urine concentration (μg/ml)			
0–4 hr			100–300
4–8 hr			10–50
Dose excreted, 8 hr (%)			85–90

sides have been injected into pleural or synovial spaces has not been determined, but we would anticipate that if large concentrations were employed, the drug absorption would be similar to that after intraperitoneal administration. The concentrations of the drugs injected intrathecally or intraventricularly are small (i.e., less than 20 mg), so systemic absorption is minimal [51].

Following intramuscular injection of an aminoglycoside, peak serum levels are reached in 30-90 min, depending upon muscle mass and circulation. Uptake is more rapid in individuals of lower body mass and delayed in those, such as paralytic individuals, who lack muscle activity. In general, peak serum levels of aminoglycosides after either bolus or slow infusion occur at the end of the infusion period, provided that there is no venous spasm.

Peak serum concentrations of gentamicin [7], sisomicin [67], netilmicin [50], tobramycin [78], and dibekacin [74,98,119] are virtually identical [62] in the same individual when identical doses are used (Table 1). Since larger concentrations of amikacin [21] and kanamycin [27] are employed (Table 2), the serum levels are four to five times those of the former agents [54]. After an initial injection over 15-30 min of 2 mg/kg of gentamicin

Table 2   Pharmacologic Properties of Amikacin and Kanamycin

Variable	Route	Dose (mg/kg body weight)	Value
Peak serum level (μg/ml)	Intramuscular	3.5	12
	Intramuscular	7.5	21
	Intramuscular (30 min)	5	20-30
	Intramuscular (2 min)	5	60
Serum level, 10 hr (μg/ml)	Intramuscular	5	2
	Intramuscular	5-7	1
Serum half-life (hr)			
Normal			2
Anuria			35-200
Volume of distribution (% body weight)			25
Protein binding (%)			0
Clearance on hemodialysis 6-8 hr (%)			50
Clearance on peritoneal dialysis, first 10 hr (mg/2L)			3
Urine concentration, 0-8 hr (μg/ml)			100-800
Dose excreted, 12 hr (%)			80-90

[42,46], sisomicin [10,28,34,41,47,79], netilmicin [48,50,97,118], or tobramycin [35,45,49,78], initial serum levels of 6–10 μg/ml are achieved, and after 1 mg/kg intramuscularly, mean peak levels of 4 μg/ml will be found in the normal 70 kg male. Rapid injection (2 min) of gentamicin or tobramycin will produce immediate serum levels of 15–25 μg/ml without toxicity [26,35,113]. Bolus injection of an amikacin will produce serum levels of 60 μg/ml [21], but serum levels of all these agents at 1 hr after the bolus injection are similar to levels after the same dose given by the intramuscular route. Furthermore, the β phase of the kinetic curves after bolus injection, infusion over 30 min, infusion over 3 hr, or intramuscular injection are the same [11,46] (Table 3). Serum levels of 20–25 μg/ml follow an intramuscular injection of 7.5 mg/kg of amikacin [8,16,43,47] or kanamycin [20,44]. The serum levels of all aminoglycosides decline exponentially, so that in normal individuals the serum levels of gentamicin, tobramycin, netilmicin, and sisomicin are less than 1 μg/ml after 8 hr [62,65,67,99,104-107].

Infusion of 1.5 mg/kg of gentamicin, tobramycin, or netilmicin over 20–30 min has yielded serum levels of 4–8 μg/ml at the end of the infusion, whereas injection of 5–7.5 mg/kg of amikacin over 20–30 min has given serum levels of 15–25 μg/ml at the end of infusion. It has been reported that 1 hr after the first injection of gentamicin (20 min), a mean serum level of only 2 μg/ml was found if the patient also received ticarcillin compared to a serum level of 3.1 μg/ml when cephalothin was given after the gentamicin [60]. Other investigators have not seen this. By 8 hr after a 30-min infusion of gentamicin, tobramycin, sisomicin, or netilmicin, mean serum levels are less than 0.5 μg/ml. The mean serum levels 8 hr after intravenous infusion of amikacin are approximately 1 μg/ml [9,66].

The precise serum level in any patient will reflect the state of renal function [52,63, 81,95]. Neither serum creatinine nor blood urea nitrogen can be used as the sole means to predict a serum level, because these tests depend not only upon renal status but upon dietary intake, muscle mass, and cardiac status. Lower serum levels have been reported during fever [66], but this is doubtful [52].

In all studies of individuals with creatinine clearances in excess of 90 ml/min per 1.73 m^2, the initial disposition rate constant is extremely rapid (Table 3). In most studies the K_{12}/K_{21} ratio is greater than 1, indicating that peripheral distribution of aminogly-

Table 3   Pharmacokinetic Parameters of Aminoglycosides in Normal Individuals

Agent	Elimination half-life (hr)	Elimination rate constant K_e (hr^{-1})	Apparent volume of distribution (% wt)	Clearance (ml/min/1.73 m^2)
Streptomycin	2–3	0.23–0.35	25	70
Kanamycin	2.1–2.4	0.29–0.33	20–25	75–85
Gentamicin	1.7–2.3	0.3 –0.41	20–25	50–90
Tobramycin	2–2.7	0.25–0.35	20–25	50–90
Sisomicin	2–2.3	0.3 –0.35	20–25	60–75
Amikacin	2.2–2.5	0.28–0.31	22–29	75–100
Netilmicin	1.9–2.3	0.29–0.31	22–26	70–85
Dibekacin	1.9–2.5	0.31	22	70–85

cosides in tissues in which they do not normally accumulate will occur after repeated dosing.

B. Distribution

Aminoglycosides are widely distributed [15,18,100], and studies in small animals using radioactive drugs show that they reach most organ systems. Tissue levels are low compared to the serum levels for all the agents, because the majority of the drugs are within the vascular compartment [18,43]. The one exception of this is the high concentrations found in renal cortical tissue, discussed in detail elsewhere. Concentrations of aminoglycosides in various body compartments are proportional to the serum levels. Hence, the levels of kanamycin and amikacin are appreciably greater than those of the other agents, since these compounds are given at much greater doses.

All the aminoglycosides reach satisfactory concentrations in noninflammed human interstitial fluid [43,114], with peak concentrations found 2 hr after the injection of the compound (Table 4). Peak interstitial concentrations range from 1.4 μg/ml with tobramycin to 4.2 μg/ml with amikacin. These levels are 17-30% of the simultaneous serum levels (Table 4). Earlier studies by Chisholm et al. [17,18] had found concentrations of gentamicin in tissues, equivalent to serum levels. In animals, interstitial fluid levels are consistently above the minimal inhibitory concentrations of susceptible organisms [15]. With repeated doses, aminoglycosides do not accumulate in tissue fluid. Tissue fluid levels are higher during the first 2 hr after bolus (2-5 min) injection than after 30-min infusion, but by 2 hr, tissue levels are identical whether bolus or infusion administration has been used [56].

None of the compounds reach adequate levels in the eye [4,37,38], and local subconjunctival instillation or direct intravitreal injection (in a small volume) are needed to achieve therapeutic concentrations [37,68]. Cerebrospinal fluid levels in the adult after either IM or IV injection are inadequate to treat meningitis due to gram-negative bacilli [13,40,44,51]. Hence, local instillation is necessary. The levels in sinus tissue have not been determined, but there is no evidence for significant lacrimal secretion or secretion by glands of the orofacial area. A salivary level of 0.4 μg/ml 4 hr after an injection of 180 mg of amikacin to an 8-year-old child has been recorded. Levels in the pleural fluid have been recorded infrequently, and often the time of serum peak has been unknown. In all reported situations the ratio of the pleural level to the serum level has been less than 1.

The aminoglycosides do enter ascitic and peritoneal fluid [31,36] after parenteral administration, and this is an important factor not only in their use in peritoneal infections, but as a potential third space which will reduce the height of the serum level [108]

Table 4   Interstitial Fluid Concentrations by the Skin-Window Technique

Agent	Fluid/serum (μg/ml)		
	1 hr	2 hr	4 hr
Gentamicin, 1.7 mg/kg	1.36-8	1.8-5.9	0.6-2.6
Tobramycin, 1.7 mg/kg	0.8-7.8	1.4-6.4	0.5-2.4
Amikacin, 7.5 mg/kg	2.2-24	4.2-25	2.1-11.6

(Table 5). Gerding et al. [36] found levels of gentamicin in ascitic fluid that were below or greater than serum levels. The ascitic fluid levels ranged from 43 to 132% of the simultaneous serum levels. Levels in serum ranged from 0.6 to 10 μg/ml with ascitic levels of 0.7–8.0 μg/ml. Ascitic levels after administration of tobramycin were similar, ranging from 1.7 to 10.8 μg/ml with serum levels of 2.5–13.6 μg/ml. Whether antibiotic concentrations in ascitic fluid can be extrapolated to patients with peritonitis and no ascitics is not known. Peritonitis following peritoneal dialysis should be best treated with instillation of the drug into the peritoneum, since the levels would be more uniform and adequate drug would be absorbed to reach other tissues.

The levels of aminoglycosides reported to occur in bile, whether with normal hepatic function or when there is obstruction to the flow of bile, are extremely variable and usually less than 1 μg/ml [87,109]. T-tube drainage has been reported to contain concentrations of aminoglycosides that are 25–90% of the serum level, whether gentamicin [87], tobramycin, or kanamycin [90] has been the agent studied. The amikacin biliary levels have ranged from 5 to 15% of the serum level and the concentrations in gallbladder wall tissue have been equally low [101]. More recent studies have refuted this and shown that amikacin produces levels in bile in concentrations equal or greater than with other agents. Blockage of the cystic duct will prevent entry of the aminoglycosides into bile, just as occurs with the antibiotics of all other classes. The effect of severe hepatic damage upon hepatic concentrations is not established. Similarly, data concerning levels in pancreatic juice are fragmentary and the methods of fluid determination often not completely adequate [90].

All the aminoglycosides reach appreciable levels in synovial fluid, so that it is unnecessary to instill the antibiotics into joints [3,19,24,70,103a]. Synovial levels of 5–8 μg/ml have followed intramuscular administration of 7.5 mg/kg of kanamycin [3] and amikacin [103a], and 2–4 μg/ml after 1.7 mg/ml of gentamicin and tobramycin [24] (Table 6). Synovial fluid concentrations of all aminoglycosides are higher than simultaneous serum concentrations at later time periods, indicating that exit from the joint is slower than from the serum. As from other body surfaces, local infusion of aminoglycosides into joints via irrigating solutions can result in systemic absorption.

Bone concentrations of aminoglycosides are less well studied in humans than in animals. Bone levels in general are lower than the serum levels. Interpretation of the actual levels in cortical or cancellous bone is made more difficult, since there usually is contamination of specimens with both blood and plasma. Bone concentrations of 0.8 and 4.3 μg/g of amikacin 1 hr after IV dose and of 8.5 μg/g 11 hr after an IM dose are re-

Table 5   Aminoglycoside Concentrations in Ascitic Fluid[a]

Agent	Serum levels (μg/ml)	Ascities (μg/ml)
Gentamicin	0.6–10	0.7–5.3
Tobramycin	2.3–13.6	1.1–10.8
Amikacin	5–25	1–20
Kanamycin	5–25	1–15

[a]Expressed as a range of values.

ported [103a]. Thus, technical problems inherent in obtaining the samples may preclude our knowing what amounts of active drug are present at the site of bone infections. Concentrations of amikacin in muscle and fat reached levels of 2.2 μg/g and 1.89 μg/g when serum levels were 14.9 μg/ml [22].

The penetration of aminoglycosides into bronchial secretions has been studied both in human and in dog models [6,80,84–86,94]. Conflicting results have been reported. Gentamicin and tobramycin have both produced concentrations in the bronchial secretions of anesthetized dogs at 2 hr of approximately 1 μg/ml, which persist for about 2 hr [84,85]. Studies in humans have produced similar levels, and the concentration of amikacin is approximately fourfold greater.

Peak sputum concentrations of 1–6 μg/ml of gentamicin have been reported by Pines [86], while Wong et al. [125] reported bronchial concentrations obtained at bronchoscopy which were 40% of the serum levels. Other investigators have found lower levels. Bergogne-Berezin et al. [6] reported levels of 3.5–4 μg/ml 1 hr after injection of amikacin. Bronchial levels of amikacin have ranged from 2.3 to 8.4 μg/ml with a mean of 5.2 μg/ml 1.5–2 hr after injection when the mean serum level was 23.7 μg/ml. Bronchial secretion levels of amikacin 7 hr after therapy were less than 1 μg/ml. Thus, bronchial secretion levels may never reach the minimal inhibitory concentration against *Pseudomonas aeruginosa* when given by parenteral injection [29]. A mean bronchial level of tobramycin of 3.75 μg/ml was found between 0.6 and 4 hr after the first dose. However, Hall et al. [43] found tobramycin levels in sputum above the minimum inhibitory concentrations (MICs) of cultured pathogens. Klastersky et al. [55] found levels of sisomicin in bronchial fluid inadequate after parenteral injection.

Authors also differ as to whether the concentration of aminoglycoside in bronchial secretions is greater or less depending upon the purulence and mucoid nature of the sputum [1,80,85,94,115].

It is not known whether the concentration of antibiotics in bronchial secretions or sputum has any effect upon clinical outcome, but studies by Klastersky's group and others have demonstrated clinical improvement when the agents are instilled into the lungs via the trachea of a tracheostomized individual. Tissue levels of aminoglycosides in pulmonary tissues, which is highly vascular, probably are similar to the levels found in the central vascular compartment.

Aminoglycosides do not penetrate prostatic tissue well, and prostatic tissue levels are inadequate to eradicate gram-negative bacteria [2]. Urinary levels of all aminoglycosides are high but depend on the degree of renal function, urine flow, and volume [2]. The urinary levels of gentamicin, tobramycin, sisomicin, and netilmicin range from 50 to 300 μg/ml, depending on volume, for the first 4 hr after a dose of 1 mg/kg. The concen-

Table 6   Synovial Penetration of Aminoglycosides

Agent	Serum levels (μg/ml)	Synovial levels (μg/ml)
Gentamicin, 1–1.5 mg/kg	2.4–6.5	2.4–5.4
Tobramycin, 1–1.5 mg/kg	1.0–5.6	1.3–4.5
Amikacin	12.3–16.2	4.1–22

trations of all aminoglycosides in urine in patients with creatinine clearances less than 30 ml/min are markedly reduced, with levels of only 10–30 μg/ml. Levels of amikacin range from 100 to 800 g in the first 4 hr after injection and are above 20 μg/ml for 24 hr. Urine levels are markedly reduced when kidney function is at a level of creatinine clearance (Ccr) of 30 ml/min.

The drugs do not enter the cerebrospinal fluid [5,10,13,76,112] (Table 7). Even with significant meningeal inflammation cisternal cerebrospinal fluid (CSF) concentrations are less than one-fourth of the serum levels, and hence below the minimal inhibitory concentrations [10,51,93]. Parenteral therapy with amikacin, gentamicin, sisomicin, netilmicin, and tobramycin produced concentrations below 1 μg/ml in lumbar, ventricular, and CSF [111]. Indeed, in 18 volunteers who underwent lumbar puncture 1–8.5 hr after 7.5 mg/kg of amikacin showed no detectable drug in the CSF. Instillation of 8–16 mg intrathecally, or preferably intraventricularly, will produce locally therapeutic concentrations for 12–18 hr [40,51,76,93,111].

Aminoglycosides do not enter human polymorphonuclear leukocytes, as several groups have shown [120]. This factor no doubt contributes to the inability of aminoglycosides to kill intraleukocytic staphylococci.

The apparent volume of distribution of aminoglycosides has varied from 20 to 30% of the weight of subjects or patients who have received the agents, but in general it averaged 25% (Table 3).

C. Elimination

Although a number of authors have speculated upon the possible metabolism of these agents, it is now known that the drugs are removed from the body by glomerular filtration [124]. The role of a possible contribution of tubular cells is discussed in later chapters. Clearly, there is evidence that all aminoglycosides are resorbed by the proximal tubule. This explains why the ratio of urinary excretion to endogenous creatinine clearance is always less than 1. Since excretion is completely dependent upon renal function, adjustment in doses must be made even when the changes of renal function (Ccr = 50–80 ml/min) are such that it would be unnecessary to adjust the dose of other antibiotics, such as penicillins and cephalosporins. Detailed methods to adjust dosage and dosing interval are given in other chapters.

Concommitant with decrease in renal function and prolongation of serum half-life there will be a decrease in urinary levels of the drugs [126]. This is extremely important

Table 7   Concentrations of Aminoglycosides in Cerebrospinal Fluid

Agent	Serum levels (μg/ml)	CSF (μg/ml)
Gentamicin	2.1–6.8	<0.1–1
Tobramycin	1.6–7.6	<0.1–1
Amikacin	5–25	<0.1–1.6

because it may be impossible in the patient with reduced renal function to achieve a urinary level that will eradicate bacteria in the bladder. Local instillation may be useful in such situations if performed by intermittent catheterization. Attempts to achieve cure of bladder bacteria will not be successful if a permanent indwelling urethral catheter is required.

A discussion of renal tissue accumulation is given in Chapter 6. There is tissue accumulation of aminoglycosides even after a single dose is given to normal subjects. This γ phase of excretion explains why the usually measured β-phase half-life of these drugs increases upon multiple-dose administration.

D. Protein Binding

Early investigators reported that aminoglycosides bound to protein. Such studies were generally based on precipitation of proteins by use of acids. When other techniques, such as ultrafiltration, have been used, no significant binding to proteins has been demonstrated [39]. However, recent ultracentrifuge methods have suggested 20% protein binding, as has equilibrium dialysis. Technical problems with all these assays precludes a definitive statement as to whether binding to protein does or does not occur. However, even if it does occur, it is not of clinical significance because at most it would be less than 20%.

III. Special Situations That Alter the Pharmacokinetics of Aminoglycosides

A. Obesity

The pharmacokinetics of gentamicin and tobramycin have been studied in subjects who ranged from 60 to 125% overweight [8,103]. If dosing is based on total body weight, excessively high serum levels will be achieved since the aminoglycosides do not penetrate adipose tissue well, owing to their high polarity. Similarly, to use only the ideal body weight rather than the total weight will produce unduly low serum levels. Thus, it is necessary to use a correction factor. Addition of a weight of 40–60% of the adipose mass to the ideal weight will result in a figure to use for calculating the initial loading dose. The subjects who are less than 80% overweight should have 40% of the adipose mass added to their lean body mass, and those who are more than 80% overweight should have a 60% correction.

B. Cystic Fibrosis

The serum levels and half-life of aminoglycosides are considerably different in young individuals, as noted in Chapter 23. However, even the young adult with cystic fibrosis has a similar problem because the extracellular space is considerably larger [104]. Thus, dosage programs of 1.5–2 mg/kg of gentamicin or tobramycin that will yield therapeutic levels in most adults will provide serum levels of less than 4 μg/ml, and doses of 2.5–3 mg/kg per dose often are necessary since the serum half-life is also shortened because of an increased clearance of the drugs. It may be necessary to administer the agents every 6 hr rather than the conventional 8 hr.

C. Burns

The half-life of aminoglycosides has been shown to be unusually short in burn patients, especially the young patient [128]. A half-life of 1.1 hr (±0.44) was found for gentamicin. Thus, larger doses and a shorter interval between doses are necessary if burn-wound sepsis is to be adequately treated.

IV. Adverse Effects of Aminoglycoside Therapy

The major toxic reactions following aminoglycoside therapy are nephrotoxicity and ototoxicity, which are considered extensively in other chapters [23]. Allergic reactions, such as eosinophilia, rash, and fever, are infrequent after aminoglycoside therapy, occurring in only 1–2% of patients. Agranulocytosis and other blood dyscrasias are reported to have occurred, albeit rarely. In each case other explanations for the reactions could be given.

Other reactions attributed to the aminoglycosides include changes in hepatic function tests, such as elevations of serum glutamic-oxaloacetic transaminase, serum glutamic-pyruvic transaminase, and alkaline phosphatase. The relation of the aminoglycoside to these changes is poorly established. Alteration of cardiac function after rapid parenteral administration of streptomycin or kanamycin has been reported [20].

Neuromuscular toxic effects of the aminoglycosides are better established [30,88, 89,122,127]. The compounds appear to impair neuromuscular transmission by a variety of mechanisms [122]. They inhibit release of acetylcholine and may affect the action of calcium at the end plate and depress sensitivity of the motor end plate to acetylcholine [73]. The ability of the various aminoglycosides to block neuromuscular transmission differs widely among the agents, with neomycin reported to be the most potent, although netilmicin is an active paralyzing agent in animals.

Patients with myasthenia gravis or severe hypocalcemia, or those who have received neuromuscular blocking agents, are particularly prone to develop this side effect. Neuromuscular blockade rarely follows the use of the agents by the IM or IV route, but has occurred after peritoneal lavage [23,31]. Use of calcium salts intravenously will reverse the neuromuscular blockade, but the cholinomimetic agents such as neostigmine or edrophonium may not always be effective [75].

Other forms of neurotoxicity, such as parathesias, pain, blurring of vision or various organic brain syndromes, and optic neuritis, have been reported, but these reports are poorly documented and quite infrequent.

Malabsorption following oral administration of neomycin, kanamycin, and paromomycin is well documented [14,28,53,64]. Several mechanisms have been postulated. The direct damage to vilus cells can be noted in the small bowel. The agents also bind to bile salt and may impair absorption by that method as they precipitate micelles [117]. Absorption of fat, protein, cholesterol, iron, and drugs such as digoxin [64] is impaired. This effect is seen only when the drugs are given by the oral route.

References

1. Alexander, R.R., Berglund, E.M., Kasik, J.E., Fox, A., and Chinn, W.M.: The concentration of tobramycin in bronchial secretions. *Chest* 75:675–678, 1979.
2. Alfthan, O., Renkonen, O.V., and Sironen, A.: Concentration of gentamicin in serum,

urine and urogenital tissue in man. *Acta Pathol. Microbiol. Scand.* [*B*] 81(Suppl. 241):92–96, 1973.

3. Baciocco, E.A., and Iles, R.I.: Ampicillin and kanamycin concentration in joint fluid. *Clin. Pharmacol. Ther.* 12:858–863, 1971.

4. Barza, M., Kane, A., and Baum, J.: Intraocular penetration of gentamicin after subconjunctival and retrobalbar injection. *Am. J. Ophthalmol.* 85:541–547, 1978.

5. Beam, T.R., and Allen, J.C.: Blood-brain and cerebrospinal fluid concentrations of several antibiotics in rabbits with intact and inflammed meninges. *Antimicrob. Agents Chemother.* 12:710–716, 1977.

6. Bergorgne-Berezin, E., Morel, C., Even, P., Benard, Y., Kaffe, H., Berthelot, G., Pierre, J., and Lambert-Zechovsky, N.: Pharmacocinétique des antibiotiques dan les voies respiratoires. *Nouv. Presse Med.* 7:2831–2836, 1978.

7. Black, J., Calesnick, B., Williams, D., and Weinstein, M.J.: Pharmacology of gentamicin, a new broad-spectrum antibiotic. *Antimicrob. Agents Chemother.* 1963: 138–147, 1964.

8. Bliovin, R.A., Mann, H.J., Griffen, W.O., Bauer, L.A., and Record, K.E.: Tobramycin pharmacokinetics in morbidly obese patients. *Clin. Pharmacol. Ther.* 26:508–512, 1979.

9. Bodey, G.P., Valdivieso, M., Feld, R., and Rodriguez, V.: Pharmacology of amikacin in humans. *Antimicrob. Agents Chemother.* 5:508–512, 1974.

10. Boger, W.P., and Gavin, J.J.: Kanamycin, its cerebrospinal fluid diffusion, renal clearance and comparison with streptomycin. *Antibiot. Annu.* 1958–59:677, 1959.

11. Bonomini, V., and Albertazzi, I.: Comparative pharmacokinetics of sisomicin after intravenous injection, intravenous infusion and intramuscular injection in man. *Infection* 4:381–385, 1976.

12. Breen, K.J., Bryant, R.E., and Levinson, J.D.: Neomycin absorption in man. *Ann. Intern. Med.* 76:211–218, 1972.

13. Briedis, D.J., and Robson, H.G.: Cerebrospinal penetration of amikacin. *Antimicrob. Agents Chemother.* 13:1042–1043, 1978.

14. Cain, G.D., Reiner, E.B., and Patterson, M.: Effects of neomycin on disaccharidase activity of the small bowel. *Arch. Intern. Med.* 122:311–314, 1968.

15. Carbon, C., Contrepois, A., and Lamotte-Barrilson, S.: Comparative distribution of gentamicin, tobramycin, sisomicin, netilmicin and amikacin in interstitial fluid in rabbits. *Antimicrob. Agents Chemother.* 13:368–371, 1978.

16. Chamberlain, G., and Needham, P.: The absorption of antibiotics from the bladder. *J. Urol.* 116:172–173, 1976.

17. Chisholm, G.D., Calman, J.S., Waterworth, P.M., and Reis, N.D.: Distribution of gentamicin in body fluids. *Br. Med. J.* 2:22–24, 1968.

18. Chisholm, G.D., Waterworth, P.M., Calman, J.S., and Garrod, L.P.: Concentration of antibacterial agents in interstitial tissue fluid. *Br. Med. J.* 1:569–573, 1973.

19. Chou, A., Hecht, R., and Winters, R.: Gentamicin and carbenicillin penetration into the septic joint. *N. Engl. J. Med.* 285:178, 1971.

20. Cohen, L.S., Wechsler, A.S., and Mitchell, J.H.: Depression of cardiac function by streptomycin and other antimicrobial agents. *Am. J. Cardiol.* 26:505–511, 1970.

21. Cubana, B.E., and Taggart, J.G.: Comparative pharmacokinetics of BB-K8 and kanamycin in dogs and humans. *Antimicrob. Agents Chemother.* 3:478–483, 1973.

22. Daschner, F., Reiss, E., and Engert, J.: Distribution of amikacin in serum, muscle, and fat in children after a single intramuscular injection. *Antimicrob. Agents Chemother.* 11:1081–1083, 1977.

23. Davia, J.E., Siemsen, A.W., and Anderson, R.W.: Uremia deafness and paralysis due to irrigating antibiotic solutions. *Arch. Intern. Med.* 125:135–139, 1970.

24. Dee, T.H., and Kozin, F.: Gentamicin and tobramycin penetration into synovial fluid. *Antimicrob. Agents Chemother.* 12:548–549, 1977.

25. Dettli, L.: Zur pharmakotherapeutischen Beeinflussung fefassloser Bezirke im Organismus im innerem gefassloser Bezirke. *Schweiz Med. Wochenschur.* 91:921–927, 1961.
26. Dobbs, S.M., and Mawer, G.E.: Intravenous injection of gentamicin and tobramycin without impairment of hearing. *J. Infect. Dis.* 134(Suppl.):114–117, 1976.
27. Doluisio, J.T., Dittert, L.W., and LaPiana, J.C.: Pharmacokinetics of kanamycin following intramuscular administration. *J. Pharmacokinet Biopharm.* 1:253–265, 1973.
28. Duenicke, A., Gruenwaldt, G., Putter, H., and Scheer, M.: Untersuchungen zur Pharmakokinetik verschiedener Sisomicin-Dosen. *Infection* 4:376–380, 1976.
29. Dull, W.L., Alexander, M.R., and Kasik, J.R.: Bronchial secretion levels of amikacin. *Antimicrob. Agents Chemother.* 16:767–771, 1979.
30. Dunkley, B., Sanghui, I., and Goldstein, G.: Characterization of neuromuscular block produced by streptomycin. *Arch. Intern. Pharmacodyn. Ther.* 201:213–223, 1973.
31. Ericsson, C.D., Duke, J.H., Jr., and Pickering, L.K.: Clinical pharmacology of intravenous and intraperitoneal aminoglycoside antibiotics in the prevention of wound infections. *Ann. Surg.* 188:66–70, 1978.
32. Erwin, F.R., Bullock, W.E., Jr., and Nattall, C.E.: Inactivation of gentamicin by penicillins in patients with renal failure. *Antimicrob. Agents Chemother.* 9:1004–1011, 1976.
33. Faloon, W.W., Paes, J.C., and Woolfolk, D.: Effect of neomycin and kanamycin upon intestinal absorption. *Ann. N.Y. Acad. Sci.* 132:879–887, 1966.
34. Feld, R., Valdivieso, M., Bodey, G.P., and Rodriguez, V.: A comparative trial of sisomicin therapy by intermittent versus continuous infusion. *Am. J. Med. Sci.* 274:179–188, 1977.
35. Gerding, D.N., Hall, W.H., and Schierl, E.A.: Antibiotic concentrations in ascitic fluid of patients with ascites and bacterial peritonitis. *Ann. Intern. Med.* 86:708–713, 1977.
36. Gillett, A.P., Falk, R.H., Andrews, J., Wise, R., and Melikian, V.: Rapid intravenous injection of tobramycin: suggested dosage schedule and concentrations in serum. *J. Infect. Dis.* 134(Suppl.):110–113, 1976.
37. Golden, B.: Subtenon injection of gentamicin for bacterial infections of the eye. *J. Infect. Dis.* 124(Suppl.):271–274, 1971.
38. Golden, B., and Coppel, S.P.: Occular tissue absorption of gentamicin. *Arch. Ophthalmol.* 84:792–796, 1970.
39. Gordon, R.C., Regamey, C., and Kirby, W.M.M.: Serum protein binding of the aminoglycoside antibiotics. *Antimicrob. Agents Chemother.* 2:214–216, 1972.
40. Graybill, J.R., Mann, J., and Charache, P.: Intrathecal gentamicin in treatment of bacterial meningitis. *Bull. Johns Hopkins Hosp.* 133:51–56, 1973.
41. Gruenwaldt, G., Doenicke, A., Scheer, M., and Fink, H.: Vergleichende Untersuchungen zur Pharmacokinetik von Sisomicin und Gentamicin. *Infection* 4:370–375, 1976.
42. Gyselynck, A.M., Forrey, A., and Cutler, R.: Pharmacokinetics of gentamicin: distribution and plasma and renal clearance. *J. Infect. Dis.* 124(Suppl.):70–76, 1971.
43. Hall, W.H., Gerding, D.N., and Schierl, E.A.: Penetration of tobramycin into infected extravascular fluids and its therapeutic effectiveness. *J. Infect. Dis.* 135:957–961, 1977.
44. Hamory, B., Ignatiaidis, P., and Sande, M.A.: Intrathecal amikacin administration. Use in the treatment of gentamicin-resistant *Klebsiella pneumoniae* meningitis. *JAMA* 236:1973–1974, 1976.
45. Horikoshi, N., Valdivieso, M., and Bodey, G.P.: Clinical pharmacology of tobramycin. *Am. J. Med. Sci.* 266:453–458, 1974.

46. Hull, J.H., and Sarubbi, F.A., Jr.: Gentamicin serum concentrations—pharmacokinetic predictions. *Ann. Intern. Med.* 85:183–189, 1976.
47. Humbert, G., Fillastre, J.P., Leroy, A., and Oksenhendler, G.: Pharmacokinetics of sisomicin in the presence of normal and impaired renal function. *Infection* 4:393–402, 1976.
48. Humbert, G., Leroy, A., Gillastre, J.P., and Oksenhendler, G.: Pharmacokinetics of netilmicin in the presence of normal and impaired renal function. *Antimicrob. Agents Chemother.* 14:40–44, 1978.
49. Jaffe, G., Meyers, B.R., and Hirschman, S.Z.: Pharmacokinetics of tobramycin in patients with stable renal impairment, patients undergoing peritoneal dialysis and patients on chronic hemodialysis. *Antimicrob. Agents Chemother.* 5:611–616, 1974.
50. Jahre, J.A., Fu, K.P., and Neu, H.C.: Kinetics of netilmicin and gentamicin. *Clin. Pharmacol. Ther.* 23:591–597, 1978.
51. Kaiser, A.B., and McGee, Z.A.: Aminoglycoside therapy of gram-negative bacillary meningitis. *N. Engl. J. Med.* 293:1215–1220, 1975.
52. Kaye, D., Levison, E., and Labovitz, E.D.: The unpredictability of serum concentrations of gentamicin: pharmacokinetics of gentamicin in patients with normal and abnormal renal function. *J. Infect. Dis.* 130:150–154, 1974.
53. Keusch, G.T., Troncale, F.J., and Buchanan, R.D.: Malabsorption due to paromomycin. *Arch. Intern. Med.* 125:273–276, 1970.
54. Kirby, W.M.M., Clarke, J.T., Libke, R.R., and Regamey, C.: Clinical pharmacology of amikacin and kanamycin. *J. Infect. Dis.* 134(Suppl.):312–315, 1976.
55. Klastersky, J., Carpentier-Meunier, F., Kahan-Coppens, L., and Thys, J.P.: Endobronchial administered antibiotics for gram-negative bronchopneumonia. *Chest* 75:586–591, 1979.
56. Kozak, A.J., Gerding, D.N., Peterson, L.R., and Hall, W.N.: Gentamicin intravenous infusion rate. Effect on interstitial fluid concentration. *Antimicrob. Agents Chemother.* 12:606–612, 1977.
57. Kroening, V., Liebig, S., and Wundschock, M.: Tobramycin—Spiegel in menschichen Lungengewebe eine Untersuchung an 30 intravital gewonnenen Lungengewebsproben. *Infection* 6:231–235, 1978.
58. Kunin, C.M.: Absorption, distribution, excretion and fate of kanamycin. *Ann. N. Y. Acad. Sci.* 132:811–818, 1966.
59. Kunin, C.M., Chalmers, T.C., and Leevy, C.M.. Absorption of orally administered neomycin and kanamycin. *N. Engl. J. Med.* 262:380–385, 1960.
60. Last, P.M., and Sherlock, S.: Systemic absorption of orally administered neomycin in liver disease. *N. Engl. J. Med.* 262:385–389, 1960.
61. LeRoy, A., Humbert, G., Oksenhendler, G., and Fillastre, J.P.: Comparative pharmacokinetics of lividomycin, amikacin and sisomicin in normal subjects and in uremic patients. *J. Antimicrob. Chemother.* 2:373–381, 1976.
62. LeRoy, A., Humbert, G., Oksenhendler, G., and Fillastre, J.P.: Pharmacokinetics of aminoglycosides in subjects with normal and impaired renal function. *Antibiot. Chemother.* 25:163–180, 1978.
63. Levy, J., and Klastersky, J.: Correlation of serum creatinine concentration and amikacin half-life. *J. Clin. Pharmacol.* 15:705–707, 1975.
64. Lindenbaum, J., Maulitz, R.M., Saha, J.R., Shea, N., and Butler, V.P.: Impairment of digoxin absorption by neomycin. *Clin. Res.* 20:410, 1972.
65. Lockwood, W.R., and Bower, J.D.: Tobramycin and gentamicin concentrations in the serum of normal and anephric patients. *Antimicrob. Agents Chemother.* 3:125–129, 1973.
66. Lode, H., Grunert, K., Koeppe, P., and Longmaach, H.: Pharmacokinetic and clinical

studies with amikacin, a new aminoglycoside antibiotic. *J. Infect. Dis.* 134(Suppl.): 316–332, 1976.

67. Lode, H., Kemmerich, B., and Koeppe, P.: Comparative clinical pharmacology of gentamicin, sisomicin and tobramycin. *Antimicrob. Agents Chemother.* 8:396–401, 1975.

68. Malthone, M.B.R., and Harden, A.: Penetration and systemic absorption of gentamicin after subconjunctival injection. *Br. J. Ophthalmol.* 56:609–612, 1972.

69. Marks, W.M., Prentice, R., Swarson, R., Cotton, E.K., and Eickhoff, T.C.: Carbenicillin and gentamicin pharmacologic studies in patients with cyctic fibrosis and pseudomonas pulmonary infections. *J. Pediatr.* 79:822–828, 1971.

70. Marsh, D.C., Jr., Matthew, E.B., and Perselin, R.A.: Transport of gentamicin into synovial fluid. *JAMA* 228:607, 1974.

71. McCrae, W.M., Raeburn, J.A., and Hanson, E.J.: Tobramycin therapy of infections due to *Pseudomonas aeruginosa* in patients with cystic fibrosis: effect of dosage and concentration of antibiotic in sputum. *J. Infect. Dis.* 134(Suppl.):191–193, 1976.

72. McLaughlin, J.E., and Reeves, D.J.: Clinical and laboratory evidence for inactivation of gentamicin by carbenicillin. *Lancet* 1:261–264, 1971.

73. McQuillen, M.P., Cantor, H.E., and O'Rourke, J.R.: Myasthenic syndrome associated with antibiotics. *Arch. Neurol.* 18:402–415, 1968.

74. Metz, H., and Tannenberg, H.: Determination of dibekacin serum levels and the antimicrobial activity of dibekacin. *Curr. Chemother.* 2:930–931, 1977.

75. Middleton, W.H., Morgan, D.D., and Moyers, J.: Neostigmine therapy for apnea occurring after administration of neomycin. *JAMA* 165:2186–2187, 1957.

76. Moellering, R.C., Jr., and Fischer, E.G.: Relationship of intraventricular gentamicin levels to cure of meningitis. *J. Pediatr.* 81:534–537, 1972.

77. Murillo, J., Standiford, H.C., Schimpff, S.C., and Tatem, B.A.: Gentamicin and ticarcillin serum levels. *JAMA* 241:2401–2403, 1979.

78. Naber, K.G., Westerfelder, S.R., and Madsen, P.O.: Pharmacokinetics of the aminoglycoside antibiotic tobramycin in humans. *Antimicrob. Agents Chemother.* 3:469–473, 1973.

79. Neu, H.C., Appel, G.B., Parry, M.F., Garvey, G.J., and Jacob, G.: Pharmacokinetics of sisomicin in normal patients and those with depressed renal function. *Infection* 4:403–405, 1976.

80. Odio, W.E., Van Laer, J., and Klastersky, J.: Concentrations of gentamicin in bronchial secretions after intramuscular and endotracheal administration. *J. Clin. Pharmacol.* 15:518–524, 1975.

81. Pechere, J.C., and Dugal, R.: Pharmacokinetics of intravenously administered tobramycin in normal volunteers and in renal impaired and hemodialyzed patients. *J. Infect. Dis.* 134(Suppl.):118–124, 1976.

82. Pennington, J.E.: Tobramycin in bronchial secretions. *Antimicrob. Agents Chemother.* 4:299–301, 1973.

83. Pennington, J.E., Dale, D.C., and Reynolds, H.Y.: Gentamicin sulfate pharmacokinetics: lower levels of gentamicin in blood during fever. *J. Infect. Dis.* 132:270–275, 1975.

84. Pennington, J.E., and Reynolds, H.Y.: Concentrations of gentamicin and carbenicillin in bronchial secretions. *J. Infect. Dis.* 128:63–68, 1973.

85. Pennington, J.E., and Reynolds, H.Y.: Pharmacokinetics of gentamicin sulfate in bronchial secretions. *J. Infect. Dis.* 131:158–162, 1975.

86. Pines, A., Raafat, H., and Plucinski, K.: Gentamicin and colistin in chronic purulent bronchial secretions. *Br. Med. J.* 2:543–545, 1967.

87. Pitt, H.A., Robert, R.B., and Johnson, W.D., Jr.: Gentamicin levels in the human biliary tract. *J. Infect. Dis.* 127:299–302, 1973.

88. Pittinger, C.B., Eryasa, Y., and Adamson, R.: Antibiotic-induced paralysis. *Anesth. Analg.* 49:487–501, 1970.

89. Pittinger, C.B., Long, J.P., and Miller, J.R.: The neuromuscular blocking action of neomycin. *Anesth. Analg.* 37:276–282, 1958.

90. Preston, F.W., Silverman, M., Henegar, G.C., and Neveril, E.: The excretion of kanamycin in bile and pancreatic fluid. *Antibiot. Annu.* 7:857–860, 1960.

91. Pridgeon, J.E.: Respiratory arrest thought to be due to intraperitoneal neomycin. *Surgery* 40:571–574, 1956.

92. Quinlan, W., and Mehigan, C.: Release of antibiotics from bone cement. *Ir. J. Med. Sci.* 147:425–429, 1978.

93. Rahal, J.J., Jr., Hyans, P.J., Simberkoff, M.S.: Combined intrathecal and intramuscular gentamicin for gram-negative meningitis. Pharmacologic study of 21 patients. *N. Engl. J. Med.* 290:1394–1398, 1974.

94. Regula, H., Wieser, O., Naumann, P., Wundt, W., and Kleinhuber, U.: Pharmakokinetische Untersuchungen über Sputum-serum, und Urinkozentration von Gentamicin nach Aerosolinhalation. *Int. J. Clin. Pharmacol.* 7:95–100, 1973.

95. Riff, L.J., and Jackson, G.G.: Pharmacology of gentamicin in man. *J. Infect. Dis.* 124(Suppl.):98–105, 1971.

96. Riff, L.J., and Jackson, G.G.: Laboratory and clinical conditions for gentamicin inactivation by carbenicillin. *Arch. Intern. Med.* 130:887–891, 1972.

97. Riff, L.J., and Moreschi, G.: Netilmicin and gentamicin comparative pharmacology in humans. *Antimicrob. Agents Chemother.* 11:609–614, 1977.

98. Rimoldi, R., Curcio, L., and Sanfilippo, A.: Pharmacokinetic study of dibekacin in humans. *Curr. Chemother.* 2:928–930, 1977.

99. Rodriguez, V., Bodey, G.P., Valdivieso, M., and Feld, R.: Clinical pharmacology sisomicin. *Antimicrob. Agents Chemother.* 7:38–41, 1975.

100. Rodriguez, V., Stewart, D., and Bodey, G.P.: Gentamicin sulfate distribution in body fluids. *Clin. Pharmacol. Ther.* 11:275–281, 1970.

101. Rubenstein, E., Beer-Gehler, G., Dubner, J., Hallynek, T., and Wolfstein, I.: Amikacin in the human biliary tract. *Curr. Chemother.* 1:18–23, 1978.

102. Schurman, D.J., Trindale, C., Hirshman, K.P., Moser, K., Kajiyama, G., and Stevens, P.: Antibiotic-acrylic bone cement composites. *J. Bone Joint Surg.* 60A:978–984, 1978.

103. Schwartz, S.N., Pazing, G.J., Lyon, J.A., Ho, M., and Pasculle, A.W.: A controlled investigation of the pharmacokinetics of gentamicin and tobramycin in obese subjects. *J. Infect. Dis.* 138:499–505, 1978.

103a. Schurman, D.J., and Wheeler, R.: Bone and joint gram-negative infection and amikacin treatment. *Am. J. Med.,* 62 (Suppl.): 160-164. 1977.

104. Setia, U., and Gross, P.A.: Administration of tobramycin and gentamicin by the intravenous route every 6 hours in patients with normal renal function. *J. Infect. Dis.* 134(Suppl.):125–129, 1976.

105. Silber, G.R., Echevaria, P., and Smith, A.L.: Pharmacokinetics of gentamicin in children and adults. *J. Infect. Dis.* 132:637–642, 1975.

106. Simon, V.C.: A comparison of sisomicin and gentamicin pharmacokinetics. *Infection* 7(Suppl. 7):268, 1979.

107. Simon, V.K., Mosinger, E.U., and Maleryzy, V.: Pharmacokinetics studies of tobramycin and gentamicin. *Antimicrob. Agents Chemother.* 3:445–450, 1973.

108. Smithivas, T., Hyams, P.J., Matalon, R., Simberkoff, M.S., and Rahal, J.J., Jr.: The use of gentamicin in peritoneal dialysis. I. Pharmacologic results. *J. Infect. Dis.* 124(Suppl.):77–83, 1973.

109. Smithivas, T., Hyams, P.J., and Rahal, J.J., Jr.: Gentamicin and ampicillin in human bile. *J. Infect. Dis.* 124(Suppl.):106–108, 1971.

110. Solberg, C.O., and Hellum, K.B.: Protection of phagocytosed bacteria against anti-microbial agents. *Scand. J. Infect. Dis.* 14(Suppl.):246–250, 1978.

111. Sood, A.F., and Farrar, W.E.J.: Intracisternal and intrathecal injections of genta-micin in *Enterobacter* meningitis. *Arch. Intern. Med.* 134:738–740, 1974.

112. Spector, R.: The transport of gentamicin in the choroid plexus and cerebrospinal fluid. *J. Pharmacol. Exp. Ther.* 194:82–88, 1975.

113. Stratford, B.C., and Dixson, S.: Serum levels of gentamicin and tobramycin after slow intravenous bolus infusion. *Lancet* 1:378, 1978.

114. Tan, T.S., and Salstrom, S.J.: Levels of carbenicillin, ticarcillin, cephalothin, cefa-zolin, cefamandole, gentamicin, tobramycin, and amikacin in human serum inter-stitial fluid. *Antimicrob. Agents Chemother.* 11:698–701, 1977.

115. Thys, J.P., and Klastersky, J.: Concentration of sisomicin in serum and in bronchial secretions after intratracheal administration. *Curr. Chemother.* 2:920–921, 1977.

116. Thys, J.P., Mouawad, E., and Klastersky, J.: Sisomicin concentrations in serum and bronchial secretions after endotracheal administration. *Infection* 7(Suppl. 3): 269–270, 1979.

117. Thompson, G.R., Barrowman, J., and Gutierrez, L.: Action of neomycin on the intraluminal phase of lipid absorption. *J. Clin. Invest.* 50:319–323, 1971.

118. Trestman, I., Parsons, J., Santoro, J., Goodhart, G., and Kaye, D.: Pharmacology and efficacy of netilmicin. *Antimicrob. Agents Chemother.* 13:832–836, 1978.

119. Ueda, Y., Saito, A., Matsumoto, F., Omori, M., Shiba, K., Yamaji, T., and Ihara, H.: Clinical studies on dibekacin. *Curr. Chemother.* 2:931–933, 1977.

120. Vaudaux, P., and Waldvogel, F.A.: Gentamicin antibacterial activity in the presence of human polymorphonuclear leukocytes. *Antimicrob. Agents Chemother.* 16: 743–749, 1979.

121. Waitz, J.A., Drube, C.G., Moss, E.L., Jr., Oden, E.M., Bailey, J.V., Wagman, G.H., and Weinstein, M.J.: Biological aspects of the interaction between gentamicin and carbenicillin. *J. Antibiot. (Tokyo)* 25:219–224, 1972.

122. Warner, W.A., and Sanders, E.: Neuromuscular blockade associated with gentamicin therapy. *JAMA* 215:1153–1154, 1974.

123. Weibert, R., Keane, W., and Shapiro, F.: Carbenicillin inactivation of aminoglyco-sides in patients with severe renal failure. *Trans. Am. Soc. Artif. Intern. Organs* 22:439–443, 1976.

124. Wilson, T.W., Mahon, W.A., Inaba, T., Johnson, G.E., and Kadar, D.: Elimination of tritiated gentamicin in normal human subjects and in patients with severely impaired renal function. *Clin. Pharmacol. Ther.* 14:815–822, 1973.

125. Wong, G.A., Pierce, T.H., Goldstein, E., and Hoeprich, P.D.: Penetration of anti-microbial agents into bronchial secretions. *Am. J. Med.* 59:219–223, 1975.

126. Wood, M.J., and Farrell, W.: Comparison of urinary excretion of tobramycin and gentamicin in adults. *J. Infect. Dis.* 124(Suppl.):133–136, 1976.

127. Wright, E.A., and McQuillen, M.P.: Antibiotic-induced neuromuscular blockade. *Ann. N.Y. Acad. Sci.* 183:358–368, 1971.

128. Zaske, D.E., Sawchuk, R.J., Gerding, D.N., and Strate, R.G.: Increased dosage requirements of gentamicin in burn patients. *J. Trauma* 16:824–828, 1976.

6

AMINOGLYCOSIDE PHARMACOKINETICS AS A GUIDE TO THERAPY AND TOXICOLOGY

JEROME J. SCHENTAG   State University of New York at
Buffalo, and Millard Fillmore Hospital, Buffalo, New York

Although strict definition limits clinical pharmacokinetics to the tracing of drug molecule movement through physiological systems, clinical medicine has rapidly incorporated these techniques because of their utility in dosage adjustment.

A well-supported pharmacokinetic model may not only describe serum concentrations, but also facilitate extrapolations from single-dose studies to multiple dosing, predict the effects of organ failure on blood levels, and potentially describe toxicity in terms of drug concentrations in blood or tissues.

It is the purpose of this chapter to integrate our present understanding of aminoglycoside pharmacokinetics with the effects of these drugs on renal physiology. Merger of these concepts is the prerequisite to an overall appreciation of two important and parallel processes—aminoglycoside renal accumulation and aminoglycoside renal toxic effects. Furthermore, better appreciation of the role of aminoglycoside pharmacokinetics in nephrotoxicity is of use in both the proper dosing and in renal monitoring during treatment with these important antibiotics.

143

I.   Pharmacokinetics of Aminoglycosides in Humans

Until very recently, aminoglycoside pharmacokinetics were considered among the least complicated of all drugs. The drugs were presumed to be excreted completely unchanged by glomerular filtration. Distribution was described in terms of extracellular water compartments, as aminoglycosides were felt to be excluded from most tissues. A one-compartment pharmacokinetic model was viewed as adequate to describe their disposition [9,11, 54,77]. Their now well-known ability to damage the kidney was considered strictly dose-related and uncommon [16,23]. Ototoxicity occurred, but only if dosing was continued after nephrotoxicity.

Several observations in the early 1970s cast doubt on the descriptive reliability of the one-compartment model. These may be summarized in four major areas. The earliest and most consistently unexplained finding was the incomplete and variable recovery of these nonmetabolized agents in the urine. Twenty-four-hour urine recovery averaged 70–90% following a single dose given to normal volunteers, difficult to explain since it should be almost 100% in a drug with a 2 hr serum half-life [9,22,54,77]. Later, Luft and Kleit [40] and others [14,15,35,87] demonstrated high and persisting concentrations of these drugs in both animal and human kidney tissue. Third, and even more convincing, was documentation of slowly rising serum peak and trough concentrations on multiple dosing, as shown in Figure 1 [12,60]. Although decreases in renal function were presumed to account for these rising concentrations, none could be clearly demonstrated. Further-

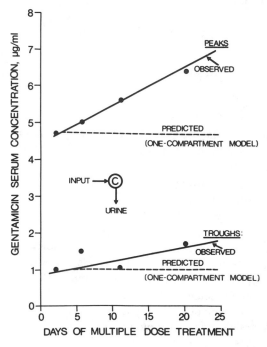

Figure 1   Measured peak and trough gentamicin serum concentrations (●) in a patient on multiple-dose therapy. Dashed lines indicate predicted values for the one-compartment model shown in the insert. (Data are modified from [12].)

more, even unmeasurable renal-function changes could not account for the biphasic decline in serum concentrations after multiple dosing [32,63].

In expanded studies employing more sensitive radioimmunoassay methods, all aminoglycosides have shown post-therapy persistence in human serum (Figure 2), and all patients have demonstrated this biphasic "washout" after the last dose. The half-life of the second phase is almost always in excess of 100 hr in both humans and animals [15, 31,66]. A two-compartment model was employed to describe these data.

In subsequent studies, the two-compartment model was shown to reliably predict aminoglycoside recovery from postmortem tissues [67,69]. In addition, cumulative urine

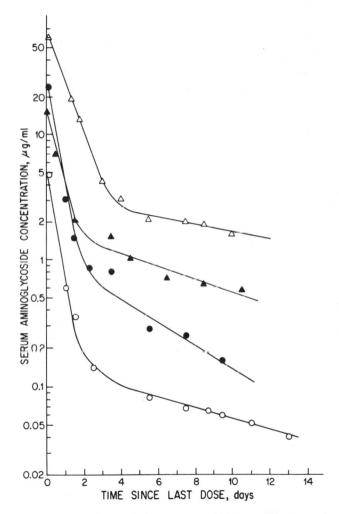

Figure 2   Serum concentrations of streptomycin (△), gentamicin (▲), amikacin (●), and tobramycin (○) at the end of a 7–25 day course of therapy. Data were obtained from separate patients with impaired renal function (creatinine clearances: 25–50 ml/min) to illustrate clearly both phases of serum decline. Curves were fitted to the experimental data using the two-compartment pharmacokinetic model. For these patients, first-phase half-life ranged from 10 to 19 hr; second phase half-lives, 82–264 hr.

collections for 10–30 days after the last dose also recovered the predicted tissue amount of drug [64], establishing another technique to verify tissue accumulation independent of pharmacokinetic modeling.

During studies of aminoglycoside tissue accumulation in both nontoxic and nephrotoxic individuals, a relationship was demonstrated between tissue amount and nephrotoxicity [61,68], as abnormally high tissue accumulation apparently preceded measurable changes in renal function. In these same studies, aminoglycoside serum concentrations were not highly predictive of nephrotoxicity [68], confirming the results of previous investigators [7,12]. It therefore appears that serum levels do not always reflect tissue levels.

The clinical utility of measured serum concentrations or predesignated "therapeutic range" is compromised when a random serum level can provide little insight into the amount of drug at the site of nephrotoxicity. Accordingly, some appreciation for nephrotoxicity in terms of tissue rather than blood concentrations must form the basis for clinical monitoring. The obvious difficulties encountered when monitoring tissue concentrations establishes the role of pharmacokinetic modeling in studies of these drugs.

II. Renal Handling of Aminoglycosides and Their Effects

A. Normal Handling of Aminoglycosides by the Kidney

Abundant evidence establishes glomerular filtration as the sole excretory route for these drugs. There is also general agreement that tubular secretion does not occur in either humans or animals. On the other hand, controversy over the extent of tubular reabsorption has continued for some time. Early studies noting almost complete excretion in normal volunteers [54] have been strongly challenged by recent clinical work establishing extremely high renal cortical tissue concentrations [14,56] and competitive antagonism of tubular reabsorption using basic amino acids such as lysine [86,88]. Investigators employing autoradiography have demonstrated proximal tubular accumulation as a result of drug reabsorpotion from urine [30]. In the face of this more direct evidence, it is reasonable to conclude that the aminoglycosides are reabsorbed from the urine filtrate. Aminoglycoside reabsorption from the peritubular capillary is also likely, but the vast surface area differences between peritubular and luminal sides of the proximal tubule (PT) establish luminal reabsorption as the primary determinant of the high kidney cortex concentrations.

Aminoglycoside proximal tubular reabsorption involves at least two steps, which differ in rate and order of magnitude. An initial binding to the PT cell brush border membrane [30] is probably a result of the electrophilic attraction between the charged amino groups and the membrane surface. This binding is rapid and apparently of both high capacity and relatively high affinity. Binding at this site is also likely to be competitive, as basic amino acids impair drug uptake, but only when given in amounts that saturate their own renal uptake [88]. The initial membrane binding correlates with the very rapid distributive phase demonstrable only after rapid intravenous injection in humans or animals, and therefore, initial renal uptake probably occurs even before the onset of renal elimination, and well before the usual 2-hr half-life begins. In kidney slices, metabolic inhibitors will not antagonize or alter the initial uptake rate [34], probably establishing that the first uptake phase in kidney slice models is also due to this competitive binding process.

Following adherence to the cell membrane, the drug is transported inside the cell [30]. This second step occurs more slowly, and is more readily saturated than the first process. Pharmacokinetic and tissue slice evidence favors saturation of this system at gentamicin urine concentrations between 5 and 25 $\mu g/ml$ [64]. Although some evidence favors the second phase as a carrier and energy-dependent membrane effect [27,34], other explanations, such as pinocytosis [84] and combined active and passive transport [34], are also offered.

After uptake by the proximal tubule (PT) cell, some drug probably diffuses back into peritubular fluids, but most remains bound intracellularly. The process of drug accumulation within PT cells occurs relatively slowly but continues throughout therapy [68].

Opinions also differ on the intracellular site damaged when toxic aminoglycoside concentrations are finally achieved. Effects on membrane ATP-mediated electrolyte transport [39], uncoupling during ATP synthesis in mitochondria [6], disruption of the lysosomal transport system [41,47,84], and nonspecific inhibition of protein synthesis are all equally plausible, and each, individually, could account for PT cell death. The relatively slow development of aminoglycoside nephrotoxicity favors lysosomal or protein synthesis damage, whereas the early loss of reabsorptive functions [70,71] and brush border membrane enzymes [46] favors disruption of energy production or its utilization. An interplay of all four mechanisms is readily conceivable.

B. Markers of Aminoglycoside Renal Effects

There has been rising interest in aminoglycoside monitoring with renal tubular (RT) indices, as there is a growing appreciation for the renal tubule as the primary site of renal damage. All these tests indicate nonspecific renal tubular damage rather than damage specific to aminoglycoside antibiotics, but all indices are useful in the early detection of changes if performed serially. Brief discussion of these indices in presented below in order of decreasing test sensitivity.

β_2-Microglobulin

An endogenously produced plasma protein (molecular weight around 11,500), this substance is freely filtered by the glomerulus, completely reabsorbed, and extensively catabolized by the PT lining cells. Normally functioning PT cells will remove almost all filtered β_2-microglobulin, resulting in negligible urinary losses [52,71,75]. Our continuing evaluation of the various urinary markers of RT damage suggests that measurement of urinary β_2-microglobulin is among the most sensitive proximal tubular markers [65,70]. In addition to high sensitivity, the test may also be highly specific to PT cell function, as distal tubule damage has not altered β_2-microglobulin excretion in preliminary studies [50].

Increased urinary β_2-microglobulin excretion is probably one of the first signs of impairment of PT cell reabsorptive functions. Urinary levels increase rapidly in response to many renal tubular insults, well in advance of demonstrable PT cell structural damage. Although this test may be too sensitive for sole use in clinical monitoring, β_2-microglobulin is a valuable research procedure in the study of mild nephrotoxins such as chronic exposure to heavy metals [75] and therapeutically administered aminoglycosides [70]; Fifty to 80% of aminoglycoside-treated patients have a rise in β_2-microglobulin excretion [71a]. However, the levels rise at least 10 times higher in aminoglycoside nephrotoxic patients than they do in nontoxics. In Figure 3 some typical patterns of β_2-microglobulin urinary excretion are presented versus time. One control patient, two aminoglycoside-treated but nontoxic individuals, and two nephrotoxic individuals are illustrated with respect to time. Although all four treated patients had an increase in urinary β_2-micro-globulin excretion, levels in the two nephrotoxic patients were at least 10 times higher than levels in the two nontoxic patients. Thus, the main differences were observed in the order of magnitude rather than excretion pattern.

Renal Enzymes

In contrast to β_2-microglobulin, which is normally produced in plasma cells, filtered, and reabsorbed, the renal enzymes are components of renal cell structure and are found in urine only when renal cell damage or death occurs [42,80]. Renal enzymes are not as sensitive as functional impairment markers such as β_2-microglobulin. This may be a dis-

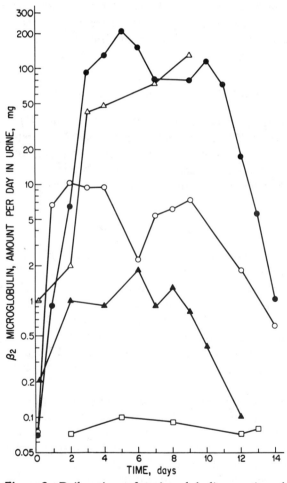

Figure 3   Daily urinary β_2-microglobulin exretion plotted versus time in five patients. One untreated control ($\square$), two treated but nonnephrotoxic patients ($\blacktriangle$,$\circ$), and two nephro-toxic patients ($\triangle$,$\bullet$) are illustrated to show typical values observed in similar populations.

tinct advantage over highly sensitive β_2-microglobulin, since there would be fewer false positives or day-to-day fluctuations. In this manner, elevations in renal enzyme excretion probably indicate more severe damage than do elevations in β_2-microglobulin. Some of the renal enzymes are excreted earlier in renal damage than others, the probable explanation being that the enzymes are found at different sites within the cell [46]. For example, alanine aminopeptidase (AAP) is a more sensitive marker than N-acetyl-β-glucosamine (NAG), because AAP is a brush border (surface)-located enzyme, whereas NAG is found within the cell itself. Thus, the different enzyme tests are unique in that they may assist in localization of the renal lesion, as well as allow the clinician to select a test with desired sensitivity for clinical use. Extensive discussions of the renal enzymes as markers of tubular injury are found in other chapters.

Renal Tubular Casts

As the PT cells undergo more extensive destruction, lysis of cell membranes occurs, and glomerular filtration rate slows and then finally ceases [81]. Each damaged nephron accumulates large amounts of cellular protein in the tubule. Distal tubular stasis of this media, rich in Tamm Horsfall protein, initiates the formation of granular casts [73]. These granular casts can be counted on a random urine using a hemocytometer and a microscope. However, a concentrated and acidic urine must be employed if the test is to be meaningful [62].

Cast counts have been very useful as an early marker of RT damage in our preliminary studies. Figure 4 shows mean cast counts for 20 nontoxic and 4 nephrotoxic patients given aminoglycosides. An increase in cast count over 500 casts/ml on 3 successive days appears to be an early warning sign of RT damage, as the excretion of significant casts precedes serum creatinine rise by 2-9 days. These results were later confirmed in a study of 154 patients [62].

Although the cast count is a less sensitive marker of PT cell damage than β_2-microglobulin, the lower sensitivity of this test may actually be considered a relative advantage, because there are fewer day-to-day fluctuations and fewer false positives [62,65].

Figure 4 also demonstrates that urinary cast counts fall to normal after serum creatinine increases, presumably because the eventual cessation of glomerular filtration in the nephron prevents their formation or excretion. Therefore, like β_2-microglobulin and enzymes, this test must be prospective, and will provide misleading information if done only after serum creatinine has already increased above baseline. Mindful of such limitations, quantitation of casts is likely to be one of the least expensive and rapid techniques available for the early identification of RT damage.

Serum Creatinine and Creatinine Clearance

The least sensitive marker of aminoglycoside-related RT damage is serum creatinine or creatinine clearance (Ccr). These tests do not determine RT function, but rather indicate that glomerular filtration has changed. Tests of glomerular filtration are of little use in monitoring for early aminoglycoside nephrotoxicity, because there may be 2-9 days of lag time between RT damage and the first rise in serum creatinine values (Figures 3 and 4). Nevertheless, the serum creatinine or Ccr can be employed to document that RT damage has occurred to such a degree that glomerular filtration has ceased. Although functional alterations in glomerular filtration occur in response to RT damage [10,13,26,37,59, 72,81], the exact nature of the control mechanism remains to be elucidated.

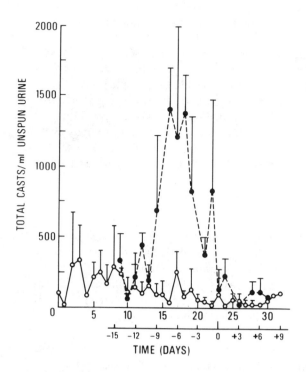

Figure 4 Mean ±1 SD urinary cast excretion in 20 nontoxic (○) and 5 nephrotoxic (●) patients treated with aminoglycosides. Daily values in the nontoxic group are illustrated with day 0 the first day of treatment. Daily values in the nephrotoxic group are illustrated on the lower scale, day 0 being the day of their first creatinine rise. The latter illustration method was used because the day of first creatinine rise was different in the five patients.

C. Features of Aminoglycoside-Related Renal Damage

Once felt to be distinctly uncommon [16,23], the reported incidence of significant amino-glycoside associated renal damage has been steadily rising. Values of 1–5% in the early 1970s rose to an average of 10% by 1977 [78,79], and have now typically been reported as 15% or more [17,49,70,83]. Although it is clear that the reported incidence is rising, this increase cannot be readily explained as due to gradual increases in dosage recommendations. However, in the face of the same dosage recommendations, surveillance of treated patients has clearly improved. Most patients are being intensely monitored before, during, and after treatment courses. In addition to a rising incidence of nephrotoxicity during treatment, improved monitoring has shown that there is also a significant incidence of post-treatment onset nephrotoxicity.

The true incidence of this effect is also controversial because there is no general agreement on a precise definition of aminoglycoside nephrotoxicity. This definition is elusive partly because of the complicated clinical setting in which the drugs are studied. It is clear that in studies with very high nephrotoxic rates, minimal rises in serum creatinine were the major criteria [78]. However, it is unknown just how many of these patients with creatinine elevations actually have aminoglycoside-related damage versus damage from other causes. Accordingly, we have employed more rigorous definitions in light of new knowledge of RT pathology and the pharmacokinetics of tissue accumulation.

The progression of aminoglycoside nephrotoxicity through time involves changes in both PT function and glomerular filtration. Although either a primary change in filtration or a severe RT insult will elevate serum creatinine, we do not attribute a serum creatinine rise to the aminoglycoside antibiotic unless the serum creatinine increase is preceded by definite evidence of RT injury. Renal tubular injury is established by a characteristic pattern of rise and fall of urinary β_2-microglobulin, enzymes, casts, and proteins prior to the increase in serum creatinine. These indices rise and fall in a reproducible and orderly fashion [70]. Aminoglycoside nephrotoxicity, as we define it, is nonoliguric except in the final stages [2,19], slow in onset (2-7 days), reversible in 10-30 days, and signs of RT damage are preceded by abnormally high tissue accumulation [61,68,70].

The sequential appearance of RT and glomerular indices in the blood and urine of a nephrotoxic patient is illustrated versus time in Figure 5. Figure 5 also presents the aminoglycoside tissue accumulation pattern and serum trough concentrations of this patient. Rapid tissue accumulation was an early event, while serum trough concentrations rose dramatically only a few days prior to serum creatinine elevation. Whereas renal tissue levels are nearly always elevated in aminoglycoside nephrotoxicity, serum levels are less predictable. Not uncommonly, patients with normal serum concentrations also suffer renal damage [7], consistent with theory, in that rapid tissue uptake can also be responsible for low serum trough concentrations.

Aminoglycosides are model compounds for the study of progressive RT damage in humans. These drugs are relatively mild nephrotoxic agents, and the renal damage progresses slowly. Therefore, combinations of RT and glomerular markets allow distinct stages of damage to be identified. However, in contrast to the slowly progressive RT injury patterns produced by aminoglycosides at recommended doses, all the RT indices rise almost simultaneously in severe RT insults such as septic shock [70]. Thus, when very large aminoglycoside doses or more potent tubular toxins are employed to study RT necrosis in animals, the renal insult may also be too severe to observe the stages that we have demonstrated in patients [70].

D. Four Stages of Aminoglycoside Damage to Human Nephrons

Aminoglycoside nephrotoxicity usually presents as an impairment of RT function progressing to RT cell necrosis if drug exposure continues [70]. A kidney slice examined by light microscopy will demonstrate nephrons in all stages, as has also been consistently reported with all RT nephrotoxic substances [18,21,25,36,55,85]. Thus, it is probable that single nephrons undergo these structural and functional changes at different rates. Since the kidney reacts as the sum of all its nephrons rather than as a single one, a critical number of nephrons must cease to function before an insensitive marker such as serum creatinine will demonstrate a rise. Fewer nephrons must be damaged to elevate urinary β_2-microglobulin or enzymes.

In a model system such as the kidney, where each nephron has internal feedback control over the cycle of its filtration, it is even possible for RT damage to occur on a continual basis without detectable changes in serum creatinine. Chronic heavy metal poisoning is probably the best example [75]. The key to this hypothesis is the balance between the rate of damage to RT lining cells and the rate at which this damage is repaired. Serum creatinine may not rise in chronic RT damage, even in the face of extensive histologic abnormalities, because even though the kidney contains some nephrons that have been

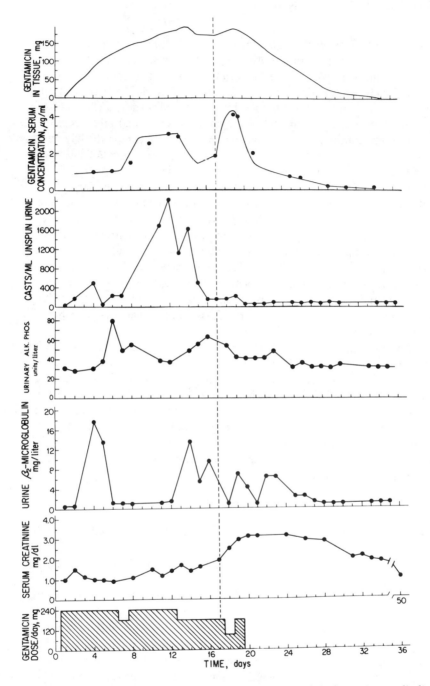

Figure 5 Dosing, serum creatinine, urinary β_2-microglobulin, urinary alkaline phosphatase, urinary cast excretion, serum trough concentrations measured (●) vs. predicted (–), and predicted tissue accumulation plotted versus time for a typical nephrotoxic patient. The dashed vertical line marks the day of first significant serum creatinine rise as a reference point. The order of elevation of these indices was tissue accumulation, β_2-microglobulin, alkaline phosphatase, casts, serum trough concentration, and serum creatinine.

152

damaged to the point where their filtration ceases, each of these damaged nephrons can regenerate its PT cells (and return to functional status) before a net decrease in filtering nephron mass occurs. Under this model, there can be a sufficient number of functional nephrons to maintain a stable and normal serum creatinine, in spite of marked RT histologic abnormalities on microscopy.

Turning back to aminoglycoside nephrotoxicity, Figure 6 presents four steps observable as the single nephron progresses from functional impairment to cell regeneration.

Functional Impairment (Stage I of Figure 6)

The rapid increase in urinary β_2-microglobulin that occurs 24-48 hr after initial exposure to aminoglycosides is the earliest index of drug effect. At this early time, no histologic changes can be detected in PT cells, but increased β_2-microglobulin excretion indicates initial failure of the PT cell to reabsorb these small proteins [45,71]. The functional nature of the early defect is also supported by the evidence that as many as 75% of aminoglycoside-treated patients demonstrate a β_2-microglobulin increase without any subsequent change in serum creatinine [71a]. However, in spite of the sensitivity of β_2-microglobulin excretion in treated patients, differences in the magnitude of β_2-microglobulin excretion can identify most of those who will later have a rise in serum creatinine. Nephrotoxic patients usually excrete over 50 mg of this protein per 24 hr [71,71a].

Structural Damage (Stage II of Figure 6)

Microscopic evidence of PT cellular impairment is usually evident by days 3 to 5 in animals given low doses of aminoglycosides [21,25]. On electron microscopy, the PT cell brush border membrane begins to slough into the luminal fluid, and cytosegresomes begin to accumulate within the cytoplasm [25,36]. The damage continues, and results in cellular swelling and mitochondrial disruption [36,85]. These changes could probably be induced earlier if overwhelming doses were given.

After demonstrating early defects in function by failure to reabsorb the filtered β_2-microglobulin, nephrons that continue to be exposed to these drugs will develop structural damage. Severe structural damage to the cell membrane and intracellular organelles is the probable explanation for the sequential appearance of urinary enzymes, then RT casts, around days 6-10 (Figures 5 and 6).

Aminoglycoside renal tissue concentration begins to decline as the nephron enters stage III, probably because the drug-laden PT cells slough off into the urine, carrying large amounts of tissue-bound drug with them. In support of this hypothesis, renal tubular casts contain far greater concentrations of gentamicin than does the surrounding urine (unpublished observations).

Clearly, nephrons with PT lining cells in stage II cannot fulfill their function to reabsorb water, electrolytes, and proteins. Continued glomerular filtration would be life-threatening. The protective mechanism linking individual nephron glomerular filtration rate with proper RT cell function must now be activated to avoid large fluid and electrolyte losses. Although the specific trigger is uncertain, this activation marks the onset of stage III.

Complete Shutdown of Glomerular Filtration (Stage III of Figure 6)

In traditional monitoring of aminoglycosides, the rise in serum creatinine is the first clinical indication that serious renal damage has occurred. Creatinine usually rises 2-9 days

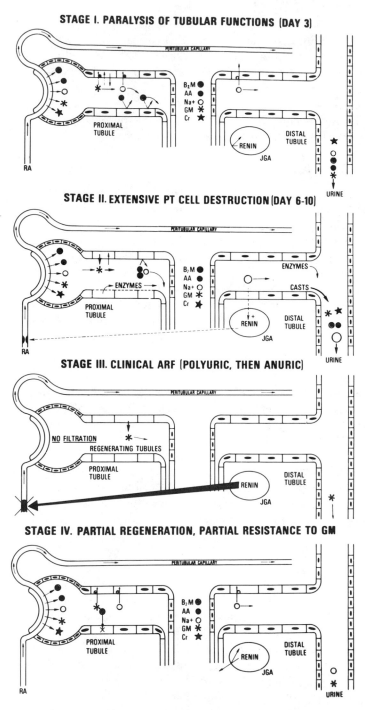

Figure 6 Stages I-IV of aminoglycoside-related damage to the single nephron. Renal handling of creatinine ($\star$), gentamicin (*), sodium ($\circ$), β_2-microglobulin ($\odot$), amino acids ($\bullet$), enzymes, and casts are shown at each of these four stages. Also illustrated is the presumed juxtoglomerular feedback control mechanism that controls filtration.

after the appearance of >500 casts/ml in urine (Figure 4). Protective in nature, but usually viewed as a definite adverse development, glomerular filtration rate must cease when a nephron loses its PT lining cells [81]. Indeed, all RT toxins produce GFR shutdown when the nephron is sufficiently damaged to become nonfunctional. A cessation of glomerular filtration is the final common pathway for all acute tubular necrosis [21]. Although the shutdown of glomerular filtration apparently occurs without changes in total renal blood flow [10,24,26,37], there is significant disagreement on this point, much of which can be attributed to differences in the timing of measurements in relation to the renal tubular insult.

During gentamicin-induced GFR cessation in rats, transmission electron microscopy (TEM) revealed no glomerular changes [5], but recent studies using scanning electron microscopy (SEM) have revealed changes in surface characteristics [14a]. The dosages of gentamicin used and the time of EM study differed in these reports. Furthermore, even if these same ultrastructural lesions are also found in humans, it has not yet been determined whether these lesions are specifically caused by aminoglycosides or whether the lesions can be observed after cessation of GFR induced by other RT toxins. The differentiation is important, because the lesions observed on SEM could be either a direct and specific aminoglycoside effect on the basement membrane structures, or they could manifest a secondary physiologic effect occurring in nonspecific response to tubular injury.

Considerable evidence also favors a role for the juxtoglomerular apparatus (JGA) in control of the shutdown of glomerular filtration, although the precise nature of the feedback signal is unknown. Anatomic arrangement of the JGA in close contact with both the distal tubule and the afferent arteriole of Bowman's capsule strongly supports this hypothesis. In addition, electron microscopy of the JGA during the very early stages of RT necrosis indicate rapid granulation and hypertrophy, and presumably JGA involvement in the formation and release of intrarenal renin [59].

Apparently, the tubuloglomerular feedback loop also functions independently of outside neuronal or hormonal influences, as it can also be demonstrated in the isolated perfused kidney [72]. Changes in distal tubular sodium concentrations appear to be intimately involved in the feedback control of nephron glomerular filtration rate [13,81]. Elevated distal sodium levels may be the initiator of the feedback loop, acting through the JGA as a sensor, and by releasing intrarenal renin to mediate the decrease in glomerular ultrafiltration coefficient.

After close inspection of the events leading up to and following stage III, the prognosis and treatment of aminoglycoside nephrotoxicity can be addressed in terms of renal pathophysiology and aminoglycoside pharmacokinetic considerations.

Renal tubular cell regeneration (stage IV) is already in progress at the time a nephron enters stage III, and the overall therapeutic goal becomes support of the patient until these new RT lining cells assume functional status and filtration resumes.

The progression of stage III cessation of GFR can be monitored only by serum creatinine. There are no casts or β_2-microglobulin in the urine of stage III patients, as shown in Figures 4 and 5. This observation is not surprising, because nonfiltering nephrons can no longer be expected to excrete β_2-microglobulin, casts, enzymes, or electrolytes.

There does not appear to be a reliable way to reverse stage III in the nephron, and this may be preferable because the return of nephron filtration before PT cells regenerate would severely disrupt electrolyte balance. Since glomerular filtration rate is now markedly reduced and no new aminoglycoside can enter the urine, the regenerating RT cells are being shielded from further exposure. Theoretically, treatment could be continued provided that dosages are reduced to prevent potentially ototoxic serum concentrations. However, not all stage II nephrons have entered stage III when creatinine first rises, and therapy should be stopped whenever possible. The drug should be stopped because rising serum concentrations induced by the failure of nephron after nephron ensure that larger amounts of drug are filtered by the remaining stage II nephrons, and places them at risk for greater damage.

Stage III can potentially be aggravated by certain therapeutic measures that may be employed as serum creatinine rises. The principal offenders can be the loop diuretics [8,29,38,89], as one of their actions is to block the feedback-mediated decrease in GFR [89]. Although in one sense the furosemide-induced block of the feedback loop will increase urine output, this also causes more severe problems in the nephron, since nephrons that continue to filter will continue to increase the drug concentration in PT lining cells, and greater overall damage would result. Short-term nephrotoxicity studies in rats [8], where exact replacement or urinary losses of water and electrolytes did not protect against renal failure, but markedly increased the aminoglycoside tissue levels, support this hypothesis. Aminoglycoside tissue levels can only continue to rise with continued treatment, if filtration does not cease after stages II and III occur.

Regeneration and Resistance to Further Damage (Stage IV of Figure 6)

As soon as a nephron enters stage III, it also apparently enters stage IV, as nephron RT cell regeneration begins as soon as exposure to the insulting substance ends at the cessation of glomerular filtration. Although it is apparent that stage IV occurs, this stage is the most difficult to study, because there are no available markers other than microscopy, and the absence of stage synchronization between tissue slice nephrons clearly complicates microscopic study. When filtration ceases, serum creatinine rises, but there is no way to predict either how long or how high. Creatinine or creatinine clearance can only confirm the eventual return of glomerular filtration, which indicates that the RT cell is again functioning.

The most striking observation on regenerated PT cells is that these cells are temporarily less sensitive to further damage by either the same substances or by other RT toxins [44]. Although these cells do not reabsorb normally during resolving tubular necrosis, they do accumulate aminoglycosides, apparently without further injury. Even more remarkably, GFR will resume in nephrotoxic rats even in the face of continued drug exposure [43]. This suggests that new PT cells transiently tolerate previously lethal aminoglycoside urine levels.

The lag time between the return of glomerular filtration and the return of normal sensitivity to aminoglycoside renal damage may be dependent on variables such as the state of nutrition [1,82] and subsequent exposure to other RT toxins. Thus, although patients with resolving ATN may not be immediately at risk for RT damage, they may become so after a variable period. The problem of altered sensitivity during RT cell regeneration deserves further intensive study, because prediction of the safe interval can be important if the patient again requires aminoglycosides.

III. Pharmacokinetic Considerations in Aminoglycoside Nephrotoxicity

An important prerequisite to the clinical utilization of multicompartment pharmacokinetic modeling is the appreciation that serum concentrations decline as a function of two relatively independent processes, tissue uptake and renal excretion. This is opposed to one-compartment-model behavior, which assumes that renal excretion is solely responsible for blood-level decline. Because of these pharmacokinetic factors, low trough concentrations could result from either very high tissue uptake or rapid renal excretion.

Because there are two forces responsible for blood-level decline after a dose, two patients with the same low serum trough concentration can differ markedly in their renal tissue concentration and their response to therapy. Two-compartment-model behavior will also explain lack of renal damage in the face of elevated trough concentrations, as these elevated trough concentrations might be due to either decreased renal excretion or abnormally low tissue uptake. As an example, two patients with similar final peak and trough serum concentrations are illustrated in Figure 7. Complete pharmacokinetic data on both of these individuals has been published [68]. In spite of similar peak and troughs by the final dose, one patient had a marked decline in renal function, whereas the other had no adverse effects. The difference between the patients was abnormally rapid tissue accumulation in the nephrotoxic patient. In this individual, abnormally rapid tissue uptake apparently preceded any detectable changes in renal function. A pharmacokinetic algorithm that further describes these two types of aminoglycoside responses is presented in Figure 8. As shown in Figure 7, uptake apparently preceded detectable changes in renal function. A pharmacokinetic algorithm that further describes these two types of aminoglycoside responses is presented as Figure 8.

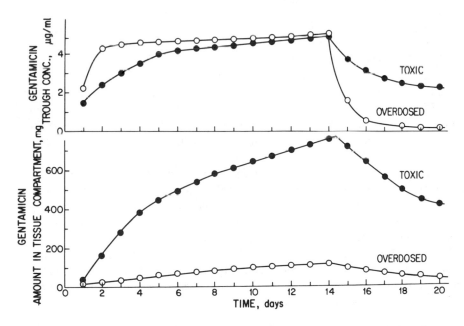

Figure 7 Serum trough concentrations and predicted tissue accumulation for a nontoxic patient who was overdosed (○) and a nephrotoxic patient (●). The data demonstrate the early and abnormally rapid tissue uptake that occurs in prenephrotoxic patients. Changes in serum trough concentrations were less marked than those in tissue accumulation. The data illustrate the differences in tissue amount that can occur in the face of similar serum trough concentrations.

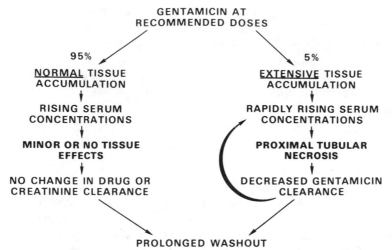

Figure 8 Proposed scheme for the two groups of appropriately dosed patients. Most patients show normal tissue uptake and washout, whereas nephrotoxic individuals begin early and rapid tissue uptake, which apparently results in nephrotoxicity.

As shown in Figures 7 and 8, aminoglycoside antibiotics accumulate in tissues of all appropriately dosed individuals. However, a small number of patients demonstrate early and rapid tissue accumulation without apparent cause. These nontoxic and prenephrotoxic patients may have similar serum levels, and may also be identical in their ability to filter aminoglycosides at the glomerulus. However, tissue accumulation rate apparently differs between these same individuals by a factor of 10-100. Renal tubular reabsorption rate markedly reflects these interindividual differences in tissue uptake.

Predisposed individuals appear to be randomly distributed and at present unidentifiable, and no clinical variables successfully identify abnormal uptake before renal function deteriorates [61,79]. Single blood levels are particularly inadequate, because underdosing, rapid tissue uptake, or normal renal excretion might yield a low trough serum concentration. Serial trough concentrations are potentially more useful, although somewhat less sensitive than desirable.

Under the dual constraints of unpredictable accumulation patterns and the lack of a prospective test to identify those patients who will abnormally accumulate, a baseline nephrotoxicity incidence around 5-10% must be anticipated even in the face of vigorous use of blood levels and dosing nomograms. However, the lack of predictable nephrotoxic effects does not mean that proper initial dosing has no value, because clearly proper dosing reduces nephrotoxicity down to this baseline. On the other hand, any reduction of significant nephrotoxicity below random events requires a method useful in the early identification of renal damage, before the damage progresses to a serum creatinine rise. This task falls upon the renal tubular indices.

IV. Monitoring Aminoglycoside Therapy—A Combined Pathophysiologic and Pharmacokinetic Approach

We have presented a complex pathogenetic scheme integrating aminoglycoside nephrotoxicity with its tissue pharmacokinetics. However, most treated patients can be monitored with a less rigorous approach, provided that the important elements of pathogenesis are considered. A series of progressively more sensitive tests is now under study, and it is likely that an individualized approach can be formulated. Those whose survival requires close monitoring must receive it, but an extensive use of resources can be avoided where the yield is small. Criteria for selecting a particular test should include an assessment of available resources and a decision regarding the affordable level of renal compromise which does not threaten patient survival. Clearly, a transient and reversible 50% decrease in renal function presents little risk to a stable ward patient, but can drastically alter prognosis in a delicately balanced intensive-care patient.

A. Use of Aminoglycoside Blood Levels

Problems with Blood Level Predictions

Perhaps no issue has become so controversial in the clinical use of these drugs as the predictability of blood levels from patient characteristics, nomograms, or pharmacokinetic models.

Two assay problems must be appreciated before blood level predictability can even be addressed. Microbiologic assays are subject to interference by unknown concurrent antibiotics [20], and carbenicillin inactivates the aminoglycosides both in vitro and in vivo [57]. Once these difficulties are controlled by use of RIA and by adding penicillinase to fresh blood before assay, respectively, the remaining problem is accurate prediction of the levels in unstable clinical settings. Here, disease-related changes in distribution volume [51,74], infusion rate fluctuations, and difficulties with changing renal function raise havoc with predictability. In our studies, blood levels of any one aminoglycoside appear no more predictable than any other, although there are occasionally different viewpoints [53].

The entire predictability controversy appears to be between those who studied small numbers of stable patients and found levels predictable [28,58] versus those who studied small numbers of unstable patients and found levels unpredictable [4,33]. Rather than taking sides on this controversy, the possibility should be considered that neither argument is very important to our patients. For proper use of blood levels, the reason why a measured level deviates from its prediction must be discovered. Since so many factors can affect blood levels, only this activity will ultimately benefit patients, and only the search will expand our knowledge. Statistically valid regression lines between measured and predicted values with good correlation but large scatter do not address the challenge of the deviant patient. Rather, careful study of these deviant blood levels must be undertaken if the complexities of aminoglycoside disposition are to be understood.

Problems with Interpretation

Blood levels rise and fall with dosing, making an appreciation of their timing in relation to drug dosing vital to their proper interpretation. A therapeutic peak concentration is dependent primarily on the minimum inhibitory concentration (MIC) of the infecting organism and the locus of infection. A nontoxic trough is mainly dependent on proper drug excretion and normal tissue uptake. In spite of this apparent simplicity, neither a therapeutic peak nor a nontoxic trough can be readily quoted. Peak levels appear most useful when there is a question of therapeutic failure, and troughs are most useful if done every 2–4 days in the quest for changes heralding abnormal tissue accumulation. Because dramatic trough level rises are soon followed by changes in creatinine, aminoglycoside serum levels have been called expensive serum creatinines [3], and in a large measure this may be correct. However, they do begin to rise before serum creatinine (Figure 5), and the rate of rise is useful in assessment of the rate of tissue uptake [61,68].

B. Role of Tissue Accumulation and Two-Compartment Modeling

Although highly useful as a research tool in the study of patients, two-compartment modeling is not likely to lend itself to widespread clinical application, because the computer time and multiple sampling schedules are expensive and time-consuming. Even more important, serial use of RT indices appears to be a more sensitive indicator of abnormal accumulation than prediction based on blood levels or urine collections. However, we continue to consider these techniques valuable basic research tools. Of available methods, only tissue accumulation will allow decisions of the cause of renal tubular damage when other possible causes are also present [70]. Second, tissue accumulation models allow the study of comparative nephrotoxicity between similar aminoglycosides [69].

C. Use of Renal Tubular Indices

There are also a number of disadvantages associated with the clinical use of the RT indices. False positives, nonspecificity, the requirement for serial use, and costs in technical time are primary concerns. Use of most tubular indices (except cast counts) requires timed urine collections, preferably daily 24-hr collections. Most of the assay procedures are complex or expensive. The RT indices are elevated by most RT insults, including aminoglycosides, and are of high value in the early identification of RT damage. However, they will not always identify the specific cause. All are prospective tests, in that they are only of value if measured before serum creatinine rises.

Several advantages offset these disadvantages. The principal value of the RT indices is that they offer the potential to completely avoid clinically significant creatinine rise if used properly. Nonspecific indicators of RT damage can be employed to identify early RT insults in intensive-care patients, and do allow intervention well in advance of severe ATN. Tests such as cast counts can potentially decrease the need for frequent blood-level determinations, and thus save considerably in the costs of monitoring these drugs.

In summary, these tests have exciting potential but also a number of technical problems. They should remain a research procedure until it is prospectively demonstrated that they can be employed to adjust the drug dose. Indeed, it remains to be prospectively determined if any of these early markers will avoid subsequent renal damage in aminoglycoside-treated patients, since their elevation indicates that some RT damage has already occurred.

V. Conclusions and Recommendations

Aminoglycoside antibiotics are among the most effective agents for gram-negative infections. However, like many of our most effective agents, they have potential side effects, in this case serious renal and ototoxicity. The renal damage occurs in the proximal tubule, is relatively mild, and will eventually heal, especially if treatment is stopped. The lesion in the renal tubule is similar to that produced by most other tubular toxins, but develops more slowly. Thus, these drugs are model compounds for the study of chronic renal tubular damage and its mechanisms. They are model compounds for clinical pharmacokinetic study as well, because of their unique disposition and single-pathway elimination. Parallel pharmacokinetic and pathophysiological study of aminoglycoside nephrotoxicity has provided much information of the mechanisms of renal damage due to these compounds, as well as a means for relative comparison.

These antibiotics will clearly cause nephrotoxicity in all patients if excessive dosages are given, but the nephrotoxic effects will also occur unpredictably among appropriately dosed individuals [61,79]. To date, the single characteristic common to all of our appropriately dosed nephrotoxic patients is excessive tissue accumulation [61]. It is probable that an individual sensitivity mediates this abnormal tissue accumulation, as accumulation begins with the first dose and appears to lead to nephrotoxicity, rather than occurring as a result of renal damage.

Proper therapeutic use of these drugs must first involve the selection of appropriate dosing schedules by use of creatinine clearance. Stable patients should then be monitored primarily with frequent serum creatinines, employing aminoglycoside serum concentrations if there is question of therapeutic failure. On the other hand, high-risk patients—especially unstable patients who cannot afford a decline in renal function—should be observed far more closely. After dosing on the basis of creatinine clearance, blood levels should be employed to maintain the desired therapeutic range. Serial use of trough concentrations (every 2-4 days) is important to identify abnormal tissue uptake. Renal tubular indices such as cast counts should be employed in selected high-risk patients to document adverse renal effects early and to determine the need for further aminoglycoside blood-level measurements or dosage adjustments.

If renal damage occurs, the drug should be stopped if possible, or dosage should be reduced. Patients should be well hydrated, and urine electrolyte losses should be quantitated and replaced. Loop diuretics should always be avoided, except if fluid overload is more life-threatening than further renal damage. We would seldom hemodialyze nephrotoxic patients unless serum trough concentrations exceed 4.0, with serum creatinine above 4.0 and rising rapidly. Hemodialysis is effective in removing the drug from blood, but after stopping dialysis in nephrotoxic patients, the blood level will often reequilibrate back to previous values (Figure 9). This occurs because of redistribution from the high tissue stores back into blood, as most of the drug is in tissues rather than in blood. On the other hand, nontoxic patients with high serum levels will have minimal tissue stores, and redistribution will be less notable after hemodialysis. Because of these differences, measurement of aminoglycoside serum levels in the first 8 hr after hemodialysis may establish a means of interpreting postdialysis tissue stores. Peritoneal dialysis, even though less efficient than hemodialysis in single-dose studies, may be more effective in nephrotoxic patients because the longer equilibration time might remove more of the tissue stores. However, this hypothesis clearly requires further study.

Finally, since the various aminoglycosides probably differ in their nephrotoxic potential, the clinical use of the least nephrotoxic aminoglycoside will clearly be the most effective means of reducing the clinical incidence of nephrotoxic reactions, all other factors being equal.

Acknowledgments

The nursing, house, and attending staffs of the Millard Fillmore and Buffalo General Hospitals referred patients and assisted with sample collections. G. Calleri, E. DeGlopper, D. Chiarmonte, J. Acquard, and D. Danner provided technical assistance. I thank Drs.

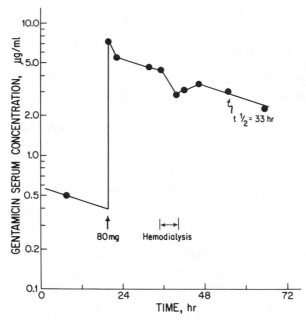

Figure 9   Gentamicin serum concentrations vs. time in a nephrotoxic patient who underwent a 4 hr hemodialysis. Overall half-life was not significantly affected because of redistribution from tissue stores.

Martin E. Plaut and William J. Jusko for their helpful suggestions throughout these studies.

Supported in part by Grant GM20852 from the National Institutes of Health.

References

1. Abel, R.M., Beck, C.H., Abbott, W.M., Ryan, J.A., Barnett, G.O., and Fischer, J.E.: Improved survival from acute renal failure after treatment with intravenous essential (L)-amino acids and glucose. *N. Engl. J. Med.* 288:695–699, 1973.

2. Anderson, R.J., Linas, S.L., Berns, A.S., Henrich, W.L., Miller, T.R., Gabow, P.A., and Schrier, R.W.: Nonoliguric acute renal failure. *N.Engl. J. Med.* 296:1134–1138, 1977.

3. Barza, M.: Gentamicin serum levels. *Clin. Pharmacokinet.* 3:202–215, 1978.

4. Barza, M., Brown, R.B., Shen, D., Gibaldi, M., and Weinstein, L.: Predictability of blood levels of gentamicin in man. *J. Infect. Dis.* 132:165–174, 1975.

5. Baylis, C., Rennke, H.R., and Brenner, B.M.: Mechanisms of the defect in glomerular ultrafiltration associated with gentamicin administration. *Kidney Int.* 12:344–353, 1977.

6. Bendirdjian, J.-P., Foucher, B., and Fillastre, J.-P.: Influence des aminoglycosides sur le métabolisme respiratoire des mitochondries isolées de foie et de rein de rat. In *Nephrotoxicity: Interaction of Drugs with Membrane Systems Mitochondria-lysosomes,* edited by Fillastre, J.-P. New York, Masson, 1978, pp. 315–332.

7. Bennett, W.M., Gilbert, D.N., Houghton, D., and Porter, G.A.: Gentamicin nephrotoxicity—morphologic and pharmacologic features. *West. J. Med.* 126:65–68, 1977.

8. Chiu, P.J.S., and Long, J.F.: Effects of hydration on gentamicin excretion and renal

accumulation in furosemide-treated rats. *Antimicrob. Agents Chemother.* 14:214–217, 1978.

9. Clarke, J.T., Libke, R.D., Regamey, C., and Kirby, W.M.M.: Comparative pharmacokinetics of amikacin and kanamycin. *Clin. Pharmacol. Ther.* 15:610–616, 1974.

10. Cox, J.W., Baehler, R.W., Sharma, H., O'Dorisio, T., Osgood, R.W., Stein, J.H., and Ferris, T.F.: Studies on the mechanism of oliguria in a model of unilateral acute renal failure. *J. Clin. Invest.* 53:1546–1558, 1974.

11. Cutler, R.E., Gyselynck, A-M., Fleet, W.P., and Forrey, A.W.: Correlation of serum creatinine concentration and gentamicin half-life. *JAMA* 219:1037–1041, 1972.

12. Dahlgren, J.G., Anderson, E.T., and Hewitt, W.L.: Gentamicin blood levels: a guide to nephrotoxicity. *Antimicrob. Agents Chemother.* 8:58–62, 1975.

13. DiBona, G.F., McDonald, F.D., Flamenbaum, W., Dammin, G.J., and Oken, D.E.: Maintenance of renal function in salt loaded rats despite severe tubular necrosis induced by $HgCl_2$. *Nephron* 8:205–220, 1971.

14. Edwards, C.Q., Smith, C.R., Baughman, K.L., Rogers, J.F., and Lietman, P.S., Concentrations of gentamicin and amikacin in human kidneys. *Antimicrob. Agents Chemother.* 9:925–927, 1976.

14a. Evan, A.P., Huser, J., Avasthi, P.S., Rankin, L.I., and Luft, F.C.: Gentamicin induced glomerular injury. Abstr. 933, Proceedings of the 11th International Congress of Chemotherapy, Boston, Mass., October 1–5, 1979.

15. Fabre, J., Rudhardt, M., Blanchard, P., and Regamey, C.: Persistence of sisomicin and gentamicin in renal cortex and medulla compared with other organs and serum of rats. *Kidney Int.* 10:444–449, 1976.

16. Falco, F.G., Smith, H.M., and Arcieri, G.M.: Nephrotoxicity of aminoglycosides and gentamicin. *J. Infect. Dis.* 119:406–409, 1969.

17. Fee, W.E., Vierra, V., and Lathrop, G.R.: Clinical evaluation of aminoglycoside toxicity: tobramycin versus gentamicin, a preliminary report. *J. Antimicrob. Chemother.* 4:31–36, 1978.

18. Ganote, C.E., Reimer, K.A., and Jennings, R.B.: Acute mercuric chloride nephrotoxicity: an electron microscopic and metabolic study. *Lab Invest.* 31:633–647, 1975.

19. Gary, N.E., Buzzeo, L., Salaki, J., and Eisinger, R.P.: Gentamicin-associated acute renal failure. *Arch. Intern. Med.* 136:1101–1104, 1976.

20. Giamarellou, H., Zimelis, V.M., Matulionis, D.O., and Jackson, G.G.: Assay of aminoglycoside antibiotics in clinical specimens. *J. Infect. Dis.* 132:399–406, 1975.

21. Gilbert, D.N., Plamp, C., Starr, P., Bennett, W.M., Houghton, D.C., and Porter, G.: Comparative nephrotoxicity of gentamicin and tobramycin in rats. *Antimicrob. Agents Chemother.* 13:34–40, 1978.

22. Gyselynck, A-M., Forrey, A., and Cutler, R.: Pharmacokinetics of gentamicin: distribution and plasma and renal clearance. *J. Infect. Dis.* 124(Suppl.):S70–S76, 1971.

23. Hewitt, W.L.: Gentamicin: toxicity in perspective. *Postgrad. Med. J.* 50(Suppl.): 55–59, 1974.

24. Hollenberg, N.K., Adams, D.G., Oken, D.E., Abrams, H.L., and Merrill, J.P.: Acute renal failure due to nephrotoxins, renal hemodynamic and angiographic studies in man. *N. Engl. J. Med.* 282:1329–1334, 1970.

25. Houghton, D.C., Hartnett, M., Campbell-Boswell, M., Porter, G., and Bennett, W.: A light and electron microscopic analysis of gentamicin nephrotoxicity in rats. *Am. J. Pathol.* 82:589–612, 1976.

26. Hsu, C.H., Kurtz, T.W., Rosenzweig, J., and Weller, J.M.: Renal hemodynamics in $HgCl_2$-induced acute renal failure. *Nephron* 18:326–332, 1977.

27. Hsu, C.H., Kurtz, T.W., and Weller, J.M.: In vitro uptake of gentamicin by rat renal cortical tissue. *Antimicrob. Agents Chemother.* 12:192–194, 1977.

28. Hull, J.H., and Sarubbi, F.A.: Gentamicin serum concentrations: pharmacokinetic predictions. *Ann. Intern. Med.* 85:183–189, 1976.

29. Inada, Y., Terashita, Z.-I., Shibouta, Y., Ando, T., Shimakawa, H., Nishikawa, K., Kikuchi, S., and Shimamoto, K.: Effect of pretreatment with 1,4-dimorpholino-7-phenylpyrido(3,4-d)pyridazine (DS-511), hydrochlorothiazide and furosemide on cephaloridine- or gentamicin-induced nephrotoxicity in the rat. *J. Takeda Res. Lab.* 37:41–48, 1978.

30. Silverblatt, F.J., and Kuehn, C.: Autoradiography of gentamicin uptake by the rat proximal tubule cell. *Kidney Int.* 15:335–345, 1979.

31. Kahlmeter, G., Jonsson, S., and Kamme, C.: Multiple-compartment pharmacokinetics of tobramycin. *J. Antimicrob. Chemother.* 4(Suppl.):5–11, 1978.

32. Kahlmeter, G., and Kamme, C.: Prolonged excretion of gentamicin in a patient with unimpaired renal function. *Lancet* 1:286, 1975.

33. Kaye, D., Levison, M.E., and Labovitz, E.D.: The unpredictability of serum concentrations of gentamicin: pharmacokinetics of gentamicin in patients with normal and abnormal renal function. *J. Infect. Dis.* 130:150–154, 1974.

34. Kluwe, W.M., and Hook, J.B.: Analysis of gentamicin uptake by rat renal cortical slices. *Toxicol. Appl. Pharmacol.* 45:531–539, 1978.

35. Kornguth, M.L., and Kunin, C.M.: Distribution of gentamicin and amikacin in rabbit tissues. *Antimicrob. Agents Chemother.* 11:974–977, 1977.

36. Kosek, J.C., Mazze, R.I., and Cousins, M.J.: Nephrotoxicity of gentamicin. *Lab. Invest.* 30:48–57, 1974.

37. Kurtz, T.W., and Hsu, C.H.: Systemic hemodynamics in nephrotoxic acute renal failure. *Nephron* 21:100–106, 1978.

38. Lawson, D.H., Macadam, R.F., Singh, H., Gavras, H., Hartz, S., Turnbull, D., and Linton, A.L.: Effect of furosemide on antibiotic-induced renal damage in rats. *J. Infect. Dis.* 126:593–600, 1972.

39. Lietman, P.S.: Aminoglycoside inhibition of a renal sodium-potassium ATP'ase: a possible model for nephrotoxicity. Abstr. 328, 18th Interscience Conference on Antimicrobial Agents and Chemotherapy, Atlanta, 1978.

40. Luft, F.C., and Kleit, S.A.: Renal parenchymal accumulation of aminoglycoside antibiotics in rats. *J. Infect. Dis.* 130:656–659, 1974.

41. Luft, F.C., and Patel, V.: Lysosomal acid hydrolases as urinary markers of aminoglycoside nephrotoxicity in the rat. In *Nephrotoxicity: Interaction of Drugs with Membrane Systems Mitochondria-lysosomes*, edited by Fillastre, J.-P. New York, Masson, 1978, pp. 127–142.

42. Luft, F.C., Patel, V., Yum, M.N., Patel, B., and Kleit, S.A.: Experimental aminoglycoside nephrotoxicity. *J. Lab. Clin. Med.* 86:213–220, 1975.

43. Luft, F.C., Rankin, L.I., Sloan, R.S., and Yum, M.N.: Recovery from aminoglycoside nephrotoxicity with continued drug administration. *Antimicrob. Agents Chemother.* 14:284–287, 1978.

44. Luft, F.C., Yum, M.N., and Kleit, S.A.: The effect of concomitant mercuric chloride and gentamicin on kidney function and structure in the rat. *J. Lab. Clin. Med.* 89:622–631, 1977.

45. Maack, T.: Renal handling of low molecular weight proteins. *Am. J. Med.* 58:57–64, 1975.

46. Mondorf, A.W., Breier, J., Hendus, J., Scherberich, J.E., Mackenrodt, G., Shah, P.M., Stille, W., and Schoeppe, W.: Effect of aminoglycosides on proximal tubule membranes of the human kidney. *Eur. J. Clin. Pharmacol.* 13:133–142, 1978.

47. Morin, J.-P., Fresel, J., Fillastre, J.-P., and Vaillant, R.: Aminoglycoside actions on rat kidney lysosomes in vivo and in vitro. In *Nephrotoxicity: Interaction of Drugs with Membrane Systems Mitochondria-lysosomes*, edited by Fillastre, J.-P. New York, Masson, 1978, pp. 253–264.

48. Nomiyama, K., Yamamoto, A., and Sato, C.: Assay of urinary enzymes in toxic nephropathy. *Toxicol. Appl. Pharmacol.* 27:484–490, 1974.

49. Panwalker, A.P., Malow, J.B, Zimelis, V.M., and Jackson, G.G.: Netilmicin: clinical efficacy, tolerance, and toxicity. *Antimicrob. Agents Chemother.* 13:170–176, 1978.

50. Pedersen, E.B., Mogensen, C.E., Solling, K., Amdisen, A., and Darling, S.: Urinary excretion of albumin, β_2-microglobulin and free light chains during lithium treatment. *Scand. J. Clin. Lab. Invest.* 38:269–272, 1978.

51. Pennington, J.E., Dale, D.C., Reynolds, H.Y., and MacLowry, J.D.: Gentamicin sulfate pharmacokinetics: lower levels of gentamicin in blood during fever. *J. Infect. Dis.* 132:270–275, 1975.

52. Peterson, P.A., Evrin, P.-E., and Berggard, I.: Differentiation of glomerular, tubular, and normal proteinuria: determinations of urinary excretion of β_2-microglobulin, albumin, and total protein. *J. Clin. Invest.* 48:1189–1198, 1969.

53. Price, K.A., and Chudzik, G.M.: Amikacin. *Lancet* 2:659–660, 1977.

54. Regamey, C., Gordon, R.C., and Kirby, W.M.M.: Comparative pharmacokinetics of tobramycin and gentamicin. *Clin. Pharmacol. Ther.* 14:396–403, 1973.

55. Reimer, K.A., Ganote, C.E., and Jennings, R.B.: Alterations in renal cortex following ischemic injury. *Lab Invest.* 26:347–363, 1972.

56. Reiner, N.E., Bloxham, D.D., and Thompson, W.L.: Nephrotoxicity of gentamicin and tobramycin given once daily or continuously in dogs. *J. Antimicrob. Chemother.* 4(Suppl.):85–101, 1978.

57. Riff, L.J., and Jackson, G.G.: Laboratory and clinical conditions for gentamicin inactivation by carbenicillin. *Arch. Intern. Med.* 130:887–891, 1972.

58. Sarubbi, F.A., and Hull, J.H.: Amikacin serum concentrations: predictions of levels and dosage guidelines. *Ann. Intern. Med.* 89:612–618, 1978.

59. Sato, T., McDowell, E.M., McNeil, J.S., Flamenbaum, W., and Trump, B.F.: Studies on the pathophysiology of acute renal failure: III. A study of the juxtaglomerular apparatus of the rat nephron following administration of mercuric chloride. *Virchows Arch. [Cell Pathol.]* 24:279–293, 1977.

60. Sawchuk, R.J., Zaske, D.E., Cipolle, R.J., and Wargin, W.A.: Kinetic model for gentamicin dosing with the use of individual patient parameters. *Clin. Pharmacol. Ther.* 21:362–369, 1977.

61. Schentag, J.J., Cumbo, T.J., Jusko, W.J., and Plaut, M.E.: Gentamicin accumulation and nephrotoxic reactions. *JAMA* 240:2067–2069, 1978.

62. Schentag, J.J., Gengo, F.M., Plaut, M.E., Danner, D., Mangione, A., and Jusko, W.J.: Urinary casts as an indicator of renal tubular damage in patients receiving aminoglycosides. *Antimicrob. Agents Chemother.* 16:468–474, 1979.

63. Schentag, J.J., and Jusko, W.J.: Gentamicin persistence in the body. *Lancet* 1:486, 1977.

64. Schentag, J.J., and Jusko, W.J.: Renal clearance and tissue accumulation of gentamicin. *Clin. Pharmacol. Ther.* 22:364–370, 1977.

65. Schentag, J.J., Jusko, W.J., and Plaut, M.E.: Comparative evaluation of tests for aminoglycoside nephrotoxicity. *Clin. Pharmacol. Ther.* 23:128, 1978.

66. Schentag, J.J., Jusko, W.J., Plaut, M.E., Cumbo, T.J., Vance, J.W., and Abrutyn, E.: Tissue persistence of gentamicin in man. *JAMA* 238:327–329, 1977.

67. Schentag, J.J., Jusko, W.J., Vance, J.W., Cumbo, T.J., Abrutyn, E., DeLattre, M., and Gerbracht, L.M.: Gentamicin disposition and tissue accumulation on multiple dosing. *J. Pharmacokinet. Biopharm.* 5:559–577, 1977.

68. Schentag, J.J., Jusko, W.J., Vance, J.W., Cumbo, T.J., Abrutyn, E., and Plaut, M.E.: Aminoglycoside nephrotoxicity in relation to tissue accumulation. In *Nephrotoxicity: Interaction of Drugs with Membrane Systems Mitochondria-lysosomes,* edited by Fillastre, J.-P. New York, Masson, 1978, pp. 63–82.

69. Schentag, J.J., Lasezkay, G., Plaut, M.E., Jusko, W.J., and Cumbo, T.J.: Comparative tissue accumulation of gentamicin and tobramycin in patients. *J. Antimicrob. Chemother.* 4(Suppl.):23–30, 1978.

70. Schentag, J.J., Plaut, M.E., Cerra, F.B., Wels, P.B., Walczak, P., and Buckley, R.J.: Aminoglycoside nephrotoxicity in critically ill surgical patients. *J. Surg. Res.* 26: 270–279, 1979.

71. Schentag, J.J., Sutfin, T.A., Plaut, M.E., and Jusko, W.J.: Early detection of aminoglycoside nephrotoxicity with urinary beta-2-microglobulin. *J. Med.* 9:201–210, 1978.

71a. Schentag, J.J., and Plaut, M.E.: Patterns of urinary β_2-microglobulin excretion in patients treated with aminoglycosides. *Kidney Int.* 17:654–661, 1980.

72. Schnermann, J., Stowe, N., Yarimizu, S., Magnusson, M., and Tingwald, G.: Feedback control of glomerular filtration rate in isolated, blood-perfused dog kidneys. *Am. J. Physiol.* 233:F217–F224, 1977.

73. Schreiner, G.E.: The identification and clinical significance of casts. *Arch. Intern. Med.* 99:356–369, 1957.

74. Schwartz, S.N., Pazin, G.J., Lyon, J.A., Ho, M., and Pasculle, A.W.: A controlled investigation of the pharmacokinetics of gentamicin and tobramycin in obese subjects. *J. Infect. Dis.* 138:499–505, 1978.

75. Shiroishi, K., Kjellstrom, T., Kubota, K., Evrin, P.-E., Anayama, M., Vesterberg, O., Shimada, T., Piscator, M., Iwata, T., and Nishino, H.: Urine analysis for detection of cadmium-induced renal changes, with special reference to β_2-microglobulin. *Environ. Res.* 13:407–424, 1977.

76. Siersbaek-Nielsen, K., Molholm Hansen, J., Kampmann, J., and Kristensen, M.: Rapid evaluation of creatinine clearance. *Lancet* 1:1133–1134, 1971.

77. Simon, V.K., Mosinger, E.U., and Malerczy, V.: Pharmacokinetic studies of tobramycin and gentamicin. *Antimicrob. Agents Chemother.* 3:445–450, 1973.

78. Smith, C.R., Baughman, K.L., Edwards, C.Q., Rogers, J.F., and Lietman, P.S.: Controlled comparison of amikacin and gentamicin. *N. Engl. J. Med.* 296:350–353, 1977.

79. Smith, C.R., Maxwell, R.R., Edwards, C.Q., Rogers, J.F., and Lietman, P.S.: Nephrotoxicity induced by gentamicin and amikacin. *Johns Hopkins Med. J.* 142:85–90, 1978.

80. Stroo, W.E., and Hook, J.B.: Enzymes of renal origin in urine as indicators of nephrotoxicity. *Toxicol. Appl. Pharmacol.* 39:423–434, 1977.

81. Thurau, K., and Boylan, J.W.: Acute renal success, the unexpected logic of oliguria in acute renal failure. *Am. J. Med.* 61:308–315, 1976.

82. Toback, F.G.: Amino acid enhancement of renal regeneration after acute tubular necrosis. *Kidney Int.* 12:193–198, 1977.

83. Trestman, I., Parsons, J., Santoro, J., Goodhart, G., and Kaye, D.: Pharmacology and efficacy of netilmicin. *Antimicrob. Agents Chemother.* 13:832–836, 1978.

84. Tulkens, P., Aubert-Tulkens, G., Van Hoff, F., and Trouet, A.: The lysosomal toxicity of aminoglycosides. In *Nephrotoxicity: Interaction of Drugs with Membrane Systems Mitochondria-lysosomes*, edited by Fillastre, J.-P. New York, Masson, 1978, pp. 231–252.

85. Vera-Roman, J., Krishnakantha, T.P., and Cuppage, F.E.: Gentamicin nephrotoxicity in rats: I. Acute biochemical and ultrastructural effects. *Lab. Invest.* 33:412–417, 1975.

86. Whelton, A.: Intrarenal antimicrobial distribution modulating factors, therapeutic and toxicologic implications. In *Nephrotoxicity: Interaction of Drugs with Membrane Systems Mitochondria-lysosomes*, edited by Fillastre, J.-P. New York, Masson, 1978, pp. 95–126.

87. Whelton, A., Carter, G.G., Bryant, H.H., Fox, L., and Walker, W.G.: Therapeutic implications of gentamicin accumulation in severely diseased kidneys. *Arch. Intern. Med.* 136:172–176, 1976.
88. Whelton, A., Carter, G.G., Craig, T.J., Bryant, H.H., Herbst, D.V., and Walker, W.G.: Comparison of the intrarenal disposition of tobramycin and gentamicin: therapeutic and toxicologic answers. *J. Antimicrob. Chemother.* 4(Suppl.):13–22, 1978.
89. Wright, F.S.: Intrarenal regulation of glomerular filtration rate. *N. Engl. J. Med.* 291:135–141, 1974.

7

ASSAYS OF AMINOGLYCOSIDE ANTIBIOTICS: PURPOSES, PROCEDURES, AND PITFALLS

MICHAEL W. LAUERMANN* and MICHAEL BARZA
Tufts–New England Medical Center Hospital,
Boston, Massachusetts

I. Factors Influencing Serum Levels of Aminoglycosides

The relative simplicity of the pharmacokinetics of the aminoglycoside antibiotics suggests that their serum levels should be quite predictable. Unfortunately, this has not proven to be the case and a wide variety of factors have been shown to influence both the peak serum level and half-life of these drugs in sick individuals [8,24,25,38,47,65,70,73, 86,92,123].

Use of the intramuscular route of administration may lead to erratic uptake, especially when repeated injections are given in the same site [44,123].

Departures from the norm in the volume into which the aminoglycoside is distributed (the "apparent volume of distribution") may have important pharmacokinetic consequences. For example, obese patients possess a large adipose compartment into which these antibiotics are only sparsely distributed; dosages based on total body weight will result in unexpectedly high serum levels [44,95]. Conversely, patients with large volumes

---

*Dr. Lauermann is now with Long Beach Memorial Hospital and St. Mary's Hospital of Long Beach, Long Beach, California.

of extracellular fluid (e.g., edema, ascites) may exhibit relatively low serum levels of aminoglycosides [101] and, in some instances, a prolonged serum half-life because of slower equilibration. The first factor, that of obesity, may be compensated for by the use of a formula ("normalized" body mass plus 40% of adipose mass [95]); the second, that of increased extracellular fluid, poses a more difficult problem for which there are no simple predictive formulas.

Several other variables may significantly alter aminoglycoside concentrations. Some authors have noted an inverse relation between the serum hematocrit and either the peak serum level [86] or the serum half-life [8] of aminoglycosides. However, others have been unable to substantiate this [21,70]. The presence of fever has been shown to reduce the peak serum concentration of these agents [79,101]. This may be related to either an increase in renal clearance of the drug or to a more rapid equilibration of gentamicin with tissue compartments [101]. Other conditions that may alter the glomerular filtration rate (GFR), and hence the rate of excretion of aminoglycosides, include pregnancy [96], juvenile diabetes [69], and severe burns [53]. Although it is not clear that the first two conditions require a significant dosage adjustment, the marked increase in GFR in burn patients, up to 200% of normal, does necessitate a substantial augmentation of dosage [37,53,126].

Because aminoglycosides are eliminated from the body almost solely by glomerular filtration, decreases in renal function have a major impact upon their pharmacokinetics. The effect is much more pronounced upon the serum half-life than upon peak serum levels of these drugs. Failure to recognize the presence of renal dysfunction may lead to drug accumulation with consequent toxicity.

Although the serum creatinine concentration is a useful guide to the presence of significant abnormalities of renal function, this is not necessarily the case in elderly patients who may have marked age-related decreases in GFR with an essentially normal serum creatinine concentration [22,81,123]. Measurements of creatinine clearance provide a more precise index of GFR but are time-consuming and subject to errors in the collection of specimens.

These logistical problems can be circumvented by determining an "estimated creatinine clearance" based upon the serum creatinine concentration and modified for age, sex, and body weight. Indeed, several nomograms [19,102] and computer formulas [44,59,91] for the dosage of aminoglycosides have been put forward to take account of these variables. Such methods clearly provide a better guide to serum levels than do simple estimates on the basis of body weight and serum creatinine concentration; however, they may still lead to errors in dosing, especially in patients with unstable renal function, fluid shifts, and other fluctuating physiological parameters [8,47,63,93].

Two other factors may make dosage adjustments particularly difficult in patients with renal insufficiency. One relates to the effects of dialysis on aminoglycoside pharmacokinetics. Hemodialysis uniformly shortens the half-life of aminoglycosides in the body. The extent of the reduction varies with the type of dialyzer coil [40], its surface area, and permeability characteristics, and is independent of blood and dialysate flow rates [20]. In contrast, peritoneal dialysis is relatively ineffective in removing aminoglycosides from the body and consequently has little effect on drug half-life [55]. A second factor arises from the ability of carbenicillin or ticarcillin to inactivate aminoglycosides within the body; the magnitude of the interaction varies from combination to combination. Although the rate of inactivation is sufficiently slow as to be of little importance in patients with normal kidneys, in those with impaired renal function and a longer serum half-life of aminoglycoside, the interaction may be clinically significant [27,31, 72,85].

There are a number of special considerations in the pediatric population which make the prediction of blood levels of aminoglycosides difficult [42,48,63,77,87,101]. The serum half-life of aminoglycosides is longer in neonates than in young children because of the immaturity of the kidney in newborns [42,43,70]. In older children, serum levels are lower than they are in young children when similar doses are given on a mg/kg basis [29,63,77]; whether this is because of differences in the apparent volume of distribution, in distribution rate constants, or in half-life has not been elucidated [29]. The presence of hypoxemia has been reported to cause a prolongation in the half-life of amikacin in neonates [70].

The purpose of reviewing the impact of these pharmacological variables upon serum levels of aminoglycosides is not to suggest that there is no value to nomograms or formulas for the determination of dosage regimens. Rather, the objective is to emphasize the striking interplay of physiological factors (summarized in Table 1) which may impair the accuracy of these guidelines. In our experience [8,9] and that of many other groups [37, 44,53,70,72,77,79,86,91] the deviations from expected results are of sufficient magnitude and frequency that it seems advisable to recommend that serum levels be measured at intervals. In relatively stable patients, one or two measurements per week may be sufficient; in others, daily studies may be necessary. An additional refinement is to obtain serum specimens at several intervals following a dose of drug; the results can be used to determine the precise half-life of the antibiotic in that individual patient and thus to permit extremely accurate dosage adjustments [93].

II. Assay Methods

There are four major categories of assays for aminoglycoside antibiotics which have undergone substantial testing; these include biological, enzymatic, radioimmunological, and chromatographic methods. Others, such as the "urease" technique, have had only

Table 1   Factors That Affect the Peak Serum Level and Half-Life of Aminoglycosides[a]

Raise peak serum level	Diminish peak serum level	Increase half-life	Decrease half-life
Obesity [44,95]	Increased extracellular fluid [101]		Burns [53]
Anemia [86]	Burns [37,126] Fever [79] Children (inverse correlation with age) [29,42]	Anemia [8] Hypoxemia in infants [70] Neonates (correlated with low gestational age and postnatal age) [42,43,63,77]	Hemodialysis [40] Concomitant carbenicillin, ticarcillin [27,28, 72,85]

[a] As applied to dosages based on body weight.

limited usage. In the present section we outline the characteristics of each type of assay with major emphasis on those which have been extensively studied.

A. Bioassays

These are the oldest and still most widely used of the assay methods. The agar-diffusion variant is clearly more precise than twofold broth-dilution methods [82]. The materials (bacteriological media, test organisms) for bioassays are readily obtainable and inexpensive. The procedures involved are straightforward (see the appendix); however, some experience is required, especially when unexpected values are found. Although results generally are obtained after 12–18 hr, a rapid (4 hr) method has been described using poured plates of Trypticase Soy Agar (BBL) with *Staphylococcus epidermidis* (*ATCC 27626*) [1], Antibiotic Medium No. 11 (Difco) added to Tryticase Soy Broth (BBL) with *Staphylococcus aureus* (*ATCC 6538P*) [119] and Heart Infusion Agar (Difco) [88] or Antibiotic Medium No. 5 (Difco) [89] with *Bacillus subtilis* (*ATCC 6633*).

The sensitivity of agar-diffusion bioassays can be increased by a variety of manipulations, including the use of thin agar layers, wells instead of filter-paper disks, and initial refrigeration of plates after application of the antibiotic. It is also possible to take advantage of certain characteristics of the medium which may affect the zone sizes. For example, the activity of aminoglycosides is influenced by the concentration of sodium chloride in the medium [97]. In one study, a threefold increase in zone diameters was noted with a test strain of *Klebsiella edwardsii* when the salt concentration in the agar was raised from 0 to 1.5%; however, further increases of salt concentration caused the zone size to be *reduced* [28]. Although the activity of aminoglycosides is diminished at acid pH, this may be more than offset by the corresponding inhibition of bacterial growth. Thus, zones in trypticase soy agar seeded with a strain of *S. epidermidis* were larger at pH 7.4 than when the pH of the medium was adjusted to 7.8 [1]. Alterations in the constituents of the medium may be usefully exploited in special circumstances when it is desirable to change the sensitivity of the bioassay.

Several rules must be carefully adhered to if bioassays are to be accurate. Most important is that the standards be dissolved in a fluid containing the same constituents as are present in the test specimens (e.g., serum, bile, or urine). In one study, assays with standards dissolved in phosphate-buffered saline resulted in overestimates of serum concentrations of as much as 100%; the use of barbitone buffer diluent for the standards produced the opposite effect [97]. In another study, the use of 7% albumin as diluent for assays of serum produced errors as great as 300% [89].

A second important consideration is to include a set of antibiotic standards on each bioassay plate [80]. This should pose no problem if large dishes are used.

A third factor relates to the test system used. Certain assay media seem to produce erratic results [99]. Table 2 summarizes a number of combinations of test organisms and agars which have been found to afford reproducible data.

A number of substances may interfere with the bioassay for aminoglycosides. The most common are other antimicrobial agents possessing activity against the test organism. One solution is to select an assay bacterium that is highly resistant to the second drug; examples of such organisms are shown in Table 2. Another approach is to inactivate the nonaminoglycoside antibiotic; in the case of penicillins and cephalosporins, this can be accomplished by adding commercially available β-lactamases [i.e., penicillinases

Table 2   Bioassay Test Systems Found to Yield Reproducible Results

Test organism	Medium	Resistant to:	Sensitive to.
S. epidermidis (*ATCC 27626*)	Trypticase soy agar (BBL) [1] Oxoid nutrient broth No. 2 with 20% horse serum [64]	Ampicillin, carbenicillin, chloramphenicol tetracycline, kanamycin, streptomycin, sulfisoxazole, lincomycin, clindamycin	Amikacin, gentamicin, vancomycin, cephalosporin (intermediate)
Klebsiella edwardsii (*NCTC 10896*)	Diagnostic sensitivity test agar (Oxoid) [99] Sensitest agar (Oxoid) [99] Mueller-Hinton (Lab M) [99] Tryptone soy agar (Lab M) [99] Antibiotic medium No. 11 (Difco) [99] Columbia agar base (Oxoid) [99]	Penicillin, ampicillin, carbenicillin, methicillin, cephalosporins, chloramphenicol, clindamycin, lincomycin, fucidin, macrolides, tetracycline, trimethoprim, sulfonamide	Gentamicin, tobramycin, kanamycin
Klebsiella pneumoniae (University of Minnesota, K-1296)	Antibiotic medium No. 11 (Difco) [54]	Ampicillin, carbenicillin, cephalothin, chloramphenicol, clindamycin, erythromycin, karamycin, methicillin, penicillin, streptomycin, tetracycline	Gentamicin
Providencia stuartii (University of Southern California, strain 73-197)	Antibiotic medium No. 11 (Difco) [58]	Ampicillin, carbenicillin, tetracycline, chloramphenicol, kanamycin, gentamicin, streptomycin, tobramycin	Amikacin
Bacillus subtilis (*ATCC 6633*)	Heart infusion agar (Difco) [89] Antibiotic medium No. 5 (Difco) [88]		Penicillins, cephalosporins, aminoglycosides, and most other antibiotics
S. aureus (*ATCC 6538P*)	Antibiotic medium No. 11 (Difco) [119]	Streptomycin, neomycin, nitrofurantoin	Gentamicin, tobramycin, amikacin, kanamycin, netilmicin, ampicillin, carbenicillin, cephalothin, penicillin

173

(BBL) or cephalosporinase (Ames or Miles Laboratories)] either to the test specimen or to the agar. Whenever this method is used, the possibility should be considered
that inactivation is not complete and separate controls may be necessary. For example,
the content of β-lactam drug in the serum may be measured and the ability of the enzyme to destroy this concentration may be ascertained. As increasingly potent antimicrobials are developed, this problem becomes progressively more difficult to overcome. Fortunately, the nonbiological assays described below appear to circumvent this
issue.

The use of heparinized tubes to collect specimens for assay may cause a reduction in
the measured concentration of gentamicin of as much as 10-50% [84]; this effect is maximal at heparin concentrations greater than 50 U/ml and is reversible with dilution of the
sample. The phenomenon appears to be the result of an ionic interaction between the
negatively charged heparin and positively charged aminoglycoside molecule, which form
an insoluble complex at certain drug concentration ratios [125]. Although this reaction
is unlikely to be of importance at the levels of heparin which are encountered in clinical
therapy (about 5 U/ml of serum), it may be significant when the anticoagulant is present
in the collection tubes where levels usually exceed 150 U/ml. Of interest is the observation that concentrations of gentamicin in serum collected from postdialysis tubing were
lower than serum collected directly from the patient. Various factors may have been
responsible for this finding, including the presence of heparin in the specimens. The
authors recommended that samples not be drawn from dialysis tubes for aminoglycoside
assays [54].

It has been reported that there may be an underreading of gentamicin levels in the
serum of jaundiced patients [88]; however, this could not be verified in a more extensive
study [64]. Furthermore, the addition of exogenous bile or bile acids to normal serum
did not interfere with the assay of gentamicin [64].

Stessman and colleagues [108] found that the recovery rate of aminoglycosides in
serum from azotemic individuals was lower than that from nonazotemic control serum
(50-70% versus 80-90%). However, this discrepancy appears to be peculiar to assays
using Difco No. 11 medium [98,109].

Carbenicillin and ticarcillin have been demonstrated to inactivate aminoglycosides
[27,31,66,118,124]. Presumably, serum specimens containing carbenicillin or ticarcillin
and aminoglycosides, if allowed to sit for prolonged periods at room temperature, might
result in some degree of degradation of the aminoglycoside. This would yield a lower
level of assayed aminoglycoside than was initially present in the serum [118,123,124].
Freezing the specimen should circumvent this problem.

The accuracy of microbiological assays varies considerably from one institution to
another. In a survey of a large number of laboratories, it was found that the best ones
produced results that were consistently accurate to within 15-20% [82]. Unfortunately,
only about 20% of laboratories using agar-diffusion bioassay were able to achieve this
level of reliability [83]. The results of bioassay have generally correlated well with those
achieved by other methods. Thus, there was a correlation coefficient of 0.96 between
bioassay and enzymatic assay in one study [26]. In another, the average difference between bioassay and chromatographic assay for gentamicin or tobramycin in paired specimens was about 20% [60]. A number of authors have found high correlation coefficients
between bioassay and radioimmunoassay results for various aminoglycosides [14-16,
50, 112-114]. In evaluating these comparisons, it must be remembered that they have
generally been performed in research laboratories and may not be indicative of the results
in everyday practice.

Despite the multiplicity of variables that may influence the results of microbiological assays, they remain the least expensive and most readily applied for general use. If properly performed, under the surveillance of an experienced supervisor, with due attention to the necessity that standards and test specimens be of similar constitution and present on the same plate, the results should be reasonably accurate.

A modification of the bioassay technique has been reported by Faine and Knight [32]. The "urease" method is based on the ability of *Proteus* to split urea, causing a rise in pH of the medium and the capacity of aminoglycosides to inhibit the expected rise. Despite modifications of the test to enhance its accuracy [12,74,75], it remains cumbersome, no more accurate than the standard agar-diffusion assay [49,74] and less accurate than enzymatic measurements [3,49,80].

B. Enzymatic Assays

Certain bacteria elaborate aminoglycoside-inactivating enzymes which function by linking the antibiotic to substances such as acetyl-coenzyme A (CoA) (acetylating enzymes) or adenosine triphosphate (ATP) (adenylylating enzymes). If the latter are tagged with radioactive labels (e.g., [14C]acetyl-CoA [13,39], [3H]acetyl-CoA [111], [14C]ATP [41,103]), the aminoglycoside becomes radiolabeled. The antibiotics, being positively charged, are able to bind to negatively charged phosphocellulose paper, which can then be "counted" [106].

The enzymes are generally not specific for one or another aminoglycoside. For example, aminoglycoside 2''-adenylyltransferase [AAD(2'') or gentamicin adenylyltransferase] will detect gentamicin, sisomicin, and tobramycin; kanamycin has a variable affinity for this enzyme [41]. Aminoglycoside 4'-adenylyltransferase [AAD(4')] may be used to detect kanamycin, amikacin, neomycin, and paromomycin as well as various aminoglycosides that are not used medically [90]. Alternatively, aminoglycoside 6'-N-acetyltransferase [AAC(6') or kanamycin acetyltransferase] may be used to detect kanamycin, amikacin, tobramycin, and sisomicin [39,111]. By the use of enzymes with differing substrate specificities, it is possible to distinguish the concentrations of pairs of aminoglycosides present in the same specimen; however, such subtle distinctions require considerable expertise. Fortunately, this problem rarely arises clinically. There appears to be no cross-reactivity of the test with other classes of antibiotics.

A number of problems noted in earlier studies with enzymatic assays appear to have been resolved successfully. For example, improved purification has resulted in greater stability of the enzymes [103]. Nonetheless, the enzyme is still degradable by repeated freeze-thawing. Nonspecific binding of the radioactive substrate to the phosphocellulose paper in the absence of antibiotic has been reduced by tromethamine (TRIS) washing [34] and by use of purified enzyme [103]. Breakdown of ATP by inherent ATPase activity of serum can be obviated by heating the serum to 60°C for 5 min [103]. Lot-to-lot variations in the binding capacity of phosphocellulose paper has been noted by some investigators [116] but not by others [103].

A number of interfering substances have been reported with enzymatic assays. An "inhibitory factor" was noted in a batch of pooled human serum; the substance, which was not identifiable, was heat-stable and had an apparent molecular weight below 30,000 [68]. Similarly, minocycline and tetracycline were found to decrease gentamicin concentrations [26]. Stevens and Young have noted a reduction of amikacin concentration by 30% when chloramphenicol was present in the serum. This may have been due to the presence of a chloramphenicol acetyltransferase in the enzymatic preparation used in their assay [110]. Spuriously high results for the assay of gentamicin have been reported in sera of patients receiving intravenous lipids; it was hypothesized that lipid might coat the cellulose paper, preventing the unreacted [^{14}C]acetyl-CoA from being adequately rinsed off [13]. ^{67}Ga, but not ^{123}I,^{99}Tc]pertechnetate, [^{99}Tc]sulfacolloid, [^{99}Tc] albumin microspheres, or [^{75}Se]methionine, was found to interfere with the enzymatic assay [10]. The suggestion was made that radioactive gallium is present as a cation which adheres to the negatively charged paper and is counted together with the tagged amino-glycoside. This phenomenon can be prevented by extraction of the sample with diiso-propyl ether before performing the assay.

The enzymatic assay is equal in accuracy to the 4 hr bioassay and may be performed more rapidly, yielding results in about 2 hr [36,104]. For gentamicin, a linear response was noted for all concentrations up to 30 μg/ml with a coefficient of variation of 3% [26]. A linear response to 50 μg/ml has been reported for amikacin [111]. The major advantages of the technique lie in its rapidity, specificity, and accuracy; its major dis-advantages arise from the need for a scintillation spectrophometer and the necessity of working with radioactive material [106].

C.   Radioimmunoassays

The production of specific antibodies to aminoglycoside antibiotics has led to the develop-ment of highly sensitive and specific tests for the assay of these compounds [14-16, 18,51,56,112,113]. These techniques are able to measure concentrations of amikacin less than 1 μg/ml [113] and of gentamicin as low as 0.1 μg/ml [67]. Comparative studies have demonstrated their sensitivity and accuracy to be at least equivalent to that of the microbiological [14-16,50,114] and enzymatic assay methods [16,50,114].

The radioimmunoassay method depends on the formation of a radiolabeled antigen-antibody complex. Aminoglycoside in the patient's serum competes with exogenous radio-labeled antibiotic (either ^{125}I or ^{3}H) for the antibody binding sites. In "solid-phase" assays, the antibody is bound to the wall of the reaction vessel and unreacted antigen can be washed off; in "liquid-phase" assays, the complex can be precipitated by reaction with a second anti-body. (Other variations of these techniques have also been used.) The bound antigen is "counted" and the results are determined by comparison with known standards.

The use of ^{125}I in place of ^{3}H as the radioactive label has reduced the cost and the time of the procedure [113] so that as many as 30 samples can be run in 3-4 hr [14]. The fact that volumes of serum as small as 5-10 μl can be used [50,56] makes the proce-dure particularly useful for pediatric work.

Thus far, the problems encountered with radioimmunoassay have been minimal. Cross-reactivity among aminoglycosides is not extensive. For example, [^{125}I]amikacin had 70% cross-reactivity with kanamycin but none with gentamicin, tobramycin, siso-micin, or netilmicin [113]. [^{125}I]tobramycin reacted to the extent of 1.5% with genta-micin; <0.8% with amikacin, sisomicin, streptomycin, or neomycin; and 5-8% with kanamycin [16,18]. [^{125}I]netilmicin exhibited 3% cross-reactivity with gentamicin and sisomicin [112]. Recently, a pooled antiserum has been described that is specific for gentamicin, tobramycin, and amikacin and which can accurately measure any of these three drugs in the presence of the others [6].

The concomitant presence of carbenicillin in serum reduced the amount of tobramycin detectable using [^{125}I]tobramycin in proportion to the concentration of carbenicillin in the sample [16]. However, in another study, using [^{3}H]gentamicin, there was no effect of carbenicillin, even in concentrations as high as 10,000 μg/ml [51]. Further experience will be necessary to determine whether other substances interfere with the radioimmunoassay.

The disadvantages of the radioimmunoassay include the necessity to work with radioactive substances and to have access to a gamma counter or liquid scintillation spectrophotometer [113]. In addition, this test is almost three times as expensive as microbiological assays [114]. Nevertheless, the development of commercial test kits which provide rapid and accurate results and are easy to use has led to a steady increase in the number of laboratories using this technique.

D. Chromatographic Assays

Chromatographic procedures have been shown to be at least as accurate as microbiological methods [5,57,60,78]. Gas-liquid chromatography is somewhat simpler than high-performance liquid chromatography and does not require fluorimetry [4,5,57,60-62,78]. It provides a linear response to concentrations of gentamicin and tobramycin ranging from 0.6 to 12 μg/ml; the coefficient of variation was 6.3-9.6% for gentamicin and 3.8-13.5% for tobramycin in one study [60].

Chromatographic assays are highly specific and free of interference from other substances, provided that appropriate internal standards are run. Although these methods appear promising, they are still primarily research tools.

E. Miscellaneous Assay Methods

For the sake of completeness, mention should be made of a number of assay methods that have undergone preliminary investigation but whose applicability is less well defined than those noted above. These include (1) luciferase assay of extracellular ATP [71]; (2) polarization fluoroimmunoassay [120], quenching fluoroimmunoassay [100], and reactant-labeled immunoassay [17]; (3) peroxidase immunoassay [107]; and (4) spectrophotometric assay, which measures the production of a chromophore from the acetylation of gentamicin [121]. Although these techniques have been favorably compared with bioassays, enzymatic assays, or radioimmunoassays, their application has been confined to the study of gentamicin and they have yet to be subjected to the rigors of clinical trials. Thus, like chromatographic assays, their use remains restricted to research facilities.

Each of the major varieties of assay described above has advantages and disadvantages, and the choice of a method must depend upon the availability of equipment, the frequency with which tests are to be done, and the degree of precision required. Microbiological assays are probably the easiest to set up initially and require the least sophisticated equipment. However, enzymatic assays and radioimmunoassays are somewhat more precise and are free of interference from other substances, especially concomitant antibiotics. They also provide information more quickly. Thus, they may be more appropriate than bioassays for busy centers in which a large number of assays are performed and where rapidity is at a premium (e.g., hospitals with busy dialysis and intensive-care units). Table 3 summarizes the characteristics of the major assay procedures.

Perhaps more important than the specific methodology of the particular test is the expertise of the medical staff in knowing when to obtain serum levels and how to interpret

Table 3 Selected Characteristics of Major Groups of Aminoglycoside Assays

Type of assay	Usual lower limit of assay for gentamicin	Specificity	Interfering substances		Accuracy (%)	Comments
			Increase apparent concentration	Decrease apparent concentration		
Agar-diffusion bioassay	0.5–1 µg/ml (may be enhanced with various manipulations)	Does not distinguish other antimicrobial substances	Other antimicrobials	Heparin ? Azotemia ? Jaundice ? Carbenicillin	±15	Test samples and standards must be in similar diluent; inactivating enzymes or test organisms "resistant" to concomitant antimicrobials may not completely vitiate their activity
Enzymatic assay	Less than 0.5 µg/ml	No cross reactivity with other antimicrobial agents; some cross reactivity within the aminoglycoside group	Concomitant administration of intralipids ^{67}Gallium	? Inhibitory "factor" in serum ? ATP'ase activity of serum Nonspecific binding of radioactive substrate to phosphocellulose paper ("high background counts") Tetracycline, minocycline Chloramphenicol	±10	Repeated freezing and thawing degrades the enzymes; may be performed more rapidly than the bioassay but requires special equipment and radioactive material
Radioimmuno- assay	0.1 µg/ml	No cross reactivity with other classes of antibiotics; little cross reactivity within the aminoglycosides		? Carbenicillin	±10	Small volumes of sample may be used; assay procedures are faster than the bioassay but require special equipment and radioactive substances

them. In one study, half the samples submitted to the laboratory for measurement of serum levels of gentamicin were improperly drawn and only 20% of the determinations were used appropriately in making dosage adjustments [2].

III. Relation of Peak and Trough Levels to Toxicity

On the basis of limited data, it appears that effective therapy of serious *Pseudomonas* infections requires that peak serum levels of gentamicin exceed 4-5 μg/ml [46,73]; for gram-negative pneumonias, 8 μg/ml may be necessary [73]. Corresponding values for other aminoglycosides are even less well defined; for amikacin, which is somewhat less active than gentamicin in vitro, peak levels > 20 μg/ml may be required [115].

The major types of toxicity due to aminoglycoside antibiotics are minifested as nephrotoxicity or ototoxicity. It is generally recommended that both peak and trough (valley) levels of aminoglycosides in the serum be determined. The utility of measuring the peak level is based upon the belief that excessive levels offer no added benefit and may be harmful; the value of determining the trough concentrations arises from the fact that several investigations have offered evidence that these, rather than peak serum levels, are the major determinants of toxicity. The results of several pertinent studies are summarized in Table 4.

Goodman et al., comparing a variable-dose with a variable-frequency regimen of gentamicin, demonstrated a significant correlation between trough levels ⩾ 4 μg/ml and a rising serum creatinine concentration; this occurred in both groups of patients [38]. Dahlgren and colleagues, in a prospective study, found renal deterioration in 8 of 21 (38%) patients with trough levels > 2 μg/ml, 2 of 6 with levels > 3 μg/ml, and 5 of 5 with levels > 4 μg/ml. In contrast, only 1 of 5 with peak levels > 10 μg/ml exhibited signs of nephrotoxicity [25]. However, it is equally plausible that the elevated trough levels were not the cause, but the earliest manifestation, of already present renal damage. In another study, no correlation was evident between nephrotoxicity and initial peak or valley levels of gentamicin or amikacin [105]. Although both peak and valley concentrations rose during therapy, more strikingly in those who developed nephrotoxicity, it was felt that this was probably the result rather than the cause of renal impairment; indeed, there was no correlation between a rising trough level of gentamicin in patients whose initial value was < 2 μg/ml and the subsequent development of renal damage. The authors felt that the likelihood of developing nephrotoxicity was least when peak and trough levels of gentamicin remained below 10 and 2 μg/ml; for amikacin the comparable values were 38.5 and 10 μg/ml.

Schentag and associates, studying the pharmacokinetics of gentamicin in 64 patients, compared the characteristics of those who subsequently developed nephrotoxicity with those who did not. They found no relation between toxicity and dosage of gentamicin, duration of therapy, initial renal function, or trough levels after the first dose. However, patients who became nephrotoxic retained three times as much gentamicin in the tissues after the first dose as those who did not [94]. This suggests that there may be some as yet unidentified factor, manifesting as tissue avidity for the antibiotic, which predisposes to toxicity.

It is equally difficult to discern the relative importance of peak and trough serum levels of aminoglycosides in terms of ototoxicity (Table 4). Some authors have reported a significant correlation with peak serum levels [11] and others with trough concentrations

Table 4 Correlations Between Peak and Trough Serum Concentrations of Aminoglycosides and Toxicity

	Nephrotoxicity		Ototoxicity		Comment
	Trough	Peak	Trough	Peak	
Goodman et al. [38]	Significant (P = 0.027)	No correlation	—	—	High trough levels could be an early indication of renal damage rather than the cause of it
Dahlgren et al. [25]	Implied correlation	No correlation	—	—	Correlation with trough levels not statistically significant, but only a small number of patients studied
Smith et al. [105]	Significant (P < 0.025)	Significant (P < 0.025)	—	—	The high trough and peak levels probably were the result and not the cause of renal damage
Schentag et al. [94]	No correlation	—	—	—	Nephrotoxic patients demonstrated a threefold greater gentamicin tissue accumulation after the first dose than nontoxic patients
Line et al. [52]	—	—	Significant (P < 0.001)	—	All patients with elevated trough levels had preexisting abnormal renal function and they were generally older
Black et al. [11]	—	—	Significant (P < 0.05)	Significant (P < 0.01)	Ototoxic patients received larger total dose (P < 0.01) and longer course of treatment (P < 0.02) than controls; patients with elevated peak *and* trough were not segregated
Nordstrom et al. [76]	—	—	Significant (P < 0.05)	No correlation	Patients with ototoxicity had significantly higher serum creatinine concentrations and longer duration of therapy
Tjernstrom et al. [117]	—	—	—	No correlation	86% of patients who developed vestibular damage had preexisting ototoxicity; none had serum peak > 10 µg/ml gentamicin
Banck et al. [7]	—	—	Implied correlation	No correlation	Studied a group of patients who had received prior ototoxic antibiotics, had renal insufficiency, were treated for a prolonged period or had received a large total dose of drug
Cox [23]	—	—	—	Implied correlation	Of the 3% who developed ototoxicity with high serum levels of gentamicin, 67% were uremic
Jackson and Arcieri [45]	—	—	—	Implied correlation	Peak levels were not determined in all patients; there was a significant correlation (P = 0.003) between ototoxicity and renal dysfunction
Gailiunas et al. [35]	—	—	No correlation	No correlation	All patients were on hemodialysis; ototoxicity was significantly (P < 0.001) correlated with total dose and duration of therapy

[11,52,76]. In one study of patients receiving long-term hemodialysis, the authors found no correlation of ototoxicity with serum levels, but an increased incidence of this adverse effect in patients who had received a cumulative gentamicin dose in excess of 17.5 mg/kg [35]. However, using extremely sensitive methods, cochlear abnormalities have been detected after as little as one dose of tobramycin given as a bolus injection [122].

It is not possible, at present, to distinguish the relative contributions of peak and valley levels of aminoglycosides to their nephro- or ototoxicity. These drugs show a marked affinity for renal tissue [30] and penetrate the otic perilymph quite readily [33]; moreover, their half-life in these sites is much longer than that in the serum. It is easy to imagine that the gradient favoring accumulation in these sites might be favored both by high peak levels and by high trough levels in the serum. Indeed, it may be the total gradient, integrated as the "area under the curve," rather than the peak or trough alone, that is the major department of accumulation and toxicity.

Lacking precise knowledge as to the most pertinent serum values to measure, most investigators recommend that both peak and valley levels be determined. Avoidance of excessive peak concentrations is easily accomplished by altering either the dose or the interval between administrations. In contrast, there is some risk in attempting to achieve trough levels below an arbitrarily chosen figure (e.g., 2 μg/ml for gentamicin), since this may expose the patient to subeffective doses for treatment of the infection. This is especially true in the very group in whom such problems most frequently arise: patients with impaired renal function in whom the serum half-life of aminoglycosides is markedly prolonged. In order to achieve sufficiently low trough levels in these individuals, it may be necessary to incur subinhibitory serum concentrations for extended periods. Although there is no obvious solution to this dilemma, knowledge of the serum levels in an individual patient permits the physician to avoid unnecessarily high values and to weigh the merits of an elevated trough concentration against the possible risks.

Appendix: Microassay for Gentamicin in the Presence of Other Antibiotics*

Materials

Plastic Petri dishes, 150 × 15 mm.
Capillary tubes, nonheparinized, 75 × 0.9 × 1.0 mm.
Thin-walled metal tube, 2 mm outside diameter with rubber tubing attachment.

Media

Modified tumor-specific antigen (TSA), 1% agar (26.66 g of TSA in 1 L of distilled water). Medium is distributed in 18 ml portions in screwcap tubes.

Organism

S. epidermidis (ATCC 27626), 4-6 hr culture on Mueller-Hinton agar. Organism is harvested in brain heart infusion and stocked in 0.5 cm^3 aliquots at -70°C. Resistant to ampicillin, carbenicillin, chloramphenicol, tetracycline, kanamycin, streptomycin, colis-

*As performed by the Infectious Disease Laboratory, New England Medical Center Hospital.

tin, sulfisoxasole, lincomycin, and clindamycin. Sensitive to gentamicin, vancomycin, amikacin, and tobramycin. Intermediate sensitivity to cephalothin.

Antibiotic Standards

Gentamicin stock standard, 3200 µg/ml, prepared in water. Working standards are made by diluting the stock solution in sterile, normal pooled human serum to final concentrations of 32:1 µg/ml.

Preparation of Plates

Agar is melted, cooled, and maintained in the liquid state at 45-50°C. 0.5 ml of 4-6 hr growth (approximately 10^8 complement-fixation units per milliliter) of the assay organism is mixed with the agar. If a penicillin or cephalosporin is present, add 0.1 ml of penicillinase and/or cephalosporinase (final concentration of the cephalosporinase depends on the directions of the suppliers). The mixture is quickly poured into a 150 mm plastic petri dish, allowed to solidify, and cooled at 4°C. Plates can be stored for at least 1 week at 4°C. (Our experience shows deterioration of the plates after 2 weeks of storage.) Wells are punched in the seeded agar with a 2 mm stainless steel cylinder with sharp edges (slightly beveled), attached to low suction. Well spacing is facilitated by using a grid system under the plate. 30-35 wells per plate can easily be applied. To remove any moisture that may accumulate in the wells, the plate is dried open and upside down at 37°C for 10-15 min.

Assay for Antibiotic

Wells are filled by using a nonheparinized capillary tube (0.5 x 0.9 x 1.0 mm). Standards and unknowns are set up in triplicate. The plate is then incubated for 4-6 hr at 37°C.

Reading and Calculation

The diameter of the zones of inhibition of growth are measured to the nearest 0.1 mm with a micrometer eyepiece. Using triple-cycle semilogarithmic graph paper, the concentration of the antibiotic standard (µg/ml) is recorded on the ordinate and the mean zone size (mm) on the abscissa. A straight line best fitting these points is drawn and the concentrations of the unknowns are then extrapolated from the graph using the mean zone size they produce.

References

1. Alcaid, D.V., and Seligman, S.J.: Simplified assay for gentamicin in the presence of other antibiotics. *Antimicrob. Agents Chemother.* 3:559–561, 1973.
2. Anderson, A.C., Hodges, G.R., and Barnes, W.G.: Determination of serum gentamicin sulfate levels. *Arch. Intern. Med.* 136:785–787, 1976.
3. Andrews, J., Gillette, P., Williams, J.D., and Mitchard, M.: Analysis of gentamicin in plasma: a comparative study of four methods. *Postgrad. Med. J.* 50(Suppl. 7):17–20, 1974.
4. Anhalt, J.P.: Assay of gentamicin in serum by high-pressure liquid chromatography. *Antimicrob. Agents Chemother.* 11:651–655, 1977.
5. Anhalt, J.P., and Brown, S.D.: High-performance liquid-chromatographic assay of aminoglycoside antibiotics in serum. *Clin. Chem.* 24:1940–1947, 1978.
6. Ashby, C.D., Lewis, J.E., and Nelson, J.C.: Measurement of three aminoglycoside antibiotics with a single radioimmunoassay system. *Clin. Chem.* 24:1734–1737, 1978.

7. Banck, G., Belfrage, S., Juhlin, I., Nordstrom, L., Tjernstrom, O., and Toremalm, N.G.: Retrospective study of the ototoxicity of gentamicin. *Acta Pathol. Microbiol. Scand.* [B]81(Suppl. 241):54–57, 1973.

8. Barza, M., Brown, R.B., Shen, D., Gibaldi, M., and Weinstein, L.: Predictability of blood levels of gentamicin in man. *J. Infect. Dis.* 132:165–174, 1975.

9. Barza, M., and Lauermann, M.: Why monitor serum levels of gentamicin? *Clin. Pharmacokinet.* 3:202–215, 1978.

10. Bhattachyarya, I., Seligsohn, R., and Lerner, S.A.: Effects of radiopharmaceuticals on radioenzymatic assays of aminoglycoside antibiotics: interference by gallium-67 and its elimination. *Antimicrob. Agents Chemother.* 14:448–453, 1978.

11. Black, R.E., Lau, W.K., Weinstein, R.J., Young, L.S., and Hewitt, W.L.: Ototoxicity of amikacin. *Antimicrob. Agents Chemother.* 9:956–961, 1976.

12. Bourne, P.R., Phillips, I., and Smith, S.E.: Modification of the urease method for gentamicin assays. *J. Clin. Pathol.* 27:168–169, 1974.

13. Broughall, J.M., and Reeves, D.S.: Properties of the gentamicin acetyltransferase enzyme and application to the assay of aminoglycoside antibiotics. *Antimicrob. Agents Chemother.* 8:222–223, 1975.

14. Broughton, A., and Strong, J.E.: Radioimmunoassay of iodinated gentamicin. *Clin. Chim. Acta* 66:125–129, 1976.

15. Broughton, A., Strong, J.E., and Bodey, G.P.: Radioimmunoassay of sisomicin. *Antimicrob. Agents Chemother.* 9:247–250, 1976.

16. Broughton, A., Strong, J.E., Pickering, L.K., and Bodey, G.P.: Radioimmunoassay of iodinated tobramycin. *Antimicrob. Agents Chemother.* 10:652–656, 1976.

17. Burd, J.F., Wong, R.C., Feeney, J.E., Carrico, R.J., and Boguslaski, R.C.: Homogeneous reactant-labelled fluorescent immunoassay for therapeutic drugs exemplified by gentamicin determination in human serum. *Clin. Chem.* 23:1402–1408, 1977.

18. Casley, D.J., Atkins, R.C., Murphy, G.F., and Johnston, C.I.: Radioimmunoassay for serum tobramycin levels using ^{125}I-labelled tobramycin. *Pathology* 10:307–315, 1978.

19. Chan, R.A., Benner, E.J., and Hoperich, P.D.: Gentamicin therapy in renal failure: a nomogram for dosage. *Ann. Intern. Med.* 76:773–778, 1972.

20. Christopher, T.G., Korn, D., Blair, A.D., Forrey, A.W., O'Neill, M.A., and Cutler, R.E.: Gentamicin pharmacokinetics during hemodialysis. *Kidney Int.* 6:38–44, 1974.

21. Churchill, D.M., McNamara, J.A., Bowmer, M.I., Ahmed, M., and Gault, M.H.: Predictability of serum gentamicin concentrations in renal failure. *Cardiovasc. Med.* 3:1307–1310, 1978.

22. Cockcroft, D.W., and Gault, M.H.: Prediction of creatinine clearance from serum creatinine. *Nephron* 16:31–41, 1976.

23. Cox, C.E.: Gentamicin, a new aminoglycoside antibiotic: clinical and laboratory studies in urinary tract infection. *J. Infect. Dis.* 119:486–491, 1969.

24. Cutler, R.E., Gyselynck, A.M., Fleet, W.P., and Forrey, A.W.: Correlation of serum creatinine concentration and gentamicin half-life. *J. Am. Med. Assoc.* 219:1037–1041, 1972.

25. Dahlgren, J.G., Anderson, E.T., and Hewitt, W.L.: Gentamicin blood levels: a guide to nephrotoxicity. *Antimicrob. Agents Chemother.* 8:58–62, 1975.

26. Daigneault, R., Gagné, M., and Brazeau, M.: A comparison of two methods of gentamicin assay: an enzymatic procedure and an agar diffusion technique. *J. Infect. Dis.* 130:642–645, 1974.

27. Davies, M., Morgan, J.R., and Anand, C.: Interactions of carbenicillin and ticarcillin with gentamicin. *Antimicrob. Agents Chemother.* 7:431–434, 1975.

28. Deacon, S.: Factors affecting the assay of gentamicin by the plate diffusion method. *J. Clin. Pathol.* 29:54–57, 1976.

29. Echeverria, P., Siber, G.R., Paisley, J., Smith, A.L., Smith, D.H., and Jaffe, N.: Age-dependent dose response to gentamicin. *J. Pediatr.* 87:805–809, 1975.

30. Edwards, C.Q., Smith, C.R., Baughman, K.L., Rogers, J.F., and Lietman, P.S.: Concentrations of gentamicin and amikacin in human kidneys. *Antimicrob. Agents Chemother.* 9:925–927, 1976.

31. Ervin, F.R., Bullock, W.E., and Nuttall, C.E.: Inactivation of gentamicin by penicillins in patients with renal failure. *Antimicrob. Agents Chemother.* 9:1004–1011, 1976.

32. Faine, S., and Knight, D.C.: Rapid microbiological assay of antibiotic in blood and other body fluids. *Lancet* 2:375–378, 1968.

33. Federspil, P., Schatzle, W., and Tiesler, E.: Pharmacokinetics and ototoxicity of gentamicin, tobramycin and amikacin. *J. Infect. Dis.* 134(Suppl.):S200–S205, 1976.

34. Forrey, A.W., Blair, A., O'Neill, M., Cutler, R.E., and Christopher, T.G.: Enzymatic assay for gentamicin. *New Engl. J. Med.* 288:108, 1973.

35. Gailiunas Jr., P., Dominguez-Moreno, M., Lazarus, J.M., Lowrie, E.G., Gottlieb, M.N., and Merrill, J.P.: Vestibular toxicity of gentamicin. *Arch. Intern. Med.* 138: 1621–1624, 1978.

36. Giamarellou, H., Zimelis, V.M., Matulionis, D.O., and Jackson, G.G.: Assay of aminoglycoside antibiotics in clinical specimens. *J. Infect. Dis.* 132:399–406, 1975.

37. Glew, R.H., Moellering, R.C., Jr., and Burke, J.F.: Gentamicin dosage in children with extensive burns. *J. Trauma* 16:819–823, 1976.

38. Goodman, E.L., Van Gelder, J., Holmes, R., Hull, A.R., and Sanford, J.P.: Prospective comparative study of variable dosage and variable frequency regimens for administration of gentamicin. *Antimicrob. Agents Chemother.* 8:434–438, 1975.

39. Haas, M.J., and Davies, J.: Enzymatic acetylation as a means of determining serum aminoglycoside concentrations. *Antimicrob. Agents Chemother.* 4:497–499, 1973.

40. Halpren, B.A., Axline, S.G., Coplon, N.S., and Brown, D.M.: Clearance of gentamicin during hemodialysis: comparison of four artificial kidneys. *J. Infect. Dis.* 133: 627–636, 1976.

41. Holmes, R.K., and Sanford, J.P.: Enzymatic assay for gentamicin and related aminoglycoside antibiotics. *J. Infect. Dis.* 129:519–527, 1974.

42. Howard, J.B., and McCracken, G.H., Jr.: Pharmacological evaluation of amikacin in neonates. *Antimicrob. Agents Chemother.* 8:86–90, 1975.

43. Howard, J.B., and McCracken, G.H., Jr.: Reappraisal of kanamycin usage in neonates. *J. Pediatr.* 86:949–956, 1975.

44. Hull, J.H., and Sarubbi, F.A.: Gentamicin serum concentrations: pharmacokinetic predictions. *Ann. Intern. Med.* 85:183–189, 1976.

45. Jackson, G.G., and Arcieri, G.: Ototoxicity of gentamicin in man: a survey and controlled analysis of clinical experience in the United States. *J. Infect. Dis.* 124(Suppl.): S130–S137, 1971.

46. Jackson, G.G., and Riff, L.J.: *Pseudomonas* bacteremia: pharmacologic and other bases for failure of treatment with gentamicin. *J. Infect. Dis.* 124(Suppl.):S185–S191, 1971.

47. Kaye, D., Levison, M.E., and Labovitz, E.D.: The unpredictability of serum concentrations of gentamicin: pharmacokinetics of gentamicin in patients with normal and abnormal renal function. *J. Infect. Dis.* 130:150–154, 1974.

48. Klein, J.O., Herschel, M., Therakan, R.M., and Ingall, D.: Gentamicin in serious neonatal infections: absorption, excretion and clinical results in 25 cases. *J. Infect. Dis.* 124(Suppl.):S224–S231, 1971.

49. Krooden, E.T., and Darrell, J.H.: Rapid gentamicin assay by enzymatic adenylylation. *J. Clin. Pathol.* 27:452–456, 1974.

50. Lewis, J.E., Nelson, J.C., and Elder, H.A.: Amikacin: a rapid and sensitive radioimmunoassay. *Antimicrob. Agents Chemother.* 7:42–45, 1975.

51. Lewis, J.E., Nelson, J.C., and Elder, H.A.: Radioimmunoassay of an antibiotic: gentamicin. *Nature New Biol.* 239:214–216, 1972.

52. Line, D.H., Poole, G.W., and Waterworth, P.M.: Serum streptomycin levels and dizziness. *Tubercle* 51:76–81, 1970.

53. Loirat, P., Rohan, J., Baillet, A., Beaufils, F., David, R., and Chapman, A.: Increased glomerular filtration rate in patients with major burns and its effect on the pharmacokinetics of tobramycin. *New Engl. J. Med.* 299:915–919, 1978.

54. Lund, M.E., Blazevic, D.J., and Matsen, J.M.: Rapid gentamicin bioassay using a multiple-antibiotic-resistant strain of *Klebsiella pneumoniae*. *Antimicrob. Agents Chemother.* 4:569–573, 1973.

55. Madhaven, T., Yaremchuk, K., Levin, N., Pohlod, K., Burch, K., Fisher, E., Cox, F., and Quinn, E.L.: Effect of renal failure and dialysis on the serum concentration of the aminoglycoside amikacin. *Antimicrob. Agents Chemother.* 10:464–466, 1976.

56. Mahon, W.A., Ezer, J., and Wilson, T.W.: Radioimmunoassay for measurement of gentamicin in blood. *Antimicrob. Agents Chemother.* 3:585–589, 1973.

57. Maitra, S.K., Yoshikawa, T.T., Hansen, J.L., Nilsson-Ehle, I., Palin, W.J., Schotz, M.C., and Guze, L.B.: Serum gentamicin assay by high-performance liquid chromatography. *Clin. Chem.* 23:2275–2278, 1977.

58. Marengo, P.B., Wilkins, J., and Overturf, G.D.: Rapid, specific microbiological assay for amikacin (BB-K8). *Antimicrob. Agents Chemother.* 6:498–500, 1974.

59. Mawer, G.E., Ahmad, R., Dobbs, S.M., McGough, J.G., Lucas, S.B., and Tooth, J.A.: Prescribing aids for gentamicin. *Br. J. Clin. Pharmacol.* 1:45–50, 1974.

60. Mayhew, J.M., and Gorbach, S.L.: Assay of gentamicin and tobramycin in sera of patients by gas-liquid chromatography. *Antimicrob. Agents Chemother.* 14:851–855, 1978.

61. Mayhew, J.M., and Gorbach, S.L.: Gas-liquid chromatographic method for the assay of aminoglycoside antibiotics in serum. *J. Chromatogr.* 151:133–146, 1978.

62. Mays, D.L., Van Apeldoorn, R.J., and Lauback, R.G.: High-performance liquid chromatographic determination of kanamycin. *J. Chromatogr.* 120:93–102, 1976.

63. McCracken, G.H., Jr.: Clinical pharmacology of gentamicin in infants 2 to 24 months of age. *Am. J. Dis. Child.* 124:884–887, 1972.

64. McDonald, D.R., and Waterworth, P.M.: Alleged effect of bile constituents on gentamicin assays. *Antimicrob. Agents Chemother.* 5:562–564, 1974.

65. McHenry, M.C., Gavan, T.L., Gifford, R.W., Jr., Geurkink, N.A., Van Ommen, R.A., Town, M.A., and Wagner, J.G.: Gentamicin dosages for renal insufficiency: adjustments based on endogenous creatinine clearance and serum creatinine concentrations. *Ann. Intern. Med.* 74:192–197, 1971.

66. McLaughlin, J.E., and Reeves, D.S.: Clinical and laboratory evidence for inactivation of gentamicin by carbenicillin. *Lancet* 1:261–264, 1971.

67. Minshew, B.H., Holmes, R.K., and Baxter, C.R.: Comparison of a radioimmunoassay with an enzymatic assay for gentamicin. *Antimicrob. Agents Chemother.* 7:107–109, 1975.

68. Minshew, B.H., Swanzy, S.R., and Schoenknecht, F.D.: Inhibition of acetylating activity in an enzymatic assay for aminoglycoside antibiotics. *Antimicrob. Agents Chemother.* 12:597–601, 1977.

69. Mogensen, C.E.: Elevated glomerular filtration rate in insulin-treated short-term diabetes. *Acta Med. Scand.* 194:559–561, 1973.

70. Myers, M.G., Roberts, R.J., and Mirhij, N.J.: Effects of gestational age, birth weight, and hypoxemia on pharmacokinetics of amikacin in serum of infants. *Antimicrob. Agents Chemother.* 11:1027–1032, 1977.

71. Nilsson, L.: New rapid bioassay of gentamicin based on luciferase assay of extracellular ATP in bacterial cultures. *Antimicrob. Agents Chemother.* 14:812–816, 1978.

72. Noone, P., Beale, D.F., Pollock, S.S., Perera, M.R., Amirak, I.D., Fernando, O.N., and Moorhead, J.F.: Monitoring aminoglycoside use in patients with severely impaired renal function. *Br. Med. J.* 2:470–473, 1978.

73. Noone, P., Parsons, T.M.C., Pattison, J.R., Slack, R.C.B., Garfield-Davies, D., and Hughes, K.: Experience in monitoring gentamicin therapy during treatment of serious gram negative sepsis. *Br. Med. J.* 1:477–481, 1974.

74. Noone, P., Pattison, J.R., and Samson, D.: Simple, rapid method for assay of aminoglycoside antibiotics. *Lancet* 2:16–19, 1971.

75. Noone, P., Pattison, J.R., and Slack, R.C.B.: Rapid assay of gentamicin. *Lancet* 2:1194–1195, 1972.

76. Nordstrom, L., Banck, G., Belfrage, S., Juhlin, I., Tjernstrom, O., and Toremalm, N.G.: Prospective study of the ototoxicity of gentamicin. *Acta Pathol. Microbiol. Scand.* [B] 81(Suppl. 241):58–61, 1973.

77. Paisley, J.W., Smith, A.L., and Smith, D.H.: Gentamicin in newborn infants. *Am. J. Dis. Child.* 126:473–477, 1973.

78. Peng, G.W., Jackson, G.G., and Chiou, W.L.: High-pressure liquid chromatographic assay of netilmicin in plasma. *Antimicrob. Agents Chemother.* 12:707–709, 1977.

79. Pennington, J.E., Dale, D.C., Reynolds, H.Y., and MacLowry, J.D.: Gentamicin sulfate pharmacokinetics: lower levels of gentamicin in blood during fever. *J. Infect. Dis.* 132:270–275, 1975.

80. Phillips, I., Warren, C., and Smith, S.E.: Serum gentamicin assay: a comparison and assessment of different methods. *J. Clin. Pathol.* 27:447–451, 1974.

81. Pijck, J., Hallynck, T., Soep, H., Baert, L., Daneels, R., and Boelaert, J.: Pharmacokinetics of amikacin in patients with renal insufficiency: relation of half-life and creatinine clearance. *J. Infect. Dis.* 134(Suppl.):S331–S341, 1976.

82. Reeves, D.S.: Accuracy of gentamicin assays. *Postgrad. Med. J.* 50(Suppl.):20–21, 1974.

83. Reeves, D.S., and Bywater, M.J.: Quality control of serum gentamicin assays— experience of national surveys. *J. Antimicrob. Chemother.* 1:103–116, 1975.

84. Regamey, C., Schaberg, D., and Kirby, W.M.M.: Inhibitory effect of heparin on gentamicin concentrations in blood. *Antimicrob. Agents Chemother.* 1:329–332, 1972.

85. Riff, L.J.: Pseudomonas bacteremia. Evaluation of factors influencing response to therapy. *Acta Pathol. Microbiol. Scand.* [B] 81(Suppl. 241):79–88, 1973.

86. Riff, L.J., and Jackson, G.G.: Pharmacology of gentamicin in man. *J. Infect. Dis.* 124(Suppl.):S98–S105, 1971.

87. Rosdahl, N., and Andersen, J.B.: Aspects of the pharmacology of gentamicin during the neonatal period. *Acta Pathol. Microbiol. Scand.* [B] 81(Suppl. 241):119–123, 1973.

88. Sabath, L.D., Casey, J.I., Ruch, P.A., Stumpf, L.L., and Finland, M.: Rapid microassay of gentamicin, kanamycin, neomycin, streptomycin, and vancomycin in serum or plasma. *J. Lab. Clin. Med.* 78:457–463, 1971.

89. Sabath, L.D., Casey, J.I., Ruch, P.A., Stumpf, L.L., and Finland, M.: Rapid microassay for circulating nephrotoxic antibiotics. In *Antimicrobial Agents and Chemotherapy 1970.* Washington, D.C., American Society for Microbiology, 1971, pp. 83–90.

90. Santanam, P., and Kayser, F.H.: Enzymatic adenylylation by aminoglycoside 4'-adenyltransferase and 2″-adenylyltransferase as a means of determining concentrations of aminoglycoside antibiotics in serum. *Antimicrob. Agents Chemother.* 10:664–667, 1976.

91. Sarubbi, F.D., Jr., and Hull, J.H.: Amikacin serum concentrations: prediction of levels and dosage guidelines. *Ann. Intern. Med.* 89:612–618, 1978.

92. Sawchuk, R.J., and Zaske, D.E.: Pharmacokinetics of dosing regimens which utilize

multiple intravenous infusions: gentamicin in burn patients. *J. Pharmcokinet. Biopharm.* 4:183-195, 1976.

93. Sawchuk, R.J., Zaske, D.E., Cipolle, R.J., Wargin, W.A., and Strate, R.G.: Kinetic model for gentamicin dosing with the use of individual patient parameters. *Clin. Pharmacol. Ther.* 21:362-369, 1977.

94. Schentag, J.J., Cumbo, T.J., Jusko, W.J., and Plaut, M.E.: Gentamicin tissue accumulation and nephrotoxic reactions. *JAMA* 240:2067-2069, 1978.

95. Schwartz, S.N., Pazin, G.J., Lyon, J.A., Ho, M., and Pasculle, A.W.: A controlled investigation of the pharmacokinetics of gentamicin and tobramycin in obese subjects. *J. Infect. Dis.* 138:499-505, 1978.

96. Semple, P.F., Carswell, W., and Boyle, J.A.: Serial studies of the renal clearance of urate and inulin during pregnancy and after the puerperium in normal women. *Clin. Sci. Mol. Med.* 47:559-565, 1974.

97. Shanson, D.C., and Daniels, J.V.: Factors affecting plate assay of gentamicin: I. Diluents. *J. Antimicrob. Chemother.* 1:219-227, 1975.

98. Shanson, D.C., Kensit, J., and Hince, C.: Uremia, gram-negative sepsis, and gentamicin assays. *Lancet* 2:875, 1975.

99. Shanson, D.C., and Hince, C.J.: Factors affecting plate assay of gentamicin: II. Media. *J. Antimicrob. Chemother.* 3:17-23, 1977.

100. Shaw, E.J., Watson, R.A.A., Landon, J., and Smith, D.S.: Estimation of serum gentamicin by quenching fluoroimmunoassay. *J. Clin. Pathol.* 30:526-531, 1977.

101. Siber, G.R., Echeverria, P., Smith, A.L., Paisley, J.W., and Smith, D.H.: Pharmacokinetics of gentamicin in children and adults. *J. Infect. Dis.* 132:637-651, 1975.

102. Siersbaek-Nielsen, K.: Rapid evaluation of creatinine clearance. (Editorial). *Lancet* 1:1133-1134, 1971.

103. Smith, A.L., and Smith, D.H.: Gentamicin: adenine mononucleotide transferase: partial purification, characterization, and use in the clinical quantitation of gentamicin. *J. Infect. Dis.* 129:391-401, 1974.

104. Smith, A.L., Waitz, J.A., Smith, D.H., Oden, E.M., and Emerson, B.B.: Comparison of enzymatic and microbiological gentamicin assays. *Antimicrob. Agents Chemother.* 6:316-319, 1974.

105. Smith, C.R., Maxwell, R.R., Edwards, C.Q., Rogers, J.F., and Lietman, P.S.: Nephrotoxicity induced by gentamicin and amikacin. *Johns Hopkins Med. J.* 142:85-90, 1978.

106. Smith, D.H., Van Otto, D., and Smith, A.L.: A rapid chemical assay for gentamicin. *New Engl. J. Med.* 286:583-586, 1972.

107. Standefer, J.C., and Saunders, G.C.: Enzyme immunoassay for gentamicin. *Clin. Chem.* 24:1903-1907, 1978.

108. Stessman, J., Michel, J., and Sacks, T.: Error in recovery rate of aminoglycosides from uraemic serum. *Chemotherapy* 23:142-148, 1977.

109. Stevens, D.L., Page, B.M.A., and Adeniyi-Jones, C.: Standardization of a rapid microbiologic assay for aminoglycosides using *Enterobacter cloacae*. *Am. J. Clin. Pathol.* 70:808-815, 1978.

110. Stevens, P., and Young, L.S.: Rapid assay of aminoglycosides by radioenzymatic techniques. In *Microbiology*, 64-72, 1975.

111. Stevens, P., Young, L.S., and Hewitt, W.L.: Improved acetylating radioenzymatic assay of amikacin, tobramycin, and sisomicin in serum. *Antimicrob. Agents Chemother.* 7:374-376, 1975.

112. Stevens, P., Young, L.S., and Hewitt, W.L.: [125]I radioimmunoassay of netilmicin. *Antimicrob. Agents Chemother.* 11:768-770, 1977.

113. Stevens, P., Young, L.S., and Hewitt, W.L.: ^{125}I-radioimmunoassay of amikacin and comparison with a microassay. *J. Antibiot. (Tokyo)* 29:829-832, 1976.

114. Stevens, P., Young, L.S., and Hewitt, W.L.: Radioimmunoassay, acetylating radio-enzymatic assay and microbioassay of gentamicin: a comparative study. *J. Lab. Clin. Med.* 86:349–359, 1975.

115. Tally, F.P., Louie, T.J., Weinstein, W.M., Bartlett, J.G., and Gorbach, S.L.: Amikacin therapy for severe gram-negative sepsis. *Ann. Intern. Med.* 83:484–488, 1975.

116. Tilton, R.C., Murphy, J.R. and Mallett, E.: Assay for gentamicin. *New Engl. J. Med.* 287:1100, 1972.

117. Tjernstrom, O., Banck, G., Belfrage, S., Juhlin, I., Nordstrom, L., and Toremalm, N.G.: The ototoxicity of gentamicin. *Acta Pathol. Microbiol. Scand.* [B] 81(Suppl. 241):73–78, 1973.

118. Waitz, J.A., Drube, C.G., Moss, E.L., Jr., Oden, E.M., Bailey, J.V., Wagman, G.H., and Weinstein, M.J.: Biological aspects of the interaction between gentamicin and carbenicillin. *J. Antibiot. (Tokyo)* 25:219–225, 1972.

119. Warren, E., Snyder, R.J., and Washington, J.A., II: Four-hour microbiological assay of gentamicin in serum. *Antimicrob. Agents Chemother.* 1:46–48, 1972.

120. Watson, R.A.A., Landon, J., Shaw, E.J., and Smith, D.S.: Polarization fluoroimmunoassay of gentamicin. *Clin. Chem. Acta* 73:51–55, 1976.

121. Williams, J.W., Langer, J.S., and Northrop, D.B.: A spectrophotometric assay for gentamicin. *J. Antibiot. (Tokyo)* 28:982–987, 1975.

122. Wilson, P., and Ramsden, R.T.: Immediate effects of tobramycin on human cochlea and correlation with serum tobramycin levels. *Br. Med. J.* 1:259–261, 1977.

123. Winters, R.E., Chow, A.W., Hecht, R.H., and Hewitt, W.L.: Combined use of gentamicin and carbenicillin. *Ann. Intern. Med.* 75:925–927, 1971.

124. Young, L.S., Decker, G., and Hewitt, W.L.: Inactivation of gentamicin by carbenicillin in the urinary tract. *Chemotherapy* 20:212–220, 1974.

125. Yourassowsky, E., de Broe, M.E., and Wieme, R.J.: Effect of heparin on gentamicin concentration in blood. *Clin. Chim. Acta* 42:189–191, 1972.

126. Zaske, D.E., Sawchuk, R.J., Gerding, D.N., and Strate, R.G.: Increased dosage requirements of gentamicin in burn patients. *J. Trauma* 16:824–828, 1976.

TOXICITY: CLINICAL AND EXPERIMENTAL EVALUATIONS AND HUMAN PATHOLOGY

8

RENAL TUBULAR TRANSPORT AND INTRARENAL AMINOGLYCOSIDE DISTRIBUTION

ANDREW WHELTON   The Johns Hopkins University
School of Medicine, Baltimore, Maryland

Studies that define the intrarenal distribution characteristics of an antibiotic provide information of both therapeutic and toxicological importance. The transport systems localized within the proximal tubular portion of the nephron can significantly modulate renal cortical drug levels while the physiologic activities of the loop of Henle and distal tubule can significantly influence inner and outer medullary drug concentrations. Hence, the study of the intrarenal gradient patterns of a given class of antibiotics can provide an important tissue end result of renal tubular active transport and physiologic activities.

This chapter is designed to provide a detailed review of the intrarenal distribution characteristics and renal cortical kinetics of aminoglycosides in health and disease. It will evaluate the modulating influence of renal physiologic parameters and various forms of renal tubular transport upon the renal tissue gradient characteristics of aminoglycosides. These data will then be assessed in terms of the clinical and toxicological implications that they provide in patient management.

I. Bacteriologic Correlations in the Study of Intrarenal Aminoglycoside Concentrations

Several years ago two important separate series of investigations indicated that it might be of clinical value to assess the factors that influence the intrarenal distribution of antibiotics since the investigations would yield information of both therapeutic and toxicologic importance. These new but widely divergent series of studies related on the one hand to the advances in our understanding of the pathophysiology of acute and chronic pyelonephritis [3,4,7,74,75,83], and on the other to the elucidation of the physiologic mechanism which either dilutes or concentrates the urine, the so-called countercurrent multiplier system [40,48,49,85].

It became apparent that the physiologic activity of the countercurrent system within the kidney might influence the intrarenal gradient pattern of many classes of antibiotics. This would be of importance since in the management of bacterial infections in humans, one of several important determinants that influences response to treatment is the concentration of the selected antibiotic produced in the parenchyma of the infected organ or biologic fluid accumulation. We therefore embarked upon a detailed series of studies that would correlate renal physiology, antibiotic pharmacokinetics, and bacterial pyelonephritis.

Several lines of investigation indicated that in the case of the pathophysiology of bacterial pyelonephritis there was an unsuspected difference in the intrarenal tissue susceptibility to infection, when investigations were designed to separately identify such a susceptibility pattern in renal cortical, medullary, or papillary tissues [3,4,74,75]. The net result of these studies was the demonstration that the renal medulla and papilla were the zones of initial activity in the acute seeding of bacterial infection in the kidney and were also the anatomic areas of continued bacterial infection in chronic active pyelonephritis. Although the reasons for this remarkable susceptibility of the inner zones of the kidney to infection are not fully resolved, some clear-cut relationships have been identified between both the structural integrity and functional activity of the kidney and its susceptibility to bacterial infections. Of foremost importance in the latter regard is the low blood flow into the renal medulla and papilla, which represents an integral part of the operation of the countercurrent multiplier system [85], since this results in a decreased delivery of leukocytes, antibody, and complement into the inner zones of the kidney. Additional factors of importance represent the reduced tissue oxygen tension and hypertonicity of the medulla and papilla, which serve to inhibit normal leukocyte function and thereby increase tissue susceptibility to infection. The anticomplementary activity of free ammonia, when the latter substance is produced in the cells of the distal and collecting duct portion of the nephron, also enhances the susceptibility of the medulla and papilla to bacterial infection [7].

Physiologic definition of the countercurrent multiplier system revealed that diffusable solutes of small molecular size such as urea were subject to dramatic changes in paren-chymal concentration in the renal medulla and papilla by the simple expedient of change in the state of systemic hydration. Production of concentrated urine dramatically increased tissue urea concentrations as such levels were progressively measured from inner cortex to papillary tip, while the production of a diuresis eliminated such a tissue gradient pat-tern with tissue urea concentrations approximating each other in all areas of the kidney. In other words, a "washout" of the solute from medulla and papilla took place during the elaboration of dilute urine. This physiologic system therefore identified a simple means by which the medullary and papillary concentrations of some appropriately sized solutes could be manipulated.

In our laboratories we then asked a question that represented the logical sequitor to the investigations described above. If the medulla and papilla are the zones of bacteriologic activity in both acute and chronic pyelonephritis and also are the anatomic locations in which various solute concentrations can be physiologically manipulated, would it be pos-sible to influence antibiotic concentrations in these therapeutically important areas simply by adjusting the state of hydration or operation of the renal countercurrent multi-plier system? This chapter provides a number of answers to these questions as they relate to the aminoglycoside group of antibiotics. In addition, the impact of adjustment of urine pH upon the intrarenal gradient patterns of aminoglycosides is reviewed, since it is known that urine pH can influence the renal clearance and tissue concentration of a number of antibiotics [45,92,93,100-104].

Several different techniques may be utilized to identify renal tissue antimicrobial concentrations; however, each of these approaches possesses limitations that complicate translation of the derived results into solid clinical therapeutic information. Details of these investigative techniques are available elsewhere.

II. Toxicologic and Pharmacokinetic Study of Intrarenal Aminoglycoside Concentrations

Morphologic evidence of drug-induced nephrotoxicity may be located in any of three zones within the nephron. These three anatomic areas are the glomerulus, the proximal tubule, and the distal tubular portion of the nephron. The toxicity may be localized to one area alone or it may be present in all or a combination of these zones. The mediation of the toxicity may be (1) dose-related (aminoglycosides, cephalosporins, amphoteracin B, and analgesics, etc.); (2) direct hypersensitivity to the glomerular and tubular basement membrane (penicillins, sulfonamides, rifampin, etc.); and (3) secondary to immune com-plex deposition within the nephron (gold salts, anticonvulsants, etc.). These mechanisms of drug-induced toxicity and their anatomic location in human kidneys are reviewed in detail in Chapter 15.

In humans and in experimental animals all morphologic evidence of aminoglycoside-induced damage in the kidney is virtually exclusively a phenomenon of the proximal tubular cells [24,34,51,56,59,80] (for additional details, see Chapter 15). There are, however, important recent data from the laboratories of Brenner and colleagues which have identified that a change in the glomerular trans-basement membrane ultrafiltration coefficient can be produced by therapeutic dosing levels of gentamicin in a rat model [6]. These functional changes have been identified in the absence of clear-cut morphologic abnormalities in the glomeruli. Similarly, a recent report from Luft and associates identified

that a change in glomerular endothelial cells can be detected by scanning electron microscopy in a gentamicin induced rat model of nephrotoxicity [28].

Aminoglycoside-induced nephrotoxicity is clinically related to (1) the total dose of
drug, and (2) the duration of therapy with the selected compound. As already mentioned,
the morphologic evidence of aminoglycoside nephrotoxicity is confined to the proximal
tubular portion of the nephron. Consequently, several investigators have evaluated the
pharmacokinetic aspects of aminoglycoside accumulation and retention within the proximal tubular portion of the nephron. The aim of such studies has been to correlate the
anatomy of the nephron with aminoglycoside pharmacokinetics and the pathophysiologic
events that lead to damage in the proximal tubular cells. A number of important recent
autoradiographic studies have confirmed that the precise cellular area of aminoglycoside
uptake within the nephron is the proximal portion of the tubule [22,42,52,81]. However, most experimental studies that assess the drug kinetic and pathophysiologic events
of nephrotoxicity have been designed to study the total concentration of aminoglycoside
found within the renal cortex since the latter area represents the intrarenal anatomic location of the proximal tubule. In general, these studies have compared the total renal cortical concentrations of one aminoglycoside versus another with the concurrent renal
functional status of the investigative animals and the pathologic tissue changes induced
by the aminoglycosides in question. The conclusions derived from these studies have
been conflicting at best. Since the renal cortical pharmacokinetics of aminoglycosides
can be modulated by several factors other than the physical chemistry of the drugs themselves, it becomes a matter of importance to evaluate the currently available data so that
we may separately review many of these modulating influences. Six clinical pathophysiologic sets of circumstances need to be evaluated as sequentially identified here:

1. Animal species selected for study
2. Dose, method of administration, and duration of therapy
3. Systemic state of hydration
4. Systemic pH and the urinary pH
5. Presence of concurrent drug therapy
6. Functional integrity of the kidney prior to therapy

A. Animal Species Selected for Study

It is axiomatic that in human clinical investigation one cannot design any drug trials in
which a toxic side effect of a drug is knowingly produced by the investigator. In patients
or human volunteers one can only look for the unsuspected or unintentional development
of a dose-related toxic side effect and then develop some conclusions based upon the
clinical sequence of events once the toxicologic side effect(s) has been recognized and the
drug discontinued. As a result, the vast majority of data pertinent to the toxic complications of the aminoglycosides have been derived from the study of animal models. In
animals it is, of course, possible to give increasing doses of the aminoglycosides over a
finite period of time, and to undertake the most detailed renal functional studies and
tissue pathologic evaluations at any given point in the course of drug therapy. The Fischer
rat model has proved particularly useful and popular in the study of aminoglycoside-
induced nephrotoxicity because of low cost, good reproducibility of results, and ease of
handling the animals. Canine models have also been used extensively and a lesser number
of investigations have been undertaken in the mouse and rabbit. This topic is reviewed in
detail in Chapter 10.

Many difficulties are inherent in the extrapolation of animal toxicologic data to human clinical practice. Suffice it to say here that all the investigative animal models show the unifying pharmacokinetic finding of significant renal cortical uptake and retention of the aminoglycosides. This is the renal "hallmark" of the aminoglycosides in both animals and humans [2,10,18,26,31,33,42,50,54,55,58,59,61,71,84,86,92,94,101]. It is reasonable to compare the toxic potential of several different aminoglycosides in the same animal model, but at the present state of the art it is unacceptable to compare the milligram per milligram toxic effects of two different aminoglycosides in two different animal models. Which animal models represent the closest approximation to human nephrotoxicity is not yet fully resolved.

B. Dose, Method of Administration, and Duration of Therapy

Since the clinical introduction of the first aminoglycoside, streptomycin, it has been recognized that all drugs in the latter class have the potential for inducing dose-related toxic side effects as manifest by nephrotoxicity, ototoxicity, and neurotoxicity. The required dose per kilogram of body weight of the patient necessary to produce nephrotoxicity varies from patient to patient and most particularly from drug to drug. A good example is the dose-toxicity relationship of streptomycin versus neomycin. In the case of streptomycin, it would be necessary to give substantially larger milligram dosing than is required for neomycin to ensure the induction of experimental (or clinical) nephrotoxicity.

In single-dose acute experiments the uptake of aminoglycoside into the renal cortex directly correlates with the dose of drug administered. An example of this is presented in Figure 1, which describes the "acute" renal cortical kinetics of gentamicin and tobramycin when graded increases in the dosage of each drug were administered to healthy canine models. Duration of drug administration was 4–5 hr and the cortical concentrations observed were those produced at a time when the serum drug levels were stable at several points throughout the therapeutic serum concentration range. As can be seen in the figure, there is a correlation between the stable serum level of aminoglycoside and the resultant cortical concentration of the drug. It is of interest to note in the case of gentamicin that a saturational effect upon renal cortical uptake appears to develop at high dosing or serum concentration values. These data tell us that the dose of aminoglycoside alone will influence the measured renal cortical concentrations of the drug and that precision in identification of the serum level of the compound is essential before deriving any conclusions about dose-concentration relationships or toxicologic associations. It is worth emphasizing here that without the detailed data base available in Figure 1, one could derive the erroneous conclusion that within the human therapeutic serum concentration range there was no difference in renal cortical levels of gentamicin and tobramycin since, for example, at a serum gentamicin level of 2 μg/ml the renal cortical level is 40 μg/g, and at a serum tobramycin level of 6 μg/ml the renal cortical concentration of tobramycin is the same as that just noted for gentamicin; however, such a conclusion is incorrect because when comparable serum values are compared throughout the therapeutic concentration range, as presented in Figure 1, gentamicin levels are always higher than those noted for tobramycin ($P < 0.01$). We will address this issue later in relation to the clinical incidence of aminoglycoside nephrotoxicity.

It is now recognized, from the precise work of Thompson and colleagues and of Bennett and associates, that the method of administration of the aminoglycosides bears a

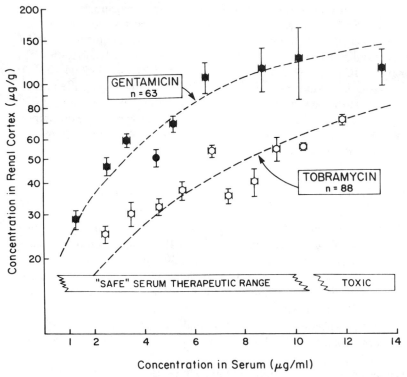

Figure 1 These data define the renal cortical concentrations of gentamicin and tobramycin in healthy canine kidneys at various stable serum concentrations of the drug. In these studies the aminoglycosides were infused intravenously for 4 hr.

direct relationship to their potential for nephrotoxicity [10,71]. The clinical importance of this issue is of sufficient magnitude that it is reviewed in detail in Chapter 20. It should be briefly noted here that Thompson and colleagues have found that when identical doses of an aminoglycoside (gentamicin or tobramycin) are given to an experimental canine model, either as a single bolus infusion or by continuous intravenous infusion, the continuous infusion of the drug leads to a greater decrease of renal function than injection once daily. Additionally, Bennett and associates have shown in a rat model that when daily doses of aminoglycoside are administered three times a day rather than once a day, the former regimen is associated with greater toxicity and a more rapid cortical accumulation of drug [10].

   The duration of therapy has a profound influence upon renal cortical uptake or accumulation of aminoglycosides. It has been recognized for quite some time that following an initial dose of gentamicin, only 40–60% of the drug is recovered in the urine, but that by the third or fourth doses of the compound, one can recover from the urine almost all of the aminoglycoside administered [39,73]. This observation correlates in part with what is happening kinetically in the renal cortex, since there is an initial rapid uptake and retention of the drug within the renal cortex with evidence of cell saturational kinetics being noted at high dosing levels, as indicated in Figure 1. As repeat administrations of the drug are given utilizing standard clinical-type dosing considerations, the renal cortex

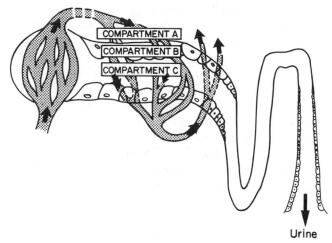

Figure 2 Schematic representation of the three separate aminoglycoside distribution compartments within the renal cortex.

continues to show uptake of the aminoglycoside for several days, followed by a plateauing of tissue drug concentration with variable cyclical changes taking place subsequently [10,11,32,50,76–78,91]. It is therefore of great interest to recognize that in clinical practice nephrotoxicity is almost always associated with therapy of more than 7–10 days' duration [82]. This suggests that maintenance of the proximal tubular cells in a setting wherein they are "fully saturated" and repetitively exposed to more aminoglycoside may be an important kinetic determinant of toxicity. It would explain the confusing clinical observation that nephrotoxicity may become manifest despite the fact that the serum levels of the drug have been carefully monitored with dosing adjustment to keep them within the accepted safe therapeutic range. This important kinetic question needs further evaluation.

C. State of Hydration

The state of hydration of a patient receiving an aminoglycoside is an important risk factor in relationship to the induction of nephrotoxicity. Specifically, dehydration and/or the use of potent diuretic agents enhance the development of aminoglycoside-induced renal toxicity [1,9,19,66].

The exact reason for the higher profile for aminoglycoside toxicity in the clinical setting of dehydration or diuretic use is not fully resolved. Three aminoglycoside distribution compartments within the renal cortex may be modulated by the state of hydration, and these locations are shown in Figure 2 as representing the intravascular and interstitial compartment of the renal cortex (compartment A), the intracellular space (compartment B), and the intraluminal glomerular ultrafiltrate (compartment C). It is probable that a combination of drug pharmacokinetic changes in each of these compartments represents the answer to the nephrotoxic potential of diuretic use of dehydration in patients receiving aminoglycoside therapy. We can subsequently examine the available data pertinent to this issue.

It is well recognized that the state of hydration has a marked influence upon the intrarenal distribution characteristics of many antibiotic chemotherapeutic compounds. However, it is typically the parenchymal drug concentrations within the inner zones of the
kidney (i.e., the medulla and papilla) which are most influenced by manipulation of the
state of hydration [92,100,101]. This is directly related to the fact that the physiologic
operation of the countercurrent multiplier system which concentrates and dilutes the
urine particularly influences the medullary and papillary tissue concentrations of diffusable solutes of small molecular size, such as urea, which has a molecular weight of 60
[85]. This information led us some years ago to question whether or not antibiotics such
as penicillins (average molecular weight 360), cephalosporins (average molecular weight
400), tetracyclines, or aminoglycosides (average molecular weight 450) could have their
tissue concentration within the inner zones of the kidney reproducibly manipulated
simply by changing the state of hydration or activity of the countercurrent multiplier
system.

Figure 3 diagrammatically represents the intrarenal location where the recycling of
urea or other molecules takes place by diffusion from the collecting duct into the interstitium of the medulla and papilla, then into the ascending limb of the loop of Henle.
It indicates that, as already known for urea, if the intrarenal gradient pattern of an antibiotic can be influenced by the state of hydration, the diffusion of such an antibiotic
across the epithelial lining of the collecting duct and ascending limb of Henle's loop
must be greater than that which takes place within the cortical distal convoluted tubule.
A good example of this physiologic influence of the countercurrent multiplier system can
be seen in Figures 4 and 5, which depict the influence of the state of hydration upon
penicillin G and two of its analogs. To put the studies in perspective, in the hydrated
or dilute urine studies, the urinary flow rates averaged approximately 2–4 ml/min per
kidney with an average osmolarity of 200–300 mosmol/L, whereas in the hydropenic or
concentrated urine studies, the urinary flow rates averaged approximately 0.1–0.2 ml/min
per kidney with an osmolarity of 800–2000 mosmol/L. It can be seen that in the case of

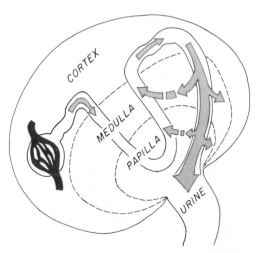

Figure 3   Diagrammatic representation of the movement of small diffusable molecules
in the distal portion of the nephron. The operation of the countercurrent multiplication
of concentrations systems influences the intrarenal parenchymal concentration of many
diffusable antibiotics.

penicillin G and carbenicillin, a significant increase in medullary and papillary drug concentrations are noted when concentrated urine (hydropenic state) is produced. Interestingly, ampicillin does not demonstrate the same medullary and papillary tissue drug kinetics; in fact, a significant difference in renal cortical concentration of ampicillin is noted between the hydrated and hydropenic states.

In the case of the aminoglycosides, the majority of available information indicates that these antibiotics are but modestly influenced by the state of hydration in terms of their renal parenchymal concentrations, despite enormous differences in urinary levels [9,19,94]. Specifically, as can be seen in Figure 6 for gentamicin, it appears that in the healthy dog kidney the drug does not undergo the countercurrent recycling already noted for many of the penicillins, cephalosporins, and tetracyclines [95,97-101]. The most striking feature of the intrarenal gradient pattern of gentamicin is the high levels of drug that are found in the renal cortex in both the hydrated and hydrapenic states. In the studies depicted in Figure 6, cortical concentrations were approximately 20-fold higher than concomitant serum levels. It is now well recognized that renal cortical accumulation of the aminoglycosides is a pharmacokinetic "hallmark" of that group of antibiotics, not only in experimental animals but also in healthy human kidneys, as clarified by Milner et al. [63], Edwards and associates [26], and Luft et al. [59]. This phenomenon is well exemplified in Figure 7, which contrasts cortical accumulation of the penicillins versus the cephalosporins versus the aminoglycosides.

Data that define the changes induced in renal cortical concentrations of the aminoglycosides by variation in the state of systemic hydration or by the concomitant administration of potent diuretics are scant. No such tissue data information are available for humans. However, in a rat model, Bennet and colleagues [9], and Chiu and Long [19], have shown that approximately a twofold change in renal cortical gentamicin concentration can be produced by changing the animals' state of hydration. The tissue level is increased by

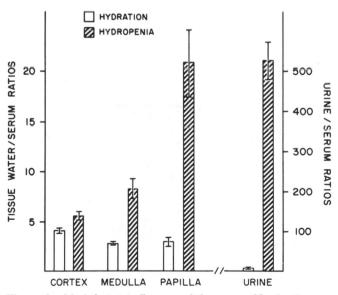

Figure 4   Modulating influence of the state of hydration upon the intrarenal tissue and urinary concentrations of penicillin G in healthy canine kidneys. See the text for a delineation of urinary flow rates and osmolality.

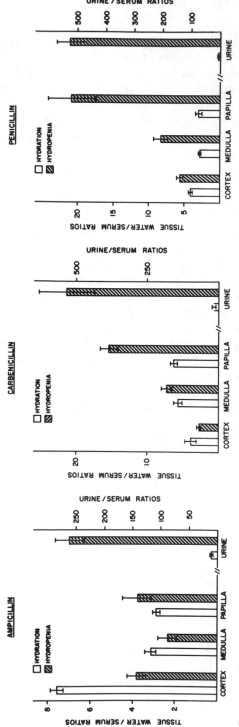

Figure 5 Comparative influence of the state of hydration, or operation of the countercurrent multiplier system, upon penicillin G versus its analogs ampicillin and carbenicillin.

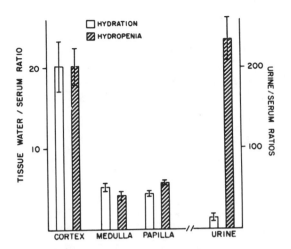

Figure 6 Intrarenal gradient pattern of gentamicin as seen in the healthy canine kidney following a 4 hr constant infusion of the drug. Note the high renal cortical concentrations of the compound. See the text for details of urinary flow rates and osmolality. The countercurrent multiplier system does not significantly modulate parenchymal concentrations of the aminoglycosides.

volume repletion. One can not extrapolate from their studies the exact renal cortical gentamicin concentration achieved at similar serum concentrations of the drug but at different states of systemic hydration. Nonetheless, these rat studies appear to have an important parallel to what is observed in clinical practice in terms of the "risk factor" role of dehydration and/or the use of potent diuretics.

In a healthy dog model we have measured renal cortical concentration of several aminoglycosides in "hydrated" and "hydropenic" animals. The average urine flow rates and urinary osmotic activity in these two classifications of the systemic state of hydration were similar to the data already described for the antibiotics detailed in Figures 4 and 5. When comparable serum concentrations are compared between these two physiological states of hydration, it is possible to see, as presented in Figure 8, that there are modest differences in the renal cortical concentrations of several aminoglycosides, in particular tobramycin, kanamycin, and amikacin. For most of these drugs, however, the available data base is insufficient to realistically evaluate the statistical significance of these differences in tissue cortical drug concentrations. The renal parenchymal concentrations of aminoglycosides, as presented in Figure 8, are derived from acute drug infusion studies. As yet, data are not available to pinpoint the tissue drug changes that occur over the framework of several days of drug administration.

The tissue morphologic consequences of the use of gentamicin alone versus gentamicin and a loop diuretic have recently been published by Adelman and colleagues [1]. In these important and carefully designed studies, they evaluated serial renal biopsies and urinary enzyme activity in dogs receiving either (1) gentamicin alone or (2) a gentamicin-furosemide combination. Equal weight-for-weight aminoglycoside dosing was given to each group of animals. In their experimental model the nephrotoxicity of gentamicin given in combination with furosemide increased dramatically. There were earlier and significantly greater changes in serum creatinine, blood urea nitrogen, and urinary enzymes in the latter group than in the dogs who received gentamicin alone. Enzymuria preceded changes in conventional renal function tests. Peak and trough serum levels of the aminoglycosides were

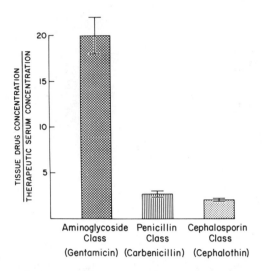

Figure 7 Difference noted in renal cortical tissue drug concentrations when three separate antibiotic classes are compared.

higher in the dogs receiving combination therapy. Tissue drug concentrations were not measured. The authors speculated that the higher serum gentamicin levels seen in their dogs receiving gentamicin and furosemide may have been secondary to increased proximal tubular reabsorption, a decreased volume of drug distribution, and/or a decreased glomerular filtration rate. They concluded that the elevated serum gentamicin levels may have resulted from or contributed to the observed increase in their toxicity.

The clinical application of the several lines of investigation pertinent to the role of the systemic state of hydration as a risk factor in the induction of nephrotoxicity is of major clinical importance. However, as so often happens in clinical practice, most patients who develop aminoglycoside-induced nephrotoxicity while receiving diuretics represent a complex dilemma in terms of interpretation of the clinical facts. Invariably, the patients also receive other drugs that influence their renal function, and indeed the underlying disease itself will frequently influence their level of renal functional activity. Nonetheless, the extant data now tell us that, extrapolating to the clinical use of aminoglycoside therapy, the presence or development of dehydration or the intercurrent use of potent diuretics will (1) increase serum concentrations of the aminoglycoside, (2) increase renal cortical levels of the drug, and (3) accelerate the development of nephrotoxicity.

III. Can Renal Cortical Drug Kinetics Predict the Clinical Nephrotoxic Potential of an Aminoglycoside?

The initial observations indicating that aminoglycosides show a predisposition for accumulation within the renal cortex [2,55,101] were surprising in view of the earlier investigations of renal lymph aminoglycoside concentrations undertaken by Chisholm and associates [16,17]. The latter investigators found that renal lymph gentamicin concentrations as identified in canulated hilar renal lymphatics of the dog approximated the levels concurrently found in the serum. Subsequent anatomic studies have indicated that lymph collected in renal/hilar lymphatics in fact reflects solute and electrolyte concentrations as typically found in the interstitial fluid of the cortex of the kidney. Therefore, the amino-

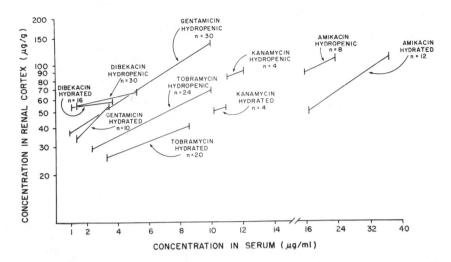

Figure 8 The systemic state-of-hydration variable influences the renal cortical concentrations of essentially all aminoglycosides. These data are derived from studies of a 4 hr infusion of aminoglycosides in healthy dogs (n = number of kidneys). See the text for details of urinary flow rates and osmolarity.

glycoside concentrations noted in the lymph kinetic studies of Chisholm and colleagues would in large part reflect cortical interstitial gentamicin concentrations and would not detect the sizable concentrations of the drug present within the cells of the renal cortex. Furthermore, it is now recognized that the aminoglycosides manifest a significant degree of renal cortical tissue binding [50,53,60,96] and retention within the cortex [29,31, 55,57,71,76-78,84,86,91]. These kinetic differences of the aminoglycosides in compartments A and B, as presented in Figure 2, resolve the difference in the renal cortical tissue concentrations of aminoglycosides as found in tissue homogenate studies versus the concentration of drug as found in renal lymph investigations. As already indicated, the renal lymph drug concentration studies will reflect only the non-tissue-bound portion of a drug. Data defining differential renal cortical concentrations for most aminoglycosides are now available and, as already noted, the majority of these studies have been performed in a dog or a rat model. Figure 9 summarizes the data for several aminoglycosides as found in a healthy canine model when various concentrations of the respective drugs were administered over a 4-5 hr infusion period [91,92,94,96,101]. It is important to emphasize that the few available reports in the literature defining renal cortical aminoglycoside concentrations as seen in "healthy" human kidneys, removed immediately upon death in patients who died from a nonrenal cause and were receiving aminoglycosides up to the time of demise, or wherein intraoperative renal tissue samples were obtained, indicate that cortical levels of gentamicin and amikacin show a trend similar to the results noted in the animal investigations [2,26,59].

The data indicate that when renal cortical concentrations of the several drugs are compared at a time when the concomitant serum values are in the usual therapeutic range, or at the serum levels seen following lavage of a surgical cavity in the case of neomycin, we can establish the following sequence of cortical drug concentration from highest to lowest: neomycin = dibekacin > gentamicin > kanamycin = amikacin > tobramycin > streptomycin. This cascade would appear to bear a remarkable similarity to the clinical

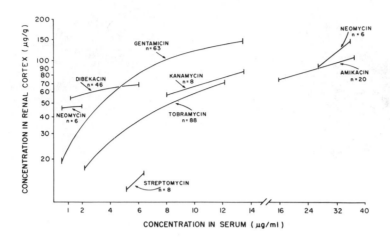

Figure 9 Renal cortical aminoglycoside concentrations as noted in healthy canine kidneys following 4 hr of drug infusion (n = number of kidneys analyzed per drug).

spectrum of greatest nephrotoxicity to least nephrotoxicity. It is clearly necessary to add that from the clinical point of view, adequate reference toxicologic data are not yet available.

It has been mentioned that continued uptake of drug into the renal cortex takes place when an experimental animal is repetitively dosed over a time frame of several days [11,50,91]. Hence, in the comparison of renal cortical drug concentrations, the duration of therapy should always be taken into consideration. The data of Bennett et al., Cuppage and associates, and Luft and colleagues indicate that following repetitive aminoglycoside therapy, cyclical changes in renal cortical drug concentration may be determined as a result of nephrotoxic cellular changes in the proximal tubule or indeed other unidentified modulating influences [11,24,57]. The quantitative aspects of drug administration must also be assessed when comparing renal cortical aminoglycoside concentration because many experimental nephrotoxic studies are undertaken at significantly higher milligram dosing values that would be used in normal clinical practice. At these dosing levels saturational kinetics may be taking place in the renal tissues, and therefore the parenchymal concentrations may not reflect what occurs in clinical practice. The high serum drug concentration may have a dramatic influence upon the measured tissue values. This is a most important issue that has not been adequately addressed in the literature.

A brief example of the complexity of this issue is presented in the two panels of Figure 10. We consider two aminoglycosides, X and Y, which are used on the same weight-for-weight basis and are known clinically and experimentally to have a significant difference in incidence of nephrotoxicity. In panel A the serum concentration for each drug is within the safe therapeutic range and is measured at 4 μg/ml. Uptake of drug X into the proximal tubular cells becomes stable or "saturated" at a level of 100 μg/g, whereas in the case of aminoglycoside Y the proximal tubular cells become stable or "saturated" at a level of 40 μg/g. The total renal cortical homogenate concentration of aminoglycoside X is 80 ± 8 μg/g, a figure that is significantly higher than the 30 ± 3 μg/g found for drug Y. However, if the serum concentration of both drugs is now markedly increased to the range frequently achieved in experimental models of nephrotoxicity, we find that the "saturated" intracellular levels of compartment 2 are completely overwhelmed

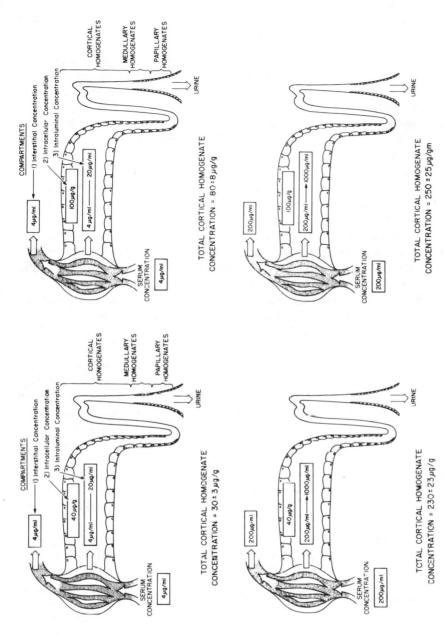

Figure 10 Renal cortical kinetics of two aminoglycosides, X and Y, which are used clinically on a similar weight-for-weight basis. Note that at low (therapeutic) serum levels renal cortical concentrations are significantly different; however, at high (toxic) serum levels no difference in renal cortical homogenate concentrations can be detected.

205

by the high concentrations in compartments 1 and 3. As a result, the total renal cortical homogenate concentration in the case of aminoglycoside X is 250 ± 25 μg/g and in the case of Y is 230 ± 23 μg/g. Therefore, in the therapeutic serum range of either of these aminoglycosides, which are used in the same weight-for-weight basis, the total cortical homogenate concentrations of the drugs are significantly different, whereas in the serum range often used for the induction of experimental nephrotoxicity, the total cortical concentration of these two drugs are not significantly different. Nonetheless, the levels within the proximal tubular cells remain essentially the same for each drug despite the major changes in the serum aminoglycoside concentration. This example underscores the difficulty inherent in the interpretation of the predictive toxicologic value of studies that identify renal cortical aminoglycoside concentration.

To return to the issue of predicting the clinic potential of aminoglycosides for the induction of nephrotoxicity, it appears that within carefully defined experimental circumstances, the renal cortical kinetics of aminoglycosides may offer a valuable predictive clue to clinical nephrotoxicity. If these data are evaluated in concert with the experimental results of morphologic and renal functional changes induced in an animal model such as the rat or dog, following repetitive high-dose administration of the test aminoglycosides, we characterize the best and most realistic combination of experimental predictive nephrotoxic information. It is appropriate to point out that there are several extant studies which do not show a good correlation between renal cortical concentrations of the aminoglycosides and the clinical and experimental nephrotoxicity of these agents. This issue is carefully addressed in Chapter 10. Similarly, the recent data of Luft and colleagues [54] and Barza and associates [5] generated in a rat model, plus the elegant studies of Thompson and colleagues in a dog model [71], have not shown a solid relationship between measured renal cortical drug levels and subsequent incidence of nephrotoxicity. In many of these studies the drug-dosing schedule utilized represented one that was substantially higher than would be used in clinical practice, but an administration schedule that would reproducibly induce nephrotoxicity. Again, it must be emphasized that before any credence can be placed on the correlation of renal cortical drug concentrations with the resultant clinical toxicity of such an agent, one must be precise in defining (1) the drug dose, (2) the duration of therapy, (3) the time relationship of removal of the renal tissue for drug analysis pruposes versus the concomitant serum drug concentration, and (4) the kinetic status of the serum drug concentration.

In the future, a much more sophisticated approach to this issue will be necessary, including (1) focusing more specifically upon the proximal tubular brush border membrane so that we can quantify the available binding sites for various aminoglycosides, (2) identifying the avidity with which these agents bind to the brush border membrane, (3) quantification of the kinetics of uptake and binding in lysosomes or other intracellular organelles, and (4) quantification of the kinetics of elution of these various drugs from the renal proximal tubular cells. A number of these issues are already under investigation in several centers.

IV. Influence of Systemic or Urinary pH upon the Renal Tissue Disposition of Aminoglycosides

The role of environmental pH, as an important factor which modulates the microbiologic activity of the aminoglycosides, has been well recognized from the earliest days of the use of streptomycin. In biologic solutions an alkaline pH enhances the antimicrobial activity of aminoglycosides. Clinically, this issue is of realistic importance only in the renal tract. The distal portion of the nephron influences final urine pH and hence within the nephron and lower urinary tract it is clinically possible to manipulate the urine pH through the physiologic range 4.5–8.5. Initial clinical experience with streptomycin indicated that alkalinization of the urine was an important additive factor in the therapeutic management of renal tuberculosis. The therapeutic advantages of this pH-related phenomenon appear to be secondary to a change in net charge in ionization of the aminoglycoside molecule. The drugs are organic bases by nature and an alkalinization of the environment decreases net molecular charge and renders the drug more capable of movement across bacterial cell wall membranes. This important phenomenon of nonionic diffusion is reviewed elsewhere [62,89,90]. As the aminoglycoside class of drugs developed and newer individual members of the group were shown to possess greater and more effective spectra of antimicrobial activity, the importance of clinically monitoring or manipulating urinary pH in the management of urinary tract infections has diminished. Nonetheless, pH remains a most important factor in any in vitro test systems used for either bacterial sensitivity testing or drug assay purposes.

Only a small number of studies have assessed the impact of systemic pH or urinary pH upon the renal tissue concentrations of the aminoglycosides. Hsu and colleagues reported in 1974 that when normal rats chronically ingested a 1% ammonium chloride solution, the resultant metabolic acidosis significantly exacerbated gentamicin nephrotoxicity in vivo [36]. In renal cortical tissue slices obtained from the latter animals there was a depression of aminohippuric acid and tetraethylammonium bromide uptake when compared to control studies. In our laboratories, using a healthy dog model, we could not detect a significant influence of urine pH upon the concommitant renal cortical, outer medullary, or papillary concentrations of gentamicin or tobramycin [94,96]. Intravenous administration of sodium bicarbonate was used to produce alkaline urine, and pretreatment with ammonium chloride was used to ensure production of acid urine. Subtle changes in renal cortical concentrations or intrarenal gradient patterns of the drug may not have been detected in the in vivo test system as used in our laboratories. Subsequently, Kluwe and Hook have reported that using an in vitro renal cortical slice preparation they could not identify an optimal pH for active renal cortical gentamicin accumulation [47]. In fact, these authors noted that under hypoxic conditions, gentamicin accumulation at both acidic and alkaline pH exceeded that noted at pH 7.4. Most recently, Chiu et al. [21] have reported that in a rat model the renal cortical uptake of gentamicin was moderately inhibited by urinary alkalinization produced by sodium bicarbonate administration, but was unaffected by acidification with ammonium chloride. These authors noted that nephrotoxicity resulting from chronic treatment of gentamicin was ameliorated by concomitant sodium bicarbonate administration. In their studies it appears that the influence of sodium bicarbonate administration upon renal cortical gentamicin uptake was independent of any hydrational influences of the salt load, and the authors speculated that the intratubular adjustment in pH changed the cationic nature of the molecule and therefore reduced its binding to and uptake by the proximal tubular cells.

In summary, the influence of systemic or urinary pH upon renal tissue concentrations of aminoglycosides remains controversial, with the majority of evidence suggesting a minor role for pH. Clearly, the data support the notion that in terms of toxic effects of the aminoglycosides, chronic acidosis is a contributing factor, but sodium bicarbonate-induced chronic alkalosis is not a risk factor. It should be emphasized that it has been reported that a physical interaction, and hence incompatibility, exists when sodium bicarbonate solutions and aminoglycoside drugs such as streptomycin are mixed in vitro [70]. This phenomenon must, therefore, be considered in assessment of the tissue slice studies already reviewed, which have shown an inhibitory effect of an alkaline pH upon aminoglycoside uptake into renal cortical tissues.

V. Concomitant Administration of Other Drugs

The incompatibility demonstrated by the direct in vitro mixing of sodium bicarbonate and streptomycin has already been noted in this chapter. Extensive lists of other drugs and solution incompatibilities are presented elsewhere [35,68,70], but the incompatible nature of the direct admixture of aminoglycosides and the penicillins or cephalosporins is of particular note insofar as these drugs are often coadministered to septicemic patients. Riff and Jackson have reported that carbenicillin and gentamicin interact to form a conjugate linked between the amino groups in the sugars of gentamicin and the β-lactam ring of the penicillin [72]. Such a conjugate inactivates the antimicrobial activity of both molecules. The reaction is now thought to occur via nucleophilic opening of the β-lactam ring, probably by the methylamino group of gentamicin with concomitant formation of a biologically inactive amide [12,27,87]. A similar reaction between any β-lactam antibiotic and any aminoglycoside can be anticipated, with the rate of inactivation varying from one drug combination to another. The full clinical significance of the in vivo consequences of the interaction of aminoglycosides with other antibiotics is not yet clearly identified, but it appears to be of particular importance in patients with severe renal impairment wherein long intervals of in vivo incubation of these drug combinations takes place before supplemental doses of the drugs are given. Ervin and colleagues carefully evaluated the inactivation of gentamicin by carbenicillin or ticarcillin and have provided dosing modification guidelines to use clinically in patients with renal failure [27]. The importance of this question in relationship to the measurement of serum drug levels in clinical practice is well identified in the studies of Hull and Sarubbi [38], Noone and Pattison [67], Weibert and Keane [88], and Kaye and colleagues [46]. They have shown that when patients receive the combined therapy of gentamicin and carbenicillin, the measured serum aminoglycoside levels are repetitively lower than the pharmacokinetically predicted values.

Only a limited number of studies have assessed the influence of concomitant antibiotic therapy upon renal aminoglycoside kinetics. Barza and colleagues [5] and Dellinger et al. [25] evaluated the influence of several different cephalosporins upon the uptake of gentamicin, tobramycin, netilmicin, or amikacin into the renal cortex of a Fischer rat model. Significant reductions in renal cortical aminoglycoside concentrations were noted during the combined therapy. Additionally, a corresponding reduction in the nephrotoxicity of the animals was noted when compared to control rats who received the aminoglycoside alone. The doses of cephalosporin used in these studies were substantially higher than would be used in clinical practice, and the obligated osmotic solute effect of the cephalosporin load upon the renal cortical aminoglycoside concentrations may in fact have played some role in the protection that combination therapy produced in these studies. Direct interaction of the aminoglycoside and cephalosporin in the tissues may also have influenced the toxicologic outcome of the investigations. In our experience, when we have acutely administered cephalothin to healthy dogs at therapeutic dosing levels, we did not find a reduction of cortical aminoglycoside levels when compared to control observations [96].

The clinical relevance of the nephrotoxicity as seen in combined aminoglycoside and cephalosporin therapy is of such importance that it is reviewed separately in Chapter 16.

We have already noted in this chapter that the nephrotoxic potential of an aminoglycoside is significantly enhanced by concomitant diuretic therapy. Mitchell and associates have recently reported that, based upon their studies of aminoglycoside kinetics in an isolated perfused rat kidney model, they noted a net secretion of potassium [64]. They suggest that diuretics which promote potassium loss might predictably increase the nephrotoxic effects of gentamicin. The report of a series of patients by Butkus and colleagues wherein clindamycin in combination with gentamicin caused acute renal failure is of interest [15]. However, the three clinical cases reported by these authors do not persuasively suggest a real synergistic nephrotoxic role for a gentamicin-clindamycin combination.

Using an in vitro preparation of proximal tubular brush border membrane, Just and Habermann reported that the binding of the basic peptide aprotinin to brush border protein could be displaced in decreasing order of magnitude by neomycin > tobramycin > gentamicin > kanamycin > streptomycin [42]. This sequence appears to relate to the number of free amino groups in the molecule of the antibiotic. This interaction of aprotinin and aminoglycoside is of particular interest in relationship to the mechanisms of uptake of these drugs through the apical border of the proximal tubular cells. An interesting sequitur to these studies are the investigations of Pastoriza-Munos et al. [69]. They have been able to demonstrate, in a rat renal cortical slice preparation, that aminoglycosides exhibit competition between one another for the same receptor sites or transport systems in the proximal tubular cell. Their data indicate that the renal cortical accumulation of gentamicin was inhibited in a dose-dependent manner by netilmicin and tobramycin. However, from the quantitative point of view, these authors noted that in their studies the affinity of gentamicin for the receptor or transport system was greater than that of tobramycin or netilmicin. This brush border binding affinity sequence is different from the cascade noted by Just and Haberman [42]. It is of interest that all published in vivo studies have shown that the total cortical homogenate concentration of gentamicin is always higher than that of tobramycin when these drugs are given on an equal weight-for-weight basis. Therefore, the conflicting data on one hand, brush border aminoglycoside binding affinity, and on the other, cortical slice uptake, pose the possibility that the quantitative kinetics of brush border interaction of the aminoglycosides is different to their transmembrane transport and retention within the proximal tubular cells; more studies are needed to clarify this issue.

It might finally be noted in this segment that aminoglycosides as a class will enhance the nephrotoxic potential of any known renal toxin.

VI. Tubular Transport of the Aminoglycosides

For many years it was not recognized that renal tubular transport systems might be involved in the renal excretory kinetics of the aminoglycosides. However, as numerous investigations began to indicate that the renal cortical accumulation and retention of the aminoglycosides represented a reproducible, oxygen-consuming, pharmacokinetic phenomenon, it became necessary to reassess the entire question of the role of renal tubular transport systems in the renal handling of aminoglycosides [18,37,41,47,92,101].

As previously noted in this chapter, all available autoradiographic data indicate that the cellular accumulation of aminoglycosides within the kidney occurs only in proximal tubular cells [22,42,52,81]. These studies also demonstrate the importance of endocytotic or pinocytotic uptake of aminoglycosides from the luminal brush border of the proximal tubular cells. However, autoradiographic data cannot quantify the fractional cellular uptake of aminoglycosides into the renal cortex.

The proximal tubule contains four separate transport systems which might modulate renal cortical aminoglycoside kinetics. These transport systems are represented diagrammatically in Figure 11. It has already been indicated that the cellular uptake mechanism(s) for the aminoglycosides is an active, oxygen-dependent phenomenon [20,37,47]. Separate secretory systems exist for organic acids (organic anionic transport) and organic bases (organic cationic transport), while separate reabsorbtive systems exist for amino acids and glucose. A brief sequential review of the available data pertinent to the interaction of each of these transport systems with the aminoglycosides will be presented. Initially, it should be clarified that there is no evidence to suggest a direct involvement of aminoglycoside transport by any of the classic electrolyte (NA^+; K^+; CA^{2+}; or Mg^{2+}) transport systems in the nephron.

Cellular uptake of the aminoglycosides in the proximal tubule could potentially take place by active or passive uptake at the antiluminal border (basolateral cell boundary) or at the luminal brush border (apical cell membrane) of these cells. The available data now indicate that reabsorption of filtered drug from the lumen of the tubule is the dominant mechanism by which the aminoglycosides enter the proximal tubular cells. Chiu and colleagues suggested from their studies of renal gentamicin extraction in the dog that as much as 20% of filter drug was reabsorbed by the proximal tubule [18]. However, their data did not take into consideration the controversial issue of the binding of gentamicin to serum proteins, if the latter exists, since the latter protein binding would reduce the calculated figure of gentamicin tubular reabsorption. More recently, Pastoriza-Munos et al. have calculated that in a rat model the fractional reabsorption of gentamicin could not have been greater than 5.4–7.9% in their studies [69]; a smaller contribution of passive or active drug uptake from the basalateral cell membrane has not been ruled out.

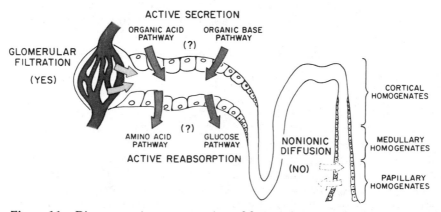

Figure 11   Diagrammatic representation of four active transport systems located within the renal proximal tubule which might modulate total renal cortical concentrations of aminoglycosides.

A. Organic Anionic (Organic Acid) Secretory Systems

Since the aminoglycosides are chemically organic bases in nature, it is not surprising that there is no evidence to suggest involvement of the anionic secretory systems in the renal excretion of aminoglycosides. Some years ago, Bergan and colleagues demonstrated that the concomitant administration of probenecid did not influence the renal excretory kinetics of gentamicin [13,14]. Tissue concentration data were not evaluated in the latter studies. Similarly, in our laboratories we have identified that at therapeutic dosing levels the coadministration of cephalothin, which is transported via the anionic secretory system, did not influence renal cortical levels of tobramycin or the renal clearance of the latter aminoglycosides [96]. In in vitro effects of the aminoglycosides upon anionic transport of p-aminohippuric acid are separately reviewed in Chapter 10. In summary, there are no human or in vitro animal data which indicate a role for anionic secretion in the renal elimination of aminoglycosides.

B. Organic Cationic (Organic Base) Secretory Systems

Once again the data pertinent to this question are scant. There was initially some preliminary but inconclusive evidence generated in our laboratories which demonstrated that in a limited number of animal investigations it was possible to produce a reduction in renal cortical gentamicin concentration by the coadministration of quinine, a competitive blocker of the cationic secretory system [101]. Subsequently, this could not be reproduced by us for tobramycin when we used a substantially larger series of animals. Although the renal cortical tobramycin concentrations were reduced during the concomitant administration of quinine, the reduction was not significant [96]. Jerauld and Silverblatt have reported rat studies in which they blocked organic cationic secretion by the administration of N'-methylnicotinamide. The latter substrate reduced renal clearance by 25%. These authors concluded that the organic base transport system does not significantly modulate renal cortical gentamicin kinetics [41]. However, in a rat model Fabre and associates have recently found that quinine administration increased renal cortical gentamicin concentrations [30]. These conflicting results in vivo may represent evidence for the activity of a dose-modulated, unidirectional or bidirectional cationic transport of the aminoglycosides. In summary, there are no human data and only minimal in vivo animal data which suggest an important modulating role for cationic secretion in the overall regulation of renal cortical aminoglycoside uptake. The available data support the possibility that cationic secretion may play a minor role in modulating renal cortical aminoglycoside levels.

C. Glucose Transport in the Proximal Tubule

Since the structural configuration of all aminoglycosides contains a number of six-carbon sugars, it is attractive to hypothesize that the proximal tubular glucose reabsorptive transport system might modulate renal cortical uptake of the aminoglycosides. Recent data from our laboratories have indicated that in a healthy canine model saturation of glucose reabsorption in the proximal tubule did not reproducibly influence the uptake of gentamicin into the renal cortex. These data therefore suggest a nonsignificant role for the glucose transport systems in the uptake of aminoglycosides into renal proximal tubular cells.

D. Amino Acid Transport in the Proximal Tubule

Data defining the interaction of proximal tubular amino acid reabsorption and aminogly-coside renal cortical uptake have only recently become available [96]. Because of the presence of unsubstituted amino side chains in the molecular configuration of all amino-glycosides, it could be anticipated that, following glomerular filtration, aminoglycosides might utilize the same proximal tubular transport systems that reabsorb filtered amino acids as the mechanism to move these antibiotics into the interior of proximal tubular cells. To test this hypothesis we infused a mixture of commercially available amino acids, as used clinically for parenteral nutrition purposes, into a healthy dog model (Freamine II, obtained from the McGaw Laboratories, Glendale, California). We noted that when saturation of proximal tubular reabsorption of amino acid transport was manifest by the appearance of amino acids in the urine, there was a significant ($P < 0.001$) reduction in the renal cortical levels of tobramycin [96].

We elected to further pursue this amino acid effect upon aminoglycoside uptake into the renal cortex since in clinical practice many of the patients who typically require hyperalimentation therapy—such as postoperative patients, nutritionally deficient patients, or the geriatric age group—are also the patients who need aminoglycoside therapy. Recognizing that the reabsorption of amino acids by the renal proximal tubule is a multi-transport and cotransport phenomenon [79], we evaluated separately the three predominant amino acid transport limbs (i.e., "base," "acid," and "neutral" amino acid transport systems). In a healthy canine model we administered L-lysine as a means of saturating base amino acid transport, L-glutamic acid in the form of sodium glutamate to saturate acid amino acid transport, and L-glycine to saturate the neutral amino acid transport limb. D-Lysine was also incorporated in our studies so that the effects of this dextro-rotary isomer of lysine, which is characteristically not transported by the proximal tubule, could be assessed [23].

The results of these studies have demonstrated that L-lysine, which can saturate and block the reabsorption of basically charged amino acids, significantly reduced the renal cortical concentrations of gentamicin ($P < 0.001$; two- to fourfold reduction) and tobramycin ($P < 0.001$; two- to threefold reduction). On the other hand, saturation of either the acid or neutral amino acid pathways of renal tubular transport, as produced in these studies, did not concomitantly influence renal cortical gentamicin levels. Since this compound is only minimally transported, if at all, by the renal proximal tubule, it suggests that the mechanism which prevents uptake of aminoglycoside into proximal tubular cells is possibly something other than saturation of a classic cellular transport system. In summary, the results of these amino acid infusion studies suggest that the "base" amino acid effect upon aminoglycoside uptake into the renal cortex is initially mediated by either (1) alteration of the proximal tubular brush border membrane charge, (2) competition for aminoglycoside receptor sites on the brush border membrane, (3) change in membrane protein binding coefficients of the aminoglycosides, (4) inhibition of brush border pino-cytotic uptake of aminoglycosides, or (5) the intraluminal formation of a nonreabsorbable complex of amino acids and aminoglycosides.

All these possibilities in concert are likely to relate to the stereochemistry of basically charged amino acids and aminoglycosides. The common primary step in amino acid and aminoglycoside transport may be the linking of a free positively charged amino group, such as is present in either L-lysine or D-lysine, with a comparable negatively charged site on the brush border membrane. Free positively charged amino groups are also present in the molecular configuration of all aminoglycosides, and it would appear from our studies that the basically charged amino acids are more avid competitors than aminoglycosides for brush border binding sites and hence transport into proximal tubular cells.

The data of Mogensen and Solling [65] are of interest in this context since they have demonstrated that the administration of several different basically charged amino acids, including L-lysine, can inhibit proximal tubular protein reabsorption. They hypothesized that the initial event in the normal reabsorption of protein in the proximal tubule is a binding between a free positive amino or guanidine group in the protein molecule and a negative site on the tubular cell surface. Just and Habermann have also demonstrated that polylysine administration will impair proximal tubular protein reabsorption [42], and as we have previously noted in this chapter, autoradiographic studies suggest a major role for pinocytosis as the mechanism by which aminoglycosides are transported into proximal tubular cells [22,81]. Our studies would suggest that before an aminoglycoside can be incorporated into an apical brush border endocytotic vacuole, a charge-mediated binding between the brush border membrane and the aminoglycoside is the initial cellular uptake or transport phenomenon.

No data are available to support the possibility that the reduction in renal cortical uptake of aminoglycosides results from the production of a nonreabsorbable complex of basically charged amino acids and aminoglycosides within the tubular lumen. In fact, this phenomenon would have been more likely to occur with the acidic amino acids. However, in these investigations L-glutamic acid did not exert a measurable influence upon the renal cortical uptake of tobramycin.

In terms of clinical relevance the amino acid studies reviewed here may bear some relationship to the clinical use of hyperalimentation solutions since the patient population wherein amino acid mixtures are administered for nutritional purposes are also the patients who frequently need aminoglycoside antibiotic therapy. The first implication is that further studies will be necessary to define the influence of a 7–14 day constant infusion of amino acid(s) upon the renal functional activity of an investigative model of aminoglycoside nephrotoxicity. The second implication is to evaluate the systemic effect of amino acid administration, if any, upon in vivo response of gram-negative pathogens to aminoglycoside administration or the aminoglycoside uptake of such bacteria in an in vivo setting.

The available transport data relative to renal cortical aminoglycoside kinetics are summarized in Table 1.

Table 1   Renal Tubular Transport of Aminoglycosides

Transport mechanism	Influence
Anion (organic acid) system	No
Cation (organic base) system	Questionable
Amino acid system	Yes
"Acid" amino acid system	No
"Base" amino acid system	Yes
"Neutral" amino acid system	No
Glucose system	No
Pinocytotic proximal uptake	Yes
Nonionic distal diffusion	No

VII. Prolonged Renal Cortical Aminoglycoside Retention

An important observation first clearly identified by Luft and Kleit in a rat model [55] was the indication that following the cessation of aminoglycoside administration, one could still identify substantial renal cortical drug accumulation several days to weeks following discontinuation of the compound. In their early studies, they demonstrated that the half-life of gentamicin in the healthy rat cortex was 109 hr, by contrast with a serum half-life of 0.5 hr. This phenomenon has also been confirmed for human patients by measurement of persistent drug concentrations in the urine [43,44,76–78] and by direct measurement of tissue aminoglycoside concentrations [26,59,63]. The pharmacokinetic data pertinent to this phenomenon are separately reviewed in Chapter 6.

In a canine model we have calculated a renal cortical $T_{1/2}$ for gentamicin and tobramycin, as presented in Figure 12, measured following 7 days of administration of 1.5 mg/kg of body weight of each drug every 12 hr. One of the most striking findings in these studies was the extent to which the drugs accumulate within the renal cortex and the remarkable duration of time they remained within the latter location. In the case of gentamicin, mean cortical concentrations of the drug 24 hr postdosing were 285 μg/g, whereas for tobramycin they were 70 μg/g.

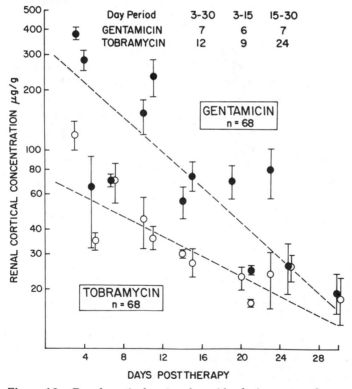

Figure 12   Renal cortical aminoglycoside elution curves for gentamicin and tobramycin as seen following 7 days of treatment of healthy dogs with 3 mg/kg of body weight of either compound. The "day period" refers to calculation of the renal cortical aminoglycoside $T_{1/2}$ as noted during the intervals days 3–30, 3–15, or 15–30 following drug administration.

The calculated cortical $T_{1/2}$ based upon all data points for gentamicin was 7 days, whereas the cortical $T_{1/2}$ for tobramycin was 12 days. The scatter of the tissue drug concentration points from 2 to 4 weeks post-therapy suggested strongly that if further tissue drug concentrations were measured during the second and third months after treatment, it might indicate a later phase of drug elution with a much longer tissue $T_{1/2}$.

The important clinical correlation in these studies relates to the issue of repetitive courses of aminoglycoside therapy. Clinical experience has shown that, when a patient receives a second course of aminoglycoside therapy within several days or a few weeks following the first course of treatment, nephrotoxicity may become manifest within a few days of the repeat course of aminoglycoside treatment. From the tissue drug point of view, Figure 12 provides a plausible answer to this particular clinical observation. The data in Figure 12 indicate that 2 or 3 weeks post-treatment, a sizable residual quantity of drug remains within the renal cortex. If additional therapy were given at such a time interval, the renal proximal tubular cells would be "preloaded" with drug, and hence this may well comprise the additional toxicologic risk factor found in patients requiring repeat aminoglycoside courses of therapy. Further studies will be necessary to determine at what point following aminoglycoside treatment the risk factor associated with pretreatment becomes negligible.

VIII. Influence of Preexisting Parenchymal Disease upon the Intrarenal Concentration Characteristics of Aminoglycosides

The presence of severe disease in the renal tissue has a profound influence upon both the renal parenchymal and urinary concentrations of virtually all antibiotics [8,94-98,101]. A number of important therapeutic implications derive from the finding that severe disease in the kidney reduces tissue aminoglycoside concentrations in all zones of the kidney, and indeed in the urine produced by such kidneys. Bennett and colleagues have evaluated the clinical implications of this question and have reported that in their experience poor antibacterial response in the renal tract of patients with significant renal impairment was in part due to low renal tissue and urinary antibiotic levels in such patients [8].

In our laboratories we have quantified the changes produced in tissue aminoglycoside levels secondary to the presence of severe renal disease. This has been done by utilizing the circumstance of patients who were on chronic dialysis and who were undergoing bilateral nephrectomy, in preparation for renal transplantation, as a source of severely diseased human kidneys. Although the drug concentration data achieved from investigations performed in such tissues reveal what happens at one end of the spectrum of renal disease, it is nonetheless a level of renal function often encountered in patients several months prior to the initiation of chronic dialysis and frequently a circumstance complicated by indolent bacterial infections.

In the presence of severe renal disease, the selection of antibiotics must of necessity be reduced and carefully monitored. For such reasons, a limited number of antibiotics tend to be used, but the use of aminoglycosides is often mandatory in renal patients with sepsis of undetermined etiology. Frequently, the renal tract is the focal origin of such bacterial infection. The clinical dosing considerations for systemic aminoglycoside use in renal impairment are reviewed in Chapter 20.

Table 2 identifies the tissue concentration levels of gentamicin, tobramycin, and kanamycin which can be produced in severely diseased kidney tissue. The table also contains the results of similar studies which assessed the influence of parenchymal disease

Table 2   Renal Tissue Antibiotic Concentrations in Health and Disease

	Healthy kidney[a] (serum/tissue)	Severely diseased kidney[a] (serum/tissue)	Urine from diseased kidney (serum/urine)
Tobramycin	1:5	1:1.2	1:8 ± 4[b]
Gentamicin	1:5	1:0.8	1:3 ± 1
Kanamycin	1:6	1:0.5	–
Carbenicillin	1:12	1:0.7	1:11 ± 2
Ampicillin	1:4	1:0.6	1:4 ± 0.8
Cephalothin	1:4	1:0.6	1:1 ± 0.2
Doxycycline	1:2	1:2	1:0.6 ± 0.08

[a]Mean value of medulla and papilla.
[b]Mean ± SE.

upon carbenicillin, ampicillin, cephalothin, and doxycycline. The details of these studies are recorded elsewhere [95,97,98,101], but it should be noted here that the patients received the usual parenteral therapeutic doses of the antibiotics listed in the table for one or more administrations prior to nephrectomy. The results are presented as tissue/serum ratios. Also, the results from healthy canine kidneys are compared with the data from the diseased human renal tract.

For each aminoglycoside it is clear that marked reductions in renal parenchymal drug concentration take place in all locations within the kidney. All zones tend to be equal in concentration and they approximate the concomitant serum values. It would therefore appear that serum antibiotic levels represent the best guide to coexisting drug levels in diseased renal tissue. For complicated renal tract infections, it may be clinically advantageous to correlate the information in Table 1 and the aminoglycoside minimal inhibitory concentration sensitivity profile of the infecting pathogen. The analysis of antibiotic concentrations in diseased tissue homogenates cannot be said with complete accuracy to represent the antibiotic concentration achievable at the "critical site" in terms of antibacterial efficacy, but the total parenchymal concentration data do suggest working guidelines of antibiotic concentration achievable in severely diseased renal parenchyma. It should be cautioned that only minimal clinical data are available to correlate these drug concentration results in severely diseased human kidneys with either short- or long-term therapeutic results.

It will be recognized from the foregoing that the available data concerning aminoglycoside concentrations found in diseased human kidneys versus healthy animal tissues represent two ends of a spectrum of renal function. In that context, it is of interest to note that in a canine model of pyelonephritis a full 80% reduction of renal function, as measured by glomerular filtration rate, was required before any change could be detected in the usual corticopapillary distribution pattern of carbenicillin [92]. This finding suggests that substantial parenchymal damage must be present before it significantly influences the tissue pharmacokinetics of an antibiotic.

In clinical practice it has been popularly suggested that patients with severe preexisting renal disease are particularly prone to the development of aminoglycoside nephrotoxicity. However, there are no clinical data to support the latter assumption. In fact, it is in patients with normal renal function at the time of initiation of aminoglycoside therapy that one sees the greatest renal cortical accumulation of the drugs and the development of delayed onset nephrotoxicity. One can only speculate that if it were possible to eliminate all other risk factors and compare 14 days of gentamicin therapy in a series of individuals with initially normal renal function versus a group of patients with severe chronic renal impairment, provided that the serum concentrations of gentamicin were maintained within the usual therapeutic range, the patients with normal renal function would be more likely to manifest drug-related nephrotoxicity. Clearly, patients with preexisting renal disease are prone to systemic gentamicin accumulation and its attendant problems of ototoxicity, neurotoxicity, and indeed dose-related nephrotoxicity. The latter circumstance is probably what has given rise to the clinical belief that preexisting renal disease enhances aminoglycoside-induced nephrotoxicity.

IX. Conclusions

In conclusion, the study of intrarenal aminoglycoside distribution in health and disease provides information of therapeutic and toxicologic value. Remarkable accumulation of aminoglycosides takes place in the renal cortex. This is the anatomic location of the proximal tubule and it is also the only site of morphologic cellular damage in aminoglycoside-induced nephrotoxicity. Manipulation of the state of hydration can significantly influence the nephrotoxic potential of an aminoglycoside. As a predictor of nephrotoxicity, renal cortical aminoglycoside kinetics are controversial. Nonetheless, with certain analytic constraints the data provide useful information when combined with the results of renal functional and morphologic changes as seen following multiple-dosing studies. Uptake of aminoglycosides into the renal cortex is predominantly mediated by the proximal tubular reabsorption of filtered drug. The initial steps in this process involve binding of the aminoglycosides to amino acid receptor sites, probably followed by pinocytotic uptake into the cell. Once the drugs are in the cell, long-range retention is characteristic, with a tissue half-life of several days. The presence of severe preexisting disease in the renal tissues markedly reduces the parenchymal and urinary concentrations of aminoglycosides. This may account in part for the clinical difficulty encountered in the eradication of bacterial pathogens from a diseased renal tract.

Acknowledgment

Supported in part by United States Public Health Service Clinical Research Centers Grant RR35.

References

1. Adelman, R.D., Spangler, W.L., Beasom, F., Ishizaki, G., and Conzelman, G.M.: Furosemide enhancement of experimental gentamicin nephrotoxicity: a comparison of functional and morphological changes with urinary enzyme activities. *J. Infect. Dis.* 140:342–352, 1979.
2. Alfthan, O., Renkonon, O.V., and Sivonen, A.: Concentration of gentamicin in serum, urine and urogenital tissue in man. *Acta Pathol. Microbiol. Scand.* 81(Suppl. 241): 92–94, 1973.

3. Andriole, V.T.: Water, acidosis and experimental pyelonephritis. *J. Clin. Invest.* 49:21–30, 1970.

4. Andriole, V.T., and Epstein, F.H.: Prevention of pyelonephritis by water diuresis: evidence of the role of medullary hypertonicity in promoting renal infection. *J. Clin. Invest.* 44:73–79, 1965.

5. Barza, M., Pinn, V., Tanguay, P., and Murray, T.: Nephrotoxicity of newer cephalosporins and aminoglycosides alone and in combination in a rat model. *J. Antimicrob. Chemother.* 4(Suppl. A):59–68, 1978.

6. Baylis, C., Rennke, H.R., and Brenner, B.M.: Mechanisms of the defect in glomerular ultrafiltration associated with gentamicin administration. *Kidney Int.* 12:344–353, 1977.

7. Beeson, P.B., and Rowley, D.: The anticomplementary effect of kidney tissue: its association with ammonia production. *J. Exp. Med.* 110:685–697, 1959.

8. Bennett, W.M., Hartnett, M.N., Craven, R., Gilbert, D.N., and Porter, G.A.: Gentamicin concentrations in blood, urine, and renal tissue of patients with end-stage renal disease. *J. Lab. Clin. Med.* 90:389–393, 1977.

9. Bennett, W.M., Hartnett, M.N., Gilbert, D., Houghton, D., and Porter, G.A.: Effect of sodium intake on gentamicin nephrotoxicity in the rat. *Proc. Soc. Exp. Biol. Med.* 151:736–738, 1976.

10. Bennett, W.M., Plamp, C.E., Gilbert, D.N., Parker, R.A., and Porter, G.A.: The influence of dosage regimen on experimental gentamicin nephrotoxicity: dissociation of peak serum levels from renal failure. *J. Infect. Dis.* 140:576–580, 1979.

11. Bennett, W.M., Reger, K., Plamp, C., Houghton, D., Porter, G.A., and Gilbert, D.N.: Reversal of gentamicin nephrotoxicity in the rat despite continued drug administration. *Clin. Res.* 26:390A, 1978.

12. Benvineste, R., and Davies, J.: Structural activity relationships among the aminoglycoside antibiotics: role of hydroxyl and amino groups. *Antimicrob. Agents Chemother.* 4:402–409, 1973.

13. Bergan, T., Brodwall, E.K., Westlie, L., and Oyri, A.: Renal excretion of gentamicin and effect of probenecid. *Acta Pathol. Microbiol. Scand.* 81(Suppl. 241):95–98, 1973.

14. Bergan, T., Westlie, L., and Brodwall, E.K.: Influence of probenecid on gentamicin pharmacokinetics. *Acta Med. Scand.* 191:221–224, 1972.

15. Butkus, D.E., deTorrente, A., and Terman, D.S.: Renal failure following gentamicin in combination with clindamycin. *Nephron* 17:307–313, 1976.

16. Chisholm, G.D.: Distribution and dosage of antibacterial agents in patients with normal and impaired renal function. *Med. J. Aust.* 1(Suppl.):25–29, 1970.

17. Chisholm, G.D., Calnan, J.S., Waterworth, P.M., and Reis, N.D.: Distribution of gentamicin in body fluids. *Br. Med. J.* 2:22–24, 1968.

18. Chiu, P.J.S., Brown, A., Miller, G., and Long, J.F.: Renal extraction of gentamicin in anesthetized dogs. *Antimicrob. Agents Chemother.* 10:277–282, 1976.

19. Chiu, P.J.S., and Long, J.F.: Effect of hydration on gentamicin excretion and renal accumulation in furosemide treated rats. *Antimicrob. Agents Chemother.* 14:214–217, 1978.

20. Chiu, P.J.S., and Long, J.F.: Urinary excretion and tissue accumulation of gentamicin and para aminohippurate in postischemic rat kidneys. *Kidney Int.* 15:618–623, 1979.

21. Chiu, P.J.S., Miller, G.H., Long, J.F., and Waitz, J.A.: Renal uptake and nephrotoxicity of gentamicin during urinary alkalinization in rats. *Clin. Exp. Pharmacol. Physiol.* 6:317–326, 1979.

22. Collier, V.U., Lietman, P.S., and Mitch, W.E.: Evidence for luminal uptake of gentamicin in the perfused rat kidney. *J. Pharmacol. Exp. Ther.* 210:247–251, 1979.

23. Crampton, R.F., and Smyth, D.H.: The excretion of the enantiomorphs of amino acids. *J. Physiol.* 122:1–10, 1953.

24. Cuppage, F.E., Setter, K., Sullivan, L.P., Reitzes, E.J., and Melnykovych, A.O.: Gentamicin nephrotoxicity. II. Physiological, biochemical and morphological effects of prolonged administration to rats. *Virchows Archiv. [Cell Pathol.]* 24:121–138, 1977.

25. Dellinger, P., Murphy, T., Barza, M., Pinn, V., and Weinstein, L.: Effects of cephalothin on renal cortical concentrations of gentamicin in rats. *Antimicrob. Agents Chemother.* 9:587–588, 1976.

26. Edwards, C.Q., Smith, C.R., Baughman, K.L., Rogers, J.F., and Leitman, P.S.: Concentrations of gentamicin and amikacin in human kidneys. *Antimicrob. Agents Chemother.* 9:925–927, 1976.

27. Ervin, F.R., Bullock, W.E., and Nuttall, C.E.: Inactivation of gentamicin by penicillins in patients with renal failure. *Antimicrob. Agents Chemother.* 9:1004–1011, 1976.

28. Evan, A.P., Huser, J., Avasthi, P.S., Rankin, L.I., and Luft, F.C.: Gentamicin induced glomerular injury. Abstr. 933 in *Proceedings of the 11th International Congress of Chemotherapy.* Washington, D.C., American Society for Microbiology, 1979.

29. Fabre, J., Fillastre, J.-P., Morin, J.-P., and Rudhardt, M.: Nephrotoxicity of gentamicin. *Contrib. Nephrol.* 10:53–62, 1978.

30. Fabre, J., Rudhardt, M., Blanchard, P., and Chauvin, P.: Kinetics of gentamicin and sisomicin in renal cortex and medulla of rats compared with organs: influence of quinine. In *Nephrotoxicity: Interaction of Drugs with Membrane Systems Mitochondria-lysosomes,* edited by Fillastre, J.-P. New York, Masson, 1978, pp. 83–94.

31. Fabre, J., Rudhardt, M., Blanchard, P., and Regamey, C.: Persistence of sisomicin and gentamicin in the renal cortex and medulla compared with other organs and serum in rats. *Kidney Int.* 10:444–449, 1976.

32. Gilbert, D.N., Houghton, D.C., Bennett, W.M., Plamp., C.E., Reger, K., and Porter, G.A.: Reversibility of gentamicin nephrotoxicity in rats: recovery during continuous drug administration. *Proc. Soc. Exp. Biol. Med.* 160:99–103, 1979.

33. Gilbert, D.N., Plamp, C., Starr, P., Bennett, W.M., Houghton, D.C., and Porter, G.: Comparative nephrotoxicity of gentamicin and tobramycin in rats. *Antimicrob. Agents Chemother.* 13:34–40, 1978.

34. Ginsburg, D.S., Quintanilla, A.P., and Levin, M.: Renal glycosuria due to gentamicin in rabbits. *J. Infect. Dis.* 134:119–122, 1976.

35. Hansten, P.D.: *Drug Interactions* (3rd ed.). Philadelphia, Lea & Febiger, 1975.

36. Hsu, C.H., Kurtz, T.W., Easterling, R.E., and Weller, J.M.: Potentiation of gentamicin nephrotoxicity by metabolic acidosis. *Proc. Soc. Exp. Biol. Med.* 146:894–897, 1974.

37. Hsu, C.H., Kurtz, T.W., and Weller, J.M.: In vitro uptake of gentamicin by rat renal cortical tissue. *Antimicrob. Agents Chemother.* 12:192–194, 1977.

38. Hull, J.H., and Sarubbi, F.A., Jr.: Gentamicin serum concentrations: pharmacokinetic predictions. *Ann. Intern. Med.* 85:183–189, 1976.

39. Jackson, G.G.: Gentamicin. *Practitioner* 198:855–866, 1967.

40. Jamison, R.L., and Robertson, C.R.: Recent formulations of the urinary concentrating mechanism: a status report. *Kidney Int.* 16:537–545, 1979.

41. Jerauld, R., and Silverblatt, F.J.: Effect of N'-methylnicotinamide on the renal accumulation and reabsorption of gentamicin in rats. *Antimicrob. Agents Chemother.* 13:893–894, 1978.

42. Just, M., and Habermann, E.: The renal handling of polybasic drugs: Parts 1 and 2. *Naunyn Schmiedebergs Arch. Pharmacol.* 300:57–66; 67–76, 1977.

43. Kahlmeter, G., Jonsson, S., and Kamme, C.: Long standing post-therapeutic gentamicin serum and urine concentrations in patients with unimpaired renal function: a pharmacokinetic evaluation. *J. Antimicrob. Chemother.* 4:143–152, 1978.

44. Kahlmeter, G., and Kamme, C.: Prolonged excretion of gentamicin in a patient with unimpaired renal function. *Lancet* 1:286, 1975.

45. Kass, E.H.: An approach to the management of resistant urinary infections. *Kidney Int.* 16:204–212, 1979.

46. Kaye, D., Levison, M.E., and Labovitz, E.D.: The unpredictability of serum concentrations of gentamicin: pharmacokinetics of gentamicin in patients with normal and abnormal renal function. *J. Infect. Dis.* 130:150–154, 1974.

47. Kluwe, W.M., and Hook, J.B.: Analysis of gentamicin uptake by rat renal cortical slices. *Toxicol. Appl. Pharmacol.* 45:531–539, 1978.

48. Kokko, J.P., and Rector, F.C., Jr.: Countercurrent multiplication system without active transport in inner medulla: new model. *Kidney Int.* 2:214–223, 1972.

49. Kokko, J.P., and Tisher, C.C.: Water movement across nephron segments involved with the countercurrent multiplication system. *Kidney Int.* 10:64–81, 1976.

50. Kornguth, M.L., and Kunin, C.M.: Distribution of gentamicin and amikacin in rabbit tissues. *Antimicrob. Agents Chemother.* 11:974–977, 1977.

51. Kosek, J.C., Mazze, R.I., and Cousins, M.J.: Nephrotoxicity of gentamicin. *Lab. Invest.* 30:48–57, 1974.

52. Kuhar, M.J., Mak, L.L., and Lietman, P.S.: Autoradiographic localization of ^{3}H gentamicin in the proximal renal tubules of mice. *Antimicrob. Agents Chemother.* 15:131–133, 1979.

53. Kunin, C.M.: Binding of antibiotics to tissue homogenates. *J. Infect. Dis.* 121:55–64, 1970.

54. Luft, F.C., Bloch, R., Sloan, R.S., Yum, M.N., Costello, R., and Maxwell, D.R.: Comparative nephrotoxicity of aminoglycoside antibiotics in rats. *J. Infect. Dis.* 138:541–545, 1978.

55. Luft, F.C., and Kleit, S.A.: Renal parenchymal accumulation of aminoglycoside antibiotics in rats. *J. Infect. Dis.* 130:656–659, 1974.

56. Luft, F.C., Patel, V., Yum, M.N., Patel, B., and Kleit, S.A.: Experimental aminoglycoside nephrotoxicity. *J. Lab. Clin. Med.* 86:213–220, 1975.

57. Luft, F.C., Rankin, L.I., Sloan, R.S., and Yum, M.N.: Recovery from aminoglycoside nephrotoxicity with continued drug administration. *Antimicrob. Agents Chemother.* 14:284–287, 1978.

58. Luft, F.C., Yum, M.N., and Kleit, S.A.: Comparative nephrotoxicities of netilmicin and gentamicin in rats. *Antimicrob. Agents Chemother.* 10:845–849, 1976.

59. Luft, F.C., Yum, M.N., Walker, P.D., and Kleit, S.A.: Gentamicin gradient patterns and morphological changes in human kidneys. *Nephron* 18:167–174, 1977.

60. Mahon, W.A., Ezer, J.I., and Inaba, T.: Renal binding of aminoglycosides. In *Assessment of Aminoglycoside Toxicity* (Symposium Proceedings, Burgenstock, Switzerland, September 24–25, 1977). Indianapolis, Ind., Eli Lilly, 1978, pp. 14–18.

61. Marre, R., Beck, H., Zullich, B., Husstedt, W., and Freiesleben, H.: Animal studies on accumulation of 11 aminoglycosides in renal tissue. In Proceedings of the 10th International Congress of Chemotherapy, Zurich, 1977. *Curr. Chemother.* 2:955–957, 1978.

62. Milne, M.D., Scribner, B.H., and Crawford, M.A.: Non-ionic diffusion and the excretion of weak acids and bases. *Am. J. Med.* 24:709–729, 1958.

63. Milner, R.D.G., Ross, J., Froud, D.J.R., and Davis, J.A.: Clinical pharmacology of gentamicin in the newborn infant. *Arch. Dis. Child.* 47:927–932, 1972.

64. Mitchell, C.J., Bullock, S., and Ross, B.D.: Renal handling of gentamicin and other antibiotics by the isolated perfused rat kidney: mechanism of nephrotoxicity. *J. Antimicrob. Chemother.* 3:593–600, 1977.

65. Mogensen, C.E., and Solling, K.: Studies on renal tubular protein reabsorption: partial and near complete inhibition by certain amino acids. *Scand. J. Clin. Lab. Invest.* 37:477–486, 1977.

66. Noel, P., and Levy, V.G.: Toxicité rénal de l'association gentamicine-furosémide. *Nouv. Presse Med.* 7:351–353, 1978.

67. Noone, P., and Pattison, J.R.: Therapeutic implications of interaction of gentamicin and penicillins. *Lancet* 2:575–578, 1971.

68. Nunning, B.C., and Granatek, A.P.: Physical compatibility and chemical stability of amikacin sulfate in combinations with non-antibiotic drugs in large-volume parenteral solutions: Parts I, II, III, and IV. *Curr. Ther. Res.* 20:352–358;379–368;369–416;417–491, 1976.

69. Pastoriza-Munoz, E., Bowman, R.L., and Kaloyanides, G.: Renal tubular transport of gentamicin in the rat. *Kidney Int.* 16:440–450, 1979.

70. Patel, J.A., and Phillips, G.: A guide to physical compatibility of intravenous drug mixtures. *Am. J. Hosp. Pharm.* 23:409–411, 1966.

71. Reiner, N.E., Bloxham, D.D., and Thompson, W.L.: Nephrotoxicity of gentamicin and tobramycin given once daily or continuously in dogs. *J. Antimicrob. Chemother.* 4(Suppl. A):85–101, 1978.

72. Riff, L.J., and Jackson, G.G.: Laboratory and clinical conditions for gentamicin inactivation by carbenicillin. *Arch. Intern. Med.* 130:887–891, 1972.

73. Riff, L.J., and Jackson, G.G.: Pharmacology of gentamicin in man. *J. Infect. Dis.* 124(Suppl.):S98–S105, 1971.

74. Rocha, H., and Fekety, F.R., Jr.: Acute inflammation in the renal cortex and medulla following thermal injury. *J. Exp. Med.* 119:131–138, 1964.

75. Rocha, H., Guze, L.B., Freedman, L.R., and Beeson, P.B.: Experimental pyelonephritis: III. The influence of localized injury in different parts of the kidney on susceptibility to bacillary infection. *Yale J. Biol. Med.* 30:341–354, 1958.

76. Schentag, J.J., Jusko, W.J., Plaut, M.E., Cumbo, T.J., Vance, J.W., and Abrutyn, E.: Tissue persistence of gentamicin in man. *JAMA* 238:327–329, 1977.

77. Schentag, J.J., Jusko, W.J., Vance, J.W., Cumbo, T.J., Abrutyn, E., DeLattre, M., and Gerbracht, L.M.: Gentamicin disposition and tissue accumulation on multiple dosing. *J. Pharmacokinet. Biopharm.* 5:559–577, 1977.

78. Schentag, J.J., Lasezkay, G., Cumbo, T.J., Plaut, M.E., and Jusko, W.J.: Accumulation pharmacokinetics of tobramycin. *Antimicrob. Agents Chemother.* 13:649–656, 1978.

79. Segal, S., and Thier, S.O.: Renal handling of amino acids. In *Handbook of Physiology, Section 8: Renal Physiology*, edited by Orloff, J., and Berliner, R.W. Washington, D.C., American Physiological Society, 1973, pp. 653–676.

80. Silverblatt, F.J.: Antibiotic nephrotoxicity: a review of pathogenesis and prevention. *Urol. Clin. North Am.* 2:557–567, 1975.

81. Silverblatt, F.J., and Kuehn, C.: Autoradiography of gentamicin uptake by the rat proximal tubule cell. *Kidney Int.* 15:335–345, 1979.

82. Smith, C.R., Maxwell, R.R., Edwards, C.Q., Rogers, J.F., and Lietman, P.S.: Nephrotoxicity induced by gentamicin and amikacin. *Johns Hopkins Med. J.* 142:85–90, 1978.

83. Stamey, T.A., Govan, D.E., and Palmer, J.M.: The location and treatment of urinary tract infections: the role of bactericidal urine levels as opposed to serum levels. *Medicine (Baltimore)* 44:1–36, 1975.

84. Trottier, S., Bergeron, M.-G., and Gauvreau, L.: Intrarenal concentration of netilmicin and gentamicin. In Proceedings of the 10th International Congress of Chemotherapy, Zurich, 1977. *Curr. Chemother.* 2:953–955, 1978.

85. Ullrich, K.J., Kramer, K., and Boylan, J.W.: Present knowledge of the countercurrent system in the mammalian kidney. *Prog. Cardiovasc. Dis.* 3:395–431, 1961.

86. Wahlig, H., Metallinos, A., Hameister, W., and Bergmann, R.: Gentamicin-konzentrationen in Geweben und Korperflussigkeiten von Versuchstieren. *Int. J. Clin. Pharmacol. Biopharm.* 10:212–229, 1974.

87. Waitz, J.A., Drube, C.G., Moss, E.L., Oden, E.M., Bailey, J.V., Wagman, G.H., and Weinstein, M.J.: Biologic aspects of the interaction between gentamicin and carbenicillin. *J. Antibiot. (Tokyo)* 25:219–225, 1972.

88. Weibert, R.T., and Keane, W.F.: Carbenicillin-gentamicin interaction in actue renal failure. *Am. J. Hosp. Pharm.* 34:1137–1139, 1977.

89. Weiner, I.M.: Transport of weak acids and bases. In *Handbook of Physiology, Section 8: Renal Physiology,* edited by Orloff, J., and Berliner, R.W. Washington, D.C., American Physiological Society, 1973, pp. 521–554.

90. Weiner, I.M., and Mudge, G.H.: Renal tubular mechanisms for excretion of organic acids and bases. *Am. J. Med.* 36:743–762, 1964.

91. Whelton, A.: Aminoglycoside renal cortical kinetics: a clue to mechanisms of nephrotoxicity. In *Antibiotics and Hospitals,* edited by Grassi, C., and Ostino, G. New York, Alan R. Liss, 1979, pp. 33–41.

92. Whelton, A.: Intrarenal antimicrobial distribution. Modulating factors, therapeutic and toxicologic implications. In *Nephrotoxicity: Interaction of Drugs with Membrane Systems Mitochondria-lysosomes,* edited by Fillaster, J.-P. New York, Masson, 1978, pp. 95–126.

93. Whelton, A.: Tetracyclines in renal insufficiency: resolution of a therapeutic dilemma. *Bull. N.Y. Acad. Med.* 54:233–236, 1978.

94. Whelton, A., Carter, G.G., Bryant, H.H., Fox, L., and Walker, W.G.: Therapeutic implications of gentamicin accumulation in severely diseased kidneys. *Arch. Intern. Med.* 136:172–176, 1976.

95. Whelton, A., Carter, G.G., Bryant, H.H., Porteous, L.A., and Walker, W.G.: Carbenicillin concentrations in normal and diseased human kidneys: a therapeutic consideration. *Ann. Intern. Med.* 78:659–662, 1973.

96. Whelton, A., Carter, G.G., Craig, T.J., Bryant, H.H., Herbst, D.V., and Walker, W.G.: Comparison of the intrarenal disposition of tobramycin and gentamicin: therapeutic and toxicologic answers. *J. Antimicrob. Chemother.* 4(Suppl. A):13–22, 1978.

97. Whelton, A., Nightingale, S.D., Carter, G.G., Gordon, L.S., Bryant, H.H., and Walker, W.G.: Pharmacokinetic characteristics of doxycycline accumulation in normal and severely diseased kidneys. *J. Infect. Dis.* 132:467–471, 1975.

98. Whelton, A., Sapir, D.G., Carter, G.G., Garth, M.A., and Walker, W.G.: Intrarenal distribution of ampicillin in the normal and diseased human kidney. *J. Infect. Dis.* 125:466–470, 1972.

99. Whelton, A., Sapir, D.G., Carter, G.G., Kramer, J., and Walker, W.G.: Intrarenal distribution of penicillin, cephalothin, ampicillin and oxytetracycline during varied states of hydration. *J. Pharmacol. Exp. Ther.* 179:419–428, 1971.

100. Whelton, A., and Walker, W.G.: An approach to the interpretation of drug concentrations in the kidney. *Johns Hopkins Med. J.* 142:8–14, 1978.

101. Whelton, A., and Walker, W.G.: Intrarenal antibiotic distribution in healthy and disease. Editorial. *Kidney Int.* 6:131–137, 1974.

102. Whelton, A., Walker, W.G., and Carter, G.G.: The influence of nonionic diffusion on renal concentration and excretion of antibiotics. *J. Clin. Invest.* 46:1132–1133, 1967.

103. Winningham, D.G., Nemoy, N.J., and Stamey, T.A.: Diffusion of antibiotics from plasma into prostatic fluid. *Nature* 219:138–143, 1968.

104. Woodruff, M.W., Malvin, R.L., and Thompson, I.M.: The renal transport of nitrofurantoin: effect of acid base balance upon its excretion. *JAMA* 175:1132–1135, 1961.

9

AUTORADIOGRAPHIC STUDIES OF INTRACELLULAR AMINOGLYCOSIDE DISPOSITION IN THE KIDNEY

FREDRIC J. SILVERBLATT   San Fernando Valley
Medical Program, University of California, Los Angeles,
California

When appropriate regard is given to potential artifacts, autoradiography can be a most powerful tool for identifying the distribution of a pharmacological compound. Coupled with electron microscopy, the technique has the unique advantage of permitting recognition of the presence of a labeled compound within individual cell organelles. Autoradiography has been used by three groups of investigators to study the intrarenal disposition of gentamicin [4,7,13]. The results have provided fresh insights into the mechanism of gentamicin transport and site of accumulation in the kidney as well as the possible pathogenesis of its nephotoxicity.

I. Technical Considerations

A. Principles of Autoradiography

Autoradiography is the process whereby the location of a source of ionizing radiation in sections of biological material is detected on a superimposed photographic emulsion. The interaction, between the ionized particles and silver halide crystals located in the film, produces a grain of metallic silver which, in the light microscope, appears as a small black dot, and by electron microscopy as a tangled string of sausages.

Optimally, grains directly overlie the biological structures from which the radiation has originated. However neither the emulsion nor tissue sections are infinitely thin, and as the β or γ rays can travel at any angle from their source, they may strike a silver halide crystal at some distance. Fortunately, the distribution of grains about a radioactive source follows the laws of probability and statistical methods can be used to determine whether the grains are associated wtih one class of biological structure or another.

B. Potential Artifacts

Gentamicin labeled with tritium (^{3}H) has been used preferentially for autoradiography because of its long half-life, its commercial availability, and the superior resolution achieved with electron microscopy. Although many labeling processes alter, to some extent, the pharmacological properties of the parent compound, tritiated gentamicin maintains its antibacterial activity [4] and behaves chromatographically as unlabeled gentamicin [13]. Hydrogen atoms are relatively labile and it is possible that as a result of in vivo metabolism the tritium may become associated with a different compound and the silver grains reflect not the position of gentamicin but of one of its metabolites. However, when the chromatographic pattern of radioactive material recovered from the kidney homogenate of rats injected with labeled gentamicin was compared with that of the original injected material, it was found that gentamicin did not undergo significant metabolism in vivo [13].

A third potential artifact is that as gentamicin is a water-soluble compound, it may shift its position during specimen processing. Fortunately, gentamicin appears to be immobilized in the tissue by conventional electron microscopic fixatives, possibly because gentamicin is cross-linked to tissue proteins by glutaraldehyde [12]. Kuhar et al. have also successfully used quick-freezing techniques to minimize diffusion [7].

II.   Distribution of Gentamicin Along the Proximal Tubules

A. Light Microscopy

Each of the three autoradiographic studies that have been done to date agree that the proximal tubule is the primary site of gentamicin accumulation. Following intravenous injection of [^{3}H]gentamicin, radioactivity appears over the luminal surface of the proximal tubules as early as 5–10 min (Figure 1). After 1 hr the grains are concentrated in the mid or basal portion of the cells (Figure 2). All proximal tubule profiles are not labeled to the same extent, and examination of serial sections has revealed that the initial portion of the *pars convoluta* has considerably fewer grains than the second portion (Figure 2). This finding suggests that the ability to reabsorb gentamicin is not distributed uniformly along the proximal tubule. It is also possible, however, that some of the heterogenous labeling is due to internephron as well as intranephron variations.

B. Electron Microscopy

Temporal studies of animals sacrificed at different times after injection indicate that gentamicin is handled primarily by the organelles of the endocytic apparatus. At 5–10 min

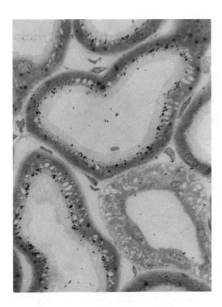

Figure 1   Light microscope autoradiograph of rat renal cortex. Animal sacrificed 10 min after intravenous injection of [^{3}H]gentamicin. Silver grains are located predominantly over the apical cytoplasm of the proximal tubules. (X630) (Reprinted from [13]. Used by permission.)

silver grains are found primarily over the apical pinocytic vesicles (Figure 3). Just et al. have reported that microvilli were also labeled at this time [4]. However, we have found that flushing the vasculature of the kidney with cold gentamicin immediately prior to fixation eliminated the labeling of the microvilli, suggesting that the gentamicin is only loosely associated with these structures. Similarly, early labeling of the basement membrane and peritubular space was virtually eliminated by a cold chase prior to fixation.

During the first hour, the label shifts progressively from the apical vesicles to large lysosomes (heterosomes) located in the mid or basal portion of cytoplasm (Figure 4). The radioactivity is not distributed homogeneously among the lysosomes; some have five or more silver grains, whereas others nearby have none.

C. Quantitative Autoradiography

Although inspection of the autoradiographs indicates that the label is largely confined to structures of the endocytic apparatus, we believed it important to confirm our visual impressions with quantitative analysis [13]. The following example illustrates why this was necessary. In Figure 3 it can be seen that the grains appear to be associated with pinocytic vesicles. However, as these organelles are also the most prevalent structures in the apical region of the cell, they would be expected, by chance alone, to have the most grains. We used two independent methods to distinguish true or specific labeling of an organelle from spurious association. One method compared the number of grains associated with a particular class of organelle, to the number that would have been expected if an equivalent area of cytoplasm were randomly labeled. In the example given, the probability

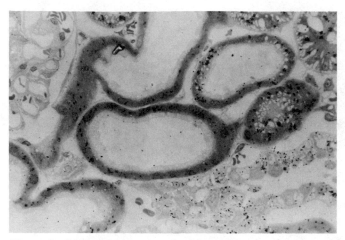

Figure 2 Light microscope autoradiograph of renal cortex of rat, sacrificed 24 hr after injection of [³H]gentamicin. In the proximal tubules, silver grains are located predominantly over mid or basal cytoplasm. Note that the initial proximal tubule segment taking off from the glomerulus is not labeled. This figure also demonstrates the presence of radioactivity in individual cells of the distal tubule. The distribution of radioactivity was similar at 1 hr and 24 hr (×630) (Reprinted from [13]. Used by permission.)

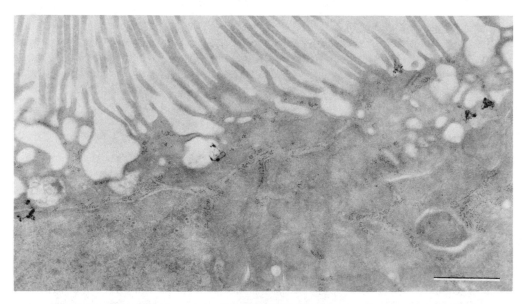

Figure 3 Electron microscope autoradiograph of proximal tubule cell from a rat renal cortex. The animal was killed 10 min after intravenous administration of [³H]gentamicin. This figure depicts a cross section of the apical cytoplasm. Except for a single grain overlying the luminal membrane, all the grains are associated with apical vesicles. Bar represents 1 μm. (×17,290) (Reprinted from [13]. Used by permission.)

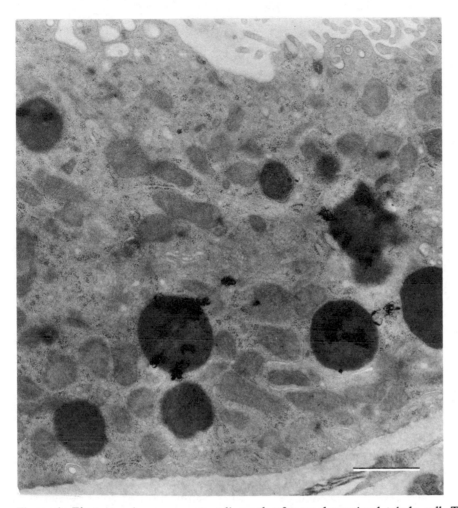

Figure 4 Electron microscope autoradiograph of a renal proximal tubule cell. The animal was killed 24 hr after administration of [^{3}H]gentamicin. Almost all the grains are associated with lysosomes. Bar represents 1 μm. (Reprinted from [13]. Used by permission.)

that the area of the cell occupied by the apical vesicles would by chance have as many grains as were observed is very small.

The second, and more exacting, method used to assess the specificity of labeling was grain density distribution analysis [12]. This method is based on the assumption that the number of grains originating from a source falls off rapidly with increasing distance from the source. The distribution of grains about apical vesicles of the animals sacrificed after 10 min demonstrated a striking exponential relationship with proximity to the vesicle membrane (Figure 5). A similar plot of the lysosomes at 24 hr confirmed the specificity of the labeling of these organelles, and indicated that the radioactivity was distributed homogeneously throughout their interior.

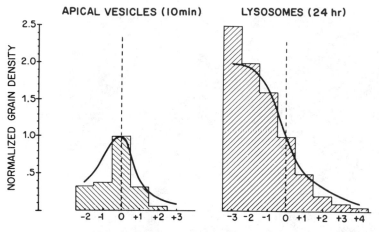

Figure 5 Distribution of grain densities about apical vesicles and lysosomes analyzed by the method of Salpeter and McHenry [12]. The numbers in the abscissa describe concentric circles without (positive numbers) and within (negative numbers) the organelles. The half-distance represents the distance from a radioactive source within which 50% of the grains emanating from that source would fall. The grain densities were normalized with respect to the membrane and therefore do not reflect the absolute amount of label associated with each organelle. The results indicate that gentamicin was associated with the membrane of the apical vesicles and was distributed homogeneously throughout the lysosomes. The observed distribution conform to theoretical distributions (superimposed solid lines) expected for a doughnut-shaped and solid disk, respectively. (Reprinted from [13]. Used by permission.)

D. Mechanism of Aminoglycoside Transport and Toxicity

Autoradiography has demonstrated that gentamicin is transported into, and accumulates within, the proximal tubule cells, and thus confirms similar results obtained by other techniques, such as micropuncture [11] and clearance studies [3]. The temporal progression of labeled gentamicin into the proximal tubule indicates that gentamicin is reabsorbed from the glomerular filtrate across the luminal membrane rather than from the tubular capillary across the basal surface. Electron microscopy shows that the label is associated with endocytic organelles, suggesting that gentamicin is transported into the proximal tubule cell by pinocytosis in a manner analogous to the mechanism responsible for reabsorption of proteins.

    Details of this latter process have been elucidated with the help of tracer proteins such as horseradish peroxidase that can be identified in tissue sections by histochemical [2] or autoradiographic techniques [10]. Uptake into the proximal tubule cell is initiated when the protein binds to specialized tubular invaginations of the surface membrane located at the base of the microvilli. These pinch off and become apical pinocytic vesicles which subsequently coalesce to form larger vacuoles. The protein-laden vacuoles undergo translocation to the mid or basal portions of the cell, where they fuse with preexisting large lysosomes. Within the lysosomes, hydrolytic enzymes degrade the proteins and the products become available for recycling within the cell or for return to the circulation.

Just and Habermann have suggested that the polycationic nature of the aminoglycosides is responsible for their pinocytosis by the proximal tubule cells [5]. They point out that polycationic polypeptides are known to be more readily reabsorbed from the glomerular filtrate than neutral or polyacidic peptides. Using isolated brush border membranes, they have demonstrated that gentamicin can competitively inhibit the binding of the cationic polypeptide aprotinin [5]. Although they failed to demonstrate any inhibitory effect of gentamicin on uptake of aprotinin in vivo, such inhibition might not be expected since aprotinin is reabsorbed in the earliest portion of the proximal tubule whereas, as mentioned above, gentamicin does not accumulate in this portion of the nephron. Just and Habermann found a good correlation between the number of amino groups of different aminoglycosides and their ability to bind to brush border: neomycin > tobramicin > gentamicin > kanamycin > streptomycin. A similar correlation has also been shown for increasing number of primary amine groups and the degree of renal accumulation of aminoglycosides [9].

The rapid appearance of gentamicin within the lysosomes may help explain the morphogenesis of the electron-dense material that accumulates in renal lysosomes of aminoglycoside-treated rats. These deposits have been called myeloid bodies because of their osmophilic lamellated appearance [6]. It has previously been suggested that myeloid bodies are the residual lipoid debris of organelles injured by gentamicin that have been ingested and degraded by the lysosomes [14]. The fact that gentamicin enters the lysosome within 20 min after administration, and is found in no other organelle other than apical vesicles prior to that time, favors a direct effect of gentamicin on the lysosomes themselves.

Accumulation of gentamicin in lysosomes might also contribute to gentamicin toxicity. Gentamicin may block an important lysosomal degradative function leading either to accumulation of a toxic digestion product or depletion of a vital metabolite. Gentamicin may disrupt the integrity of the lysosomal membrane, releasing potentially harmful enzymes into the cytoplasm. The interaction of aminoglycosides with lysosomes is discussed in detail in Chapter 13. On the other hand, accumulation of gentamicin in lysosomes and the formation of myeloid bodies may be unrelated to the mechanism that produces cell necrosis. Although autoradiography suggests that endocytosis is the predominant pathway by which gentamicin is transported by the proximal tubule cell, the inherent lack of sensitivity of this technique cannot exclude the possibility that a small amount of gentamicin enters the cell by one or more additional mechanisms.

III. Labeling of the Distal Nephron

Although labeling is, for the most part, confined to the proximal tubule, in our study we noted a considerable number of silver grains over some distal elements as well. Ten minutes after injection, cells of papillary collecting ducts are labeled, some very intensively (Figure 6). By 1 hr, however, collecting ducts are free of silver grains. A different pattern is seen over the distal convoluted tubules, where silver grains appear over the cytoplasm of individual cells at 1 hr and 24 hr (Figure 2).

By electron microscopy the pattern of labeling differs from that of proximal tubule cells. In the distal convoluted tubule cells, silver grains are associated predominantly with the plasma membrane (Figure 7). Silver grains overlying the papillary collecting ducts did not appear to be associated with any particular structure.

The significance of the labeling of the distal nephron is not obvious at this time, as

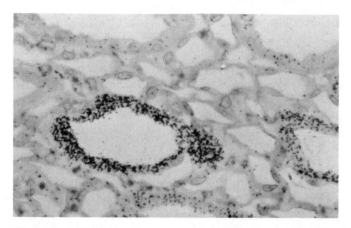

Figure 6 Light microscope autoradiograph of a renal papilla of a rat sacrificed 10 min after injection of [³H]gentamicin. Collecting ducts are labeled, some very heavily. (×630)

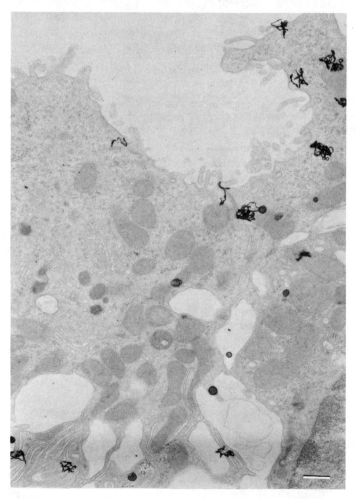

Figure 7 Electron microscope autoradiograph of a distal tubule cell of a rat sacrificed 24 hr after injection of [³H]gentamicin. The silver grains overlie, or are adjacent to, the plasma membrane. Bar represents 1 μm. (×7500)

micropuncture and clearance studies have indicated that gentamicin is neither reabsorbed nor secreted by the distal nephron [1,11]. The transient association of silver grains with papillary collecting ducts shortly after administration suggests that these cells may be permeable to the exceptionally high concentration of the drug present in the papilla at that time. Why gentamicin binds persistently to the plasma membrane of some distal convoluted tubule cells is not known but may reflect the affinity of aminoglycosides to kidney proteins [8].

IV. Autoradiography of Kidney Slices

Sections of kidney cut by the hand-held Stadie-Riggs microtome have been widely used to study the mechanism of renal transport of a number of compounds, including aminoglycosides [3,11]. Gentamicin has been reported to be concentrated by renal slices up to fourfold over the medium [11]. Accumulation could not be prevented by inhibitors of the organic base or organic acid transport system, but was prevented by inhibitors of oxidative phosphorulation [3]. An autoradiographic study of the distribution of tritiated gentamicin in cortical slices casts doubt, however, on the validity of using kidney slices for studying gentamicin transport. Figure 8 is an autoradiograph of a rat cortical slice

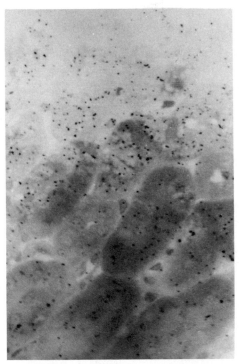

Figure 8 Light microscope autoradiograph of a renal cortical slice incubated in [³H]gentamicin for 20 min. Most of the silver grains overlie cells along the cut edge. These cells appear pale and swollen, suggesting that they have been injured by the microtome blade. (×630)

incubated for 20 min in [³H]gentamicin. As can be seen, the majority of grains are associated with the cytoplasm of cells along the cut edge of the slice and, in particular, with cells that have morphological evidence of injury. A very small proportion of the total number of grains appear in the lumen of healthy appearing proximal tubule segments on the nontraumatized capsular side and may represent some secretory activity. The preferential labeling of injured cells may be due to the propensity of aminoglycosides to bind (nonspecifically) to renal cell proteins [8].

V. Conclusions

Autoradiography has confirmed that gentamicin and presumably other aminoglycosides are reabsorbed from the glomerular filtrate by the proximal tubule. This appears to be accomplished by pinocytosis in a manner analogous to the process whereby proteins are reabsorbed from the glomerular filtrate. Reabsorbed gentamicin is sequestered in lysosomes, where it may be responsible for the formation of myeloid bodies. Whether the accumulation of gentamicin within the lysosomes of the proximal tubules is also responsible for the toxic effects of the drug on these cells is still a matter of conjecture.

Autoradiography has also shown that the renal slice technique is probably unsuitable for studying gentamicin transport, as most of the drug that becomes associated with cell appears to bind to the cytoplasm of cells traumatized by the microtome knife.

Finally, autoradiography has provided unexpected, and at present unaccountable, evidence that elements of the distal nephron may also participate in the handling of gentamicin by the kidney.

References

1. Chu, P.H., Brown, A, Miller, G., and Long, J.F.: Renal extraction of gentamicin in anesthetized dogs. *Antimicrob. Agents Chemother.* 10:277–282, 1976.
2. Graham, R.C., Jr., and Karnovsky, M.J.: The early stages of absorption of injected horseradish peroxidase in the proximal tubules of mouse kidney: ultrastructural cytochemistry by a new technique. *J. Histochem. Cytochem.* 14:291–302, 1966.
3. Hsu, C.H., Kurtz, T.W., and Weller, J.M.: In vitro uptake of gentamicin by rat renal cortical tissue. *Antimicrob. Agents Chemother.* 12:192–194, 1977.
4. Just, M., Erdman, G., and Habermann, E.: Renal handling of polybasic drugs. 1. Gentamicin and aprotinin in intact animals. *Naunyn-Schmiedeberg's Arch. Pharmacol.* 300:57–66, 1977.
5. Just, M., and Habermann, E.: The renal handling of polybasic drugs. 2. In vitro studies with brush border and lysosomal preparations. *Naunyn-Schmiedeberg's Arch. Pharmacol.* 300:67–76, 1977.
6. Kosek, J.C., Mazze, R.I., and Cousins, M.J.: Nephrotoxicity of gentamicin. *Lab. Invest.* 30:48–57, 1974.
7. Kuhar, M.J., Mak, L., and Lietman, P.S.: Autoradiographic location of [³H]gentamicin in the proximal renal tubules of mice. *Antimicrob. Agents Chemother.* 15:131–133, 1979.

8. Kunin, C.: Binding of antibiotics to tissue homogenates. *J. Infect. Dis.* 121:55–64, 1970.

9. Luft, F.C., and Kleit, S.A.: Renal parenchymal accumulation of aminoglycoside antibiotics in rats. *J. Infect. Dis.* 130:656–659, 1974.

10. Maunsbach, A.B.: Absorption of I^{125}-labeled homologous albumin by rat kidney proximal tubule cells: a study of micro-perfused single proximal tubules by electron microscopic autoradiography and histochemistry. *J. Ultrastruct. Res.* 15:197–241, 1966.

11. Pastoriza-Munos, E., Bowman, R.L., and Kaloyanides, G.J.: Renal transport of gentamicin in the rat. *Kidney Int.* 16:440–450, 1979.

12. Salpeter, M.M., and McHenry, F.A.: Electron microscopic autoradiography: analysis of autoradiograms. In *Advanced Techniques in Biological Electron Microscopy,* edited by Koebler, J.K. New York, Springer-Verlag, 1973, p. 113.

13. Silverblatt, F.J., and Kuehn, C.: Autoradiography of gentamicin uptake by the rat proximal tubule cell. *Kidney Int.* 15:335–345, 1979.

14. Vera-Roman J., Krishnakantha, T.P., and Cuppage, F.E.: Gentamicin nephrotoxicity in rats: I. Acute biochemical and ultrastructural effects. *Lab. Invest.* 3:412–417. 1975.

10

ANIMAL MODELS IN THE STUDY OF AMINOGLYCOSIDE NEPHROTOXICITY

RICHARD A. PARKER, WILLIAM M. BENNETT, and GEORGE A. PORTER   University of Oregon Health Sciences Center, Portland, Oregon

The nephrotoxic potential of aminoglycoside antibiotics has undergone extensive evaluation in a number of experimental animals. Although the drugs have been examined in many species, the bulk of the data have been generated employing the rat model.

This chapter reviews the experimental data available for each of the drugs. To the extent possible, it has been organized by drug and by structural or functional defect induced by the particular aminoglycoside. The studies evaluating gentamicin will be examined first, in part because the nephrotoxic potential of the newer aminoglycosides is referenced to that of gentamicin.

I. Gentamicin

At present, gentamicin is the most widely used aminoglycoside antibiotic. The frequency with which nephrotoxic reactions associated with gentamicin administration have been noted seems to be increasing. This is due, in part, to an increased clinical awareness of the problem. Because of its clinical efficacy, the drug has been the subject of extensive laboratory investigation, both in vitro and in vivo. The bulk of the in vivo studies has been carried out in rats. However, several studies employing dogs, cats, guinea pigs, mice, and monkeys have also been completed. As is the case with many drugs, the results of these investigations are not directly comparable because of a marked species and strain variability in susceptibility to the nephrotoxic potential of gentamicin. In addition, it is now clear that variables such as the route and frequency of drug administration, the volume status of the animal, its age and sex, as well as concomitant drug administration must all be taken into account when comparing the results of studies.

Unfortunately, in spite of the extensive interest in gentamicin from a laboratory investigative standpoint, the mechanism by which it induces renal insufficiency remains unknown. Consequently, the preponderance of investigative work with the drug has been descriptive of the histological and renal functional changes that are consequent to its administration. Out of this has developed a characteristic histological picture and functional profile of gentamicin nephrotoxicity.

A. Histologic Alterations

Studies evaluating both the acute and chronic administration of the drug have been performed. Vera-Roman and associates [51] assessed the ultrastructural alterations in Sprague-Dawley rats following the administration of a single dose of gentamicin in the range 10–160 mg/kg of body weight. The changes noted were dose-related and confined primarily to the proximal tubule; 160 mg/kg produced apical vacuolization primarily in the pars convulata. Cytosegresomes were also prominent, some containing altered cytoplasmic membranes, producing the so-called myelin bodies. The fact that these changes represent injury, but not cellular death, was suggested by a complete resoltuion of the findings 48 hr following the dose of gentamicin. The changes at lower doses were qualitatively the same but quantitatively not as significant. At 10 mg/kg no histologic alterations were noted.

Several authors have evaluated the renal morphologic consequences of prolonged administration of gentamicin. Kosek et al. [31] treated Fischer 344 rats continuously for 28 days with doses ranging from 1 to 40 mg/kg per day (single daily dose). At 1 mg/kg, electron microscopic (EM) changes were noted as early as 48 hr following the first dose.

These consisted primarily of increased numbers of cytosegresomes, many of which contained prominent myeloid bodies. Changes were confined to the proximal tubules and were present in EM but not light microscopic (LM) sections. As the dose was progressively increased, the appearance of focal followed by diffuse proximal tubular necrosis was noted. In addition, the number of cytosegresomes increased commensurate with the increasing dose of the drug. Animals sacrificed 8 days following the last dose of gentamicin were found to have essentially the same histologic picture as that described earlier. These authors did not comment upon the presence of any regenerative activity in the animals studied 8 days following treatment.

Houghton and co-workers examined the effect of 14 continuous days of gentamicin administration [21]. In this study employing Fischer 344 rats, the renal morphologic consequences of 40 mg/kg per day of gentamicin (administered in two daily doses) were assessed at various times during the treatment period and for 4 weeks following discontinuation of the drug. Seven continuous days of therapy produced minimal LM alterations. Ultrastructural changes were, however, evident. These consisted of marked increases in the number of cytosomes and cytosegresomes, within which numerous myeloid bodies were noted (Figure 1). The mitochondria were swollen and the endoplasmic reticulum dilated. Ten days of treatment produced progressive damage. Proximal tubular necrosis and desquamation were evident in approximately one-half of the tubules in the outer cortex. The changes on EM were consistent with the tubular necrosis noted by LM. However, the cytosegresomes and myeloid bodies, although still numerous, were less conspicuous that at 7 days. In addition, EM evidence of regenerating cells was seen. By 14 days the proximal tubular epithelium had undergone nearly complete alteration. Electron microscopically, focal areas of necrosis were present; however, regeneration was extensive (Figure 2). In the recovery period, the return of histologic integrity proceeded such that regenerating proximal tubular epithelial cells were regaining normal height by 7 days after the last dose (Figure 3). Following 4 weeks of recovery, the LM and EM findings were comparable to controls. Scattered areas of prominent residual scarring were, however, present.

The data from these and other studies provide a picture of the characteristic renal morphologic response to gentamicin administration. Histologic changes are dose-related and dependent on the duration of drug administration. The earliest changes are ultrastructural with the development of increased numbers of lysosomes and cytosegresomes in association with myeloid bodies. Shortly thereafter, dependent on the dose of gentamicin, proximal tubular necrosis develops and progresses to involve essentially all of the cortex. The glomeruli are spared significant LM and EM changes. Coexistent with necrosis, however, is active tubular epithelial regeneration which proceeds despite continued drug administration. Following the discontinuation of the drug, normal tubular architecture is reestablished fairly rapidly.

The origin and significance of the cytosegresomes and myeloid bodies noted in the previous studies remains conjectural. Cytosomes and cytosegresomes are components of the digestive system of the cell. As such, they are involved in the process of autophagocytosis. The proximal tubule of control rats contains a small number of cytosegresomes, probably reflecting the steady state of cell organelle disposal. The number of cytosegresomes are increased in states of sublethal cell injury, suggesting that perhaps these organelles provide a means of clearing toxins or injured cytoplasmic constituents. Gentamicin may enhance the appearance of these autophagic vacuoles by its ability to induce focal cytoplasmic injury. Additionally, the increased number of cytosegresomes may represent a cellular

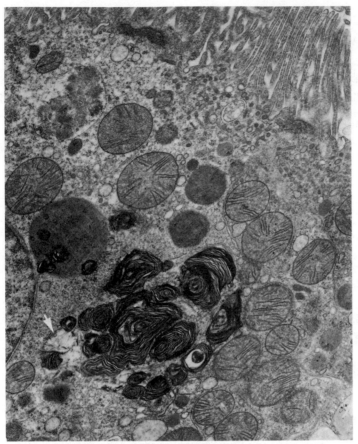

Figure 1 Electron micrograph of a Fischer rat proximal tubule cell after the animal had received gentamicin, 40 mg/kg per day, for 3 days. A cytosegresome is distended by numerous myeloid bodies and other material, including a mitochondrial remnant (arrow). The cytosomal membrane is ill defined, possibly disrupted. Small myeloid bodies are forming in one of several other smaller, round cytosomes. (×24,000) (Reprinted from [21]. Used by permission.)

attempt to digest and exclude the gentamicin. Similarly, the genesis and significance of the myeloid bodies in gentamicin-treated kidneys is not known. It is possible that the myeloid bodies are a result of the interaction of polycationic gentamicin with the acidic phospholipid membranes of lysosomes. This possibility is supported by autoradiographic data demonstrating preferential gentamicin localization in the lysosomal fraction of the cytoplasm [25].

It is not possible to correlate the appearance of increased numbers of cytosegresomes and myeloid bodies with the toxic effect of gentamicin. Kosek's [31] data demonstrating their appearance in the tubules of rats treated with extremely low, apparently nontoxic dosages of the drug would argue against a direct relationship between their appearance and the onset of gentamicin-induced nephrotoxicity. In addition, other drugs are known to induce myeloid body formation seemingly independent of a significant nephrotoxic

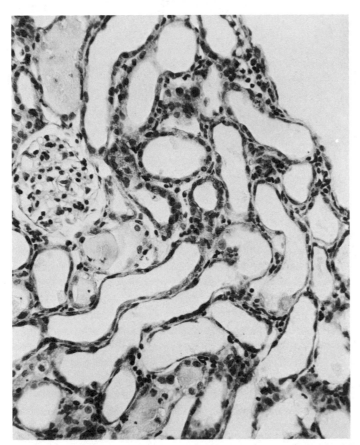

Figure 2 Photomicrograph of the outer renal cortex of a Fischer rat that had received gentamicin, 40 mg/kg per day, for 14 days. Proximal tubules are lined by squamoid regenerating cells. Necrotic debris remain in the lumina of a few tubules. (×300) (Reprinted from [16]. Used by permission.)

potential. Thus, it would appear that the myeloid bodies serve only as a marker of gentamicin treatment and not as an indicator of nephrotoxicity.

Of much interest is the appearance of regenerating cells in the presence of continued gentamicin administration. This will be addressed more fully below.

B. Renal Function Alterations

The effects of gentamicin on the functional characteristics of the kidney are extensive. Alterations in excretory function, concentrating ability, organic ion transport, oxidative metabolism, tubular function, urinary enzyme excretion, and renal blood flow have all been identified and evaluated.

Excretory Function and Concentrating Ability

The nature and degree of dysfunction induced by gentamicin is dose-dependent. Characteristically, the drug produces nonoliguric acute renal failure (ARF) that is associated with polyuria, urinary hypoosmolarity and a progressive but reversible decline in

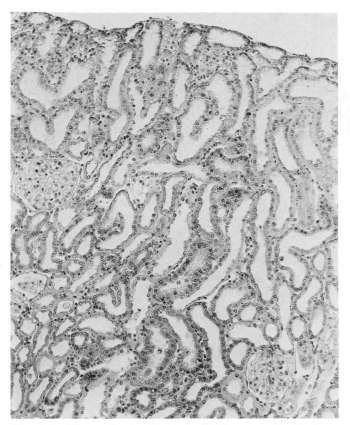

Figure 3   Photomicrograph of the renal cortex of a Fischer rat 1 week after the end of a 10 day course of gentamicin, 40 mg/kg per day. Proximal tubules are repopulated by normal-appearing epithelial cells. (×130) (Reprinted from [21]. Used by permission.)

glomerular filtration rate (GFR). Gentamicin-induced renal excretory dysfunction has been evaluated in several species with various dosage regimens [8,13,18,46,48]. The time course over which the development of excretory dysfunction occurs has been best demonstrated in the rat [17,21,31,36]. In Fischer 344 rats, treatment with 40 mg/kg per day of gentamicin (given twice daily) produces urinary hypoosmolality early in the course of drug administration. Urinary osmolalities significantly lower than controls may be seen within 3 days of instituting gentamicin. The serum creatinine rises shortly thereafter, reaching a peak value at approximately day 14 of continuous drug administration (Figure 4). The rise in serum creatinine is coincident with the onset of tubular necrosis at 7–10 days of treatment. The subsequent course of the renal insufficiency as measured by excretory function is independent of the continued administration of the drug at this dose (see below).

Polyuria and urinary hypoosmolality seen in the foregoing model is related to a gentamicin-induced resistance to antidiuretic hormone (ADH) [43]. Administration of exogenous ADH to these animals fails to produce a rise in urinary osmolality or the expected increase in urinary cyclicAMP excretion. Based upon studies assessing the effect of gentamicin on ADH-induced water flow in the toad urinary bladder, gentamicin appears

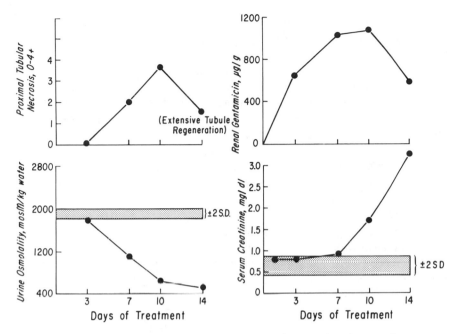

Figure 4   Renal changes after gentamicin, 40 mg/kg per day, in two doses.

to impair the ADH-stimulated generation of cyclicAMP [41,45]. This resistance to ADH
induced by gentamicin may, in part, be related to enhancement of prostaglandin modulation of water flow.

Organic Ion Transport

The histologic damage resulting from gentamicin administration is confined almost
exclusively to the renal cortex and specifically to the proximal tubule. As has been noted
above, the histologic changes and renal excretory dysfunction associated with gentamicin
occur relatively late in the course of drug administration. In an attempt to localize the
earliest site of injury, studies assessing the tubular transport of various organic ions have
been carried out by several groups of investigators. The majority of these studies have
evaluated the transport in renal cortical slices of the organic acid, p-aminohippurate (PAH),
and the organic base, N-methylnicotinamide (NMN). PAH accumulation in rat renal
cortical slices may be stimulated or suppressed by gentamicin, depending on the dosage
and duration of therapy. Cohen et al., employing gentamicin in a dose of 100 mg/kg per
day for 3 days, demonstrated an increased PAH uptake in cortical slices [7]. Lapkin and
associates, using the same dose for 2 days, also demonstrated augmented PAH uptake in
slices [32]. Conversely, the uptake of the organic base NMN has been shown to be impaired by gentamicin. Bennett and co-workers [2] studied the characteristics of organic
cation and anion transport during the course of 14 days of continuous gentamicin administration (40 mg/kg per day, given twice daily) to Fischer 344 rats. PAH uptake was
enhanced as early as day 1 of treatment. This stimulation progressed and was maximal
at day 7. Enhanced PAH uptake was then followed by an abrupt decline that was coincident with the onset of azotemia and proximal tubular necrosis (see above). On the

other hand, the accumulation of the organic base NMN was impaired by day 3 and remained suppressed throughout the treatment period (Table 1).

The mechanism of augmented PAH uptake is not known. Kinetic studies by Lapkin [32] are consistent with the hypothesis that gentamicin stimulates the production of a PAH carrier protein. In addition, he also noted that gentamicin impaired PAH runout from cortical slices. It has been suggested that gentamicin initiates cellular damage by altering renal tubular membrane permeability. This might lead to a compensatory increase in sodium-potassium pump activity in an attempt to maintain normal ion distribution. Because PAH transport has been linked to sodium-potassium exchange, enhancement of the pump activity might secondarily increase PAH uptake.

The mechanism of gentamicin's impairment of organic base transport (NMN) also remains to be defined. Gentamicin is an organic base. As such, it may decrease NMN transport by binding to or altering the base transport proteins in the renal cortex.

It appears that alterations in organic ion transport represent subtle manifestations of nephrotoxicity. It is clear from the data describing the time course of development of gentamicin nephrotoxicity that elevations in serum creatinine and the onset of tubular necrosis occur relatively late in the course of exposure to the drug. Organic ion transport changes, however, are evident much earlier, occurring independent of significant histologic changes.

Aminoglycoside-induced alterations in organic ion transport may have utility in the investigative assessment of nephrotoxic potential. As will be described later in this chapter, early stimulation of PAH transport serves as a predictor of the nephrotoxicity of a particular aminoglycoside at a given dose.

Renal Metabolism

The metabolic alterations induced by gentamicin have been assessed in both acute and chronic studies. Information is available only in the rat. Vera-Roman and associates [51] evaluated the effect of gentamicin on protein synthesis in the renal cortex. Following a single injection of gentamicin into Sprague-Dawley rats in a dosage range of 10–160 mg/kg, they were unable to demonstrate any affect on the renal cortical uptake of [^{3}H]leucine. They concluded from this that the mechanism of nephrotoxicity was not likely impairment of protein synthesis within the kidney.

Gentamicin has been shown to impair tissue respiration in renal cortical slices. Kluwe and Hook have demonstrated a gentamicin-induced concentration-dependent impairment of tissue respiration that correlated well with a decreased ability of cortical slices to accumulate PAH [28]. Cuppage et al. evaluated the effect of 28 days of gentamicin administration on oxygen uptake and phosphorus/oxygen ratios in renal cortical homogenates [9]. Low-dose (1–10 mg/kg per day) gentamicin-stimulated oxygen uptake suggesting an uncoupling of oxidative phosphorylation while higher doses (20–40 mg/kg per day) significantly decreased oxygen consumption as compared to controls. Kluwe and Hook determined the formation of glucose and ammonia and the rate of slice respiration in renal cortical slices taken from Sprague-Dawley rats treated for 2–4 days with 100 mg/kg of gentamicin [29]. Ammoniagenesis was impaired by approximately 40% following 2 days of treatment and remained suppressed at 4 days. Gluconeogenesis was suppressed by the fourth day. In addition, oxygen consumption was depressed by 35% in animals receiving 4 days of the drug. The slice accumulation of amino acids and sugars was not altered.

The significance of these gentamicin-induced alterations in renal metabolism,

Table 1 Effect of Gentamicin (40 mg/kg per day) on PAH and NMN Transport[a]

Slice/medium ratios (mean ± SEM)	Days of treatment							
	1	3	4	5	7	10	14
Control							
PAH	8.58 ± 0.27	8.61 ± 0.22	9.06 ± 0.32	8.67 ± 0.32	8.69 ± 0.22	8.41 ± 0.32	8.54 ± 0.41
NMN	6.25 ± 0.23	5.95 ± 0.12	6.39 ± 0.10	5.63 ± 0.17	6.07 ± 0.13	5.77 ± 0.26	5.82 ± 0.13
Gentamicin, 40 mg/kg per day							
PAH	10.58 ± 0.61[b]	8.69 ± 0.32	11.38 ± 0.43[b]	10.20 ± 025[b]	11.48 ± 0.26[b]	2.88 ± 0.36[b]	1.36 ± 0.14[b]
NMN	5.60 ± 0.30[b]	5.11 ± 0.10[b]	4.61 ± 0.12[b]	4.09 ± 0.07[b]	4.34 ± 0.10[b]	1.70 ± 0.18[b]	1.40 ± 0.05[b]

[a]Each number represents the mean of at least six experiments.
[b]Significantly different from control.

243

particularly as they relate to mechanisms of nephrotoxicity, are not clear. The disruption of ammoniagenesis and the impairment of gluconeogenesis suggests a toxic alteration of mitochondrial function. The oxygen consumption data further suggests an overall impairment by gentamicin of cellular respiration. All of which could be manifest by renal functional impairment and, ultimately, cell death.

Tubular Function

Consistent with its role as a proximal toxin, gentamicin alters the fractional excretion of sodium and potassium and the tubular reabsorption of phosphate and glucose. Sprague-Dawley rats treated with 112.5 mg/kg per day of gentamicin for 15 days (one daily injection) increase their fractional excretion of sodium from 0.66 to 1.23 mEq/L and that of potassium from 3.7 to 8.7 mEq/L [34]. The same pattern of increased fractional sodium excretion occurs in the Fischer 344 rat receiving 40 mg/kg per day of gentamicin for 14 days. In the Sprague-Dawley model noted above, the tubular reabsorption of phosphate is markedly impaired by 15 days of gentamicin (92.1% to 35.9%).

Several investigators have evaluated the effect of gentamicin on the renal tubular handling of glucose. Kluwe and Hook noted significant glucosuria on each of 4 days of gentamicin administration (100 mg/kg) in Sprague-Dawleys [29]. Luft observed a decrease in tubular reabsorption of glucose (99% to 85%) in Sprague-Dawley rats receiving gentamicin in a dose of 112.5 mg/kg per day for 15 days [34]. No data relative to the time course of development of glucosuria were presented in this study. The renal handling of glucose by dogs receiving gentamicin has been evaluated by Cronin [8]. He treated dogs for 10 days with either 7 mg/kg per day or 30 mg/kg per day of the drug. The animals receiving the lower dose failed to develop a decrease in creatinine clearance or glucosuria over a 31 day observation period. In contrast, all the dogs receiving the higher dose of gentamicin developed significant fasting glucosuria and decreased tubular reabsorption of glucose by 8 days of treatment. These changes paralleled the decrease in creatinine clearance noted in this treatment group. The glucosuria and decreased GFR were maximal at 15 days and demonstrated a return to near normal by 31 days.

Similar findings of gentamicin-induced glucosuria have been recorded in the rabbit [18]. In rabbits receiving either 4 or 8 mg/kg per day of gentamicin, a direct correlation was observed between the ensuing degree of impaired glomerular filtration and the quantity of glucosuria.

In none of the studies noted above have gentamicin-induced changes in serum glucose been noted. Similarly, significant alterations in serum potassium or sodium do not occur as a result of gentamicin administration.

Renal Blood Flow and Urinary Enzyme Excretion

Gentamicin's effect on renal blood flow has recently been characterized by Appel and associates. The characteristic pattern of urinary enzyme excretion associated with gentamicin administration has been noted by several investigators. Both of these topics are dealt with in detail in Chapters 11 and 12, respectively.

C. Renal Parenchymal Accumulation

Gentamicin selectively and preferentially accumulates within the renal parenchyma. Following a single subcutaneous injection of 10 mg/kg in Sprague-Dawley rats, Luft and

Kleit demonstrated a gentamicin half-life of 109 hr in renal tissue [35]. This compared with a serum half-life of approximately 30–35 min. In the same study, these workers noted gentamicin half-lives in lung and hepatic tissue not significantly different from that in the serum.

The renal parenchymal distribution of gentamicin is not uniform. Rather, there is marked preferential accumulation of the drug in the renal cortex as opposed to the medulla. Luft and Kleit noted that 80% of the renal gentamicin concentration was confined to the cortex, with 10% in the juxtamedullary area and only 5% in the papillary region. Similar findings have been described by other investigators administering gentamicin to various animals over a more prolonged time period [12,29]. When administered to the Fischer 344 rat at a dose of 40 mg/kg per day (given twice daily) for 14 days, the drug progressively accumulates in the renal cortex. A maximal concentration of 1200 μg/g of tissue is noted at 10 days of treatment. By day 14, the concentration decreases to 570 μg/g. As described in detail below, if administration of the drug is continued, the renal cortical concentrations undergo cyclical changes independent of renal function or histology.

The renal handling of gentamicin remains a controversial topic. As discussed in Chapter 8, the bulk of the data suggest that gentamicin enters the renal parenchyma after undergoing filtration and proximal tubular reabsorption. However, important data also exist ascribing a significant role to peritubular transport of the drug. In any case, the drug appears to accumulate intracellularly. This is supported by autoradiographic studies (see Chapter 9) as well as the studies of Szwed and associates [47]. These investigators measured the concentration and half-life of gentamicin simultaneously in serum, renal parenchyma, and renal lymph. The drug concentrations and half-lives in the lymph and serum were the same, whereas those in the parenchyma were as noted above. Because renal lymph concentrations are thought to be satisfactory indicators of those present in renal interstitial fluid, the high and persistent concentrations of gentamicin in the renal parenchyma suggest intracellular accumulation or binding to cell membranes.

The significance of the renal cortical accumulation of gentamicin as it relates to nephrotoxicity is not known. It is attractive to suggest that a cause-and-effect relationship exists between the cortical accumulation and the proximal tubular toxicity. However, as will be seen when the various aminoglycosides are compared in terms of their nephrotoxic potential, there is a poor correlation between the absolute amount of drug present in the cortex and the resulting renal dysfunction. Further confusion arises when the data dissociating tissue concentration and renal function/histology (see below) are examined.

D. Modifiers of Toxicity

The nephrotoxic potential of gentamicin has been modified in the laboratory animal by manipulation of several factors. It is clear that important determinants of nephrotoxicity include (1) volume and (2) acid/base status of the animal, (3) frequency of administration of the drug, (4) age of the animal, and (5) sex of the animal, and (6) concomitantly administered drugs.

Volume

Sodium restriction preceding the administration of gentamicin markedly enhances nephrotoxicity. Bennett et al. [1] administered gentamicin in a dose of 40 mg/kg per day (two daily doses) for 10 days to Fischer 344 rats. These animals had either been deprived of dietary sodium for the 2 weeks preceding the institution of drug treatment or had supplemental sodium provided as 1% saline substituted for their drinking water. The low-sodium diet markedly potentiated nephrotoxic effects of the drug as evidenced by animal mortality, renal failure, pathological changes, and increased renal cortical concentrations of the drug (Table 2). The high-sodium diet resulted in significantly lower cortical concentrations of tentamicin but did not protect the animals from the nephrotoxic effects of the drug.

In an acute study, Chiu and Long explored the effects of furosemide-induced volume depletion and decreased GFR on gentamicin nephrotoxicity in Sprague-Dawley rats [5]. In animals receiving a continuous infusion of gentamicin for 3 hr, they noted that concominant furosemide induced a gentamicin-independent fall in GFR. In this setting, the gentamicin tissue concentrations were significantly increased and urinary excretion of the drug impaired. Replacement of the furosemide-induced volume loss ameliorated the increased accumulation of gentamicin in the renal cortex as well as returning urinary excretion of the drug to levels comparable to non-furosemide-treated animals.

The mechanism of volume depletion enhancement of nephrotoxicity is speculative. Presumably, increased proximal reabsorption of the drug in association with the volume-induced stimulus to sodium reabsorption results in increased cortical exposure to gentamicin. Activation of the renal pressor system has also been suggested as a possible mechanism for volume-related enhancement of toxic insults. However, the failure of renin suppression (saline supplement) to protect against gentamicin nephrotoxicity does not support the possibility of an activated renin system playing a primary role.

Acid-Base

Controlled pertubations of acid-base homeostasis and the resulting effect on gentamicin nephrotoxicity have been explored in rats only. Hsu and associates substituted 1% NH_4Cl for the drinking water of Sprague-Dawley rats, producing a significant metabolic acidosis [23]. The animals and nonacidotic controls were then treated with gentamicin, 20 mg/kg per day for 16 days. By 12 days of therapy, the serum urea nitrogen of acidotic animals was significantly greater than the controls. This difference persisted throughout

Table 2   Renal Function and Cortical Gentamicin Concentrations in Rats with Varying Sodium Intake[a]

	Group 1[b]	Group 2[c]	Group 3[d]
Serum creatinine (mg/dl)	1.2 ± 0.1	1.2 ± 0.1	9.8 ± 3[e]
Cortical gentamicin concentration (μg/g tissue)	381 ± 22	278 ± 21[e]	900 ± 123[e]

[a]Gentamicin, 40 mg/kg per day (two divided doses), for 10 days. Sacrifice 7 days after treatment.
[b]Standard diet.
[c]One percent saline substituted for drinking water.
[d]Sodium-deficient diet.
[e]Significantly different from group 1.

the treatment period. At sacrifice, the histologic changes in the acidotic animals were markedly more severe than those noted in the controls. In addition, PAH transport in renal cortical slices was profoundly depressed in kidneys taken from the acidotic animals. Data with regard to tissue concentration of gentamicin were not reported. It is not known why or how acidosis potentiates gentamicin toxicity.

Dosage Frequency

The relationship between the frequency with which a daily dose of gentamicin is administered and the subsequent development of nephrotoxicity has been evaluated in rats, rabbits, monkeys, and dogs. Plamp et al. administered a total daily dose of gentamicin (40 mg/kg) to Fischer 344 rats as one single daily injection, or as two equal injections eight hours apart, or as three equal injections given every 8 hr [42]. The animals were sacrificed after 2 or 10 days of therapy such that sacrifice bloods provided peak serum level data for the gentamicin. Following 2 days of treatment, the peak serum drug concentrations were highest in the rats receiving a single injection daily and lowest in those receiving three separate injections.

Renal tissue concentrations at 2 days were highest in the animals receiving three doses daily (1105 μg/g). The serum creatinines were no different between the various groups or when compared to controls. Following 10 days of treatment, the once-daily regimen still produced the highest serum drug levels; those receiving gentamicin three times daily had serum levels greater than the twice-daily group. The tissue concentrations of the drug revealed the highest value (1372 μg/g) in the once-daily group and the lowest (666 μg/g) in those animals receiving three doses daily (Figure 5). When the serum creatinines were analyzed at the end of 10 days of treatment, the greatest increase occurred in those animals treated three times daily (Figure 6).

Similar results were obtained by Frame et al. in rabbits [13]. Rabbits given gentamicin every 8 hr for 3 weeks developed renal insufficiency, whereas those treated with the same total dose in a single injection failed to demonstrate deterioration in their renal function. Similarly, Thompson and colleagues compared once-daily versus continuous infusions of gentamicin in mongrel dogs [49]. Following 10 days of treatment, all the dogs suffered a decline in renal function. However, the magnitude of the decline was significantly greater in the dogs given gentamicin by continuous infusion.

Currently available data do not demonstrate a correlation between nephrotoxicity and peak or trough levels of gentamicin. The frequency with which the drug is administered seems to be of greater importance in modulating toxicity than do the serum levels.

Age

Karniski and colleagues evaluated age as a modifier of gentamicin nephrotoxicity [27]. These investigators treated sexually immature and sexually mature rabbits for up to 28 days with a total dose of 15 mg/kg of gentamicin administered daily in two equal injections. None of the sexually immature group developed renal insufficiency, whereas greater than 50% of the sexually mature rabbits incurred a serum creatinine of more than 3 mg/dl. Cytosegresomes were present in the kidneys from all animals; however, tubular necrosis was noted only in those with creatinines greater than 3 mg/dl. There were no significant differences in the serum or renal tissue concentrations of gentamicin between the two groups.

It would appear that age is a risk factor for the development of gentamicin nephro-

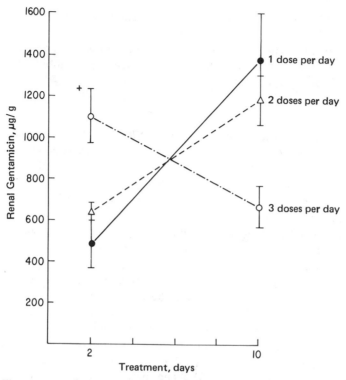

Figure 5 Influence of gentamicin (40 mg/kg per day) dosage schedule on renal gentamicin concentrations. (+) Each data point represents mean ± 2SE of four rats.

toxicity. The reason for this is not clear. It may be, however, that in immature animals
drug transport systems have not yet achieved a sufficiently sophisticated state to allow
gentamicin access to the critical intracellular sites necessary to induce nephrotoxicity.
Other workers have shown a similar dependence of PAH transport on the age of the
animal. This failure of immature cells to transport gentamicin may in part account for
the presence of regenerating tubular cells that are noted during continued gentamicin
administration (see below). Age-dependent susceptibility to gentamicin is correlated with
the clinical observation that gentamicin nephrotoxicity is rarely noted in the pediatric
age group.

Sex

A predilection to the nephrotoxic effect of gentamicin is dependent on the sex of the
Fischer 344 rat. When the drug is administered twice daily in a total dose of 40 mg/kg,
female rats show a marked resistance to gentamicin-induced nephrotoxicity when compared to their male counterparts (Table 3). Following 14 days of treatment the male
animals achieved a mean serum creatinine of 3.5 mg/dl compared to the female mean of
0.8 mg/dl. The PAH transport in renal cortical slices taken from these animals showed
a pattern of toxicity only in the males (Table 3). Gentamicin concentration in the renal
cortex was the same in males and females at 7 and 10 days of treatment (1000 µg/g of
tissue). However, at 14 days the drug concentration in males had fallen to 500 µg/g of
tissue, whereas that in the females remained at 1000 µg/g.

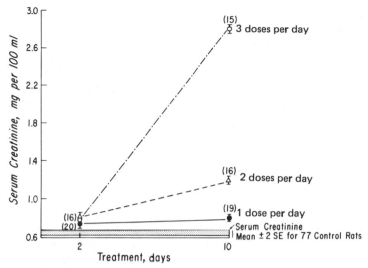

Figure 6 Influence of gentamicin (40 mg/kg per day) dosage schedules on serum creatinine (mean ± 2SE). Number of animals given in parentheses.

Similar patterns of gentamicin nephrotoxicity based upon sex have not been reported in other animal models. However, other workers have noted that female mice are much less susceptible to the nephrotoxic effect of chloroform than are males. The basis for this difference appeared to be the presence or absence of testosterone. Whether a similar mechanism is operative in the Fischer rats noted above remains to be determined.

Concominant Drugs

As might be expected from the extensive data on drug-drug interactions, the nephrotoxicity of gentamicin may be modified by the concurrent administration of other

Table 3 Effect of Gentamicin (40 mg/kg per day) on Male (M) and Female (F) Fischer 344 Rats[a]

| | | Sex | Days of treatment | | |
			7	10	14
Serum		M	0.80	1.70	3.50
creatinine (mg/dl)		F	0.78	0.65	0.80
Slice/medium		M	12.0	2.88	1.36
	Control	M	8.69	8.41	8.54
PAH ratio		F	9.30	9.00	9.80
	Control	F	8.21	7.81	8.77
Renal cortical gentamicin		M	1132	1252	568
concentration (μg/g)		F	1492	1354	1370

[a] Data = means; n = minimum of 4.

compounds. Elsewhere in this volume (Chapter 16) Luft describes the effects of cephalosporins on the toxicity of gentamicin. Furosemide has been shown to enhance the nephrotoxicity of gentamicin, presumably via a volume-depletion mechanism [5,33].

On the other hand, the concurrent administration of carbenicillin appears to protect
the Sprague-Dawley rat from gentamicin nephrotoxicity [3]. This protective effect of carbenicillin is present not only in relation to renal function but also in the form of less
severe structural changes. In this instance, carbenicillin may function as a nonresorbable
anion in exerting its protective effect. In addition, a significant sodium load is provided
by the administration of carbenicillin, perhaps providing protection by a volume-related
mechanism.

E. Prolonged Gentamicin Administration

The histologic and metabolic consequences of the prolonged administration of gentamicin
has been examined in three experimental protocols. Cuppage et al. administered the drug
(3–40 mg/kg per day, given twice daily) to both Fischer 344 and Sprague-Dawley rats for
a period of 28 days [9]. When these animals were sacrificed at the conclusion of the treatment period, there was no elevation in BUN or serum creatinine in any of the treatment
groups. At the higher dose levels, only focal evidence of proximal tubular injury was
present. In addition, slight impairment of proximal function as determined by PAH
secretion and oxidative metabolism was noted. These findings are in marked contrast to
the structural and functional alterations noted following 14 days of gentamicin (40 mg/kg)
to Fischer 344 rats. However, of interest is the fact that significant proximal tubular
regenerative activity is present at the conclusion of 14 days of treatment in the Fischer
model.

Gilbert and colleagues [16] further explored the significance of regeneration at 14
days and focal histologic changes at 28 days by administering gentamicin (40 mg/kg per
day in two divided doses) for 42 consecutive days to Fischer 344 rats. Groups of animals
were sacrificed at intervals throughout the treatment period in order to observe the time
course of the gentamicin-induced lesions. At the conclusion of 21 days of continuous
treatment, there was a return to nearly normal renal cortical histology. This was in contrast to the necrosis of 90% of the proximal tubules noted at the tenth day of treatment.
Following 28 days of continuous treatment, treated and control animals could not be
distinguished by light microscopy. The serum creatinine reached a peak at day 14, returned to baseline by day 21, and remained there throughout the remainder of the
42 day treatment period (Figure 7). Of great interest were the marked cyclical changes
that the renal cortical gentamicin concentration underwent during the study (Figure 7).
When compared to the changes in serum creatinine, it is evident that no direct relationship exists between gentamicin-induced renal dysfunction and the associated tissue levels
of the drug. The same is true for the relationship of tissue concentration and histology.
Although serum creatinine and histology returned to normal by 21 days, the renal cortical slice transport of PAH remained suppressed from day 10 on, confirming a persistent
subtle toxic effect of gentamicin. The gentamicin-induced urinary hypoosmolality did not
return to baseline until day 35 of treatment, lagging significantly behind the normalization of histology and serum creatinine. None of the animals in this study failed to survive
the treatment period.

It is evident that the rat kidney is capable of regeneration and a return to essentially

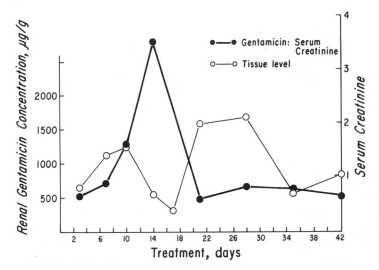

Figure 7 Sacrifice cortical gentamicin concentrations and serum creatinines in animals treated continuously with gentamicin, 40 mg/kg per day, for up to 42 days. Each point represents a mean of four animals.

normal function in spite of the continuous presence of a toxin that had previously induced severe structural and functional damage. The mechanisms for the resistance displayed by newly regenerated tubular cells to the toxic effects of gentamicin are not known. It has been previously shown that rat kidneys in the recovery phase from a previous acute insult may be resistant to a new acute challenge. However, these studies did not evaluate the continued presence of the toxin.

Data from this study confirm the disassociation of renal function and histology from concomitant renal tissue concentrations of gentamicin. The significance of gentamicin's reaccumulation in the renal cortex during the treatment interval is not clear. The fact that the drug is present and yet unable to induce a significant toxic effect suggests the possibility that newly regenerated cells do not possess sufficiently matured drug transport systems to allow gentamicin access to critical intracellular sites. On the other hand, it is possible that only a certain population of cells is susceptible to gentamicin-induced necrosis. Once these cells have undergone necrosis and subsequent regeneration, renal function and histology return to normal in spite of the continued presence of the toxin. The presence of the drug in the renal cortex in these situations probably represents nonspecific binding of gentamicin to the cortical tissue.

The duration of resistance of these newly regenerated cells to gentamicin is not known. Of great interest will be the results of rechallenge experiments designed to evaluate the susceptibility of regenerated renal tubular epithelia to gentamicin at a time remote from an initial necrosis producing gentamicin insult.

Luft and colleagues have demonstrated a similar recovery of function and histology following 21 days of continuous gentamicin administration to Sprague-Dawley rats (67.5 mg/kg per day) [37]. None of the animals in this study died from progressive uremia. In addition, during the last 7 days of treatment, these workers substituted netilmicin or tobramycin for gentamicin and found the same degree of renal structural and functional recovery at 21 days.

II. Tobramycin

The nephrotoxicity of tobramycin has been evaluated in much the same fashion as gentamicin. Animals that have been employed in these studies include rats, dogs, and guinea pigs. Although the quantity of data relative to nephrotoxicity is less for tobramycin as compared to gentamicin, sufficient information is available to characterize the nephrotoxic effects of tobramycin in the various animal models.

A. Histologic Alterations

Houghton et al. assessed the morphologic sequelae of tobramycin administration in Fischer 344 rats [22]. The drug was given in two divided doses totaling 40 or 120 mg/kg per day for up to 14 days. In these studies, the renal injury induced by tobramycin was found to be similar to that previously described for gentamicin but significantly less severe. Changes were confined primarily to the proximal tubules. Initial alterations at 3 days included the appearance of cytosomes and cytosegresomes, the number of which increased with continued drug administration (Figure 8). Light microscopy failed to demonstrate differences from control animals with either dose of tobramycin. By 7 days of treatment, rare foci of proximal tubular necrosis was noted by light but not by electron microscopy in both dosage groups. After ten days of 40 mg/kg per day of tobramycin, proximal tubular necrosis was only slightly more extensive than that noted earlier. No ultrastructural evidence of cellular necrosis was apparent. In comparison, the 120 mg/kg per day group displayed more extensive necrosis at 10 days; however, the total proportion of necrotic tubules was not in excess of 5% of the total proximal tubular area in any section. No further necrosis was evident at the conclusion of 14 days of treatment with 120 mg/kg per day. However, substantial proximal tubular regenerative activity was present (Figure 9).

Very similar results have been recorded by Fry et al. in studies utilizing guinea pigs [15]. Following 10 days of 25 mg/kg per day of tobramycin (once daily or in three divided doses) only rare areas of focal proximal necrosis were present by light microscopy. Electron microscopy displayed the typical lysosomal changes and myeloid bodies.

B. Renal Function Alterations

Excretory Function and Concentrating Ability

The functional sequelae of tobramycin has been evaluated by Houghton in the Fischer 344 rat as noted previously [22]. Following 3, 7, 10, and 14 days of tobramycin at 40 and 120 mg/kg per day (two divided doses), the blood urea nitrogen (BUN) and serum creatinine were modestly but significantly elevated (Table 4). However, there was no significant difference between animals in the 40 and 120 mg/kg per day groups. In contradistinction to the relatively minimal toxicity noted at these dose levels, tobramycin in a total daily dose of 180 mg/kg per day produces significant elevations in BUN and serum creatinine following only 3 days of administration. At the conclusion of 14 days of treatment, marked elevations of BUN and creatinine (means of 158 mg/dl and 3.2 mg/dl, respectively) are noted.

Mongrel dogs treated with tobramycin, 45 mg/kg per day, demonstrated a 10–40% reduction in creatinine clearance over a 10-day treatment period [44]. The degree of impairment was dependent on the frequency of drug administration (see below).

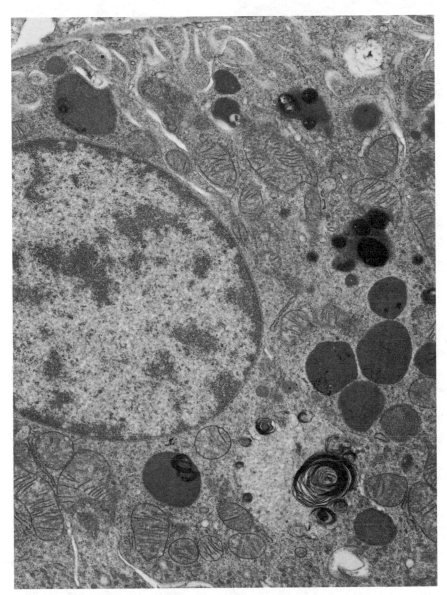

Figure 8 Electron micrograph of a Fischer rat proximal tubule cell after the animal had received 3 days of tobramycin, 120 mg/kg per day. The matrix of the large cytosome (lower right) has a more lucent, floccular appearance than the others and contains several myeloid bodies. Its limiting membrane is indistinct. Dense inclusions, many of them myeloid bodies, are also present in other cytosomes. (×14,600)

Engle and Abt administered tobramycin to dogs in doses of 25, 50, and 75 mg/kg per day (once daily intramuscularly (IM) for 14 days) [11]. A significant decline in renal function was noted only in the animals receiving the 75 mg/kg per day dose. Serum creatinine rose from 1.0 to 3.6 mg/dl, whereas the creatinine clearance fell from 42.5 to 7.8 ml/min. Glycosuria, proteinuria, and polyuria were noted in this treatment group, as were three deaths prior to the completion of treatment.

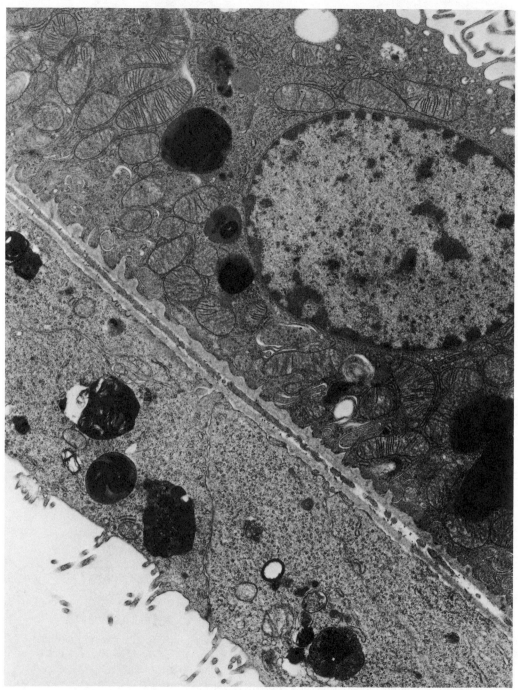

Figure 9 Electron micrograph of adjacent proximal tubules from a Fischer rat after it had received tobramycin, 120 mg/kg per day, for 14 days. The cells at the lower left are regenerative. Their cytoplasm is simplified, containing only a few mitochondria and cytosomes with myeloid bodies and having only a few microvilli. Polyribosomes are particularly numerous throughout the cytoplasm. Except for the presence of myeloid bodies, the cell at the upper right shows no significant alteration. (×14,600) (Reprinted from [22]. Used by permission.)

254

Table 4   Effect of Tobramycin (40, 120, and 180 mg/kg per day) on Renal Function[a]

Tobramycin	Days of treatment			
	3	7	10	14
40 mg/kg per day				
Serum creatinine (mg/dl)	0.67	0.86	0.74	—
Urinary osmolarity	1600	1550	1400	1400
120 mg/kg per day				
Serum creatinine (mg/dl)	0.80	0.75	0.70	0.73
Urinary osmolarity	2000	1500	1575	1400
180 mg/kg per day				
Serum creatinine (mg/dl)	0.77	1.66	5.25[b]	3.2
Urinary osmolarity	1800	940	426[b]	492

[a] Data = means.
[b] One animal.

Urinary osmolality decreases in response to tobramycin administration [17]. Following 7 of 14 days of treatment, Fischer 344 rats receiving 40 or 120 mg/kg per day of the drug exhibit significant decreases in their urinary osmolality (Table 4). Seven additional days of treatment result in little additional change. At no time during the 14 day treatment period were significant differences between the 40 mg/kg and 120 mg/kg dosage groups noted; 180 mg/kg per day, however, produced marked impairment in urinary concentrating ability when compared to the lower doses of tobramycin (Table 4).

The functional and histologic changes noted in Fischer rats receiving tobramycin, 40 and 120 mg/kg per day (two equal daily doses), are very similar in spite of marked differences in the renal tissue concentrations of the drug during the treatment period (Table 5). This disassociation of functional and structural changes from drug tissue concentrations is similar to that noted wtih gentamicin.

Organic Ion Transport

Tobramycin-induced alterations in renal cortical slice transport of organic ions have been evaluated in Fischer 344 rats only. Pretreatment of rats with tobramycin, 40 mg/kg per day (two divided doses), failed to induce significant changes in cortical slice PAH transport at any time during a 14 day experimental period (Table 6) [2]. However, treatment with 180 mg/kg per day induced significant changes in PAH uptake consisting of stimulation at 3 days of treatment followed by profound impairment at 7 days (Table 6). This inhibition of PAH transport by 180 mg/kg per day of tobramycin was coincident with the onset of marked azotemia (see above).

Slice transport of the organic base NMN was suppressed by both doses of tobramycin. At 40 mg/kg per day, impairment was significant at day 7 of treatment, whereas only three days of treatment were needed to suppress NMN transport with the 180 mg/kg per day dose.

The characteristics of organic ion transport seen with tobramycin pretreatment fit the

Table 5 Renal Cortical Concentrations of Tobramycin ($\mu g/g$)

Tobramycin	Days of treatment			
	3	7	10	14
40 mg/kg per day	505	590	624	480
120 mg/kg per day	838	1278	1404	1103

general pattern of aminoglycoside-induced changes in transport noted in the discussion of gentamicin. At a nonnephrotoxic dose (40 mg/kg) no initial stimulation of PAH transport is noted. However, a dose producing azotemia results in an initial stimulation followed by marked suppression of organic acid uptake by slices. Thus, the potential nephrotoxicity of a given aminoglycoside at a given dose seems to be predicted by the pattern of PAH transport.

C. Renal Parenchymal Accumulation

Tobramycin is selectively concentrated in the kidney. Following a single intravenous injection, Luft and Kleit found the half-life of tobramycin in renal tissue to be 74 hr [35]. As was the case for gentamicin, these workers found that 85% of the tobramycin present in renal tissue was located within the cortex.

In Fischer rats, Gilbert et al. noted that tobramycin given in a dose of 120 mg/kg per day (two divided doses) produced peak tissue levels approximately three times that seen with 40 mg/kg per day (Table 5). In spite of this concentration disparity, there was no significant difference in the functional alteration induced by the two dosage regimens.

D. Modifiers of Toxicity

In contradistinction to gentamicin, very little information is available with regard to the modification of tobramycin nephrotoxicity. Reiner and associates examined the effect of dosage frequency on toxicity in mongrel dogs [44]. They compared once daily intravenous (IV) injections with a continual IV infusion of tobramycin administered for 10 days in a total daily dose of 45 mg/kg.

In spite of the fact that the once-daily injections resulted in the highest peak serum

Table 6 Effect of Tobramycin on PAH Transport (Slice/Medium Ratios)

	Days of treatment			
	3	7	10	14
Control	8.61	8.69	8.41	8.54
Tobramycin, 40 mg/kg per day	9.16	8.89	9.70	9.74
Tobramycin, 180 mg/kg per day	9.40[a]	5.30[a]	1.53[a]	2.06[a]

[a]Significantly different from control.

concentrations, the maximum nephrotoxicity occurred in those animals receiving the drug on a continual basis. They were unable to correlate the degree of nephrotoxicity with either the renal tissue concentration or serum level of the drug at the time of death. As with gentamicin, it appears to be the total exposure time to elevated serum levels rather than the peak of the serum level that is important in determining the nephrotoxicity of a given dose of tobramycin.

III. Netilmicin

Netilmicin as a nephrotoxin has been evaluated in several animal models. The data accumulated suggest that its nephrotoxic potential is less than both gentamicin and tobramycin.

A. Histologic Alterations

The renal structural consequences of netilmicin administration, particularly those seen on light microscopy, are very similar qualitatively to those noted following gentamicin exposure. The changes induced by netilmicin do not demonstrate the same degree of dose dependency that has been observed with gentamicin.

The effect of netilmicin, 40 and 120 mg/kg per day (two divided doses) for 14 days to Fischer 344 rats, was examined by Ormsby and associates [40]. Following 3 days of treatment, the histologic changes were primarily confined to the cytosol and consisted of cytosegresomes with myeloid body inclusions. At 10 days of treatment rare foci of proximal tubule necrosis was present. Little deterioration in the histologic picture was present at the conclusion of 14 days of drug administration. Similar results have been obtained by Frame in rabbits [14] and Luft [38] and Bowman [4] in Sprague-Dawley rats.

B. Renal Function Alterations

The effect of netilmicin on excretory function, concentrating ability, organic ion transport, and tubular function has been examined in several studies.

Excretory Function and Concentrating Ability

Fischer 344 rats treated with 40 or 120 mg/kg per day (two divided doses) required 14 days of treatment to demonstrate a rise in serum creatinine when compared to controls [40]. A decrease in urine osmolality was noted at 3 days of treatment in the rats receiving 120 mg/kg per day and at 7 days in those receiving 40 mg/kg per day of the drug.

Increasing the dose of netilmicin to 180 mg/kg per day in the Fischer 344 rat model produces surprisingly little further deterioration in the serum creatinine. Following 3 days of treatment, no significant elevation of the serum creatinine is noted. At 7 and 14 days of treatment the serum creatinine is statistically significantly elevated compared to controls; however, the absolute changes are minimal (0.66 versus 0.50 mg/dl at 7 days, and 0.67 versus 0.49 mg/dl at 14 days). It can be seen that there is no progression of renal insufficiency between days 7 and 14.

Luft examined the consequences of daily netilmicin in a dosage range of 30–120 mg/kg per day (once daily) in Sprague-Dawley rats [38]. Treatment for 14 days failed to produce a significant decline in renal function as measured by creatinine clearance. Urinary osmolality was, however, decreased in a dose-dependent manner by 3 days of treatment in all the treatment groups. Very similar results have been recorded by Bowman et al. in Sprague-Dawley's [4]. In the rat it appears that the limiting factor in netilmicin dosage determinations is its relatively high acute toxicity. In Sprague-Dawley rats, doses in excess of 160 mg/kg per day produce mortality very early in the treatment period [6]. Similarly, when given to Fischer 344 rats at a dose of 240 mg/kg per day, death occurs during the first 4 days of the treatment period.

Frame and colleagues treated rabbits with netilmicin (60 mg/kg or 30 mg/kg every 8 hours) for 21 and 11 days, respectively [14]. With the exception of one animal that died at the end of 3 days of treatment (30 mg/kg), no significant elevations in serum creatinine, when compared to pretreatment values, were noted. These data are made difficult to interpret, however, because of the absence of control animals in the treatment groups.

In two separate studies, dogs receiving netilmicin for 14 and 28 days in doses ranging from 20 to 80 mg/kg per day failed to develop changes in serum creatinine or creatinine clearance [11,52].

Organic Ion Transport

There is a paucity of published data regarding the effects of netilmicin pretreatment on organic ion transport in renal cortical slices. Bowman et al. examined PAH transport in cortical slices obtained from Sprague-Dawley rats pretreated with 100 mg/kg of netilmicin daily for 2 days [4]. They demonstrated a marked stimulation of PAH uptake compared to saline controls (slice/medium ratios of 18.1 versus 10.5, respectively) 24 hr following the last dose of the drug. (The absolute values of the slice/medium ratios in this study are, in part, explained by the use of acetate in the incubation media.) The sequential effect of netilmicin on PAH or NMN transport during the course of a prolonged treatment period was not reported in this study.

Netilmicin-induced alterations in organic ion transport have been examined in greater detail in the Fischer 344 rat. When treated with netilmicin (40 mg/kg per day in two divided doses), the cortical slice transport of PAH is unaffected up to day 7 of treatment. At this point, a slight but significant stimulation of PAH uptake is noted. With continued treatment, there is a gradual decline in uptake such that at 10 days, treated and controls are no different, whereas at 14 days, PAH uptake in the treated animals is significantly depressed (Table 7). Increasing the dose of netilmicin to 180 mg/kg per day results in no significant change in PAH slice transport until day 14, at which time impairment of transport is noted (Table 7).

Netilmicin-induced PAH transport alterations are in marked contrast to those seen in Fischer rats receiving 40 mg/kg per day of gentamicin, 180 mg/kg per day of tobramycin, and 150 mg/kg per day of neomycin. The latter regimens are associated with early, very significant stimulation of PAH uptake, followed by profound transport impairment and tubular necrosis.

The significance of the slight stimulation in PAH transport noted at 7 days of treatment with 40 mg/kg per day of netilmicin is not known. However, it is clear that the degree of stimulation of PAH uptake induced by nonnephrotoxic doses of netilmicin is much less and occurs much later than that seen with other aminoglycosides given in doses

Table 7 Effect of Netilmicin on PAH Transport (Slice/Medium Ratios)

| | Days of treatment | | | |
	3	7	10	14
Control	8.10	7.80	8.30	9.5
Netilmicin, 40 mg/kg per day	9.04	9.52[a]	8.13[a]	7.9[a]
Netilmicin, 180 mg/kg per day	8.30	7.60	8.38	6.80[a]

[a]Significantly different from control.

resulting in significant renal functional and structural alterations. These data would support the utility of aminoglycoside-induced alterations in organic anion transport as predictors of nephrotoxic potential in the Fischer rat. The conflicting data presented by Bowman in Sprague-Dawley rats may be explained by strain variability and/or the utilization of acetate in the incubation media.

Tubular Function

Chiu examined the effect of netilmicin on urinary glucose and fractional excretion of potassium [6]. Sprague-Dawley rats failed to develop glucosuria when treated with netilmicin in doses up to 170 mg/kg per day for 7 days. The fractional excretion of potassium was, however, increased in these treatment groups. When compared to gentamicin-treated rats, the increase in fractional potassium excretion was much less dramatic.

C.   Renal Parenchymal Accumulation

As is the case with gentamicin, netilmicin is preferentially accumulated in renal cortical tissue. Multiple investigators have observed renal cortical concentrations of netilmicin in excess of those found in the medulla [4,38,40].

Luft noted no significant differences in cortical concentrations of netilmicin in Sprague-Dawley rats receiving 30–120 mg/kg per day for 15 days. Ormsby, however, observed cortical concentrations of netilmicin three times greater at 14 days in Fischer rats receiving 120 mg/kg per day compared to those receiving 40 mg/kg per day. In spite of this substantial concentration difference, there were no renal functional differences between these treatment groups. Doses of 180 mg/kg per day of netilmicin given to Fischer rats produce cortical concentrations of the drug at 14 days that are comparable to those seen with 120 mg/kg per day.

It is clear from the tissue concentrations of netilmicin in these several studies that there is no direct relationship between the quantity of drug present in the renal tissue and subsequent nephrotoxicity. In spite of achieving peak tissue concentrations very similar to those seen with nephrotoxic doses of gentamicin, netilmicin fails to produce significant renal insufficiency with the doses employed in the various animal models.

D.   Modifiers of Toxicity

In contrast to gentamicin, only one study has been designed to evaluate modifiers of netilmicin's effects on renal function. Luft and associates explored the effects of concomitantly administered nonaminoglycoside antibiotics on the potential nephrotoxicity of netilmicin [19]. Ampicillin, carbenicillin, methicillin, cefamadole, and clindamycin all failed to enhance the nephrotoxic effects of either 30 or 60 mg/kg per day of netilmicin given to Sprague-Dawley rats.

IV. Amikacin

Data on the nephrotoxicity of amikacin generated from animal studies are scant. Information is available for the cat, rat, and rabbit.

A. Histologic Alterations

As is the case with gentamicin, amikacin produces dose-dependent ultrastructural and light microscopic alterations in the renal cortex of the Fischer 344 rat. Changes are limited to the convoluted portion of the proximal tubules. When given twice daily for 14 days in a total dose of 120 mg/kg per day, distinctive morphologic changes, primarily ultrastructural, are noted in proximal tubules. However, there is neither light nor electron microscopic evidence of severe cell injury. When the dose is increased to 360 mg/kg per day, progressive, severe proximal tubular injury results. By 10 days of treatment, tubular necrosis is noted in 10–50% of the outer cortical tubules (Figure 10). In addition, active tubular epithelial regeneration is present. At the conclusion of 14 days of treatment, acute necrotic activity has generally resolved, replaced in large part by epithelial regeneration.

Ultrastructural changes noted at both 120 mg/kg per day and 360 mg/kg per day consist of cytosomes, cytosegresomes, and myeloid bodies. At the higher dose, all stages of cell injury are seen, including mitochondrial disruption, dilitation of the endoplasmic reticulum, and cellular edema.

Similar findings were noted by Hottendorf in Sprague-Dawley rats treated with amikacin (50 or 100 mg/kg per day, once daily) for 28 days [20]. The lower-dose animals displayed no tubular necrosis, whereas at the higher dose, scattered, focal areas of proximal tubular necrosis were noted. In the same study, cats receiving either 45 or 90 mg/kg per day of amikacin did not develop significant light microscopic changes at the conclusion of 28 days of therapy.

B. Renal Function Alterations

Amikacin-induced functional changes follow a pattern similar to that seen with the structural consequences noted above. The degree of functional alteration is dose-dependent.

Excretory Function

Amikacin induces no change in BUN or serum creatinine when administered to Fischer 344 rats in a total daily dose of 120 mg/kg (two divided doses) for 14 days. Tripling the dose to 360 mg/kg, however, results in profound derangements in excretory function. Significant rises in serum creatinine do not occur until after day 7 of treatment. By 10 days the serum creatinine has risen to 1.91 mg/dl (baseline 0.5 mg/dl) and rises to a peak of 4.25 mg/dl at the conclusion of 14 days of drug administration. In spite of the significant azotemia (BUN 156 mg/dl) induced by this dose of amikacin, no mortality is noted

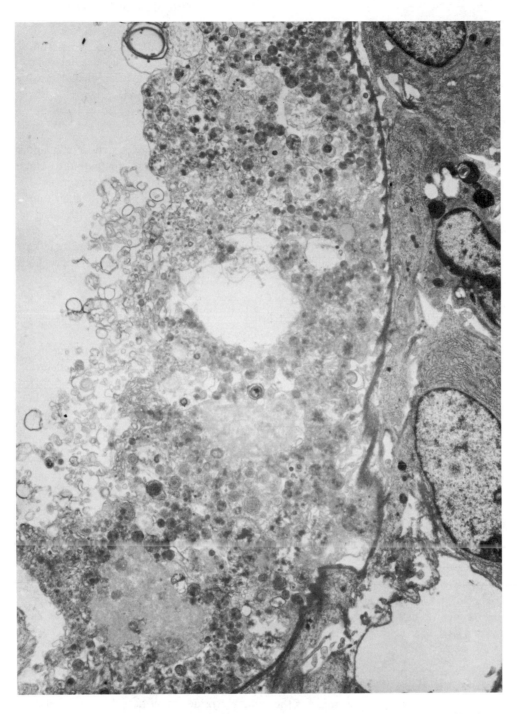

Figure 10 Electron micrograph of a Fischer rat proximal tubule after the animal had received amikacin, 360 mg/kg per day, for 10 days. A severely injured cell consists of a chaotic mixture of disrupted organelles. Attachment to the basal lamina appears tenuous. (×3400)

with 14 continuous days of treatment. This is compatible with the histologic picture described earlier of resolving necrosis and active regeneration. Luft et al. have noticed a similar dose dependency on alterations in renal function and histology in Sprague-Dawley rats treated with 150–375 mg/kg per day of amikacin for 15 days [34].

Organic Ion Transport

As is the case with other nonnephrotoxic doses of aminoglycosides, amikacin in a dose of 120 mg/kg per day (Fischer 344 rats, two divided doses) failed to stimulate the uptake of PAH by renal cortical slices. By day 7 of treatment, PAH transport was suppressed relative to controls. In contrast, 360 mg/kg per day of amikacin produced significant stimulation in PAH uptake by day 7. This stimulation persisted through day 10, being followed by marked impairment in transport on day 14 of treatment.

Transport of the organic base NMN is impaired by the third day of either 120 mg/kg or 360 mg/kg per day of amikacin. This suppression of base transport persists throughout 14 days of treatment with the drug.

C. Renal Parenchymal Accumulation

Amikacin is concentrated in the renal cortex in much the same fashion as gentamicin. Peak cortical concentrations are dose-dependent, with a level of 5394 μg/g of wet tissue weight occurring following 10 days of 360 mg/kg per day in Fischer rats.

In rabbits, the kidney is the only organ to accumulate amikacin following a single injection of the drug [30]. At the conclusion of 7 days of treatment, the kidney remains the prime site of drug accumulation, although detectable levels are present in other tissues.

V. Sisomicin

There is little published information evaluating the nephrotoxic potential of sisomicin in animal models. Luft and colleagues have examined the nephrotoxicity of the drug in Sprague-Dawley rats [34]. Treatment with 30–75 mg/kg per day for 15 days failed to produce significant changes in renal tubular handling of sodium, potassium, glucose, or phosphate. However, significant elevations of BUN and serum creatinine, as well as evidence of drug-induced histological changes, were present at all dosages of sisomicin. Fabre and associates have demonstrated the preferential concentration of the drug in the renal cortex as well as a prolonged tissue half-life [12].

VI. Neomycin

Neomycin is the most nephrotoxic of the aminoglycoside antibiotics.

A. Histologic Alterations

In doses ranging from 25 to 100 mg/kg per day, dose-dependent renal tubular damage and/or necrosis has been noted in rabbits, rats, dogs, and guinea pigs [10,26,39]. The histologic changes are consistent from species to species with epithelial injury and necrosis being confined to the proximal tubule. Typical changes include foamy vacuolization of the proximal tubule epithelium. The nuclei are displaced by the vacuoles and are frequently pyknotic or frankly necrotic. Glomerular changes are not noted by light microscopy. In animals that survive the neomycin-induced acute renal failure, regeneration of the necrotic tubular epithelium is observed.

B. Renal Function Alterations

The functional sequelae of neomycin administered in a total daily dose of 150 mg/kg (two divided doses) to Fischer 344 rats has been examined by Bennett and co-workers [2]. The serum creatinine was significantly elevated by day 7 of treatment, reaching a level of 9.1 mg/dl on day 10 of drug administration. Death from uremia was noted in all rats treated beyond 10 days with this dose of neomycin.

Organic ion transport in renal cortical slices obtained from these animals was characteristic of that seen following exposure to nephrotoxic aminoglycosides. PAH uptake was significantly enhanced following 1 day of treatment. By day 7, however, this enhancement had been converted to a significant depression that was most profound on day 10. The impaired PAH transport was coincident with histologic evidence of nearly total proximal tubular necrosis. Studies beyond 10 days were not possible, owing to animal attrition.

VII. Kanamycin

Kanamycin-induced nephrotoxicity has been evaluated in mice, rats, cats, and dogs. The bulk of data relate to histologic alterations with little information relative to renal functional impairment.

A. Histologic Alterations

No histologic changes were noted by Tisch and associates when kanamycin was administered in doses up to 100 mg/kg per day to dogs, rats, and cats for time periods ranging from 40 days to 9 months [50]. When these authors increased the dosage to 100–200 mg/kg per day, cloudy swelling of the proximal tubular cells together with focal areas of tubular necrosis became evident. Animals exposed to doses in excess of 200 mg/kg per day developed generalized proximal tubular necrosis, the severity of which was dose-related. Histologic alterations in the glomeruli were not noted.

Rats employed in this study were both males and females. It is interesting that the authors noted more severe necrosis of the renal tubular epithelium in the males as compared to the females at any given dose of the drug.

B. Renal Function Alterations

In the dog, doses of kanamycin in excess of 200 mg/kg per day result in albuminuria, hematuria, altered PSP excretion, and an elevated BUN [55]. Anuria ensues if the drug is continued in this dosage range.

VIII. Streptomycin

Streptomycin is the least nephrotoxic of the aminoglycosides. Early reports of nephrotoxic potential have since been discounted on the basis of impurities in the drug preparations employed in the initial studies. Recent work by Luft and associates support this conclusion [34]. These workers were unable to demonstrate significant changes in serum creatinine, tubular function, or renal histology as a consequence of treating Sprague-Dawley rats with 150-375 mg/kg per day of streptomycin for 15 consecutive days.

In a separate study, Luft and Kleit evaluated the renal parenchymal accumulation characteristics of streptomycin [35]. Unlike the other nephrotoxic aminoglycosides, streptomycin was not found to be concentrated in the renal cortex. Rather, the drug was evenly dispersed between cortex and medulla. In addition, the renal tissue half-life of streptomycin was only 4.6 hr, significantly shorter than the 109 hr recorded for gentamicin. The relationship, if any, between streptomycin's apparent lack of nephrotoxicity and its failure to concentrate in the renal cortex remains to be determined.

IX. Comparative Nephrotoxicity

Several studies have been published comparing the nephrotoxic potentials of two or more of the aminoglycosides. These studies are important because, in large measure, they minimize the risk of comparing results that have been obtained in different species or strains using varying doses or frequency of administration.

In the rat, the rank order of nephrotoxicity (most to least) is (1) neomycin [34]; (2) gentamicin [40]; (3) sisomicin, amikacin, and kanamycin [2]; (4) tobramycin [15]; (5) netilmicin [22]; and (6) streptomicin [4]. The basis for this differential in nephrotoxicity is not known. However, it is clear from these studies that with the possible exception of streptomycin, the degree of drug accumulation in the renal cortex does not correlate with the differences in nephrotoxic potential.

One comparative study is available utilizing the guinea pig [60]. In this model, gentamicin produced significantly greater elevations of BUN as compared to tobramycin. Following 10 days of treatment (25 mg/kg per day), light microscopy was unable to differentiate gentamicin-treated animals from those receiving tobramycin. Electron microscopic evidence of proximal tubular necrosis was present, however, in the gentamicin animals only. Both drugs induced characteristic lysosomal changes and the development of myeloid bodies. These ultrastructural changes of cytosomes and cytosegresomes are expected sequelae of all aminoglycosides and are consistently observed in each of the various animal models. They function not as predictors of toxicity but rather as markers of exposure to aminoglycosides.

Igarashi and colleagues compared netilmicin and gentamicin in squirrel monkeys [24]. Although the dosages of netilmicin (50-100 mg/kg per day) were somewhat greater than those of gentamicin (25-75 mg/kg per day), the nephrotoxicity of gentamicin as measured by morbidity and serum creatinine was consistently greater than netilmicin. Treatment in this study was continued for 21 days in surviving animals.

Acknowledgments

Some of the studies reported in this chapter were supported, in part, by grants from the National Institutes of Health and the Oregon Heart Association.

References

1. Bennett, W.M., Hartnett, M.N., Gilbert D., Houghton, D., and Porter, G.A.: Effect of sodium intake on gentamicin nephrotoxicity in the rat. *Proc. Soc. Exp. Biol. Med.* 151:736–738, 1976.

2. Bennett, W.M., Plamp, C.E., Parker, R.A., Gilbert, D.N., Houghton, D.C., and Porter, G.A.: Renal transport of organic acids and bases in aminoglycoside nephrotoxicity. *Antimicrob. Agents Chemother.*, 1979, in press.

3. Bloch, R., Luft, F.C., Rankin, L.I., Sloan, R.S., Yum, M.N., and Maxwell, D.R.: Protection from gentamicin nephrotoxicity by cephalothin and carbenicillin. *Antimicrob. Agents Chemother.* 15:46–49, 1979.

4. Bowman, R.L., Silverblatt, F.J., and Kaloyanides, G.J.: Comparison of nephrotoxicity of netilmicin and gentamicin in rats. *Antimicrob. Agents Chemother.* 12:474–478, 1977.

5. Chiu, P.J.S., and Long, J.F.: Effect of hydration of gentamicin excretion and renal accumulation in furosemide-treated rats. *Antimicrob. Agents Chemother.* 14:214–217, 1978.

6. Chiu, P.J., Miller, G.H., Brown, A.D., Long, J.F., and Waitz, J.A.: Renal pharmacology of netilmicin. *Antimicrob. Agents Chemother.* 11:821–825, 1977.

7. Cohen, L., Lapkin, R., and Kaloynides, G.J.: Effect of gentamicin on renal function in the rat. *J. Pharmacol. Exp. Ther.* 193:264–273, 1975.

8. Cronin, R.E.: Aminoglycoside induced glycosuria in the dog. *Clin. Res.* 26:461A, 1978.

9. Cuppage, F.E., Setter, K., Sullivan, L.P., Reitzes, E.J., and Melnykovych, A.O.: Gentamicin nephrotoxicity. *Virchows Arch [Cell Pathol.]* 24:121–138, 1977.

10. Einspruch, B.C., and Gonzalez, V.V.: Clinical and experimental nephropathy resulting from use of neomycin sulfate. *JAMA* 173:809–811, 1960.

11. Engle, J.E., and Abt, A.: Comparison of nephrotoxicity of netilmicin and tobramycin in dogs. *Am. Soc. Nephrol.*, 14a, 1977.

12. Fabre, J., Rudhardt, M., Blanchard, P., and Regamey, C.: Persistence of sisomicin and gentamicin in renal cortex and medulla compared with other organs and serum of rats. *Kidney Int.* 10:444–449, 1976.

13. Frame, P.T., Phair, J.P., Watanakunakorn, C., and Bannister, T.W.P.: Pharmacologic factors associated with gentamicin nephrotoxicity in rabbits. *J. Infect. Dis.* 135:952–956, 1977.

14. Frame, P.T., Watanakunakorn, C., and Cluxton, R.J.: Comparative nephrotoxicity of netilmicin and gentamicin in rabbits. *Curr. Ther. Res.* 24:344–351, 1978.

15. Fry, T.L., Fried, F.A., and Goven, B.A.: Renal toxicity: tobramycin and gentamicin. *Invest. Urol.* 15:100–103, 1977.

16. Gilbert, D.N., Houghton, D.C., Bennett, W.M., Plamp, C.E., Reger, K., and Porter, G.A.: Reversibility of gentamicin nephrotoxicity in rats: recovery during continuous drug administration. *Proc. Soc. Exp. Biol. Med.* 160:99–103, 1979.

17. Gilbert, D.N., Plamp, C.E., Starr, P., Bennett, W.M., Houghton, D.C., and Porter, G.A.: Comparative nephrotoxicity of gentamicin and tobramycin in rats. *Antimicrob. Agents Chemother.* 13:34–40, 1978.

18. Ginsburg, D.S., Quintanilla, A.P., and Levin, M.: Renal glycosuria due to gentamicin in rabbits. *J. Infect. Dis.* 134:119–122, 1976.

19. Hagstrom, G.L., Luft, F.C., Yum, M.N., Sloan, R.S., and Maxwell, D.R.: Nephrotoxicity of netilmicin in combination with non-aminoglycoside antibiotics. *Antimicrob. Agents Chemother.* 13:490–493, 1978.

20. Hottendorf, G.H.: Comparative ototoxicity (cats) and nephrotoxicity (rats) of amikacin and gentamicin. U.S. Amikacin Symposium, University of California Medical School, Los Angeles, Calif., Nov. 9–10, 1976.

21. Houghton, D.C., Hartnett, M., Campbell-Boswell, M., Porter, G.A., and Bennett, W.M.: A light and electron microscopic analysis of gentamicin nephrotoxicity in rats. *Am. J. Pathol.* 82:589–599, 1976.

22. Houghton, D.C., Plamp, C.E., DeFehr, J.M., Bennett, W.M., Porter, G.A., and Gilbert, D.N.: Gentamicin and tobramycin nephrotoxicity. *Am. J. Pathol.* 93:137–147, 1978.

23. Hsu, C.H., Kurtz, T.W., Easterling, R.E., and Weller, J.M.: Potentiation of gentamicin nephrotoxicity by metabolic acidosis. *Proc. Soc. Exp. Biol. Med.* 146:894–897, 1974.

24. Igarashi, M., Levy, J.K., and Jerger, J.: Comparative toxicity of netilmicin and gentamicin in squirrel monkeys. *J. Infect. Dis.* 137:476–480, 1978.

25. Just, M., Erdmann, G., and Habermann, E.: The renal handling of polybasic drugs. *Arch. Pharmacol.* 300:57–66, 1977.

26. Karlson, A.G., Gainer, J.H., and Feldman, W.H.: The effect of neomycin on tuberculosis in guinea pigs infected with streptomicin-resistant tuburcle bacilli. *Am. Rev. Tuberc.* 62:345, 1950.

27. Karniski, L., Chonko, A., Steven, Y.R., Cuppage, F., and Hodges, G.: The effects of gentamicin on renal function in the sexually mature versus sexually immature rabbit. *Am. Soc. Nephrol.*, 92A (abstr.), 1978.

28. Kluwe, W.M., and Hook, J.B.: Analysis of gentamicin uptake by rat renal cortical slices. *Toxicol. Appl. Pharmacol.* 45:531–539, 1978.

29. Kluwe, W.M., and Hook, J.B.: Functional nephrotoxicity of gentamicin in the rat. *Toxicol. Appl. Pharmacol.* 45:163–175, 1978.

30. Kornguth, M.L., and Kunin, C.M.: Distribution of gentamicin and amikacin in rabbit tissues. *Antimicrob. Agents Chemother.* 11:974–977, 1977.

31. Kosek, J.C., Mazze, R.I., and Cousins, M.J.: Nephrotoxicity of gentamicin. *Lab. Invest.* 30:48–57, 1974.

32. Lapkin, R., Bowman, R., and Kaloyanides, G.J.: Effect of gentamicin on p-aminohippurate metabolism and transport in rat kidney slices. *J. Pharmacol. Exp. Ther.* 201:233–242, 1977.

33. Lawson, D.H., Macadam, R.F., Singh, H., Gavras, H., Hartz, S., Turnbull, D., and Linton, A.L.: Effect of furosemide on antibiotic-induced renal damage in rats. *J. Infect. Dis.* 126:593–600, 1972.

34. Luft, F.C., Bloch, R., Sloan, R.S., Yum, M.N., Costello, R., and Maxwell, D.R.: Comparative nephrotoxicity of aminoglycoside antibiotics. *J. Infect. Dis.* 138:541–545, 1978.

35. Luft, F.C., and Kleit, S.A.: Renal parenchymal accumulation of aminoglycoside antibiotics in rats. *J. Infect. Dis.* 130:656–659, 1974.

36. Luft, F.C., Patel, V., Yum, M.N., Patel, B., and Kleit, S.A.: Experimental aminoglycoside nephrotoxicity. *J. Lab. Clin. Med.* 86:213–220, 1975.

37. Luft, F.C., Rankin, L.I., Sloan, R.S., and Yum, M.N.: Recovery from aminoglycoside nephrotoxicity with continued drug administration. *Antimicrob. Agents Chemother.* 3:284–287, 1978.

38. Luft, F.C., Yum, M.N., and Kleit, S.A.: Comparative nephrotoxicities of netilmicin and gentamicin. *Antimicrob. Agents Chemother.* 10:845–849, 1976.

39. Nelson, A.A., Radomski, J.L., and Hagan, E.C.: Renal and other lesions in dogs and rats from intramuscular injection of neomycin. *Fed. Proc.* 10:366–367, 1951.

40. Ormsby, A.M., Parker, R.A., Plamp, C.E., Stevens, P., Houghton, D.C., Gilbert, D.N., and Bennett, W.M.: Comparison of the nephrotoxic potential of gentamicin, tobramycin and netilmicin in the rat. *Curr. Ther. Res.* 25:335–343, 1979.

41. Parker, R.A., Porter, G.A., and Bennett, W.M.: Gentamicin induced resistance to ADH-stimulated water flow in the toad urinary bladder: partial correction by prostaglandin inhibition. *Clin. Res.* 27:64A, 1979.

42. Plamp, C.E., Bennett, W.M., Gilbert, D.N., Parker, R.A., and Porter, G.A.: The influence of dosage regimen on experimental gentamicin nephrotoxicity: dissociation of peak and serum levels from renal failure. *J. Infect. Dis.,* 1979, in press.
43. Plamp, C.E., Reger, K., Bennett, W.M., McClung, M., and Porter, G.A.: Vasopressin-resistant polyuria in gentamicin nephrotoxicity. *Clin. Res.* 26:151A, 1978.
44. Reiner, N.E., Bloxham, D.D., and Thompson, W.L.: Nephrotoxicity of gentamicin and tobramycin given once daily or continuously in dogs. *J. Antimicrob. Chemother.* 4:85–101, 1978.
45. Souliere, C.R., Goodman, D.B.P., Appel, G.B., and Forrest, J.N.: Gentamicin selectively inhibits antidiuretic hormone induced water flow in the toad urinary bladder. *Am. Soc. Nephrol.,* 99A (abstr.), 1978.
46. Sugarman, A., Brown, R.S., and Rosen, S.: Features of gentamicin nephrotoxicity in the rat. *Clin. Res.* 25:449A, 1977.
47. Szwed, J.J., Luft, F.C., Black, H.R., Elliott, R.A., and Kleit, S.A.: Comparison of the distribution of tobramycin and gentamicin in body fluids of dogs. *Antimicrob. Agents Chemother.* 5:444–446, 1974.
48. Thiel, G., de Rougemont, D., Konrad, L., Oeschger, A., Torhorst, J., and Brunner, F.: Gentamicin induced acute renal failure in the rat. 7th International Congress of Nephrology, D45 (abstr.), 1978.
49. Thompson, W.L., Reiner, N.E., and Bloxham, D.D.: Gentamicin and tobramycin nephrotoxicity in dogs on continuous or once daily intravenous injection. In *Proceedings of the 10th International Congress of Chemotherapy.* Washington, D.C., American Society for Microbiology, 1977, p. 207.
50. Tisch, D.E., Huftalen, J.B., and Dickinson, H.L.: Pharmacological studies with kanamycin. *Ann. N.Y. Acad. Sci.* 76:44–64, 1958.
51. Vera-Roman, J., Krishnakantha, T.P., and Cuppage, F.E.: Gentamicin nephrotoxicity in rats. *Lab. Invest.* 33:412–417, 1975.
52. Weinberg, E., Szot, R., McCormick, G., Christie, B., and Schwartz, E.: Comparative nephro- and ototoxic effects of netilmicin and tobramycin given intramuscularly to dogs. Abstr., *18th Interscience Conference on Antimicrobial Agents and Chemotherapy.* Washington, D.C., American Society for Microbiology, 1978, p. 329.

11

AMINOGLYCOSIDE NEPHROTOXICITY: PHYSIOLOGIC STUDIES OF THE SITES OF NEPHRON DAMAGE

GERALD B. APPEL   Columbia University College of Physicians and Surgeons, Columbia Presbyterian Medical Center, New York, New York

The nephrotoxic potential of the aminoglycoside antibiotics has been amply confirmed in both clinical studies and in animal models [2,3]. Although the pattern of renal injury has been well described both morphologically and in terms of commonly employed measures of kidney function, few studies have focused on the site(s) of physiologic damage within the nephron. Fewer still have been studies that attempt to relate such physiologic alterations to the mechanisms of the acute renal failure produced by the aminoglycoside antibiotics. Knowledge of the sequence and severity of physiologic damage to various nephron sites may be of considerable value to the clinician searching for signs of early aminoglycoside nephrotoxicity. Physiologic studies of the sites of nephrotoxicity may lead to more accurate experimental comparative studies of nephrotoxicity among the aminoglycoside antibiotics and therefore help to produce less toxic newer agents. Such studies may also prove of value in studying therapeutic methods of preventing aminoglycoside-induced renal damage. Finally, because the aminoglycoside antibiotics induce a nonoliguric pattern of acute renal failure, they provide a clinically relevant model for the study of the pathophysiologic mechanisms of this pattern of renal damage, which is being encountered more and more frequently.

A major problem in the study of aminoglycoside, or indeed any drug-induced, renal damage lies with the definition of "nephrotoxicity." To the pathologist nephrotoxicity may include a spectrum of morphologic alterations ranging from minor abnormalities of the lysosomes and mitochondria of the proximal tubular cells as seen by electron microscopy to frank tubular cell necrosis apparent under low-power light microscopy. To the research investigator or renal physiologist, subtle tubular defects such as increased urinary excretion of proximal tubular cell lysosomal enzymes or an impairment in maximum concentrating ability may signify renal damage. To the clinician nephrotoxicity virtually always implies a decrease in glomerular filtration rate (GFR) as noted by a rise in serum creatinine or blood urea nitrogen (BUN). It is clear from numerous investigations that many morphologic changes, as well as abnormalities in sensitive indicators of renal tubular function, do not always correlate with alterations in the GFR [17,31,34].

Although the physiologist, morphologist, and research investigator are all clearly describing evidence of renal damage induced by the aminoglycoside antibiotics, these changes may have little immediate relevance to the clinician unless good correlations are established with the clinical occurrence of nephrotoxicity. Thus, despite their value in understanding the mechanisms and patterns of drug toxicity, it is best to be cautious in translating the results of these investigations into clinical practice. This is certainly true of in vitro techniques as well as investigations in animal models, where methods of drug administration, species-strain differences, and experimental conditions can have profound effects upon the results of such studies [3].

This chapter reviews physiologic studies related to aminoglycoside nephrotoxicity and attempts to define the functional sites of damage within the nephron. These findings are related to morphologic and biochemical studies as well as other measures of renal function and dysfunction whenever possible in an attempt to clarify the mechanism(s) of renal failure in aminoglycoside nephrotoxicity. The relationship of these findings to the clinical pattern of renal damage seen in humans is also discussed.

I. Pattern of Aminoglycoside Nephrotoxicity

The pattern of aminoglycoside nephrotoxicity has been well delineated in animal models [2,3]. Initially, there is increased excretion of lysosomal enzymes in the urine [32,42, 59], followed by polyuria and a urinary concentrating defect [4,33,59]. Mild proteinuria and granular cylindruria subsequently herald the decline in glomerular filtration rate, as manifested by an elevation of serum creatinine and BUN [29,32,33]. The renal failure is progressive and nonoliguric in nature [22,33,59].

The most striking morphologic correlates of this nephrotoxicity occur predominantly in the pars convoluta of the proximal tubule [29,32,58]. Here, in animals treated with as low as 1 mg/kg of body weight of gentamicin, significant electron microscopic alterations are noted within 48 hr of drug administration [29]. These consist of the formation of lysosomal cytosegrosomes containing dense lamellated myeloid bodies [7,22,29,33]. Gentamicin has been well localized in these lysosomal bodies [24], which may represent either phagocytozed residues of cellular organelles or the aminoglycosides' cationic binding to anionic lipoproteins of the lysosomal plasma membrane [2]. With further administration of the aminoglycoside, these myeloid bodies increase in number and spill into the tubular lumens and the urinary sediment [22]. Ultimately, damage to other cell organelles appear and the proximal tubular cells undergo cell necrosis [22,28]. There is only scant evidence for any morphologic changes in the glomeruli in this spectrum of tubular damage [34,45].

II. Sodium Reabsorption

With such impressive alterations in the proximal tubular cells, it is not surprising that attention has been focused on the physiologic correlates of these morphologic abnormalities. A number of studies have focused on the ability of the renal tubule to perform its normal physiologic functions during aminoglycoside-induced nephrotoxicity. Of these functions the reabsorption of Na^+ is paramount. It is estimated that between 90 and 95% of renal tubular oxygen consumption is utilized merely in the reabsorption of filtered sodium. In both human and animal models of acute tubular necrosis (ATN), sodium reabsorption is impaired and the measurement of urinary Na^+ concentration in the presence of azotemia will reveal relatively high values. At the same time the urinary creatinine concentration, being diluted by a large volume of unreabsorbed "water," remains relatively low. These values contrast with the findings in "prerenal" azotemia, where the nephron being underperfused avidly reabsorbs "salt" and water, leading to a low urinary sodium concentration and a high urinary creatinine concentration [15,21]. These physiologic relationships are exploited by clinical and research nephrologists to determine whether the tubule is adequately performing its function of NaCl reabsorption in the azotemic state. The fractional excretion of sodium (F_ENa^+), the ratio of the amount of Na^+ excreted in the urine to the amount filtered by the glomerulus, is an accurate indicator of NaCl reabsorption by the nephron [15]. In acute azotemia it is high in states of tubular dysfunction (>1%) and low in conditions of prerenal azotemia (<1%) [1]. In humans the F_ENa^+ has been elevated in aminoglycoside nephrotoxicity, confirming the state of tubular dysfunction [19]. Table 1 shows representative values for F_ENa^+ in Sprague-Dawley rats treated with gentamicin for varying periods of time. It can be seen that with nephrotoxicity and decreasing GFR, the F_ENa^+ dramatically rises to levels that are very high for the experimental conditions of this study.

The elevated F_ENa^+ neither localizes the site of tubular damage along the nephron, nor allows further insight into the mechanism of aminoglycoside nephrotoxicity. It does, however, allow the clinician to distinguish true nephrotoxicity from prerenal azotemia in the seriously ill patient, and confirms the classic pattern of "ATN-type" renal failure that occurs with drugs of this class.

Table 1 GFR, Urinary Flow Rate, and F_ENa^+ in Rats treated with 80 mg/kg of Gentamicin Body Weight for 5, 10, and 12 Days

	GFR (% of control value)	Urine flow rate (% of control volue)	F_ENa^+ (%)
Control	100	100	0.04
5 Days	98	165	0.13
10 Day	52	249	2.16
12 Days	15	380	2.75

III. Enzymuria

The release of lysosomal enzymes and brush border enzymes has been extensively studied during aminoglycoside nephrotoxicity. Lysosomal enzymuria occurs early in the course of aminoglycoside administration [33,42,59] and a dramatic rise in the urinary levels of these enzymes has been claimed to correlate with the induction of renal damage [42,60]. Although such studies on uninary enzymes will be reviewed elsewhere in this text, certain words of caution are warranted.

We have recently reviewed the lysosomal enzyme excretion of 35 children with "minimal-change" nephrotic syndrome (A. S. Appel, J. P. Hayslett, G. B. Appel, and N. J. Siegel, unpublished observation). These children have a normal glomerular filtration rate (by normal serum creatinines and creatinine clearances) but suffer recurrent bouts of heavy proteinuria and the nephrotic syndrome. The proteinuria remits either spontaneously or with corticosteroid therapy. We have found a good correlation between high levels of the urinary excretion of the lysosomal enzyme N-acetyl-β-glucosaminidase (NAG) and heavy proteinuria in this population. With remission of proteinuria, enzymuria slowly declines to control levels, only to rise up again as relapse of the nephrotic syndrome and proteinuria occur. Since the addition of purified serum albumin does not interfere with the chemical assay for NAG, and since the serum level of NAG is too low to produce such high urinary NAG levels through glomerular filtration, the enzymuria is probably the result of albumin "leakage" through the glomeruli, with subsequent reabsorption by the proximal tubular cells leading to lysosomal enzyme leakage. Significant levels of urinary enzymuria can occur without any associated deleterious effect on glomerular filtration rate; therefore, caution must be exercised in interpreting lysosomal enzymuria as a marker of significant aminoglycoside nephrotoxicity. Others have also found lysosomal enzymuria in a variety of clinical states other than those associated with known renal damage [36], and in some animal studies it has been possible to dissociate lysosomal enzymuria from nephrotoxicity as indicated by a decreased GFR [34].

Clearly, lysosomal enzymuria is a very sensitive indicator of alterations within the proximal tubular cells. Whether this sensitivity and a lack of specificity will vitiate attempts to use this enzyme excretion as a marker of aminoglycoside nephrotoxicity is unclear. The use of brush border enzymuria as a marker of renal damage also awaits further definition [39]. It is clear at present, however, that urinary enzyme excretion alone does not mandate a future reduction in nephron GFR.

IV. Glucose Reabsorption

Virtually all of the glucose filtered by the glomeruli is reabsorbed in the proximal tubule. Glycosuria at normal serum glucose concentrations localizes a defect in tubular function to this particular segment of the nephron. Although there are rare case reports in humans of renal glycosuria as part of the Fanconi syndrome following aminoglycoside therapy [38,48], the association remains to be proven. More convincing data are available from animal models of aminoglycoside-induced renal damage.

In a study of albino rabbits receiving gentamicin sulfate in doses only slightly greater than those used clinically in humans (4–8 mg/kg twice a day subcutaneously), Ginsburg et al. were able to demonstrate glycosuria [20]. None of the animals developed significant hyperglycemia, and the glycosuria was correlated with the extent of the renal damage.

Those animals developing major renal dysfunction developed the heaviest glycosuria. Of interest, urinary amino acid excretion, another indicator of proximal tubular reabsorptive capacity, was not consistently impaired in these studies. This is in contrast to an early study of neomycin nephrotoxicity in rats, in which both glycosuria and amino aciduria were felt to be important indicators of aminoglycoside nephrotoxicity [14].

We and others have checked for renal glycosuria in gentamicin-treated Sprague-Dawley rats (40–80 mg/kg per day for 10–15 days). Despite the development of nonoliguric renal failure with manifestations similar to those found in humans with gentamicin nephrotoxicity, we did not detect significant glycosuria [4]. Moreover, detailed investigations of the reabsorptive capacity of the tubule for glucose ($T_{m\,glucose}$) did not show any impairment over a range of serum glucose concentrations from normal to over 600 mg/dl. Others, however, have noted mild glycosuria with gentamicin administration, especially if the rats were acidotic at the time [28]. Of interest, these same investigators found no defect in the renal cortical slice accumulation of a labeled glucose analog in this model.

In humans suffering from aminoglycoside nephrotoxicity, glycosuria is not a major manifestation of their renal dysfunction. Whether these varying results represent species variations in the threshold or pattern of glucose handling or variations in study technique are unclear. They may represent a changing balance between the preservation of GFR and the filtration of glucose, and the damaged tubules ability to reabsorb this load.

V. Reabsorption of Low-Molecular-Weight Proteins

Another physiologic function of the proximal tubular cells involves the reabsorption and catabolism of low-molecular-weight proteins which are freely filtered by the glomeruli. Insulin and growth hormone, as well as many other small protein molecules, are handled by the kidney in this manner [35]. β-2-Microglobulin, an endogenous protein of low molecular weight, is easily measured in the serum and urine. Under normal conditions the protein is freely filtered and completely reabsorbed by the proximal tubules. The appearance in the urine of β-2-microglobulin is an indication of "tubular" proteinuria and incomplete reabsorption by the proximal tubule [43]. In aminoglycoside nephrotoxicity, excessive urinary excretion of this molecule has been documented [47]. Unfortunately, like urinary enzymuria, this is an extremely sensitive marker of tubular function, and abnormal excretion of β-2-microglobulin may be present without alterations in the GFR. Moreover, the serum levels of β-2-microglobulin are elevated in certain disease states, making interpretation of urinary excretion rates problematic [37]. As mentioned previously, there have been only sporadic cases of Fanconi syndrome or aminoaciduria attributed to aminoglycoside therapy in humans [48]. Clearly, both further experimental work and studies in humans are needed to define whether the excretion of β-2-microglobulin can be utilized clinically as a predictor of impending aminoglycoside nephrotoxicity as measured by a change in GFR.

VI. Secretory Pathways for Organic Anions and Bases

The interaction of aminoglycoside antibiotics with secretory pathways for organic acid and base transport has been extensively studied by a number of investigators. In vivo studies in rats show that gentamicin-treated animals secrete p-aminohippurate (PAH), an organic anion, into the proximal tubule at a higher rate than control animals [11]. Studies investigating the uptake of organic acids and bases in vitro by kidney slices have produced conflicting results. Thus, the addition of gentamicin directly to the transport media surrounding the kidney slice has been reported to have no influence on organic acid and base transport in one study [23], whereas it has caused an inhibition of such transport in another study [13]. Several investigators have noted an enhancement of the organic anion transport system, as measured by slice PAH uptake, early in the course of aminoglycoside nephrotoxicity [9,30]. This is followed later in the course of renal damage by an abrupt decline in slice uptake (i.e., PAH transport) [9]. Based on kinetic evaluation of the PAH uptake by slices from gentamicin-treated animals, it has been proposed that gentamicin increases the amount or availability of a carrier protein involved in PAH transport, at least early in the course of toxicity [30].

No enhancement and in some instances an impairment in the transport of the organic base system has been documented by measurement of N-methylnicotinamide (NMN) uptake by kidney slices from aminoglycoside-treated rats [9,11,28,30]. It has been proposed that early stimulation and later suppression of organic acid transport systems is a characteristic of all aminoglycosides with substantial nephrotoxic potential, whereas suppression of the organic base transport system occurs with all aminoglycoside antibiotics whether they are nephrotoxic or not [9]. It should be stressed, however, that there are a number of possible pathophysiologic mechanisms for these findings and that the specificity and sensitivity of these techniques in studying comparative aminoglycoside nephrotoxicity awaits further clarification of the comparative toxicity of a number of these agents in humans.

VII. Alterations at Other Nephron Sites

Although a large amount of effort has focused on the physiologic consequences of the morphologic changes noted in the proximal tubule of the nephron, it is clear that defects in function of other nephron segments exists. The acute renal failure associated with aminoglycoside administration is not only nonoliguric in nature, but actually associated with polyuria in the early stages of drug-induced damage [29,33]. Several studies in animal models have demonstrated that the polyuria is associated with a decreased urinary osmolality [11,29,33]. The decreased urinary osmolality has been shown to result from a defect in the ability of the kidneys of aminoglycoside-treated animals to concentrate their urine rather than being secondary to polydipsia [44]. When vasopressin was administered after a period of water deprivation, the defect in maximum urinary concentrating ability persisted, and therefore is not due to a lack of endogenous vasopressin [4,44].

An early study of gentamicin nephrotoxicity in rats found no defect in the tubular reabsorption of solute-free water ($TC H_2O$) in clearance studies over a range of solute excretion [11]. This occurred at a time when the GFR was decreased. Thus, it was argued that there could be no major depression of sodium chloride transport along the loop of Henle to cause the impairment in urinary concentrating ability that was noted. This impairment might reflect the loss of functioning nephrons, resulting in an increased non-reabsorbable solute load per remaining nephron, thus restricting the back diffusion of water from the collecting duct during antidiuresis. It was also suggested that gentamicin might cause a structural defect along the distal nephron, impairing the back diffusion of water. Although this latter mechanism would explain the decreased urinary concentrating ability noted, it is difficult to reconcile with the normal values for $TC H_2O$.

In detailed investigations designed to localize the intrarenal site(s) of the urinary concentrating defect, we studied the Sprague-Dawley rat model of aminoglycoside nephrotoxicity (80 mg/kg per day). Urinary concentrating ability was impaired by day 5 of drug administration before alterations in GFR, renal blood flow, or renal blood flow distribution were noted. Through studies of free water clearance and distal solute delivery, we confirmed the adequate delivery of proximal solute to Henle's loop and integrity of the loop's "Cl⁻ pump" [4]. Since the aminoglycoside-treated animals exhibiting this defect in concentrating ability continued to eat and gain weight normally, it is unlikely that relative starvation led to medullary urea depletion and impaired concentrating ability. We thus excluded inadequate solute or water delivery out of the proximal nephron, medullary washout of solute, defective sodium chloride reabsorption by the ascending limb of Henle's loop, and medullary urea depletion as causes of altered urinary concentrating ability. We concluded that in gentamicin nephrotoxicity there is an early defect in the sensitivity of the collecting duct epithelia to endogenous vasopressin. Of interest, the cells of this part of the nephron show only a few myeloidlike inclusions within the lysosomes, in contrast to the marked changes within the proximal tubular cells.

That this is, indeed, the mechanism of the early polyuria noted with aminoglycoside therapy is supported by a number of in vitro studies using the toad bladder model of vasopressin-mediated water flow [41,52]. The bilobed urinary bladder of *Buffo marinus* is the structural and functional analog of the vasopressin-sensitive collecting duct epithelia of the mammalian kidney. The addition of vasopressin to the bathing fluid surrounding a suspended water-filled hemibladder causes an efflux of water from the "bag," leading to a gravimetrically noticeable decrease in the bag's weight. The other hemibladder may serve as a control for alterations in vasopressin-mediated water flow induced by the addition of a drug to the serosal bathing medium of one member of the bladder pair. Drugs that clinically induce a nephrogenic (vasopressin-resistant) diabetes insipidus have been amply shown to inhibit vasopressin-mediated water flow in this model [51].

Table 2 shows the percent inhibition of vasopressin-mediated water flow induced by the addition of gentamicin to the serosal bathing medium of this preparation. Neomycin caused greater inhibition of water flow than gentamicin, which caused greater inhibition than streptomycin [52]. Similarly, studies have established a dose-response curve with increasing inhibition of the action of vasopressin by increasing concentrations of gentamicin [52]. Moreover, through studies using 8-CPT cyclic AMP rather than vasopressin to stimulate the water flow, it is clear that aminoglycosides produce a dose-dependent, reversible inhibition of vasopressin-mediated hydroosmotic water flow at a step proximal

Table 2   Effect of 3 mM Gentamicin on Serosal Surface of Vasopressin-Treated Toad Bladder

Amount vasopressin present (mU/ml)	Percent inhibition of water flow
0 (basal)	No effect
0.5	89
3.0	45
30	23

to the formation of cyclin AMP [52]. More recent data suggest that the results of in vitro toad bladder studies may be a pH-related phenomenon, rather than truly related to inhibition of the action of vasopressin by the aminoglycoside. While further studies will be needed to clarify this point, the in vivo observation of a vasopressin-resistant defect in urinary-concentrating ability stands.

Other areas of tubular function have been less well studied in aminoglycoside-induced renal damage. Detailed investigations in gentamicin-treated rats into the pattern and time course of tubular defects in K^+ excretion following an exogenous KCl load did not reveal an early defect in K^+ secretion [4]. Indeed, later defects in K^+ handling may be attributed equally to alterations in GFR and extensive tubular damage. On the other hand, following an oral NH_4^+Cl load, gentamicin-treated Sprague-Dawley rats display a clear defect in net acid excretion secondary to impaired NH_4^+ excretion [4]. This occurred at a time prior to alterations in GFR or renal blood flow (80 mg/kg times 5 days). Again, the exact localization of such a defect along the nephron is impossible without micropuncture studies. However, data obtained from in vitro kidney slice preparation techniques support such a defect in ammoniagenesis [28].

Finally, there have been a number of clinical reports of disturbances in calcium magnesium, and potassium balance in patients receiving aminoglycoside antibiotics [6,27]. The pathogenesis of these defects are unclear, although hypoparathyroidism, intrinsic tubular defects leading to the wasting of these cations, and the effects of hyperaldosteronism are all possible contributions. The fact that most of the patients studies were severely ill, and were receiving multiple other drugs makes interpretation of these results difficult. Certainly the counterpart of these defects has not been described in animal models of aminoglycoside nephrotoxicity.

VIII. Alterations in Renal Hemodynamics and GFR

By definition, all forms of renal failure are associated with a reduction in the glomerular filtration rate. Clinically, this is manifested by the retention of the nitrogenous end products of metabolism, and is monitored by a rise in serum creatinine and blood urea nitrogen together with a fall in the creatinine clearance. The aminoglycosides have been well demonstrated to be capable of causing such a decline in GFR in humans [19,25,61]. Whether by measurement of rising BUN and serum creatinine concentration and declining creatinine or inulin clearance in intact animal studies [4,29,33] or through micropuncture studies evaluating single-nephron GFR [8,46], a decline in GFR has been equally well documented in experimental models of aminoglycoside nephrotoxicity. The pathogenetic relationship between the alterations in tubular morphology and function and the abnormalities of GFR are unclear in aminoglycoside-induced renal failure. This is not surprising, because in no form of acute renal failure has the mechanism of the decline in GFR been proven to the satisfaction of all investigators in the field [31,54]. Nevertheless, a number of hypotheses have been addressed to explain the mechanism of the altered GFR noted in the many experimental models of acute renal failure of the ATN type. A review of these theories in relation to what is known clinically and experimentally about aminoglycoside nephrotoxicity is of considerable interest.

Obstruction of the tubular lumens by casts of cellular debris and associated interstitial edema have long been advanced as a cause of the diminished GFR of acute renal failure (ARF). The obstruction would lead to increased back pressure, and a decrease in the net driving pressure for ultrafiltration within the glomeruli. In some models of ARF there is considerable evidence for a major role of obstruction in the pathogenesis of the altered GFR [16,31], whereas in other models pressure measurements would relegate such a hypothesis to a minor or insignificant role [18,40]. In human and experimental amino-glycoside nephrotoxicity, the appearance of significant cylindruria does antedate the fall in GFR [2]. Pathology studies often show debris within the tubular lumens, cast forma-tion, and interstitial edema [22,58]. Micropuncture studies have found some collapsed tubular lumens on the surface of the kidney with other nephrons with widely patent lumena [46]. However, direct measurements of intratubular pressures in single nephrons have not revealed an increased tubular pressure at a time when single nephron GFR was markedly depressed [8]. Moreover, the pressures recorded in the patent tubules are not in excess of the pressure that would be expected to be necessary to cause an increase in the pressure in Bowman's capsule that would be sufficient to stop glomerular ultrafiltra-tion [8]. Thus, as in many other models of ARF, obstruction does not seem to be the major factor involved in the decreased GFR of aminoglycoside nephrotoxicity.

Passive back flow or reabsorption of the glomerular filtrate across the damaged proxi-mal tubular cells has also been proposed as a factor in the decreased GFR in experimental models of ATN. In such models the reabsorption of creatinine, insulin, or other usually nonreabsorbed markers of renal GFR occurs despite the presence of continued filtration by the glomeruli. There are experimental models of ATN both in favor [5,31,53,55] and against [18,40] such a hypothesis. In gentamicin-induced renal failure micropuncture studies have shown the microinjection of radiactive-labeled inulin into the proximal tubular lumens leads to relatively high recovery rates (97%) of the marker substance in the effluent from the tubules [8]. This occurs despite a markedly depressed GFR, making it unlikely that a significant percentage of the marker of glomerular filtration is being passively back-absorbed across damaged proximal tubular epithelium [8]. Although a second study found a lower percentage recovery of labeled inulin (63%) in tubules with a normal single-nephron GFR, it should be noted that many tubules on the surface of this kidney model were collapsed and nonfiltering. Indeed, in this same study the urine flow rate in the late proximal tubule was actually greater than in control animals, clearly imply-ing no back leak of fluid in those glomeruli still filtering [46].

Alterations in renal hemodynamics have been proposed as another factor that is re-lated to the diminished GFR of ATN [31,40,54]. Essential to these findings is a shunting of blood flow from the renal cortex to the renal medulla and an overall vasoconstriction of the vasculature of the kidney. Whether this is in response to alterations in the amount of sodium chloride allowed to be delivered to the distal tubules by the damaged more proximal tubular segments, and subsequent release of "intrarenal" renin with local angio-tensin generation [56], or whether due to alterations in renal prostaglandins or another mechanism is unclear at present [57]. Similarly, whether decreased glomerular filtration is a direct result of afferent arteriolar constriction, efferent arteriolar dilataion, or a com-bination of these is unclear.

In gentamicin-induced acute renal failure in rats we have documented a decrease in renal blood flow by both the inulin extraction method and through the use of Sr^{85}-labeled microspheres [4]. In addition, the microsphere method allows one to study the distribution of renal blood flow to various cortical segments of the kidney [26]. In con-trast to many oliguric forms of ATN, renal blood flow was not shifted from the cortex to the juxtamedullary regions of the kidney [31,40,54]. This relatively preserved outer cortical blood flow may be related to the nonoliguric nature of gentamicin-induced ATN.

In several other models of polyuric renal acute failure, such a preservation of outer cortical renal blood flow has been documented [26,49,50]. Persistent perfusion of outer cortical nephron may lead to glomerulotubular imbalance in a small number of filtering nephrons and the continued high urine volumes noted at a time when GFR is markedly depressed. Other studies have documented by micropuncture technique that the reduction in blood flow to the nephron is probably not the major contribution to the altered GFR of the single nephrons micropunctured [8]. Similarly, in our studies in intact rats renal blood flow was reduced to two-thirds normal in both animals with moderate and severe degrees of renal insufficiency (GFR 50% and 15% of control). Thus, it is unlikely that decreased renal blood flow per se is the major cause of the decreased GFR in gentamicin-induced renal failure. This does not exclude the possibility that alterations in blood flow trigger other mechanisms which lead to the alterations in GFR.

Finally an alteration in the glomerular permeability to ultrafiltration has been proposed as a mechanism for the diminished GFR seen in ATN [10,12]. Whether this is due to anatomic destruction or obliteration of the capillary loops of the glomeruli leading to a decreased surface area for filtration or due to an intrinsic abnormality of the capillary wall preventing filtration, the results would be a diminished GFR. In a micropuncture study of Munich Wistar rats receiving either 4 or 40 mg/kg per day of gentamicin for 10 days, the major cause for the decrease in single-nephron GFR was found to be a decrease in the glomerular capillary ultrafiltration coefficient [8]. Since the ultrafiltration coefficient is defined as the product of the hydraulic permeability of the glomerular capillary wall and the total surface area available for filtration, detailed studies were performed to detect any alteration by electron microscopy in the glomerular capillary wall. In agreement with previous studies of biopsy and autopsy material in humans and animal models, there were no major glomerular abnormalities noted [8]. Although this does not exclude a subtle morphologic abnormality beyond our current level of detection or a biochemical defect in parts of the glomerular capillary wall, it does make an anatomic decrease in the glomerular surface area improbable. Thus, we are left with an intrinsic defect in the filtration property of the glomerulus as the cause of the diminished GFR noted. Several problems appear in examining the results of this study: for example, why did the rats receiving only 4 mg/kg per day have as significant a decrease in GFR as those receiving 40 mg/kg per day? Was nephron sampling homogeneous? Do the results apply equally to aminoglycoside nephrotoxicity earlier and later in the course of the damage? Nevertheless, the results are intriguing.

Whether this abnormality or some combination of the many mechanisms proposed above for the reduction of GFR noted in aminoglycoside nephrotoxicity is truly the major factor remains to be shown. Moreover, the crucial link between the induction of tubular damage and the ultimate processes that result in a decrease in GFR (whether through changes in renal blood flow, altered glomerular ultrafiltration, backflow of ultrafiltrate, etc.) remains unknown, as it does for most models of ARF. Future investigations on the mechanisms of cellular damage by the aminoglycoside antibiotics as well as further investigations into the patterns of physiologic damage induced by these agents may provide suggestions as to the nature of this crucial step in the induction of acute renal failure of the ATN type.

IX. Conclusions

The clinical pattern of aminoglycoside-induced acute renal failure has been well described in both animal models and humans. Extensive investigations of animal models of aminoglycoside nephrotoxicity by both in vivo and in vitro techniques have defined some of the mechanisms behind this pattern of renal damage. Indeed, a number of studies pointing to alterations in glomerular capillary wall permeability to ultrafiltration, alterations of cortical blood flow in this nonoliguric model of ARF, and studies of the early concentrating defect noted may have significance far beyond the aminoglycoside model of acute renal failure alone. Similarly, development of sensitive and specific tubular markers of nephrotoxicity and detection of early, clinically relevant, defects in tubular function may prove of great future value to the clinician as well as the investigator developing less nephrotoxic aminoglycoside analogs. Indeed, physiologic studies of the aminoglycoside nephrotoxicity may force nephrologists and clinicians alike to reevaluate their definition of nephrotoxicity.

References

1. Anderson, R.J., Linas, S.L., Berns, A.S., Henrich, W.L., Miller, J.R., Gabow, P.A. and Schrier, R.W.: Nonoliguric acute renal failure. *N. Engl. J. Med.* 296:1134–1138, 1977.
2. Appel, G.B., and Neu, H.C.: Gentamicin in 1978. *Ann. Intern. Med.* 89:528–538, 1978.
3. Appel, G.B., and Neu, H.C.: The nephrotoxicity of antimicrobial agents. *N. Engl. J. Med.* 296:663–670, 722–728, 784–787, 1977.
4. Appel, G.B., Siegel, N.J., Appel, A.S., and Hayslett, J.P.: Hemodynamic and tubular functional alterations in gentamicin induced experimental acute renal failure. *Kidney Int.* 14:721, 1978.
5. Bank, N., Multy, B.F., and Aynedjian, H.S.: Role of "leakage" of tubular fluid in uremia due to mercury poisoning. *J. Clin. Invest.* 48:695, 1968.
6. Bar, R.S., Wilson, H.E., and Mazzaferri, E.L.: Hypomagnesemic hypocalcemia seconding to renal magnesium wasting: a possible consequence of high dose gentamicin therapy. *Ann. Intern. Med.* 82:646–649, 1975.
7. Barr, G.A., Mazze, R.I., Cousins, M.J., and Kosek, J.C.: An animal model for combined methoxyfluorane and gentamicin nephrotoxicity. *Br. J. Anaesth.* 45:306–311, 1973.
8. Baylis, C., Rennke, H.R., and Brenner, B.M.: Mechanism of the defect in glomerular ultrafiltration associated with gentamicin administration. *Kidney Int.* 12:344–353, 1977.
9. Bennett, W.M., Plamp, C.E., Gilbert, D.N., and Porter, G.A.: Kinetics of gentamicin uptake in rat cortical slices and the effect of aminoglycoside pre-treatment on the transport of PAH and N-methyl nicotinomide. In *Nephrotoxicity: Interaction of Drugs with Membrane System Mitochondria-lysosomes,* edited by Fillastre, J.P. New York, Masson, 1978, pp. 143–156.
10. Blantz, R.C.: Mechanism of acute renal failure after uranyl nitrate. *J. Clin. Invest.* 55:621, 1975.
11. Cohen, L., Lapkin, R., and Kaloyanides, G.: Effect of gentamicin on renal function in the rat. *J. Pharmacol. Exp. Ther.* 193:265–273, 1975.
12. Cox, J.W., Bachler, R.W., Sharmar, H., O'Dorisio, T., Osgood, R.W., Stein, J.H., and Ferris, T.F.: Studies on the mechanism of oliguria in a model of unilateral acute renal failure. *J. Clin. Invest.* 53:1546, 1974.
13. Dahlager, J., and Milman, N.: Aminoglycoside nephrotoxicity: I. Effects of aminoglycoside antibiotics on iodohippurate accumulation in rabbit renal cortical slices. *J. Antibiot. (Tokyo)* 30:597, 1977.
14. Emerson, B., and Pryse-Davies, J.: Studies on the nephrotoxic effect of neomycin. *Aust. Ann. Med.* 13:149–156, 1964.
15. Espinel, C.H.: The Fe Na^+ test. *JAMA* 236:579–581, 1976.

16. Finn, W.F., Arendshorst, W.J., and Gottschalk, C.W.: Pathogenesis of oliguria in acute renal failure. *Circ. Res.* 36:675–681, 1975.

17. Finck, R.E.S., Jeremy, D., and Whyte, H.W.: Structural renal damage and its relation to clinical features in acute oliguric renal failure. *Q. J. Med.* 31:429, 1962.

18. Flamenbaum, W.: Pathophysiology of acute renal failure. *Arch. Intern. Med.* 131: 911–927, 1973.

19. Gary, N.E., Buzzeo, L., Salaki, J., and Eisinger, R.: Gentamicin associated acute renal failure. *Arch. Intern. Med.* 136:1101–1104, 1976.

20. Ginsburg, D.S., Quintanilla, A.P., and Levin, M.: Renal glycosuria due to gentamicin in rabbits. *J. Infect. Dis.* 134:119–122, 1976.

21. Handa, S.P., and Morrin, P.A.F.: Diagnostic indices in acute renal failure. *Can. Med. Assoc. J.* 96:78–82, 1967.

22. Houghton, D.C., Hartnett, M., and Campbell-Boswell, P., Porter, G.A., and Bennett, W.M. A light and electron microscopic analysis of gentamicin nephrotoxicity in rats. *Am. J. Pathol.* 82:589–612, 1976.

23. Hsu, C.H., Kurtz, T., and Weller, J.: In vitro uptake of gentamicin by rat renal cortical tissue. *Antimicrob. Agents Chemother.* 12:192–194, 1977.

24. Just, M., Erdmann, G., and Haberman, E.: The renal handling of polybasic drugs, I. *Naunyn Schmiedebergs Arch. Pharmacol.* 300:5–66, 1977.

25. Kahn, T., and Stein, R.M': Gentamicin and renal failure. *Lancet* 1:498, 1972.

26. Kashgarian, M., Siegel, N.J., Ries, A.L., DiMeola, H.J., and Hayslett, J.P.: Hemodynamic aspects in the development and recovery phases of experimental postischemic acute renal failure. *Kidney Int.* 10:S160–S168, 1976.

27. Keating, M.J., Sethi, M.R., Bodey, G.P., and Samaan, N.A.: Hypocalcemia with hypoparathyroidism and renal tubular dysfunction associated with aminoglycoside therapy. *Cancer* 39:1410–1414, 1977.

28. Kluwe, W.M., and Hook, J.B.: Analysis of gentamicin uptake by rat renal cortical slices. *Toxicol. Appl. Pharmacol.* 45:531–539, 1978.

29. Kosek, J.D., Mazze, R.I., and Cousins, M.J.: Nephrotoxicity of gentamicin. *Lab. Invest.* 30:48–57, 1974.

30. Lapkin, R., Bowman, R., and Kaloyanides, G.: Effect of gentamicin on p-aminohippurate metabolism and transport in rat kidney slices. *J. Pharmacol. Exp. Ther.* 201:233–242, 1977.

31. Levinsky, N.G.: Pathophysiology of acute renal failure. *N. Engl. J. Med.* 296:1453–1458, 1977.

32. Luft, F.C., Patel, V., Yum, M.N., and Kleit, S.A.: Nephrotoxicity of cephalosporin-gentamicin combination in rats. *Antimicrob. Agents Chemother.* 9:831–839, 1976.

33. Luft, F.C., Patel, V., Yum, M.N., Patel, G., and Kleit, S.A.: Experimental aminoglycoside nephrotoxicity. *J. Lab. Clin. Med.* 86:213–220, 1976.

34. Luft, F.C., Yum, M.N., and Kleits, A.: The effect of concomitant mercuric chloride and gentamicin on kidney function and structure in the rat. *J. Lab. Clin. Med.* 89:622–631, 1977.

35. Maack, T.: Renal handling of low molecular weight proteins. *Am. J. Med.* 58:57–64, 1975.

36. Mamadani, B.H., Muni, H.M., Subbarayudu, K., and Durea, G.: Urinary N-acetyl β-D glucosaminidase (NAG) in health and disease. *Kidney Int.* 12(A):470, 1977.

37. Michalski, J.P., Daniels, T.E., Talal, N., and Grey, H.M.: Beta-2-microglobulin and lymphocytic infiltration in Sjogren's syndrome. *N. Engl. J. Med.* 293:1228–1231, 1975.

38. Milne, M.D.: Disorders of amino acid transport. *Br. Med. J.* 1:327, 1964.

39. Mondorf, A.W., Hendas, J., Beir, J., Scherberich, J.E., and Schoeppe, W.: Tubular

toxicity induced by aminoglycosides in human kidneys. In *Nephrotoxicity: Inter-action of Drugs with Membrane System Mitochondria-lysosomes,* edited by Fillastre, J.P. New York: Masson, 1978, pp. 167–174.

40. Oken, D.E.: On the passive back flow theory of acute renal failure. *Am. J. Med.* 58:77–82, 197.

41. Parker, R.A., Porter, G.A., and Bennett, W.M.: Gentamicin induced resistance to ADH stimulated water flow in the toad urinary bladder: partial correction by pros-taglandin inhibition. *Clin. Res.* 27:64, 1979.

42. Patel, V., Luft, F.C., Yum, M.N., Patel, B., Zeman, W., and Kleit, S.A.: Enyzmuria in gentamicin-induced kidney damage. *Antimicrob. Agents Chemother.* 7:364–369, 1975.

43. Peterson, P.A., Evrin, P.E., and Berggard, I.: Differentiation of glomerular, tubular and normal proteinuria: determinations of urinary excretion of beta-2-microglobulin, albumin, and total protein. *J. Clin. Invest.* 48:1187–1198, 1969.

44. Plamp, C.E., Reger, K., Bennett, W.M., McClung, M., and Porter, G.: Vasopressin resistant polyuria and gentamicin nephrotoxicity. *Clin. Res.* 26:151A, 1978.

45. Sack, K., and Freisleben, D.: Experimental studies on the functional nephrotoxicity of tobramycin, gentamicin, and sisomicin. *Infection* 3(Suppl.):540–549, 1975.

46. Safirstein, R., and Kahn, T.: Micropuncture study of the polyuric phase of gentami-cin nephrotoxicity. *Kidney Int.* 12(A):534, 1977.

47. Schentag, J.J., Sutfin, T.A., Plaut, M.E., and Jusko, W.J.: Early detection of amino-glycoside nephrotoxicity with urinary beta-2-microglobulin. *J. Med.* 9:201–210, 1978.

48. Schwartz, J.H., and Schein, P.: Fanconi syndrome associated with cephalothin and gentamicin therapy. *Cancer* 41:769–772, 1978.

49. Siegel, N.J., Feldman, R.A., Lytton, B., Hayslett, J.P., and Kashgarian, M.: Renal cortical blood flow distribution in obstructive nephropathy in rats. *Circ. Res.* 40: 379–384, 1977.

50. Siegel, N.J., Gunstream, S.K., Handler, R.I., and Kashgarian, M.: Renal function and cortical blood flow during recovery phase of acute renal failure. *Kidney Int.* 12: 199–204, 1977.

51. Singer, I., and Forrest, J.N., Jr.: Drug induced states of nephrogenic diabetes in-sipidus. *Kidney Int.* 10:82–95, 1976.

52. Souliere, C.R., Goodman, D.B.P., Appel, G.B., and Forrest, J.N.: Gentamicin selec-tively inhibits antidiuretic hormone induced water flow in the toad urinary bladder. *Kidney Int.* 14:733, 1978.

53. Stein, J.H., Gottschalk, J., Osgood, R.W., and Ferris, T.: Pathophysiology of a neph-rotoxic model of acute renal failure. *Kidney Int.* 8:27, 1975.

54. Stein, J.H., Patak, R.V., and Lifschitz, M.D.: Acute renal failure: clinical aspects and pathophysiology. *Contrib. Nephrol.* 14:118–141, 1978.

55. Steinhauser, M., Eisenbach, G.M., and Helmstadter, V.: Concentration of lissamine green in proximal tubule of antidiuretic and mercury poisoned rats and the per-meability of tubules. *Pfluegers Arch.* 3:11, 1969.

56. Thurau K., and Boylan, J.W.: Acute renal success. *Am. J. Med.* 61:308–315, 1976.

57. Torres, V.E., Strong, C.G., Romero, J.C., and Wilson, D.M.: Indomethacin enhance-ment of glyceral-induced acute renal failure in rabbits. *Kidney Int.* 7:170, 1975.

58. Vera-Roman, J., Krishnakantha, T.P., and Cuppage, F.E.: Gentamicin neophrotoxicity in rats. *Lab. Invest.* 33:412–417, 1975.

59. Wellwood, J.M., Lovell, D., Thompson, A.E., and Tighe, J.R.: Renal damage caused by gentamicin: a study of the effects of renal morphology and urinary enzyme excre-tion. *J. Pathol.* 118:171–182, 1976.

60. Wellwood, J.M., Simpson, P.M., Tighe, J.R., and Thompson, A.E.: Evidence of gentamicin nephrotoxicity in patients with renal allografts. *Br. Med. J.* 3:278–281, 1975.

61. Wilfert, J.N., Burke, J.P., Bloomer, H.A., and Simth, C.B.: Renal insufficiency associated with gentamicin therapy. *J. Infect. Dis.* 124(Suppl.):148–152, 1971.

12

URINARY ENZYMATIC MARKERS OF RENAL DAMAGE

A. WERNER MONDORF   University of Frankfurt Center of Internal
Medicine, Frankfurt, Federal Republic of Germany

The use of urine enzymes of renal origin for the detection of inflammatory and toxic lesions has not yet been established clinically, for several reasons. On the one hand, there is a lack of detailed knowledge regarding the origin of the enzymes; on the other hand, an increase in enzyme level is not specific and can have multiple causes. The daily oscillations in the normal output of these enzymes makes it difficult to establish a schedule of collection periods. Methodological difficulties are also created by the presence of inhibitors and the very low concentration of the enzymes in urine. The concentration and milieu of urine is subject to wide fluctuations.

A urine enzyme that would allow diagnostic conclusions should in both normal and pathological conditions of the kidney originate only in the kidney; in fact, it should come from a specific section of the nephron and from only a definite part of the cell. Such an enzyme should possess high stability and antigenicity for exact determination of enzymatic and immunologic characteristics. The daily enzyme output must not fluctuate over too wide a range, and the day-to-day elimination must be constant in healthy subjects.

Of all the enzymes that have been researched to date, alanine aminopeptidase (AAP), an enzyme produced in the brush border membranes of the proximal tubule, best fulfills these requirements. This section of the nephron has a huge surface area and is densely covered with AAP. For this reason the turnover processes in this section are very constant, resulting in a constant daily and day-to-day output [7,29]. Also, because of the high molecular weight of the AAP, glomerular filtration is not possible. AAP is very sensitive to deep freezing but very stable at room temperature. It is a glycoprotein and is therefore a potent antigen of the kidney cortex. AAP is directly correlated with the duration and strength of toxic alterations of the proximal tubule and is therefore a very sensitive parameter for cell membrane damage in this area.

Clinical and animal studies have shown that aminoglycosides have a nephrotoxic effect [2,4,8,14,20-22,29,38,43]. Other studies report nephrotoxic effects of cephalosporins [3,10,12,34,37,40,42,46] and their combination with aminoglycosides [6,11,15-17, 24,26,35,36]. Indifferent [47], and in animal studies even protective [13] behavior of the cephalosporins has been reported. For this reason it was evident that the potentially toxic characteristics of antibiotics and chemotherapeutics should be investigated using the AAP parameter [19,29,33,41]. The broadest studies, with both volunteers and patients, have investigated the effect of aminoglycosides, cephalosporins, and combinations thereof [19,28-30,41].

I. Characteristics of Alanine Aminopeptidase

The alanine aminopeptidase (AAP), substrate alanine p-nitranilide, in human urine is identical to the arylamidase, leucin aminopeptidase, clycinaminopeptidase, and cystin aminopeptidase. In the literature the molecular weight of this glycoprotein AAP varies between 157,000 [45] and 300,000 [28]. It is obvious from these studies that the AAP is not filtered by the glomerula.

It has been proven that AAP originates in the brush border membrane of the proximal tubule (Figure 1) [4,31,45]. This membrane-bound part of the AAP constitutes at least

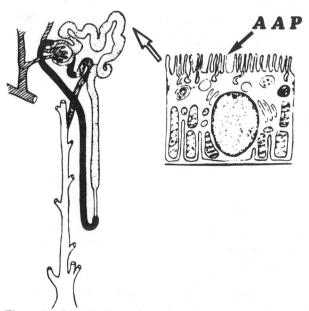

Figure 1   Model of a nephron from the human kidney with a cell of the proximale tubule. On top of the cell is the brush border membrane, which is closely covered with the enzyme alanine aminopeptidase (AAP).

95% of the total AAP in the urine [8]; the other 5% possibly originates in the proximal tubular cell. It has been proven immunologically that the isolated membrane enzyme is identical to the corresponding enzyme in the urine [31,39,45].

The normal values in the 24 hr urine differ for men and women; the values for women are lower than those for men [9,23,25,28,29,38,43]. The values vary from 2000 to 6000 mU/24 hr for men and 1500 to 3700 mU/24 hr for women. No correlation with age or sex has been observed in children [1]. The normal values of the 24 hr urine of survey groups according to researchers statistically follows a normal [29] or logarithmic [38] distribution. There is a circadian rhythm with a lower output at night. In women it appears that AAP output increases with age [38]. AAP activity in the urine is correlated to diuresis [29].

II. Diagnostic Value of AAP Elimination in Urine

An increase of AAP activity in urine indicates an unspecific response to renal or extrarenal inflammatory or toxic lesions to the proximal tubular cells. In this case AAP elimination is no longer correlated only with diuresis. The peak outputs of AAP are observed in acute tubular necrosis and toxic lesions of the tubular membrane [1,4,8,14,19,22,23, 25,27,28,32,33,39], whereby the amount of enzyme output is correlated with the extent of the damage. The predamaged kidney reacts more sensitively to toxic noxae than does the healthy organ; that is, the elimination of AAP is greater than would normally be expected under pathological conditions [1,4,8,19,25,39,41]. Inflammatory injuries of the kidney result in a varying production of AAP.

The amount of AAP in the urine depends upon the extent of proximal tubular participation. In acute glomerular diseases the AAP concentration in the urine is fundamentally higher than it is in acute pyelonephritis. On the other hand, the AAP levels are higher in cases of acute diseases than in chronic diseases. In chronic diseases, AAP levels vary to the extent that [4,5,19] the absorbing area of the proximal tubular cells is destroyed. No relation has been found between proteinuria and bacteriuria on the one side, and AAP elimination on the other [1]. In cases of kidney tumors and tumors outside the kidney, elevated AAP values have been observed [1,25,28]. Similar AAP levels are observed with icteric patients [4,23,27]. During rejection episodes after kidney transplantation the AAP values are elevated; the values decrease both after normalization of kidney function and after failure of the transplanted kidney [19,33]. Short-term toxic kidney damage without a decrease in kidney function reveals an increase in AAP levels in the urine and characterizes the sensitivity of this parameter [25,29].

III. Effect of Aminoglycosides on Proximal Tubular
Membrane of the Human Kidney

The secretion behavior of AAP is characterized by narrow day-to-day fluctuations (Figure 2). In 24 hr urine, AAP production for women is statistically significantly lower than for men, with an average of 800 ml/24 hr (Table 1). It is probable that if a larger sample group is examined, both groups will follow a logarithmic distribution (Figure 3).

The effect of aminoglycoside on AAP elimination from the proximal tubule was clearly demonstrated in studies with informed volunteers using the applied model [29]. After a collection period of 1 or 2 days (-2, -1), aminoglycosides were applied on

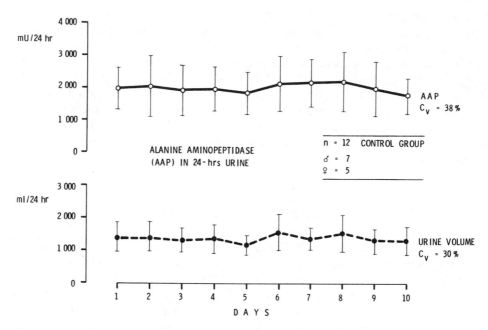

Figure 2 AAP activity in 24 hr urines and the volumes of the 24 hr urines over a 10 day period in healthy subjects. C_v, coefficient of variability.

three consecutive days (1, 2, 3). Data were also collected for a period of 3–6 days (4, 5, 7, 8, 9) after administration of the drugs. The initial activities were very uniform in all groups and ranged from 1500 to 3000 mU/24 hr. All aminoglycosides examined in clinically comparable daily dosages (amikacin, 10 mg/kg; gentamicin, 3 mg/kg; netilmicin, 3 mg/kg; ribostamycin, 15 mg/kg; sisomicin, 3 mg/kg; tobramycin, 3 mg/kg) showed a statistically significant increase on the third and fourth days after administration of the various drugs. In all cases a cumulative increase in AAP elimination, which was no longer correlated only with diuresis, could be observed. The highest values of AAP activities in urine could, in accordance with the accumulation of the aminoglycosides in kidney tissue, be measured 1 day after the last application (Figure 4). On the seventh day after the beginning of drug administration, normal values could again be measured. At the point of the greatest AAP elimination, the individual aminoglycosides showed statistically significant differences

Table 1 Mean Values of AAP Activity in 24-hr Urines[a]

	Male	Female
n	88	36
$\bar{x}$ (mU/24 hr)	2302	1490
s	±931	±530
CV (%)	40	36

[a]With the Wilcoxon test both groups are highly significant different. There was no influence of age ($\chi^2 = 5.797$). $2\alpha < 0.001$.

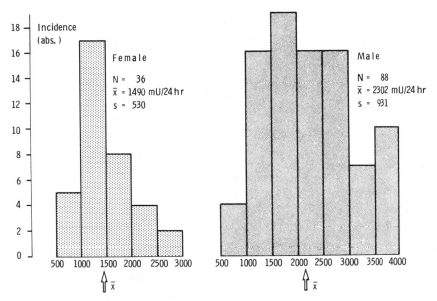

Figure 3   Frequency distribution of AAP activity in 24 hr urines in female and male.

among themselves: tobramycin differentiated itself from all others, with the exception of netilmicin, to a statistically significant degree (Figure 5).

Increased aminoglycoside doses produced correspondingly high elimination rates (Figure 6). It was thus also observable that the individual reactions to these higher dosages were clearly different. Doubling the dose of netilmicin from 3 to 6 mg/kg per day

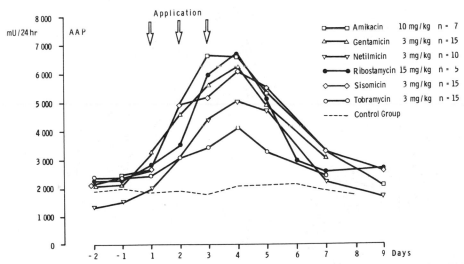

Figure 4   Mean levels of AAP activity in 24 hr urines before, during, and after IM administration of amikacin (10 mg/kg), ribostamycin (15 mg/kg), gentamicin, netilmicin, sisomicin, and tobramycin (3 mg/kg) on three consecutive days compared with an untreated control group.

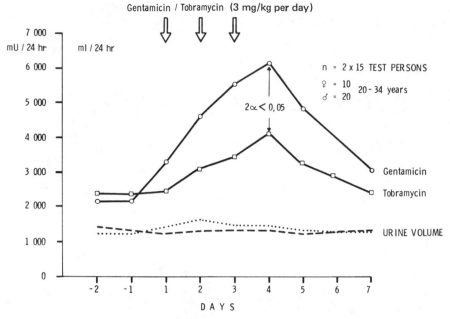

Figure 5 Mean levels of AAP activity after administration of gentamicin and tobramycin. On day 4, 1 day after the last dose of the aminoglycosides, the levels for gentamicin (3 mg/kg) were significantly higher than were those for tobramycin. The urine volumes of the two test groups were identical.

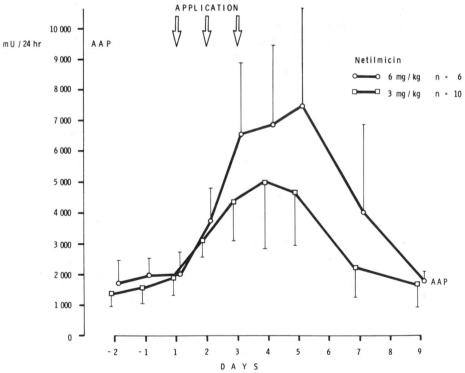

Figure 6 Mean levels and standard deviations of AAP activity in 24 hr urines after netilmicin (3 and 6 mg/kg of body weight) was administered once daily on days 1, 2, and 3.

288

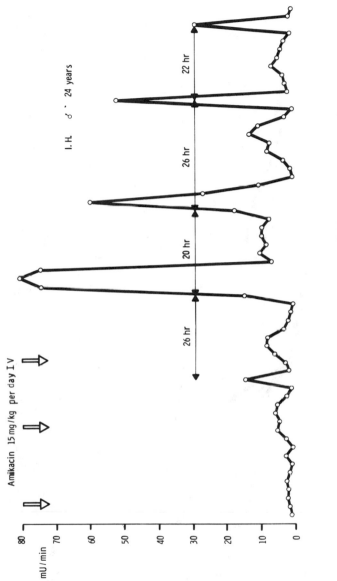

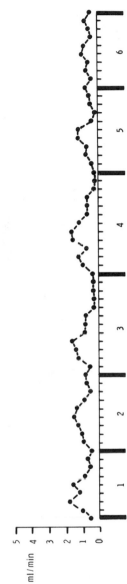

Figure 7 AAP activity in urine measured at 2 hr intervals. Collection periods 1 and 2 cover the time from 7 a.m. in the morning up to 9 p.m. in the evening; collection periods 3 and 4 represent the full 24 hr and collection periods 5 and 6 cover 20 hr on the corresponding days. The urine volumes in ml/min have been plotted for comparison with the AAP activity. Amikacin (7.5 mg/kg) was given IV twice a day.

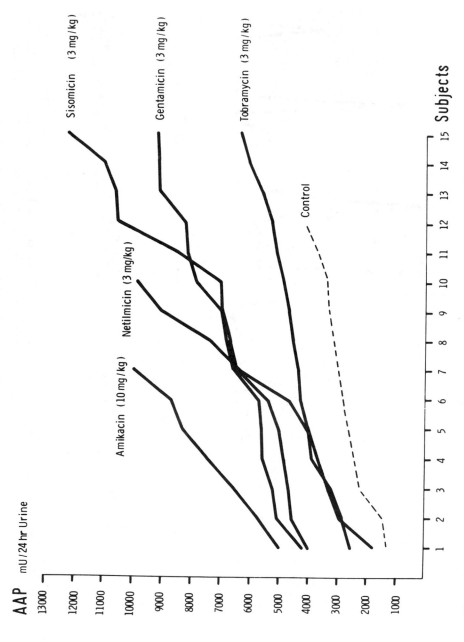

Figure 8 Maximum peak of AAP activity, plotted in increasing amounts of each subject in a test group, was reached after the onset of the application of different aminoglycosides. For example, in the test group of amikacin, subject 1 reached the maximum with 5000 mU/24 hr, and subject 7 had the highest peak, 10,000 mU/24 hr.

290

nearly doubled the AAP activity in the 24 hr urines based on the mean values. The test persons reacted as high and low responders. A larger number of subjects would make the difference even more evident. With the higher dosage, the cumulative effect also became more obvious. The maximum AAP output could be seen on day 5 instead of day 4. The β phase of the aminoglycosides could easily be represented in a single trial, in that the enzyme activities were measured in 2 hr intervals over a period of 6 days following administration of amikacin 15 mg/kg of body weight on three consecutive days (Figure 7). Nightly peaks were observed which were diuresis-independent and exhibited an approximately linear reduction in the course of the days following application. Every test subject reached an individual maximum of AAP elimination, in most cases on the third or fourth day after application. Comparison of graphs of these maxima provides very good information as to the different effects of aminoglycosides (Figure 8). This information is much more useful than that obtained from plotting the mean values. From these illustrations it is evident that the test subjects reacted very uniformly to tobramycin, gentamicin, and amikacin, whereas the sensitivity to sisomicin and netilmicin varied greatly. The kidney function parameters did not change pathologically at any time in any experiment of this investigation. It must be noted that the numerically small sample size limits the conclusions that can be drawn from these studies.

IV. Effect of Cephalosporins on Proximal Tubular Membrane of the Human Kidney

The cephalosporins (cefamandole, cefazedon, cefazolin, cefoxitin, cefuroxim, cephacetril, cephalothin) have, with the exception of a slight effect from cephacetril, no influence on the membranes of the proximal tubules [30]. Tested in the same model as the aminoglycosides, it could be demonstrated that after the last application of cephacetril, AAP values return directly to the normal range (Figure 9). Cephacetril had both responder and nonresponder groups [30]. During and after application of any of the

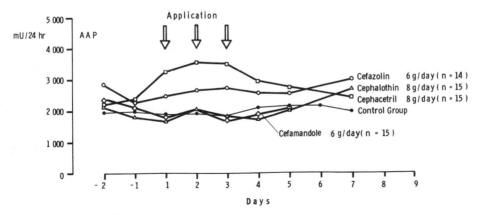

Figure 9   Mean values for the cephalosporin group. On days -2 and -1, the means of enzyme determinations in 24 hr urines before treatment are shown. Administration of the cephalosporins was as single infusions on days 1, 2, and 3. Days 4, 5, and 7 are control days after cessation of treatment.

cephalosporins, creatinine in serum and creatinine clearance showed no pathological deviations.

V. Effect of the Combination of Aminoglycosides with Cephalosporins on Proximal Tubular Membrane of the Human Kidney

It was evident from studies with test subjects that the toxicity of the combination of aminoglycosides and cephalosporins is almost exclusively bound to the potential toxicity of aminoglycosides (Tables 2 and 3). The mean values in the tables demonstrate the homogeneous manner in which different test groups react to damaging noxae on proximal tubular membranes. The combinations of gentamicin (3 mg/kg) with cefazedon (6 g), cefazolin (6 g), cefuroxim (6 g), and cephalothin (8 g) and those of tobramycin (3 mg/kg) with cefamandole (6 g), cefoxitin (6 g), cefuroxim (6 g), and cephalothin (8 g) order themselves after the potential toxicity of the respective aminoglycosides in the combination. The appearance of slight additive effects after administration of the combinations compared to that of the single drug is probably caused by differences in the sensitivity of the test groups or by the number of high responders in such a group (Figures 10 and 11). The mean value curve of the combinations shows a higher standard deviation than that for any of the single drugs.

In the study using the combination of cefuroxim (6 g) plus gentamicin (3 mg/kg) and cefuroxim (6 g) plus tobramycin (3 mg/kg), no definite additive effects could be observed (Figures 12 and 13). Except for a slight decrease in the mean values during the first 3 days of application, the curves follow those of the aminoglycosides.

The difference between gentamicin and tobramycin is also evident to a statistically significant degree in combination with cephalosporins. A study of the same test group with identical persons in a 4 month interval made this clear. The combination of cefuroxim and tobramycin produced, to a statistically significant degree, less enzyme elimination than did the combination of cefuroxim with gentamicin (Figure 14). No protective or additive effects of the cephalosporins were observable in these studies. At no point during or after the application of the drug combinations were kidney function parameters changed pathologically.

Table 2 Mean Values of AAP Activity in 24 hr Urines[a]

| | Days | | Gentamicin, 3 mg/kg per 24 hr, plus: | | | | Gentamicin 3 mg/kg per 24 hr (n = 15) |
			Cephalothin, 8 g/24 hr (n = 10)	Cefuroxim, 6 g/24 hr (n = 10)	Cefazolin, 6 g/24 hr (n = 10)	Cefazedon, 6 g/24 hr (n = 10)	
Before application	−1		2377	1514	1773	2104	2149
Application	1		2098	1793	2369	2516	3274
	2		3162	3027	3662	3697	4575
	3		3518	4928	5980	6370	5537
After application	4		6321	6803	8068	8046	6154
	5		7073	5178	6318	7937	4826
	7		4687[b]	2719	2796[b]	4698	3042
$\bar{x}$			4176	3708	4423	5052	4422
s			1926	1964	2372	2453	1445

[a]The cephalosporins were administered IV 30 min before gentamicin was injected IM.
[b]n = 6.

Table 3   Mean Values of AAP Activity in 24 hr Urine

| | Days | Cephalothin, 8 g/24 hr (n = 15) | Tobramycin, 3 mg/kg per 24 hr, plus: | | | |
			Cefuroxim, 6 g/24 hr (n = 6)	Cefoxitin, 6 g/24 hr (n = 10)	Cefamandole, 6 g/24 hr (n = 15)	Tobramycin, 3 mg/kg per 24 hr (n = 15)
Before application	-1	2210	1576	2652	2198	2401
Application	1	2185	2018	1846	2318	2440
	2	2559	2478	2559	2458	3063
	3	4241	3077	2968	3657	3149
After application	4	4921	4789	4822	4719	4109
	5	3498	3492	3480	3402	3256
	7	2707	1941	2857	2612	2407
$\bar{x}$		3188	2767	3026	3052	2975
s		1064	1114	932	920	624

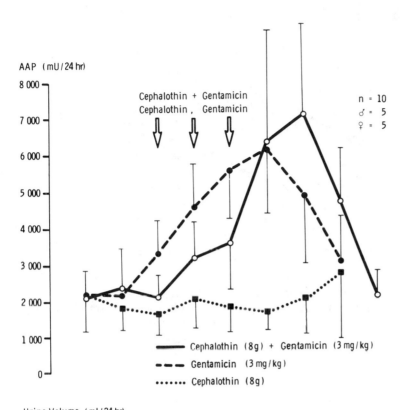

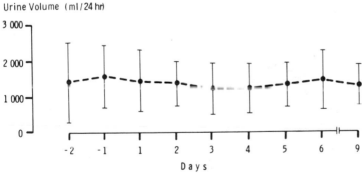

Figure 10 Mean values and standard deviations of AAP activity in 24 hr urines before, during, and after administration of gentamicin (3 mg/kg), the combination of gentamicin (3 mg/kg) and cephalothin (8.0 g), and cephalothin (8.0 g) alone. Also given are the mean values and standard deviations of the urine volume from the test group with the combination of cephalothin and gentamicin.

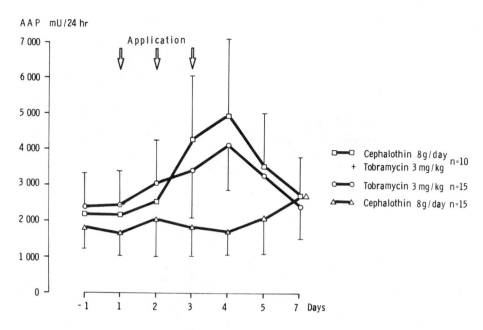

Figure 11 Mean values and standard deviations of AAP activity in 24 hr urines of different test groups with cephalothin (8 g), tobramycin (3 mg/kg), and the combination of cephalothin (8.0 g) and tobramycin (3 mg/kg) using the same model as in Figure 9.

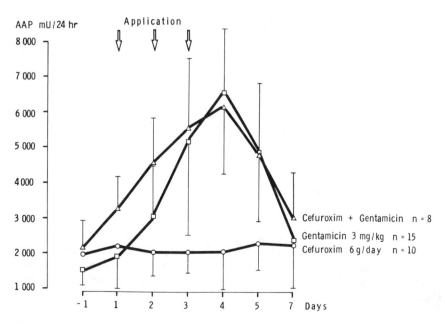

Figure 12 Mean values and standard deviations of AAP activity in 24 hr urines before, during, and after administration of cefuroxim (6 g), gentamicin (3 mg/kg), and the combination of cefuroxim (6 g) with gentamicin (3 mg/kg) in different groups of test subjects.

296

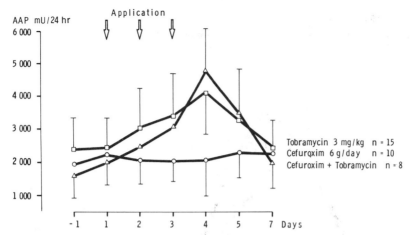

Figure 13 Mean values and standard deviations of AAP activity in 24 hr urines before, during, and after administration of cefuroxim (6 g), tobramycin (3 mg/kg), and cefuroxim (6 g) plus tobramycin (3 mg/kg) in groups with different test subjects.

VI. Conclusions

1. Aminoglycosides alter the brush border membrane of the proximal tubular cell, which could be measured by AAP elimination, a constituent of the brush border membrane itself. The AAP elimination in 24 hr urines showed a cumulative increase after onset of treatment based on the accumulation of aminoglycosides in the kidney cortex.
2. With increasing doses of aminoglycosides, the accumulation in the kidney seemed to be more rapid and was followed by a sharp increase of AAP elimination, which returned to normal later and reached normal values after a longer time.
3. There are statistically significant differences in the potential nephrotoxicity between aminoglycosides.
4. Even in high doses, cephalosporins have almost no effect on the brush border membrane.
5. It was shown that the combination of cephalosporins with aminoglycosides had only slight additive effects on their nephrotoxicity. Comparing the same test subjects for single aminoglycosides and months later in combination with a cephalosporin, no additive or protective effects could be seen. Therefore, it seems clear that the potential nephrotoxic effect of the combination of cephalosporins with aminoglycosides was related only to the nephrotoxicity of the aminoglycosides.
6. Comparing the AAP activities in 24 hr urines in two test groups with identical subjects, the combination cefuroxim plus gentamicin and cefuroxim plus tobramycin was statistically significantly different and verified investigations with single aminoglycosides in different test groups.

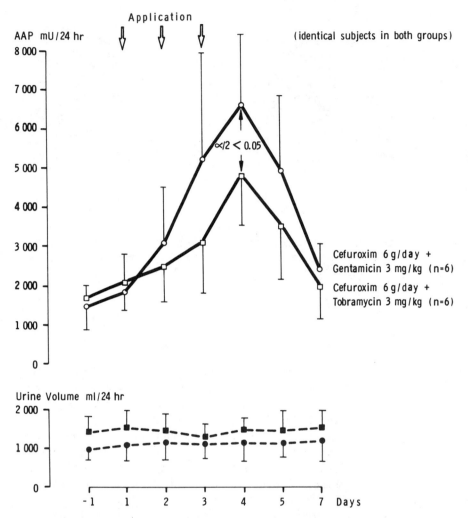

Figure 14 Mean values and standard deviations of AAP activity in 24 hr urines and of urine volumes before, during, and after administration of cefuroxim (6 g), gentamicin (3 mg/kg), and the combination gentamicin (3 mg/kg) and cefuroxim (6 g) to the same test subject group investigated after an interval of 4 months. On day 4, one day after the last application, the difference between both groups is statistically significant. The devided dose of cefuroxim was IV administered 30–60 min before the IM administration of gentamicin.

298

References

1. Appel, W., Schilling, W., and Huth, E.: Zur Enzymdiagnostik im Harn bei Kindern. *Z. Kinderheilkd.* 114:233–248, 1973.
2. Baylis, C., Rennke, H.R., and Brenner, B.M.: Mechanisms of the defect in glomerular ultrafiltration associated with gentamicin administration. *Kidney Int.* 12:344–353, 1977.
3. Benner, J.: Renal damage associated with prolonged administration of ampicillin, cephaloridine and cephalothin. *Antimicrob. Agents Chemother.*, 417–420, 1969.
4. Bergmann, H., and Scheler, F.: Der Nachweis tubulärer Funktionsstörungen der Niere durch Bestimmung der Aminopeptidase-Aktivität im Harn. *Klin. Wochenschr.* 42:275–279, 1964.
5. Bergmann, H., and Truss, F.: Harnenzyme als Indikator tubulärer Nierenschäden. *Med. Welt* 34:1760–1764, 1964.
6. Bobrow, S.N., Jaffe, E., and Young, R.C.: Anuria and acute tubular necrosis associated with gentamicin and cephalothin. *JAMA* 222:1546–1547, 1972.
7. Burchardt, U., Hempel, A., Höpfner, J.B., and Hempel, R.D.: Alaninaminopeptidase- und Kreatininausscheidung unter verschiedenen Diuresezuständen beim Menschen. *Z. Inn. Med.* 31:550–552, 1976.
8. Burchardt, U.: Alaninaminopeptidaseausscheidung mit dem Harn und osmotische Nephropathie. *Z. Inn. Med.* 30:65–69, 1975.
9. Burchardt, U., Peters, J.E., and Mampel, E.: Zum gegenwärtigen Stand der Harnenzymdiagnostik. *Z. Aerztl. Fortbild.* 69:619–623, 1975.
10. Burton, J.R., Lichtenstein, N.S., Colvin, R.B., and Hyslop, N.E.: Acute renal failure during cephalothin therapy. *JAMA* 229:679–682, 1974.
11. Cabanillas, F., Burgos, R.C., Rodriguez, R.C., and Baldizón, C.: Nephrotoxicity of combined cephalothin-gentamicin regimen. *Arch. Intern. Med.* 135:850–852, 1975.
12. Carling, C.P., Idelson, B.A., Casano, A.A., Alexander, E.A., and McCabe, W.R.: Nephrotoxicity associated with cephalothin administration. *Arch. Intern. Med.* 135:797–801, 1975.
13. Dellinger, P., Murphy, T., Pinn, V. Barza, M., and Weinstein, L.: Protective effect of cephalothin against gentamicin-induced nephrotoxicity in rats. *Antimicrob. Agents Chemother.* 9:172–178, 1976.
14. Egger, E., Jung, K., and Mallmann, B.: Die Arylamidase im Harn von nierentransplantierten Patienten. *Z. Klin. Chem. Klin. Biochem.* 11:565–566, 1973.
15. Falge, P., Hofstetter, A., and Staehler, G.: Kombinierte Cephalothin-Gentamicin-Behandlung schwerster urologischer Infektionen. *Med. Welt* 24:368–371, 1973.
16. Fillastre, J.P., Laumonier, R., Humbert, G., Dubois, D., Metayer, J., and Delpech, A.: Insuffisances rénales aiguës après administration de céphalotine et de gentamycine. *Nouv. Presse Med.* 2:444, 1973.
17. Fillastre, J.P., Laumonier, R., Humbert, G., Dubois, D., Metayer, J., Delpech, A., Leroy, J., and Robert, M.: Acute renal failure associated with combined gentamicin and cephalothin therapy. *Br. Med. J.* 2:396–397, 1973.
18. Harrison, W.O., Silverblatt, F.J., and Turck, M.: Gentamycin nephrotoxicity: failure of three cephalosporins to potentiate injury in rats. *Antimicrob. Agents Chemother.* 8:202–215, 1975.
19. Horpácsy, G., Dutz, W., May, G., Zinsmeyer, J., and Mebel, M.: Die differential-diagnostische Bedeutung der Bestimmung der Alanin-Aminopeptidase-Aktivitäten im Harn nach der Nierentransplantation. *Z. Urol. Bd.* 66:365–366, 1973.
20. Horpácsy, G., Zinsmeyer, J., Schröder, K., and Mebel, M.: Changes in serum and urine lysozyme activity after kidney transplantation: influence of graft function and therapy with azathioprine. *Clin. Chem.* 24:74–79, 1978.

21. Horpacsy, G., Zinsmeyer, J., and Mebel, M.: Continuous determination of various enzymes and sodium concentration in urine. *Eur. Urol.* 4:334–337, 1978.
22. Klaus, D.: Über Aminopeptidasen im menschlichen Organimus. *Aerztl. Forsch.* 16:9–17, 1962.
23. Klaus, D.: Über Aminopeptidasen im menschlichen Organismus. *Aerztl. Forsch.* 16:18–23, 1962.
24. Kleinknecht, D., Ganeval, D., and Droz, D.: Acute renal failure after high doses of gentamicin and cephalothin. *Lancet* 1:1129, 1973.
25. Luckmann, E., and Körner, F.: Die Leucin-Aminopeptidase im Harn bei Nierenerkrankungen. *Klin. Wochenschr.* 50:1003–1008, 1972.
26. Luft, F.C., Patel, V., Yum, M.N., and Kleit, S.A.: Nephrotoxicity of cephalosporin-gentamicin combinations in rats. *Antimicrob. Agents Chemother.* 9:831, 1976.
27. Mason, E.E., Chernigoy, F.A., Gulesserian, H.P., and Tector, A.J.: Enzymuria in incteric surgical patients. *Surg. Gynecol. Obstet.* 122:333–339, 1966.
28. Mattenheimer, H.: Enzymes in the urine. *Med. Clin. Am.* 55:1493–1508, 1971.
29. Mondorf, A.W., Breier, J., Hendus, J., Scherberich, J.E., Mackenrodt, G., Shah, P.M., Stille, W., and Schoeppe, W.: Effect of aminoglycosides on proximal tubular membranes of the human kidney. *Eur. J. Clin. Pharmacol.* 13:133–142, 1978.
30. Mondorf, A.W., Zegelmann, M., Klose, J., Maske, L., Scherberich, J.E., Stefanescu, T., Müller, H., and Schoeppe, W.: Effect of various cephalosporins on the proximal tubule of human kidney. *Eur. J. Clin. Pharmacol.* 13:357–363, 1978.
31. Mondorf, A.W., Kinne, R., Scherberich, J.E., and Falkenberg, F.: Isolierung, enzymatische und immunologische Charakterisierung einer Plasmamembranfraktion vom proximalen Tubulus der menschlichen Niere. *Clin. Chim. Acta* 37:25–32, 1972.
32. Mondorf, A.W.: Die Bedeutung lokalisationsspezifischer Enzyme im Harn für toxikologische und klinisch-pharmakologische Untersuchungen. *Arzneim.-Forsch. (Drug Res.)* 23:1634–1637, 1973.
33. Mondorf, A.W., Carpenter, C.B., Scherberich, J.E., and Merill, J.: Brush border proteins in urine after human kidney transplantation and acute renal failure. 5th International Congress of Nephrology, 1972.
34. Müller, S., Wiesendanger, W., Wegmann, W., Scheitlin, W., and Moccetti, T.: Akutes Nierenversagen durch Cephalothin? *Schweiz. Med. Wochenschr.* 103:889–894, 1973.
35. Noone, P., Pattison, J.R., and Shafi, M.S.: Acute renal failure after high doses of gentamicin and cephalothin. *Lancet* 1:1387–1388, 1973.
36. Opitz, H., v. Herrath, D., and Schaefer, K.: Akute Niereninsuffizienz nach Gentamicin-Cephalosporin-Kombinationstherapie. *Med. Welt* 22:434–438, 1971.
37. Pasternak, D.P., and Stephens, B.G.: Reversible nephrotoxicity associated with cephalothin therapy. *Arch. Intern. Med.* 135:599–602, 1975.
38. Peters, J.E., Schneider, I., and Haschen, R.J.: *Clin. Chim. Acta* 36:289–301, 1972.
39. Peters, J.E., Mampel, E., Schneider, I., Burchardt, U., Fukala, E., Ahrens, I., and Haschen., R.J.: Alanin-Aminopeptidase (L-Alanyl-Peptihydrolase) im Harn bei Nierenerkrankungen. *Clin. Chim. Acta* 37:213–223, 1972.
40. Pickering, M.J., Spooner, G.R., Quesada, A., and Cade, J.R.: Declining renal function associated with administration of cephalothin. *South. Med. J.* 63:426–428, 1970.
41. Raab, W.P.: Diagnostic value of urinary enzyme determinations. *Clin. Chem.* 18:5–25, 1972.
42. Rahal, J.J., Meyers, B.R., and Weinstein, K.: Treatment of bacterial endocarditis with cephalothin. *N. Engl. J. Med.* 279:1305, 1968.
43. Roth, M.: Une Ultramicro-méthode fluorimétrique pour le dosage de la leucine-aminopeptidase dans les liquides biologiques. *Clin. Chim. Acta* 9:448–453, 1964.
44. Scherberich, J.E., and Mondorf, A.W.: Immunochemical studies on membrane sur-

face proteins of human kidney by Con-A-affinity chromatography. *Protides Biol. Fluids* 23:575, 1975.

45. Scherberich, J.E., Falkenberg, F., Mondorf, A.W., Müller, H., and Pfleiderer, G.: Biochemical and immunological studies on isolated brush border membranes of human kidney and their membrane surface proteins. *Clin. Chim. Acta* 55:179–197, 1976.

46. Schulte-Lippern, M., Freyland, M.D., Frotscher, U., Jenett, G., Messerschmidt, W., Richter, R., Zschaege, B., and Wilbrand, R.: Akutes Nierenversagen nach hochdosierter Cephalothinbehandlung. *Med. Klin.* 68:202–206, 1973.

47. Stille, W., and Arndt, J.: Argumente gegen eine Nephrotoxizität von Cephalothin und Gentamycin. *Med. Welt* 23:1603–1605, 1972.

13

AMINOGLYCOSIDE-INDUCED LYSOSOMAL DYSFUNCTIONS IN KIDNEY

J. P. MORIN, and J. P. FILLASTRE   Université de
Rouen, Hôpital de Boisguillaume, Bois Guillaume, France

Aminoglycoside antibiotics are widely used for the treatment of severe gram-negative bacterial infections. All these compounds are basic organic molecules of low molecular weight (about 500) with pK values between 7.5 and 8.5. These molecules are largely water-soluble, not metabolized in the body, lightly bound to plasma proteins, and almost exclusively eliminated from the body by glomerular filtration under active form. In human therapy, the incidence of renal complications is estimated to be between 4 and 9% [30,37], depending on the drug. The clinical and pathological manifestations of these renal complications have been described [1,5,49,54,56,82].

Kidney dysfunction is preceded and accompanied by the development of selective lesions of the lysosomes in the proximal tubule cells [34,40,41,53]. These lesions consist of the intralysosomal accumulation of electron-dense multilamellar structures (myeloid bodies), which strongly suggests the deposition of polar lipids organized in closely apposed layers. Koseck et al. [53] showed that these lesions are demonstrable in Fisher rats treated with 1 mg of gentamicin per kilogram per day for only 2 days, a dosage level at which kidney function is still unaffected. Alterations of the other subcellular structures are observed only at larger doses (above 10 mg/kg for 7 days) and reflect the necrosis of the tubular cells [94].

The demonstration of specific lysosomal lesions at concentrations similar to those used in human chemotherapy led us to investigate in closer detail the interaction between lysosomes and aminoglycosides. In this chapter we review our observations and those of other authors in this domain, as obtained from in vitro and in vivo studies.

I. Lysosomes: General Properties

A. Definitions and Main Characteristics

The name lysosomes (lytic bodies) was coined in 1955 to designate a new group of cytoplasmic organelles identified in rat liver and characterized by the presence of several acid hydrolases in a latent form [21]. Since then, lysosomes have been observed in all nucleated cells from animal or vegetal organisms where they have been looked for. The properties of lysosomes and their physiological functions have been examined in great detail in a number of reviews and treatises (e.g., Refs. 18, 22, and 39).

In brief, lysosomes are small intracellular organelles of about 0.5–2 μm in diameter and characterized by:

1. Enzymatic equipment, consisting of a large variety of acid hydrolases (proteases, glycosidases, esterases, lipases, nucleases) capable by cooperative action of breaking down almost all biological constituents [8,93]. The majority of these enzymes have an acid optimum of pH. Oxidoreductases have not been demonstrated.
2. A semipermeable membrane, which restricts the action of the acid hydrolases on the molecules present in lysosomes and prevents the destructive action of these enzymes on the other intracellular components. It is this membrane that is responsible for the "latency" of lysosomal hydrolases when assayed in homogenates prepared under conditions that preserve the integrity of the subcellular particles. The exclusion limit of this lysosomal membrane is about 200 MW for saccharides and peptides. Thus, the products of the digestion of the corresponding macromolecules may freely diffuse out of lysosomes. For other constituents, including ions, the permeability of the lysosomal membrane depends largely on their size and lipophilicity, but most substances above 1000 MW do not part across lysosomal membranes.
3. An acid pH which is at least 1 unit lower than the pH of the suspension medium for isolated lysosomes in vitro [78] and probably 2–3 units lower than the culture fluid in living cells [75] (see also Ref. 45). How this low pH is maintained is not fully understood. It probably depends in part on a Donnan-type equilibrium [38,88] which is due to the large amounts of fixed acidic charges on the lysosomal membrane or in the lysosomal matrix [35], but perhaps also on the existence of some sort of proton pump [20,75].
4. A great heterogeneity of their contents, as seen in the electron microscope. This reflects the various stages of digestive activities in which lysosomes are engaged, which vary greatly from one cell type to another and even within a given cell population. Lysosomes have often been characterized by their cytochemical reaction for acid phosphatase, which can be assayed for by both light and electron microscopy.

B. Role in the Handling of Intracellular and Extracellular Substances

Although surrounded by a semipermeable membrane, lysosomes are unique among intracellular organelles in that they are readily accessible in vivo to both extracellular and
intracellular constituents, through the route of heterophagy and autophagy (for nomenclature, see Ref. 22).

Autophagy

During autophagy, portions of the cytoplasm (including organelles such as mitochondria, parts of endoplasmic reticulum, etc.) are sequestered within vacuoles that eventually fuse with lysosomes. This process occurs physiologically in all cell types and is
believed to play a role in the turnover of the cell constituents. In kidney, intralysosomal
segregation and breakdown of mitochondria have been clearly demonstrated [27]. The
rate of their segregation would account for the reported half-life of mitochondria in this
tissue (about 8.5 days [10]). An indirect but striking proof of the importance of autophagy
in all cell types is given by the study of the lysosomal disorders caused by the genetic
defect of one lysosomal enzyme [36]. Lysosomes become overloaded with undigested
material which often is undoubtedly of cellular origin: for instance, glycogen in the
α-glucosidase deficiency. Another proof is also given by the observation that chloroquin,
a compound that accumulates in lysosomes, inhibiting the lysosomal protease cathepsin
B and slowing the cell protein turnover [100]. Also, stimulation of the protein turnover
in hepatocytes, which results from the administration of large doses of insulin, is accompanied by an increase in the number of autophagic vacuoles [77].

Under normal circumstances, autophagy does not provoke long-term overloading of
lysosomes, because the material that is being segregated is susceptible to the acid hydrolases. Lipofuscin could be an exception in this respect [12]. Alterations of the cell
constituents, by drugs for instance, could render them undigestible by lysosomal enzymes; this would then lead to their intralysosomal storage. An example is the lysosomal
phospholipidoses induced by a variety of cationic amphiphilic drugs which bind to
phospholipids of subcellular membranes and are thought to prevent their further degradation by lysosomal phospholipases [62].

Heterophagy (Endocytosis)

Heterophagy allows the intralysosomal accumulation of extracellular macromolecules
and other nonpermeating molecules. This process, better known under the name of endocytosis (for reviews, see Refs. 44 and 83), occurs through the formation of intracellular
vesicles (endocytic vesicles or phagosomes), which are derived from invaginations of the
pericellular membrane. Endocytosis is a widespread property of all eukaryotic cells.
Although there is a fair degree of specialization with respect to the prehension of large
particles (phagocytosis), the uptake of small droplets of fluid (pinocytosis) is observed
in almost all cell types, but is particularly efficient in intestinal or tubular epithelia. There
is theoretically no substance that cannot be internalized in that way. As proposed by
Jacques [44] and verified experimentally in a number of instances [32,80,101], substances that bind to the pericellular membrane will, however, be taken up preferentially.
The amount of these substances that is internalized in endocytic vesicles will indeed be
the sum of the amount present in the fluid phase (i.e., the volume endocytosed times the

extracellular concentration—"fluid pinocytosis") plus the amount bound to the membrane ("adsorptive pinocytosis"). Since pinocytotic vesicles are small (diameter of 0.5 μm or less), the ratio of their surface to their volume is high, which allows considerable transport to occur by the membrane phase. Thus, pinocytosis of denatured albumin by yolk sac cells is up to 80-fold faster than the uptake of nonbinding substances such as sucrose or polyvinylpyrrolidone [68].

Figure 1 shows a computer calculation of the rate of fluid and adsorptive pinocytosis made on the basis of the model of Jacques [44] and de Duve et al. [19] (see also Ref. 101) and applied to cultured fibroblasts such as L cells for which morphometric data on pinocytosis are available [85]. A substance that binds to the pericellular membrane with a dissociation constant of only 2×10^{-5} M and with a binding capacity of 4×10^{-7} mol/mg of cell protein (which are the parameters calculated for gentamicin in kidney brush border, see below) will be taken up about 30-fold faster than sucrose or inulin, which do not bind to the membrane.

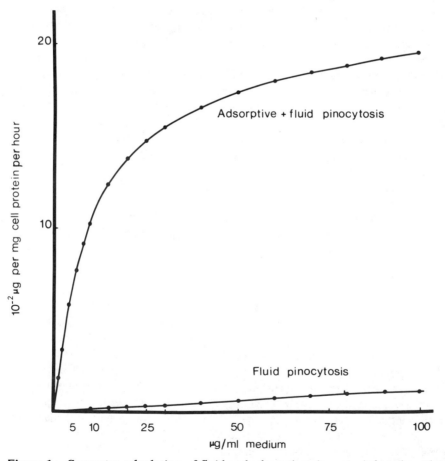

Figure 1   Computer calculation of fluid and adsorptive pinocytosis based on the L-cell morphometric data of Steinman [85] or endocytosis in L cells, using the dissociation constant and binding capacity of the membrane for gentamicin, reported by Just et al. [46].

Endocytic vesicles quickly fuse with lysosomes, where their content is brought into contact with lysosomal hydrolases. This transfer from cell surface to lysosomes is very rapid. Straus and Oliver [86], Novikoff [72,73], and Graham and Karnovsky [33] showed that horseradish peroxidase (an enzyme with a molecular weight of about 40,000 commonly used as a cytochemical marker of endocytosis) reaches lysosomes only a few minutes after it is detected in endocytic vacuoles in the apical region of the proximal tubular cells. Christensen [13] showed that 3 min after an intravenous (IV) injection of cytochrome c, 72% of the material was detected in the endocytic region, whereas only 3% was inside the lysosomes. Thirty minutes later these values became 15 and 52% respectively.

Similar observations have been published on the adsorption of hemoglobin [64] or of [^{125}I]albumin by the proximal tubule cells [63]. The appearance of these molecules in large phagosomes close to the luminal surface of the tubule cells a few minutes after the injection may be taken to indicate that the protein was absorbed from the lumen by a process related to pinocytosis.

It seems well established that the renal handling of the plasma proteins that are filtered through the renal glomeruli involves proximal tubule cell reabsorption by pinocytosis. Pinocytosed proteins are catabolized by lysosomal enzymes [14,17]. Thus, the kidney is an important organ for the catabolism of various plasma proteins; Katz et al. [48], Schultze and Heremans [81], and Strober and Waldmann [87] showed that 10–15% of whole albumin catabolism takes place in the kidney.

Pinocytosis is also responsible for the intrarenal transport of nonprotein substances. Light microscopic studies have demonstrated that a large number of carbohydrates and other nonprotein substances, if administered in large doses, cause renal tubular changes which are frequently referred to as osmotic nephrosis. Following the injection of sucrose, mannitol, or dextran (which are not degraded by the organism except in the intestinal brush border), the initial change in ultrastructure is an increase in the frequency and size of the apical vacuoles close to brush border. The content of these vacuoles is transferred to lysosomes. Endocytosis appears to be the major route of entry for these compounds.

In addition to proteins, polysaccharides, and so on, lysosomes accumulate a variety of basic organic compounds of low molecular weight, many of which are dyes and/or drugs, for instance acridine, quinine, chloroquine, and neutral red [4,16,52]. The rate of accumulation of these compounds is often too fast to be accounted for by endocytosis. Furthermore, this accumulation is reversible and is also observed in vitro, although often to a lower extent than in vivo. de Duve et al. [19] and Reijngoud et al. [78] have proposed that organic bases tend to accumulate in lysosomes by ion trapping because of the difference in pH between lysosomes and the extracellular fluid, and of the higher permeability coefficients through biological membranes of the unprotonated form than of the protonated form of these bases [65,95]. Dingle and Barrett [26] have also shown that lysosomes contain acid glycolipids which bind organic bases such as acridine. As far as we know, there is at the present time no direct evidence for specific transport systems in the lysosomal membrane.

In view of these findings, de Duve et al. [19] have introduced the term "lysosomotropic agents," which designates all the compounds that are taken up in lysosomes either by pinocytosis (as such, or linked to an endocytosed macromolecule) or by any other mechanism, including ion trapping. These agents may stay within the lysosomes and act either by altering the lysosomal content or the lysosomal ability to degrade engulfed substances, thus inducing lysosomal storage disorders. The agent may also modify the properties of the lysosomal membrane, inducing an increase or a decrease of its permeability toward exogenous substrates or even to lysosomal enzymes themselves, which could escape lysosomes and provoke cell injury (see also Refs. 3 and 18).

II. Gentamicin Uptake and Localization in the Experimental Animal

A. Whole-Body Tissular Localization

Aminoglycosides are concentrated within the renal cortex [2,57,99], but their localization in other tissues has not been well documented. In this study, adult Wistar rats weighing 250–300 g were given a single intramuscular (IM) injection of gentamicin at the dose of 4 mg/kg. ^{14}C-labeled gentamicin (generously provided by the Schering Co., Kenilworth, N.J.) was used as a tracer at the dose of 2 μCi per rat. Animals were sacrificed by immersing them into liquid nitrogen after light ether anesthesia and maintained at –20°C until 100 μm slices were cut under congelation and dried at –20°C according to the technique of Ullberg [92] as modified by Cohen and Delassue [15]. The slices were put in contact with x–ray Kodirex films for 3 weeks.

As shown in Figure 2, gentamicin is almost exclusively localized within the kidneys and cartilages. No labeling was found in the muscle, liver, brain, or lung 2 hr after the injection. Association of gentamicin with cartilage had not yet been reported, to our knowledge, and may be related to the presence of large amounts of chondroitin sulfate (a polyanionic compound) in the cartilages.

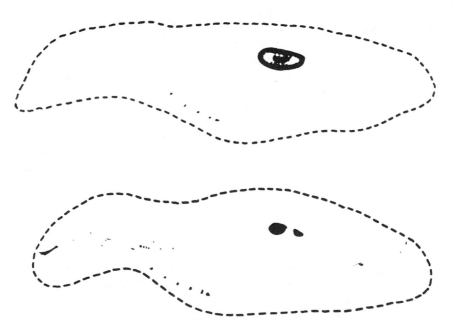

Figure 2 Whole-body autoradiogram of rat injected with ^{14}C-labeled gentamicin. Two hours after the injection, dashed lines represent body limits. Top: Sagittal section through the kidney which is highly labeled, primarily at the level of the cortex. Bottom: Median sagittal section displaying intense labeling of cartilages: nasal, tracheal, and costal cartilages are clearly seen. In this slice, kidney was cut tangentially through cortex.

B. Kidney Cellular Localization

Most of the intrarenal gentamicin appears to be intracellular, because the concentration of the drug in the interstitial fluid never exceeds that in the plasma, whereas the tissue/plasma peak concentration ratio reaches values of 10-12. Just et al. [46] reported the localization of tritiated gentamicin within the proximal tubule cells of the rat kidney by an autoradiographic method (see also Ref. 55). Using a technique of microdissection of rabbit kidney nephron [69], we have been able to pinpoint the gentamicin distribution along a single nephron from the glomerulus to the collecting duct (unpublished data). No detectable amount of labeled product was found within the glomeruli, the descending and ascending loop of Henle, and the distal tubular cells. Low amounts of silver grains were seen in the epithelial cells of the macula densa area. Large amounts were found within the proximal tubular cells. The gentamicin distribution was not homogenous all along the proximal tubular cells: an increasing gradient of intracellular gentamicin concentration was recorded from the initial portion to the distal portion of the proximal convoluted tubule. This gradient seems to be in correlation with the concentration gradient of primary urine solutes observed along the proximal tubule, suggesting that gentamicin uptake is dependent on tubular fluid gentamicin concentration.

C. Subcellular Localization in Tubular Cells

Subcellular localization of a constituent can be approached by two different ways: cell fractionation or morphology. Although the former approach is very quantitative, it necessitates the disruption of cells. When doing so, one must keep in mind the danger of disrupting the subcellular organelles as well, thus releasing the constituent and allowing its readsorption on various structures. Morphology is rarely quantitative and raises the problem of the positive identification of the constituent investigated.

Sherman strain rats (250-300 g) were injected with ^{14}C-labeled gentamicin and sacrificed 6 and 24 hr after the injection. Kidney cortices were homogenized in ice-cold 0.25 M sucrose and the postnuclear supernatant subjected to centrifugation in a linear sucrose gradient. As shown in Figure 3, the distribution pattern of gentamicin resembles closely those of two marker enzymes of lysosomes: N-acetyl-β-D-glucosaminidase and cathepsin B, either 6 hr or 24 hr after the injection of the drug. These results strongly suggest that gentamicin is associated with lysosomes. In contrast, when gentamicin is added to a cortex homogenate rather than being injected to the animal, distribution of the antibiotic and of the lysosomal enzymes is clearly dissociated. Thus, the association of gentamicin to lysosomes shown in Figure 3 is unlikely to result from a fractionation artifact. Similar results were obtained by Just et al. [46], who showed the importance of preserving the integrity of lysosomes during homogenization and centrifugation procedures. When lysosomes were purposedly damaged, gentamicin was partially solubilized and partially readsorbed on structures distinct from lysosomes.

Association of gentamicin with lysosomes was confirmed by autoradiographic studies at the ultrastructural level [46], which showed a heavy accumulation of radioactive label in "dense bodies" within 20 min after the injection of the tritiated drug.

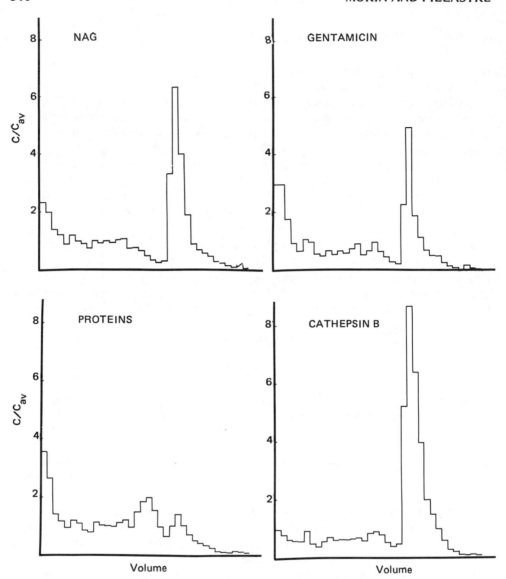

Figure 3 Distribution patterns of protein, gentamicin, cathepsin B, and N-acetyl-β-D-glucosaminidase (NAG) in a postnuclear supernatant of kidney cortex homogenate from rat injected 6 hr previously with ^{14}C-labeled gentamicin, as analyzed by sucrose gradient centrifugation. The ordinate is C/C_{av}, where C is the concentration of the compound in each fraction and C_{av} is the theoretical concentration of the compound if it were uniformly distributed all along the gradient.

D. Mechanism of Gentamicin Uptake by Proximal Tubular Cells

Specific Pathways

The mechanism of gentamicin uptake by kidney proximal tubule cell has been previously approached from the physiological point of view by monitoring the competition of gentamicin with substrates of the active transport mechanisms described in kidney cells: organic anions and cations, glucose, and amino acids. Gentamicin does not interfere with the uptake of p-aminohippurate (PAH), α-methyl-D-glucopyranoiside (a sugar analog), and α-aminoisobutyrate (an amino acid analog); this suggests that the corresponding specific transport pathways are not involved in gentamicin uptake [50,51]. However, Whelton and co-workers [97,98] have reported that the infusion of large amounts of amino acids (at a rate that results in amino aciduria) significantly lowers the cortical uptake of gentamicin and tobramycin.

Even though transport by specific pathways cannot be ruled out, it is unlikely that it plays a major role in the handling of aminoglycosides. Molecules transported by these systems are indeed delivered in the cytosol. How, then, would aminoglycosides cross the lysosomal membrane?

Nonionic Diffusion

This pathway seems unlikely because the aminoglycosides are probably much too hydrophilic to diffuse easily through biological membranes. Alkalinization of the urine has been reported to lower the nephrotoxicity of gentamicin [43], which suggests a reduced uptake of the drug, at a time when the concentration of its un-ionized form is increased. Also, nonionic diffusion takes place mostly in the distal tubule and in the collecting duct, where little gentamicin accumulates.

Endocytosis

As discussed in Section I, proximal tubule cells display very high and constant endocytic activity, which should enable them to take up rapidly the compounds present in the urinary fluid. In addition, the process of concentration of the urinary fluid, which takes place along the proximal tubule, will enhance the efficiency of the uptake.

Just et al. [46] have reported that the kidney brush border membrane displays binding sites toward a variety of polybasic substances, including gentamicin. In vitro studies showed a dissociation constant of about 2×10^{-5} M and a binding capacity of about 18 nmol/mg of membrane protein. Thus, at the concentrations of gentamicin prevailing in the proximal tubular fluid [about 50–100 μg/ml (i.e., 0.1–0.2 mmol)], the membrane of the tubular cells is largely covered by bound gentamicin. As a result, the uptake of gentamicin by endocytosis will be very efficient (see our model in Figure 1); detailed calculations for kidney cells are not yet possible, because of the lack of morphometric data, such as area of brush border membrane, size of endocytic vesicles, and so on.

This endocytosis hypothesis is substantiated by the observation that 5 min following the injection of radiolabeled gentamicin, most of the label is located over the brush border or at the apical pole of the cell in close association with endocytic vesicles. Ten to 15 min later, lysosomes become labeled, whereas the apical region and the brush border loose their radioactivity.

Furthermore, it should be pointed out that accumulation of aminoglycosides in lysosomes is not a property unique to proximal tubular cells. Under appropriate conditions, Tulkens and Trouet [89–91] (see also Tulkens et al. [89] showed that aminoglycosides readily enter cultured fibroblasts and accumulate in lysosomes, where they reach concentrations 20- to 100-fold larger than in the culture fluid.

A point of interest is to know whether this intracellular transport of gentamicin by endocytosis could be inhibited, hence reducing the renal toxicity of the drug. As far as endocytosis itself is concerned, we know no means to lower its rate except by severely inhibiting or poisoning the cell energy production. Competition of the association of gentamicin with its brush border binding sites could theoretically reduce its endocytic uptake. However, this binding seems relatively unspecific and involves a large number of sites [47]. Thus, competing agents will need to be given in large doses, which may raise some problems. We suggest that the inhibition of the aminoglycoside uptake following the infusion of large doses of amino acids [98] (see above) actually reflects a competition between lysine or other basic amino acids and aminoglycosides for these binding sites, rather than for the specific amino acid transport system. Aprotinin, a basic polypeptide that competes efficiently with gentamicin for brush border membrane binding in vitro, was, however, ineffective in vivo at pharmacologically acceptable doses. Alkalinization of the urine could be effective as discussed above [43], since binding of gentamicin to brush border requires ionization of the drug. This requires more detailed investigation, however.

III. Aminoglycoside Effects on Kidney Lysosomes

A. In Vitro Studies

Lysosomal Membrane Integrity

According to de Duve et al. [18,23], free/total and unsedimentable/total activity ratios of lysosomal enzymes are presumed to reflect the lysosomal membrane integrity: as long as the lysosomal membrane remains intact, the enzymes present inside the lysosomes cannot act upon their substrates present in the incubation medium. Once the membrane is injured, the enzymatic activities are released and cause a rapid lysis of the substrates present in the incubation medium. A fraction enriched in lysosomes and prepared from rat kidney was incubated for 60 min at 37°C; free, total, and unsedimentable activities were assayed before and after the incubation at 37°C in the presence or in the absence of antibiotic.

Aminoglycosides labilize the lysosomes of rat kidney in vitro at concentrations ranging from 5 to 50 μg/ml of standard lysosomal suspension (Figure 4). A statistical study using the Student's t test for mean and variation allowed us to classify these molecules in three groups in term of the intensity of their labilizing effect. A highly active class includes sisomicin, tobramycin, and gentamicin, an intermediary-active class includes neomycin and lividomicin, and a low-active class includes kanamycin, amikacin, netilmicin, streptomycin, paromomycin, and kasugamycin. Except for neomycin, a good correlation is seen between the relative intensity of aminoglycoside labilizing effect and the relative nephrotoxicity of these molecules as reported by Appel et al. [5].

Cephalosporins induce little alteration of the lysosomal membrane. Except for cephalothin, the labilizing effect of which is similar to that of kanamycin, concentrations higher than 50 μg/ml must be reached to observe a significant effect of these molecules (Table 1).

In the case of cephalosporin-aminoglycoside combinations, three types of protocols have been followed, the simultaneous presence of both compounds, the aminoglycoside prior to the cephalosporin, or the reverse within a time interval of 30 min. This mimics what can happen in vivo during therapy.

As shown in Figure 5, the addition sequence has to be considered with regard to the result: when the cephalosporin was added first and the aminoglycoside 30 min later, a potentiation of the aminoglycoside labilizing effect is evidenced. When the aminoglycoside was added first and the cephalosporin 30 min later, a protective effect of cephalosporin-against aminoglycoside-induced labilization is seen during the simultaneous contact of

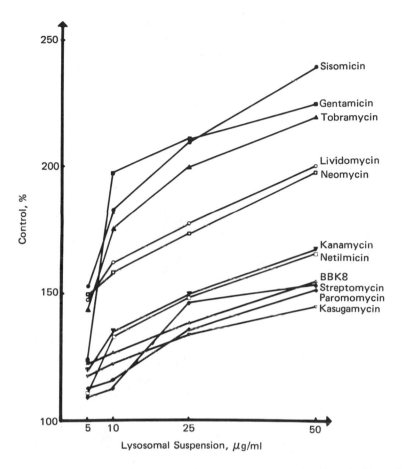

Figure 4 Structural latency of a lysosomal preparation of rat kidney. Comparative effect of the addition of 11 aminoglycosides at different concentrations. Results are expressed in comparison with the control. The lysosomal suspension, about 0.2 mg/ml of protein, was incubated for 1 hr at 37°C, and the free and total activities of N-acetyl-β-hexosaminidase were then assayed.

both drugs, except in the case of kanamycin. In the case of the simultaneous addition of aminoglycoside and cephalosporin to the incubation medium, an important protective effect of cephalosporin against aminoglycoside effect is seen [71]. In the case of amikacin and streptomycin only a partial protection was seen, whereas in the case of kanamycin no protection could be evidenced (Figure 6).

These results fit well with the in vivo observations of Dellinger et al. [24], Luft et al. [58], and Barza et al. [9], who described a lower nephrotoxicity for the aminoglycoside-cephalosporin combination that for the aminoglycoside alone. Furthermore, Dellinger et al. [24] reported the loss of the cephalosporin effect against gentamicin nephrotoxicity when both compounds of the association are injected within a time interval of 6 hr to rats. In this case, no effect of the injection sequence was recorded in vivo. We suggest that this is due to the short half-life of aminoglycoside in rats (30 min [29]), which does not allow the simultaneous presence of both drugs in the serum.

Table 1 Comparative Study of Lysosomal Membrane Stability in the Presence of Cephalosporins[a]

	5	10	25	50	100	250
Cephalotin	100.86 ± 0.63	113.55 ± 1.76	135.34 ± 2.30	133.38 ± 1.33	138.27 ± 1.49	142.53 ± 1.68
Cephradin	101.80 ± 1.11	101.01 ± 1.31	101.30 ± 1.30	96.37 ± 1.74	95.82 ± 2.44	115.19 ± 2.45
Cefazolin	103.48 ± 1.95	101.01 ± 2.18	101.17 ± 2.11	98.47 ± 2.25	102.15 ± 1.98	99.37 ± 1.75
Cephapirin	101.63 ± 2.27	101.09 ± 1.92	98.74 ± 3.07	98.72 ± 3.18	81.40 ± 1.71	69.72 ± 2.13
Cefuroxime	98.25 ± 0.76	100.90 ± 1.28	97.76 ± 2.86	96.93 ± 2.76	92.14 ± 1.93	86.71 ± 2.55
Cefamandole	99.67 ± 0.83	98.91 ± 1.339	105.02 ± 2.94	98.28 ± 2.53	93.04 ± 1.82	81.51 ± 2.26
Cephaloridine	104.63 ± 1.21	102.24 ± 1.37	104.39 ± 1.22	105.21 ± 1.08	87.56 ± 2.60	69.22 ± 2.46
Cefoxitin	96.75 ± 1.76	98.04 ± 1.55	99.37 ± 1.45	81.35 ± 2.09	71.36 ± 1.95	63.18 ± 1.37

[a] Results are expressed in percent of the nondrug control value incubated 60 min at 37°C.

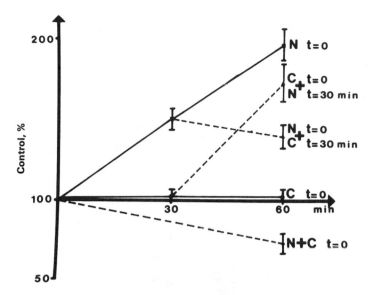

Figure 5 Suppression of the structural latency of N-acetyl-β-glucosaminidase of a lysosomal preparation by the addition of netilmicin (N) and/or cefazolin (C), 25 μg/ml each, added in various sequences.

As previously described [70], aminoglycosides seem to induce a modification of the lysosomal membrane permeability to the enzyme substrates rather than a membrane disruption, which seems to be induced by thermal shock or mechanical disruptions.

B. Influence of Gentamicin on Tubular Cells in Animals

Morphology

As mentioned in our introduction, lysosomes show prominent alterations after treatment with gentamicin. As clearly shown by Koseck et al. [53] and confirmed by Houghton et al. [40], Wellwood et al. [96], and Luft et al. [60], these alterations consist mainly of (1) the intralysosomal deposition of osmophilic material with a lamellar aspect and a concentric deposition of osmophilic material with a lamellar aspect and a concentric deposition (myeloid bodies); and (2) some degree of enlargement of the lysosomes, which, however, do not appear swollen or overloaded by other material than myeloid bodies. These modified lysosomes were described as "cytosegresomes" or autophagic vacuoles. It should, however, be stressed that they do not resemble vacuoles that develop after stimulation of autophagy, as observed in liver after glucagon administration [25] (see also Ref. 28). In the latter situation, lysosomes are overloaded with an heterogenous material, in which fragments of mitochondria and other subcellular structures are still easily recognizable, obviously as a result of a failure of the lysosomal digestion to keep pace with a temporary enhancement of the autophagic sequestration. With gentamicin, the picture is much more homogeneous and recalls that observed in lysosomal enzyme defects, for instance in Niemann-Pick or Tay-Sachs diseases (for reviews, see Refs. 11 and 74). The lesions also resemble those induced by treatment with amphiphilic cationic drugs [61,62] (see also Ref. 42). Thus, we are probably dealing with a delay in the

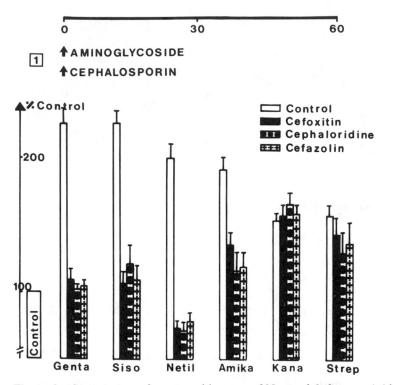

Figure 6 Suppression of structural latency of N-acetyl-β-glucosaminidase in a lysosomal preparation from rat kidney incubated 60 min in the presence of aminoglycoside alone or of aminoglycoside + cefoxitin, + cephaloridine, or + cefazolin, added simultaneously with the aminoglycoside.

degradation of one or a few specific constituents rather than with a general stimulation of the autophagic activity of the renal cell.

Only at large concentrations (higher than 10 mg/kg) for 7 days or more do light and electron microscopy reveal pyknosis and general damage of the tubular cells. The autophagic images reported at those concentrations [94] are difficult to interpret in view of the extensive tubular necrosis which develops and which allows many cell fragments to be endocytosed by neighboring cells.

In view of the effects of gentamicin on the lysosomal membrane as reported above and of the striking effects exerted by gentamicin on the lysosomes of cultured cells [6] (see Tulkens et al. [89], we have investigated the biochemical alterations of the lysosomes of the kidney cortex after administration of the antibiotic in vivo.

Lysosomal Membrane Stability

As shown in Table 2, lysosomes prepared from rats injected with 3 or 6 mg of gentamicin per kilogram for 7 days show no change in the ratio of free/total activity of lysosomal enzymes. Lysosomal membrane alteration is only evidenced above 20 mg/kg for 7 days. In addition, gentamicin therapy induces a significant enhancement of the lysosomal susceptibility to thermal shock, suggesting a modification of the membrane properties.

After 25 and 75 mg/kg therapy for 7 days, a loss of sensitivity of the lysosomal membrane to an added dose in vitro was recorded during the thermal shock. Gentamicin added

Table 2   Effect of Gentamicin Therapy on Lysosomal Structural Latency and Sensitivity
to Thermal Shock[a]

	Without incubation	Incubation 60 min at 37°C
Control (n = 11)	57.00	100
3 mg/kg, 8 days	58.8	111.2
6 mg/kg, 8 days	60.7	126.4
20 mg/kg, 8 days (n = 4)	77.57	137.70
50 mg/kg, 13 h 30 (n = 4)	80.32	125.57
50 mg/kg, 2 days (n = 4)	88.85	137.60
50 mg/kg, 4 days (n = 4)	89.47	134.63
50 mg/kg, 6 days (n = 4)	96.00	143.27
50 mg/kg 8 days (n = 4)	109.78	142.10
100 mg/kg, 8 days (n = 2)	105.50	141.70

[a] Results are expressed in percent of the issued lysosomal suspension from control rats,
incubated 60 min at 37°C.

to control lysosomal suspensions brought to a similar initial free/total activity ratio as
preparations from treated rats showed a pattern strictly parallel to that of control prepara-
plus gentamicin (Figure 7). To explain this fact, we speculate that the intralysosomal
gentamicin concentrations in treated rat preparations might conceal the effect of the
gentamicin dose added in vitro [70].

Lysosomal Enzyme Activities

The activity of lysosomal enzymes has been studied in kidney cortices of Sprague-
Dawley rats treated with 10 and 20 mg of gentamicin per kilogram per day for 2-24 days.
Most of these enzymes, which included several glycosidases and a protease, cathepsin D,
showed no significant change in their specific activity, except for sphingomyelinase, the
activity of which was lowered to 63% of the controls.

In parallel with this effect on sphingomyelinase, gentamicin treatment increased by
25% the cellular content in phospholipids, which is highly significant. Although we do
not know for certain where this excess of phospholipids is located, there could be little
doubt that lysosomes is their major site of deposition, in view of typical images of myeloid bodies, which are known to be made largely, if not exclusively, of complex lipids
(see, for instance, Refs. 7 and 84). However, morphometric and chemical studies need to
be made to confirm this point.

It is thus remarkable that the lesions induced by gentamicin in fibroblasts and kidney
cortex are both associated with similar biochemical alterations of the lysosomes. However, the degree of the decrease of enzyme activity and of phospholipid overloading is
less pronounced in kidney than in fibroblasts. The reasons for this are not yet clear, but
it may be speculated that this reflects the heterogeneity of the kidney tissue, as compared with homogenous cultures of fibroblasts. Not all tubular cells may be affected
similarly by gentamicin treatment. In this connection, we showed (see above) that the
cellular gentamicin distribution is not constant along the tubule, and we know that the
phospholipidosis and the sphingomyelinase deficiency are dose-dependent [6]. In addition, the most affected cells may quickly undergo necrosis and be replaced by younger
cells, which, according to Gilbert et al. [31], Houghton et al. [41], and Luft et al. [59],

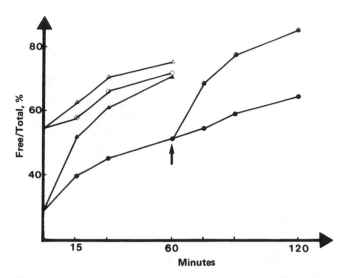

Figure 7 Release from lysosomal preparations issued from untreated (solid symbols) and gentamicin-treated (open symbols) animals, 75 mg/kg daily for 21 days. Results are expressed as free/total activity ratio of N-acetyl-β-D-glucosaminidase during incubation at 37°C in the absence (round symbols) or the presence (triangular symbols) of gentamicin, 25 μg/ml added in vitro at the beginning of the incubation or 60 min later (arrow).

are less susceptible to gentamicin injury. Finally, a nonnegligible part of the sphingomyelinase activity and of the phospholipids in the kidney cortex may be associated with other cells than the proximal tubule cells and may therefore not be involved in the disorders described.

Extralysosomal Effects

In addition to the effects described for lysosomes, we have noted an inhibitory action of gentamicin toward two enzymes of the kidney brush border, alkaline phosphodiesterase I and alanine aminopeptidase, the activity of which were reduced to 50% of that of controls. The inhibition of alanine aminopeptidase can be reproduced in vitro, although large concentrations (100–200 μg/ml) of gentamicin are required.

Alanine aminopeptidase activity is present in very high amounts in kidney brush border (about 300–350 U/g of kidney cortex protein). Mondorf [66] has shown this enzyme to be located on the surface of brush border membrane. The significance of such high amounts of alanine aminopeptidase in the brush border membrane of proximal tubule cells is not known. In our experiments, values of both V_{max} and K_m were affected, suggesting a loss of enzymatic activity accompanied by a competitive inhibition. Mondorf et al. [66,67] reported an increased urinary excretion of alanine aminopeptidase during aminoglycoside therapy.

The specific localization of this enzyme in the kidney cells and its early increase of activity in the urine after gentamicin treatment can be related with the binding of the antibiotic to the brush border (see above). Thus, we suggest that alanine aminopeptidase urinary levels are an early and valuable index of gentamicin toxicity.

Other enzymes, including lysosomal hydrolases, are also excreted in the urine. These enzymes may have various origins, such as the plasma (through glomerular filtration), prostatic secretion, desquamation, or necrosis of tubular cells. It is probably the latter phenomenon that accounts for the increased enzymuria of lysosomal enzymes reported by Patel et al. [76] after gentamicin treatment.

IV. Conclusions

Aminoglycosides can be added to the long list of lysosomotropic agents. Lots of compounds may reach lysosomes of kidney proximal cell and be concentrated there. But this property is not by itself sufficient to induce nephrotoxic syndromes.

The aminoglycoside-induced storage pathology, as shown in fibroblasts [6], as well as in the rat kidney cortex, is characterized by a decrease in sphingomyelinase activity and the accumulation of phospholipids. Further studies are, however, necessary to ascertain the extent of the enzymatic lesions induced by aminoglycosides and the pathogeny of the lysosome overloading. Enzyme inhibition is most probably responsible for the formation of myeloid bodies. Lysosomal alterations represent a very early stage of kidney dysfunction, as they are already demonstrable at a time when no other subcellular changes can be detected.

Furthermore, aminoglycosides labilize the lysosomes of the kidney proximal tubule cells in vitro as in vivo. The pathological significance of the correlation between the in vitro and the in vivo findings is, however, speculative, for it must be emphasized that in vitro aminoglycosides interact with the lysosomal membrane from the outside, whereas in vivo they interact from the inside of the lysosomes.

At the beginning of the chapter, we pointed out two mechanisms by which lysosomotropic agents may cause lysosomal toxicity, that is, lysosome overloading and alteration of the lysosomal membrane.

The relative contribution of lysosomal membrane damage and of lysosomal overloading cannot be ascertained at this stage. Lysosomal disruption probably occurs in vivo after high doses of drug, as shown in the recent study of Watanabe [95a]. At lower doses, or at an early stage of chronic administration, lysosomal overloading seems more prominent.

Acknowledgments

This work was supported in part by INSERM Grant 76-1-176-5 and in part by Fondation de la Recherche Médicale Française. We gratefully acknowledge the excellent technical assistance of Mrs. G. Viotte and Mr. J. Fresel, and thank Mrs. Falize for secretarial help. We also thank Drs. F. Van Hoof, G. Aubert-Tulkens, and P. Tulkens (International Institute of Cellular and Molecular Pathology and Université Catholique de Louvain, Brussels, Belgium) for communicating to us a manuscript in press (Ref. 6), for allowing us to use unpublished data on the action of gentamicin on kidney enzymes, and for helpful discussions during the writing of this chapter.

References

1. Abramowicz, M., and Edelmann, C.M.: Nephrotoxicity of antiinfective drugs. *Clin. Pediatr.* 7:389, 1968.

2. Alfthan, O., Renkonen, O.V., and Sivonen, A.: Concentration of gentamicin in serum, urine and urogenital tissue in man. *Acta Pathol. Microbiol. Scand. [B],* 81(Suppl. 241):92, 1973.

3. Allison, A.C., and Davies, P.: Mechanisms of endocytosis and exocytosis. *Symp. Soc. Exp. Biol.* 28:419–446, 1974.

4. Allison, A.C., and Young, M.R.: Uptake of dyes and drugs by living cells in culture. *Life Sci.* 3:1407–1414, 1964.

5. Appel, G.B., and Neu, H.C.: The nephrotoxicity of antimicrobial agents. *N. Engl. J. Med.* 269:663–670, 722–728, 784–787, 1977.

6. Aubert-Tulkens, G., Van Hoof, F., and Tulkens, P.: Gentamicin induces a lysosomal phospholipidosis in cultured rat fibroblasts. Quantitative ultrastructural and biochemical study. *Lab. Invest.* 40(4):481–491, 1979.

7. Bangham, A.D.: Physical structure and behavior of lipids and lipid enzymes. *Adv. Lipid Res.* 1:65–104, 1963.

8. Barrett, A.J., and Heath, M.F.: Lysosomal enzymes. In *Lysosomes: A Laboratory Handbook,* 2nd ed., edited by Dingle, J.T. Amsterdam, Elsevier/North-Holland, 1977, pp. 19–145.

9. Barza, M., Pinn, V., Tanguay, P., and Murray, T.: Nephrotoxicity of newer cephalosporins and aminoglycosides alone and in combination in a rat model. *J. Antimicrob. Chemother.* 4(Suppl.):59–68, 1978.

10. Beattie, D.S., Basford, R.E., and Koritz, S.B.: The turnover of the protein components of mitochondria from rat liver, kidney and brain. *J. Biol. Chem.* 242:4584–4586, 1967.

11. Brady, O.R., and King, F.M.: Niemann-Pick's disease. In *Lysosomes and Storage Diseases,* edited by Hers, H.G., and Van Hoof, F. New York, Academic Press, 1973, pp. 439–452.

12. Chio, K.S., Reiss, U., Fletcher, B., and Tappel, A.L.: Peroxidation of subcellular organelles formation of lipofucin-like fluorescent pigments. *Science* 166:1535–1536, 1969.

13. Christensen, E.I.: Rapid protein uptake and digestion in proximal tubule lysosomes. *Kidney Int.* 10:301–310, 1976.

14. Christensen, E.I., and Maunsbach, A.B.: Intralysosomal digestion of lysozymes in renal proximal tubule cells. *Kidney Int.* 6:396–407, 1974.

15. Cohen, Y., and Delassue, H.: Modifications de la méthode d'autoradiographie de S. Ullbert sur coupes de souris entières. *C.R. Soc. Biol.* 153:300–304, 1959.

16. Cohn, Z., and Wiener, E.: The particulate hydrolases of macrophages: I. Comparative enzymology isolation and properties. *J. Exp. Med.* 118:991–1008, 1963.

17. Davidson, S.J.: Protein absorption by renal cells: II. Very rapid lysosomal digestion of exogenous ribonuclease in vitro. *J. Cell Biol.* 59:213–222, 1973.

18. de Duve, C.: *The Lysosome Concept* (Ciba Foundation Symposium on Lysosomes), edited by de Reuck, A.V., and Cameron, M.P. London, Churchill, 1963, pp. 1–35.

19. de Duve, C., de Barsy, T., Poole, B., Trouet, A., Tulkens, P., and Van Hoof, F.: Lysosomotropic agents. *Biochem. Pharmacol.* 23:2495–2531, 1974.

20. de Duve, C., Okuma, S., Poole, B., and Tulkens, P.: The intralysosomal pH. In *Microenvironments and Metabolic Compartmentation,* edited by Stere, P.A., and Eastbrook, R.W. New York, Academic Press, 1978, pp. 371–380.

21. de Duve, C., Pressman, B.C., Gianetto, R., Wattiaux, R., and Appelmans, F.: Tissue fractionation studies: 6. Intracellular distribution patterns of enzymes in rat liver tissue. *Biochem. J.* 60:604–617, 1955.

22. de Duve, C., and Wattiaux, R.: Functions of lysosomes. *Ann. Rev. Physiol.* 28: 435–492, 1966.

23. de Duve, C., Wattiaux, R., and Wibo, M.: Effects of fat soluble compounds on lysosomes in vitro. *Biochem. Pharmacol.* 9:97–116, 1962.

24. Dellinger, P., Murphy, T., Pinn, V., Barza, M., and Weinstein, L.: Protective effect of cephalothin against gentamicin-induced nephrotoxicity in rats. *Antimicrob. Agents Chemother.* 9:172–178, 1976.

25. Deter, R.L., and de Duve, C.: Analog modeling of glucagon induced autophagy in rat liver. *Exp. Cell Res.* 94:127–139, 1975.

26. Dingle, J.T., and Barrett, A.J.: The uptake of biologically active substances by lysosomes. *Biochem. J.* 109:19, 1968.

27. Ericsson, J.L.E.: Transport and digestion of hemoglobin in the proximal tubule. *Lab. Invest.* 14:16–39, 1965.

28. Ericsson, J.L.E.: Mechanism of cellular autophagy. In *Lysosomes in Biology and Pathology,* Vol. 2, edited by Dingle, J.T., and Fell, H.B. Amsterdam, North-Holland, 1969, pp. 345–394.

29. Fabre, J., Rudhardt, M., Blanchard, D.P., and Regamey, C.: Persistence of sisomicin and gentamicin in renal cortex and medulla compared with other organs and serum of rats. *Kidney Int.* 10:444–449, 1976.

30. Falco, F.G., Smith, H.M., and Acieri, G.M.: Nephrotoxicity of aminoglycosides and gentamicin. *J. Infect. Dis.* 119:406–409, 1969.

31. Gilbert, D.N., Plamp, C., Starr, P., Bennett, W.M., Houghton, D.C., and Porter, G.: Comparative nephrotoxicity of gentamicin and tobramycin in rats. *Curr. Chemother.* 2:936–937, 1978.

32. Goldstein, J.L., and Brown, M.S.: The low-density lipoprotein pathway and its relation to atherosclerosis. *Annu. Rev. Biochem.* 46:897–930, 1977.

33. Graham, R.C., and Karnovsky, M.J.: The early stages of absorption of injected horse radish peroxydase in the tubule of mouse kidney. Ultrastructural cytochemistry by a new technique. *J. Histochem. Cytochem.* 14:291–302, 1966.

34. Harrison, W., Silverblatt, F., and Turck, M.: Experimental gentamicin nephrotoxicity. Abstr. 74, 14th Interscience Conference on Antimicrobial Agents and Chemotherapy, San Francisco, 1974.

35. Henning, R., Plattner, H., and Stoffel, W.: Nature and localization of acidic groups on lysosomal membranes. *Biochim. Biophys. Acta* 330:61–75, 1973.

36. Hers, H.G., and Van Hoof, F., eds.: *Lysosomes and Storage Disease.* New York, Academic Press, 1973.

37. Hewitt, W.L.: Gentamicin, toxicity in perspective. *Postgrad. Med. J.* 505:55–59, 1974.

38. Hollemans, M., Rejngoud, D.J., and Tager, J.M.: Evidence against a Mg-ATP-dependent proton pump in rat liver lysosomes. *Biochim. Biophys. Acta* 551:55–66, 1979.

39. Holtzman, E.: Lysosomes. A survey. *Cell Biology Monographs,* Vol. 3. New York, Springer-Verlag, 1976.

40. Houghton, D.C., Hartnett, M., Campbell Boswell, M., Porter, G., and Bennett, W.M.: A light and electron microscopic analysis of gentamicin nephrotoxicity in rats. *Am. J. Pathol.* 82:589–599, 1976.

41. Houghton, D.C., Plamp, E.C., Defehr, J.M., Bennett, W.M. Porter, G., and Gilbert, D.: Gentamicin and tobramycin nephrotoxicity. *Am. J. Pathol.* 93:137–150, 1978.

42. Hruban, Z., Slesers, A., and Hopkins, E.: Drug induced and naturally occurring myeloid bodies. *Lab. Invest.* 27:62–70, 1972.

43. Hsu, C.H., Kurtz, T.W., and Weller, J.M.: Potentiation of gentamicin nephrotoxicity by metabolic acidosis. *Antimicrob. Agents Chemother.* 12:132–194, 1977.

44. Jacques, P.: Endocytosis. In *Lysosomes in Biology and Pathology,* Vol. 2, edited by Dingle, J.T., and Fell, H.B. Amsterdam, North-Holland, 1969, pp. 395–420.

45. Jensen, M.S., and Bainton, D.S.: Temporal changes in pH within phagocytes vacuole of the polymorphonuclear neutrophilic leucocytes. *J. Cell Biol.* 59:379–388, 1973.

46. Just, M., Erdman, G., and Habermann, E.: The renal handling of polybasic drugs: 1. Gentamicin and aprotinin in intact animals. *Naunyn-Schmiedebergs Arch. Pharmacol.* 300:57–66, 1977.

47. Just, M., and Habermann, E.: The renal handling of polybasic drugs: 2. In vitro Studies with brush border and lysosomal preparations. *Naunyn-Schmiedebergs Arch. Pharmacol.* 300:67–76, 1977.

48. Katz, J.A., Sellers, A.L., and Bonorris, G.: Effect of nephrectomy on plasma albumin catabolism in experimental nephrosis. *J. Lab. Clin. Med.* 63:680–686, 1974.

49. Kleinknecht, D., and Fillastre, J.P.: *La Tolérance rénale des antibiotiques.* Paris, Masson, 1973, p. 89.

50. Kluwe, W.M., and Hook, J.B.: Analysis of gentamicin uptake by rat renal cortical slices. *Toxicol. Appl. Pharmacol.* 45:531–539, 1978.

51. Kluwe, W.M., and Hook, J.B.: Functional nephrotoxicity of gentamicin in the rat. *Toxicol. Appl. Pharmacol.* 45:163–175, 1978.

52. Koenig, H.: Intravital staining of lysosomes by basic dyes and metallic ions. *J. Histochem. Cytochem.* 2:120–121, 1963.

53. Kosek, J.D., Mazze, R.I., and Cousins, M.J.: Nephrotoxicity of gentamicin. *Lab. Invest.* 30:48–57, 1974.

54. Kovnat, P., Labowitz, E., and Levison, S.P.: Antibiotics and the kidney. *Med. Clin. North Am.* 57:1045–1064, 1973.

55. Kuhar, M.J., Mak, L.L., and Lietman, P.S.: Autoradiographic localization of (^{3}H) gentamicin in the proximal renal tubules of mice. *Antimicrob. Agents Chemother.* 15:131–133, 1979.

56. Kunin, C.M.: Nephrotoxicity of antibiotics. *JAMA* 202:204–208, 1967.

57. Luft, F.C., and Kleit, S.A.: Renal parenchymal accumulation of aminoglycoside antibiotics in rats. *J. Infect. Dis.* 130:656–659, 1974.

58. Luft, F.C., Patel, V., Yum, M.N., and Kleit, S.A.: Nephrotoxicity of cephalosporin-gentamicin combinations in rats. *Antimicrob. Agents Chemother.* 9:831–839, 1976.

59. Luft, F.C., Rankin, L.I., Sloan, S.A., and Yum, M.N.: Recovery from aminoglycoside nephrotoxicity with continued drug administration. *Antimicrob. Agents Chemother.* 14:284–287, 1978.

60. Luft, F.C., Yum, M.N., Walker, P.D., and Kleit, S.A.: Gentamicin gradient patterns and morphological changes in human kidneys. *Nephron* 18:167–174, 1977.

61. Lullmann, H., Lullmann-Rauch, R., and Wasserman, O.: Drug induced phospholipidosis. *CRC Crit. Rev. Toxicol.* 4:185–218, 1975.

62. Lullmann, H., Lullmann-Rauch, R., and Wasserman, O.: Lipidosis induced by amphiphilic cationic drugs. *Biochem. Pharmacol.* 27:1103–1108, 1978.

63. Maunsbach, A.B., and Neustein, H.B.: Auto-radiographic demonstration of hemoglobin in the lysosomes of rabbit proximal tubule cells during experimental hemoglobinemia. *J. Ultrastruct. Res.* 25:183, 1968.

64. Miller, F., and Palade, G.E.: Lytic activities in renal protein absorption droplets. *J. Cell Biol.* 23:519, 1964.

65. Milne, M.D., Scribner, B.H., and Crawford, M.A.: Non-ionic diffusion and the excretion of weak acids and bases. *Am. J. Med.* 24:709–729, 1958.

66. Mondorf, A.W., Hendus, J., Beier, J., Scherberich, J.E., and Scoeppe, W.: Tubular toxicity induced by aminoglycosides in human kidneys. In *Nephrotoxicity: Interaction of Drugs with Membrane Systems Mitochondria-lysosomes,* edited by Fillastre, J.P. New York, Masson, 1978, pp. 167–174.

67. Mondorf, A.W., Zegelman, M., Klose, J., Hendus, J., and Breier, J.: Comparative studies on the action of aminoglycosides and cephalosporins on the proximal tubule of the human kidney. *J. Antimicrob. Chemother.* 4(Suppl.):53–57, 1978.

68. Moore, A.T., Williams, K.E., and Lloyd, J.G.: The effect of chemical treatment of albumin and orosomucoid on rate of clearance from the rat blood stream and rate of pinocytosis capture by rat yolk sac cultured in vitro. *Biochem. J.* 164:607–616, 1977.

69. Morel, F., Chabardes, D., and Imbert, M.: Functional segmentation of the rabbit

distal tubule by microdetermination of hormone dependent adenylate cyclase activity. *Kidney Int.* 9:264–277, 1976.

70. Morin, J.P., Fillastre, J.P., and Vaillant, R.: Prediction of aminoglycoside-cephalosiporin nephrotoxicity. In *Current Chemotherapy. Proceedings of the 10th International Congress of Chemotherapy* (Zurich, September 18–23, 1977), Vol. 2. Washington, D.C., American Society for Microbiology, 1978, pp. 960–962.

71. Morin, J.P., Fresel, J., Fillastre, J.P., and Vaillant, R.: Aminoglycoside actions on rat kidney lysosomes in vivo and in vitro. In *Nephrotoxicity: Interaction of Drugs with Membrane Systems Mitochondria-lysosomes,* edited by Fillastre, J.P. New York, Masson, 1978, pp. 253–263.

72. Novikoff, A.B.: Lysosomes and related particles. In *The Cell,* Vol. 2, edited by Brachet, J., and Mirsky, A.E. New York, Academic Press, 1961, pp. 432–488.

73. Novikoff, A.B.: Lysosomes in the physiology and pathology of cells. Contribution of staining methods. Ciba Foundation Symposium Lysosomes, 1963, pp. 36–77.

74. O'Brien, J.S.: Tay Sachs' disease and juvenile GM_2-gangliosidosis. In *Lysosomes and Storage Diseases,* edited by Hers, H.G., and Van Hoof, F. New York, Academic Press, 1973, pp. 323–345.

75. Okuma, S., and Poole, B.: Fluorescence probe measurement of the intralysosomal pH in living cells and the perturbation of pH by various agents. *Proc. Natl. Acad. Sci. USA* 75:3327–3331, 1978.

76. Patel, V., Luft, F.C., Yum, M.N., Patel, B., Zeman, W., and Kleit, S.A.: Enzymuria in gentamicin induced kidney damage. *Antimicrob. Agents Chemother.* 7:364–369, 1975.

77. Pfeiffer, U.: Inhibition by insulin of the formation of autophagic vacuoles in rat liver or morphometric approach to the kinetics of intracellular degradation by autophagy. *J. Cell Biol.* 78:152–167, 1978.

78. Reijngoud, D.J., Oud, P.S., Kas, J., and Tager, J.M.: Relationship between medium pH and that of the lysosomal matrix as studied by two independent methods. *Biochim. Biophys. Acta* 448:290–302, 1976.

79. Reijngoud, D.J., Oud, P.S., and Tager, J.M.: Effects of ionophores on intralysosomal pH. *Biochim. Biophys. Acta* 448:303–313, 1976.

80. Roberts, A.V.S., Williams, K.E., and Lloyd, J.B.: The pinocytosis of ^{125}I labelled polyvinyl pyrrolidone, ^{14}C sucrose and colloidal, [198]Au gold by rat yolk sac cultured in vitro. *Biochem. J.* 168:239–244, 1977.

81. Schultze, H.E., and Heremans, J.F.: *Molecular Biology of Human Proteins.* Vol. 1. Amsterdam, Elsevier, 1966.

82. Silverblatt, F.J.: Antibiotic nephrotoxicity. A review of pathogenesis and prevention. *Urol. Clin. North Am.* 2:557–567, 1975.

83. Silverstein, S.C., Steinman, R.M., and Cohn, Z.A.: Endocytosis. *Annu. Rev. Biochem.* 46:669–722, 1977.

84. Staubli, W., Scheizer, W., and Suter, J.: Some properties of myeloid bodies induced in rat liver by antidepressant drug maprodilene. *Exp. Mol. Pathol.* 28:177–195, 1978.

85. Steinmann, R.M., Silver, J.M., and Cohn, Z.A.: Pinocytosis in fibroblasts. Quantitative studies in vitro. *J. Cell Biol.* 63:949–969, 1974.

86. Straus, W., and Oliver, J.: Cellular mechanisms of protein metabolism in the nephron: VI. The immunological demonstration of egg white in droplets and other cellular fractions of the rat kidney after intraperitoneal injections. *J. Exp. Med.* 102:1, 1955.

87. Strober, W., and Waldmann, T.A.: The role of the kidney in the metabolism of plasma proteins. *Nephron* 13:35–66, 1974.

88. Tager, J.M., and Reijngoud, D.J.: Mechanism of accumulation of basic compounds in lysosomes. In *Nephrotoxicity: Interaction of Drugs with Membrane Systems Mitochondria-lysosomes,* edited by Fillastre, J.P. New York, Masson, 1978, pp. 265–280.

89. Tulkens, P., Aubert-Tulkens, G., Van Hoof, F., and Trouet, A.: The lysosomal toxicity of aminoglycosides. In *Nephrotoxicity: Interaction of Drugs with Membrane Systems Mitochondria-lysosomes,* edited by Fillastre, J.P. New York, Masson, 1978, pp. 231–251.

90. Tulkens, P., and Trouet, A.: Uptake and intracellular localization of streptomycin in the lysosomes of cultured fibroblasts. *Arch. Int. Physiol. Biochim.* 80:623–624, 1972.

91. Tulkens, P., and Trouet, A.: Uptake and intracellular localization of kanamycin and gentamicin in the lysosomes of cultured fibroblasts. *Arch. Int. Physiol. Biochim.* 82:1018–1019, 1974.

92. Ullberg, S.: Studies on the distribution and fate of ^{35}S labelled benzylpenicillin in the body. *Acta Radiol.,* Suppl. 118, 1–110, 1954.

93. Vaes, G.: Digestive capacities of lysosomes. In *Lysosomes and Storage Diseases,* edited by Hers, H.G., and Van Hoof, F. New York, Academic Press, 1973, pp. 43–78.

94. Vera-Roman, J., Krishnakantha, T.P., and Cuppage, F.E.: Gentamicin nephrotoxicity in rats: I. Acute biochemical and ultrastructural effects. *Lab. Invest.* 33: 412–417, 1975.

95. Waddell, W.J., and Bates, R.G.: Intracellular pH. *Physiol. Rev.* 49:285–329, 1969.

95.a Watanabe, H.: Drug-induced lysosomal changes and nephrotoxicity in rats. *Acta Pathol. Japan* 28(6):867–889, 1978.

96. Wellwoud, J.M., Lovell, D., Thompson, A.E., and Tighe, J.R.: Renal damage caused by gentamicin: a study of the effects on renal morphology and urinary enzyme secretion. *J. Pathol.* 118:171–182, 1976.

97. Whelton, A.: Intrarenal antimicrobial distribution modulating factors, therapeutic and toxicologic implications. In *Nephrotoxicity: Interaction of Drugs with Membrane Systems Mitochondria-lysosomes,* edited by Fillastre, J.P. New York, Masson, 1978, pp. 95–126.

98. Whelton, A., Carter, G.G., Craig, T.J., Bryant, H.H., Herbst, D.V., and Walker, W.G.: Comparison of the intrarenal disposition of tobramycin and gentamicin: therapeutic and toxicological answers. *J. Antimicrob. Chemother.* 4:13–22, 1978.

99. Whelton, A., and Walker, W.G.: Intrarenal antibiotic distribution in health and disease. *Kidney Int.* 6:131–137, 1974.

100. Wibo, M., and Poole, B.: Protein degradation in cultured cells: II. Uptake of chloroquine by rat fibroblasts and the inhibition of cellular protein degradation and cathepsin B. *J. Cell Biol.* 63:430–440, 1974.

101. Williams, K.E., Kidston, E.M., Beck, F., and Lloyd, J.B.: Quantitative studies of pinocytosis: I. Kinetics of uptake of ^{125}I polyvinylpyrrolidone by rat yolk sak cultured in vitro. *J. Cell Biol.* 64:113–122, 1975.

102. Morin, J. P., Viotte, G., Vandwalle, A., Van Hoof, F., Tulkens, P., and Fillastre, J. P.: Gentamicin induced nephrotoxicity: A cell biology approach. *Kid. Int.* 18:583–590, 1980.

103. Vandewalle, A., Farman, N., Morin, J. P., Fillastre, J. P., Hatt, P. Y., and Bonvalet, J. P.: Gentamicin incorporation along the nephron: Autoradiographic study on isolated tubules. *Kid. Int.,* May/June, 1981.

14
MITOCHONDRIA MODIFICATIONS WITH THE AMINOGLYCOSIDES

J. P. BENDIRDJIAN, J. P. FILLASTRE, and B. FOUCHER
Université de Rouen, Hôpital de Boisguillaume,
Bois Guillaume, France

The nephrotoxicity of aminoglycosides has been well established. Its degree has been assessed as between 4 and 10%, according to various authors. The mechanism of renal involvement is not yet clear. The concentration of antibiotics in the kidney has been demonstrated, and it has been seen to be greater in the cortex than in the medulla. In the animal, renal concentration attains levels tenfold that of the serum concentration peak after a single intramuscular or intraperitoneal injection. Comparative serum half-life and renal lymph half-life studies, autoradiograms on thin slices or on isolated nephrons have eliminated the possibility of aminoside accumulation in the renal interstitium. Thus, renal sequestration of aminosides is intracellular and is greater in the proximal than in the distal part of the tubule [17,18,26,46,53,54].

Ultrastructural renal alterations occur early, within 48 hr following even a low dose of aminosides [27]. Lysosomes are the first affected, transformed into cytosegresomes having high acid phosphatase activity and containing myeloid bodies. Then the Golgi apparatus becomes hypertrophied and the number of ribosomes increases. Later, the mitochondria are affected [27,49]. These data suggested the interest of studying in vitro lysosome and mitochondria activity in the presence of increasing concentrations of aminoglycosides.

The energy freed by metabolite oxidation in the Krebs cycle serves to synthesize adenosine triphosphate (ATP) from its precursor, adenosine diphosphate (ADP), through coupled reactions. This process, termed oxidative phosphorylation, occurs through the action of the mitochondrial respiratory chain in which electrons are transported from nicotinamide-adenine dinucleotide hydrogenase (NADH) to oxygen.

Chance and Williams measured the redox states of all the carriers within the respiratory chain under various experimental conditions [8,9]. Five metabolic states of the mitochondria were described. They are shown on Table 1. The sequence of respiratory chain carriers was established, as well as the sites of oxidative phosphorylation. Oxidative phosphorylation can easily be demonstrated by comparing the rates of mitochondrial oxygen consumption between states 3 and 4. Figure 1 shows that adding ADP to a mitochondrial suspension in physiological medium corresponding to Chance and Williams's state 4 greatly increases oxygen consumption, which is accompanied by ADP phosphorylation. When ADP has been completely transformed into adenosine triphospate (ATP), the rate of oxygen consumption returns to its initial level. It is now known that certain substances can inhibit this system. Antimycin and mucidine block electron transport between cytochrome b and cytochrome c, inhibiting respiration [43–45]. Oligomycin inhibits the increase in oxygen consumption usually provoked by ADP and ADP phosphorylation [43,44]. Other examples could also be given for uncoupling agents [23, 29,32,43] or ionophores [34–36]. We studied the in vitro activity of aminosides on mitochondrial respiration of rat livers and kidneys.

Table 1   Various Metabolic States of Mitochondria Isolated in Physiological Medium

State	Concentration			Respiration	Limiting factor	Oxidation step of the respiratory chain
	Oxygen	ADP	Substrate			
1	>0	Low	Low	Slow	ADP	Increases from NADH ($<10\%$) to cytochrome a (100%)
2	>0	High	Low	Slow	Substrate	Whole chain in oxidized
3	>0	High	High	Fast	Respiratory chain	Increases from NADH (50%) to cytochrome a (96–100%)
4	>0	Low	High	Slow	ADP	Increases from NADH (0%) to cytochrome a (100%)
5	0	High	High	Nil	Oxygen	Whole chain is reduced

Source: Data derived from [g].

I. Material and Methods

A. Mitochondria Isolation

Male and female Sherman rats weighing 200–400 g were sacrificed by decapitation. The kidneys and liver were quickly harvested and plunged into an isolating solution maintained at 0–4°C. The medium was composed of 0.33 M sucrose, 1 mM ethylenediaminetetraacetic acid (EDTA)-Na_2, and 5 mM TRIS-HCL buffer pH 7.4 [33]. After rough teasing, the tissues were ground in a Potter homogenizer in the presence of the isolating medium. The ratio of tissue weight to medium volume was approximately 1:10. The mitochondria were isolated by differential centrifugation at 2°C. The first centrifugation at 900 g for 10 min eliminates intact cells, large cellular debris, and nuclei. The supernatant is then centrifuged at 9000 g for 10 min. The resulting pellet of mitochondria then undergoes two washings at 9000 g for 10 min in each medium. The final pellet is suspended in a small quantity of isolating medium to attain 60 mg protein per milliliter. Protein concentration is determined by a modified biuret method in the presence of sodium chelate. Bovine serum albumin is used as reference.

B. Measurement of Oxygen Consumption

Oxygen consumption by mitochondria is determined by oxypolarography in an incubation medium maintained at 25°C (composition: 80 mM KCl, 10 mM K_2HPO_4, 20 mM TRIS-HCl pH 7.2 [25]. The medium is air-saturated by magnetic agitation for approximately 1 hr at room temperature. The resulting oxygen concentration is 240 mM. A quantity of mitochondria corresponding to 2.5 mg/ml protein is added to 1.5 ml of incubation medium. Variations in oxygen concentration are automatically recorded.

The effects of various aminoside antibiotics tested on mitochondria in this study are always compared to those of gentamicin C (mixture of forms C_1, C_{1a}, and C_2), which is used as a reference antibiotic.

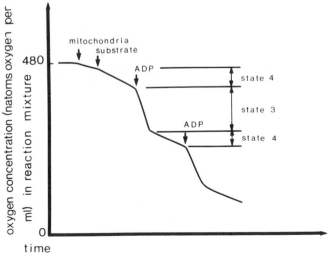

Figure 1   Example of polarographic recording of substrate oxidation by rat liver mitochondria, showing states 3 and 4 described by Chance and Williams.

C. NADH Redox State Variations in the Respiratory Chain

NADH becomes fluorescent after irradiation at 366 nm. Variations in the level of oxido-reduction in the respiratory chain can be qualitatively determined by this fluorescence. The determinations are made at 400–3000 nm in a fluorescence cuvette on a Eppendorf 1101 M spectrophotometer. The incubation medium is identical to that used to measure oxygen consumption. Mitochondrial protein concentration is 2.5 mg/ml. Results are expressed in arbitrary units.

D. Variations in Magnesium and Calcium Content of the
Mitochondrial Membrane

Variations in the magnesium and the calcium content of the mitochondrial membrane are qualitatively evaluated using chlortetracycline (or aureomycin). This compound, when in the presence of calcium or magnesium, emits fluorescence when submitted to irradiation at 366 nm. The maximum intensity of this fluorescence is 530 nm. It particularly appears when divalent cations are in an apolar medium. Thus, chlortetracycline can serve as a probe within the mitochondrial membrane [7]. When it is added (in an alcohol solution at a concentration of 5 μM) to a mitochondrial suspension submitted to irradiation at 366 nM, a fluorescence induced by the presence of calcium and magnesium immediately appears. This fluorescence stabilizes in 1–2 min. It is measured the whole length of a spectrum, from 530 to 3000 nm, in a fluorescence cuvette on an Eppendorf 1101 M spectrophotometer. The incubation medium contains 80 mM KCl and 20 mM TRIS-HCl pH 7.4. Mitochondrial protein concentration is 2.5 mg/ml. Results are expressed in arbitrary units. It was shown that chlortetracycline, at a concentration of 5 μM, does not modify mitochondrial respiratory activity.

E. Measurement of Proton Ejection and ATPase Activity
by pH Assessment

Substrate oxydation by the mitochondria is accompanied by proton ejection. Various compounds can modify this phenomenon to greater or lesser degrees. Ejection can be assessed easily by measuring the pH in the incubation medium with a pH meter attached to a recording apparatus. ATPase activity can be similarly measured, since ATP hydrolysis liberates protons which would thus lower the exterior medium pH. This measurement is made in the presence of antimycin in order to eliminate interference of any protons coming from oxidation of the endogenous substrate.

F. Study of Mitochondrial Membrane Permeability

Variations in the permeability of the mitochondrial membrane can be studied by an indirect method consisting of measuring the absorbance of a mitochondrial suspension. In certain conditions, mitochondria can accumulate various substances or ions. A simultaneous increase in water occurs in order to maintain osmotic balance. The resulting increase in organelle volume is reflected in a decreased absorbance of the suspension. Variations in absorbance are measured at 546 nm.

II. Results

A. Effect of Aminoglycosides on the Respiratory Activity of Rat Liver and Kidney Mitochondria

Liver Mitochondria

The following aminoside antibiotics were studied: neomycin B, gentamicin, tobramycin, sisomicin, dibekacin, netilmicin, kanamycin A, amikacin, and sorbistin A_1. These antibiotics share similar activity (i.e., activation of state 4 and inhibition of state 3). The degree of these two effects varies greatly among them; however, the intensity of the activating effect of state 4 is linked to the intensity of the inhibitory effect of state 3. An aminoside that induces strong activation in state 4 will also induce strong inhibition in state 3. At a concentration of 50 μg/ml (at which the differences in activity among antibiotics appear most clearly), we found it possible to classify aminosides according to the intensity of respiratory modifications which they induced. In decreasing activity, we found neomycin B, sisomicin, tobramycin, dibekacin, gentamicin C, netilmicin, kanamycin A, amikacin, and sorbistin A_1. Our study included aminoside concentrations from 10 to 1000 μg/ml. We have attempted to establish an equation of antibiotic activity based on the percentage of inhibition in state 3 according to the aminoside concentration. We did, in fact, note that the effect of aminosides on liver mitochondrial respiration varies linearly with the inverse of concentrations from 0 to 500 μg/ml (Figure 2). A simple equation is obtained: $Y = (a/c) + b$, where Y is the percent of mitochondrial respiration inhibition in state 3, c is the aminoside concentration in μg/ml, and a and b represent characteristics of the aminoside. For each antibiotic, these two terms are defined by performing a linear regression from experimental points on curves representing inhibition of respiratory activity in rat liver mitochondria in state 3. Correlation coefficient values (r) range from -0.94 to -0.99 and are therefore highly significant. The values of a and of b, of correlation coefficient r, as well as of the degree of significance of the coefficient r,

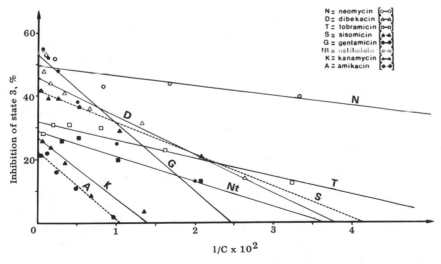

Figure 2 Inhibition curves of succinate oxidation by rat liver mitochondria in state 3 as an inverse function of the concentration of various aminoglycosides.

Table 2   Values of Coefficients a and b in the Relationship y = a/c + b, Illustrating the Effect of a Number of Aminosides on Rat Liver Mitochondria in State 3

	a	b	r	Signification limit of r for P = 0.95[a]
Neomycin B	- 295	49	-0.94	-0.75 (7)
Gentamicin C	-2170	53	-0.97	-0.75 (7)
Tobramycin	- 574	32	-0.96	-0.81 (6)
Sisomicin	-1011	42	-0.99	-0.81 (6)
Dibekacin	-1230	46	-0.99	-0.81 (6)
Netilmicin	- 798	29	-0.98	-0.95 (4)
Kanamycin A	-1966	27	-0.94	-0.88 (5)
Amikacin	-2248	23	-0.99	-0.88 (5)

[a]Numbers in parentheses indicate number of experimental points.

are given in Table 2 for eight of the aminoglycosides studied. The constant b corresponds to the theoretical maximum percent of inhibition of state 3 that an aminoside can induce. It follows that the higher the value of b, the greater the degree of intracellular damage. Gentamicin C, neomycin B, and dibekacin have very similar high potential toxicities. Netilmicin, kanamycin A, and amikacin, on the other hand, have low potential activities. Tobramycin and sisomicin occupy an intermediate position. The coefficient a indicates the large degree to which the mitochondrial effect of aminosides can vary. The coefficient of neomycin B is clearly lower than that of the other aminosides, indicating that the toxic potential of neomycin B remains high even at low concentrations. In contrast, for amikacin, gentamicin, and kanamycin A, the coefficient is high, reflecting their weak activity at low concentrations, which increases rapidly with the concentration (Figure 2).

Kidney Mitochondria

A study was also made using rat kidney mitochondria. As seen in Figure 3, similar results were obtained with both kidney and liver mitochondria. Only the effects of dibekacin and netilmicin appear weaker on kidney mitochondria than on liver mitochondria.

Discussion

Our initial goal was to attempt to assess the possible harmful effects of aminoside antibiotics. To date, this toxicity has been well established for renal tubule cells. The results of this study were compared to known data concerning the nephrotoxicity of these antibiotics. It was seen that our results parallel experimental data demonstrating their nephrotoxicity. Neomycin B, which appears to be the most active on mitochondrial respiration, is without a doubt the most nephrotoxic of these aminosides [1,2,16,28]. Sisomicin is slightly more toxic than tobramicin [6], which itself has been reported to be of closely similar [19,51,52] or of lesser toxicity than gentamicin [3,40,48]. Netilmicin and amikacin are less toxic than gentamicin [3,12,13,24,30,31]. In Table 3, these aminosides are classified according to the degree of their effect on mitochondrial respiration and according to data in the literature with regard to their nephrotoxicity. A good correlation is seen between nephrotoxicity and the degree to which the intensity of mitochondrial

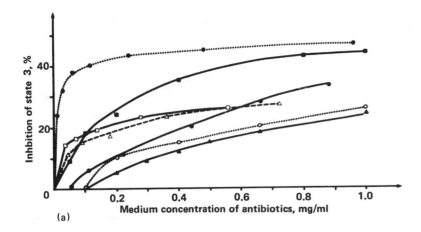

(a)

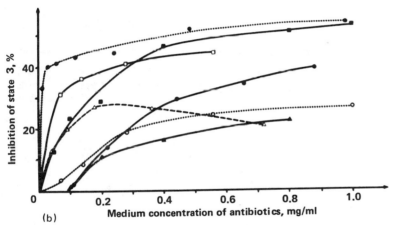

(b)

Figure 3   In vitro effects of some aminoglycoside antibiotics on state 3 respiration of mitochondria from rat kidney (a) and rat liver (b). (---●---) neomycin B; (—■—) gentamicin C; (—□—) dibekacin; (—△—) netilmicin; (—●—) ribostamycin; (---○---) kanamycin A; (—▲—) amikacin. Medium: KCl, 80 mM; K_2HPO_4, 10 mM; TRIS-HCl, pH 7.2, 20 mM; rotenone, 1.4 µg/ml; succinate, 10 mM; ADP, 0.33 mM. The values are means of 12 experiments for gentamicin C and 6 experiments for other antibiotics. In both cases no effect is seen with sorbistin A_1.

respiration is altered. It would appear that the effect of aminoside antibiotics on mitochondrial respiratory metabolism is an interesting and practical model for assessing the possible nephrotoxicity of new antibiotics in this group. Their influence on mitochondrial respiration appears to be well correlated with the degree of clinical manifestations of renal toxicity after a long course of these antibiotics. Mitochondria are easily obtainable and only 20 mg of antibiotics suffices for testing. This in vitro study of the action of aminosides on rat mitochondria is proposed as a model that may provide an accurate estimation of the toxicity of antibiotics in this group. Based on these results, we do not

Table 3   Classification of Aminoglycosides According to Their Nephrotoxicity and to Their Effect on Mitochondrial Respiration

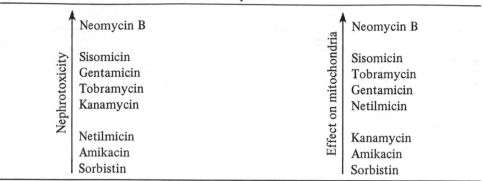

Nephrotoxicity	Effect on mitochondria
Neomycin B	Neomycin B
Sisomicin	Sisomicin
Gentamicin	Tobramycin
Tobramycin	Gentamicin
Kanamycin	Netilmicin
Netilmicin	Kanamycin
Amikacin	Amikacin
Sorbistin	Sorbistin

intend to state that aminosides actually have an in vivo effect on mitochondrial respiratory metabolism.

B. Mechanism of Action of Gentamicin on Rat Liver Mitochondria

In the preceding section we showed that aminosides activate state 4 and inhibit state 3 of rat liver and kidney mitochondria when succinate is the respiratory substrate. Identical results were found using malate-glutamate as substrate (Figure 4). We attempted to further

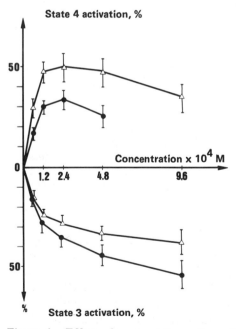

Figure 4   Effect of gentamicin on the oxidation of malate-glutamate ($\triangle$——$\triangle$) and of succinate + rotenone ($\bullet$——$\bullet$) by rat liver mitochondria. Proteins, 2.5 mg/ml; malate-glutamate, 5 + 5 mM; succinate, 10 mM; rotenone, 2 μg; ADP, 0.33 mM.

elucidate the mechanism of gentamicin activity on mitochondria by studying the effect of adding magnesium, EDTA, or potassium, to the incubation medium.

Modification of the Effect of Gentamicin on Mitochondrial Respiration
Incidental to Increasing Concentrations of Magnesium

A progressive inhibition of the effect of gentamicin on mitochondrial respiratory activity in state 3 or 4 is noted when increasing concentrations (from 0.030 to 0.50 mM) of Mg^{2+} are added to the incubation medium (Figure 5). This observation could suggest competition between magnesium and aminosides. Many interactions between magnesium and membrane amino compounds have been observed. Membrane polyamines freely exchange with magnesium in solution [38,47], and Silver et al. showed that magnesium protects *Escherichia coli* against the effects of steroidamine [42]. A number of authors have observed that bacteria become less sensitive to the effect of aminoglycosides in the presence of magnesium or various other cations [14,15,21,41,50]. Zimelis and Jackson demonstrated that this effect resulted from increased resistance of the bacterial membrane to the activity of the aminoside rather than from an inactivation of the antibiotic [55]. It is likely that the phenomenon of interaction that has been observed stems from the fact that the base organic characteristics of aminoglycosides [39] in certain conditions induce competition with magnesium.

Based on our findings, the effect of gentamicin on mitochondria in the presence of magnesium can be interpreted in either of two ways: (1) keeping in mind the important role played by magnesium in maintaining the integrity of the mitochondrial membrane [4,5], it could result from competition between the aminoside and the cation at the mitochondrial membrane; or (2) it could result from the inactivation of gentamicin by magnesium through chelate formation [11,12]. Neither are these two hypotheses mutually exclusive, and the two types of activity might coexist. EDTA can chelate a part of membrane magnesium [4,33], freeing sites that could be occupied by the cationic form of gentamicin. In an attempt to more precisely define the mechanism by which gentamicin

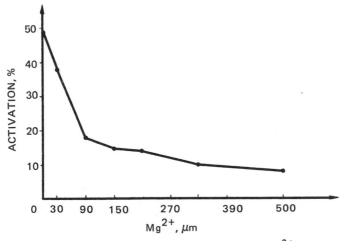

Figure 5   Effect of increasing concentration of Mg^{2+} on sensitivity of rat liver mitochondria in state 4 to gentamicin. Proteins, 2.5 mg/ml; succinate, 10 mM; rotenone, 2 μg; gentamicin, 2.4×10^{-4} M.

affects the mitochondrial membrane, we studied its activity on divalent membrane cations by using a fluorescent probe: chlortetracyclin.

Variations in Mitochondrial Membrane Content of Divalent Cations in the Hydrophobic Phase and in the Presence of Gentamicin

This study was performed on mitochondria isolated in the presence and in the absence of EDTA.

In the Presence of EDTA. When added to a suspension of mitochondria isolated in the presence of EDTA, chlortetracycline exhibits a fluorescence. The subsequent addition of gentamicin induces a decrease in fluorescence. This decrease correlates with the amount of aminosides that is added. The addition of 0.25 mM of Mg^{2+} 2 min after gentamicin induces an increase in fluorescence (Figure 6A). The addition of Mg^{2+} to a mitochondrial

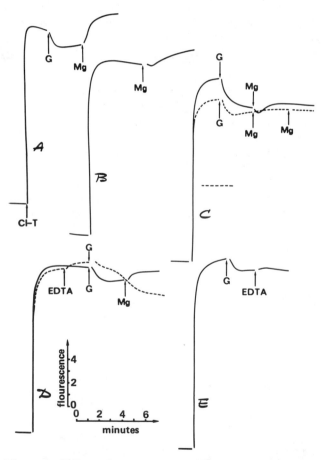

Figure 6 Effect of gentamicin, magnesium, and EDTA on chlortetracyclin fluorescence in the presence of rat liver mitochondria isolated in the presence of EDTA. Proteins, 2.5 mg/ml; chlortetracycline (Cl-T), 5 μM. Gentamicin (G): A, 6 $\times$ 10^{-5} M; D, E, 2.4 x 10^{-4} M; C, 9.6 x 10^{-4} M. Mg^{2+}, 0.25 mM; EDTA-Na_2, 0.5 mM. C: dotted line, medium containing 0.25 mM Mg^{2+}; solid line, control (medium without Mg^{2+}). D: dotted line, effect of gentamicin added after EDTA; solid line, control (without addition of EDTA).

suspension without gentamicin also induces a slight increase in fluorescence (Figure 6B). When magnesium is previously added to the incubation medium, the later addition of gentamicin still reduces fluorescence, but much less (two times less) that in the absence of magnesium. Adding the cation again after the aminoside results in no further modification of fluorescence (Figure 6C). The addition of 0.5 mM of EDTA-Na$_2$ to the mitochondrial suspension in the presence of chlortetracyclin induces almost no change in fluorescence. Subsequent addition of gentamicin results in a much greater drop in fluorescence when gentamicin is added alone (Figure 6D). The degree of fluorescence is not altered when EDTA is added after gentamicin; it remains at the level reached after the addition of the aminoside (Figure 6E).

In the Absence of EDTA. Isolated, in the absence of EDTA, mitochondria are affected much less by gentamicin. The addition of gentamicin to a suspension of mitochondria in the presence of chlortetracyclin induces a decrease in fluorescence. This decrease is less than that induced in mitochondria isolated in the presence of EDTA. The subsequent addition of magnesium has no affect. When EDTA is added to the mitochondrial suspension in the presence of chlortetracyclin, no effect is observed and the addition of gentamicin has virtually no effect (Figure 7).

In conclusion, it may be stated that the presence of EDTA in the isolating medium frees certain sites initially occupied by magnesium because the addition of the cation to these mitochondria induces an increase in fluorescence, whereas this result is not observed in mitochondria isolated in the absence of EDTA. In addition, gentamicin seems to act by chelating a part of the magnesium or of the membrane calcium, or by replacing these cations. However, it is surprising to find that this effect is greater in mitochondria where cation content was previously diminished by EDTA than in those having retained all their membrane cations. It would thus seem that EDTA sensitizes the membrane to gentamicin by chelating a part of the membrane cations, with gentamicin therefore affecting only certain divalent cations of the membrane. These cations would be unaffected by EDTA activity. We have shown that gentamicin added to mitochondria in a medium containing

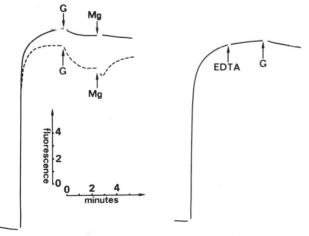

Figure 7 Effect of gentamicin, magnesium, and EDTA on chlortetracyclin fluorescence in the presence of rat liver mitochondria isolated in the absence of EDTA. Proteins, 2.5 mg/ml; gentamicin, 2.4 × 10^{-4} M; Mg^{2+}, 0.25 mM; EDTA-Na$_2$, 0.5 mM. Dotted line: control (mitochondria isolated in the presence of EDTA).

magnesium diminishes the fluorescence of chlortetracyclin and that further addition of Mg^{2+} after the antibiotic has no effect. Then magnesium added to the suspension medium does not appear to be able to inactivate gentamicin (at least not totally), because in this case gentamicin can still decrease the fluorescence of chlortetracyclin, although to a lesser degree. Thus, a part of the mitochondrial membrane cations would be affected by gentamicin without magnesium being able to inhibit this action. In sum, the presence of EDTA in the isolating medium of the mitochondria would sensitive the mitochondrial membrane to gentamicin by chelating part of the membrane cations. The antibiotic would then be able to chelate or to replace a part of the membrane cations that are not affected by EDTA through two different mechanisms, one of competition with magnesium, and the other noncompetitive. These two types of phenomena could correspond to the replacement and to the chelation of membrane cations, respectively. Suarez and Kurtz, using the aminoside neomycin B, showed that this antibiotic does not alter the Ca^{2+} ionic concentration of a solution and that the solution did not contain antibiotic-Ca^{2+} complex. Extrapolating to gentamicin, it could be suggested that extramitochondrial magnesium does not act by inactivating gentamicin but rather by modifying the membrane. The chelating capacity of gentamicin would appear only in conditions involving a role of the medium polarity and, probably, its ionic strength. The latter appears to alter the behavior of EDTA with regard to mitochondria. The mitochondria are affected by gentamicin only when EDTA is added to the organelle isolating medium (medium of low ionic strength), whereas its addition only to the incubation medium (medium of high ionic strength) no longer sensitizes the mitochondrial membrane to the antibiotic. The fact that different conditions are required to induce EDTA and gentamicin chelating capacity could result from a difference in hydrophilic characteristics between these two compounds, resulting in different sites of activity in the mitochondrial membrane.

This interaction between gentamicin and Mg^{2+} led us to search for possible effects of gentamicin on the permeability of the mitochondrial membrane.

Effect of Gentamicin on the Permeability of the Mitochondrial Membrane

The addition of gentamicin to a mitochondrial suspension in a KCl 80 mM medium induces only a very slight swelling of the mitochondria. In contrast, a clear swelling appears when a substrate such as β-hydroxybutyrate or malate-glutamate is added. The swelling is prevented by addition of an uncoupler such as carbonyl cyanide m-chlorophenylhydrazone (ClCCP). When antimycin (0.4 μg/mg protein) is previously added to the incubation medium, gentamicin induces no swelling of the mitochondria in the presence of the respiratory substrate (Figure 8). Thus, the mitochondrial swelling induced by gentamicin requires a source of energy. This energy is provided by respiratory substrate oxidation, since antimycin, which inhibits electron transport, prevents the occurrence of this swelling. By lowering the amount of energy available in the mitochondrial membrane, the uncoupler removes the energy required for swelling. The increase in mitochondrial volume implies an active penetration of a substance into the mitochondria. This substance may be either gentamicin or potassium. Evaluations made in medium without potassium showed that gentamicin did not induce swelling in these conditions. Swelling appeared only when potassium was added to the medium (Figure 9). The phenomenon thus appears due to active K^+ penetration into the mitochondria induced by gentamicin. Swelling does not appear in the presence of Mg^{2+}, similar to observations above, because

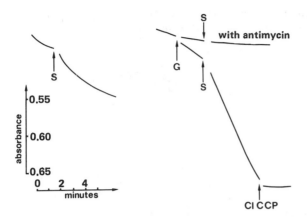

Figure 8 Effect of gentamicin on the absorbance of a mitochondrial suspension of rat liver mitochondria in a KCl medium. Proteins, 1 mg/ml; substrate (S), malate-glutamate, 5 + 5 mM; gentamicin (G), 2.4 x 10^{-4} M; ClCCP, 0.2 μM; antimycin, 0.4 μg/mg of proteins.

the cation appears to act as an antagonist of gentamicin. It is well known that certain compounds, such as H_2O, CH_3COOH, and NH_3, can passively penetrate the internal mitochondrial membrane. For a salt to penetrate the mitochondria passively, the anion and the cation must be able to penetrate simultaneously. Mitochondria in an isotonic medium of potassium acetate do not swell because K^+ cannot enter. In contrast, swelling appears when valinomycin, an ionophorus antibiotic, and an uncoupler are added; valinomycin renders the membrane permeable to K^+, while the uncoupler induces proton ejection. The following cycle is established: uptake of K^+, ejection of H^+, and equilibration

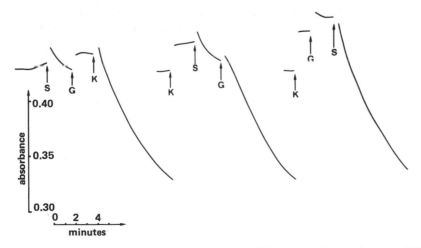

Figure 9 Effect of gentamicin on rat liver swelling in a medium of sucrose 0.20 M + 10 mM TRIS-HCl pH 7.2. Proteins, 2.5 mg/ml; substrate (S), TRIS-succinate, 5 mM; gentamicin (G), 2.4 x 10^{-4} M; KCl (K), 15 mM; Mg^{2+}, 0.5 mM. Evidence that the presence of both K^+ and of a substrate is necessary for gentamicin to induce mitochondrial swelling.

of the electrical charge of the mitochondria, permitting the acetate ion in the form of CH_3COOH to penetrate. The following diagram summarizes this hypothesis.

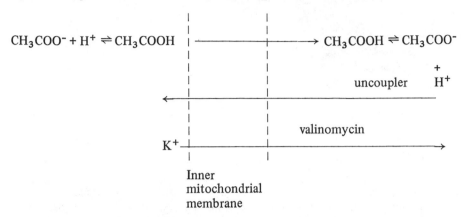

After creating identical conditions, but using gentamicin in the place of valinomycin, we observed no swelling of the mitochondria (Figure 10).

Genatmicin thus appears to have no effect on passive transport, and its activity appears linked only to energy. As shown, this energy is provided by substrate oxidation. It can also originate, although to a lesser degree, in ATP hydrolysis. We observed that in the presence of antimycin and rotenone, which inhibits the energy from substrate oxidation, a weak but clear mitochondrial swelling occurred. This would indicate that another source of energy was used. It is most likely that this source is ATP. We verified this hypothesis by measuring ATPase activity induced by gentamicin (Figure 11).

Effect of Gentamicin on the Electron Transport Chain

In earlier studies we showed that gentamicin alters the oxidation rate of malate + glutamate and of succinate + rotenone [20]. The site of action of gentamicin on the electron

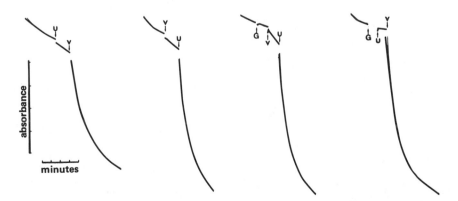

Figure 10   Effect of gentamicin on passive swelling. Medium: potassium acetate, 100 mM; rotenone, 2 μg; gentamicin (G), 2.4×10^{-4} M; uncoupling agent, ClCCP (U), 0.015 μM; valinomycin (V), 0.25 μg/ml.

transport chain might have been located between flavin adenine dinucleotide (FAD) and oxygen. Using ascorbate + N,N,N′,N′-tetramethyl-p-phenylenediamine (TMPD) as substrate, we observed that gentamicin at a concentration of 2.4×10^{-4} M inhibits state 3 by $22.5 \pm 6.3\%$, without a clear effect on state 4. Thus, gentamicin acts on the whole of the electron transport chain. We studied the effect of gentamicin on the oxidation step of NAD in the respiratory chain. The oxidation of NAD is approximately 50% for mitochondria in state 3 and nearly nil in mitochondria in state 4, where NAD is in the NADH form [9]. When a respiratory substrate is added to a mitochondrial suspension, NAD is greatly reduced. This reduction occurs very rapidly with malate-glutamate at a concentration of 5 + 5 mM. It reaches its maximum, then diminishes very slightly with time, and stabilizes after 3 or 4 min (Figure 12A). In the presence of an electron transport inhibitor such as rotenone or antimycin, NAD is totally reduced (Figure 12D). In contrast, in the presence of an uncoupler such as 2,4-dinitrophenol (2,4-DNP), which accelerates the oxidation process, NAD is maximally oxidized (Figure 12E). The addition of gentamicin (2.4×10^{-4} M) after the respiratory substrate induces oxidation back of NADH, attaining a stable maximum within a few minutes (Figure 12C). The addition of ADP (0.4 mM) after gentamicin induces another transient oxidation of NADH; the recorded variation in

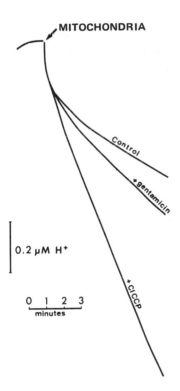

Figure 11 pH measurement of ATPase activity induced by gentamicin and ClCCP in rat liver mitochondria. Proteins, 2.5 mg/ml; antimycin, 0.1 μg/ml of proteins; ATP, 1 mM; gentamicin, 9.6×10^{-4} M; ClCCP, 2 μM. Incubation medium: KCl, 80 mM; TRIS-HCl, pH 7.2, 20 mM. Final volume, 3 ml. Verification was made that according to the concentration of H^+ ions, the incubation medium response is linear throughout the pH range used.

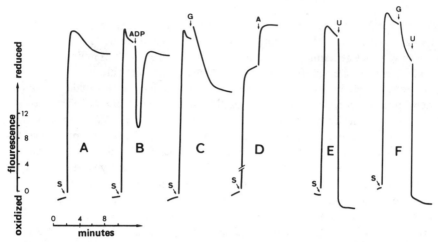

Figure 12 Variation of NAD oxidation step in rat liver mitochondria in the presence of various compounds. Measurements made in arbitrary units of fluorescence. S: substrate (malate-glutamate), 7.5 mM; ADP, 0.20 mM. G: gentamicin, 2.4×10^{-4} M; A, antimycin: 0.1 μg/mg of proteins; U: uncoupler (2,4-DNP), 5×10^{-5} M; proteins, 2.5 mg/ml.

fluorescence is greater as the interval between addition of gentamicin and ADP is shorter (Figure 13). Studying mitochondrial oxygen consumption with regard to the time of gentamicin addition showed that as long as the antibiotic is added to the incubation medium before the substrate, the percentage of inhibition remains the same. In contrast, when gentamicin is added at various times between the substrate and the ADP, the in-

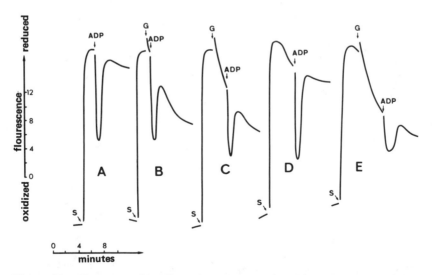

Figure 13 Variation of NAD oxidation step induced by addition of ADP to a rat liver mitochondrial suspension in the presence of gentamicin. ADP is added at various intervals after the addition of antibiotic: B, 30 sec; C, 2 min; E, 4 min. A and D are controls. Measurements made in arbitrary units of fluorescence. S: substrate (malate-glutamate), 7.5 mM; ADP, 0.20 mM; gentamicin (G), 2.4×10^{-4} M.

Table 4 Effect on State 3 at the Time Gentamicin Is Added to Incubation Medium[a]

t = <0	0	45	60	90	120	140	150	State 4	State 3	$\%\triangle_3$
Control	M		S				ADP	36	91	
G	M		S				ADP	52	53	–42
	M	G	S				ADP		55	–40
	M		S	G			ADP		65	–28
	M		S		G		ADP		73.5	–19
	M		S			G	ADP		91	0

[a]t, time, in seconds, as of addition of mitochondria; M, mitochondria: 2.5 mg/ml protein; S, succinate: 10 mM; G, gentamicin: $2-4 \times 10^{-4}$ M; oxygen consumption in states 3 and 4 is expressed in n atoms oxygen: mn/mg proteins; $\%\triangle_3$, percent of state 3 inhibition.

hibitory effect of the aminoside on state 3 is weaker the closer its time of addition is to that of ADP (Table 4).

Thus, maximum inhibition of state 3 is obtained when the addition of ADP to the incubation medium follows that of gentamicin with a sufficiently long interval so that the alteration in the NAD oxidation step induced by the antibiotic has stabilized. This indicates that the oxidation rate of substrate by mitochondria in state 3 is all the greater as the initial NAD oxidation step is weaker, or that this oxidation rate is in proportion to the difference between the initial and the final states of NAD oxidation-reduction, $\triangle E$. If E_i represents initial oxidation-reduction potential and E_f final oxidation-reduction potential of the reaction

$$NADH + H^+ \rightleftharpoons NAD^+ + 2H$$

the free enthalpy change $\triangle G = -nF \triangle E$, proportional to $\triangle E = E_f - E_i$, has a higher absolute value as E_i is weaker. This is true because E_f, which corresponds to the oxidation step of NAD in the mitochondria in state 3, is largely constant. By raising E_i, gentamicin reduces the absolute value of $\triangle G$. However, we observed that the P/O ratio was only very slightly altered by the presence of gentamicin (in relation to the ATPase activity weakly induced by the antibiotic) (Table 5), indicating virtually total phosphorylation of ADP. The energy necessary for ATP synthesis from ADP and inorganic phosphate (P_i) is $\triangle G_0$ = 7.3 kcal (corresponding to $\triangle E = 0.16$ V). It is therefore evident that despite the increase in the NAD oxidation step by gentamicin, $\triangle E = E_f - E_i$ remains above or equal to 0.16 V because phosphorylation remains possible in the presence of gentamicin. Only the rate of the reaction is diminished. Gentamicin thus does not directly affect the

Table 5 Effect of Gentamicin on the P/O Ratio of Liver Mitochondria[a]

Gentamicin in 10^{-4} M	0	0.6	1.2	2.4	4.8
P/O	2.8 ± 0.06	2.78 ± 0.11	2.66 ± 0.07	2.64 ± 0.10	2.58 ± 0.09

[a]Proteins, 2.5 mg/ml; substrate, malate-glutamate 5 + 5 mM, ADP 0.330 mM.

coupling process of oxidative phosphorylation as would an actual uncoupler such as 2,4-DNP. Moreover, gentamicin does not alter the final degree of NAD oxidoreduction of mitochondria uncoupled by 2,4-DNP (Figure 12F). In this case also the increased initial oxidation step of the nucleotide induced by gentamicin is reflected by a decrease in the oxidation rate of the substrate (Figure 14). It is our opinion that the interaction that we demonstrated between gentamicin and magnesium should be associated with the decreased oxidation rate induced by the antibiotic. When the formula giving the constant of the enzymatic oxidation rate of the substrates is considered:

$$k = \text{constant} \times e - \Delta H_a/RT \cdot e - \Delta S_a/RT$$

where

k = the constant of the enzymatic reaction rate
ΔH_a = activation energy
ΔS_a = activation entropy

it is seen that this rate is lower with higher ΔH_a. The high catalytic effect of magnesium in many enzymatic reactions is well known. It consists of a reduction in ΔH_a, activation energy. In competing with magnesium on certain membrane sites, gentamicin would reduce the catalytic effect of the cation, increasing ΔH_a and thus decreasing the oxidation rate.

In conclusion, the mechanism of the effect of gentamicin can be summarized as follows. Initially, because of the amine basic characteristics of the amino groups, gentamicin would be fixed on the mitochondrial membrane sites freed by the effect of EDTA during the isolation of the mitochondria. This fixation would provide access for the antibiotic to the divalent cations Ca^{2+} or Mg^{2+} situated in a hydrophobic environment and not accessible to EDTA. These cations, without being released from the mitochondrial membrane, would simply be moved over the polar phase (protein phase of the membrane). The properties of the membrane would thus be altered, particularly its energy level diminished, implying oxidation of the transport chain. This freed energy would activate

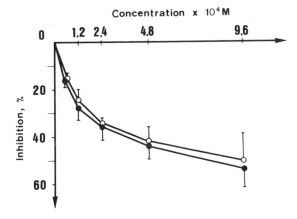

Figure 14   Compared effects of gentamicin on oxidation of succinate by rat liver mitochondria in state 3 (●——●) and after uncoupling by 2,4-DNP (o——o). Proteins, 2.5 mg/ml; succinate, 10 mM; rotenone, 2 μg; ADP, 0.33 mM; 2,4-DNP, 3 × 10^{-2} mM.

the H^+ pump of the mitochondrial membrane [10], which would expulse the protons produced by the substrate oxidation. This expulsion of H^+ (which we demonstrated by pH measurement, see Figure 15), would be compensated by the penetration of K^+. In the presence of Mg^{2+}, the sites that were previously freed by the EDTA would be preferentially occupied by the cation, and gentamicin would no longer be effective. We demonstrated that when EDTA is added to the incubation medium before gentamicin, the effect of the antibiotic is increased, whereas it has no effect when added subsequently. It thus appears that in these experimental conditions (of high ionic strength) EDTA further sensitizes the membrane to the action of gentamicin, for instance by capturing a part of the polar Mg^{2+} of the membrane which was not chelated in the isolating medium (of low ionic strength). In contrast, gentamicin would protect the membrane against the subsequent effect of EDTA. Thus, according to the experimental conditions (ionic strength, presence or absence of Mg^{2+}), EDTA and gentamicin would mutually alter their affinity for membrane divalent cations, which under the influence of external factors might occupy different positions within the membrane. This proposed possible mechanism of the effect of gentamicin is illustrated in Figure 16.

C. Structure–Activity Relationship of Aminoglycosides on the Respiratory Metabolism of Rat Liver Mitochondria

In order to study aminoglycosides having the closest possible structures, we selected derivatives of 4,6-disubstituted 2-deoxystreptamine. This group includes most of the aminoglycosides used in clinical treatment. We compared the effects of 10 antibiotics: gentamicin C_1, A, and B; kanamycin A and B; tobramycin; sisomicin; netilmicin; dibekacin; and amikacin. Two distinct groups of antibiotics clearly emerge from the comparison

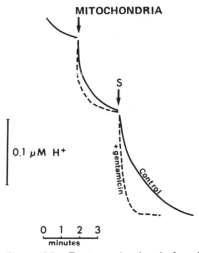

Figure 15 Protons ejection induced by gentamicin on rat liver mitochondria. Proteins, 2.5 mg/ml; succinate (S), 5 mM; rotenone, 3 μg; gentamicin, 2.4×10^{-3} M. Incubation medium: KCl, 80 mM; TRIS-HCl, pH 7.2, 10 mM; K_2HPO_4, 5 mM. Final volume, 3 ml. Verification was made that according to the concentration of H^+ ions, the incubation medium response is linear throughout the pH range used.

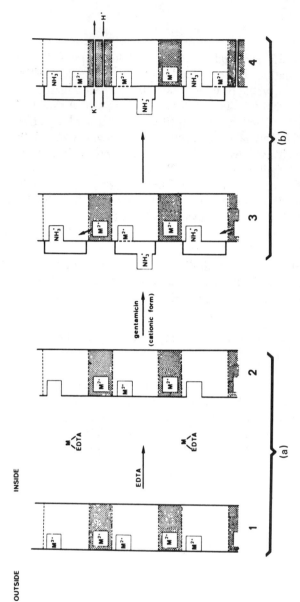

Figure 16 Suggested mechanism of action of gentamicin on isolated rat liver mitochondria; (a) sucrose medium; (b) KCl medium. 1, Schematic representation of the inner membrane of mitochondria; spotted surface; apolar (or hydrophobic phase; white surface, polar phase; M^{2+}, divalent cations. 2, Chelating effect of EDTA on M^{2+} in polar phase. 3, Scheme describing the replacement of M^{2+} by gentamicin in its cationic form and the protective effect of the antibiotic against subsequent action of EDTA on the membrane. M^{2+} displaced from the hydrophobic phase into the polar phase. 4, Permeability characteristics of the membrane resulting from the action of gentamicin.

of the effects of these drugs on mitochondrial respiratory intensity in state 3 or 4 up to a concentration of 4.8×10^{-4} M (Figure 17). The first group includes gentamicin C_1, dibekacin, sisomicin, tobramycin, kanamycin B, and netilmicin; the effect of these drugs is clear as of a concentration of 6×10^{-5} M. The second group is composed of kanamycin A, gentamicin A and B, and amikacin; the effect of these drugs appears as of concentrations ranging from 1.2 to 2.4×10^{-4} M. Netilmicin has a special effect, decreasing with increased concentrations as of 4.8×10^{-4} M, when it joins the second group of antibiotics.

When considering the chemical structure of the two groups (Figure 18 and Table 6), it is immediately seen that all the aminoglycosides of the first group (the most active) have five primary or secondary amino groups, whereas those of the second group have only four. However, in addition to the four amino groups, amikacin has an amide radical resulting from the transformation of the amino group into C_1 by the N substituent: γ-amino-α-hydroxybutyryl (AHBA). It thus appears that the number of amino groups is the first most important factor. It is likely that the highly nucleophilic characteristics of the amino groups is the reason. The existence of a lone pair of electrons on nitrogen permits the fixation of a proton. This proton acceptance depends on the nature of the N substituent and on the pH.

Role of the Nature of the N Substituent

The various aminoglycosides studied herein contain only primary or secondary amino groups, and in this case the N substituent is most often a methyl group. The donor inductive effect of the methyls will tend to increase the nucleophilic characteristics of the secondary amino groups; however, the simultaneous increase in stereohindrance prevents access to the lone pair of electrons of nitrogen, which diminishes the nucleophilic effect

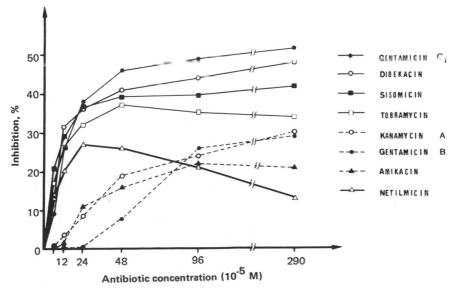

Figure 17   Effect of 4,6-disubstituted 2-deoxystreptamine derivatives on oxidation of succinate by rat liver mitochondria in state 3.

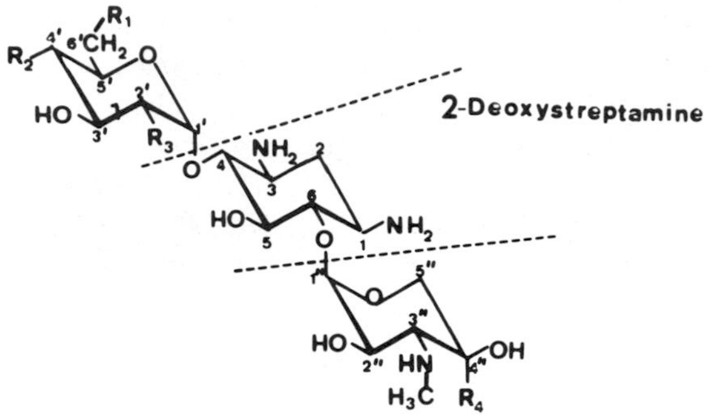

	R_1	R_2	R_3	R_4
GENTAMICIN A:	OH	OH	NH_2	H
GENTAMICIN B:	NH_2	H	OH	CH_3

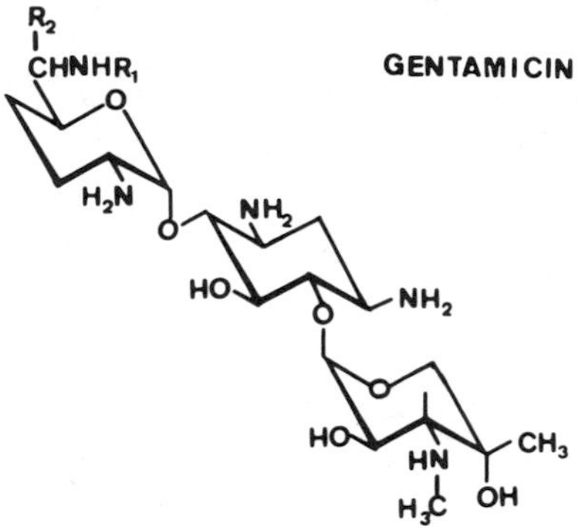

	R_1	R_2
GENTAMICIN C_1:	CH_3	CH_3
GENTAMICIN C_2:	H	CH_3
GENTAMICIN C_{1a}:	H	H

Figure 18 Structure of some of the naturally occurring gentamicins.

of the function. Results obtained with gentamicin C_1 and sisomicin, which contain an aminomethyl group, are similar to those obtained with tobramycin, kanamicin B, or dibekacin, which have only primary amino groups. It can thus be concluded that the steric factor is not sufficiently important to counter the donor inductive effect of $-CH_3$ and that as a result the acceptance capacity of protons of $-NH_2$ or of $-NH-CH_3$ will be closely similar. However, this steric factor appears more important in the case of the N substituent of amikacin C_1. This factor could well explain the behavior of this molecule with regard to other aminoglycosides having five amino groups. In amikacin C_1, the N substituent (AHBA) is in fact an important steric hindrance and renders access to the lone pair of electrons of nitrogen difficult, if not impossible. In addition, the presence of a highly electron-attracting carbonyl lowers the electron cloud of nitrogen and further reduces its nucleophilic characteristics. This confirms that the nucleophilic characteristics of the amino group are important as well as the presence of a nitrogenous

Table 6 Position of Substituents of Some 4, 6-Disubstituted 2-Deoxystreptamine-Containing Aminoglycosides

| | 2-Deoxystreptamine | | | | | | | | | | |
	1	3	5	2'	3'	4'	5'	2"	3"	4"	5"
Kanamycin B	NH_2	NH_2	OH	NH_2	OH	OH	$-CH_2NH_2$	OH	NH_2	OH	$-CH_2OH$
Tobramycin	NH_2	NH_2	OH	NH_2		OH	$-CH_2NH_2$	OH	NH_2	OH	$-CH_2OH$
Gentamicin C_1	NH_2	NH_2	OH	NH_2			$-CH-NH-CH_3$	OH	NH_2	OH / CH_3	$-CH_2OH$
Sisomicin	NH_2	NH_2	OH	NH_2			$-CH_2NH_2$	OH	$-NH-CH_3$	OH / CH_3	
Netilmicin	$-NH-CH_2-CH_3$	NH_2	OH	NH_2			$-CH_2NH_2$	OH	$-NH-CH_3$	O— / CH_3	
Gentamicin B	NH_2	NH_2	OH	OH	OH		$-CH_2NH_2$	OH	$-NH-CH_3$	OH / CH_3	
Gentamicin A	NH_2	NH_2	OH	NH_2	OH	OH	$-CH_2OH$	OH	$-NH-CH_3$	OH	
Dibekacin	NH_2	NH_2	OH	NH_2			$-CH_2NH_2$	OH	NH_2	OH	$-CH_2OH$
Amikacin	NH-L-AHBA[a]	NH_2	OH	OH	OH	OH	$-CH_2NH_2$	OH	NH_2	OH	$-CH_2OH$
Kanamycin A	NH_2	NH_2	OH	OH	OH	OH	$-CH_2NH_2$	OH	NH_2	OH	$-CH_2OH$

[a]L-AHBA, $-CO-CH-CH_2-CH_2NH_2$.
$\quad\quad\quad\quad$ OH

group. It is also noted that netilmicin C_1 contains an N substituent ($-CH_2-CH_3$) with a considerable blocking effect, which probably affects access to the lone pair of electron of nitrogen to a certain degree. This special structure of netilmicin may have some relationship to the fact that its effect resembles that of molecules with five amino groups at low concentrations and that of those with four amino groups at higher concentrations. The equilibrium among the forms of netilmicin having more or less protons would be altered based on antibiotic concentration: at low concentrations, the forms with the most protons would dominate, and at high concentrations only the form with the least protons would remain.

We studied the effects of N-acetylgentamicin and of penta-N-benzyloxycarbonyl-sisomicin, which were seen to be totally ineffective on mitochondria. This finding is compatible with the characteristic of highly electron-attracting N substituents, which suppresses any nucleophilic nature of the amino groups.

Influence of pH on the Effect of Gentamicin on Mitochondria

We measured the pK of various 4,6-disubstituted 2-deoxystreptamine-containing aminoglycosides as shown in Figure 19. Considering the mean pK of all amino groups of each antibiotic, it can be seen that this pK is approximately 8 for all the aminosides. This level is close to the pH, at which the various antibiotics were tested on mitochondria (pH 7.2), and it is likely that at this pH the proportion of cationic forms of aminoglycosides is rather low. However, when the effect of gentamicin on mitochondria was studied at pH 8.2, it was seen that in these conditions its effect was much lower (Table 7). We thus conclude that protonation of aminoglycoside amino groups is required for their action on mitochondria, and that it is sufficient at pH 7.2. It should be noted that the

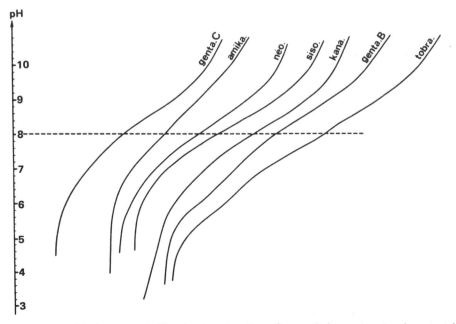

Figure 19   Evaluation of pK value by titration of some 2-deoxystreptamine-containing aminoglycosides.

effect of the antibiotic on the mitochondria depends on the number of amino groups and not on whether the molecule is basic, since the pK's are nearly identical for all the aminoglycosides, whether they are weakly or highly active.

Relationship between the Number and the Location of Hydroxyls of
4,6-Disubstituted 2-Deoxystreptamine Containing Aminoglycosides
and Their Effect on Mitochondrial Respiratory Activity

We studied the possible effects of hydroxylated groups on molecular activity. The less active group of antibiotics contains from 5 to 7 -OH, and the more active from 4 to 6. Thus, there appears to be no clear relationship between the effect of the antibiotic on the mitochondria and its number hydroxyls. The molecule's toxic effect seems slightly less with an increased number of hydroxyls. Among aminoglycosides of the first group (the most active) it is seen that kanamycin B, tobramycin, and dibekacin, whose molecular structures differ only by the presence or absence of hydroxyls on carbons 3' and 4', contain 6, 5, and 4 OH, respectively. However, kanamycin B and tobramycin have identical effects on mitochondria (Figure 20), whereas dibekacin is slightly more toxic.

It appears that no precise relationship can be demonstrated between the effect on mitochondria of the aminoglycoside chemical subgroup studied and the number or location of the hydroxyl radicals of their molecule. Such a relationship is evident, however, with regard to the number of amino groups. Moreover, although neither are derivatives of 4,6-disubstituted 2-deoxystreptamine, we noted that neomycine B and sorbistin A_1 have 6 and 3 primary or secondary amino groups, respectively. The former has the greatest effect on mitochondria, whereas the latter has practically none. In addition, one of the sorbistine A_1 secondary functions was highly hindered and deactivated by the N substituent (propionyl). Similarly, ribostamicin, which has four amino groups, has an activity very close to that of kanamycin A or amikacin. The role of the number of amino groups of aminoside molecules appears rather important with regard to antibacterial activity. For example, gentamicin A and B have a very low antibacterial activity compared to gentamicin C, which is very high [37]. Gentamicin A and B have only four amino groups, whereas gentamicin C has five. N acetylation is one of the mechanisms whereby bacteria inactivate aminoglycosides. However, a direct relationship between antibacterial activity and the number of amino groups has not been established. Amikacin, with a wide antimicrobial spectrum, and netilmicin have only four free amino groups. They alter mitochondrial respiratory activity only slightly while being very active toward bacteria. In contrast, the number and location of hydroxyl functions

Table 7   Effect of Gentamicin on the Oxidation of Malate-Glutamate by Rat Liver Mitochondria at pH 7.2 and pH 8.2[a]

State	pH 7.2			pH 8.2		
	Control	+ Gentamicin	%△	Control	+ Gentamicin	%△
3	62 ± 6	48 ± 9	−23	73 ± 8	77 ± 15	+5
4	12.5 ± 0.8	17.5 ± 3.3	+40	15.3 ± 1.1	15.8 ± 3.6	+3

[a]Malate-glutamate, $5 + 5$ mM; ADP, 0.33 mM. Proteins, 2.5 mg/ml; gentamicin, 1.2×10^{-4} M. Oxygen consumption expressed in n atoms/mn per mg of proteins.

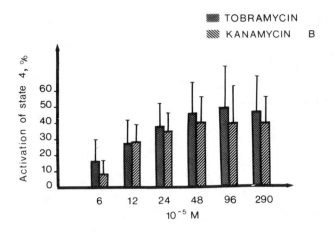

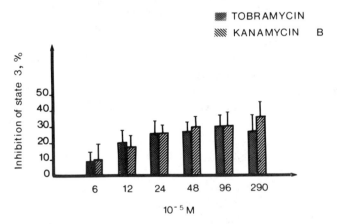

Figure 20   Comparison of the effect of kanamycin B and tobramycin on succinate oxidation by rat liver mitochondria. Proteins, 2.5 mg/ml; succinate, 10 mM; rotenone, 2 μg; ADP, 0.33 mM.

strongly influence the antibacterial effect of a drug. Better knowledge of these relationships might permit future partial synthesis of drugs with high antibacterial activity without harmful effects on intracellular organelles.

III.   Conclusions

1. Aminosides alter rat liver and kidney mitochondrial respiration. They stimulate state 4 and inhibit state 3. It is our opinion that the study of the aminoglycoside effect on mitochondrial respiration may be a rapid and efficient method of assessing the possible harmful effect of these antibiotics on (the behavior of) certain subcellular elements and may indicate the nephrotoxicity of this group of antibiotics. This conclusion stems from our observation of the close correlation between their effects on mitochondria and clinical observations of their nephrotoxicity.

The degree of these alterations grows with the number of free amino groups of the molecules. Neomycin B, known to be the most nephrotoxic of these antibiotics, has the greatest effect on the mitochondrial system and has the largest number of free amino groups. Antibiotics having average nephrotoxicity have five free amino groups, and those with lower nephrotoxicity have only four. Finally, sorbistin A_1, one of the most recently discovered molecules in this group, has only three free amino groups and this antibiotic showed low order of toxicity in mice. The relationship between structure and activity appears of value in assessing possible nephrotoxic potential and should be studied further by evaluating a larger number of aminosides.

2. Aminosides appear to have an effect on certain membranes, as suggested by our results with gentamicin on the mitochondrial membrane. Gentamicin is fixed on the membrane structure, and this fixation depends on the pH. The result of the activity of gentamicin on the mitochondrial membrane seems to be an alteration in permeability of K^+ ions and active transport of this ion within the mitochondria. The energy required is provided by the electron transport chain, which is partially oxidized, after which the ejection of protons is compensated by the entrance of K^+.

Acknowledgments

This work was supported by INSERM Grant 76-1-176-5. We are grateful to Mrs. D. Falize for preparation of the manuscript and to Mrs. D. Brooner for the translation. We also wish to thank the various laboratories that donated the antibiotics.

References

1. Appel, G.B., and Neu, H.C.: The nephrotoxicity of antimicrobial agents [three parts]. *N. Engl. J. Med.* 296:663–670, 722–728, 784–787, 1977.
2. Asselineau, J., and Zalta, J.: Antibiotiques de nature osidique. In *Les Antibiotiques— structure et exemples de mode d'action.* Paris, Hermann, 1973, pp. 41–64.
3. Bennet, W., Plamp, C., Gilbert, D., and Porter, G.: Experimental aminoglycoside nephrotoxicity in the rat: comparison of gentamicin, tobramycin and netilmicin. *Clin. Pharmacol. Ther.* 23:109, 1978.
4. Binet, A.: Les Ions Ca^{2+} et Mg^{2+} dans les mitochondries; leur régulation par un facteur cytoplasmique. Thèse Docteur ès Sciences Naturelles, Université de Paris-Sud, Centre d'Orsay, 1975.
5. Binet, A., and Volfin, P.: Regulation by Mg^{2+} and Ca^{2+} of mitochondrial membrane integrity; study of the effects of a cytosolic molecule and Ca^{2+} antagonists. *Arch. Biochem. Biophys.* 170:576–586, 1975.
6. Brummett, R.E., Fox, K.E., Bendrick, T.W., and Himes, D.L.: Comparative ototoxicity of tobramycin, gentamicin, amikacin and sisomycin. In *Current Chemotherapy. Proceedings of the 10th International Congress of Chemotherapy,* Vol. 2. Washington, D.C., American Society for Microbiology, 1978, pp. 939–941.
7. Caswell, A.H.: The migration of divalent cations in mitochondria visualized by a fluorescent chelate probe. *J. Membr. Biol.* 7:345–364, 1972.
8. Chance, B., and Williams, G.R.: Respiratory enzymes in oxidative phosphorylation. *J. Biol. Chem.* 217:383, 1955.
9. Chance, B., and Williams, G.R.: The respiratory chain and oxidative phosphorylation. *Adv. Enzymol.* 17:65–134, 1956.

10. Chappell, J.B., and Crofts, A.R.: Ion transport and reversible volume changes of isolated mitochondria. In *Regulation of Metabolic Processes in Mitochondria*. Amsterdam, Elsevier, 1966, pp. 293–316.

11. Chenoweth, M.B.: Chelation as a mechanism of pharmacological action. *Pharmacol. Rev.* 8:57–87, 1956.

12. Chiu, J.S., Miller, G.H., Brown, A.D., Long, J.F., and Waitz, J.A.: Renal pharmacology of netilmicin. *Antimicrob. Agents Chemother.* 11:821–825, 1977.

13. Christensen, E.F., Reiffenstein, J.C., and Madissoo, H.: Comparative ototoxicity of amikacin and gentamicin in cats. *Antimicrob. Agents Chemother.* 12:178–184, 1977.

14. d'Amato, R.F., Thornsberry, C., Baker, C.N., and Kirven, L.A.: Effect of calcium and magnesium ions on the susceptibility of *Pseudomonas* species to tetracyclin, gentamicin, polymyxin B and carbenicillin. *Antimicrob. Agents Chemother.* 7:596–600, 1975.

15. Davis, D.S., and Iannetta, A.: Relative antagonism in vitro of calcium in serum to the bactericidal activities of gentamicin and tobramicin on *Pseudomonas aeruginosa*. *Chemotherapy* 19:243–253, 1973.

16. Einspruch, B.G., and Gonzalev, U.V.: Clinical and experimental nephropathy resulting from use of neomycin sulfate. *JAMA* 173:809, 1960.

17. Fabre, J., Rudhardt, M., Blanchard, P., and Regamey, C.: Persistence of sisomicin and gentamicin in renal cortex and medulla compared with other organs and serum of rats. *Kidney Int.* 10:444–449, 1976.

18. Fabre, J., Rudhardt, M., Blanchard, P., and Chauvin, P.: Kinetics of gentamicin and sisomicin in renal cortex and medulla of rats compared with other organs. Influence of quinine. In *Interactions of Drugs with Membrane Systems: Mitochondria-lysosomes*, Rouen, October 27–28, 1977.

19. Federspil, P., Schatzle, W., and Tiesler, E.: Pharmacokinetics and ototoxicity of gentamicin, tobramycin and amikacin. *J. Infect. Dis.* 134:200–205, 1976.

20. Fillastre, J.P., Kuhn, J.M., Bendirdjian, J.P., Foucher, B., Leseur, J.P., Rollin, P., and Vaillant, R.: Prediction of antibiotic nephrotoxicity. *Adv. Nephrol.* 6:343–370, 1976.

21. Gilbert, D.N., Kutscher, E., Ireland, P., Barnett, J.A., and Sanford, J.P.: Effect of the concentrations of magnesium and calcium on the in vitro susceptibility of *Pseudomonas aeruginosa* to gentamicin. *J. Infect. Dis.* 124:537–545, 1971.

22. Hava, M., Sobek, V., and Mikulaskova, J.: On the role of calcium ions in the toxic neomycin action. *Biochem. Pharmacol.* 8:76, 1961.

23. Hemker, H.C., and Hulsmann, W.C.: Dinitrophenol-induced ATPase of rat-liver mitochondria. *Biochim. Biophys. Acta* 48:221–223, 1961.

24. Hoyme, U., and Madsen, P.O.: Netilmicin and tobramycin in the therapy of complicated urinary tract infections. In *Current Chemotherapy. Proceedings of the 10th International Congress of Chemotherapy*, Vol. 2. Washington, D.C., American Society for Microbiology, 1978, pp. 987–989.

25. Inouye, B., Uchinomi, Y., and Wachi, T.: In vitro effect of rifampicin and its derivatives on energy transfer reactions in mitochondria. *J. Antibiot.* 27:192–198, 1974.

26. Kornguth, M.L., and Kunin, C.M.: Distribution of gentamicin and amikacin in rabbit tissues. *Antimicrob. Agents Chemother.* 11:974–977, 1977.

27. Kosek, J.C., Mazze, R.I., and Cousins, M.J.: Nephrotoxicity of gentamicin. *Lab. Invest.* 30:48–57, 1974.

28. Kovnat, P., Labovitz, E., and Levison, S.P.: Antibiotics and the kidney. *Med. Clin. North Am.* 57:1045, 1973.

29. Lee, C.P., and Ernster, L.: The energy-linked nicotinamide nucleotide transhydrogenase reaction: its characteristics and its use as a tool for the study of oxidative

phosphorylation. In *Regulation of Metabolics Processes in Mitochondria*. Amsterdam, Elsevier, 1966, pp. 218–234.

30. Lerner, S.A., Seligsohn, R., Rudd, R., and Matz, G.J.: Comparative clinical studies of ototoxicity and nephrotoxicity of amikacin and gentamicin. In *Current Chemotherapy. Proceedings of the 10th International Congress of Chemotherapy*, Vol. 2. Washington, D.C., American Society for Microbiology, 1978, pp. 945–946.

31. Luft, F.C., Yum, M.N., and Kleit, S.A.: Comparative nephrotoxicities of netilmicin and gentamicin in rats. *Antimicrob. Agents Chemother.* 10:845–849, 1976.

32. Mahler, H.R., and Cordes, E.H.: Biological oxidations. In *Biological Chemistry*. New York: Harper & Row, 1966, pp. 554–623.

33. Packer, L., Utsumi, K., and Mustafa, M.G.: Oscillatory states of mitochondria: I. Electron and energy transfer pathways. *Arch. Biochem. Biophys.* 117:381–393, 1966.

34. Pressmann, B.C.: Mechanism of action of transport mediating antibiotics. *Ann. N.Y. Acad. Sci.* 147:829–841, 1969.

35. Pressmann, B.C.: Biological applications of ionophores. *Ann. Rev. Biochem.* 45:501–530, 1976.

36. Pressmann, B.C. In Van Dam, K., and Myer, A.J.: Oxidation and energy conservation by mitochondria. *Ann. Rev. Biochem.* 40:115–160, 1971.

37. Price, K.E., Godfrey, J.C., and Kawaguchi, H.: Effect of structural modifications on the biological properties of aminoglycoside antibiotic containing 2 deoxystreptamine. *Adv. Appl. Microbiol.* 18:191–307, 1974.

38. Quigley, J.W., and Cohen, S.S.: The isolation of envelopes of *Escherichia coli* spheroplasts in the zonal ultracentrifuge. *J. Biol. Chem.* 244:245P–2458, 1969.

39. Rinehart, K.L.: *The Neomycins and Related Antibiotics*. New York: Wiley, 1964, p. 137.

40. Rosa, F. de, Buoncristiani, U., Capitanucci, P., and Frongillo, R.F.: Tobramycin toxicological and pharmacological studies in animals and pharmacokinetic research in patients with varying degrees of renal impairment. *J. Int. Med. Res.* 2:100–114, 1974.

41. Schoenknecht,   : Aminoglycosides from the clinical microbiologist's viewpoint. *Int. J. Clin. Pharmacol.* 11:103–114, 1975.

42. Silver, S., Wendt, L., Bhattacharyya, P., and Beauchamp, R.S.: Effects of polyamines on membrane permeability. *Ann. N.Y. Acad. Sci.* 197:838–862, 1975.

43. Slater, E.C.: Mechanism of energy conservation in mitochondrial oxido-reductions. In *Regulation of Metabolic Processes in Mitochondria*. Amsterdam, Elsevier, 1966, pp. 166–179.

44. Slater, E.C., Berden, J.A., Bertina, R.M., and Albracht, S.P.J.: Antimycin and oligomycin as probes of the energy state of mitochondrial membranes. In *Energy Transduction in Respiration and Photosynthesis*. Bari, Adriatica Editrice, 1971, pp. 35–51.

45. Subik, J., Behun, M., and Musilek, V.: Antibiotic mudicin, a new antimycin, a like inhibitor of electron transport in rat liver mitochondria. *Biochem. Biophys. Res. Commun.* 57:17–22, 1974.

46. Szwed, J.J., Luft, F.C., Black, H.R., Elliott, R.A., and Kleit, S.A.; Comparison of the distribution of tobramycin and gentamicin in body fluids of the dog. *Antimicrob. Agents Chemother.* 5:444–446, 1974.

47. Tabor, C.W., and Kellogg, P.D.: The effect of isolation conditions on the polyamine content of *Escherichia coli* ribosomes. *J. Biol. Chem.* 242:1044–1052, 1967.

48. Valdivieso, M., Horikoshi, N., Rodriguez, U., and Bodey, G.P.: Therapeutic trials with tobramycin. *Am. J. Med. Sci.* 268:149–156, 1974.

49. Vera-Roman, J., Krishnakantha, T.P., and Cuppage, F.E.: Gentamicin nephrotoxicity in rats. I. Acute biochemical and ultrastructural effects. *Lab. Invest.* 33:412–417, 1975.

50. Washington, J.A., Snyder, R.J., Kohner, P.C., Wilte, C.G., Ilstrup, D.M., and McCall, J.T.: Effect of cation content of agar on the activity of gentamicin, tobramycin and amikacin against *Pseudomonas aeruginosa. J. Infect. Dis.* 137:103–110, 1978.

51. Welles, J.S., Emmerson, J.L., Gibson, W.R., Nickander, R., Owen, N.V., and Anderson, R.C.: Preclinical toxicologic studies with tobramycin. *Toxicol. Appl. Pharmacol.* 22:332, 1972.

52. Welles, J.S., Emmerson, J.L., Gibson, W.R., Nickander, R., Owen, N.V., and Anderson, R.C.: Preclinical toxicology studies with tobramycin. *Toxicol. Appl. Pharmacol.* 25:398–409, 1973.

53. Whelton, A., Carter, G.G., Bryant, H.H., Cody, T.S., and Walker, W.G.: Tobramycin and gentamicin intrarenal kinetic comparisons: therapeutic and toxicological answers. In *Current Chemotherapy. Proceedings of the 10th International Congress of Chemotherapy,* Vol. 2. Washington, D.C., American Society for Microbiology, 1978, pp. 951–953.

54. Whelton, A., and Walker, W.G.: Intrarenal antibiotic distribution in health and disease. *Kidney Int.* 6:131–137, 1974.

55. Zimelis, V.M., and Jackson, G.G.: Activity of aminoglycoside antibiotics against *Pseudomonas aeruginosa:* specificity and site of calcium and magnesium antagonism. *J. Infect. Dis.* 127:663–669, 1973.

15

PATHOLOGY OF DRUG NEPHROTOXICITY IN HUMANS

STEEN OLSEN   University Institute of Pathology, Kommunehospitalet, Aarhus, Denmark

KIM SOLEZ   The Johns Hopkins University School of Medicine, Baltimore, Maryland

The kidney is very sensitive to the action of toxic agents. This is also true of other parenchymal organs, the liver being the best example, but in the kidney the lesions induced by drugs and toxins are extremely diverse clinically and pathophysiologically as well as morphologically. This diversity may be related to the complicated structure and function of the kidney.

The main purpose of this chapter is to describe the histopathology of renal disease induced by antibiotics with special reference to the aminoglycosides. We have found it desirable, however, to facilitate the understanding of nephrotoxicity of this group of drugs by presenting first an overview of the subject of drug nephrotoxicity in general. Pathogenesis is not well understood in many types of drug nephrotoxicity, but to the extent that we have knowledge of the mechanisms responsible for drug-induced renal diseases, they seem to express themselves in several rather clear-cut and uniform morphological patterns [1,86,115]. For this reason, we have chosen to give a description of the various histopathological categories that may be seen in drug-induced renal disease. Following this, the antibiotic-induced nephropathies will be described and discussed in more detail.

Before entering into the substance of this chapter, we feel it important to stress some factors that have impact on the reliability of our current knowledge of causal relationships between a drug and the renal pathological lesions ascribed to it. Several sources of error may affect the evaluation of the pertinent literature. The most important of these are the following:

1. *Lack of morphological studies.* This source of error applies particularly to cases with rather mild symptoms which are reversible following withdrawal of the drug. The clinician will often hesitate to perform renal biopsies in such cases.
2. *In cases that have been investigated solely by autopsy,* postmortem autolytic changes may be misinterpreted as vital lesions.
3. Many reports present *incomplete histopathological data* and the description is sometimes restricted to a brief claim of the presence of "degenerative" lesions of the tubules, and so on.
4. *Nomenclature* may offer an unsurmountable obstacle to the understanding of the types of damage inflicted by a given drug. Thus, the term "acute tubular necrosis," which is appropriate in cases with extensive tubular cell necrosis, is also widely used when such necroses are few and small or not present at all (see later). To add to the confusion, the term "acute tubular necrosis" has also gained popularity as denoting a clinical state and is used in many situations in which morphological studies have not been performed.
5. *Uncertainty as to the responsible agent.* Many patients are treated with several drugs in the period preceding the onset of clinical renal disease. All too often adverse reactions against specific drugs are postulated and reported without due reference to other possible causal factors, such as other nephrotoxic drugs, trauma, hypotension, or dehydration.
6. *The disease for which the drug is given may itself be responsible* for the renal disease, or it may have modified the histological picture of a lesion induced primarily by a drug. It is important to keep this in mind, for example, when ascribing an interstitial nephritis to the action of a drug, because sepsis alone may give rise to interstitial nephritis, a fact that has been known for many years [148]. Patients suffering from systemic lupus erythematosus (SLE) or periarteritis have often been treated with antibiotics when signs and symptoms of renal disease appear. Arteritis or glomerulitis may then erroneously be interpreted as being due to the drug.
7. *Kidney lesions may have been present before the drug was given.* Age-related interstitial fibrosis and tubular atrophy, arteriolar sclerosis, and lymphocytic interstitial inflammation may all exist unnoticed without giving rise to overt clinical symptoms. If a patient with one of these lesions undergoes renal biopsy because of suspicion of

drug-induced nephropathy, the preexisting lesions may be misinterpreted as being due to the drug in question.

8. The difficulty in obtaining precise knowledge of the causal relationship between administration of drugs and renal lesions in clinical situations leads investigators to conduct experimental studies in animals. It is incontestable that *animal models* have contributed greatly to our understanding of drug-induced kidney disease, but it should be remembered that this approach is associated with other sources of error because of the different sensitivity of various species to the action of specific drugs. When evaluating experimental studies it should furthermore be kept in mind that doses (per kilogram of body weight) used in animal experiments are often much higher than those used in humans.

In the following survey we have tried to avoid misinterpretations due to the factors listed above as far as possible, and reports which in our view are too much biased by these factors have been omitted. On the other hand, we had to take into consideration the fact that well-performed morphological studies using modern methods are relatively scarce in this field, and some observations are cited that are less well substantiated than we might like.

I. Typical Histological Patterns

A. Drug-Induced Glomerulonephritis

Glomerulonephritis can be elicited by at least two different immunopathogenetic mechanisms. One of these is dependent on the presence of circulating antibodies reacting against normal glomerular basement membrane. The other mechanism, which is by far the most common in human glomerular disease, is triggered when circulating, soluble antigen–antibody complexes are deposited in or along the glomerular basement membrane. The last type may be due to bacterial antigens, but the antigen is often unknown. By immunofluorescence a granular pattern is seen due to the deposition of immunoglobulins and complement in glomeruli.

Glomerulonephritis of the so-called membranous type may be elicited by gold (sodium aurothiomalate and sodium aurothiosulfate) administered to patients with rheumatoid arthritis [18,78,119,124,132,141] and by mercury, which is contained in skin-lightening creams [69]. It can also follow administration of penicillamine [4,54,64,76,77], a compound related to penicillin which is used to treat Wilson's disease and cystinuria.

By light microscopy there may be no abnormalities or the glomerular capillary loops may be thickened. Subepithelial "spikes" are often seen using methenamine silver stains. Electron microscopically there are discrete subepithelial dense deposits with intervening processes of basement membrane that project outward between deposits and correspond to the spikes seen by light microscopy. These changes, as well as the fine granular pattern of immunoglobulin and complement seen by immunofluorescence, are identical with those seen in idiopathic membranous glomerulonephritis (Figure 1).

Clinically, the nephrotic syndrome is observed. The symptoms and glomerular lesions tend to disappear following withdrawal of the drug.

The ultrastructural and immunopathological picture is (by analogy with experimental models) thought to indicate a pathogenesis involving deposition of antigen–antibody complexes. It has been proposed that the responsible antigen in the case of gold-induced proteinuria might be the metal acting as an antigen or as a hapten, but gold cannot be

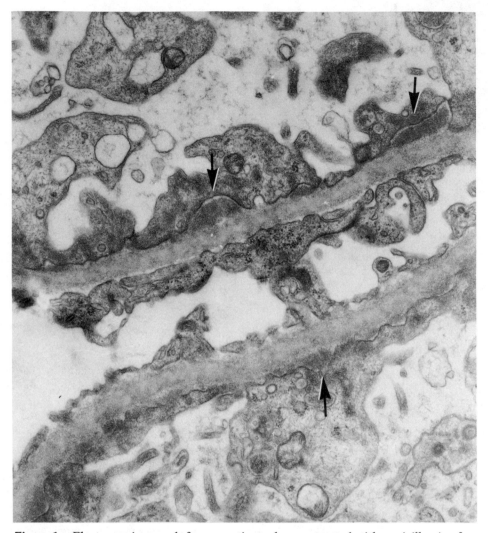

Figure 1   Electron micrograph from a patient who was treated with penicillamine for rheumatoid arthritis. After 16 weeks severe proteinuria developed. The renal biopsy showed normal glomeruli by light microscopy. By electron microscopy multiple discrete subepithelial deposits were seen (arrows), interpreted as state I of a membranous glomerulonephritis. Withdrawal of the drug was followed by permanent remission and a control biopsy showed normal glomerular ultrastructure. (×15,000) (Courtesy of G. Ditscherlein, from G. Ditscherlein, and W. Schneider, *Dtsch. Gesundheitswes* 30:2401, 1975, VEB Verlag Volk und Gesundheit, Berlin.)

detected in the glomerular deposits, and other mechanisms may therefore be operative. By electron microscopy gold can be demonstrated in tubular lysosomes (irrespective of the presence of proteinuria), and hypothetically the gold salts might cause tubular damage with release of tubular epithelial antigens. In experimental models using another heavy metal, mercuric chloride, anti-GBM antibodies as well as immune complexes have been reported [108] and the presence of antinuclear antibodies in another study points toward a possible role of nuclear antigens released by the action of mercuric chloride [144].

A case of trimethadione-induced nephrotic syndrome reported by Bar-Khayim and co-workers probably also can be categorized as a membranous glomerulonephritis, although the authors felt that its histopathology was unique [7]. The published electron micrographs seem to be compatible with a late stage of membranous glomerulonephritis, although the conspicuous infiltration of eosinophils these authors observed in the glomerular tuft is not usually present in other types of membranous glomerulonephritis.

The very rare occurrence of crescentic glomerulonephritis associated with arteritis reported as a complication to treatment with penicillin is discussed under the heading "Arteritis."

Glomerulonephritis with morphological lesions characteristic for SLE has been reported following the anticonvulsants trimethadione and diphenylhydantoin [136]. The glomerular alterations may be similar to the membranous or the proliferative types of lupus nephritis. So-called virus-like inclusions, characteristic of (but not specific for) SLE, have also been observed in this drug-induced nephropathy. Patients with drug-induced SLE may also show lupus-like lesions in other organs, and the serologic tests for antinuclear antibodies may be positive.

B. Arteritis

Arteritis associated with crescentic glomerulonephritis has been reported following treatment with penicillin [117] either as a pure renal disease or as a generalized hypersensitivity reaction also affecting arteries in other parts of the body. The apparent relation between the treatment and the disease could be coincidental or the clinical signs and symptoms for which the patients were treated with penicillin could conceivably be due to preexisting polyarteritis [86]. Sulfa compounds have also been reported able to induce arteritis [45].

C. Acute Interstitial Nephritis

An acute nephropathy characterized morphologically by interstitial infiltration of leukocytes occurs sometimes after the administration of drugs, most frequently following methicillin [5,16,56,65,85,96,113]. The clinical presentation is typically that of acute renal failure occurring after 1–2 weeks of administration of the responsible drug. Proteinuria is frequent and hematuria as well as leukocyturia may also occur. Fever and skin rash as well as a blood eosinophilia are present in some patients. Most cases recover some weeks after withdrawal of the drug. The renal function may, however, be persistently reduced in a minority of patients.

Renal biopsies performed during the acute phase have revealed a patchy but severe interstitial infiltration by inflammatory cells (Figure 2). The infiltration is dominated by mononuclear cells (large and small lymphocytes, plasma cells, and histiocytes). There is also interstitial edema. A focal accumulation of histiocytes may form small granulomas without necroses. Méry and Morel-Maroger have described the presence of multinuclear giant cells [85]. Some biopsies may show a mixed infiltrate of polymorphs and eosinophils. Interstitial infiltration entirely dominated by eosinophils can also be seen occasionally. Associated with the interstitial inflammatory infiltration there is tubular damage, which typically affects small foci of distal tubules. There may be mild focal epithelial

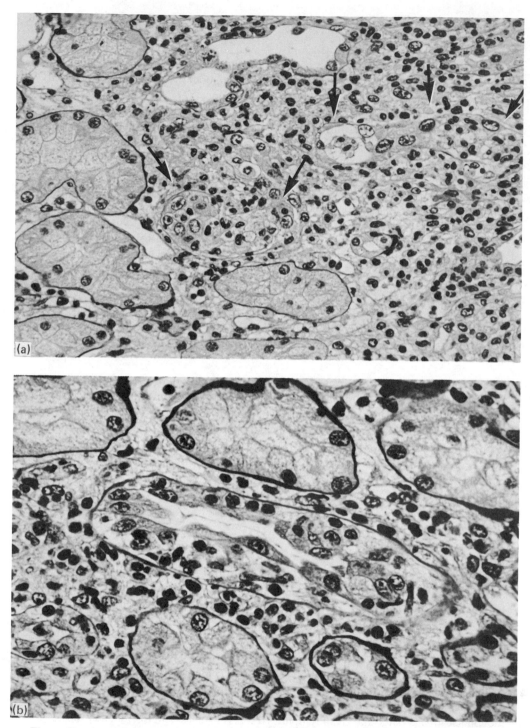

Figure 2 Acute interstitial nephritis in a patient who was treated prophylactically with methicillin following cardiac surgery. Acute renal failure appeared 2 weeks after the operation. (a) Severe interstitial infiltration with lymphocytes and histiocytes. Some tubules show disappearance of the basement membrane, epithelial degeneration, and mononuclear cell invasion (arrows). (×450) (b) Higher magnification from the same biopsy. Another tubule with penetration of mononuclear cells and epithelial changes. (Periodic acid-silver methenamine, ×800)

necrosis, but extensive necrosis is not seen, even in those cases that show considerable tubular loss. The tubular basement membrane may show defects or be totally lacking in foci of necrosis. Compared with biopsies from patients suffering from acute renal failure following shock, hemolysis, and so on, the acute interstitial nephritis differs by lacking the tendency to severe distal tubular dilatation and by the much more conspicuous interstitial cellular infiltration with a tendency for tubules to disappear in areas of intense inflammation. Glomeruli and vessels are normal. Immunofluorescence in most cases is negative [96], but some cases have been reported in which the tubular basement membranes showed the presence of immunoglobulins [16,56].

It has long been known that interstitial nephritis may be caused by sepsis [148]. The observation that it may follow a prophylactic administration of methicillin to patients without infection [96] indicates, however, that methicillin alone, without the action of infection, may give rise to interstitial nephritis.

Acute drug-induced interstitial nephritis is not dose-dependent and there is characteristically a time lag between the start of the treatment and the onset of the disease. These circumstances, together with the blood eosinophilia and skin rash which are often present, point strongly toward an allergic mechanism.

Penicilloyl hapten has been demonstrated together with IgG and C_3 along the tubular basement membrane [5], and antitubular antibodies have been detected in the blood of one patient [16]. Immunoglobulin deposits in the tubular basement membrane are, however, lacking in most cases [96]. The dense mononuclear infiltrate around damaged tubules might indicate that the disease is due to a cellular immune reaction rather than to a humoral pathogenesis involving circulatory antibodies.

Penicillin, ampicillin, rifampicin, allopurinol, polymyxin B-sulfate, and antibacterial sulfonamides are other drugs that may give rise to interstitial nephritis [5,13,85,93,111, 135,139]. The same applies to acetazolamide (Diamox) [50], a diuretic related to the sulfonamides. The diuretic agents furosemide and dihydrochlorthiazide may also elicit a hypersensitivity reaction with acute interstitial nephritis, but a definite etiologic relation was not established [80].

D. Chronic Interstitial Nephritis

Interstitial mononuclear cell infiltration associated with progressive interstitial fibrosis and tubular atrophy is a histological picture which may have several causes. One of them is long-term administration of phenacetin and possibly other analgesics. This important renal disease, which was initially discovered in Switzerland [131], accounts for a relatively large proportion of all cases of chronic terminal uremia, at least in Europe and Australia [29,51,57,70,116]. In the Aarhus series of 400 patients treated with renal allotransplantation, the renal failure was due to analgesic nephropathy in 18%. Morphologically, this particular form of chronic interstitial nephritis is characterized by its obligatory association with papillary necrosis, which is probably primary, constituting a kind of intraparenchymal "obstruction," leading to chronic inflammation, fibrosis, and nephron atrophy (Figure 3).

Another renal disease, which may be regarded as a chronic interstitial nephritis (although not associated with papillary necrosis), has recently been reported to occur in patients treated with lithium salts for endogenous manic-depressive disease [60]. After several years of treatment, some patients develop polyuria, with decreasing concentration

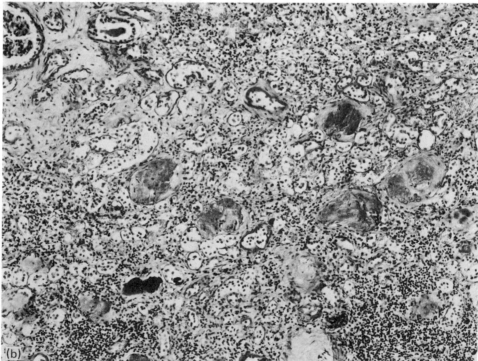

Figure 3 Chronic interstitial nephritis and multiple papillary necrosis. The patient had used phenacetin for many years and developed chronic uremia. (a) Calcified necrotic papilla. (b) The renal cortex shows glomerular obsolescence and tubular atrophy as well as severe diffuse interstitial lymphocytic infiltration and fibrosis. (Hematoxylin and eosin, ×100)

ability as well as a mild impairment of renal function. The kidneys of patients with these functional alterations show interstitial fibrosis, tubular atrophy, and glomerular sclerosis. The lesions are typically focal in distribution. Progression toward uremia is rare and the lesions seem to be very slowly developing. Macroscopically, this nephropathy is characterized by the presence of multiple, small cortical cysts.

E. Extensive Tubular Necrosis ("Toxic Acute Tubular Necrosis")

Renal tubular epithelium is particularly vulnerable to the action of several toxic agents. Poisoning with mercuric chloride is one of the earliest known examples of extensive tubular necrosis. The necrotic cells are desquamated following ingestion of the poison and the basement membrane subsequently becomes covered by flat basophilic regenerating epithelium. Other heavy metals and some organic compounds, mainly solvents, may also give rise to extensive tubular necrosis. This type of renal damage is rather seldom caused by drugs in the clinical situation, although very high doses of some drugs (e.g., gentamicin; see later) used in animal experiments have given rise to this extreme destruction of renal tissue.

When used clinically in high doses (6 g/day or more) cephaloridine may give rise to extensive necrosis of the proximal tubules. It is true that most reports are based on autopsies, but one "biopsy" performed 1 hr postmortem has shown the same changes [109] and the dose-dependent toxic action of this drug, particularly on proximal tubular epithelium, has been amply demonstrated in animal experiments [120]. The degree of toxicity in animals is dependent not only on dose but also on the degree of polymerization of the compound [17]. Lower doses comparable to those recommended for clinical use cause only mild tubular damage.

The potent nephrotoxin bacitracin has (in autopsy studies) been shown to give rise to extensive tubular necroses. Calcification of the necrotic cells indicates that the "necrosis" was real and not just due to postmortem autolytic changes [47]. The aminoglycosides cause dose-related patchy necrosis as described in more detail below.

Calcification of tubular epithelial cells as a sequel to necrosis is a well-known late stage of the mercuric chloride lesion. As mentioned above, it may also take place in tubular necrosis following the action of drugs. We have seen this lesion in a patient who was treated for metastasizing B-cell tumor with streptozotocin, a B-cell poison used to generate experimental diabetes mellitus in animals (Figure 4). It has also been described as the characteristic morphologic picture in amphotericin nephrotoxicity [134].

F. Acute Tubular Nephropathy ("Vasomotor Nephropathy," "Shock Kidney")

The degree of severity of tubular necrosis may vary tremendously, but the term "tubular necrosis" has been used in this chapter only for renal conditions in which at least whole tubular cross sections, if not whole tubular segments, undergo necrosis. Renal diseases in which only a few individual cells in a given tubular cross section become necrotic (e.g., acute renal failure following shock or hemolysis) will not be referred to as "acute tubular necrosis" but rather, since the presence of occasional missing tubular cells rather than obviously necrotic ones is the predominant histologic feature [127], as "acute tubular nephropathy." In the normal kidney, tubular cells die, are promptly discharged into the

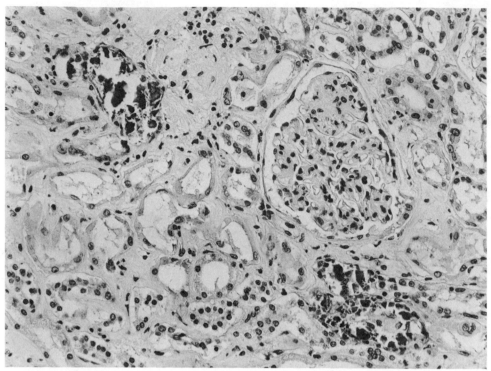

Figure 4   This picture is from an autopsy specimen of the kidney from a patient who developed acute renal failure following treatment with streptozotocin (a potent B-cell poison) for a metastasizing B-cell tumor of the pancreas. The patient died eventually from her malignant disease. There is widespread calcification of the tubular epithelium, probably following necrosis of the tubular epithelium. (Hematoxylin and eosin, ×800)

tubular lumen, and are replaced by new cells. The average lifetime of a tubular cell in the mouse is approximately 190 days [24]. Animal studies suggest that cell division in the normal kidney is not constantly occurring but rather occurs in cycles with peaks of mitotic activity occurring every 7 days [63]. It is likely that cell death is similarly cyclical in the normal kidney. This normal process of cell death probably accounts for the sloughed tubular cells with pyknotic nuclei seen occasionally in tubular lumens in normal human kidneys. It appears that in the sort of acute renal failure that follows shock or hemolysis this cell death is accelerated and cell replacement does not keep pace with cell loss. Obviously necrotic cells are seldom seen lining tubules in acute tubular nephropathy, presumably because they are expelled into the lumen as they die. Loss of individual tubular epithelial cells is significantly more prominent in biopsies from patients who have acute renal failure at the time of biopsy than it is in biopsies from patients who are biopsied shortly after recovery from acute renal failure, suggesting that this lesion plays an important role in the pathophysiology of acute renal failure [127].

Acute tubular nephropathy is the most frequent morphologic type of acute renal failure [95]. Clinically, it appears as acute anuric or oliguric renal failure following volume depletion with or without clinically overt circulatory shock. Etiologically, a multitude of causes can be responsible [6]. Some of these are: episodes of hypotension,

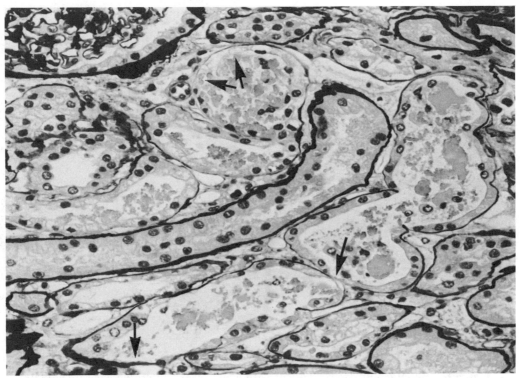

Figure 5 Acute renal failure following injection of sodium aurothiosulfate as a treatment for rheumatoid arthritis. There are heme-stained casts in the dilated distal tubules. Focally, there are defects of the epithelium (arrows), probably following shedding of epithelial cells without replacement. Figures 5 and 6 are examples of the same renal disease. There is no widespread tubular necrosis. (Periodic acid–silver methenamine, ×400)

mismatched transfusions, hemolysis following malaria or hemolytic blood diseases, and myolysis due to surgical or other trauma. This type of acute renal failure is reversible if the patient survives.

The histopathology is characterized by rather mild lesions. There is a tendency to dilatation of the distal tubules, which may be empty or contain hyaline, granular, or pigmented casts. There is a patchy mild interstitial infiltration with mononuclear cells, predominantly in the inner cortex and the outer medulla. Necrosis or desquamation of single distal tubular cells or of small groups of tubular cells may occasionally occur (Figure 5), but the picture is very distinct from the extreme necrosis that characterizes heavy metal-induced tubular necrosis (see discussion above). This misleading nomenclature has led some investigators [among them one of us (St.O)] to prefer the more appropriate name "acute tubulointerstitial nephropathy" for this entity, but we have used the term "acute tubular nephropathy" in this chapter to avoid confusion with other entities characterized by a combination of interstitial inflammation and tubular changes (e.g., acute interstitial nephritis following methicillin).

Several drugs may give rise to acute renal failure with this histological picture. Among them are barbiturates (Figure 6) and some antibiotics.

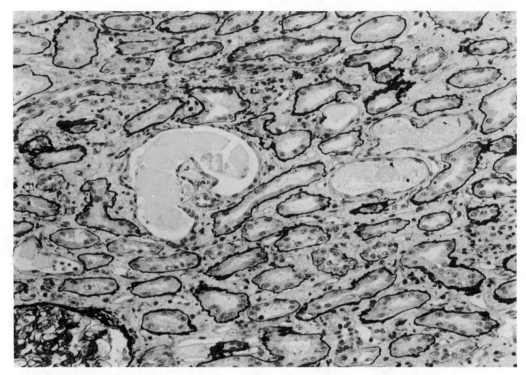

Figure 6   Renal biopsy from a patient with acute renal failure following barbiturate poisoning with shock. The most impressive lesion is a marked distal tubular dilatation. The dilated tubules are covered by flat, but in this location not necrotic epithelium. This renal disease is often erroneously called "acute tubular necrosis." Compare with Figure 5. (Periodic acid–silver methenamine, ×200)

The pathogenesis of this disorder is still not known despite extensive clinical and experimental investigations. One hypothesis, which has recently gained some popularity, explains the oliguria as being due to a severe decrease in glomerular filtration caused by afferent arteriolar constriction. The immediate cause of the vasoconstriction is thought to be a local release of renin brought about by a lowered proximal sodium resorption with a resulting high sodium concentration in the macula densa. Other hormones, such as vasopressin and thromboxane A_2, may also play a role in this type of acute renal failure [126,129]. A decrease in glomerular permeability or tubular obstruction by hyaline casts may also contribute to renal insufficiency in this disorder [128,129].

G.   Tubular Crystalline Deposits

Patchy deposition of crystalline material is very frequent in various kidney diseases, especially in acute renal failure. It can be detected easily by polarization microscopy. The presence of scattered intratubular crystals has no diagnostic importance. A few diseases are morphologically characterized by extensive deposition of crystals. The most remarkable disorders of this type are the nephropathy of gout, some inborn errors of amino acid metabolism, primary oxalosis, glycol poisoning, and the acute methoxyfluorane nephropathy [3,44,55]. Methoxyfluorane is a general anesthetic which was formerly widely used. Mild polyuria is very frequent in patients receiving this agent, and a severe polyuria associated with acute renal failure is occasionally observed. Acute oliguria or anuria is another clinical presentation. Recovery is the usual outcome, but persisting or progressive deterioration of the renal function may occur. The early phase is characterized morphologically by extensive deposition of calcium oxalate crystals in the tubules. If the patient develops progressive renal disease, interstitial fibrosis and tubular atrophy may be found.

The acute oliguric or polyuric renal failure may be due to the intratubular crystalline deposition, but the action of fluoride may also be of importance. Oxalate as well as fluoride are metabolites of methoxyfluorane.

Halothane, another general anesthetic, has also been reported to be able to induce acute renal failure, with a renal histology like that reported from methoxyfluorane-induced cases [32]. It may also indirectly generate an acute renal failure of the "hepatorenal" type when halothane-induced hepatitis occurs.

Extensive deposits in tubular and collecting ducts of uric acid crystals associated with acute renal failure may occur in patients with leukemia or malignant lymphomas treated with cytotoxic drugs. It is still uncertain if these crystalline casts are responsible for the renal failure, which may also be due to other mechanisms (e.g., an action of the elevated serum concentration of uric acid on some receptor responsible for the maintenance of normal tubuloglomerular balance).

Reaction to sulfonamides is associated with deposition of crystals in the distal part of the nephron [79]. There is also a severe acute interstitial nephritis (described above), and it may be this interstitial reaction rather than the crystal deposition per se which is the cause of the acute renal failure that may follow sulfonamide administration. Crystal deposition is rare with the newer highly soluble sulfa compounds.

H. Osmotic Nephrosis

Swelling and vacuolization of the proximal tubular cells, a morphologic pattern often termed "osmotic nephrosis," may be produced by sucrose or other hypertonic solutions administered parenterally (glucose, dextran, mannitol, etc.) [31,38,82,137]. Osmotic nephrosis was formerly a very frequent additional feature in biopsies from patients suffering from acute renal failure because hyperosmotic solutions were widely used in the treatment of anuria. During recent years it has become relatively rare, although it is occasionally still observed in patients treated with one of the above-mentioned solutions.

Light microscopically, there is a severe hydropic change and swelling of the proximal epithelium due to accumulation of numerous vacuoles, and the cytoplasm is very faintly stained or watery clear (Figure 7). The epithelial cells are *not* necrotic, the nuclei and the brush border appearing quite normal. The other parts of the nephron are usually spared, but the severity of the alteration is dose-dependent and in patients who have been treated with large doses of the responsible solution, there may also occur some hydropic degeneration in the distal tubules and in the epithelium of Bowman's capsule.

Ultrastructural studies in the rat [137] and in humans [94] have shown that the vacuoles are dilated lysosomes. The other cytoplasmic organelles and the nuclei are normal.

The alteration is reversible, and it is questionable whether it has any harmful effect on renal function [82]. This opinion is not in conflict with the fact that acute oliguric renal

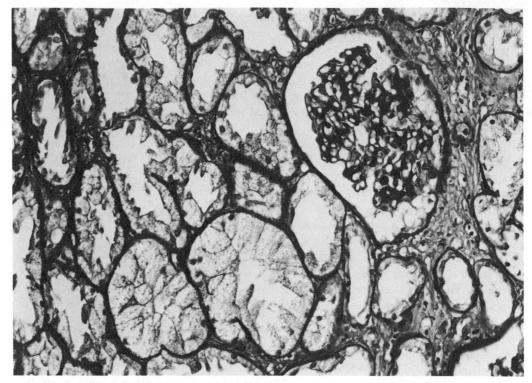

Figure 7 Severe hydropic change of proximal epithelium in a patient treated with infusion of low-molecular-weight dextran. (Periodic acid–Schiff, ×350)

failure may follow infusions of Rheomacrodex (low-molecular-weight dextran 10%, mean MW 40,000), because this complication may be unrelated to the histologic tubular lesion.

I. Neoplasia

Thorotrast, a radioactive contrast medium widely used for angiography from 1930 until 1955 has been demonstrated to give rise to malignant tumors several years after the administration. Most Thorotrast-induced tumors have been situated in the liver. Several cases of kidney tumors (mainly urothelial carcinomas in the renal pelvis) have also been reported [53,140].

Patients who have used the analgesic phenacetin for a long period of time seem to have an increased incidence of carcinoma of the renal pelvis [14].

Myleran and cis-platinum compounds used for leukemia and other malignancies may induce severe dysplasia of the epithelium in the papillary ducts and collecting tubules, but invasive neoplasms have not been reported [52].

J. Miscellaneous Histologic Lesions

Polyvinylpyrrolidone (a polymer of vinylpyrrolidone, MW 10,000–700,000) is used as a vehicle for retarding the resorption of various drugs intended for subcutaneous injection (e.g., *Insipidin retard,* a sustained-release preparation of vasopressin). It is not metabolized in the body, and following long-term administration it is stored in the reticuloendothelial system in many organs [106], including the kidneys, where it can be detected in interstitial macrophages and in mesangial cells. The foreign material is stored as cytoplasmic granules which stain gray-blue with hematoxylin and eosin and brown-red with Congo red (although without green dichroism). It has not been determined whether these deposits have any deleterious action on the renal tissue.

The characteristic lysosomal changes (myeloid bodies) in the proximal tubular epithelium brought about by gentamicin are described in detail below.

K. Acute Renal Failure with Normal Histology

Some drugs may give rise to acute renal failure without any lesions detectable by light microscopy. We have seen this in some cases of acute lithium intoxication and in one case of poisoning with phenindione.

II. Histopathology of Antibiotic Nephrotoxicity

A. Penicillins

The nephrotoxicity of penicillin is low. The few cases of toxicity that have been reported are of two types: (1) an acute generalized hypersensitivity reaction with arthritis and glomerulonephritis, or (2) an acute interstitial nephritis of the methicillin type [5,117]. Whereas recent reports seem to indicate a relatively high incidence of interstitial nephritis following methicillin, it remains the rare exception following penicillin. The nephrotoxic sequelae following treatment with semisynthetic penicillins is dominated by methicillin nephropathy. The hypersensitivity-related acute interstitial nephritis following methicillin was described above. Acute interstitial nephritis has also been reported following administration of ampicillin, oxacillin, and other drugs belonging to this group. Direct dose-related toxicity or tubular necrosis has not been reported following penicillins (natural or semisynthetic) even after high doses or sustained treatment.

B. Aminoglycosides

The morphologic changes in aminoglycoside nephrotoxicity are quite different from the hypersensitivity-induced interstitial nephritis that characterizes penicillin and methicillin nephrotoxicity. The aminoglycosides cause dose-related focal necrosis of the proximal tubules. Interstitial chronic inflammation at the corticomedullary junction occurs in the later stages of the disease, as it does in other types of primary tubular damage, but the picture is not that of a primary interstitial nephritis.

On the other hand, the tubular changes brought about by aminoglycosides are different from the confluent necrosis of lengthy segments of the proximal tubule seen in heavy metal-induced acute renal failure. Both in humans and in experimental animals the tubular necrosis caused by aminoglycosides is focal and patchy except when very high doses (or very sensitive species) are used (Figure 8). Commonly, most tubules show no necrosis. In those tubular cross sections that show changes, some cells are heavily vacuolated, flattened, or missing altogether, whereas others appear relatively normal (Figure 9). Except for the cytoplasmic vacuolization, most cases of gentamicin toxicity show a picture similar to the acute tubular nephropathy observed following shock or crush injury.

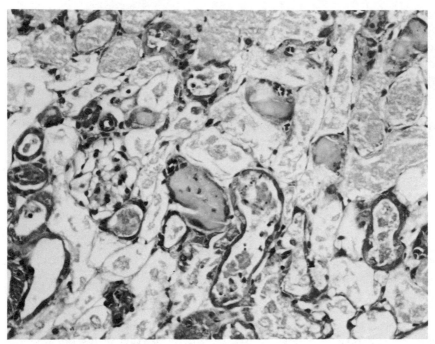

Figure 8   Extensive tubular necrosis in a Fisher 344 rat given gentamicin in a dose of 40 mg/kg (eight times the maximum therapeutic dose in humans) IM for 10 days. This particular rat strain, which is a popular model for gentamicin toxicity, is particularly sensitive to gentamicin. Other strains and other species show only very patchy changes with comparable doses. (Hematoxylin and eosin, ×300)

The erroneous belief that aminoglycosides in therapeutic doses commonly cause confluent necrosis of entire tubular segments in humans has its origin in early autopsy studies of neomycin-induced acute renal failure [102]. Neomycin, discovered in 1949, was the first aminoglycoside antibiotic to be used widely in septic patients. By the end of the 1950s, however, the parenteral use of this agent had been largely abandoned because of reports of severe nephrotoxicity and ototoxicity. Neomycin is poorly absorbed from the gastrointestinal tract and now is used primarily in the management of hepatic encephalopathy and for sterilization of the bowel prior to colonic surgery. Neomycin is also used in irrigating solutions by surgeons and still occasionally causes deafness and acute renal failure when used for this purpose [81]. Because systemic therapy with neomycin was discontinued before the advent of the renal biopsy, only autopsy studies of neomycin-induced renal changes are available. These tend to overestimate the tubular lesions produced. Nuclear pyknosis, nuclear loss, loss of brush border, separation of tubular cells from the basement membrane, and interstitial edema all can be found in normal autopsy kidneys as a result of postmortem autolysis. Evidence for tubular necrosis is thus easily found in postmortem kidneys. Animal studies of neomycin nephrotoxicity show very focal necrotic lesions and vacuolization of proximal tubular cytoplasm similar to that observed in renal biopsies from patients who have been treated with other aminoglycosides (see below) [36].

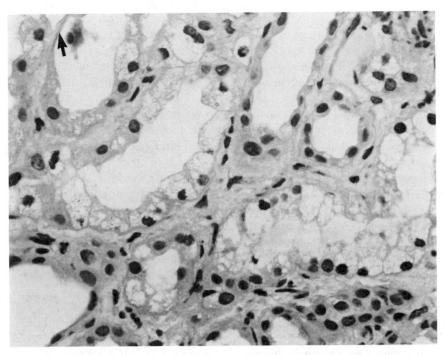

Figure 9   Vacuolation of proximal tubular cells from a relatively pure case of gentamicin toxicity in humans. At the arrow there is an apparently bare segment of tubular basement membrane where an epithelial cell has been sloughed. This is a selected field. Most tubules in the biopsy showed no conspicuous changes except for moderate dilatation. (Hematoxylin and eosin, ×600) (From Ref. 15; slide supplied by Dr. Donald Houghton.)

The aminoglycoside that is most widely used today is gentamicin. The frequency of nephrotoxicity is approximately 10% in patients treated with this drug [125]. Most animal studies of gentamicin nephrotoxicity have employed doses 10-40 times higher than those used in humans. Two studies that have used doses in the therapeutic range (1-5 mg/kg per day) are of particular interest. Kosek et al. showed that ultrastructural changes (formation of "myeloid bodies") occurred in the proximal tubules of rats receiving gentamicin in doses as low as 1 mg/kg [74]. Baylis et al. [10] observed these same ultrastructural changes in the tubules and a 48% decrease in glomerular filtration rate in rats receiving 4 mg/kg. A substantial decrease in the glomerular capillary ultrafiltration coefficient was observed in these rats, even though no morphologic abnormalities were observed in the glomerular capillaries using transmission electron microscopy. The authors state that no tubular necrosis was observed.

In acute renal failure secondary to gentamicin in humans, recovery almost invariably follows discontinuation of drug therapy. Renal biopsy provides neither therapeutic nor prognostic information in a typical case and thus few biopsies are performed. The biopsies that have been done have demonstrated that the presence of myeloid bodies in proximal tubular cells is a marker for gentamicin therapy rather than gentamicin nephrotoxicity. In a study of 109 nephrectomies and renal biopsies performed over a 2 year period at the University of Oregon Health Sciences Center, Houghton et al. [61] found myeloid bodies in the proximal tubules of 19. Fifteen of these were from patients who had received gentamicin therapy. Only one of these had shown clinical signs of nephrotoxicity. None of the patients without myeloid bodies had received gentamicin. The one patient that showed clinical nephrotoxicity showed focal necrosis and desquamation of proximal tubular cells by light microscopy and myeloid bodies in proximal *and distal* tubules by electron microscopy (Figure 10) [15,61].

The nature of the myeloid bodies is unclear. However, most authors agree that they represent modified lysosomes. It has been suggested that they form through a process of focal cytoplasmic degradation and persist because of an interference with the normal ability of lysosomal enzymes, particularly sphingomyelinase, to digest membranes [2,62]. Myeloid bodies are also seen in the kidneys of rats given other polycationic substances, such as neutral red or chloroquine [121]. Two recent autoradiographic studies by Just et al. [66] and Silverblatt and Kuehn [123] suggest that gentamicin is transported into the proximal tubular cell by pinocytosis and becomes sequestered in lysosomes. Houghton et al. have suggested that by accumulating gentamicin lysosomes may protect other cell constituents from the drug's toxic effects [61]. Following this line of reasoning, clinical nephrotoxicity would ensue when the capacity of lysosomes to accumulate gentamicin is overloaded.

Since myeloid bodies are probably present in some proximal tubular cells of all patients who receive gentamicin, the presence of small numbers of myeloid bodies in the urine cannot be used as an indication of gentamicin nephrotoxicity. Any stimulus that increases the rate of turnover of proximal tubular cells will increase the rate of appearance of myeloid bodies in the urine of the gentamicin-treated patient. Similar nonspecificity is found when the presence of brush border enzymes in the urine is used as an early indication of nephrotoxicity. Brush border loss can be demonstrated histologically in animals and humans receiving gentamicin [10,127]. Mondorf et al. have demonstrated a rise in brush border enzymes in the urine of normal human volunteers given therapeutic doses of various aminoglycoside antibiotics (including gentamicin) for 3 days [90]. However, in sick patients a variety of other minor insults to the kidney may produce similar degrees of enzymuria, and the variation in renal functional reserve from patient to patient makes it impossible to predict from patterns of enzyme excretion which gentamicin-treated patients will develop clinical nephrotoxicity. This does not mean that brush border loss is unimportant in the production of clinical acute renal failure. In a recent study of 57 biopsies from patients with acute renal failure, many of whom had received gentamicin, Solez et al. [127] found that thinning or absence of brush border was one of the two histologic changes that was significantly more prominent in biopsies obtained while the patient had renal failure than it was in biopsies obtained after recovery.

Mondorf et al. have demonstrated that there are three principal enzymes in the human proximal tubule brush border [89,91,114]. Alanine aminopeptidase is apparently the most superficially located. This brush border enzyme is the first to appear in the urine after gentamicin therapy. Next most superficial is γ-glutamyltranspeptidase, the enzyme whose presence in the urine has proved useful in the diagnosis of renal allograft rejection. The most deeply situated enzyme in the brush border and the one most closely associated with the basement membrane is alkaline phosphatase. A similar sequence probably occurs in the rabbit, with maltase being the superficially situated enzyme in the tubular brush border and alkaline phosphatase being the deeply situated, tightly bound enzyme. In work reported in abstract form, Solez et al. [130] found a significant negative correlation between renal cortical alkaline phosphatase activity and serum creatinine in rabbits with

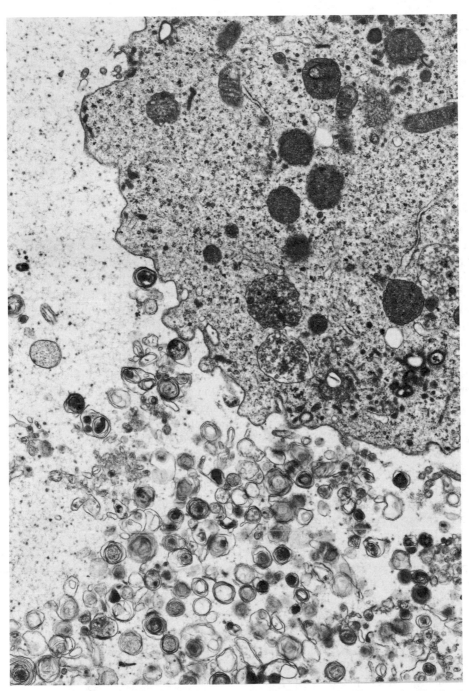

Figure 10 Electron micrograph showing myeloid bodies in the lumen of a distal tubule in a patient with gentamicin-induced acute renal failure. (×24,000) (From Ref. 15; used with permission.)

acute renal failure caused by tobramycin, an aminoglycoside similar to gentamicin. No correlation between renal cortical maltase and serum creatinine was observed. These results suggest that the loss of the more deeply situated components of the tubular brush border may contribute to impairment of renal function in aminoglycoside-induced acute renal failure.

C. Cephalosporins

In discussing the nephrotoxicity of the cephalosporin antibiotics, one must immediately distinguish between nephrotoxicity due to cephaloridine and nephrotoxicity due to cephalothin and the other newer cephalosporins.

Cephaloridine causes dose-related proximal tubular damage similar to that brought about by the aminoglycosides in both humans and experimental animals [40,100,122]. No glomerular changes are observed and although interstitial chronic inflammation may be found secondary to tubular damage, there is no evidence that this drug causes a primary interstitial nephritis. Commercial preparations of cephaloridine contain a mixture of polymers of differing molecular weights. Boyd et al. have shown that those constituents with a molecular weight less than 1000 are more nephrotoxic in rats than higher molecular-weight constituents [17]. Electron microscopy of cephaloridine-induced acute renal failure in rabbits shows loss of apical microvilli from proximal tubular cells and simplification of the lateral interdigitations [120]. Small vesicles are seen surrounding the nucleus, but there are no myeloid bodies. No electron microscopic studies of cephaloridine toxicity in humans have been published. Inhibition of organic anion transport by probenecid in the rabbit prevents uptake of cephaloridine by proximal tubular cells and tubular necrosis [138]. Inhibition of renal cation transport increases the nephrotoxicity of cephaloridine, suggesting that this transport system prevents the accumulation of drug in the kidney and is therefore important in the excretion of the drug [147].

The nephrotoxicity caused by cephalothin and the other new cephalosporins is quite different from that observed with cephaloridine. There is little evidence that cephalothin by itself is directly toxic to the kidney in therapeutic doses [100,122]. Mondorf et al. observed no increased excretion of the brush border enzyme alanine aminopeptidase in human volunteers given therapeutic doses of cephalothin, cephamandole, or cepharolin for 3 days [92]. Barza has recently reviewed the two patterns of nephrotoxicity which are observed with cephalothin [11]. The first pattern is that of non-dose-related allergic or hypersensitivity reaction. In this "hypersensitivity" group the kidney shows a primary interstitial nephritis. Clinically, a rash and eosinophilia is often observed. In the second pattern of nephrotoxicity tubular necrosis is observed, suggesting a direct toxic effect, but in these cases other nephrotoxic drugs (usually gentamicin) have usually been given with cephalothin, or the dose of cephalothin has been inappropriately high for the level of renal function. Barrientos et al. have reported the only relatively pure case of the "toxic" type of cephalothin-induced acute renal failure recorded in the literature in which a renal biopsy was performed [9].

Cephalothin and gentamicin are often used together. The possible synergistic toxicity of these two drugs is a subject of great complexity. Animal studies, summarized by Barza [11] tend to suggest either that cephalothin protects against gentamicin-induced acute renal failure or that there is no interaction. However, most clinical studies suggest that concurrent cephalothin therapy increases the nephrotoxicity of gentamicin [23,41, 71,142]. The histologic picture in these clinical studies has usually been one of acute tubular nephropathy (Figure 11) (see the earlier discussion).

One experimental study of gentamicin-cephalothin combination therapy not cited by Barza [11] is worthy of note. Hautmann et al. [58] studied the nephrotoxicity of these two drugs separately and in combination in rabbits that had undergone unilateral nephrectomy 20 days before. Under these circumstances cephalothin in a dose of 1000 mg/kg per day did not cause renal failure by itself but significantly increased the severity of renal failure caused by gentamicin 5 mg/kg per day. Glomerular changes in the form of mesangial proliferation were described in the cephalothin-treated rabbits in this study. It is of interest that mild mesangial proliferation has been recorded in two clinical studies of cephalothin nephrotoxicity [99,101].

D. Tetracyclines

Tetracyclines have little direct nephrotoxicity. The systemic toxicity of these drugs in patients with preexisting renal impairment [48,146] is beyond the scope of this discussion. Tapp and Lowe [133] reported that necrosis of the proximal tubules in rats could be produced by giving the animals tetracycline (600 mg/kg) and hemoglobin (0.3 g/kg). Neither agent given alone produced similar changes. There have been no reports of tetracycline nephrotoxicity in human patients with hemoglobinuria. Demethylchlortetracy-

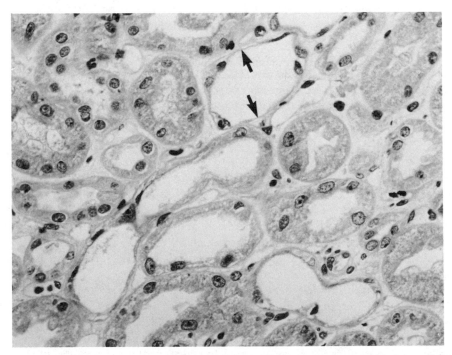

Figure 11   Representative light microscopic field from biopsy of case 3 in reference 71, a 55-year-old woman who developed acute renal failure after treatment with gentamicin and cephalothin and was biopsied on day 26 when serum creatinine was 13 mg/dl. Mild tubular lesions with focal absence of tubular cells (arrows) are seen. (Hematoxylin and eosin, ×480) (Slide supplied by Dr. Dominique Droz.)

cline (demeclocycline) produces a vasopressin-resistant form of nephrogenic diabetes insipidus. The usefulness of this agent in states of water retention is limited by the fact that it causes a natriuresis and reduces glomerular filtration rate [25,33,34,97,98]. No studies of changes in renal morphology produced by this agent have been reported.

A single case of acute interstitial nephritis induced by minocycline has been reported [143]. Immunofluorescence studies in this case showed arteriolar staining for IgA and C_3, staining of glomerular capillary loops for C_3, and staining of the proximal tubular brush border for IgA.

Tetracyclines may reduce the excretion of lithium carbonate when the two agents are given concurrently, leading to signs of lithium toxicity [84] (see earlier).

E. Amphotericin B

Amphotericin B is probably the only widely used antibiotic that causes clinical nephrotoxicity in over half of the patients treated with the drug [75]. Amphotericin appears to damage the kidney through a combination of renal vasoconstriction and direct tubular damage [21,22]. The renal morphologic changes in patients with amphotericin nephrotoxicity consist of extensive calcification of the proximal and distal tubules, sometimes associated with disruption of the tubular basement membrane, focal tubular necrosis, and tubular dilatation [145]. A striking vacuolization of the media of small arteries and arterioles has been described in the kidneys of patients who have received amphotericin [21]. It has been suggested that the vacuolization is related to the intense renal vasoconstriction and increased solute permeability of plasma membranes brought about by this drug. The change is not specific for amphotericin nephrotoxicity. We have observed such vacuolization in biopsies from a number of patients with acute renal failure who did not receive amphotericin (Figure 12). Some of these patients had received gentamicin.

Amphotericin has been reported to cause vasopressin-resistant nephrogenic diabetes insipidus [8]. It has also been suggested that amphotericin and gentamicin have synergistic renal toxicity [27].

F. Rifampicin

Rifampicin nephrotoxicity, like cephalothin nephrotoxicity, is characterized by two different histologic patterns. However, unlike the situation with cephalothin, rifampicin nephrotoxicity does not appear to be dose-related, and both histologic patterns appear to result directly or indirectly from immunologic events. Rifampicin nephrotoxicity appears to occur only when the drug is given intermittently [43,49,110]. In many cases antirifampicin antibodies can be demonstrated in the patients serum.

The first pattern of nephrotoxicity seen with rifampicin is that of a severe systemic reaction to reinstitution of rifampicin therapy, often accompanied by fever, intravascular hemolysis, shock, nausea and vomiting, and diarrhea [24a,26,43,72,88]. In this type of rifampicin toxicity the damage to the kidney is probably secondary to dehydration, shock, and/or hemolysis and is therefore analogous to that seen after incompatible blood transfusions. Pigmented casts, tubular necrosis, interstitial edema, and mild chronic interstitial inflammation are seen on renal biopsy. Patchy cortical necrosis is occasionally observed [28].

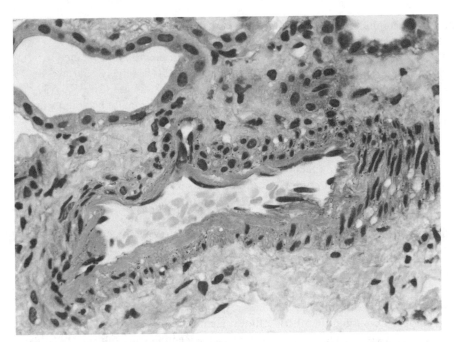

Figure 12   Medial vacuolization in an artery from a renal biopsy from the same patient as in Figure 9. The patient had received gentamicin but not amphotericin. This change is not specific for amphotericin nephrotoxicity. (Hematoxylin and eosin, ×550) (Slide supplied by Dr. Donald Houghton.)

The second pattern of nephrotoxicity seen with rifampicin is an acute interstitial nephritis. Signs of systemic toxicity are often less abrupt and severe than those described in the preceding paragraph. The histologic picture is one of a striking interstitial infiltrate of lymphocytes, plasma cells, eosinophils, and neutrophils, with or without tubular damage [30,46,83,87,104,107]. Sometimes the infiltrate may surround hyalinized arterioles [87]. A mild proliferative glomerulonephritis has been described in some cases [46,87]. In many cases immunofluorescence studies are completely negative. Two cases have been recorded in which IgG and C_3 were demonstrated along tubular basement membranes in a granular [46] or linear [87] pattern, suggesting the possibility of antitubular basement membrane antibody-mediated interstitial nephritis.

In patients with preexisting glomerulonephritis that is being treated with steroids, rifampicin therapy may be associated with another type of nephrotoxicity: rifampicin shortens the half-life of prednisolone and other steroids by inducing hepatic microsomal enzymes that increase the rate of steroid degradation [59].

G.   Polypeptide Antibiotics

Polymyxins are moderately nephrotoxic. Some patients treated with polymyxin B-sulfate suffer from a mild, reversible acute renal failure. The typical lesion is an acute interstitial nephritis which may be of granulomatous character [13]. Blood eosinophilia may be present and eosinophils participate in the interstitial infiltrate. These features suggest a hypersensitivity reaction rather than direct tubular toxicity.

Colistin (= polymyxin E) causes a reversible reduction in creatinine clearance in normal human volunteers [20]. However, clinical nephrotoxicity is largely confined to patients receiving large doses or patients with preexisting renal disease (20,35,37,39, 73,103,105,112]. Renal biopsies or autopsy specimens have occasionally been reported to show no significant morphologic abnormalities [39]. Most reports have described a picture of tubular necrosis [37,73,103,112], although one case is recorded in which an interstitial nephritis with acute and chronic inflammatory cells was found [105].

Another antibiotic in this group, bacitracin, is a potent nephrotoxic agent and for that reason it was withdrawn from parenteral use soon after its discovery. Bacitracin nephrotoxicity is clinically characterized by proteinuria and decreased renal function. Histologically, there is severe tubular degeneration and, in advanced cases, extensive tubular necrosis with subsequent calcification [47]. Its action seems to be a direct dose-related tubulonecrotic one.

H. Bactrim

The nephrotoxicity of the sulfonamides has been reviewed recently by Appel and Neu [1]. The currently popular combination of trimethoprim and sulfamethoxazole, also known as co-trimoxazole, causes a reduction in creatinine clearance in normal subjects, but this may be accounted for in part by the fact that it reduces tubular secretion of creatinine [12,118]. Clinical nephrotoxicity has been described by Kalowski et al. in 16 patients treated with bactrim [67,68]. The renal morphologic picture described was one of "tubular necrosis with associated cellular infiltration" [6768]. It is difficult to judge whether the predominant injury was tubular or interstitial since no light micrographs were included in these publications. Two cases of an acute allergic reaction to bactrim involving the skin, lung, kidney, liver, and nervous system have recently been described [19,42].

Acknowledgment

This work was supported in part by United States Public Health Service Grants HL-07835 and AM26809. Dr. Solez is the recipient of USPHS Research Career Development Award AM00835.

References

1. Appel, G.B., and Neu, H.C.: The nephrotoxicity of antimicrobial agents. *N. Engl. J. Med.* 296:663–670, 722–729, 784–787, 1977.
2. Aubert-Tulkens, G., Van Hoof, F., and Tulkens, P.: Gentamicin-induced lysosomal phospholipidosis in cultured rat fibroblasts: quantitative ultrastructural and biochemical study. *Lab. Invest.* 40:481–491, 1979.
3. Aufderheide, A.C.: Renal tubular calcium oxalate crystal deposition. *Arch. Pathol.* 92:162–166, 1971.
4. Bacon, P.A., Tribe, C.R., Mackenzie, J.C., Verrier Jones, J., Cumming, R.H., and Amer, B.: Penicillamine nephropathy in rheumatoid arthritis. *Q. J. Med.* N.S. 45: 661–684, 1976.
5. Baldwin, D.S., Levine, B.B., McCluskey, R.T., and Gallo, G.R.: Renal failure and interstitial nephritis due to penicillin and methicillin. *N. Engl. J. Med.* 279:1245–1252, 1968.

6. Balslфv, J.T., and Jфrgensen, H.E.: A survey of 499 patients with acute anuric renal insufficiency. *Am. J. Med.* 34:753–764, 1963.

7. Bar-Khayim, Y., Teplitz, C., Garella, S., and Chazan, J.A.: Trimethadione (Tridione(R))-induced nephrotic syndrome. *Am. J. Med.* 54:272–280, 1973.

8. Barbour, G.L., Straub, K.D., O'Neal, B.L., and Leatherman, J.W.: Vasopressin-resistant nephrogenic diabetes insipidus. A result of amphotericin B therapy. *Arch. Intern. Med.* 139:86–88, 1979.

9. Barrientos, A., Bello, I., and Gutierrez-Millet, V.: Renal failure and cephalothin. *Ann. Intern. Med.* 84:612, 1976.

10. Baylis, C., Rennke, H.R., and Brenner, B.M.: Mechanisms of the defect in glomerular ultrafiltration associated with gentamicin administration. *Kidney Int.* 12:344–353, 1977.

11. Barza, M.: The nephrotoxicity of cephalosporins: an overview. *J. Infect. Dis.* 137: S60–S73, 1978.

12. Berglund, F., Killander, J., and Pompeius, R.: Effect of trimethoprimsulfamethoxazole on the renal excretion of creatinine in man. *J. Urol.* 114:802–808, 1975.

13. Beirne, G.J., Hansing, C.E., Octaviano, G.N., and Burns, R.O.: Acute renal failure caused by hypersensitivity to polymyxin B sulfate. *JAMA* 202:156–158, 1967.

14. Bengtsson, U., Angervall, L., Ekman, H., and Lehmann, L.: Transitional cell tumors of the renal pelvis in analgesic abusers. *Scand. J. Urol. Nephrol.* 2:145–150, 1968.

15. Bennett, W.M., Gilbert, D.N., Houghton, D., and Porter, G.A.: Gentamicin nephrotoxicity—morphologic and pharmacologic features. *West. J. Med.* 126:65–68, 1977.

16. Border, W.A., Lehman, D.H., Egan, J.D., Sass, H.J., Glode, J.E., and Wilson, C.B.: Antitubular basement-membrane antibodies in methicillin-associated interstitial nephritis. *N. Engl. J. Med.* 291:381–384, 1974.

17. Boyd, J.F., Butcher, B.T., and Stewart, G.T.: The nephrotoxicity and histology of cephaloridine and its polymers in rats. *Br. J. Exp. Pathol.* 52:503–516, 1971.

18. Brass, II., and Lapp, II.: Nephropathie durch Goldtherapie. Klinische und elektronen-optische Untersuchungen. *Arch. Klin. Med.* 216:371–383, 1969.

19. Brockner, J., and Boisen, E.: Fatal multisystem toxicity after co-trimoxazole. *Lancet* 1:831, 1978.

20. Brumfitt, W., Black, M., and Williams, J.D.: Colistin in *Pseudomonas pyocyanea* infections and its effect on renal function. *Br. J. Urol.* 38:495–500, 1966.

21. Bullock, W.E., Luke, R.G., Nuttall, C.E., and Bhathena, D.: Can mannitol reduce amphotericin B nephrotoxicity? Doubleblind study and description of a new vascular lesion. *Antimicrob. Agents Chemother.* 10:555–563, 1976.

22. Butler, W.T., Hill, G.J., II, Szwed, C.F., and Knight, V.: Amphotericin B renal toxicity in the dog. *J. Pharmacol. Exp. Ther.* 143:47–56, 1964.

23. Cabanillas, F., Burgos, R.C., Rodriquez, R.C., and Baldizon, C.: Nephrotoxicity of combined cephalothin-gentamicin regimen. *Arch. Intern. Med.* 135:850–852, 1975.

24. Cameron, I.L.: Cell renewal in the organs and tissues of the nongrowing adult mouse. *Tex. Rep. Biol. Med.*, 203–248, 1970.

24a. Campese, V.M., Marzullo, F., Schena, F.P., and Coratelli, P.: Acute renal failure during intermittent rifampicin therapy. *Tubercle* 56:191–198, 1975.

25. Carrilho, F., Bosch, J., Arroyo, A., Mas, A., Viver, J., and Rodes, J.: Renal failure associated with demeclocycline in cirrhosis. *Ann. Intern. Med.* 87:195, 1977.

26. Chan, W.C., O'Mahoney, Sister M.G., Yu, D.Y.C., and Yu, R.Y.H.: Renal failure during intermittent rifampicin therapy. *Tubercle* 56:191–1975.

27. Churchill, D.N., and Seely, J.: Nephrotoxicity associated with combined gentamicin-amphotericin B therapy. *Nephron* 19:176–181, 1977.

28. Cochran, M., Moorhead, P.J., and Platts, M.: Permanent renal damage with rifampicin. *Lancet* 1:1428, 1975.

29. Clausen, E., and Jensen, K.: Renal biopsies from patients with a high analgesic intake.

Histological and bacteriological study. *Acta Pathol. Microbiol. Scand.* 72:219–232, 1968.

30. Cordonnier, D., and Muller, J.M.: Acute renal failure after rifampicin. *Lancet* 2: 1364, 1972.

31. Cornu, P., Caulin, C., Marsan, C., Ertel, M., Manicacci, M., Sonigo, R., and Lamotte, M.: La Néphropathie osmotique. Maladie iatrogène spécifique ou réaction aspécifique du tubule? *Sem. Hop. Paris* 49(20):1429–1440, 1973.

32. Cotton, J.R., Jr., Schwartz, M.M., Lindley, J.D., and Hunsicker, L.G.: Acute renal failure following halothane anesthesia. *Arch. Pathol. Lab. Med.* 100:628–629, 1976.

33. Cox, M., Shook, A., and Singer, I.: Demeclocycline-induced azotemia, natriuresis and anti-kaliuresis in congestive heart failure. *Clin. Res.* 27:495A, 1979 (abstr.).

34. Danovitch, G.M., Le Roth, D., and Glick, S.: Renal function during treatment of inappropriate secretion of antidiuretic hormone with demeclocycline. *Isr. J. Med. Sci.* 14:852–857, 1978.

35. Duncan, D.A.: Colistin toxicity. Neuromuscular and renal manifestations. *Minn. Med.* 56:31–35, 1973.

36. Einspruch, B.C., and Gonzalez, V.V.: Clinical and experimental nephropathy resulting from use of neomycin sulfate. *JAMA* 173:809–811, 1960.

37. Elwood, C.M., Lucas, G.D., and Muehrcke, R.C.A.: Acute renal failure associated with sodium colistimethate treatment. *Arch. Intern. Med.* 118:326–334, 1966.

38. Evans, W.A., and Wong, D.: Renal changes following the administration of low molecular weight dextran. *Aust. N.Z. J. Surg.* 63:69–73, 1966.

39. Fekety, F.R., Norman, P.S., and Cluff, L.E.: The treatment of gram-negative bacillary infections with colistin. The toxicity and efficacy of large doses in forty-eight patients. *Ann. Intern. Med.* 57:214–229, 1962.

40. Fillastre, J.P., and Kleinknecht, D.: Acute renal failure associated with cephalosporin therapy. *Am. Heart J.* 89:809–810, 1975.

41. Fillastre, J.P., Laumonier, R., Humbert, G., Dubois, D., Metayer, J., Delpech, A., Leroy, J., and Robert, M.: Acute renal failure associated with combined gentamicin and cephalothin therapy. *Br. Med. J.* 2:396–397, 1973.

42. Finlayson, W.B., and Johnson, G.: Multisystem toxicity after co-trimoxazole. *Lancet* 2:682–683, 1978.

43. Flynn, C.T., Rainford, D.J., and Hope, E.: Acute renal failure and rifampicin: danger of unsuspected intermittent dosage. *Br. Med. J.* 1:482, 1974.

44. Frascino, J.A., Vanamee, P., and Rosen, P.P.: Renal oxalosis and azotemia after methoxyflurane anesthesia. *N. Engl. J. Med.* 283:676–679, 1970.

45. French, A.J.: Hypersensitivity in the pathogenesis of the histopathologic changes associated with sulfonamide chemotherapy. *Am. J. Pathol.* 22:679–701, 1946.

46. Gabow, P.A., Lacher, J.W., and Neff, T.A.: Tubulointestinal and glomerular nephritis associated with rifampin. *JAMA* 235:2517–2518, 1976.

47. Genkins, G., Uhr, J.W., and Bryer, M.S.: Bacitracin nephropathy. Report of a case of acute renal failure and death. *JAMA* 155:894–897, 1954.

48. George, C.R.P., and Evans, R.A.: Tetracycline toxicity in renal failure. *Med. J. Aust.* 1:1271–1273, 1971.

49. Girling, D.J., and Michison, D.A.: Sensitivity to rifampicin. *Br. Med. J.* 3:114, 1974.

50. Glushien, A.S., and Fisher, E.R.: Renal lesions of sulfonamide type after treatment with acetazolamide (Diamox). *JAMA* 160:204–206, 1956.

51. Goldberg, M., and Murray, T.G.: Analgesic-associated nephropathy. An important cause of renal disease in the United States? *N. Engl. J. Med.* 299:716–717, 1978.

52. Gonzalez-Vitale, J.C., Hayes, D.M., Cvitkovic, E., and Sternberg, S.S.: The renal pathology in clinical trials of cis-platinum(II) diamminedichloride. *Cancer* 39:1362–1371, 1977.

53. Griffiths, M.H., Thomas, D.P., Xipell, J.M., and Hope, R.N.: Thorotrast-induced bilateral carcinoma of the kidney. *Pathology* 9:43–48, 1977.

54. Gärtner, H.V., Neild, G.H., Bohle, A., Hallauer, W., Hoppe-Seyler, G., Lüttgen, F.M., and Schollmeyer, P.: Perimembranöse Glomerulonephritis nach Penicillaminetherapie. Bericht über 31 Fälle. *Klin. Wochenschr.* 53:835–837, 1975.

55. Halpren, B.A., Kempson, R.L., and Coplon, N.S.: Interstitial fibrosis and chronic renal failure following methoxyflurane anesthesia. *JAMA* 223:1239–1242, 1973.

56. Hansen, E.S., and Tauris, P.: Methicillin-induced nephropathy. A case with linear deposition of IgG and C_3 on the tubular-basement-membrane. *Acta Pathol. Microbiol. Scand. [A]* 84:440–442, 1976.

57. Harvald, B.: Renal papillary necrosis. A clinical survey of sixty-six cases. *Am. J. Med.* 35:481–486, 1963.

58. Hautmann, R., Kurth, M., Buss, H., and Lutzeyer, W.: Nephrotoxicity of gentamicin-cephalothin combination therapy. *Med. Welt* 28:1617, 1977.

59. Hendrickse, W., and McKiernan, J.: Rifampicin-induced non-responsiveness to corticosteroid treatment in nephrotic syndrome. *Br. Med. J.* 1:306, 1979.

60. Hestbech, J., Hansen, H.E., Amdisen, A., and Olsen, S.: Chronic renal lesions following long-term treatment with lithium. *Kidney Int.* 12:205–213, 1977.

61. Houghton, D.C., Campbell-Boswell, M.V., Bennett, W.M., Porter, G.A., and Brooks, R.E.: Myeloid bodies in the renal tubules of humans: relationship to gentamicin therapy. *Clin. Nephrol.* 10:140–145, 1978.

62. Hruban, Z., Slesers, A., and Hopkins, E.: Drug-induced and naturally occurring myeloid bodies. *Lab. Invest.* 27:62–70, 1972.

63. Hubner, K.: Rhythmic growth and its regulation factors. *Tokai J. Exp. Clin. Med.* 2:77–89, 1977.

64. Jaffe, I.A., Treser, G., Suzuki, Y., and Ehrenreich, T.: Nephropathy induced by D-penicillamine. *Ann. Intern. Med.* 69:549–556, 1968.

65. Jensen, H. AErenlund, Halveg, A.B., and Saunamäki, K.I.: Permanent impairment of renal function after methicillin nephropathy. *Br. Med. J.* 4:406–408, 1971.

66. Just, M., Erdmann, G., and Habermann, E.: The renal handling of polybasic drugs. 1. Gentamicin and aprotinin in intact animals. *Naunyn Schmiedebergs Arch. Pharmacol.* 300:57–66, 1977.

67. Kalowski, S., Nanra, R.S., Mathew, T.H., and Kincaid-Smith, P.: *Prog. Biochem. Pharmacol.* 9:129–140, 1974.

68. Kalowski, S., Nanra, R.S., Mathew, T.H., and Kincaid-Smith, P.: Deterioration in renal function in association with co-trimaxazole therapy. *Lancet* 1:394–397, 1973.

69. Kibukamusoke, J.W., Davies, D.R., and Hutt, M.S.R.: Membranous nephropathy due to skin-lightening cream. *Br. Med. J.* 2:646–647, 1974.

70. Kincaid-Smith, P.: Analgesic nephropathy. *Kidney Int.* 13:1, 1978.

71. Kleinknecht, D., Ganeval, D., and Droz, D.: Acute renal failure after high doses of gentamicin and cephalothin. *Lancet* 1:1129, 1973.

72. Kleinknecht, D., Homberg, J.C., and Decrois, G.: Acute renal failure after rifampicin. *Lancet* 1:1238–1239, 1972.

73. Koch-Weser, J., Sidel, V.W., Federman, E.B., Kanarek, P., Finer, D.C., and Eaton, A.E.: Adverse effects of sodium colistimethate. Manifestations and specific reaction rates during 317 courses of therapy. *Ann. Intern. Med.* 72:857–868, 1970.

74. Kosek, J.C., Mazze, R.I., and Cousins, M.J.: Nephrotoxicity of gentamicin. *Lab. Invest.* 30:48–57, 1974.

75. Kovnat, P., Labovitz, E., and Levision, S.P.: Antibiotics and the kidney. *Med. Clin. North Am.* 57:1045–1063, 1973.

76. Lachmann, P.J.: Nephrotic syndrome from penicillamine. *Postgrad. Med. J.* 44(Suppl.): 23–27, 1968.

77. Lange, K.: Nephropathy induced by D-penicillamine. *Contrib. Nephrol.* 10:63–74, 1978.

78. Lee, J.C., Dushkin, M., Eyring, E.J., Engleman, E.P., and Hopper, J.: Renal lesions associated with gold therapy. Light and electron microscopic studies. *Arthritis Rheum.* 8:1–13, 1965.

79. Lehr, D., and Antopol., W.: Specific morphology of crystals appearing in the urine during administration of sulfanilamide derivatives. *Am. J. Clin. Pathol.* 12:200–209, 1942.

80. Lyons, H., Pinn, V.W., Cortell, S., Cohen, J.J., and Harrington, J.T.: Allergic interstitial nephritis causing reversible renal failure in four patients with idiopathic nephrotic syndrome. *N. Engl. J. Med.* 288:124–128, 1973.

81. Masur, H., Whelton, P.K., and Whelton, A.: Neomycin toxicity revisited. *Arch. Surg.* 111:822–825, 1976.

82. Matheson, N.A.: Renal effects of low molecular weight dextran. *Monogr. Surg. Sci.* 3:303–364, 1966.

83. Mattson, K., Riska, H., Forsstrom, J., and Kock, B.: Acute renal failure following rifampicin administration. *Scand. J. Resp. Dis.* 55:291–297, 1974.

84. McGennis, A.J.: Lithium carbonate and tetracycline interaction. *Br. Med. J.* 1:1183, 1978.

85. Méry, J.-P., and Morel-Maroger, L.: Acute interstitial nephritis. A hypersensitivity reaction to drugs. *Proceedings of the 6th International Congress of Nephrology, Florence, 1975.* Basel, Karger, 1976, pp. 524–529.

86. Michielsen, P., and De Schepper, P.J.: Renal diseases due to drugs. In *Drug-Induced Diseases,* Vol. 4, edited by Meyler, L., and Peck, H.M. Amsterdam, Excerpta Medica, 1972, p. 261.

87. Minetti, L., di Belgioioso, G.B., and Busnach, G.: Immunohistological diagnosis of drug-induced hypersensitivity nephritis. *Contrib. Nephrol.* 10:15–29, 1978.

88. Mohring, K., Asbach, H.W., Schubothe, H., and Weber, S.: Hamolytische Krise mit akutem Nierenversagen unter Rifampicin-behandlung. *Dtsch. Med. Wochenschr.* 99:1458–1462, 1974.

89. Mondorf, A.W.: Die Bedeutung lokalisationsspezifischer Enzyme in Harn für toxikologische und klinisch-pharmakologische Untersuchungen. *Arzneim. Forsch.* 23:1634–1637, 1973.

90. Mondorf, A.W., Breier, J., Hendus, J., Scherberich, J.E., Mackenrodt, G., Shah, P.M., Stille, W., and Schoeppe, W.: Effect of aminoglycosides on proximal tubular membranes of the human kidney. *Eur. J. Clin. Pharmacol.* 13:133–142, 1978.

91. Mondorf, A.W., Kinne, R., Scherberich, J.E., and Falkenberg, F.: Isolierung, enzymatische und immunologische Charakterisierung einer Plasmamembranfraktion vom proximalen Tubulus der Menschliche Niere. *Clin. Chim. Acta* 37:25–32, 1972.

92. Mondorf, A.W., Zegelman, M., Klose, J., Maske, L., Scherberich, J.E., Stefanescu, T., Muller, H., and Schoeppe, W.: Effects of various cephalosporins on the proximal tubule of the human kidney. *Eur. J. Clin. Pharmacol.* 13:357–363, 1978.

93. Murphy, F.D., Kuzma, J.F., Polley, T.Z., and Grill, J.: Clinicopathologic studies of renal damage due to sulfonamide compounds. A report of fourteen cases. *Arch. Intern. Med.* 73:433–443, 1944.

94. Olsen, S.: Ultrastructure of the renal tubules in acute renal insufficiency. *Acta Pathol. Microbiol. Scand.* 71:203–218, 1967.

95. Olsen, S.: Renal histopathology in various forms of acute anuria in man. *Kidney Int.* 10:S2–S8, 1976.

96. Olsen, S., and Asklund, M.: Interstitial nephritis with acute renal failure following cardiac surgery and treatment with methicillin. *Acta Med. Scand.* 199:305–310, 1976.

97. Oster, J.R., Epstein, M., and Ulano, H.B.: Deterioration of renal function with demeclocycline administration. *Curr. Ther. Res.* 20:794–801, 1976.

98. Padfield, P.L., Hodsman, G.P., and Morton, J.J.: Demeclocycline in the treatment of the syndrome of inappropriate antidiuretic hormone release: with measurement of plasma ADH. *Postgrad. Med. J.* 54:623–627, 1978.

99. Pasternak, D.P., and Stephens, B.G.: Reversible nephrotoxicity associated with cephalothin therapy. *Arch. Intern. Med.* 135:599–602, 1975.

100. Perkins, R.L., Apicella, M.A., Lee, I.-S., Cuppage, F.E., and Saslaw, S.: Cephaloridine and cephalothin: comparative studies of potential nephrotoxicity. *J. Lab. Clin. Med.* 71:75–84, 1968.

101. Pickering, M.J., Spooner, G.R., de Quesada, A., and Cade, J.R.: Declining renal function associated with administration of cephalothin. *South. Med. J.* 63:426–428, 1970.

102. Powell, L.W., Jr., and Hooker, J.W.: Neomycin nephropathy. *JAMA* 160:557–560, 1956.

103. Price, D.J.E., and Graham, D.I.: Effects of large doses of colistin sulphomethate sodium on renal function. *Br. Med. J.* 4:525–527, 1970.

104. Ramgopal, V., Leonard, C., and Bhathena, D.: Acute renal failure associated with rifampicin. *Lancet* 1:1195–1196, 1973.

105. Randall, R.E., Jr., Bridi, G.S., Setter, J.G., and Brackett, N.C., Jr.: Recovery from colistimethate nephrotoxicity. *Ann. Intern. Med.* 73:491–492, 1970.

106. Reske-Nielsen, E., Bojsen-Møller, M., Vetner, M., and Hansen, J.C.: Polyvinyl-pyrrolidone-storage disease. Light microscopical, ultrastructural and chemical verification. *Acta Pathol. Microbiol. Scand. [A]* 84:397–405, 1976.

107. Riska, H., Mattson, K., Kock, B., and Forsstrom, J.: Acute renal failure subsequent to the administration of rifampicin. *Scand. J. Resp. Dis.* 57:183–189, 1976.

108. Roman-Franco, A.A., Turellio, M., Albini, B., Ossi, E., Millgrom, F., and Andres, G.A.: Anti-basement membrane antibodies and antigen–antibody complexes in rabbits injected with mercuric chloride. *Clin. Immunol. Immunopathol.* 9:464, 1978.

109. Rosenthal, T., and Boichis, H.: Nephrotoxicity of cephaloridine. *Br. Med. J.* 4:115, 1971.

110. Rothwell, D.L., and Richmond, D.E.: Hepatorenal failure with self-initiated intermittent rifampicin therapy. *Br. Med. J.* 1:481, 1974.

111. Ruley, E.J., and Lisi, L.M.: Interstitial nephritis and renal failure due to ampicillin. *J. Pediatr.* 84:878–881, 1974.

112. Ryan, K.J., Schainuck, L.I., Hickman, R.O., and Striker, G.E.: Colistimethate toxicity. Report of a fatal case in a previously healthy child. *JAMA* 207:2099–2101, 1969.

113. Sanjad, S.A., Haddad, G.G., and Nassar, V.H.: Nephropathy, an underestimated complication of methicillin therapy. *J. Pediatr.* 84:873–877, 1974.

114. Scherberich, J.E., Faldenberg, F.W., Mondorf, A.W., Muller, H., and Pfleiderer, G.: Biochemical and immunological studies on isolated brush border membranes of human kidney cortex and their membrane surface proteins. *Clin. Chim. Acta* 55:179–197, 1974.

115. Schreiner, G.E., and Maher, J.F.: Toxic nephropathy. *Am. J. Med.* 38:409–449, 1965.

116. Schreiner, G.E., Maher, J.F., and Golden, A.: Nephrotoxicity of analgesic abuse. *Proceedings of the 2nd International Congress of Nephrology*, 1963. Amsterdam, Excerpta Medica, 1964, pp. 789–792.

117. Schrier, R.W., Bulger, R.J., and Van Arsdel, P.P., Jr.: Nephropathy associated with penicillin and homologues. *Ann. Intern. Med.* 64:116–127, 1966.

118. Shouval, D., Ligumsky, M., and Ben-Ishay, D.: Effect of co-trimoxazole on normal creatinine clearance. *Lancet* 1:244–245, 1978.
119. Silverberg, D.S., Kidd, E.G., Shnitka, T.K., and Ulan, R.A.: Gold nephropathy. A clinical and pathologic study. *Arthritis Rheum.* 13:812–825, 1970.
120. Silverblatt, F., Turck, M., and Bulger, R.: Nephrotoxicity due to cephaloridine: a light- and electron-microscopic study in rabbits. *J. Infect. Dis.* 122:33–44, 1970.
121. Silverblatt, F.J.: Antibiotic nephrotoxicity: a review of pathogenesis and prevention. *Urol. Clin. North Am.* 2:557–567, 1975.
122. Silverblatt, F., Harrison, W.O., and Turck, M.: Nephrotoxicity of cephalosporin antibiotics in experimental animals. *J. Infect. Dis.* 128:S367–S372, 1973.
123. Silverblatt, F.J., and Kuehn, D.: Autoradiography of gentamicin uptake by the rat proximal tubule cell. *Kidney Int.* 15:335–345, 1979.
124. Skrifvars, B. V., Tornroth, R. S., and Tallqvist, G. N.: Gold-induced immune complex nephritis in seronegative rheumatoid arthritis. *Ann. Rheum. Dis.* 36:549–556, 1977.
125. Smith, C.R., Baughman, K.L., Edwards, C.Q., Rogers, J.F., and Lietman, P.S.: Controlled comparison of amikacin and gentamicin. *N. Engl. J. Med.* 296:349–353, 1977.
126. Solez, K., Ideura, T., and Saito, H.: Role of thromboxane and outer medullar microvascular injury in post-ischemic acute renal failure. *Clin. Res.* 28:461A, 1980.
127. Solez, K., Morel-Maroger, L., and Sraer, J.-D.: The morphology of "acute tubular necrosis" in man: analysis of 57 renal biopsies and comparison with the glycerol model. *Medicine* 58:362–376, 1979.
128. Solez, K., Racusen, L.C., and Whelton, A.: Glomerular epithelial cell in early post-ischemic acute renal failure in rabbits and man. *Am. J. Path.* 103:163–173, 1981.
129. Solez, K., Silvia, C.B., Ideura, T., Hamilton, B., and Saito, H.: Clonidine after renal ischemia to lessen acute renal failure and microvascular injury. *Kidney Int.* 18:309–322, 1980.
130. Solez, K., Silvia, C.B., and Whelton, A.: Tubular brush border enzymes, tubular casts, and tobramycin-freamine II nephrotoxicity. *Kidney Int.* 14:719, 1978 (abstr.).
131. Spühler, O., and Zollinger, H.V.: Die chronisch-interstitielle Nephritis. *Z. Klin. Med.* 151:1, 1953.
132. Strunk, S.W., and Ziff, M.: Ultrastructural studies of the passage of gold thiomalate across the renal glomerular capillary wall. *Arthritis Rheum.* 13:39–52, 1970.
133. Tapp, E., and Lowe, B.: Tetracycline toxicity in haemoglobinuria. *Br. Med. J.* 1:143–144, 1966.
134. Takacs, F.J., Tomkiewicz, Z.M., and Merrill, J.P.: Amphotericin B nephrotoxicity with irreversible renal failure. *Ann. Intern. Med.* 59:716–724, 1963.
135. Tannenberg, A.M., Wicher, K.J., and Rose, N.R.: Ampicillin nephropathy. *JAMA* 218:449, 1971.
136. Thoenes, W., Thoenes, G., and Ansorge, R.: Drug-induced lupus nephritis. Light-, electron- and immufluorescence-microscopic investigations on kidney biopsies. *Verh. Dtsch. Ges. Pathol.* 56:346–352, 1972.
137. Trump, B.F., and Janigan, D.T.: The pathogenesis of cytologic vacuolization in sucrose nephrosis. An electron microscopic and histochemical study. *Lab. Invest.* 11:395–411, 1962.
138. Tune, B.M.: Relationship between the transport and toxicity of cephalosporins in the kidney. *J. Infect. Dis.* 132:189–194, 1975.
139. Velentzas, C., Oreopoulos, D.G., Brandes, L., Wilson, D.R., and Marquez-Julio, A.: Allopurinol-induced interstitial nephritis. *Ann. Intern. Med.* 86:196–198, 1977.

140. Verhaak, R.L.O.M., Harmsen, A.E., and van Unnik, A.J.M.: On the frequency of tumour induction in a thorotrast kidney. *Cancer* 34:2061–2068, 1974.
141. Viol, G.W., Minielly, J.A., and Bistricki, T.: Gold nephropathy. Tissue analysis by X-ray fluorescent spectroscopy. *Arch Pathol. Lab. Med.* 101:635–640, 1977.
142. Wade, J.C., Petty, B.G., Conrad, G., Smith, C.R., Lipsky, J.J., Ellner, J., and Lietman, P.S.: Cephalothin plus an aminoglycoside is more nephrotoxic than methicillin plus an aminoglycoside. *Lancet* 2:604–606, 1978.
143. Walker, R.B., Thomson, N.M., Dowling, J.P., and Ogg, C.S.: Minocycline-induced acute interstitial nephritis. *Br. Med. J.* 1:524, 1979.
144. Weening, J.J., Fleuren, G.J., and Hoedemaeker, J.: Demonstration of antinuclear antibodies in mercuric chloride-induced glomerulopathy in the rat. *Lab. Invest.* 39:405–411, 1978.
145. Wertlake, P.T., Butler, W.T., Hill, G.J., II, and Utz, J.P.: Nephrotoxic tubular damage and calcium deposition following amphotericin B therapy. *Am. J. Pathol.* 43:449–457, 1963.
146. Whelton, A.: Tetracyclines in renal insufficiency: resolution of a therapeutic dilemma. *Bull. N.Y. Acad. Med.* 54:223–236, 1978.
147. Wold, J.S., Turnispeed, S.A., and Miller, B.L.: The effect of renal cation transport inhibition on cephaloridine nephrotoxicity. *Toxicol. Appl. Pharmacol.* 47:115–122, 1979.
148. Zollinger, H.V.: Interstitial nephritis. In *The Kidney,* edited by Mostofi, F.K., and Smith, D.E. Baltimore, Md., Williams & Wilkins, 1966, pp. 269–281.

16

CEPHALOSPORIN AND AMINOGLYCOSIDE INTERACTIONS: CLINICAL AND TOXICOLOGIC IMPLICATIONS

FRIEDRICH C. LUFT Indiana University School of Medicine, Indianapolis, Indiana

The potential nephrotoxicity of the aminoglycoside antibiotics is well established [2]. In several earlier chapters in this volume, the features and proposed mechanisms of this nephrotoxicity are discussed in detail. The effect of the cephalosporin antibiotics on renal function and structure has been the topic of an excellent recent review by Barza [5]. With one notable exception, this class of β-latam antibiotics causes renal damage only rarely in humans [29]. The cephalosporins resemble the penicillins. As organic anions, they are actively secreted into the lumen of the proximal tubule by the proximal tubular cells. Cephalosporins are actively transported from the capillaries into the cells of the renal cortex via an anionic transport system. The drugs then diffuse down their respective concentration gradients into the tubular lumen.

Cephaloridine is the only consistently nephrotoxic cephalosporin. Its effects upon the kidney have been studied in great detail in experimental animals [17,64]. The diffusion of this drug down its concentration gradient into the tubular lumen is relatively impaired perhaps because of precipitation or polymerization of the drug within the tubular cells [12]. Cephaloridine accumulates within the proximal tubular cells and causes tubular damage proportional to the degree of accumulation within the cells [52,64]. The role of the anionic transport system in the transport of cephaloridine is supported by the observation that cephaloridine nephrotoxicity can be ameliorated by the prior administration of probenecid [65]. In addition, Wold et al. [70] have shown that in newborn rabbits, which are ordinarily resistant to the toxic effects of cephaloridine [28], the susceptibility to nephrotoxicity parallels the maturation of the renal anionic transport system. Substrate stimulation of the anionic transport system by p-aminohippurate or penicillin increases the nephrotoxicity of cephaloridine in newborn rabbits. The nephrotoxicity of cephaloridine in humans [7] and in experimental animals [52] is a dose-related phenomenon. Cefazolin resembles cephaloridine in its effect upon the nephron; however, it is only one-third to one-fourth as toxic [9,59,71].

In contrast to cephaloridine and cefazolin, cephalothin, cephalexin, cefamandole, and cephapirin are only minimally nephrotoxic [43,59,66]. A dose-related nephrotoxic response relationship for these drugs is difficult to establish. Isolated reports of cephalothin toxicity in humans suggest that at least some of the cases are the result of an allergic interstitial nephritis, as observed with methicillin, rather than a direct toxic effect on renal tubular epithelium [5].

A number of agents in combination with the aminoglycoside gentamicin have been reported to result in increased nephrotoxicity. These agents include methicillin [72], methoxyflurane [4], amphotericin B [16], cis-platinum [20], clindamycin [14], and furosemide [49], as well as cephaloridine and cephalothin [5,29]. In addition, the state of sodium balance and the presence of acidosis have been shown to modulate the nephrotoxic potential of gentamicin in experimental animals [8,38]. The nature of the clinical situations in which aminoglycosides are administered, the presence of coexisting disease, other drugs, or the infectious process itself greatly complicate the interpretation of any resulting nephrotoxicity.

Barza [5] has emphasized that since the combination of cephalosporins and aminoglycosides provides excellent antibacterial activity in vivo [31,39], and may provide synergism for certain infections [1,49], the question of whether or not such combinations involve increased risk of renal injury assumes significant clinical importance. In addition to numerous case reports and brief clinical series, several prospective randomized trials, one of which was also conducted in a double-blind fashion, suggest that in humans the combination of aminoglycoside plus cephalothin is more nephrotoxic than aminoglycoside plus a penicillin derivative. Studies in experimental animals have produced conflicting results and in general have failed to verify the human experience. It is the purpose of the present discussion to review briefly the previously reported human and animal studies. Unfortunately, a reconciliation of all the reported findings cannot be made.

I. Observations in Humans

The widespread use of the combination of cephalothin and gentamicin has been followed by numerous reports suggesting an unexpectedly high rate of nephrotoxicity in patients receiving this combination. In 1971, Opitz et al. [51] reported that 10 of 14 patients with bronchopneumonia receiving the combination of cephalothin and gentamicin developed renal insufficiency, compared to none of 30 patients receiving either agent alone. Subsequently, isolated case reports from Sweden [24], the United States [11,57], Puerto Rico [15], France [26,32,40], and New Zealand [3], some but not all of which were documented histologically, have also implicated the combination of cephalothin and gentamicin as being particularly nephrotoxic. Moreover, this combination has recently been implicated in the development of the Fanconi syndrome [58].

Plager [53] found a high incidence of renal tubular damage among 119 patients with malignant disease, 69 of whom died following recent treatment with the combination of cephalothin and gentamicin. All had normal blood urea nitrogen (BUN) values prior to therapy. In this series, tubular abnormalities were found in 33 (48%) patients who received that combination prior to death, as opposed to only 3% in a control population. Seven of the 33 patients, who showed either hydropic degeneration or acute tubular necrosis, had normal BUN values within the last 3 days of life. Seventeen patients had increasing BUN values (52–98 mg/dl). Death in these 17 patients was attributed to complications of their underlying disease. Sixteen patients showed initial evidence of renal insufficiency during the period of cephalothin-gentamicin administration and died in severe renal failure. Changes of tubular injury at postmortem examination correlated with the clinical evidence of renal failure. In addition, Plager observed that patients receiving the combination appeared particularly susceptible to other nephrotoxic insults such as blood loss or ongoing bacterial infection.

Hansen and Kaaber [34] treated 26 patients with 32 courses of cephalothin-gentamicin combination therapy and observed an increase in serum creatinine during six courses of treatment. A transient decrease in serum potassium occurred in 11 patients. Burck and Sörgel [13] examined the records of 12,000 patients, 355 of whom received 418 courses of gentamicin and cephalothin. The combined application of cephalothin and gentamicin was implicated in the development of acute renal failure in 14% of cases. The acute renal failure was generally reversible. The authors concluded that the combination should be administered only when clearly superior to other regimens and that close monitoring of renal function was mandatory. The studies previously discussed are retrospective surveys. Interpretation must be limited because of the nature of the controls provided. Some of the surveys failed to provide control groups.

Three prospective randomized trials have been performed addressing the question of possible enhanced risk from nephrotoxicity when cephalothin and an aminoglycoside are given in combination. All three yielded affirmative answers. Klastersky and associates [39] randomly administered the combinations of cephalothin-tobramycin, cephalothin-ticarcillin, and ticarcillin-tobramycin to 186 patients with cancer and suspected life-threatening bacterial infection. Although the three combinations were similarly effective, the administration of the cephalothin-tobramycin combination was associated with a significantly higher frequency of nephrotoxicity than were the other two regimens.

The European Organization for Research on Treatment of Cancer reported the results of a prospective randomized trial of three antibiotic combinations: carbenicillin plus cephalothin, carbenicillin plus gentamicin, and cephalothin plus gentamicin [22]. These regimens were employed in 625 trials with granulocytopenic cancer patients suspected of having gram-negative bacteremia. Although the combinations were equally effective in the treatment of infection, the combination of cephalothin and gentamicin was found to be substantially more nephrotoxic than the other regimens. Severe renal dysfunction was observed in 12% of patients receiving that combination, whereas renal dysfunction occurred in only 4% of patients receiving carbenicillin plus cephalothin and 2% of patients receiving carbenicillin plus gentamicin. Older patients, whose creatinine levels were initially in the high normal range, exhibited a 26% incidence of renal dysfunction related to treatment with cephalothin plus gentamicin.

Wade et al. [67] performed a prospective randomized, double-blind trial to determine if cephalothin plus an aminoglycoside is more nephrotoxic than methicillin plus an aminoglycoside. Patients were assigned to one of four treatment groups: cephalothin and gentamicin, cephalothin and tobramycin, methicillin and gentamicin, or methicillin and tobramycin. Nephrotoxicity was defined as a rise in serum creatinine of over 0.4 mg/dl if the initial level was less than 3.0 mg/dl, or a rise of over 0.9 mg/dl if the initial creatinine was 3.0 mg/dl or above. Definite nephrotoxicity was assumed when no other cause for decreasing renal function could be identified in the 72 hr before the rise in serum creatinine. There was no significant difference in nephrotoxicity between the combined gentamicin and tobramycin groups. However, definite nephrotoxicity developed in 12 of 47

(26%) of the combined cephalothin groups and in only 3 of 43 (7%) of the combined methicillin groups ($P < 0.05$).

Although the studies described above support a strong indictment against the combination of cephalothin and an aminoglycoside, other studies have reached opposite conclusions. When Fanning et al. [25] reported the findings of the Boston Collaborative Drug Surveillance Program, they were unable to document an increase in nephrotoxicity when gentamicin and cephalothin were given in combination, compared to either drug alone. They reviewed the records of 1073 patients, 334 of whom received gentamicin, 492 of whom received cephalothin, and 247 of whom received the combination. The unadjusted frequency in rise of BUN was 6.6% with gentamicin, 2.0% with cephalothin, and 9.3% with the combination. Increases in the BUN values were evaluated only if the attending physician felt that antibiotics were responsible for their elevation. The mean increase in BUN was 59 mg/dl with gentamicin, 76 mg/dl with cephalothin, and 46 mg/dl with the combination. The increase was not correlated with age, sex, first-discharge diagnosis, diuretic administration, or drug dosage, either total or daily; however, it was correlated with mortality. Patients who died had a frequency of rising BUN values of 13.6%, as opposed to a frequency of 4.9% in survivors. When the statistical analysis was adjusted for mortality, no significant difference in the frequency of rising BUN between groups receiving gentamicin alone or the combination could be shown.

Stille and Arndt [60] reviewed their results obtained from 74 immunosuppressed patients who received 85 courses of the combination cephalothin plus gentamicin. Renal insufficiency implicating the combination was identified in 6 patients, all in whom septic shock may have also played a role. They concluded that, in contrast to the finding of Opitz et al. [51], an added nephrotoxicity engendered by the combination could not be documented. Wellwood and colleagues [68] assessed renal damage by measuring urinary enzyme excretion in 36 patients who received 53 courses of antimicrobial treatment. Only gentamicin caused enzymuria. The enzymuria observed in patients receiving cephalothin and gentamicin was similar to those receiving only gentamicin. Mondorf et al. [47] observed the effect of aminoglycosides, cephalosporins, and their combination on the urinary excretion of alanine aminopeptidase in healthy adults. They found that the aminoglycosides all caused increased excretion of this enzyme. Cephalothin and cefazolin given immediately prior to gentamicin prevented enzymuria.

The interpretation of studies examining the possibility of an enhanced nephrotoxicity with the combination cephalothin plus aminoglycoside is complicated by the recent observation that cephalosporin antibiotics interfere with the Jaffé reaction, which is universally used in the determination of creatinine in both manual and automated techniques. Swain and Briggs [62] found that cephaloglycin, cephalothin, cephaloridine, cefoxitin, and cephacetrile all reacted with alkaline picrate solution to give a creatinine-like response. Carbenicillin and methicillin specifically do not interfere with the Jaffé reaction (R. Swain, personal communication). Rankin et al. [55] examined the potential implication of these findings in human volunteers with normal renal function. Creatinine concentrations in urine and plasma were measured by two automated techniques, the Programachem P1040 method and the AutoAnalyzer technique. Glomerular filtration rate was monitored by the clearance of [^{99m}Tc]pentetic acid (DTPA). Following a 4 g intravenous dose of cephalothin, the plasma creatinine concentration was significantly elevated as performed by the Programachem technique. Plasma creatinine increased from 1.0 ± 0.1 to 1.26 ± 0.15 mg/dl. Urine creatinine excretion was significantly elevated by either method. The clearance of [^{99m}Tc]DTPA was unchanged. By either method, plasma and urine creatinine concentrations were directly correlated with plasma and urinary cephalothin values (r = 0.35, P < 0.05).

Whether or not these in vitro and in vivo findings have particular relevance to studies indicating that the combination of cephalothin and aminoglycoside is nephrotoxic is unclear. The study by Klastersky et al. [39] and the EORTC study [22] each included groups receiving cephalothin combined with another β-lactam antibiotic. This combination did not lead to an increased incidence of diminished renal function as determined by plasma creatinine concentrations in either study. In addition, Plager based his conclusions on histological evidence rather than on renal function studies [53]. Nevertheless, the fact that cephalothin can interfere with the Jaffé reaction underscores the care that must be taken in interpreting the studies that have appeared in the literature.

II. Observations in Experimental Animals

In order to clarify the implication that cephalosporin-aminoglycoside combinations are particularly nephrotoxic, investigators have turned to animal models. Such models present a number of obstacles in scientific investigations. First, it is by no means certain that human kidneys behave similarly to those of any particular animal species with respect to susceptibility to nephrotoxic agents. For instance, it is quite well established that the rat, rabbit, and dog, all commonly employed as animal models, differ in their susceptibility to aminoglycoside nephrotoxicity [30,45,63]. Even within animal species, differences may exist in susceptibility to renal injury. The Wistar rat, studied by Flandre and Damon [27], appeared considerably more resistant to the nephrotoxic effects of gentamicin than the Fischer rat employed by other investigators [41]. Even within the same strain, differences may be encountered. We have observed considerable variability in susceptibility of the Sprague-Dawley rat to the nephrotoxic effects of gentamicin, when switching from the colony of one commercial supplier to another.

Another problem is the question of dose. Drugs are conveniently administered on the basis of weight; however, glomerular filtration rate correlates more realistically with total body surface area than with weight. A 200 g rat has only 1/350 the mass of its prototype 70 kg human counterpart, yet it has approximately 1/70 the surface area [21]. Its glomerular filtration rate also approximates 1/70 that of humans. Dosage interval provides an additional obstacle. Although animals are conveniently dosed once or twice daily, patients are given drugs to assure adequate concentrations, which generally requires more frequent drug administration. It is therefore apparent that dosage interval is quite important in the study of antibiotic nephrotoxicity. Whereas aminoglycosides are more toxic in experimental animals when given in smaller divided daily doses [54] or when infused continuously [63], cephaloridine appears more toxic when given as a single daily injection [5]. Yet another problem arises because of possible interactions between certain agents. It is well established that β-lactam antibiotics may inactivate aminoglycosides under certain circumstances in vivo and in vitro [23,50]. An additional consideration is the complexity of the acute renal failure model provided by the aminoglycosides. In many models of acute renal failure, those involving ischemia or those in which specific tubular toxins such as mercuric chloride or uranyl nitrate are given, the renal failure occurs abruptly. The administration of aminoglycosides, on the other hand, causes a relatively gradual acute renal failure which becomes established over a period of days to weeks [45]. As tubular necrosis progresses, active regeneration occurs, and the animal may recover normal renal function even with continued administration of the drug [37,46].

The course of acute renal failure may be modulated by the state of sodium balance, as well as by the administration of agents that promote an osmotic diuresis [42]. Since such influences may be imposed by the administration of a β-lactam antibiotic concurrently with an aminoglycoside, the interpretation of results obtained under such circumstances becomes highly complicated.

The results of animal studies examining nephrotoxicity following the concurrent administration of β-lactam antibiotics and aminoglycosides are outlined in Table 1. Hautmann et al. [36] studied the effects of the cephalothin-gentamicin combination on rabbits. The doses outlined on the table were given as two equally divided intramuscular (IM) injections daily over 14 days. Urinary output, plasma urea, and plasma creatinine concentrations were monitored and renal tissue was obtained prior to and at the end of the experimental period. The authors' data indicate that the addition of cephalothin in increasing doses resulted in increased nephrotoxicity as compared to animals that received only gentamicin. Interpretation of their data is complicated by the fact that conventional statistical analysis was not applied, and by the fact that the authors do not provide a discrete grading scale of the pathologic changes observed.

Harrison et al. [35] studied the effects of increasing doses of cephaloridine, cephalothin, cefazolin, and gentamicin, either alone or in combination, on the renal morphology of Sprague-Dawley rats. None of the combination regimens produced more renal injury than did gentamicin alone. Plasma constituents reflecting glomerular filtration rate were not examined by these investigators.

Hagstrom et al. [33] gave Sprague-Dawley rats netilmicin 30 or 60 mg/kg per day for 14 days either separately or combined with cefamandole 400 mg/kg per day, a variety of penicillins, or clindamycin. None of the combinations resulted in increased nephrotoxicity.

Dellinger et al. [19] studied the effects of 6, 12, 25, or 50 mg/kg per day gentamicin either alone or combined with 200, 400, or 800 mg/kg per day cephalothin, on the renal function and structure of the Fischer 344 rat. They found that the administration of the two drugs simultaneously resulted in a significant protective effect of cephalothin against gentamicin-related nephrotoxicity. They also observed that an equiosmolar amount of sodium sulfate provided a similar degree of protection and suggested that the phenomenon might be related to the presence of a nonresorbable anion in the urine. In additional studies [18], these same investigators noted that renal cortical concentrations of gentamicin were significantly lower in rats given gentamicin and cephalothin simultaneously than in animals given gentamicin alone. They suggested that the reduction in nephrotoxicity observed in their initial studies may have been related to a lower cortical concentration of gentamicin in rats receiving cephalothin.

That cephalosporins given to rats receiving gentamicin could ameliorate gentamicin-induced nephrotoxicity was confirmed by observations from our laboratory [44] and by Sugarman and associates [61], who also found that an equivalent improvement was provided by either sodium chloride or aminohippurate. Subsequently, Barza and associates [6] have studied the comparative nephrotoxicity of newer cephalosporins and aminoglycosides alone and in combination in the Fischer 344 rat. Cephalothin, cefamandole, and cefazolin were given concomitantly with gentamicin, tobramycin, netilmicin, or amikacin. Their results confirmed their previous observations with cephalothin and gentamicin [19], and extended their findings to other cephalosporin and aminoglycoside congeners. In addition, they observed that cortical concentrations of aminoglycosides were generally decreased when cephalosporins were given concomitantly.

Roos and Jackson [56] gave rats gentamicin 12 mg/kg alone, concomitantly with

Table 1 Nephrotoxicity of Cephalosporins Administered with Aminoglycosides to Animals

Animal	Cephalosporin (dose)	Aminoglycoside (dose)	Result	Reference
Rabbits	Cephalothin (100, 500, 1000 mg/kg per day)	Gentamicin (5, 25 mg/kg per day)	Synergistic nephrotoxicity (function, histology)	62
Rats	Cephaloridine (500, 1100 mg/kg per day) Cephalothin (1100 mg/kg per day) Cefazolin (1100 mg/kg per day)	Gentamicin (15, 20, 60, 120 mg/kg per day)	Cephalosporins failed to potentiate renal injury	63
Rats	Cefamandole (500 mg/kg per day)	Netilmicin (30, 60 mg/kg per day)	Cefamandole failed to potentiate renal injury	64
Rats	Cephalothin (200, 400, 800 mg/kg per day)	Gentamicin (6, 12, 25, 50 mg/kg per day)	Simultaneous administration had protective effect; separation of doses: no interaction	65
Rats	Cephalothin (400 mg/kg)	Gentamicin (12 mg/kg)	Concomitant cephalothin reduced renal cortical concentration of gentamicin	66
Rats	Cephalothin (100 mg/kg every 4 hr) Cefazolin (20, 50 mg/kg every 4 hr) Cephaloridine (50 mg/kg every 4 hr)	Gentamicin (5 mg/kg every 4 hr)	Cefazolin 50 mg/kg and cephaloridine had protective effect	67
Rats	Cephalothin (500 mg/kg per day)	Gentamicin (10, 100 mg/kg per day)	Simultaneous administration had protective effect	68
Rats	Cephalothin (400 mg/kg per day) Cefamandole (400 mg/kg per day) Cefazolin (200 mg/kg per day)	Gentamicin (25 mg/kg per day) Tobramycin (25 mg/kg per day) Netilmicin (25 mg/kg per day) Amikacin (25 mg/kg per day)	Protective effect; greatest reduction in creatinine observed with gentamicin and tobramycin	69
Rats	Cephalothin (200, 400, 800 mg/kg per day)	Gentamicin (12 mg/kg per day)	Dose-dependent protective effect; greatest when cephalothin was given before or concomitant with gentamicin	70
Rats	Cephalothin (100, 500, 1000 mg/kg per day)	Gentamicin (60 mg/kg per day)	Dose-dependent protective effect; also observed with carbenicillin at a lower dose	71

393

cephalothin 200, 400, or 800 mg/kg per day, or with sodium chloride 32, 64, or 129 mg/kg per day. Cephalothin 800 mg/kg per day was also given either 4 hr before or 4 hr after gentamicin 12 mg/kg per day. The experiments were conducted for either 6 or 15 days. The investigators monitored the renal concentrations of gentamicin at sacrifice as well as the total excretion of gentamicin in the urine. Histologic injury was computed via a discrete scoring system. Sodium chloride provided no protection. The protective effect of cephalothin was dose-dependent and required a molar ratio of cephalothin to gentamicin of at least 500:1 when the drugs were given concomitantly. The renal cortical concentration of gentamicin was reduced by concomitant cephalothin administration, but not by sodium chloride. The protective effect of cephalothin was enhanced when it was given 4 hr prior to gentamicin, and the urinary excretion of the drug was significantly increased as well. However, under these circumstances the cortical concentration of gentamicin was not reduced. Cephalothin given after gentamicin was also protective; however, neither the cortical concentration of gentamicin nor its urinary excretion was significantly altered. The authors suggested that the cephalothin protective effect may in part be related to a cephalothin-induced increase in gentamicin excretion.

An additional study from our laboratory [10] examined the possibility that a non-cephalosporin β-lactam antibiotic, carbenicillin, may also ameliorate gentamicin-induced nephrotoxicity in the Sprague-Dawley rat. Studies in humans have compared the cephalothin-gentamicin combination to gentamicin in combination with a penicillin derivative rather than to gentamicin alone. The possibility thus existed that cephalothin was not necessarily contributing an additional nephrotoxic insult, but rather that β-lactam antibiotics provided variable degrees of protection from gentamicin nephrotoxicity. Cephalothin was given at 100, 500, or 1000 mg/kg per day. Carbenicillin was given at 50, 100, 250, 500, or 1000 mg/kg per day. The gentamicin dose was 60 mg/kg per day. Five percent glucose solution was used as a diluent. The drugs were given concomitantly, but were injected at separate sites for 14 days. According to creatinine clearance measurements, cephalothin provided protection at 500 mg/kg per day, whereas carbenicillin provided protection at 100 mg/kg per day. Increasing the doses of either drug failed to provide additional protection. Amelioration of the histological damage did not occur until cephalothin 1000 mg/kg per day or carbenicillin 500 mg/kg per day were given. Although the doses of cephalothin and carbenicillin employed in the study were extremely large when compared to humans on a weight basis, the findings may have clinical relevance. Maximum doses of carbenicillin and cephalothin in humans approach 30 and 16 g/day, respectively. A 30 g/day carbenicillin dose in humans provides 167 mg of carbenicillin per liter of glomerular filtrate per day, assuming a normal glomerular filtration rate of 125 ml/min. A 16 g/day cephalothin dose provides 88 mg of cephalothin per liter of glomerular filtrate. In the rat, protection was observed at a dose of 8.3 mg of carbenicillin per liter of filtrate per day and 42 mg of cephalothin per liter of filtrate per day. When compared on the basis of glomerular filtration rate, doses of carbenicillin and cephalothin employed in humans assume a magnitude that could conceivably modulate the effects of gentamicin upon the kidney. Since we could not demonstrate that the β-lactam antibiotics employed in our study lowered the renal parenchymal concentration of gentamicin, we were unable to confirm the observations of previous investigators suggesting that the protective effect is in part related to an enhanced excretion of gentamicin. It is possible that the bioassay employed in our study failed to provide the necessary sensitivity to show subtle changes in renal parenchymal gentamicin concentrations. The use of radiolabeled aminoglycoside would circumvent such problems and would provide a direct assessment of aminoglycoside concentrations and excretion.

Whelton and associates [69] performed a systematic series of studies to identify active secretory or reabsorptive transport pathways that might play a role in the production of high intracellular aminoglycoside concentrations in the proximal tubular epithelium of the canine renal cortex. Quinine, a potent competitor for base transport, failed to significantly influence renal cortical tobramycin accumulation. Cephalothin, which competes for the acid secretory pathway of the renal proximal tubule, also failed to influence the renal cortical concentration of tobramycin in 12 acute experiments. A solution of mixed amino acids was used to inhibit the active proximal tubular absorption of amino acids. At a relatively high rate of infusion, the mixed amino acid solution caused a significant reduction in renal cortical tobramycin uptake.

An additional study pertinent to the question of a possible influence by cephalosporins on the renal cortical accumulation of aminoglycosides was performed by Morin et al. [48]. These investigators observed that aminoglycoside antibiotics affect rat renal-cell lysosomes in vitro. The toxic effect of the aminoglycosides was assessed by incubating each drug at concentrations from 5 to 50 μg/ml with lysosomal suspensions, and measuring the release of N-acetyl-β-D-glucosaminidase after 60 min of incubation at 37°C. The addition of cefoxitin, cephaloridine, or cefazolin inhibited N-acetyl-β-D-glucosaminidase release to a significant degree in lysosomal suspensions containing 50 μg/ml gentamicin, sisomicin, netilmicin, or amikacin. No protective effect was observed for kanamycin or streptomycin. In additional studies, the investigators gave rats ^{14}C-labeled gentamicin 50 mg/kg alone, or in combination with cefazolin 500 mg/kg. The animals were sacrificed at 6 and 24 hr after injection. Both regimens resulted in high renal cortical concentrations of gentamicin at 6 and 24 hr; however, the concentrations in animals receiving cefazolin were not significantly different from those receiving gentamicin alone. The authors concluded that cephalosporins probably do not interfere with the renal concentration and distribution of gentamicin.

III. Conclusions

Whereas the study by Hautmann et al. [36] provides evidence that the combination of cephalothin and gentamicin is more toxic than either drug alone in the rabbit, studies in rats suggest either no added toxicity [33,35] or demonstrate an ameliorative effect [18, 19,44,56,61]. The relevance of these findings to the human situation remains unclear. Species differences may account for the discrepancy between the results obtained in rabbits and rats. Additional studies in rabbits and other experimental animal models are necessary to clarify the issue. Although the animal studies have not verified the results reported in many of the human studies, they have offered additional insight into the mechanism of aminoglycoside nephrotoxicity and drug interaction within the renal cortex. Particularly intriguing are the findings of Dellinger [18], Barza [6], and Roos and Jackson [56] indicating that cephalosporins lower the concentration of aminoglycoside within the renal cortex and may facilitate urinary aminoglycoside excretion. This area warrants particular additional investigation.

Although the information obtained in humans remains controversial, three prospective studies, one of which was conducted in a double-blind fashion, indicate that the combination of cephalothin plus aminoglycoside is more nephrotoxic than penicillin derivatives plus aminoglycoside. The data from human studies thus far address only cephalothin. Whether or not other cephalosporins should engender similar concern is presently unknown. The prudent physician should consider the data at hand. It would appear reasonable to reserve the combination cephalothin plus aminoglycoside to those circumstances in which it provides a particular therapeutic advantage.

References

1. Anderson, E.T., Young, L.S., and Hewitt, W.L.: Antimicrobial synergism in the therapy of gram-negative rod bacteremia. *Chemotherapy* 24:45–54, 1978.
2. Apell, G.B., and Neu, H.C.: The nephrotoxicity of antimicrobial agents. *N. Engl. J. Med.* 296:663–670, 1970.
3. Bailey, R.R.: Renal failure in combined gentamicin and cephalothin therapy [letter to the editor]. *Br. Med. J.* 2:777, 1973.
4. Barr, G.A., Mazze, R.I., Cousins, M.J., and Kosek, J.C.: An animal model for combined methoxyflurane and gentamicin nephrotoxicity. *Br. J. Anaesth.* 45:306–312, 1973.
5. Barza, M.: The nephrotoxicity of cephalosporins: an overview. *J. Infect. Dis.* 137 (Suppl.):S60–S73, 1978.
6. Barza, M., Pinn, V., Tanguay, P., and Murray, T.: Comparative nephrotoxicity of newer cephalosporins and aminoglycosides alone and in combination in a rat model. In *Current Chemotherapy*. Washington, D.C., American Society for Microbiology, 1978, pp. 964–966.
7. Benner, E.J.: Renal damage associated with prolonged administration of ampicillin, cephaloridine, and cephalothin. *Antimicrob. Agents Chemother.*, 417–420, 1969.
8. Bennett, W.M., Hartnett, M.N., Gilbert, D., Houghton, D., and Porter, G.A.: Effect of sodium intake on gentamicin nephrotoxicity in the rat. *Proc. Soc. Exp. Biol. Med.* 151:736–738, 1976.
9. Birkhead, H.A., Briggs, G.B., and Saunders, L.Q.: Toxicology of cefazolin in animals. *J. Infect. Dis.* 128(Suppl.):S379–S381, 1973.
10. Bloch, R., Luft, F.C., Rankin, L.I., Sloan, R.S., Yum, M.N., and Maxwell, D.R.: Protection from gentamicin nephrotoxicity by cephalothin and carbenicillin. *Antimicrob. Agents Chemother.* 15:46–49, 1979.
11. Bobrow, S.N., Jaffe, E., and Young, R.C.: Anuria and acute tubular necrosis associated with gentamicin and cephalothin. *JAMA* 222:1546–1547, 1972.
12. Boyd, J.F., Butcher, B.T., and Stewart, G.T.: The nephrotoxic effect of cephaloridine and its polymers. *Int. J. Clin. Pharmacol.* 7:307–315, 1973.
13. Burck, H.C., and Sörgel, G.: Nephrotoxicity of the combined application of cephalothin and gentamicin. In *The Proceedings of the 6th International Congress of Nephrology*. Florence, International Congress of Nephrology, 1975, Abstr. 780.
14. Butkus, D.E., de Torrente, A., and Terman, D.S.: Renal failure following gentamicin in combination with clindamycin. *Nephron* 17:307–313, 1976.
15. Cabanillas, F., Burgos, R.C., Rodriguez, R.C., and Baldizon, C.: Nephrotoxicity of combined cephalothin-gentamicin regimen. *Arch. Intern. Med.* 135:850–852, 1975.
16. Churchill, D.N., and Seely, J.: Nephrotoxicity associated with combined gentamicin-amphotericin B therapy. *Nephron* 19:176–181, 1977.
17. Currie, G.A., Little, P.J., and McDonald, S.J.: The localization of cephaloridine and nitrofuranotin in the kidney. *Nephron* 3:282–288, 1966.
18. Dellinger, P., Murphy, T., Barza, M., Pinn, V., and Weinstein, L.: Effects of cephalothin on renal cortical concentrations of gentamicin in rats. *Antimicrob. Agents Chemother.* 9:587–588, 1976.

19. Dellinger, P., Murphy, T., Pinn, V., Barza, M., and Weinstein, L.: The protective effect of cephalothin against gentamicin-induced nephrotoxicity in rats. *Antimicrob. Agents Chemother.* 9:172–178, 1976.

20. Dentino, M.E., Luft, F.C., Yum, M.N., and Einhorn, L.H.: Long term effect of cis-diamminedichloride platinum on renal function and structure in man. *Cancer* 41: 1274–1281, 1978.

21. Diack, S.L.: The determination of the surface area of the white rat. *J. Nutr.* 3:289–296, 1930.

22. The EORTC International Antimicrobial Therapy Project Group: Three antibiotic regimens in the treatment of infection in febrile granulocytopenic patients with cancer. *J. Infect. Dis.* 137:14–29, 1978.

23. Ervin, F.R., Bullock, W.E., Jr., and Nuttall, C.E.: Inactivation of gentamicin by penicillins in patients with renal failure. *Antimicrob. Agents Chemother.* 9:1004–1011, 1976.

24. Ewald, U., and Jernelius, H.: Njurinsufficiens vid behandling med antibiotikombinationen cefalothin-gentamicin. *Lakartidningen* 71:2595–2597, 1974.

25. Fanning, W.L., Gump, D., and Jick, H.: Gentamicin- and cephalothin-associated rises in blood urea nitrogen. *Antimicrob. Agents Chemother.* 10:80–82, 1976.

26. Fillastre, J.-P., Laumonier, R., Humbert, G., and Dubois, D.: Insuffisances rénales aiguës après administration de céphalotine et de gentamycine. *Br. Med. J.* 2:396–397, 1973.

27. Flandre, O., and Damon, M.: Experimental study of the nephrotoxicity of gentamicin in rats. In *Gentamicin, First International Symposium.* Basel, Schwabe & Co., 1967, pp. 47–61.

28. Fleming, P.C., and Jaffé, D.: The nephrotoxic effect of cephaloridine. *Postgrad. Med. J.* 43:89–90, 1967.

29. Foord, R.D.: Cephaloridine, cephalothin and the kidney. *J. Antimicrob. Chemother.* 1(Suppl.):S119–S133, 1975.

30. Frame, P.T., Phair, J.P., Watanakunakorn, C., and Bannister, T.W.P.: Pharmacologic factors associated with gentamicin nephrotoxicity in rabbits. *J. Infect. Dis.* 135: 952–956, 1977.

31. Gaya, H., Klastersky, J., and Schimpff, S.C.: Nephrotoxicity during combined therapy with gentamicin and cephalothin. In *Program and Abstracts of the 15th Interscience Conference on Antimicrobial Agents and Chemotherapy.* Washington, D.C., American Society for Microbiology, 1975, Abstr. 401.

32. Giudicelli, C.I., Allard, P., Kumaludin, T., and Masbernard, A.: Insuffisance rénale aiguë due a l'association céphalothine-gentamycin. *Lyon Med* 231:1171–1177, 1974.

33. Hagstrom, G.L., Luft, F.C., Yum, M.N., Sloan, R.S., and Maxwell, D.R.: Nephrotoxicity of netilmicin in combination with nonaminoglycoside antibiotics. *Antimicrob. Agents Chemother.* 13:490–493, 1978.

34. Hansen, M.M., and Kaaber, K.: Nephrotoxicity in combined cephalothin and gentamicin therapy. *Acta Med. Scand.* 201:463–467, 1977.

35. Harrison, W.O., Silverblatt, F.J., and Turck, M.: Gentamicin nephrotoxicity: failure of three cephalosporins to potentiate injury in rats. *Antimicrob. Agents Chemother.* 8:209–215, 1975.

36. Hautmann, R., Kurth, M., Buss, H., and Lutzeyer, W.: Zur Nephrotoxizität der Gentamycin-Cephalotin Kombinationstherapie. *Med. Welt* 28:1617–1620, 1977.

37. Houghton, D.C., Plamp, C.E., DeFehr, J.M., Bennett, W.M., Porter, G., and Gilbert, D.: Gentamicin and tobramycin nephrotoxicity. *Am. J. Pathol.* 93:137–152, 1978.

38. Hsu, C.H., Kurtz, T.W., Easterling, R.E., and Weller, J.M.: Potentiation of gentamicin

nephrotoxicity by metabolic acidosis. *Proc. Soc. Exp. Biol. Med.* 146:894–897, 1974.

39. Klastersky, J., Hensgens, C., and Debusscher, L.: Empiric therapy for cancer patients: comparative study of ticarcillin-tobramycin, ticarcillin-cephalothin, and cephalothin-tobramycin. *Antimicrob. Agents Chemother.* 7:640–645, 1975.

40. Kleinknecht, D., Ganeval, D., and Droz, D.: ARF after high doses of gentamicin and cephalothin. *Lancet* 1:1129, 1973.

41. Kosek, J.C., Mazze, R.I., and Cousins, M.J.: Nephrotoxicity of gentamicin. *Lab. Invest.* 30:48–57, 1974.

42. Levinsky, N.G., and Alexander, E.A.: Acute renal failure. In *The Kidney,* edited by Brenner, B., and Rector, F. Philadelphia, W.B. Saunders, 1976, p. 806.

43. Linton, A.L., Bailey, R.R., and Turnbull, D.I.: Relative nephrotoxicity of cephalosporin antibiotics in an animal model. *Can. Med. Assoc. J.* 107:414–416, 1972.

44. Luft, F.C., Patel, V., Yum, M.N., and Kleit, S.A.: The nephrotoxicity of cephalosporin-gentamicin combinations in rats. *Antimicrob. Agents Chemother.* 9:831–839, 1976.

45. Luft, F.C., Patel, V., Yum, M.N., Patel, B., and Kleit, S.A.: Experimental aminoglycoside nephrotoxicity. *J. Lab. Clin. Med.* 86:213–220, 1975.

46. Luft, F.C., Rankin, L.I., Sloan, R.S., and Yum, M.N.: Recovery during aminoglycoside administration. *Antimicrob. Agents Chemother.* 14:284–287, 1978.

47. Mondorf, A.W., Klose, J., Breier, J., Hendus, J., and Schoeppe, W.: The effect of cephalosporins and aminoglycosides on proximal tubular membrane of human kidney. In *Program and Abstracts of the 17th Interscience Conference on Antimicrobial Agents and Chemotherapy.* Washington, D.C., American Society for Microbiology, 1977, Abstr. 201.

48. Morin, J.P., Fillastre, J.-P., and Vallant, R.: Prediction of aminoglycoside-cephalosporin nephrotoxicity. In *Current Chemotherapy,* Vol. 2. Washington, D.C., American Society for Microbiology, 1978, pp. 960–962.

49. Noël, P., and Levy, V.G.: Toxicité rénale de l'association gentamicin-furosémide. *Nouv. Presse Med.* 7:351–353, 1978.

50. Noone, P., and Pattison, J.R.: Therapeutic implications of interaction of gentamicin and penicillins. *Lancet* 2:575–578, 1971.

51. Opitz, A., Herrmann, I., Herrath, D.V., and Schaefer, K.: Akute Niereninsuffizienz nach Gentamycin-Cephalosporin-Kombinationstherapie. *Med. Welt* 22:434–438, 1971.

52. Perkins, R.L., Apicella, M.A., Lee, I.S., Cuppage, F.E., and Saslaw, S.: Cephaloridine and cephalothin: comparative studies of potential nephrotoxicity. *J. Lab. Clin. Med.* 71:75–84, 1968.

53. Plager, J.E.: Association of renal injury with combined cephalothin-gentamicin therapy among patients severely ill with malignant disease. *Cancer* 37:1937–1943, 1976.

54. Plamp, C., Bennett, W., Gilbert, D., and Porter, G.: The effect of dosage regimen on experimental gentamicin nephrotoxicity: dissociation of peak serum levels from renal failure. *Clin. Res.* 26:125A, 1978.

55. Rankin, L.I., Swain, R.R., and Luft, F.C.: The effect of cephalothin on creatinine measurements in man. *Clin. Res.* 26:652A, 1978.

56. Roos, R., and Jackson, G.G.: Protective effect of cephalothin on gentamicin nephrotoxicity: effect of cephalothin anion, not sodium cation. In *Current Chemotherapy.* Washington, D.C., American Society for Microbiology, 1978, pp. 962–964.

57. Schultze, R.G., Winters, R.E., and Kauffman, H.: Possible nephrotoxicity of gentamicin. *J. Infect. Dis.* 124(Suppl.):S145–S147, 1971.

58. Schwartz, J.H., and Schein, P.: Fanconi syndrome associated with cephalothin and gentamicin nephrotoxicity. *Cancer* 41:769–772, 1978.

59. Silverblatt, F., Harrison, W.E., and Turck, M.: Nephrotoxicity of cephalosporin antibiotics in experimental animals. *J. Infect. Dis.* 128(Suppl.):S367–S372, 1973.

60. Stille, W., and Arndt, I.: Argumente gegen eine Nephrotoxizität von Cephalothin und Gentamycin. *Med. Welt* 23:1603–1605, 1972.

61. Sugarman, A., Brown, R.S., and Rosen, S.: Gentamicin nephrotoxicity and the beneficial effect of simultaneous administration of cephalothin or other sodium salts. *Proc. Am. Soc. Nephrol.* 10:80A, 1976.

62. Swain, R.R., and Briggs, S.L.: Positive interference with the Jaffe reaction by cephalosporin antibiotics. *Clin. Chem.* 23:1340–1342, 1977.

63. Thompson, W.L., Reiner, N.R., Bloxham, D.D.: Gentamicin and tobramycin nephrotoxicity in dogs given continuous or once daily intravenous injections. In *Current Chemotherapy*, Vol. 2. Washington, D.C., American Society for Microbiology, 1978, pp. 941–942.

64. Tune, B.M.: Relationship between the transport and toxicity of cephalosporins in the kidney. *J. Infect. Dis.* 132:189–194, 1975.

65. Tune, B.M., and Kempson, R.L.: Nephrotoxic drugs [letter to the editor]. *Br. Med. J.* 3:635, 1973.

66. Venuto, R.C., Stein, J.H., and Ferris, T.F.: Failure to demonstrate nephrotoxicity of cephalothin in rabbits with reduced renal function. *Proc. Soc. Exp. Biol. Med.* 139:1065–1067, 1972.

67. Wade, J.C., Smith, C.R., Petty, B.G., Lipsky, J.J., Conrad, G., Ellner, J., and Lietman, P.S.: Cephalothin plus an aminoglycoside is more nephrotoxic than methicillin plus an aminoglycoside. *Lancet* 3:604–606, 1978.

68. Wellwood, J.M., Simpson, P.M., Tighe, J.R., and Thompson, A.E.: Evidence of gentamicin nephrotoxicity in patients with renal allografts. *Br. Med. J.* 3:278–281, 1975.

69. Whelton, A., Carter, G.G., Bryant, H.H., Cody, T.S., Craig, T.J., and Walker, W.G.: Tobramycin and gentamicin intrarenal kinetic comparisons: therapeutic and toxicologic answers. In *Current Chemotherapy*, Vol. 2. Washington, D.C., American Society for Microbiology, 1978, pp. 951–953.

70. Wold, J.S., Joost, R.R., and Owen, N.V.: Nephrotoxicity of cephaloridine in newborn rabbits: rate of the renal anionic transport system. *J. Pharmacol. Exp. Ther.* 201:778–785, 1977.

71. Wold, J.S., Joost, R.R., and Owen, N.V.: The nephrotoxicity of cephaloridine, cefazolin, and mercuric chloride in the newborn rabbit. *Toxicol. Appl. Pharmacol.* 37:103(Abstr. 27), 1976.

72. Yver, L., Becq-Giraudan, B., Pourrat, O., and Sudre, Y.: La Néphrotoxicité de l'association méthicilline-gentamycine. *Sem. Hop. Paris* 52:1903–1907, 1976.

17

RISK FACTORS FOR AMINOGLYCOSIDE NEPHROTOXICITY

THOMAS R. MOENCH   The Johns Hopkins Hospital, Baltimore, Maryland

CRAIG R. SMITH   The Johns Hopkins University School of Medicine, Baltimore, Maryland

Aminoglycosides are frequently used to treat patients with serious gram-negative infections. Because nephrotoxicity may limit the clinical utility of aminoglycosides [3,13,40,68], recognition of factors associated with a greater incidence of renal damage is important. These risk factors may include characteristics of the patient, or characteristics of the drug and its administration (Table 1).

Information about potential risk factors is derived from studies in experimental animals, case reports, and prospective clinical trials of aminoglycosides. Because species variability in susceptibility to nephrotoxicity is great, the results from studies done in

Table 1   Potential Risk Factors for Aminoglycoside Toxicity

Patient factors	Drug factors
Age	Dose
Sex	Duration of therapy
Race	Frequency of administration
Prior renal insufficiency	Serum levels
Sepsis	Prior aminoglycoside exposure
Acidosis	Prior aminoglycoside toxicity
Sodium volume depletion	Tissue binding
Potassium depletion	Drug interactions
Renal transplants	

experimental animals should be confirmed in clinical trials. However, studies in experimental animals are useful to identify potentially significant factors.

The methodology used in clinical trials to gather information about potential risk factors is important. For example, the frequency of observed nephrotoxicity will vary inversely with the amount of renal dysfunction required to define nephrotoxicity. The interval at which measurements of renal function are made may be important since it is possible that more toxicity will be found if it is looked for more often. Other important points include the need for a definition of (1) the patient population studied, (2) the dose and means of dose adjustment, and (3) the control group used for comparison. Because of the multiple potential variables, conclusions drawn from comparisons of various studies are difficult. The purpose of this chapter is to examine currently available studies providing information about risk factors for nephrotoxicity and to suggest areas for further research.

I. Patient Factors

A. Age

Advanced age has often been suggested as a risk factor for aminoglycoside nephrotoxicity. Gary et al. [27] reported a retrospective analysis of five patients with gentamicin nephrotoxicity, all of whom were greater than 45 years old. However, no control group was reported. Since older people are more likely to have disease states requiring aminoglycoside antibiotics, the lack of a control group may bias the conclusion. The EORTC group [21] reported increased nephrotoxicity with age greater than 40 in a subgroup of patients treated with gentamicin and cephalothin. However, drug dosage was fixed. Thus, older patients with decreased creatinine clearance (Ccr) due to age or renal disease may have been relatively overdosed. A large retrospective review of amikacin treatment found patients with nephrotoxicity were significantly older (mean age 55 years vs. 42 years) than those without nephrotoxicity. However, the method of adjusting dosage for renal failure was not stated and aminoglycoside levels were not monitored [43]. On the contrary, a larger retrospective study of 3506 tobramycin courses found no significant relationship between nephrotoxicity and age [6]. Several other studies [13,44,60] all using aminoglycoside serum levels to guide dosage and two of them prospective [44,60], have found no significant relationship between nephrotoxicity and age.

Use of an incremental rise in the serum creatinine (Scr) to define nephrotoxicity may bias studies toward finding age as a risk factor. Because the relationship between Scr and creatinine clearance (Ccr) is not linear [14], a renal insult will result in a much larger increase in Scr if renal function is already impaired than if it is normal. In fact, a Ccr of 120 ml/min may fall as much as 30% without a detectable change in Scr [19]. Because older patients have a lower Ccr for each value of serum creatinine, they will be more likely to show an increase in Scr after a renal insult than will younger patients. Varying the increment Scr required to qualify as nephrotoxicity according to the magnitude of initial Scr (e.g., $\geqslant$0.5 mg/dl if initial Scr $<$ 3, and $\geqslant$1.0 mg/dl if initial Scr $\geqslant$ 3.0) is of some help but only partially corrects this bias. In addition, the difference in the relationship of Scr and Ccr with age may result in overdosing of aged patients if the serum creatinine is used to calculate the dose.

We conclude there is no convincing evidence that age is a risk factor for nephrotoxicity. Findings to the contrary are probably explained by (1) failure to use an appropriate control group, (2) inappropriate dosing of older patients based on Scr and failure to measure serum aminoglycoside levels, or (3) bias in ascribing nephrotoxicity caused by use of Scr rise rather than percent change of Ccr as the definition of toxicity.

B. Race and Sex

Race and sex have not been frequently suggested as risk factors for aminoglycoside nephrotoxicity. Keating et al. [39] reported nephrotoxicity to be more common in males, but the difference was not statistically significant. Smith et al. [60] found no statistically significant difference in the sex of their nephrotoxic and nonnephrotoxic patients. They also found no significant difference in the risk of nephrotoxicity between blacks and whites.

C. Prior Renal Insufficiency

Assessment of the importance of prior renal insufficiency as a risk factor has proved difficult. Patients with renal disease may develop intercurrent complications, such as dehydration or hypotension, that may exacerbate their renal dysfunction unrelated to aminoglycoside exposure. Unless aminoglycoside serum levels are used to adjust dosage, patients with renal failure are more likely to receive excessive doses. Most important, the preceding discussion of the nonlinear relationship of Scr to Ccr obviously applies directly to the study of prior renal insufficiency as a potential risk factor. Studies using an incremental rise in Scr as the criterion for nephrotoxicity will overestimate nephrotoxicity in patients with preexisting renal insufficiency and underestimate it in patients with normal Ccr.

Thus, there are many studies that show a greater likelihood of an incremental rise in Scr during aminoglycoside therapy in patients with an elevated initial Scr [21,27,43,47]. Lane et al. [43] attempted to make a correction by using a graded criteria of nephrotoxicity according to initial Scr (nephrotoxicity defined as rise in Scr $\geqslant$ 0.4 if initial Scr $<$ 3, or rise in Scr $\geqslant$ 0.9 if initial Scr $\geqslant$ 3). However, this study was retrospective, and gave no indication of the dosing method used or whether serum aminoglycoside levels were employed in dosing. Smith et al. [60] could not demonstrate a greater risk of nephrotoxicity in patients with prior renal insufficiency using a similar differential definition of nephrotoxicity. Their study was prospective and they frequently measured serum levels and dosed patients to keep these levels in a predefinied range. Reyman et al. [57], in another well-designed prospective study, could not find an increased risk with prior renal insufficiency. Two studies regularly monitored Ccr as well as Scr. Fee et al. [24] measured Ccr frequently and concluded that prior renal insufficiency was indeed a risk factor. However, they defined nephrotoxicity as an increase in Scr of $\geqslant 0.5$ mg/dl for patients with abnormal initial Scr rather than analyzing the change in Ccr. The only study to analyze Ccr data showed that although nephrotoxic patients with low initial Ccr showed greater increases in Scr than did nephrotoxic patients with normal initial Ccr, the former group did hot have any greater reduction in Ccr [37].

Thus, the evidence favoring prior renal insufficiency as a risk factor is flawed by methodologic problems, and the better designed studies do not demonstrate an increased risk. We conclude that if any difference in risk does exist, it is small. To demonstrate such a difference would require a large prospective study evaluating nephrotoxicity using a direct measure of the glomerular filtration rate, and dosing patients according to serum aminoglycoside levels.

D. Sepsis

Little has been written about the presence of infection as a risk factor for nephrotoxicity. Two studies address this question. Keating et al. [39] found a greater incidence of major azotemia in patients with documented infection than in patients with no infectious source ultimately demonstrated. However, the difference between the two groups did not reach statistical significance. No information is given regarding other differences in the two groups, including the duration of therapy. In an autopsy study of cancer patients receiving gentamicin and cephalothin, Plager noted a significantly higher incidence of tubular injury in patients with bacterial infection than in those without infection [55]. He postulated that the antibiotic placed the patient at risk for severe renal tubular injury when other nephrotoxic insults were superimposed. Clearly, similar data on patients treated with aminoglycosides alone would be of interest.

It may indeed be true that various superimposed nephrotoxic insults make aminoglycoside nephrotoxicity more likely. However, it is another matter to sort out one from another since patients with sepsis are likely to have other physiologic abnormalities, such as hypotension, sodium and volume depletion, acidosis, decreased baseline Ccr, and exposure to other drugs. In fact, gram-negative sepsis can itself cause an acute glomerulonephritis [71]. Thus, sepsis remains a risk factor of uncertain significance.

E. Acidosis

Acidosis is a common disturbance in severely ill patients receiving aminoglycosides and is therefore of interest as a potential risk factor for nephrotoxicity. There are no clinical studies of this factor. However, Hsu et al. [34] found significant azotemia and tubular necrosis in gentamicin-treated rats with metabolic acidosis, whereas gentamicin-treated controls without acidosis showed no nephrotoxicity.

F. Sodium Volume Depletion

Sodium depletion has been studied as a risk factor. Bennett et al. studied sodium depletion in rats induced by dietary restriction [7]. He found that a low-sodium diet markedly potentiated the nephrotoxicity of gentamicin, but that an abnormally high sodium diet was not protective. However, volume change may have also produced these results. Kahn [38] found that a low-sodium but normal-chloride diet did not cause nephrotoxicity in gentamicin-treated rats and postulated that it may be chloride rather than sodium depletion that is of importance.

In a prospective clinical trial, Reymann et al. [57] found a 46% incidence of aminoglycoside-induced nephrotoxicity in patients with volume depletion but only a 6% incidence in patients without volume depletion (P < 0.005). The analysis attempted to exclude patients with renal deterioration attributable directly to the volume depletion itself (e.g., hypotension or prerenal azotemia).

These studies suggest that volume depletion may be a risk factor. Confirmation by additional clinical studies is necessary. Careful attention to the volume status of patients receiving aminoglycosides seems warranted.

G. Potassium Depletion

Potassium homeostasis may be altered by antibiotics. For example, gentamicin and carbenicillin can cause hypokalemia. Brinker et al. [11] studied gentamicin nephrotoxicity in dogs with potassium depletion and found increased nephrotoxicity when compared to normokalemic controls. Increased potassium wasting was noted during the gentamicin-treatment period. These interesting observations should be followed with clinical studies since potassium depletion may be common in patients given aminoglycosides.

H. Renal Transplantation

Initial reports of aminoglycoside therapy in renal transplant patients [46,61] did not suggest an increased risk of nephrotoxicity but gave minimal information about how or if it was looked for. Wellwood et al. [67] then reported that renal transplant recipients were at high risk for gentamicin nephrotoxicity. They found that transplant patients treated with gentamicin universally had a rise in urinary enzymes (N-acetyl-β-D-glucosaminidase and β-galactosidase), which returned to baseline after treatment. These changes were not observed in transplant patients treated with other antibiotics. Serum creatinine rose by >0.4 ml/dl (average 2.8 mg/dl) in 15 of 20 patients treated with gentamicin and significant proteinuria developed in 11 of 17. No other cause for the renal dysfunction was identified except possible acute rejection. However, three of the patients had transplant biopsies, and no evidence for rejection was found. Renal pathology was not described in two patients who required nephrectomy. Serum gentamicin levels were apparently obtained in some patients, but were apparently not routine or the basis for dosing.

A much more optimistic view of the safety of gentamicin in renal transplant patients is presented by Anderton and Raeburn [2]. They treated 29 transplant patients dosing by means of a nomogram and frequent serum aminoglycoside levels. Only two patients have a decrease in Ccr during therapy, both of whom had biopsy proven severe rejection.

The findings of Wellwood et al. are disturbing, but the clinical significance of the abnormalities of the sensitive urinary enzyme measure of renal dysfunction are uncertain and controversial [1,28,69]. Scr deterioration appeared to indicate a significant impact on renal function, but there was no Ccr data, and acute rejection was not excluded in most of the patients. The Ccr data from Anderton and Raeburn are reassuring. Although renal transplant patients may be at some degree of increased risk for aminoglycoside nephrotoxicity, this relationship remains unproven.

II. Drug Factors

A. Dose

A direct relationship of aminoglycoside nephrotoxicity to daily and total dose has been easily and consistently found in experimental animals [20,48,50,66]. Dose relationships in clinical studies have been more difficult to demonstrate. Dahlgren et al. [16] and Bendush and Weber [6] found less toxicity with gentamicin dosage $\leq$ mg/kg per day. However, in the dose range from 4.5 to 7.5 mg/kg per day the correlation between dose and toxicity has not been strong.

Total dose has correlated in some clinical studies [6,13,24] but not others [47,60]. However, both negative studies dosed according to frequent serum levels, thus decreasing the dose in those who developed rising serum levels as a result of nephrotoxicity. On the basis of the results obtained in experimental animals and the experience in clinical trials with daily doses less than 3 mg/kg per day, it seems clear that aminoglycoside nephrotoxicity is dose-related. However, in the dosage range usually used to treat patients with severe gram-negative infections, the correlation is weak. This weak relationship may exist because the dose-response curve for toxicity is relatively flat in this dosage range or because other confounding clinical events (e.g., hypotension) make a relationship to dose difficult to demonstrate in clinical studies.

B. Duration of Therapy

Except for Falco's review of patients treated with gentamicin prior to 1969 [22], clinical studies have consistently demonstrated a relationship between aminoglycoside nephrotoxicity and duration of therapy [6,13,60]. Most of these studies have related nephrotoxicity with a duration of therapy of greater or less than 10–11 days. A threshold duration of therapy before which nephrotoxicity is not seen has not been clearly identified, but development of nephrotoxicity before 5–6 days of therapy appears to be unusual. The consistency of these findings strongly suggests that duration of therapy is a risk factor for aminoglycoside nephrotoxicity.

C. Serum Aminoglycoside Levels

On an intermittent dose schedule, valley (predose) serum levels greater than 2 μg/ml of gentamicin or greater than 10 μg/ml of amikacin have been consistently correlated with nephrotoxicity [29,35,37,47,60]. Is this rise in valley levels the cause, or the result of the nephrotoxicity? This remains a difficult question. The data most relevant have been reported by Kahlmeter [37]. He found that the majority of his nephrotoxic patients had valley levels below 2.0 μg/ml when the Ccr started to decrease. Thus, it seems more likely that elevated or rising valley levels are an early sign of nephrotoxicity rather than its cause.

"Peak" levels have similarly increased in patients with nephrotoxicity [47,60]. Eighty-three percent of Smith's [60] patients with toxicity had 1 hr postinfusion "peak" gentamicin levels greater than 10 μg/ml, whereas only 22% without toxicity had such a level. These findings do not define a particular "toxic level" that causes nephrotoxicity. It is probable that rising "peak" serum levels are due to decreased renal clearance of the aminoglycosides. An observation that gives evidence against the toxicity of even extremely high "peak" levels (at least if exposure is brief) was reported by Ho et al. [32]. They reported a patient who inadvertently received 9 g of amikacin IV within 4 hr but developed no nephrotoxicity or ototoxicity. The 4 hr postdose serum level was 300 μg/ml, and the "peak" level was estimated to be 600–1000 μg/ml.

The integral of the serum aminoglycoside concentration as a function of time—the "area under the curve"—has also been correlated with nephtoxicity. Fee et al. [24] found a significant correlation. However, since the area under the curve is also a measure of drug clearance, this relationship may be caused by nephrotoxicity rather than indicating a pharmacokinetic factor in the etiology of the nephrotoxicity. To look for the latter effect would require a study able to demonstrate that a correlation between the area under the curve and toxicity preceded any change in glomerular filtration rate, as measured by the most sensitive means available.

We conclude that aminoglycoside levels, like total or daily dose, correlate roughly with nephrotoxicity risk. However, in the "therapeutic range" commonly employed, the slope of the dose-response curve for toxicity may be rather flat. Much of the rise in serum levels in patients experiencing toxicity is probably due to nephrotoxicity rather than being its cause.

D. Frequency of Administration

The frequency of aminoglycoside administration may be important to the toxicity and efficacy of these agents since the frequency of administration determines the time course of aminoglycoside serum and tissue levels. The extremes of frequency that have thus far been studied range from a single daily dose to continuous infusion; the most commonly used dosing interval has been 8 hr.

Daily and 8 hr dose intervals were compared by Labovitz et al. [42] in a randomized study of patients with urinary tract infections. Eleven patients receiving 160 mg of gentamicin daily had a 14% mean drop in Ccr during treatment (P < 0.01), whereas 10 patients receiving 60 to 80 mg of gentamicin every 8 hr had no significant change in Ccr. This study has been criticized because the groups varied in mean age, and although the problem caused by their differing initial Ccr may have been overcome by using change in Ccr as the criteria for nephrotoxicity, there may have been other significant age-related differences in the two groups. Frame et al. [26] obtained opposing results in rabbits. Nine rabbits given 7.5 mg/g daily for 21 days had no change in renal function, whereas nine given 2.5 mg of gentamicin/g every 8 hr showed rising Scr. Goodman et al. [29] compared variable-dose and variable-frequency gentamicin-dosing regimens in 20 seriously ill patients, many of whom had preexisting renal dysfunction. Dosing alterations appropriate for renal impairment were individualized by varying either the gentamicin dose or the frequency of administration. Further impairment in renal function was common during treatment, but no significant difference could be detected between the two regimens with this small sample size. Setia and Gross [59] treated 18 patients with tobramycin or gentamicin 1 mg/kg intravenously (IV) every 6 hr for 3 to 19 days. Valley aminoglycoside serum levels were consistently less than 2 μg/ml. One episode of nephrotoxicity (possibly due to Bence Jones proteinuria) was observed with this more frequent dosing regimen.

Three studies have treated cancer patients with aminoglycosides by continuous intervenous infusion [10,39,64]. Azotemia occurred in 10–13% of patients. However, no concurrent control group was reported. Although this toxicity does not appear excessive as judged by historical controls, the wide variability in the incidence of nephrotoxicity in past studies makes the use of concurrent controls extremely important when making any statement about relative risk.

Reiner et al. [56] compared continuous IV infusion vs. once-daily IV injection of gentamicin or tobramycin in dogs treated for 10 days. The continuous-infusion regimen was significantly more nephrotoxic than was the single-daily-dose regimen.

Thus, the relationship of frequency of administration to risk of nephrotoxicity has not yet been adequately studied. Human studies have not been well controlled. Animal studies show increased toxicity with frequent or continuous administration. Human prospective comparative trials will be needed to establish the relative safety and efficacy of the various frequency regimens.

E. Prior Aminoglycoside Exposure/Toxicity

Only a few reports have mentioned the incidence of prior aminoglycoside therapy in patients with and without aminoglycoside nephrotoxicity. Meyer et al. [52] observed six cases of possible nephrotoxicity during amikacin therapy, and noted that all six had prior aminoglycoside therapy. However, five of the six also had severe underlying disease.

Valdivieso et al. [64] found ototoxicity but not nephrotoxicity associated with prior aminoglycoside use. Lau et al. [44] in a trial of amikacin-carbenicillin vs. gentamicin-carbenicillin in leukopenic patients state that previous aminoglycoside exposure was significantly (P < 0.05) associated with amikacin but not with gentamicin therapy. Lane et al. [43], in a review of 1548 patients treated in phase II and III trials of amikacin, found a significantly greater percentage of patients who had received prior aminoglycoside in those who developed nephrotoxicity than in those who did not. However, this difference disappeared if patients who had also previously received other potentially nephrotoxic antibiotics were excluded. Thus, nephrotoxicity was not more frequent in patients previously receiving only aminoglycosides. We conclude that there is currently no convincing evidence that prior aminoglycoside exposure or nephrotoxicity is a risk factor for subsequent nephrotoxicity. There is no information regarding the possibility that prior aminoglycoside nephrotoxicity might make subsequent toxicity more likely.

F. Tissue Binding

Schentag et al. [58] postulated that aminoglycoside nephrotoxicity might be related to unusually high tissue accumulation of aminoglycosides. Thus, patients with greater propensity to tissue accumulation were at greater risk of nephrotoxicity. They studied the washout phase of gentamicin elimination after 64 courses of gentamicin therapy, calculating the rate of elimination from the tissue compartment. Using a two-compartment pharmacokinetic model, they then calculated the rate of tissue accumulation. They found that in patients developing nephrotoxicity the rate of tissue accumulation was greater than in patients without nephrotoxicity, even before any measurable change in renal function was observed. However, renal function was monitored with the Scr and extrapo-lated by nomogram to calculate Ccr. Also, the possibility that the washout curve was altered by nephrotoxicity was not excluded, and the extrapolated early tissue accumula-tion was not confirmed with tissue levels in any of the nephrotoxic patients. On the contrary, Kahlmeter [37] found no correlation between percent reduction of Ccr and post-therapeutic recovery of gentamicin in the urine. Furthermore, Reiner et al. [56] found no difference in renal aminoglycoside concentration between dogs with and without aminoglycoside nephrotoxicity. The hypothesis that differences in tissue ac-cumulation of aminoglycosides account for individual patient differences in suscepti-bility to aminoglycoside nephrotoxicity is an important but as yet unproven possibility.

G. Drug Interactions

Various drugs have been suspected to increase the nephrotoxicity of aminoglycosides. These include furosemide, several antibiotics, and methyoxyflurane.

Aminoglycoside-Furosemide Interaction

Multiple studies in experimental animals have shown an increase in nephrotoxicity when furosemide was given with aminoglycosides compared to that observed with amino-glycosides alone [1,36,38,45]. Kahn [38] found this effect in rats could be prevented by a diet high in NaCl and KCl, but could not be reproduced by a low-sodium but normal chloride diet. He postulated that the enhanced nephrotoxicity could be due to chloride rather than sodium depletion.

In humans, Tilstone et al. [62] found healthy volunteers to have a 22% fall in gentami-cin clearance 6 hr after a 20 mg IV bolus of furosemide. Several case reports [54] have described nephrotoxicity with the use of this combination, but Smith et al. [60] could not detect such an effect during a prospective comparison of amikacin and gentamicin. Reyman similarly could not demonstrate increased nephrotoxicity with the combination [57]. At present, an insufficient amount of information is available to exclude a possible increase in toxicity with this combination.

Aminoglycoside-Cephalothin Interaction

As the antibiotic combination of an aminoglycoside and cephalothin became popular, multiple case reports of nephrotoxicity associated with such treatment appeared [9,25, 30,41,63]. Since the nephrotoxicity could have resulted from either agent alone or from other concurrent nephrotoxic events in these severely ill patients, confirmation of the purported synergistic toxicity was sought in experimental animals. However, in experi-mental animals, cephalothin was found either to protect against [5,8,17,49] or to not influence [31] aminoglycoside nephrotoxicity.

Clinical studies of this problem have arrived at conflicting conclusions. Klastersky et al. [40] in a prospective randomized trial in cancer patients comparing tobramycin-cephalothin and tobramycin-ticarcillin, found nephrotoxicity in 21% and 6% respectively (P < 0.05). The EORTC group [21] compared gentamicin-cephalothin and gentamicin-carbenicillin in a prospective randomized trial in febrile granulocytopenic patients with cancer, finding nephrotoxicity in 12% and 2%, respectively (P < 0.05). Wade et al. [65] in a prospective randomized and double-blind trial of general medical patients with suspected sepsis, found 25.5% incidence of nephrotoxicity in the group treated with amino-glycoside and cephalothin compared to only 7% with an aminoglycoside and methicillin (P = 0.017).

In contrast are the results of two other studies. Fanning et al. [23] retrospectively analyzed data from the Boston Collaborative Drug Surveillance Program to determine the frequency of azotemia attributed to antibiotic therapy in general medical patients receiving gentamicin and cephalothin, alone or in combination. The incidence was 8.6% with gentamicin alone, 7.9% with cephalothin alone, and 9.3% with gentamicin and cephalothin used together. They concluded that substantial synergistic toxicity was unlikely. Giamarellou et al. [28] prospectively administered either gentamicin or gentamicin and cephalothin to alternate patients with mild infections. They found a 0.3 mg/dl or greater rise in Scr in 21% of the 33 gentamicin-treated and 27% of the 34 (gentamicin + cephalothin)-treated patients, but this difference was not statistically significant with this number of patients.

There are some important differences between the studies which appear to indicate increased nephrotoxicity with the combination (positive studies), and those which do not (negative studies). First, the experimental design of the positive studies was more rigorous and less prone to bias. Treatment assignment was truly random in all positive studies, and double-blind in that of Wade et al. Neither negative study was randomized or double-blind. Nephrotoxicity was prospectively defined and monitored in two of the three positive studies, and in one of two negative studies. The retrospective design of Fanning's study is also a disadvantage. The small numbers in Giamarellou's study result in a large beta error, and the inability to find a significant difference between the two groups may well result from lack of power of the study rather than the absence of a true difference. Second, the patient populations studied in the positive studies were probably more severely ill (i.e., cancer patients with and without granulocytopenia [21,40] or medical patients with suspected sepsis [65]). The patients in the negative studies had "mild infections" [28], or a wide variety of presentations in the retrospective study [23]. This raises the possibility of whether a subgroup of patients with severe disease causing superimposed nephrotoxic insults might be sensitive to a synergistic nephrotoxic effect of an aminoglycoside and cephalothin as postulated by Plager [55]. Third, it is important to note that in the positive studies, the control group was gentamicin plus a penicillin (carbenicillin, ticarcillin, or methicillin). This is potentially important in light of a recent animal study by Bloch et al. [8] showing that rats are protected from gentamicin nephrotoxicity not only by cephalothin, but also by carbenicillin, and that carbenicillin gave protection at a lower dose than cephalothin. Thus, they hypothesize that perhaps "beta lactam antibiotics do not contribute to aminoglycoside nephrotoxicity but rather that they afford various degrees of protection" [8]. If true this would reconcile the animal studies showing protection by cephalothin with the human studies which showed increased toxicity. This explanation does, however, require the assumption that the rate of nephrotoxicity attributed to the aminoglycoside alone must be greater than 20%, which seems unlikely. In addition, methicillin and ticarcillin have not yet been studied and shown to have any protective effect against aminoglycoside nephrotoxicity.

We conclude that there is convincing evidence that the combination of cephalothin with an aminoglycoside causes nephrotoxicity more frequently than does the combination of aminoglycosides with other β-lactam antibiotics. Whether the observed greater nephrotoxicity with the aminoglycoside-cephalothin combination is due to an intrinsic additive or synergistic toxicity of the combination, to some special characteristic of these ill patients, or to failure of the cephalosporin to protect against the nephrotoxicity of the aminoglycoside as effectively as the other β-lactam antibiotics is uncertain. We do not know if other cephalosporins interact with aminoglycosides in the same manner as that of cephalothin.

Aminoglycoside-Methicillin Interaction

Methicillin is itself a potential nephrotoxin [18]. However, case reports of nephrotoxicity in patients treated with gentamicin and methicillin together have raised the question of synergistic toxicity. Milman [53] reported six cases of renal failure associated with gentamicin, two of whom were also receiving methicillin. Yver et al. [70] reported acute renal failure in five cases out of six treated with both methicillin and gentamicin for greater than 6 days. The cases were collected in a 30 month period, but the total number of cases exposed to the combination without nephrotoxicity is unknown. The prospective study of Wade et al. [65], described above, found nephrotoxicity in only 4 of 43 patients (7%) treated with methicillin and an aminoglycoside. The duration of therapy was 7 ± 2.8 days (mean ± SD). This study is strong evidence against any increased nephrotoxicity with the combination of methicillin and an aminoglycoside given for the time period stated. It cannot provide evidence against toxicity of the combination if it was administered for more prolonged periods. However, it should be emphasized that neither is there any convincing evidence favoring such toxicity.

Aminoglycoside-Clindamycin Interaction

Butkus et al. [12] raised the question of a possible synergistic nephrotoxicity between clindamycin and gentamicin when they reported three cases of acute renal failure during therapy with this combination. Duration of gantamicin therapy was long (13–17 days), no control group was reported, and no estimate of the frequency of cases treated with the combination without toxicity was given. We could not find similar case reports appearing subsequently. The toxicity observed in the Butkus report is entirely compatible with prolonged gentamicin use itself. In addition, in our studies we have not found clindamycin to increase aminoglycoside-induced nephrotoxicity. There is no evidence for a synergistic toxicity when clindamycin is also given.

Aminoglycoside-Amphotericin B Interaction

Churchill and Seely [15] reported deterioration of renal function after administering a low total dose of amphotericin B to each of four consecutive patients already receiving gentamicin. The gentamicin therapy was prolonged and/or repetitive, and although the nephrotoxicity seemed to correlate with the administration of amphotericin B, it cannot be proven that amphotericin B had any causative relationship to the nephrotoxicity. Thus far there has been no further information published providing evidence for or against this interaction.

Aminoglycoside-Methoxyflurane Interaction

Mazze and Cousins [51] reported a case of polyuric renal insufficiency following methoxyflurane anesthesia and gentamicin therapy. The rise in Scr occurred 5 days postoperatively. The authors state that usually the greatest change in renal function in patients with nephrotoxicity due to methoxyflurane alone occurs within 3 days postoperatively. However, the time course of azotemia following methoxyflurane is not well documented. They report the observation of several other cases of nephrotoxicity following the combination of methoxyflurane and an aminoglycoside, but only one of them has been published [33]. Although Barr et al. [4] showed synergistic toxicity in rats treated with methoxyflurane and gentamicin, we have not found subsequent case reports or any controlled studies in humans to support this association.

III. Conclusions

Many factors have received attention as potentially increasing the risk of developing aminoglycoside nephrotoxicity. Because of confounding variables, study design, and number of patients studied, currently available studies could be expected to detect only relatively large effects. There is reasonably good evidence that long duration of therapy, high dose, volume depletion, and concurrent administration of cephalothin are significant risk factors, whereas advanced age and prior renal insufficiency probably are not. The importance of the other factors reviewed remains uncertain. To clarify the importance of these remaining factors or to detect small effects of factors presently thought not to be significant will require large carefully designed studies using multivariate analysis. Further understanding of the factors associated with nephrotoxicity might result in safer use of aminoglycosides and add to our understanding of the pathophysiology of their nephrotoxicity.

References

1. Adelman, R.D., Spangler, W.L., Beasom, F., Ishizaki, G., and Conzelman, G.M.: Furosemide enhancement of experimental gentamicin nephrotoxicity: comparison of functional and morphological changes with activities of urinary enzymes. *J. Infect. Dis.* 140:342-352, 1979.
2. Anderton, J.L., and Raeburn, J.A.: Gentamicin nephrotoxicity in patients with renal allografts. *Br. Med. J.* 4:165-166, 1975.
3. Appel, G.B., and Neu, H.C.: Gentamicin in 1978. *Ann. Intern. Med.* 89:528-538, 1978.
4. Barr, G.A., Mazze, R.I., Cousins, M.J., and Kosek, J.C.: An animal model for combined methoxyflurane and gentamicin nephrotoxicity. *Br. J. Anaesth.* 45:306-311, 1973.
5. Barza, M., Pinn, V., Tanguay, P., and Murray, T.: Nephrotoxicity of newer cephalosporins and aminoglycosides alone and in combination in a rat model. *J. Antimicrob. Chemother.* 4(Suppl. A):59-68, 1978.
6. Bendush, C.L., and Weber, R.: Tobramycin in sulfate: a summary of worldwide experience from clinical trials. *J. Infect. Dis.* 134(Suppl.):S219-S234, 1976.
7. Bennett, W.M., Hartnett, M.N., Gilbert, D., Houghton, D., and Porter, G.A.: Effect of sodium intake on gentamicin nephrotoxicity in the rat. *Proc. Soc. Exp. Biol. Med.* 151:736-738, 1976.
8. Bloch, R., Luft, F.C., Rankin, L.I., Sloan, R.S., Yum, M.N., and Maxwell, D.R.: Protection from gentamicin nephrotoxicity by cephalothin and carbenicillin. *Antimicrob. Agents Chemother.* 15:46-49, 1979.
9. Bobrow, S.N., Jaffe, E., and Young, R.C.: Anuria and acute tubular necrosis associated with gentamicin and cephalothin. *JAMA* 222:1546-1547, 1972.

10. Bodey, G.P., Chang, H.Y., Rodriguez, V., and Stewart, D.: Feasibility of administering aminoglycoside antibiotics by continuous intravenous infusion. *Antimicrob. Agents Chemother.* 8:328–333, 1975.

11. Brinker, K., Bronin, R., Bulger, R., Southern, P., and Henrich, W.: Potassium depletion: risk factors for and consequence of gentamicin nephrotoxicity. *Clin. Res.* 27: 764A, 1979.

12. Butkus, D.E., de Torrente, A., and Terman, D.S.: Renal failure following gentamicin in combination with clindamycin. *Nephron* 17:307–313, 1976.

13. Bygbjerg, I.C., and Moller, R.: Gentamicin-induced nephropathy. *Scand. J. Infect. Dis.* 8:203–208, 1976.

14. Carlon, G.C., Scheiner, E., Colaco, F.M., Howland, W.S., and Goldiner, P.L.: Nephrotoxic antibiotics in patients with renal failure. *Crit. Care Med.* 7:1–4, 1979.

15. Churchill, D.N., and Seely, J.: Nephrotoxicity associated with combined gentamicin-amphotericin B therapy. *Nephron* 19:176–181, 1977.

16. Dahlgren, J.C., Anderson, E.T., and Hewitt, W.L.: Gentamicin blood levels: a guide to nephrotoxicity. *Antimicrob. Agents Chemother.* 8:58–62, 1975.

17. Dellinger, P., Murphy, T., Pinn, V., Barza, M., and Weinstein, L.: Protective effects of cephalothin against gentamicin induced nephrotoxicity in rats. *Antimicrob. Agents Chemother.* 9:172–178, 1976.

18. Ditlove, J., Weidmann, P., Bernstein, M., and Massey, S.G.: Methicillin nephritis. *Medicine* 56:483–491, 1977.

19. Enger, E., and Bhegen, E.M.: The relationship between endogenous creatinine clearance, and serum creatinine in renal failure. *Scand. J. Clin. Lab. Invest.* 16:273–280, 1964.

20. Engle, J.E., Abt, A.B., Schneck, D.W., and Schoilwerth, A.C.: Netilmicin and tobramycin: comparison of nephrotoxicity in dogs. *Invest. Urol.* 17:98–102, 1979.

21. The EORTC International Antimicrobial Therapy Project Group: Three antibiotic regimens in the treatment of infection in febrile granulocytopenic patients with cancer. *J. Infect. Dis.* 137:14–29, 1978.

22. Falco, F.G., Smith, H.M., and Arcieri, G.M.: Nephrotoxicity of aminoglycosides and gentamicin. *J. Infect. Dis.* 119:406–409, 1969.

23. Fanning, W.L., Gump, D., and Jick, H.: Gentamicin- and cephalothin-associated rises in blood urea nitrogen. *Antimicrob. Agents Chemother.* 10:80–82, 1976.

24. Fee, W.E., Vierra, V., and Lathrop, G.R.: Clinical evaluation of aminoglycoside toxicity: tobramycin versus gentamicin, a preliminary report. *J. Antimicrob. Chemother.* 4(Suppl. A):31–36, 1978.

25. Fillastre, J.P., Laumonier, R., Humbert, G., Dubois, D., Metayer, J., Delpech, A., Leroy, J., and Robert, M.: Acute renal failure associated with combined gentamicin and cephalothin therapy. *Br. Med. J.* 2:396–397, 1973.

26. Frame, P.T., Phair, J.P., Watanakunakorn, C., and Bannister, T.W.P.: Pharmacologic factors associated with gentamicin nephrotoxicity in rabbits. *J. Infect. Dis.* 135: 952–956, 1977.

27. Gary, N.E., Buzzeo, L., Salaki, J., and Eisinger, R.P.: Gentamicin-associated acute renal failure. *Arch. Intern. Med.* 136:1101–1104, 1976.

28. Giamarellou, H., Metzikoff, C., Papachristophorou, A.S., Dontas, A.S., and Daikos, G.K.: Prospective comparative evaluation of gentamicin or gentamicin plus cephalothin in the production of nephrotoxicity in man. *J. Antimicrob. Chemother.* 5: 581–590, 1979.

29. Goodman, E.L., VanGelder, J., Holmes, R., Hull, A.R., and Sanford, J.P.: Prospective comparative study of variable dosage and variable frequency regimens for administration of gentamicin. *Antimicrob. Agents Chemother.* 8:434–438, 1975.

30. Hansen, M.M., and Kaaber, K.: Nephrotoxicity in combined cephalothin and gentamicin therapy. *Acta Med. Scand.* 201:463–467, 1977.

31. Harrison, W.O., Silverblatt, F.J., and Turck, M.: Gentamicin nephrotoxicity: failure of three cephalosporins to potentiate injury in rate. *Antimicrob. Agnets Chemother.* 8:209–215, 1975.

32. Ho, P.W.L., Pien, F.D., and Namiko, K.: Massive amikacin "overdose." *Ann. Intern. Med.* 91:227–228, 1979.

33. Hollenberg, N.K., McDonald, F.D., Gotran, R., Galvaner, E.G., Warhol, M., Vandam, L.D., and Merrill, J.P.: Irreversible acute oliguric renal failure. *N. Engl. J. Med.* 286:877–879, 1972.

34. Hsu, C.H., Kurtz, T.W., Easterling, R.E., and Weller, J.M.: Potentiation of gentamicin nephrotoxicity by metabolic acidosis. *Proc. Soc. Exp. Biol. Med.* 146:894–897, 1974.

35. Hewitt, W.L.: Gentamicin: Toxicity in perspective. *Postgrad. Med. J.* 50(Suppl. 7): 55–59, 1974.

36. James, G., Smith, W., Bryant, H., and Balazs, T.: Enhancement of antibiotic nephrotoxicity by furosemide: studies in mice. *Toxicol. Appl. Pharmacol.* 33:199, 1975.

37. Kahlmeter, G.: Nephrotoxicity of gentamicin and tobramycin. *Scand. J. Infect. Dis.* 18(Suppl.):15–40, 1979.

38. Kahn, T.: Effect of furosemide on gentamicin and netilmicin nephrotoxicity. *Kidney Int.* 12:527, 1977.

39. Keating, M.J., Bodey, G.P., Valdivieso, M., and Rodriguez, V.: A randomized comparative trial of three aminoglycosides—comparison of continuous infusions of gentamicin, amikacin, and sisomicin combined with carbenicillin in the treatment of infections in neutropenic patients with malignancy. *Medicine* 58:159–170, 1979.

40. Klastersky, J., Hensgens, C., and Debusscher, L.: Empiric therapy for cancer patients: comparative study of ticarcillin-tobramycin, ticarcillin-cephalothin, and cephalothin-tobramycin. *Antimicrob. Agents Chemother.* 7:640–645, 1975.

41. Kleinknecht, D., Ganeval, D., and Droz, D.: Acute renal failure after high doses of gentamicin and cephalothin. *Lancet,* 1129, 1973.

42. Labovitz, E., Levison, M.E., and Kaye, D.: Single-dose daily gentamicin therapy in urinary tract infection. *Antimicrob. Agents Chemother.* 6:465–470, 1974.

43. Lane, A.Z., Wright, G.E., and Blair, D.C.: Ototoxicity and nephrotoxicity of amikacin. *Am. J. Med.* 62:911–918, 1977.

44. Lau, W.K., Young, L.S., Black, R.E., Winston, D.J., Linne, S.R., Weinstein, R.J., and Hewitt, W.L.: Comparative efficacy and toxicity of amikacin/carbenicillin versus gentamicin/carbenicillin in leukopenic patients. *Am. J. Med.* 62:959–966, 1977.

45. Lawson, D.H., Macadam, R.F., Singh, H., Garras, H., Hartz, S., Turnbull, D., and Linton, A.L.: Effect of furosemide on antibiotic-induced renal damage in rats. *J. Infect. Dis.* 126:593–599, 1972.

46. Leigh, D.A.: Treatment of gram-negative infections in patients before and after renal transplantation. *Postgrad. Med. J.* 45:254–260, 1969.

47. Lerner, S.A., Seligsohn, R., and Matz, G.J.: Comparative clinical studies of ototoxicity and nephrotoxicity of amikacin and gentamicin. *Am. J. Med.* 62:919–923, 1977.

48. Luft, F.C., Bloch, R., Sloan, R.S., Yum, M.N., Costello, R., and Maxwell, D.R.: Comparative nephrotoxicity of aminoglycoside antibiotics in rats. *J. Infect. Dis.* 138:541–545, 1978.

49. Luft, F.C., Patel, V., Yum, M.N., and Kleit, S.A.: Nephrotoxicity of cephalosporin-gentamicin combinations in rats. *Antimicrob. Agents Chemother.* 9:831–839, 1976.

50. Luft, F.C., Yum, M.N., and Kleit, S.A.: Comparative nephrotoxicity of netilmicin and gentamicin in rats. *Antimicrob. Agents Chemother.* 10:845–849, 1976.

51. Mazze, R.I., and Cousins, M.J.: Combined nephrotoxicity of gentamicin and methoxyflurane anesthesia in man. *Br. J. Anaesth.* 45:394–398, 1973.

52. Meyer, R.D., Lewis, R.P., Carmalt, E.D., and Finegold, S.M.: Amikacin for serious gram-negative bacillary infections. *Ann. Intern. Med.* 83:790–800, 1975.
53. Milman, N.: Renal failure associated with gentamicin therapy. *Acta Med. Scand.* 196:87–91, 1974.
54. Noel, P., and Levy, V.G.: Toxicité rénale de l'association gentamicine-furosémide. *Nouv. Presse Med.* 7:351–353, 1978.
55. Plager, J.E.: Association of renal injury with combined cephalothin-gentamicin therapy among patients severely ill with malignant disease. *Cancer* 37:1937–1943, 1976.
56. Reiner, N.E., Bloxham, D.D., and Thompson, W.L.: Nephrotoxicity of gentamicin and tobramycin given once daily or continuously in dogs. *J. Antimicrob. Chemother.* 4(Suppl. A):85–101, 1978.
57. Reymann, M.T., Bradac, J.A., Cobbs, C.G., and Dismukes, W.E.: Correlation of aminoglycoside dosage with serum concentration during therapy of serious gram-negative bacillary disease. *Antimicrob. Agents Chemother.* 16:353–361, 1979.
58. Schentag, J.J., Cumbo, T.J., Jusko, W.J., and Plaut, M.E.: Gentamicin tissue accumulation and nephrotoxic reactions. *JAMA* 240:2067–2069, 1978.
59. Setia, U., and Gross, P.A.: Administration of tobramycin and gentamicin by the intravenous route every 4 hr. in patients with normal renal function. *J. Infect. Dis.* 134 (Suppl.):S125–S129, 1976.
60. Smith, C.R., Maxwell, R.R., Edwards, C.Q., Rogers, J.F., and Lietman, P.S.: Nephrotoxicity induced by gentamicin and amikacin. *Johns Hopkins Med. J.* 142:85–90, 1978.
61. Taguchi, Y., and Siddiquie, N.: Gentamicin and kidney transplantation: an assessment in vitro and in vivo. *J. Urol.* 107:911–916, 1972.
62. Tilstone, W.J., Semple, P.F., Lawson, D.H., and Boyle, J.A.: Effects of furosemide on glomerular filtration rate and clearance of practolol, digoxin, cephaloridine, and gentamicin. *Clin. Pharmacol. Ther.* 22:389–394, 1977.
63. Tobias, J.S., Whitchouse, J.M., and Wrigley, P.F.M.: Severe renal dysfunction after tobramycin/cephalothin therapy. *Lancet,* 425, 1976.
64. Valdivieso, M., Feld, R., Rodriguez, V., and Bodey, G.P.: Amikacin therapy of infections in neutropenic patients. *Am. J. Med. Sci.* 270:453–463, 1975.
65. Wade, J.C., Smith, C.R., Petty, B.G., Lipsky, J.J., Conrad, G., Ellner, J., and Lietman, P.S.: Cephalothin plus an aminoglycoside is more nephrotoxic than methicillin plus an aminoglycoside. *Lancet,* 604–606, 1978.
66. Welles, J.S., Emmerson, J.L., Gibson, W.R., Nickander, R., Owen, N.V., and Anderson, R.C.: Preclinical toxicology studies with tobramycin. *Toxicol. Appl. Pharmacol.* 25:398–409, 1973.
67. Wellwood, J.M., Simpson, P.M., Tighe, J.R., and Thompson, A.E.: Evidence of gentamicin nephrotoxicity in patients with renal allografts. *Br. Med. J.* 3:278–281, 1975.
68. Wilfert, J.N., Burke, J.P., Bloomer, H.A., and Smith, C.B.: Renal insufficiency associated with gentamicin therapy. *J. Infect. Dis.* 124(Suppl.):S148–S155, 1971.
69. Wilson, D., Weinstein, A., Hall, P., and Vidt, D.: Early indicators of aminoglycoside nephrotoxicity. *Clin. Pharmacol. Ther.* 25:253–254, 1979.
70. Yver, L., Becq-Giraudon, B., Pourrat, O., and Sudre, Y.: La Néphrotoxicité de l'association méthicilline-gentamycine. *Sem. Hop. Paris* 52:1903–1907, 1976.
71. Zappacosta, A.R., and Ashby, B.L.: Gram-negative sepsis with acute renal failure: occurrence from acute glomerulonephritis. *JAMA* 238:1389–1390, 1977.

OTOTOXICITY

18

STUDIES OF AMINOGLYCOSIDE OTOTOXICITY IN ANIMAL MODELS

ROBERT E. BRUMMETT and KAYE E. FOX
University of Oregon Health Sciences Center
School of Medicine, Portland, Oregon

Drugs that damage the ear are referred to as being *ototoxic*. The aminoglycoside antibiotics are ototoxic drugs of special clinical significance because their ototoxic effects result in *permanent* loss of inner ear function. Furthermore, because a narrow margin of safety exists between therapeutic and ototoxic doses of the aminoglycosides, ototoxicity is a major limiting factor in the clinical use of these antibiotics in humans.

The irreversible nature of the ototoxic effect of the aminoglycoside antibiotics precludes, for ethical reasons, testing these drugs in human volunteers. Therefore, the only human data available on their ototoxicity come from patients receiving the drugs for therapeutic reasons. Such data are confounded by the effects of other drugs these patients are invariably receiving together with the antibiotics and by the questionable value of measurements of inner ear function in these patients, since most of them are seriously ill. For these reasons animal models have become extremely valuable for assessing the ototoxicity of newly developed aminoglycoside antibiotics during their preclinical testing stage. It is now apparent that the results from animal studies agree well with the clinical experience in humans for the relative ototoxic potencies of these drugs. The demonstration of the relevance of animal data to humans has further enhanced the value of animal studies for preclinical testing of the aminoglycosides. Furthermore, comparative information on the ototoxicity of the aminoglycoside antibiotics, which takes years to accumulate from patient studies, can be obtained in a few months by animal studies. Also, factors attending ototoxicity, such as interactions with other drugs, can be studied easily in these models. Finally, animal models are also necessary for studies of the mechanism of the ototoxic effect of these drugs. Knowledge of the mechanism of ototoxicity may ultimately prove useful clinically, for it may reveal the means of reducing or eliminating the risk to humans of ototoxicity from these important antibiotics.

I. Animal Models

The guinea pig is the most commonly used animal for ototoxicity studies. Guinea pigs are particularly suited for these investigations for several reasons. Of special importance is the fact that the inner ear of the guinea pig is easily accessible for study because the guinea pig cochlea can be exposed by making an opening into the middle ear (Figure 1). This facilitates the placement of electrodes for electrophysiological studies and the withdrawal of perilymph for drug assays. In other species the cochlea is encased in dense bone and is very difficult to expose surgically.

Unfortunately, the guinea pig is not useful for studies of the vestibular portion of the inner ear. Even when one inner ear is surgically destroyed, the guinea pig does not have difficulty in maintaining balance. Furthermore, the corneal-retinal potential that is used as a measure of vestibular function in electronystagmography (ENG) is so small in guinea pigs that it is difficult to separate it from the electrical potentials produced by spontaneous muscle activity. Therefore, the cat is often employed for investigations of the vestibular aspects of ototoxicity. The cat relies heavily on the vestibular system for maintenance of balance, and early signs of vestibular toxicity can be detected by simply observing for ataxia. In addition, the corneal-retinal potential of the cat is of suitable magnitude for ENG recordings. However, cats are very expensive to maintain in the laboratory; therefore, rabbits are often used as an alternative to cats for vestibular toxicity tests.

Primates are ideal animals in which to study auditory and vestibular toxicity because of their close evolutionary relationship to humans. The rhesus and pig-tailed macaca have been studied for auditory toxicity and the squirrel monkey for vestibular toxicity. The main limiting factors to investigating ototoxicity in primates are that primate auditory systems are difficult to study because they are relatively inaccessible and primates are very expensive to purchase and care for in the laboratory.

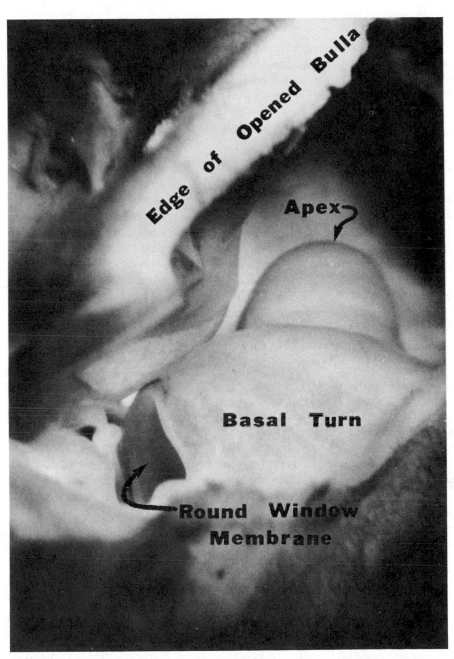

Figure 1 Photomicrograph of a guinea pig cochlea exposed through an opening made into the middle ear.

II. Experimental Techniques

Auditory function is studied using techniques that are often not well known to investigators working outside this field. So that these special tests can be appreciated, we feel
it is appropriate to present the following brief descriptions of these techniques. For more
detailed information see Vernon and Smith [56].

A. Preyer Pinna Reflex

The Preyer pinna reflex is elicited in rodents when they are exposed to loud sound bursts.
The reflex presents as a brief twitch of the pinna when the sound is presented. The reflex
is similar to the middle ear muscle contraction that occurs reflexly in humans when they
are exposed to loud sounds. It should not be confused with the searching movements of
the pinna of cats, dogs, and so on, that occur when these animals are trying to locate the
source of sounds. The Preyer pinna reflex should also not be confused with hearing per se,
as hearing is a perceptive process. However, the Preyer pinna reflex has proven to be a
reliable, albeit crude, index of auditory function in animals such as the guinea pig. It is
very useful for periodic monitoring of the auditory system during long-term ototoxicity
experiments involving chronic drug administration. In addition, testing for the presence
of the Preyer pinna reflex is an effective method of screening out animals with questionable auditory function before starting an experiment.

B. AC Cochlear Potential

The ac cochlear potential is an electrical analog of sound entering the ear (Figure 2).
When a tone is presented to the ear, the ear generates an ac potential of the same frequency. When someone speaks a person's name into an ear, the ear generates an electrical
analog of their voice. If this electrical activity is monitored from the ear, appropriately
amplified, and played through a speaker, the listener will be able to recognize the name.
In fact, the fidelity of the process is so good that if the listener is familiar with the voice
of the person speaking, he will recognize it as well.
    The ac cochlear potential is generated by the hair cells in response to sound. Sound
energy entering the ear is transferred through the ossicles of the middle ear to the perilymphatic fluid of the cochlea in the inner ear. Movement of the perilymph causes
motion of the basilar membrane. The hair cells are attached at their bases to the basilar
membrane, while their apical hairs or cilia are attached to a relatively rigid structure
called the tectorial membrane. Therefore, movement of the basilar membrane results
in distortion or bending of the cilia. It is thought that distortion of the cilia causes ionic
changes in the hair cell membrane, giving rise to the ac cochlear potential and at the
same time initiating discharge spikes in the cochlear nerve. The ac cochlear potential is
not transmitted by the cochlear nerve. Evidence for the origin and physiological relevance
of the ac cochlear potential comes from the fact that the potential as well as the ability
to hear are absent when the hair cells are destroyed as a result of excessive sound or
aminoglycoside antibiotic ototoxicity.
    This potential was originally termed the ac cochlear potential by Wever and Bray
[55]. It is often referred to in the literature as the cochlear microphonic (CM), because
Adrian thought it was an artifact of recording similar to the microphonic obtained from

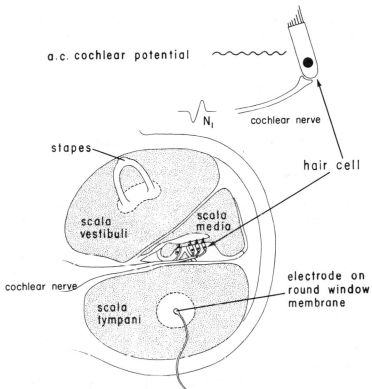

a.c. cochlear potential

N_1

cochlear nerve

stapes

hair cell

scala vestibuli

scala media

cochlear nerve

scala tympani

electrode on round window membrane

Figure 2   Schematic cross section of the cochlea, depicting the sources of the ac cochlear potential and the N_1 nerve action potential with the recording site on the round window membrane. (Reprinted from [4], p. 33, by courtesy of Marcel Dekker, Inc.)

vacuum tubes [1]. The term "cochlear microphonic" has persisted despite the fact that Adrian retracted his original statement [2]. The reason for this persistence may be because it is easier to say "cochlear microphonic" than "ac cochlear potential."

The ac cochlear potential can be recorded from electrodes placed within the cochlea or on the round window membrane. The latter site is most frequently used for ototoxicity studies because it is experimentally convenient and, more important, noninvasive to the cochlea.

C.   Cochlear Nerve Action Potential

The cochlear nerve action potential or N_1 potential, as it is generally referred to in the literature, is the first negative wave occurring in the summed action potential of the cochlear nerve (Figure 2). It is generated by using either a click or a tone pulse with a rapid onset. Either of these stimuli initiates synchronous discharges in a limited number of cochlear nerve fibers. The nerve discharges add together to produce a potential with sufficient magnitude to be recorded. N_1 potentials are measures of electrical activity generated in the cochlear nerve by sound and therefore represent a step further along in the process of hearing than does the ac cochlear potential. These potentials are recorded from electrodes placed either within the cochlea or on the round window membrane.

The N_1 threshold is generally determined by inspection of the N_1 waveform visualized on an oscilloscope while the sound intensity to the ear is varied. It can also be measured using computer techniques in which the background activity is reduced by averaging over a number of recorded nerve potentials.

D. Electronystagmography

Electronystagmography (ENG) is an electrophysiological measurement of nystagmus. Nystagmus produced in the vestibular portion of the inner ear can be used as a measure of vestibular function. The electrical activity measured is the potential difference between the cornea and the retina of the eye. The retina is about 300 μV positive to the cornea. As the eye is rotated to the left by an animal or human which has electrodes applied to the outer canthi, the electrode at the left canthes becomes more positive relative to the electrode at the right canthes. As the eye is rotated to the right, the opposite reaction occurs. Nystagmus results when the eye is deviated in one direction and then rapidly returns to the front. A recording from an electronystagmograph is shown in Figure 3. In this figure, movement of the eye to the right causes the line to move up and movement to the left causes the line to move down. In this example the eye is slowly moving to the right and quickly returns to the left. These eye movements are generated when the fluid in the semicircular canals is induced to move. This can be done experimentally by introducing water that is either colder or warmer than body temperature into the external auditory canal. These temperature changes generate convection currents. The currents cause displacement of the sensory hair cells in the ampullar crista, resulting in stimulation

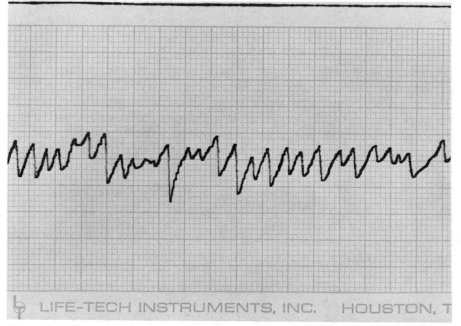

Figure 3   Electronystagmograph (ENG) from a patient. The response was elicited by instillation of cold (30°) water into the right ear. The eyes deviate to the right (slow phase—recording movement upward) and quickly return to the left (fast phase—recording movement downward).

of the vestibular nerve. After the vestibular nerve activity has been processed by the central vestibular projections in the brain, impulses are relayed to the extraocular muscles that control the movement of the eyes. ENG recordings are subject to errors due to changes in the corneal-retinal potential. For example, the potential can be lowered merely by exposure of the eyes to light. In addition, the degree of mental alertness of the test subject is an important factor; therefore, central nervous system depressant drugs may impair the measurement of nystagmus. Thus, the ENG is not only a very indirect measure of vestibular activity, but it can be easily affected by many factors not involving vestibular function.

E. Morphological

In addition to standard methods of cross-sectional light and electronmicroscopy for studying the structural effects of ototoxic drugs, there is a special morphologic preparation unique to the cochlea. This is known as a surface preparation of the organ of Corti. The technique involves fixation of the cochlea by perfusion with an osmium-containing fixative, removal of strips of the organ of Corti from the turns of the cochlea by dissection, and mounting these strips on microscope slides for viewing [30]. With this preparation it is possible, by light microscopy, to visualize rows of inner and outer hair cells. Lost cells are easily detected and counted (Figure 4). The data are often displayed as cochleograms showing missing hair cells in various turns of the cochlea. This analysis allows recognition of the pattern of hair cell loss as well as quantitation of the amount of damage produced by ototoxic agents such as the aminoglycoside antibiotics (Figure 5).

F. Behavioral

Behavioral studies are of value in investigating ototoxicity because they monitor function in an animal's entire afferent auditory pathway as well as the central auditory connections in the brain and the efferent motor system required for the animal to make a given response to a perceived sound stimulus. The technique of classical operant conditioning is employed in these studies. An animal is trained to give a response (e.g., press a bar) for a reward when it is presented with a specified sound frequency. The sound intensity is varied at the given frequency to establish an auditory threshold.

Behavioral studies of aminoglycoside antibiotic ototoxicity are valuable for several reasons, despite the fact that these drugs are known to have an end-organ effect (i.e., destruction of hair cells). For example, behavioral studies require no invasive techniques and permit monitoring auditory function in chronic treatment experiments without exposing the animals to anesthesia or trauma.

G. Other

The most commonly applied methods for studying aminoglycoside antibiotic ototoxicity have been described above. In addition, techniques are available for studying single unit nerve activity in the cochlear and vestibular nerves and their radiations to the cerebral cortex and cerebellum. Also, there are techniques for studying evoked potentials that occur in response to sound stimuli along the pathways for audition and balance. These methods are described in Vernon and Smith [56].

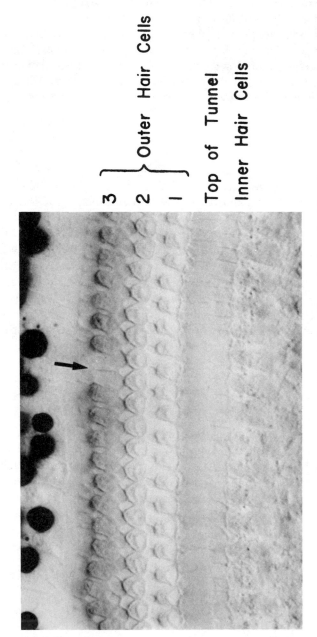

3
2 } Outer Hair Cells
1

Top of Tunnel

Inner Hair Cells

Figure 4   Surface preparation of the organ of Corti of a guinea pig viewed with a phase-contrast microscope. The location of a missing outer hair cell is shown.

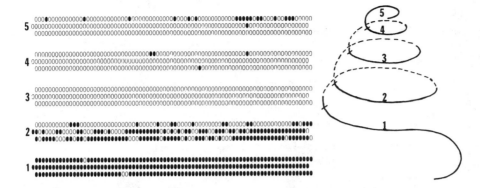

Figure 5 Cochleogram of a guinea pig treated with gentamicin, showing locations of missing outer hair cells (filled circles). Sections shown are from the correspondingly numbered turns of the cochlea schematically depicted in the figure.

III. Results of Ototoxicity Studies in Animals

A.   Morphology

Light-microscopic observations of the ototoxic lesion produced by the aminoglycoside antibiotics in the guinea pig cochlea reveal a selective destruction of the end-organ sensory hair cells [6,12,13,21–23,26,30]. Figure 6 shows a cross section through the basal turn of the cochlea of an untreated guinea pig and another that was chronically treated with kanamycin. It is very clear from the photomicrographs that the two outer hair cells of the organ of Corti have been completely destroyed following drug treatment, whereas the supporting cells of the basilar membrane and cells of the adjoining stria vascularis are unaffected. Characteristically, hair cell destruction begins in the basal turn of the cochlea and proceeds toward the apex (Figure 5). Furthermore, hair cell destruction begins in the innermost row of outer hair cells and proceeds to the outermost row. The inner hair cells are the most resistant to the effects of the drugs and are the last to be destroyed. The extent of the cochlear lesion is determined by the dose and frequency of administration of the drugs and may occur following a latent period after drug treatment has stopped.

Electron microscopic studies of the ototoxic lesion due to aminoglycoside antibiotics reveal damage at the ultrastructural level [38,41,53]. Early changes are seen in the mitochondria of the hair cells and appear as vesiculations of the mitochondrial cristae. Other early changes are the clumping of nuclear chromatin, nuclear swelling, and loss of ribosomes from the cytoplasmic matrix. In addition, it has been observed that the lysosomes are converted into dense bodies and that the plasma membranes develop blebs and lose their subsurface cisternae. However, it is not possible to determine the initial site of action of the drugs from these studies because by the time the cochleae were examined, extensive damage had already occurred and involved multiple structural sites. From the standpoint of determining the mechanism of action of these drugs, it is important to be able to distinguish primary sites of drug effects from secondary sites, since it is reasonable to assume that the latter effects are nonspecific degenerative changes.

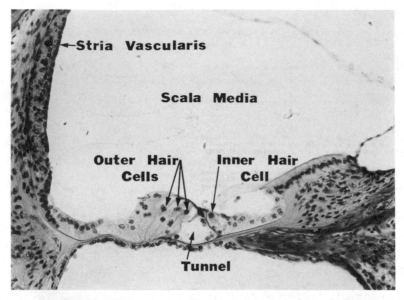

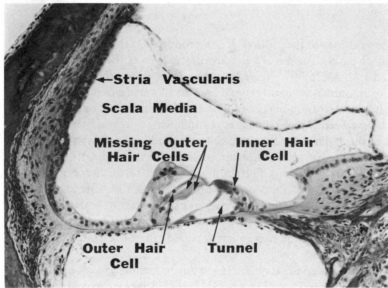

Figure 6 Photomicrographs of cross sections of the cochleae of a normal (upper) and a kanamycin treated (lower) guinea pig.

B. Electrophysiology

The effects of the aminoglycoside antibiotics on cochlear function correlate directly with the morphologic lesion. That is, sensitivity to high-frequency sound is lost first, followed by progressive losses toward lower frequencies, until complete deafness occurs [25,34, 35]. This pattern parallels the well-known frequency-specific locations of the hair cells in the cochlea. High frequencies are processed in the basal turn with a progression to lower frequencies toward the apex. Figure 7 shows a graph of the 1 μV isopotential func-

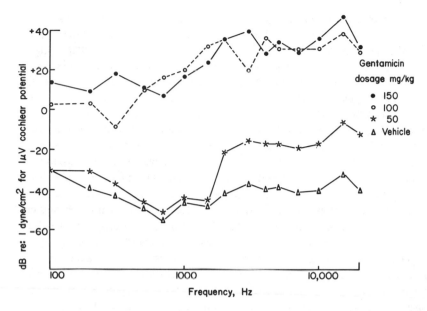

Figure 7   One microvolt isopotential functions of the ac cochlear potentials of guinea pigs treated 4 weeks with the indicated daily subcutaneous doses of gentamicin. Sound pressure increases upward along the vertical axis.

tions of the ac cochlear potential in guinea pigs treated chronically with varying doses of gentamicin. The amount of sound necessary to produce 1 μV of ac cochlear potential is plotted for the different sound frequencies. The graph shows that with increasing doses of gentamicin the high-frequency function of the cochlea is lost first, followed by losses at lower frequencies until the ear is completely destroyed. There is no recovery of function in the ears as badly damaged as these. However, there is evidence that early functional losses in sensitivity at very high frequencies may be reversible if drug administration is stopped (unpublished observations). This issue deserves further study, for if it is true, it may be possible to avoid deafness in humans from these drugs by testing patients at frequencies above the 8000 Hz routinely used as the limit for human audiometry. Then, when an increase in hearing threshold to high-frequency sound is first detected, the patient's therapy can be altered by reducing the dosage of the aminoglycoside being used or by switching to an alternative drug that is not ototoxic. Even if the patient's hearing loss for high-frequency sound is not reversed, prompt change of therapy will probably spare the patient's hearing for the important human speech communication frequencies between 300 and 3000 Hz.

Similar patterns of sound-frequency sensitivity loss due to aminoglycoside antibiotic ototoxicity can be demonstrated by measuring the N_1 cochlear nerve action potential as well as auditory thresholds.

C.   Dose Effect

Animal models of drug ototoxicity have been of major importance for the precise quantitation of dose-effect relationships of the aminoglycoside antibiotics. In particular, new drugs have been compared with older drugs in these systems to evaluate their toxic potential in phase 1 preclinical testing studies. When the antibacterial potencies of two aminoglycoside antibiotics are considered in the comparison of their ototoxic potencies, it is possible to decide whether or not one drug is superior to the other in terms of relative risks or, in other words, to determine which drug has the better therapeutic safety margin.

Because the morphological and electrophysiological effects of aminoglycoside ototoxicity parallel one another, data from either can be used to construct dose-effect curves for these drugs [5,8–11,14,33]. Figures 8 and 9 show an example of each approach for a number of drugs using the guinea pig model. It is clear from the figures that the aminoglycoside antibiotics exhibit a broad range of ototoxic potencies from the markedly ototoxic gentamicin to the apparently nonototoxic netilmicin. Two interesting observations about the structure–activity relationships of these drugs emerge from the data. First, gentamicin C_1, which is less potent than gentamicin, is in fact a constituent of gentamicin. Gentamicin is composed of three structurally distinct but very closely related antibiotics: gentamicin C_1, C_2, and C_{1a}. It is obvious that minor changes in chemical structure, in the case of the gentamicins shifting a methyl group, can result in marked changes in ototoxic potency. This is even more strikingly demonstrated in the second case. Netilmicin is N-ethyl-substituted sisomicin. Netilmicin is nonototoxic at the highest doses tested, whereas the parent drug, sisomicin, is nearly as ototoxic as gentamicin. Following is a rank ordering of a number of aminoglycoside antibiotics in terms of ototoxic potency in guinea pigs: gentamicin > sisomicin >> tobramycin = amikacin > *Sch 21420* > gentamicin C_1 >> netilmicin (ototoxic dose unknown).

When the relative antibacterial potencies of the aminoglycoside antibiotics are taken into consideration, the ranking of the drugs according to therapeutic ratio is different from the simple ranking according to ototoxic potency. Table 1 shows the relative ototoxic potencies of the drugs determined from experiments in guinea pigs. All potencies are expressed in comparison with gentamicin, which is one of the most potent ototoxic

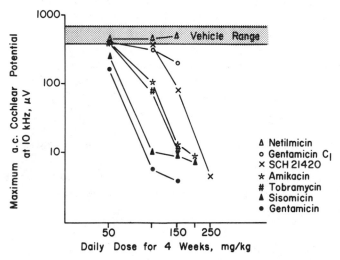

Figure 8   Maximum outputs of the ac cochlear potentials at 10 kHz of guinea pigs treated daily for 4 weeks with subcutaneous injections of the indicated doses of aminoglycoside antibiotics.

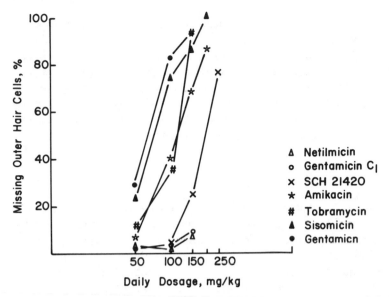

Figure 9 Percent of missing outer hair cells of guinea pigs treated daily for 4 weeks with subcutaneous injections of the indicated doses of aminoglycoside antibiotics.

aminoglycoside antibiotics. The potencies of the same drugs with regard to their antibiotic properties are also shown. Again, these are relative values using gentamicin as the reference and either the recommended daily dose for humans or, where this information is not available, the relative potencies computed from comparative tests of the drug and gentamicin in in vitro bacterial systems or in animals infected with bacteria. The therapeutic ratio is then computed as the ratio of the relative amount of drug compared to gentamicin that produces a similar degree of ototoxicity to that amount required to produce an antibiotic effect comparable to gentamicin. The greater the therapeutic ratio, the safer the drug in terms of avoiding ototoxicity at therapeutic doses comparable to gentamicin. Netilmicin, not surprisingly, heads the list. Gentamicin C_1, because of lack of antibiotic potency, falls lower than expected on the basis of its ototoxic potency. Tobramycin appears superior to amikacin for the same reason, although both drugs are equipotent

Table 1 Relative Potencies of Several Aminoglycoside Antibiotics Compared with Gentamicin

| Drug | Ratio of drug/gentamicin | | Therapeutic ratio (1)/(2) |
	Ototoxic (1)	Antibiotic (2)	
Netilmicin	>3.0	1.0[a]	>3.0
Gentamicin C_1	3.0	2.0[a]	1.5
Tobramycin	1.6	1.0[b]	1.6
Amikacin	1.9	5.0[b]	0.34

[a] Based on in vitro and animal data.
[b] Based on recommended daily dose for humans.

with regard to ototoxicity. Needless to say, this analysis is an oversimplification. Strains of bacteria have emerged clinically that are resistant to gentamicin, negating the value of a comparison between gentamicin and a drug that is an effective antibiotic for that strain. Also, sensitive species of bacteria show large variations in the effective doses of the drugs required to control their growth. However, although the analysis presented here is oversimplified, it does support the conclusion that there are large variations in the relative safety of these drugs and that changes in the chemical structure of the aminoglycosides can reduce ototoxic potency without reducing antibiotic potency.

D. Time of Onset

It is an old observation that the ototoxic effects of the aminoglycoside antibiotics become apparent only after repeated doses of drug and may be delayed for some time after drug administration has stopped [4,43,47,48]. The reason that it is not possible to damage hearing with a single dose of these drugs is that they produce lethal effects acutely at plasma concentrations below those necessary to produce ototoxicity. These lethal effects are respiratory arrest due to neuromuscular blockade and cardiac arrest due to cardiotoxic effects. The fact that doses of the drugs below those which produce acute death can be given repeatedly to produce ototoxicity leads to the conclusion that during chronic administration there is preferential accumulation of the drugs in the inner ear (see Section III. E). However, the fact that cochlear damage may not become apparent for some time after drug administration has been stopped leads to the additional conclusion that even after an ototoxic concentration of drug has been attained in the cochlea, the pathological changes in the hair cells proceed through a latent period druing which no changes in cochlear function can be detected. Hair cell destruction would appear to be at the end of a series of molecular events initiated by the drugs but during which time the cells are functionally intact. For these reasons most dose-response studies of aminoglycoside antibiotic ototoxicity in animals include daily drug administration for periods of 2–4 weeks followed by a period of several weeks during which the animals receive no drugs. At the end of this time the ototoxic lesions are permanently fixed and the electrophysiological and morphological studies can be done with the assurance that no further changes would have occurred in the cochlea.

Two experiments illustrate the latency to onset of aminoglycoside antibiotic-induced ototoxicity. In one experiment a guinea pig was anesthetized with pentobarbital and the ability of the left cochlea to generate an ac cochlear potential determined. The surgical wound was closed and the animal allowed to recover. Ten days later the guinea pig was given 400 mg/kg of kanamycin subcutaneously every 6 hr for a total of five doses. Two hours after the last dose the ability of the right cochlea to generate an ac cochlear potential was determined and the animal allowed to recover from the surgery and anesthetic as before. Two weeks later both cochleae were reevaluated. The results are shown in Figure 10. It is clear from the data that at the end of 2 weeks following treatment with kanamycin there was marked functional damage to both cochleae, whereas there was no effect 2 hr following the last dose of drug 13 days earlier.

In another experiment guinea pigs were given the same course of kanamycin treatment as described above and then tested for their ability to elicit a Preyer pinna reflex in response to different sound frequencies. Figure 11 shows the results. The stippled area indicates the range of thresholds obtained during daily testing for 3 days prior to treatment

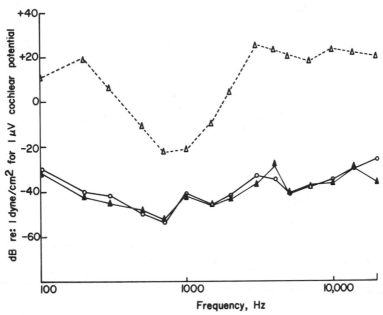

Figure 10 One microvolt isopotential functions of the ac cochlear potential from a guinea pig before and after treatment with kanamycin (400 mg/kg subcutaneously every 6 hr for five doses). △, Left ear, 2 weeks postdrug; ▲, left ear before drug; ○, right ear 2 hr postdrug.

with kanamycin. The curve labeled day 1 shows the thresholds determined two hours after the last dose of drug. Thresholds were then determined daily, but to simplify the graph, not all are shown. By day 3 there is a suggestion of an increase in threshold at 15,000 Hz. A clear effect occurred on day 6 and progressed to day 12. No further changes occurred after day 12 (data not shown). These results demonstrate the latency to onset of the ototoxic effect of the aminoglycoside antibiotics. It was further learned in these experiments that if the dose or frequency of administration of kanamycin were increased, animals began to die of respiratory arrest at progressively earlier times and with increasing incidence. Therefore, this dosing schedule for kanamycin (400 mg/kg every 6 hr for 24 hr) is the maximum rate of kanamycin administration possible for producing an ototoxic effect without lethality. The mechanism of the latency between the time drug administration is stopped and auditory changes are detected remains unknown.

E. Drug Distribution

One explanation for the ototoxic potency differences between the different aminoglycoside antibiotics is that the drugs may distribute to the cochlea in different amounts. This hypothesis is particularly attractive because it is known that these drugs accumulate in cochlear perilymph [8,12,13,33,49,58]. To study this question the concentration of a number of aminoglycoside antibiotics was determined in plasma and perilymph following acute and chronic administration. Figures 12 and 13 show the results following subcutaneous administration of a single dose of 150 mg/kg of drug in guinea pigs. With the

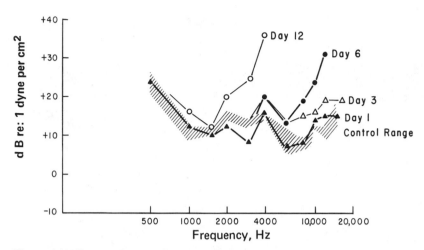

Figure 11 Preyer pinna reflex thresholds obtained for 3 consecutive days before (control range) and on the indicated days following treatment of a guinea pig with kanamycin (400 mg/kg subcutaneously every 6 hr for five doses). (Reprinted from [4], p. 39, by courtesy of Marcel Dekker, Inc.)

exception of tobramycin, the drugs behave in an identical manner. Tobramycin appears to be absorbed slightly slower than the others, but otherwise follows the same pattern of elimination. Plasma drug concentrations peak soon after injection and the plasma half-life is between 1 and 1.5 hr. Perilymph peak drug concentrations are delayed until 2–4 hr after drug injection. By 6 hr perilymph concentrations exceed those in plasma because perilymph half-lives of the drugs are much longer (5 or more hours) than plasma half-lives. Similar results have been obtained for kanamycin and neomycin. Therefore, there are no differences between drugs in distribution to plasma or perilymph following acute administration. However, the differences in plasma and perilymph half-lives of the drugs suggest a mechanism to explain why repeated doses of drugs are required to produce ototoxicity. The limiting factor in the maximum kanamycin dosing schedule described in Section III. D was acute lethality from systemic effects (respiratory arrest) of the drugs. From the values for half-lives of the drugs in plasma and perilymph it is possible to show how a maximally tolerated dose of drug given at 6 hr intervals does not accumulate in plasma to lethal levels while accumulating in perilymph to ototoxic levels. Table 2 shows the results of plasma and perilymph drug analysis 4 hr after the first dose of kanamycin in such a dosing schedule and 4 hr after the last (fifth) dose. The plasma levels show no drug accumulation, whereas kanamycin accumulated in perilymph to more than twice its original concentration.

   In a second series of experiments, drug concentrations were determined in plasma and perilymph following chronic drug treatment for 4 weeks. The samples for analysis were taken either 2 or 4 hr after the last dose so that peak perilymph concentrations could be determined for comparison with concentrations observed after a single dose of drug. Table 3 summarizes the findings. The results show that the various drugs accumulate differently in these fluids but not necessarily in relation to their relative ototoxic potencies (Table 1).

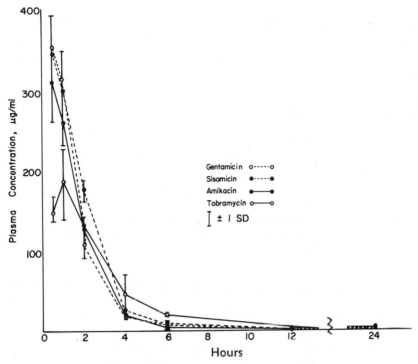

Figure 12   Guinea pig plasma concentrations of the indicated aminoglycoside antibiotics following a single 150 mg/kg subcutaneous dose of each drug. (Reprinted from [12], by permission of Academic Press, Inc.)

F. Mechanism of Action

The mechanism of the ototoxic action of the aminoglycoside antibiotics is unknown. In particular it is not known how the aminoglycosides cause the preferential death of hair cells in the organ of Corti. The highly selective nature of this toxic effect must be accounted for in any hypothesis related to the mechanism of the ototoxicity of these drugs. In a recent report [32] we have discussed the relationship between the concentration of aminoglycoside antibiotics found in perilymph of ears that have had hair cells destroyed following chronic drug administration and the concentration of these drugs required to produce cytotoxic effects in a mammalian cell culture in vitro. We found that the concentration of kanamycin in perilymph associated with complete hair cell destruction in the organ of Corti was five-thousandths or less than the concentration required to stop growth of a lymphoma cell line in vitro. Reports of similar dose-effect relationships for other in vitro cell lines and other aminoglycosides [16,31,42,46] support our conclusion that the cytotoxic effect of the aminoglycosides on the hair cells is highly selective and not simply a result of preferential accumulation of drug in perilymph.

Hair cells are difficult to study biochemically, and as a consequence there have not been definitive studies relating specific biochemical effects of the aminoglycosides on

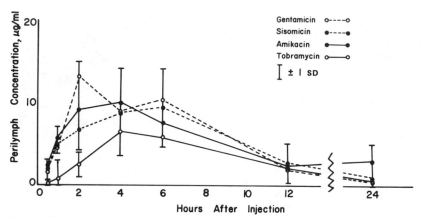

Figure 13 Guinea pig perilymph concentrations of the indicated aminoglycoside antibiotics following a single 150 mg/kg subcutaneous dose of each drug. (Reprinted from [12], by permission of Academic Press, Inc.)

hair cells to their ototoxicity. This is despite the important contributions of Thalmann et al. [50–52] to our understanding of the biochemistry of the hair cells as well as other structures in the cochlea. Recently, Schacht [45] has reported results of the effects of neomycin on phospholipid metabolism and calcium binding in the inner ear. The entire organ of Corti was studied in his experiments, so that analysis was not selective, in that it included all the supporting cells of the organ of Corti as well as the hair cells. Also, because of the 3 week daily dosing schedule followed prior to the biochemical measurements, it would appear that the animals' cochleae were quite extensively damaged. There was no assessment made of cochlear function or hair cell condition. Despite these reservations, the model presented by Schacht for the mechanism of the ototoxic action of the aminoglycosides is intriguing. Schacht proposes that the aminoglycosides disrupt phospholipid metabolism in the hair cell plasma membranes by binding between calcium sites in the membranes and the membrane phospholipids.

Table 2 Kanamycin Concentrations in Guinea Pig Plasma and Perilymph

Time (hr) after first injection[a]	Number of injections	Plasma[b] (μg/ml)	Perilymph[b] (μg/ml)
4	1	79 ± 28	24.0 ± 9.6
28	5	79 ± 37	53.9 ± 17.0

[a]Guinea pigs injected subcutaneously every 6 hr with 400 mg/kg of kanamycin.
[b]±SD.

Table 3 Aminoglycoside Concentrations in Guinea Pig Plasma and Perilymph Following Acute and Chronic Drug Administration[a]

Drug	Plasma (μg/ml) Acute	Chronic	Chronic/ acute	Perilymph (μg/ml) Acute	Chronic	Chronic/ acute
Amikacin	128	152	1.2	9	16	1.8
Gentamicin	142	402	2.8	20	33	1.6
Netilmicin	148	98	0.7	10	12	1.2
Sch 21420	13	24	1.8	9	24	2.7
Sisomicin	41	49	1.2	7	21	3.0
Tobramycin	129	152	1.2	9	16	1.8

[a]Chronic treatment was 150 mg/kg per day subcutaneously for 4 weeks except sisomicin, which was 100 mg/kg. Acute treatment was a single subcutaneous injection of the same dose as the chronic treatment. Samples in both acute and chronic were taken 2 hr following the last dose of drug except for Sch 21420, which were taken at 4 hr. Plasma concentrations are the mean of three animals and perilymph concentrations are the mean of six ears.

IV. Results of Animal Studies of the Ototoxic Interaction of the Aminoglycoside Antibiotics with "Loop" Diuretics and Sound

A. "Loop" Diuretics

The "loop" or "loop-inhibiting" diuretics are so named because they are potent inhibitors of chloride and therefore sodium transport in the loop of Henle of the nephron. Ethacrynic acid was the first loop diuretic introduced into therapy and was followed shortly by furosemide. Bumetanide and piretanide are recent additions to these therapeutically important drugs. These drugs not only share common effects on the kidney but on the ear as well. Of special interest is the fact that ethacrynic acid and furosemide are distinctly different in chemical structure, which makes it difficult to account for their similar sites of action.

The loop diuretics produce a dose-related, reversible loss of hearing in humans and animals. Indeed, the time course of depression of the ac cochlear potential by ethacrynic acid in the guinea pig (Figure 14) is identical to that described for the loss of hearing reported by patients [37]. As shown in Figure 14, following intravenous injection of 40 mg/kg of ethacrynic acid there is a rapid decrease in the ac cochlear potential which reaches a minimum by 10–15 min and then recovers completely during the following several hours. If an aminoglycoside antibiotic such as kanamycin has also been injected, the recovery phase following the initial fall of the ac cochlear potential seen with the diuretic alone is interrupted by a second fall to near-zero values (Figure 15). The ac cochlear potential never recovers after the second fall and by 24 hr the hair cells of the cochleae are destroyed [44]. The same course of events occurs with regard to hearing in humans treated with drugs from both groups [37]. This unique interaction between the aminoglycoside antibiotics and the loop diuretics was first described in humans and later studied in animals. The antibiotic given alone in a single dose has no effect [44]. As a matter of fact, this interaction is the only case in which it is clearly established that a

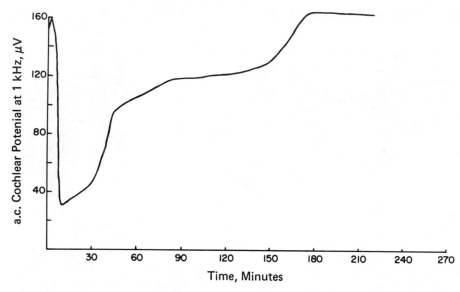

Figure 14   Ac cochlear potential of a guinea pig following a single 40 mg/kg intravenous dose of ethacrynic acid. (Reprinted from [54], p. 33. Copyright 1973, American Medical Association.)

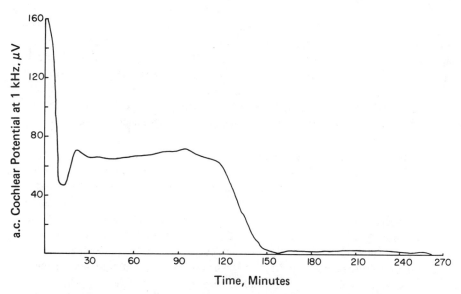

Figure 15   Ac cochlear potential of a guinea pig following a single 40 mg/kg intravenous dose of ethacrynic acid that was preceded 2 hr earlier by a single 400 mg/kg subcutaneous dose of kanamycin. (Reprinted from [54], p. 33. Copyright 1973, American Medical Association.)

438

single dose of an aminoglycoside antibiotic can be followed by permanent cochlear damage associated with hair cell destruction.

The guinea pig model has been most useful for studying the dose-effect and time-effect relationships of the interaction between the aminoglycoside antibiotics and loop diuretics [54]. Figure 16 shows the dose-effect curve obtained when the dose of kanamycin was varied while the dose of ethacrynic acid was kept at 40 mg/kg. The averaged sensitivity function for untreated guinea pigs under the conditions of this experiment was 40 dB; therefore, it is apparent that kanamycin was ototoxic even at doses lower than 50 mg/kg and was approaching maximum effect around 400 mg/kg. Figure 17 shows the dose-effect curve for varying doses of ethacrynic acid in the presence of a constant, 400 mg/kg, dose of kanamycin. The slope of the curve is very steep and suggestive of an all-or-none type of response. From a theoretical standpoint the distinct difference in the shapes of the dose-effect curves for the two drugs implies that the two drugs are acting by different mechanisms [18]; therefore, they are not producing the interaction by simple additive effects at the same site in the cochlea.

Figure 18 shows a study of the relationship between the time of administration of kanamycin and to the time of administration of ethacrynic acid [15]. Time 0 in the figure is the time of administration of ethacrynic acid. Kanamycin administration was varied between 8 hr before ethacrynic acid and up to 2 hr after. The results are striking and show that a single dose of kanamycin given as early as 8 hr before administration of ethacrynic acid will result in cochlear damage. This result is in agreement with case reports of the interaction in humans where the aminoglycoside was given many hours before the diuretic [37]. In Figure 18 maximum damage occurred when kanamycin was given simultaneously with ethacrynic acid or up to 4 hr before ethacrynic acid. The magnitude of the ototoxic effect of the interaction progressively decreased with increasing delay in the administration of kanamycin following ethacrynic acid. No effect occurred

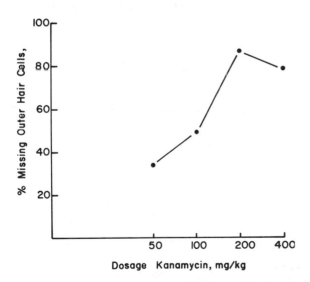

Figure 16   Percent of missing outer hair cells of guinea pigs that were treated with a single 40 mg/kg intravenous dose of ethacrynic acid 2 hr after the indicated subcutaneous doses of kanamycin. (Reprinted from [15], p. 243. Copyright 1979, American Medical Association.

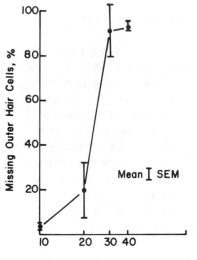

Dosage ethacrynic acid,(mg/kg)

Figure 17   Percent of missing outer hair cells of guinea pigs that were treated with a single 400 mg/kg subcutaneous dose of kanamycin followed in 2 hr by the indicated intravenous doses of ethacrynic acid. (Reprinted from [15], p. 244. Copyright 1979, American Medical Association.)

if kanamycin injection was delayed for 2 hr after administration of ethacrynic acid. The curve in Figure 18 bears a striking resemblance to the curve for aminoglycoside perilymph concentrations following a single dose of drug (Figure 19) when the latter is reversed with respect to time. This suggests that the basis of the time effect of the interaction lies in the pharmacokinetics of the aminoglycoside antibiotic in the perilymph. This hypoth-

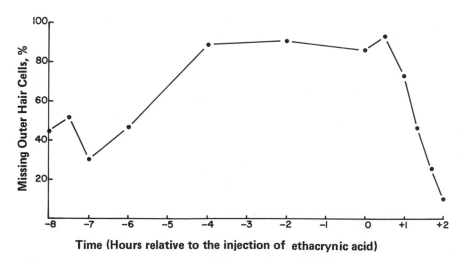

Time (Hours relative to the injection of ethacrynic acid)

Figure 18   Percent of missing outer hair cells of guinea pigs that were given a single 40 mg/kg intravenous dose of ethacrynic acid at time 0 and a single 400 mg/kg subcutaneous dose of kanamycin at the indicated times relative to the injection of ethacrynic acid; (●) mean. (Reprinted from [15], p. 245. Copyright 1979, American Medical Association.)

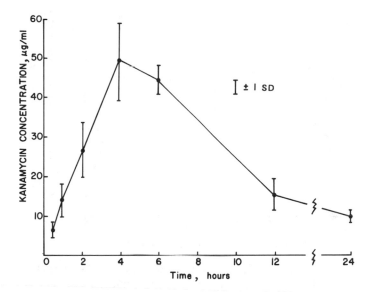

Figure 19 Guinea pig perilymph concentrations of kanamycin following a single 400 mg/kg subcutaneous dose of kanamycin.

esis rests on several assumptions. First, the effect of ethacrynic acid is to trigger the oto-toxic action of kanamycin; that is, a minimum concentration of ethacrynic acid at the site of action is required to initiate the ototoxic effect. Second, the amount of damage to the cochlea as a result of the interaction is proportional to the concentration of kana-mycin in perilymph at the time ethacrynic acid triggers the ototoxic effect. These first two assumptions are supported by the dose-effect data presented in Figures 16 and 17. Those data showed that the dose-effect relationship for ethacrynic acid was all-or-none, whereas for kanamycin it was graded. However, the interaction also occurs, albeit with diminishing magnitude, even when kanamycin is given simultaneously with ethacrynic acid or up to about 1.5 hr later. At these times the kanamycin concentration in peri-lymph at the time the ethacrynic acid was injected was obviously zero; therefore, the triggering effect of ethacrynic acid must last for nearly 2 hr following its intravenous in-jection. A third assumption is that equal ototoxic effects are produced by equal concen-trations of kanamycin in perilymph at the time ethacrynic acid triggers the response. By examination of Figure 18 it is noted that the magnitude of the interaction effect pro-duced when kanamycin was given 8 hr before ethacrynic acid (-8 hr) is equal to that when kanamycin was given 1.5 hr after ethacrynic acid (+1.5 hr). Therefore, the concen-tration of kanamycin in perilymph is assumed to be equal at these two times. Now, if the perilymph concentration curve following a single dose of 400 mg/kg of kanamycin (Figure 19) is reflected about its ordinate, superimposed on the time coordinate (abcissa) of Figure 18, and aligned such that the perilymph kanamycin concentrations are equal at -8 and +1.5 hr, we obtain the result shown in Figure 20. The parallelism between the time-effect curve and the theoretical perilymph kanamycin curve is striking, including the fact that both curves peak at about the same time.

The preceding analysis supports the hypothesis that the concentration of kanamycin present in the cochlea at the time ethacrynic acid is injected is the major determinant

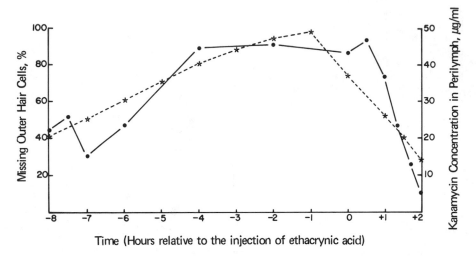

Figure 20   Relationship between the percent of missing outer hair cells (●, mean) resulting from the combined administration of kanamycin and ethacrynic acid to guinea pigs (Figure 18) and the theoretical concentration of kanamycin (☆) in perilymph at the time of ethacrynic acid injection.

of the magnitude of the ototoxic effect due to the interaction between the two drugs. In addition, the pharmacokinetics of kanamycin in the cochlea account for the time effect of the interaction.

The time course of the effect on cochlear ultrastructure of the interaction between kanamycin and ethacrynic acid has also been studied in the guinea pig [44]. Changes were seen in the outer hair cells within 3 hr following ethacrynic acid when 400 mg/kg of kanamycin was given subcutaneously 2 hr before an intravenous dose of 40 mg/kg of ethacrynic acid. These changes appear in electronmicrographs of outer hair cells as clumping of nuclear chromatin and loss of the subsurface cisternae from the plasma membranes (Figure 21). No effects were seen in the mitochondria, although mitochondrial changes have been reported following chronic exposure to the aminoglycoside antibiotics alone. Also, only the outer hair cells of the organ of Corti were affected. By 6 hr the hair cells and their nuclei were markedly swollen (Figure 22). Mitochondria were also swollen and distorted at this time. By 24 hr the outer hair cells were completely destroyed (Figure 23). Interestingly, the inner hair cells still appeared normal at 24 hr although it is known that they too eventually die. The outer hair cells were unchanged following injection of either drug alone. Ethacrynic acid alone, however, caused a marked edema of the stria vascularis which, like the effects of this drug alone on cochlear function, was completely reversible [44]. The edema of the stria vascularis also occurred in the interaction, and it appeared that kanamycin increased the magnitude and duration of this effect and very possibly caused death of some cells in the stria vascularis. Kanamycin alone at this dosage had no effect on the stria vascularis.

One possible mechanism proposed to explain the interaction was that ethacrynic acid altered the permeability of the hair cells to kanamycin, causing those cells to be exposed rapidly to ototoxic concentrations of the aminoglycoside. However, studies of the effect of ethacrynic acid on the concentration of kanamycin in perilymph during the interaction

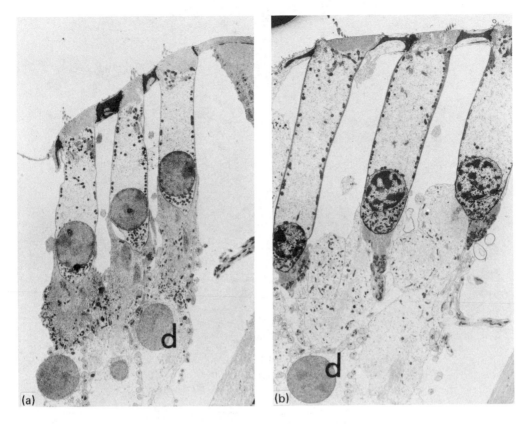

Figure 21   Electronmicrographs of outer hair cells of the basal turns of guinea pig coch-leae. Guinea pigs were injected with either saline subcutaneously followed in 2 hr by saline intravenously (SAL/SAL) or with 400 mg/kg of kanamycin subcutaneously followed in 2 hr by 40 mg/kg of ethacrynic acid intravenously (KAN/EA). Cochleae were fixed for electron microscopy 3 hr after the second injection. (a) SAL/SAL control. Hair cells are elongated and nuclei are similar in appearance to those of the supporting Deiter's cells (d) beneath. (×1800) (b) KAN/EA treated. Nuclei are crenated and contain clumped chromatin. Note normal appearance of Deiter cell nucleus. (×1800) (Reprinted from [44], by permission of *Acta Oto-Laryngologica*.)

do not support this hypothesis if it is assumed the perilymph levels of kanamycin reflect the concentration of drug in the hair cells [44]. Although the mechanism of the inter-action is unknown, it is clear that it is selective for hair cells since although both kanamy-cin and ethacrynic acid are cytotoxic for mammalian cells growing in vitro, their combined effects on these cells are simply additive and show no potentiation, as observed in their ototoxic interaction on hair cells in vivo [32].

   The other loop diuretics, furosemide, bumetanide, and piretanide, have also been shown to produce an ototoxic interaction with kanamycin in the guinea pig (unpublished observations). In the case of furosemide a detailed dose-effect study was performed. The dose-effect relationships for ethacrynic acid and furosemide are shown in Figure 24. Ethacrynic acid is more potent that furosemide despite the fact that their diuretic potencies are about equal. Therefore, it is possible to separate the ototoxic interaction potency of

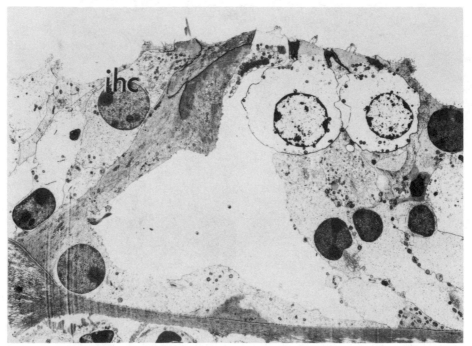

Figure 22 Electronmicrograph of organ of Corti of a guinea pig 6 hr after second injection following KAN/EA (see legend, Figure 21). Outer hair cells show severe damage, whereas the inner hair cell (ihc) looks normal. (×1800) (Reprinted from [44], by permission of *Acta Oto-Laryngologica.*)

the loop diuretics from their diuretic potency. Other types of diuretics do not produce this interaction. They are: mannitol, hydrochlorothiazide, mercuhydrin, and acetazolamide [7].

All the aminoglycoside antibiotics interact with the loop diuretics. This includes netilmicin, which is not ototoxic at the highest doses tested (Figure 8). Although definitive dose-effect data are lacking for the aminoglycosides in the interaction, it appears that they do not differ greatly in their potency. In addition, three nonaminoglycoside drugs have been shown to interact with the loop diuretics in a manner identical to kanamycin. These are polymyxin B, viomycin, and capreomycin [27]. Polymyxin E and vancomycin were also tested and found to be ineffective [27].

B. Sound

It is well known that high-intensity sound can produce damage to the cochlea resembling that produced by the aminoglycoside antibiotics [20]. Indeed, when broad-band sound is used as the damaging agent, the damage begins at the basal end of the cochlea, affecting hearing in the high frequencies first just as with the aminoglycoside antibiotics. Therefore, it is not surprising that investigators considered the possibility that sound and the aminoglycosides might interact. The results of early experiments by Vernon et al. (unpub-

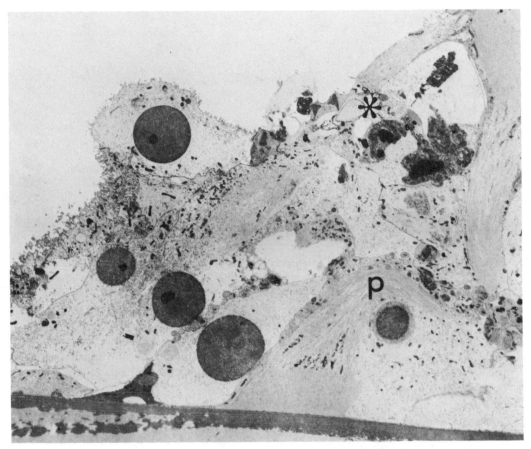

Figure 23   Electronmicrograph of organ of Corti of guinea pig 24 hr after second injection following KAN/EA (see legend, Figure 21). All three rows of outer hair cells have been selectively destroyed (*). The pillar cells (p) have collapsed and the tunnel filled in by adjacent cells. (×2400) (Reprinted from [44], by permission of *Acta Oto-Laryngologica*.)

lished observations) using guinea pigs treated with 400 mg/kg of kanamycin and exposed for a short duration to 120 dB sound gave no evidence of an interaction between the two agents. However, later reports by Dayal et al. [28,29] and Hawkins et al. [24], in which longer exposures to sound were used, suggested that an interaction did occur. Jauhiainen et al. [36] clearly showed an interaction between neomycin and sound when both agents were given for long periods of time. Brown et al. [3] repeated their experiment with neomycin and in addition studied kanamycin. The results of the experiment with neomycin are shown in Figure 25. Both sound and neomycin administered alone to guinea pigs produced a small ototoxic effect, as shown by the slight shifts to the right and reductions of the maxima of the intensity-function curves. However, both agents administered together produced a marked ototoxic effect that was greater than the sum of their individual effects. Concentrations of neomycin were determined in perilymph during the interaction but were found to be unchanged by the presence of sound. Therefore, it does not appear

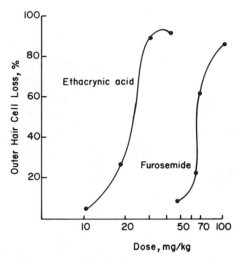

Figure 24   Dose-effect curves of the percent of outer hair cell loss in guinea pigs resulting from the administration of a single 400 mg/kg subcutaneous dose of kanamycin followed in 2 hr by the indicated intravenous doses of ethacrynic acid or furosemide.

that redistribution of drug due to an effect of sound can explain the interaction. From these studies in animals it would appear that there is a risk to humans of ototoxicity from this interaction if humans are exposed to *very high* levels of sound for a prolonged time during treatment with aminoglycoside antibiotics.

Vernon et al. [57] looked for an ototoxic interaction between the loop diuretics ethacrynic acid and furosemide, and sound. They were unable to demonstrate such an interaction even after several days of exposure of guinea pigs to drug and sound combined.

V.   Relationship of Animal Studies to Humans

Since there are numerous instances in which the qualitative or quantitative effects of drugs in various species of animals differ markedly from those found in humans, animal data are always suspect with respect to their relevance for human beings. Therefore, it is necessary to ask how well the animal models of aminoglycoside antibiotic ototoxicity fit the experience with these drugs when used in humans. In the case of the guinea pig, for which the greatest amount of data exists, it can be said that the model fits humans very well. First, as pointed out in our discussion of ototoxic pathology, the site and pattern of the cochlear lesion due to the aminoglycosides is identical in humans and guinea pigs. Second, the functional effects of these drugs on the cochlea in both species is identical. Humans and guinea pigs both lose their ability to hear high-frequency sound first. Furthermore, it can be shown that changes in the ac cochlear potential, so often used in animal studies of drug ototoxicity, reflect changes in sound perception by animals. For example, loss of the ac cochlear potential is correlated with loss of behavioral hearing in guinea pigs, as shown in Figure 26. This figure shows the correlation between changes due to kanamycin in behaviorally determined hearing thresholds with ability to generate

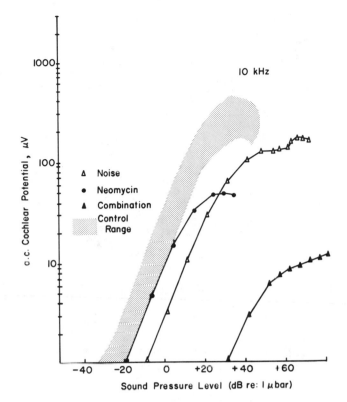

Figure 25 Maximum output of the ac cochlear potentials at 10 kHz from guinea pigs treated with noise alone (115 dB white noise, 10 hr/day for 7 days), neomycin alone (200 mg/kg subcutaneously daily for 7 days), and the combination of noise and neomycin. These data were obtained by increasing the sound pressure at 10 kHz in 5 dB increments until no further increase occurs in the ac cochlear potential. Sound pressure increases to the right along the horizontal axis. (Reprinted from [3], by permission of *Acta Oto-Laryngologica.*)

the ac cochlear potential. The guinea pigs used in the experiment had been made monaural by the destruction of one ear and then trained to respond to a given frequency of sound. Third, the time-effect relationships of the drugs are similar between guinea pigs and humans. In both species ototoxicity occurs only after chronic drug administration and can be demonstrated, using the appropriate dosing schedule, to occur with a latency after drug administration has been stopped. Fourth, although ototoxic doses of aminoglycosides on a milligram per kilogram basis are greater in the guinea pig than in humans, it can be argued that this is an artifact of size, since for some drugs the magnitude of the effect is related to dose based on body surface area rather than body weight [17,40]. For example, the average daily dose of gentamicin recommended for humans is 3 mg/kg, which may be ototoxic, depending on duration of therapy and ability of the patient to eliminate the drug. If this dose is computed in terms of body surface area of a 70 kg human and then this new dosing unit is applied to a 500 g guinea pig and converted back to a weight-based dose, the result is that a 15.0 mg/kg dose in the guinea pig is equal in

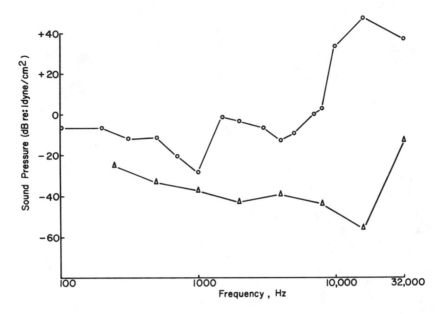

Figure 26   One microvolt isopotential functions of the ac cochlear potential (0) and the auditory thresholds (△) of a guinea pig treated with daily subcutaneous doses of kana-mycin (200 mg/kg) for 2 weeks. Both measures of auditory function show a decrease at the high frequencies.

magnitude to a 3 mg/kg dose in a human. A dose of 50 mg/kg of gentamicin is ototoxic in the guinea pig when given daily for 2 weeks. Therefore, it is clear that the doses used in guinea pig studies, although purposely high enough to be ototoxic, are not greatly out of line with those known to be ototoxic in humans. Fifth, the biological half-lives of the aminoglycoside antibiotics in the blood of humans and guinea pigs are similar [8,12,13, 39]. The pharmacokinetics of the aminoglycosides in perilymph have not been exten-sively studied in humans for obvious reasons, but zum Gottesberge and Stupp [19] found that streptomycin assayed in human perilymph 2 and 5 hr after a therapeutic dose of drug appeared to reach concentrations similar to those found in guinea pigs. Sixth, in the section on interactions with loop diuretics it was pointed out that the guinea pig and human react to the drugs in a similar way.

In summary, the guinea pig has proven to be an excellent model for humans with regard to the ototoxic effects of the aminoglycoside antibiotics. From what is known of the response of other animals such as the cat and monkey to the ototoxicity of these drugs, it can be assumed that they are also good models for humans. However, guinea pigs are less expensive and easier to handle experimentally; therefore, the guinea pig is an ideal animal for the preclinical testing of new aminoglycoside antibiotics and hopefully will aid in eventually determining the mechanism of the ototoxic effect of these drugs.

Acknowledgments

Support for research reported in this chapter came from the following sources: NINCDS Grant NS 12808, NINCDS Grant NS 09889-0452, Schering Corporation, and Eli Lilly and Company.

References

1. Adrian, E.D.: The microphonic action of the cochlea; and interpretation of Wever and Bray's experiments. *J. Physiol.* 71:28–29, 1931.
2. Adrian, E.D., Bronk, D.W., and Phillips, G.: The nervous origin of the Wever and Bray effect. *J. Physiol.* 73:2P–3P, 1931.
3. Brown, J.J., Brummett, R.E., Meikle, M.B., and Vernon, J.A.: Combined effects of noise and neomycin. *Acta Otolaryngol. (Stockh.)* 86:394–400, 1978.
4. Brummett, R.E.: Ototoxity: its detection and prevention. In *Horizons in Clinical Pharmacology*, edited by Palmer, R.F. New York, Marcel Dekker, 1976, pp. 31–45.
5. Brummett, R.E.: Comparative ototoxicity of aminoglycoside antibiotics in a guinea pig model. In *Assessment of Aminoglycoside Toxicity*. Burgenstock, Eli Lilly, 1978, pp. 56–61.
6. Brummett, R.E., Himes, D.L., Saine, B., and Vernon, J.A.: A comparative study of the ototoxicity of tobramycin and gentamicin. *Arch. Otolaryngol.* 96:505–512, 1972.
7. Brummett, R.E., West, B.A., Traynor, J., and Manor, N.: Ototoxic interaction between aminoglycoside antibiotics and diuretics. *Toxicol. Appl. Pharmacol.* 29:45, 1974.
8. Brummett, R.E., and Fox, K.E.: Comparative ototoxicity of gentamicin and gentamicin C_1. *Pharmacologist* 17:248, 1975.
9. Brummett, R.E., and Brown, R.T.: Tobramycin ototoxicity: a second look. *Arch. Otolaryngol.* 101:540–543, 1975.
10. Brummett, R.E., and Fox, K.E.: Comparative ototoxic liability of *Sch 20569* and gentamicin—electrophysiology and morphology. *Fed. Proc.* 35:621, 1976.
11. Brummett, R.E., and Fox, K.E.: Comparative ototoxicity of BB-K8 (amikacin), gentamicin, sisomicin and tobramycin in the guinea pig. *Proc. West. Pharmacol. Soc.* 20:449–453, 1977.
12. Brummett, R.E., Fox, K.E., Bendrick, T.W., and Himes, D.L.: Ototoxicity of tobramycin, gentamicin, amikacin and sisomicin in the guinea pig. *J. Antimicrob. Chemother.* 4(Suppl. A):73–83, 1978.
13. Brummett, R.E., Fox, K.E., Brown, R.T., and Himes, D.L.: Comparative ototoxic liability of netilmicin and gentamicin. *Arch. Otolaryngol.* 104:579–584, 1978.
14. Brummett, R.E., Fox, K.E., Bendrick, T.W., and Himes, D.L.: Comparative ototoxicity of tobramycin, gentamicin, amikacin and sisomicin. *Curr. Chemother.* 2:939–941, 1978.
15. Brummett, R.E., Brown, R.T., and Himes, D.L.: Quantitative relationships of the ototoxic interaction of kanamycin and ethacrynic acid. *Arch. Otolaryngol.* 105:240–246, 1979.
16. Casemore, D.: Gentamicin as a bactericidal agent in virological tissue culture. *J. Clin. Pathol.* 20:298–299, 1967.
17. Clark, A.J.: *Handbuch der Experimentellen Pharmakologie*, Vol. 4: *General Pharmacology*, edited by Heubner, W., and Schuller, J. Berlin, Julius Springer, 1937, pp. 166–169.
18. Goldstein, A., Aronow, L., and Kalman, S.M.: *Principles of Drug Action* (2nd ed.). New York, Wiley, 1974, p. 91.

19. zum Gottesberge, A.M., and Stupp, H.F.: Streptomycinspiegel in der Perilymphe des Menschen. *Acta Otolaryngol. (Stockh.)* 67:171–176, 1969.

20. Hawkins, J.E.: Comparative otopathology: aging, noise, and ototoxic drugs. *Adv. Otorhinolaryngol.* 20:125–141, 1973.

21. Hawkins, J.E., Rahway, N.J., and Lurie, M.H.: The ototoxicity of streptomycin. *Ann. Otol. Rhinol. Laryngol.* 61:789–809, 1952.

22. Hawkins, J.E., Berger, V., and Aran, J.M.: Antibiotic insults to Corti's organ. In *Sensorineural Hearing Processes and Disorders*, edited by Graham, A.D., Boston, Little, Brown, 1967, pp. 411–425.

23. Hawkins, J.E., Johnsson, L.G., and Aran, J.M.: Comparative tests of gentamicin ototoxicity. *J. Infect. Dis.* 119:417–426, 1969.

24. Hawkins, J.E., Marques, D.M., and Clark, C.S.: Noise and kanamycin interaction in the guinea pig cochlea. *J. Acoust. Soc. Am.* 58(Suppl. 1):(Abstr.), 1975.

25. Hawkins, J.E., Stebbins, W.C., Johnsson, L.G., Moody, D.B., and Muraski, A.: The patas monkey as a model for dihydrostreptomycin ototoxicity. *Acta Otolaryngol. (Stockh.)* 83:123–129, 1977.

26. Davis, H., Deatherage, B.H., Rosenblut, B., Fernandez, C., Kimura, R., and Smith, C.A.: Modification of cochlear potentials produced by streptomycin poisoning and by extensive venous obstruction. *Laryngoscope* 68:596–629, 1958.

27. Davis, R.R., and Brummett, R.E.: Cochlear destruction resulting from an interaction of some non-aminoglycoside antibiotics and ethacrynic acid. *Pharmacologist* 21:144, 1979.

28. Dayal, V.S., Kokshanian, A., and Mitchell, D.P.: Combined effects of noise and kanamycin. *Ann. Otol. Rhinol. Laryngol.* 80:897–902, 1971.

29. Dayal, V.S., and Barek, W.G.: Cochlear changes from noise, kanamycin and aging: Part II. Potentiating effects of noise and kanamycin. *Laryngoscope* 85(Suppl. 1): 8–11, 1975.

30. Engstrom, H., Ades, H.W., and Anderson, A.: *Structural Pattern of the Organ of Corti.* Stockholm, Almqvist & Wiksell, 1966.

31. Fischer, A.: Gentamicin as a bactericidal antibiotic in tissue culture. *Med. Microbiol. (Berl.)* 161:23–39, 1975.

32. Fox, K.E., and Brummett, R.E.: The relationship between the cytotoxicity of kanamycin and ethacrynic acid for mammalian cells in vitro and their ototoxicity in vivo. *Acta Otolaryngol. (Stockh.)* 87:72–78, 1979.

33. Fox, K.E., Brummett, R.E., Brown, R., and Himes, D.: A comparative study of the ototoxicity of gentamicin and gentamicin C_1. *Arch. Otolaryngol.* 106:744–750, 1980.

34. Frost, J.O., Daly, J.F., and Hawkins, J.E.: The ototoxicity of kanamycin in man. *Antibiot. Annu.* 1958–1959:700–707, 1959.

35. Frost, J.O., Hawkins, J.E., and Daly, J.F.: Kanamycin: II. Ototoxicity. *Annu. Rev. Resp. Dis.* 82:23–30, 1960.

36. Jauhiainen, T., Kohonen, A., and Jauhianinen, M.: Combined effects of noise and neomycin on the cochlea. *Acta Otolaryngol. (Stockh.)* 73:387–390, 1972.

37. Johnson, A.H., and Hamilton, C.H.: Kanamycin ototoxicity—possible potentiation by other drugs. *South. Med. J.* 63:511–513, 1970.

38. Kaneko, Y., Nakagawa, T., and Tanaka, K.: Reissner's membrane after kanamycin administration. *Arch. Otolaryngol.* 92:457–462, 1970.

39. Leroy, A., Humbert, G., Oksenhendler, G., and Fillastre, J.-P.: Pharmacokinetics of aminoglycosides in subjects with normal and impaired renal function. In *Antibiotics and Chemotherapy*, Vol. 25: *Pharmacokinetics*, edited by Schonfield, H. Basel, S. Karger, 1978, pp. 163–180.

40. Levine, R.R.: *Pharmacology: Drug Actions and Reactions* (2nd ed.). Boston, Little, brown, 1978, pp. 242–243.

41. Lundquist, P.G., and Wersall, J.: Kanamycin-induced changes in cochlear hair cells of the guinea pig. *Z. Zellforsch.* 72:543–561, 1966.

42. Maskowitz, M., and Kelkar, N.: Sensitivity of cultured mammalian cells to streptomycin and dihydrostreptomycin. *Science* 141:647–648, 1963.

43. Owada, K.: Experimental studies on the toxicity of kanamycin, its hydrolyzed products and neomycin. *Chemotherapia* 5:277–293, 1962.

44. Russell, N.J., Fox, K.E., and Brummett, R.E.: Ototoxic effects of the interaction between kanamycin and ethacrynic acid: cochlear ultrastructure correlated with cochlear potentials and kanamycin levels. *Acta Otolaryngol.,* accepted for publication, 1979.

45. Schacht, J.: Biochemistry of neomycin ototoxicity. *J. Acoust. Soc. Am.* 59:940–944, 1977.

46. Schafter, T., Pascale, A., Shimonaski, G., and Came, P.: Evaluation of gentamicin for use in virology and tissue culture. *Appl. Microbiol.* 23:565–570, 1972.

47. Shambaugh, G.E., Derlacki, E.L., Harrison, W.H., House, H., House, W., Hildyard, V., Schuknecht, H., and Shea, J.J.: Dihydrostreptomycin deafness. *JAMA* 170:1657–1660, 1959.

48. Simmons, F.B., Galambas, R., and Albrite, J.P.: Serial studies of the onset and progression of drug-induced cochlear damage in cats. *Arch. Otolaryngol.* 102:233–239, 1960.

49. Stupp, H., Rauch, S., Sous, H., Braun, J.P., and Lagler, F.: Kanamycin dosage and levels in ear and other organs. *Arch. Otolaryngol.* 86:515–521, 1967.

50. Thalmann, R., Thalmann, I., and Comegys, T.H.: Dissection and chemical analysis of substructures of the organ of Corti. *Laryngoscope* 80:1619–1645, 1970.

51. Thalmann, R., Thalmann, I., and Comegys, T.H.: Quantitative cytochemistry of the organ of Corti. Dissection, weight determination and analysis of single outer hair cells. *Laryngoscope* 82:2059–2078, 1972.

52. Thalmann, R., Ise, I., Bohne, B., and Thalmann, I.: Actions of "loop" diuretics and mercurials upon the cochlea. *Acta Otolaryngol. (Stockh.)* 83:221–232, 1977.

53. Theopold, H.M.: Comparative surface studies of ototoxic effects of various aminoglycoside antibiotics on the organ of Corti in the guinea pig. *Acta Otolaryngol. (Stockh.)* 84:57–64, 1977.

54. West, B.A., Brummett, R.E., and Himes, D.L.: Interaction of kanamycin and ethacrynic acid—severe cochlear damage in guinea pigs. *Arch. Otolaryngol.* 98:32–37, 1973.

55. Wever, E.G., and Bray, C.W.: Auditory nerve impulses. *Science* 71:215, 1930.

56. Vernon, J.A., and Smith, C.A., eds.: *Handbook of Auditory and Vestibular Research Methods.* Springfield, Ill. Charles C Thomas, 1976.

57. Vernon, J.A., Brummett, R.E., and Brown, R.T.: Noise trauma induced in the presence of loop-inhibiting diuretics. *Trans. Am. Acad. Opthalmol. Otolaryngol.* 84:407–413, 1977.

58. Voldrich, L.: The kinetics of streptomycin, kanamycin and neomycin in the inner ear. *Acta Otolaryngol. (Stockh.)* 60:243–248, 1965.

19

OTOTOXICITY: CLINICAL CONSIDERATIONS AND COMPARATIVE INFORMATION

C. L. BENDUSH   Eli Lilly and Company,
Indianapolis, Indiana

Aminoglycoside drugs have proven to be ototoxic when used parenterally, orally, or topically. Members of this class are similar with respect to their histopathologic and pathophysiologic effects. However, they have shown significant differences in their quantitative toxic effects and in their tendency to adversely affect either the auditory or the vestibular apparatus. Minimizing the risk of ototoxicity for the patient depends upon an appreciation of the innate toxic characteristics of the individual drugs and the potentially aggravating effects of certain clinical risk factors, the most important of which is probably impaired renal function. Therefore, monitoring renal function and making appropriate dosage adjustments are of paramount importance in minimizing ototoxicity. The measurement of peak and trough serum aminoglycoside levels indicates the extent to which the drug is being excreted by the kidneys. Consequently, impaired renal excretion of the aminoglycoside predisposes the patient to ototoxicity unless the serum levels are maintained within a predetermined range, in which case ototoxicity and renal dysfunction are independent of each other [112]. Although maintaining peak and trough levels below those recommended will minimize the risk of ototoxicity, reports have appeared in the literature of such patients who have sustained ototoxicity.

The current state of knowledge regarding the occurrence of aminoglycoside ototoxicity in the clinic, or indeed comparisons of ototoxicity between various aminoglycosides, is limited because:

1. Baseline function tests are frequently difficult, if not impossible, to obtain because patients are too ill to cooperate or occasionally the necessary facilities are not readily available.
2. Aminoglycoside ototoxicity is nonspecific; therefore, the problem of distinguishing it from auditory nerve dysfunction due to other causes is often difficult, if not impossible.
3. There are no standardized definitions for cochlear or vestibular impairment.
4. Little is known about the possible accelerating effect of infectious illnesses or effects of drugs used in their treatment on preexisting "normal" hearing loss (presbycusis).
5. Standardized clinical trials have not been undertaken frequently enough to compile prospective comparative clinical data for the various aminoglycosides.

Occasionally, patients are started on aminoglycoside therapy without baseline auditory function studies, then after several days of treatment tests are done that show cochlear or vestibular dysfunction. Under such circumstances, it is impossible to determine whether the impairment is due to the aminoglycoside or to other causes. Such circumstances not only complicate establishment of a causal relationship but also create a dilemma for the clinician regarding continued therapy.

Evaluation of various aminoglycosides in animal models has allowed a comparison of their ototoxic potential when the drugs are given in equal amounts. However, extrapolation of such data to the clinical setting is complicated by the fact that these drugs are usually given to patients at much lower doses and for shorter periods of time than those used in animal studies. Also, they are given according to different dosage schedules; in humans the dosage of amikacin and kanamycin is three to five times higher (15 mg/kg per day) than that of tobramycin or gentamicin (3 to 5 mg/kg per day).

Aminoglycosides in current use are generally well tolerated when given according to the currently accepted guidelines. Nevertheless, it is important to select patients for whom the benefits will outweigh the potential risk of eighth-nerve impairment, since aminoglycoside damage to cells of the inner ear is permanent, although the symptoms of ototoxicity, excluding hearing loss, usually show improvement or abate entirely.

I. Historical Overview

The pressing needs of World War II and the ineffectiveness of penicillin against most gram-negative organisms encouraged research by both government and private industry for new antibiotics. The advent of the aminoglycosides was recorded in 1944 with the announcement by Schatz et al. [107] of their discovery of streptomycin from a strain of *Streptomyces griseus*. The isolation of this antibiotic resulted from a well-planned systematic search of soil actinomyces. This approach to the development of antibiotics contrasts significantly with the more or less serendipitous discovery of penicillin and opened the door for future development of important new antibiotics.

Early in the clinical experience with streptomycin, Brown and Hinshaw [17] reported the occurrence of vestibular disturbance. Later, in 1949, dihydrostreptomycin, an analog of streptomycin, became available for therapeutic use, primarily because it was less vestibulotoxic [60,61]. Unfortunately, deleterious effects on one apparatus were replaced by those on another; it was soon apparent that dihydrostreptomycin, although not as vestibulotoxic as streptomycin, was more cochleotoxic and resulted in a higher incidence of hearing loss. Because the clinical use of dihydrostreptomycin was characterized by a significant risk of cochlear impairment, including latent irreversible deafness [32], and because its therapeutic effectiveness was apparently no greater than that of streptomycin, approval for its continued medical use was withdrawn by the United States Food and Drug Administration.

Eventually, the inherent ototoxic characteristics of the aminoglycosides were to be confirmed for all members of this antibiotic class. Clearly, ototoxicity now became a major clinical concern in the use of these antibiotics.

The published literature contains a number of reports describing attempts to minimize or prevent the toxicity of streptomycin and dihydrostreptomycin [33,56,76,89,101]. Studies ranged from aminoglycoside combinations with pantothenic acid and its calcium salts to every-other-day dosage regimens [72,88]. One such attempt, based on the theory that streptomycin and dihydrostreptomycin were specifically vestibular- or cochlear-toxic, suggested that administering half the usual dose of streptomycin combined with half the dose of dihydrostreptomycin would reduce the risk of ototoxicity [97]. These attempts did not significantly reduce the occurrence of ototoxicity.

The need for antibiotics with a broader spectrum of activity against pathogens not susceptible to streptomycin, together with the premise that molecular changes might result in less toxicity, led to the development of a number of other aminoglycosides.

Neomycin and the more recently introduced kanamycin, gentamicin, tobramycin, and amikacin can be classified by their predisposition for toxicity to either the vestibular or cochlear functions. Whereas neomycin, kanamycin, and amikacin may be characterized by their predominantly cochlear toxicity, gentamicin and tobramycin are capable of producing both kinds of toxicity; however, in the case of gentamicin, vestibular toxicity clearly predominates over cochlear toxicity.

Numerous clinical and experimental animal studies of aminoglycoside ototoxicity have been published in the last 25 years. These studies have confirmed the ototoxic potential for all aminoglycoside antibiotics. Although the data from animal investigations are not directly applicable to humans, such studies have increased our understanding of this important problem. They have facilitated detailed comparative studies of aminoglycosides, so that it is possible to rank them according to their ototoxic potential.

Although the incidence of ototoxicity reported in the clinical literature is usually low, especially with newer aminoglycosides, when damage does occur it is irreversible. For proper assessment of this potential risk in the use of aminoglycosides, it is important to appreciate the information currently available on the mechanisms of toxicity as they relate to the anatomy and physiology of the ear and the pharmacokinetics of these drugs in the inner ear.

II. Anatomy and Functions of the Ear

The human ear, as the sense organ of hearing, is a highly differentiated and specialized receptor that receives and transmits to the central nervous system information about the acoustic environment.

Anatomically, the ear is divided into three parts: the external ear, the middle ear, and the inner ear (Figure 1).

A. External and Middle Ear

Anatomy

The external ear consists of a pinna, auditory canal, and supporting structures. It is bounded on its inward side by the tympanic membrane.

The middle ear, or tympanic cavity, is an air-filled chamber in the temporal bone that opens via the eustachian tube into the nasopharynx. It contains a bridge with three auditory ossicles: the malleus, incus, and the stapes. The malleus forms one end of the bridge and is attached to the inner surface of the tympanic membrane; the opposite end of the bridge attaches to the oval window of the cochlea via the foot plate of the stapes (Figure 1).

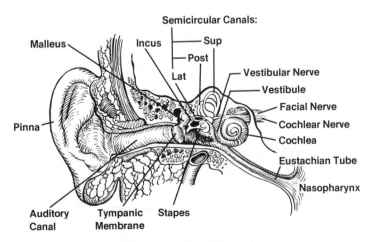

Figure 1   Anatomical relationships of the human ear. To make the relationships clear, the cochlea has been turned slightly and the middle ear muscles have been omitted. Sup, superior; Post, posterior; Lat, lateral. (Modified and redrawn from Brödel: *Three Unpublished Drawings of the Anatomy of the Human Ear,* Philadelphia, W.B. Saunders, 1946. (Reproduced by permission of W.B. Saunders Company.)

Function

Sound waves are directed into the auditory canal by the pinna of the external ear. They are funneled down the auditory canal, where they strike, and set in motion, the tympanic membrane. Sound-wave energy is thus converted to mechanical energy at the tympanic membrane and is transmitted by the bridge of auditory ossicles to the oval window of the cochlear duct (Figures 1 and 3).

B. Inner Ear

Anatomy

The inner ear (Figure 2) consists of two parts: the membranous labyrinth inside the osseous labyrinth. The space between them is filled with a limpid fluid, the perilymph.

The osseous labyrinth consists of the cochlea, vestibule, and three bony semicircular canals (Figure 2). The entire cavity is lined with a thin fibroserous membrane. The membrane forms a narrow tunnel, the perilymphatic duct, that connects to the subarachnoid space.

The membranous labyrinth consists of two sacs, the utricle and the saccule, as well as three semicircular canals and the cochlear duct. The membranous semicircular ducts open at both ends into the utricle. The utriculosaccularis duct connects the utricle with the saccule via a union with the endolymphatic duct. The endolymphatic duct originates in the saccule and ends as a blind pouch (endolymphatic sac) in the subdural space.

The saccule communicates with the cochlear duct via a tiny tube, the canalis reuniens. The membranous apparatus contains endolymph that is separate from and does not mix with perilymph.

The inner ear is innervated by the auditory nerve (eighth cranial nerve), which is sensory only and contains two branches. These differ in their function, origin, and destination. One branch is known as the cochlear nerve, and the other as the vestibular nerve (Figure 1).

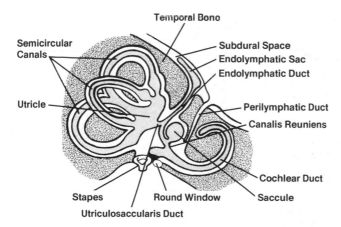

Figure 2 Relationship between the membranous and osseous labyrinths. [Modified from W.F. Ganong: *Review of Medical Physiology* (9th ed.), Los Altos, Calif., Lange Medical Publications, 1979. Reproduced by permission of the publisher.]

Function

The membranous labyrinth is a dynamic organ that is concerned with two separate sensory functions: (1) as a receiver and transducer of sound wave energy via the cochlear duct, it is the organ of hearing; and (2) as a sensor for positional changes in fluid pressure via the utricle, saccule, and semicircular canals, it is the organ of equilibrium.

The cochlear and vestibular apparatuses perform their functions as a result of excitation of specialized neuroepithelial hair cells located within discrete structures in each organ.

C. Cochlea

Anatomy

The cochlea forms the anterior part of the osseous labyrinth and is positioned horizontally in front of the vestibule (Figure 1). Resembling a snail shell, it consists of a spiral osseous canal of 2¾ turns around a hollow, conical central pillar called the modiolus. The osseous spiral lamina projects into the canal coiling around the modiolus.

The basilar membrane (Figure 3) stretches from the spiral lamina to the spiral ligament or outer wall of the bony cochlea and completely divides its cavities into two passages, or scalae, which communicate with each other at the apex of the modiolus by a small opening. The upper passage is the scala vestibuli, which terminates at the oval window, and the lower is the scala tympani, terminating at the round window (Figure 4). A second partition, Reissner's membrane, extends from the spiral lamina to the spiral ligament. These membranes enclose a triangular area called the scala media or cochlear duct, which is located between the scala vestibuli above and the scala tympani below and contains endolymph (Figure 3).

The cochlear duct (scala media) ends blindly and is connected near its basal end to the saccule. It contains the organ of Corti, also referred to as the spiral organ of Corti, which is located along the inner part of the basilar membrane, and conforms to the spiral shape

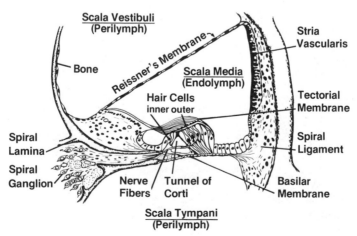

Figure 3   Cross section of one turn of the cochlea of a guinea pig. (Modified; reproduced with permission from Davis et al.: Trauma in the guinea pig, *J. Acoust. Soc. Am.* 25: 1180, 1953.)

of the cochlea. The organ of Corti with the overlying tectorial membrane projects into the cochlear duct. The organ of Corti, which is the organ of hearing, contains the auditory neuroepithelium (i.e., sensory hair cells) together with various supporting cells and nerve fibers (Figure 3). On the medial side of the tunnel of Corti, a single row of inner hair cells is orderly arranged. On the lateral side are three more parallel rows of outer hair cells. The outer hair cells number about 12,000 and the inner cells only 3500.

Function

Mechanical sound energy is transmitted from the middle ear to the cochlea via the foot plate of the stapes. The foot plate is attached to the oval window of the cochlea, where movement increases fluid pressure in the scala vestibuli. This increased fluid pressure displaces Reissner's membrane into the cochlear duct, which in turn stimulates the hair cells in the organ of Corti. Thus, neuroimpulses of electrochemical energy are initiated by movement of the hair cells in the neuroepithelium and travel via the auditory nerve to the cochlear nuclei of the brain stem with final connecting pathways to the auditory cerebral cortex.

The function of hearing is differentiated into three major categories. First, as a detector, the ear is capable of comprehending mechanical energy in the form of vibrations. Second, as an analyzer, the ear provides information about the intensity and frequency of vibrations. Third, the ear functions as a transducer, converting the mechanical energy of sound into electrochemical energy to activate the auditory nerve.

D. Vestibule and Semicircular Canals

Anatomy

The bony vestibule contains the membranous sacs, the utricle, and the saccule; it is situated behind the cochlea and in front of the osseous semicircular canals (Figures 1 and 2).

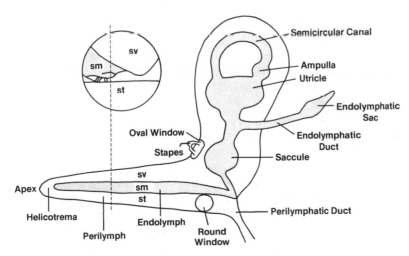

Figure 4 Anatomy of the inner ear (diagrammatic). The cochlea is shown uncoiled to demonstrate the anatomic relationships (stippling = endolymphatic space; clear = perilymphatic space). sv = scala vestibuli; sm = scala media; st = scala tympani.

A small island of neuroepithelial tissue (the macula), containing hair cells and supporting cells, is located in the utricle and saccule of the membranous labyrinth. A similar receptor structure, the crista ampullaris, is located in the expanded end or ampulla of each of the membranous semicircular canals.

Small crystals of calcium carbonate, the otoliths, are located on the hair cells of the maculae. They give weight to the hair cells and make them more sensitive to change in position. In a similar manner, the hair cells in the ampullae of the semicircular canals contain, in a sparsely scattered manner, small amounts of calcite material.

The vestibular branch of the auditory nerve arises from bipolar cells in the vestibular ganglion situated in the internal acoustic meatus. The peripheral fibers divide into three branches, which are distributed around the hair cells of the saccule, the utricle, and the ampullae of the semicircular canals. The central fibers, forming part of the vestibular branch, terminate in the fourth ventricle of the brain stem.

Function

The vestibular neuroepithelial tissues of the inner ear function to maintain equilibrium and reflex control of the eyes.

The three semicircular canals in each ear are arranged at right angles to one another, so that any movement of the head affects neurosensory cells of the canals. The neural impulses traveling along the vestibular nerve set up reflexes to eye muscles and to muscles of the body to compensate changes in position.

E. Inner Ear Fluids

Perilymph fills the space between the membranous structures of the inner ear and the bony labyrinth. The exact origin and fate of perilymph are still unsettled. Since the perilymphatic duct provides a direct communication between the subarachnoid and perilymphatic spaces (Figure 2), it would seem that perilymph could be derived from the cerebrospinal fluid. However, the perilymphatic duct is not always patent, and various experiments have failed to support such a conclusion. Other possible sources of perilymph are (1) as an ultrafiltrate of blood derived from the vascular bed located superior to the outer attachment of Reissner's membrane in the scala vestibuli, (2) as a product of filtration of endolymph through Reissner's membrane with resultant changes in composition due to selective filtration, and (3) as a secretion of the membrane lining of the osseous labyrinth. The perilymph is presumably resorbed in the spiral ligament near the basilar membrane in the scala tympani, although it may contribute to endolymph by selective passage through Reissner's membrane.

Endolymph fills the membranous labyrinth: the cochlear duct, saccule, utricle, and the semicircular canals. It is thought to be secreted by the stria vascularis in the cochlear duct and by the secretory cells associated with the vestibular neuroepithelia. However, the exact origin and fate of endolymph are still not completely resolved. Indeed, endolymph may be, at least in part, derived from perilymph as fluid and electrolytes pass across Reissner's membrane (Figure 5). It is probable that endolymph is resorbed via the endolymphatic sac in the subdural space (Figure 2). The stria vascularis may act in a manner similar to kidney tubular cells by extracting sodium ions and exchanging potassium ions.

In classic studies using the guinea pig, Smith and co-workers [111] found the concentrations of potassium ions and sodium ions in endolymph to be similar to those in intra-

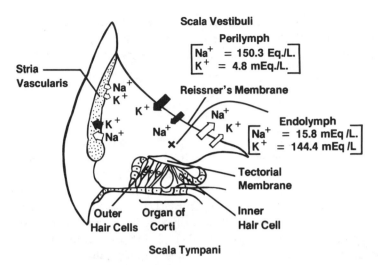

Figure 5 Cross section of the cochlea of a guinea pig, showing Na^+ and K^+ kinetics. Values for Na^+ and K^+ in endolymph and perilymph are taken from Table 1. (Arrows indicate directional flow of Na^+ and K^+; clear arrows = postulated flow; black arrows = flow based on experimental evidence.)

cellular fluid (Table 1). Therefore, endolymph is an extracellular fluid with an intracellular distribution of ions (Figure 5). In this regard, endolymph is quite different from cerebrospinal fluid or perilymph, since the latter fluids have a typical extracellular ionic distribution (i.e., high in sodium and low in potassium).

Endolymph has a high potassium content that may be related to the maintenance of the electrical endocochlear potential. Loss of endocochlear potential is one of the earliest findings of aminoglycoside ototoxicity. The ionic distribution of the endolymph, therefore the endocochlear potential, is probably dependent upon the "Na-K ATPase pump" in the stria vascularis and other secretory tissues. Membrane adenosine triphosphatase (ATPase) is thought to be closely related to the active transport of monovalent cations

Table 1 Comparison of Electrolytes (and Protein) in Labyrinthine Fluids and Spinal Fluid of the Guinea Pig[a]

	Spinal fluid (mEq/L)	Perilymph (mEq/L)	Endolymph (mEq/L)
Potassium	4.2 ± 0.5 (11)	4.8 ± 0.4 (13)	144.4 ± 4.0 (10)
Sodium	152.0 ± 1.8 (17)	150.3 ± 2.1 (18)	15.8 ± 1.6 (12)
Chloride	122.4 ± 1.0 (19)	121.5 ± 1.2 (17)	107.1 ± 1.4 (14)
	(mg %)	(mg %)	(mg %)
Protein	21.0 ± 2.0 (8)	50.0 ± 5.0 (10)	15.0 ± 2.0 (9)

[a]Values are shown with standard error of the mean (in parentheses) equal to the number of animals tested.

Source: Ref. 111.

across biological membranes [70,139]. This pump is known to be inhibited by ouabain and also by aminoglycosides [62]. In the case of aminoglycosides, the mechanism may be the result of its interaction with magnesium ions that are necessary for activation of ATPase.

III. Pathology

Histopathologic findings have demonstrated that aminoglycosides are toxic initially to the hair cells and then to the supporting cells of the neuroepithelium and secretory tissues of the vestibular and cochlear apparatuses of the inner ear. Evidence of this toxicity due to streptomycin was reported by Causse and co-workers [21,22] in 1948 and 1949 and later confirmed by Hawkins and Lurie [52,53] and Tyberghein [127]. Subsequently, animal investigations have demonstrated that although all aminoglycosides can produce ototoxic effects, the extent of damage differs from one to another [2,9,18,19,36,135]. For example, in guinea pigs given equal doses of tobramycin or gentamicin [18,35,36], the tobramycin group showed less outer hair cell loss than did the gentamicin group in both histopathologic [35,36] and electrophysiologic [18] evaluations. Investigators [54,103, 118] had earlier suggested that streptomycin was primarily toxic to the neurons of the eighth nerve or its nuclei in the brain stem, but this postulate has been disproven.

In animal models, a quantitative approximation of cochlear damage can be made by counting the missing hair cells and plotting the percentage of hair cells remaining on successive portions of the basilar membrane (cytocochleogram) [31]. Stebbins and co-workers [117] correlated hearing impairment with cochlear hair cell damage (cyto-cochleogram) in monkeys treated with aminoglycoside.

Light microscopy in animals revealed that selective histological changes in the cellular constituents of the organ of Corti initially are most marked in the basal turn of the cochlea, then progress apically [10,99,140]. Because hair cells in the basal turn are concerned with high-frequency sound perception, the initial sign of aminoglycoside cochleotoxicity is loss of high-frequency hearing.

Electron microscopy studies of cochlear hair cells following streptomycin therapy have demonstrated early changes in the mitochondria followed by merging and swelling of sensory cells and finally hair cell destruction, scarring, and dissolution of the basilar membrane [34,136]. The decrease of mitochondrion and endoplasmic membrane population suggests a metabolic disturbance caused by toxic effects of the drug. Outer hair cells in the organ of Corti appear to be more susceptible to destruction than do inner hair cells. This selective pattern of hair cell loss suggests the possibility of differences in membrane characteristics or in nutritional and metabolic requirements of inner and outer hair cells at various levels in the cochlea.

Aminoglycoside toxicity to the inner ear is not confined to the neuroepithelial cells. Changes in the stria vascularis of the cochlear duct in guinea pigs and humans have been reported following streptomycin therapy [46,102].

Histochemical experiments also demonstrate depletion of succinic acid dehydrogenase and ATPase in the cells of the stria vascularis before any changes are apparent in the organ of Corti. Johnsson and Hawkins [69] described changes in the stria vascularis: atrophy of marginal cells, loss of mitochondria, and reduction in stria length. Streptomycin injury to the secretory tissue of the vestibular system, corresponding to the stria vascularis of the cochlea, has also been reported [114,115].

Aminoglycoside toxicity of the neurosensory organs in the vestibular apparatus and semicircular canals has features similar to those found in the cochlear duct. Histopathological changes are characterized by changes in the orderly arrangement of hair cells in the ampullar cristac of the semicircular canals and the maculae of the utricle and saccule. Although the cells do not completely disappear, scarring is apparent and function is lost or impaired.

Figure 6 shows a typical patient audiogram consistent with the initial high-frequency hearing loss of aminoglycoside toxicity. This audiogram is indistinguishable from those seen frequently in older patients with presbycusis, those with congenital hearing impairment, or in patients exposed to excessive noise trauma. As the toxic effects of the aminoglycoside progress to involve hair cells toward the apex of the cochlear duct, the speech frequencies become involved and deafness results. Usually, early hearing impairment, as determined by audiogram, is not accompanied by clinical deafness since speech-frequency tones are not involved, although the patient may complain of tinnitus or "fullness" in the ears.

The reason why a particular aminoglycoside may predominately affect the vestibular or cochlear apparatus has not been explained. It is possible that the selectivity may represent a primary toxicity not to the sensory hair cells in the vestibule or the cochlea, but to the respective secretory areas that control the ionic environment [57].

IV. Mechanisms of Ototoxicity

Aminoglycosides are absorbed only minimally from the intestinal tract. Following parenteral administration, they are distributed mainly in the extracellular fluid compartment, with only trace quantities in intracellular fluids. They also poorly cross the blood–brain barrier, a fact that argues against the cerebrospinal fluid (CSF) as a significant source of

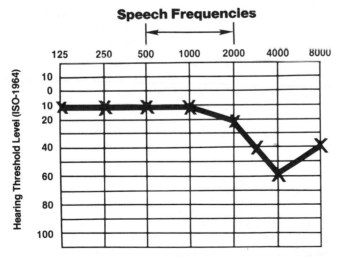

Figure 6 Audiogram showing high-frequency hearing loss for patient treated with aminoglycoside: Note that speech frequencies show a near-normal response. (Supplied by W.E. Fee, Jr., M.D., Division of Otolaryngology, Stanford University Medical Center, Stanford, Calif.)

perilymph. In animal studies, surprisingly high concentrations have been found in the perilymph and endolymph. Since aminoglycosides are not found in significant amounts in CSF, it is probable that they find their way to the inner ear fluid by the vascular route. Additionally, an extremely long perilymph half-life of 15 hr with detectable aminoglycoside levels at 25 hr has been found [90,132,142]. Thus ototoxicity may, to a large extent, be explained by the high concentration and persistence of these antibiotics in the inner ear fluids.

Stupp and co-investigators [121] compared perilymph concentrations of neomycin with those of dihydrostreptomycin, kanamycin, and streptomycin in animal studies. They found the levels of neomycin to be the highest, followed by those of dihydrostreptomycin and finally kanamycin and streptomycin; the latter two were nearly identical to each other. Such findings correlate with the known cochleotoxic potential for these agents. The local application of cochleotoxic aminoglycosides to the inner ear also shows a similar relationship between toxic cellular damage and the concentration of the antibiotic in the perilymph [38,86,109,116,137].

The following hypothesis may explain the mechanism of ototoxicity. The ionic composition of the endolymph is similar to that of intracellular fluid, in that both fluids contain more potassium ions than sodium ions. This ionic distribution in endolymph presupposes the existence of an active-transport mechanism, such as a sodium-potassium pump, to maintain ionic differences between the endolymph and the serum or perilymph. It is possible that an initial toxic effect of the aminoglycosides is an interruption of the active-transport system, leading to alterations in the normal ionic concentrations of the labyrinthine fluids. These alterations in turn impair electrical activity to the point where nerve conduction is impaired. Injury of the hair cells and other neural elements may stem from changes in the osmotic gradient of the endolymph, from direct toxic effects of the aminoglycosides, or from both. Initial toxic effects on the cell membrane or alterations of the mechanism that controls the osmotic gradient are apparently reversible; this finding would explain why the resulting ototoxicity also sometimes appears to improve or be reversible. However, as the deleterious effects progress, there is a point at which damage to the hair cells is so severe that it is irreparable, resulting in irreversible cochlear or vestibular dysfunction.

A theory proposed by Hawkins and Lurie [52] suggests that strongly basic antibiotics may gradually decrease the polarization of the organ of Corti, reducing the strong endocochlear potential inside the cochlear partition between the organ of Corti and the endolymph. Stupp et al. [121] discount this theory that ototoxicity is related to the basic character of the aminoglycoside, because the inner ear, as far as its content of acid substances is concerned, is not different from other comparable tissues; therefore, ototoxicity cannot be explained on this basis.

The effects of aminoglycosides on the cell membrane itself have been the subject of investigation [105,106]. Neomycin has been shown to inhibit the metabolism of phosphoinositide, which is thought to be essential for the regulation of membrane structure and permeability. Cell membrane lipids such as phosphoinositol bind with the cationic aminoglycoside. Such binding could disrupt the cell membranes of the neuroepithelial or secretory tissues and allow the antibiotic to enter the cell and produce cell death due to interference with intracellular protein metabolism. Since cell membrane lipid content may differ between cochlear and vestibular hair cells, a possible explanation for the selective toxicity of some aminoglycosides could be a greater affinity for the membrane lipid of one tissue over the other.

For technical reasons the effects of antibiotics at the cellular level have usually been studied in bacterial systems rather than in mammalian systems. Microorganisms are useful for the study of the intracellular effects induced by aminoglycosides.

As with most clinically useful antibiotics, with the exception of those that act on the cell wall, the aminoglycosides act primarily by inhibiting bacterial protein synthesis. Although early biochemical studies by Umbreit [128] indicated that streptomycin interfered with the microbial citric acid cycle, subsequent studies demonstrated that their antimicrobial effect was dependent upon inhibition of protein synthesis. Specifically, aminoglycosides have been shown in vitro to inhibit microbial protein synthesis through an interaction with the 30S ribosomal subunit [11]. Interestingly, aminoglycosides also have been shown to exert an inhibitory effect on mammalian protein synthesis [8]. The propensity of aminoglycosides to damage neuroepithelial cells of the inner ear and proximal renal tubular cells is probably due, in part, to (1) a unique susceptibility of the enzymes and ribosomes of these cells to aminoglycoside-induced inhibition [131], and (2) the relatively high concentrations of aminoglycosides that may occur in the kidney and inner ear.

In summary, aminoglycosides reach high levels in the inner ear, where they are removed slowly. The initial manifestation of toxicity is probably an alteration in the ionic balance of the endolymph. This, in turn, leads to alterations of the endocochlear potential and of the osmotic gradient of the fluids of the inner ear. Simultaneously, an interaction between the aminoglycoside and the cell membrane lipids probably alters permeability and subsequent respiratory functions. Some recovery of function is possible if further exposure to the aminoglycoside is stopped. The loss of endocochlear potential may be reflected initially by changes in the audiogram for high-frequency responses, and at the same time the patient may complain of fullness in the ear, tinnitus, or dizziness. Continued exposure to the offending drug will lead to alteration in the secretory and neuroepithelial cell wall membranes, either directly or as a result of osmotic changes, with consequent damage to internal cellular structures. The result is permanent vestibular or cochlear dysfunction, with impaired equilibrium and progression of impairment of high-frequency hearing to a state of hearing loss involving conversational tones.

V. Relationship to Renal Impairment

Of clinical significance is the frequent observation that ototoxicity due to aminoglycosides is closely associated with renal impairment [64]. This is due to failure to properly reduce the dosage, since it has been shown that if the serum aminoglycoside levels for gentamicin or tobramycin are maintained within a predetermined range, ototoxicity and nephrotoxicity are independent events [112]. Therefore, whether the renal insufficiency is due to the aminoglycoside or to other causes, it is important to monitor serum antibiotic levels in such patients and to adjust the dosage according to the extent of renal impairment. Because aminoglycosides are excreted entirely by glomerular filtration, serum antibiotic concentrations rise in the presence of renal dysfunction, and increased concentrations apparently lead in turn to elevated levels in the inner ear fluids, resulting in ototoxicity.

Peak serum concentrations exceeding 12 μg/ml for gentamicin [58,64] or tobramycin [91] have been suggested as the critical point at which the clinician may predict an increased risk of ototoxicity. Corresponding values for kanamycin and amikacin [3] are in the range 30-35 μg/ml.

Trough serum antibiotic concentrations probably are more closely related to oto-toxicity than are peak serum concentrations [81,92]. Trough levels are those measured just before a dose, usually 8 hr after the preceding dose. The concentrations of antibiotic should be at, or nearly at, zero with normal renal function. In any event, trough levels for tobramycin [91] or gentamicin [92] should not exceed 2 μg/ml. Levels higher than this indicate decreased glomerular filtrate of the aminoglycoside, and rising levels are indicative of progressive renal dysfunction.

Pharmacokinetically, if serum peak levels and trough levels of the aminoglycoside increase above those expected, the area under the curve of serum antibiotic levels will increase. This, in turn, apparently will lead to an increased diffusion of drug into the labyrinthine fluid, where its slow removal results in progressive accumulation.

The risk of increased peak levels accompanied by increased trough levels is in contrast to the situation where the peak "critical" levels may be exceeded following rapid intra-venous injection or infusion. Because the trough levels remain at or near zero, the increased peaks apparently do not imply a significantly increased risk of clinical ototoxicity. How-ever, even in this circumstance subtle events may be occurring in the inner ear. Wilson and Ramsden [139] observed an immediate dramatic reduction in cochlear electrical activity when peak serum concentrations of tobramycin exceeded 8 to 10 μg/ml. The changes were promptly reversed when the serum levels declined, and no sequelae were noted.

Gailiunas and co-workers [42] reviewed the problem of vestibular toxicity of genta-micin for patients on long-term hemodialysis. Signs and symptoms associated with vestibu-lar toxicity developed in an unexpectedly high percentage of patients as compared to the overall reported incidence of ototoxicity reported in patients without renal impairment. It therefore seems that patients undergoing long-term hemodialysis are at significantly higher risk of developing aminoglycoside toxicity. This review identified cumulative dose as the greatest risk factor; the toxic effect was unrelated to serum antibiotic concentrations.

Tjernstrom and co-workers [124] concluded that persisting renal impairment in patients doubles the risk of vestibular damage following gentamicin therapy, even though peak levels never exceeded 10 μg/ml. This conclusion is indirectly in conflict with the more recent findings of Smith and co-investigators [112] that nephrotoxicity and auditory ototoxicity are independent events.

An interesting relationship exists between the epithelial tissues of the kidney (the glomerulus and tubular cells) and the inner ear (the stria vascularis). Both of these tissues are similar histologically, and both are involved in the transport of fluids and electrolytes. They regulate the composition of the urinary ultrafiltrate and labyrinthine fluids, respec-tively. Patients with certain hereditary diseases (e.g., Alport's disease) have nephritis associated with high-frequency sensorineural hearing loss. Consequently, they may be more susceptible to aminoglycoside-induced ototoxicity because of altered permeability of the membranes of the inner ear, or the risk of ototoxicity may be related to altered renal function that results in elevated serum aminoglycoside levels followed by increased levels of drug in the inner ear.

VI.   Relationship of Ototoxicity to Chemical Structure

Each of the aminoglycosides possesses a unique toxicity profile. Examination of individual toxicity indices reveals that some aminoglycosides have greater potential to cause vestibular than auditory damage, and vice versa. Table 2 presents a relative ranking of aminoglycosides by Hawkins et al. [55] based on human and laboratory animal experiences.

Several investigators [15, 24, 93] have shown that the inositolic (cyclitolic) portion of the aminoglycoside molecule is responsible for toxic effects [i.e., streptidine for streptomycin and dihydrostreptomycin and 2-deoxystreptamine (2-DOS) for the other aminoglycosides discussed in this chapter]. These investigators were further able to demonstrate that the general toxic effects were antagonized by the administration of calcium. The ototoxicity appears to be related to the presence of amino or methyl amine groups in the glycosidic moiety. Those aminoglycosides with free amino groups linked to the glycosidic portions are predominantly cochleotoxic, the degree of toxicity being proportional to the total number of amino groups, whereas those containing methyl amine functions ($-NHCH_3$) are mainly vestibulotoxic. If both radicals are present, the drug would be expected to have both vestibular and auditory toxicity.

A relationship between a compound's chemical structure and its inherent potential for ototoxicity is apparent; however, more data are clearly necessary before making accurate predictions concerning an aminoglycoside's principal type of ototoxicity based on chemical structure.

VII. Diuretic Agents: Potentiation of Ototoxicity

Evidence indicates that potent "loop" diuretic agents such as ethacrynic acid and furosemide enhance the ototoxicity of aminoglycosides. The effect of concurrent administration of these agents is particularly prominent in patients with renal impairment [67,85]. Bilateral sensorineural hearing loss, transient and permanent, has been reported following oral and parenteral administration of these diuretics [48,82,96,108,110].

West et al. [138] found in the laboratory that kanamycin and ethacrynic acid in combination enhanced ototoxicity over that produced by either agent alone when the drugs were administered within 1 hr of each other.

Table 2   Qualitative Assessment of Relative Vestibular and Auditory Toxicity for Various Aminoglycosides Exhibited in Humans and Laboratory Animals[a]

Antibiotic	Relative Toxicity[b]	
	Vestibular	Auditory
Streptomycin	+++	+
Gentamicin	++	+
Tobramycin	+	+
Amikacin	+	+++
Kanamycin	+	+++
Dihydrostreptomycin	+	++
Neomycin	+	++++

[a]Values for tobramycin, amikacin, and gentamicin were added as estimates by the author.
[b]++++ equals a maximum effect.

Source: Adapted from J. E. Hawkins, Jr.: Antibiotics and the inner ear, *Trans. Am. Acad. Ophthalmol. Otolaryngol.*, 63:206–218, 1959, Table III.

Quick and Duvall [98] have described the primary effect of ethacrynic acid as atrophy of the intermediate cell layer of the stria vascularis; active ion transport in the cochlear duct is inhibited, causing an imbalance in the K^+ and Na^+ concentrations in the endolymph. Others [45,74] have shown a direct toxic effect on the sensory hair cells of the inner ear due to inhibition of glycolytic metabolism. Ethacrynic acid has also been shown to inhibit (sodium and potassium)-activated ATPase, which is associated with active ion transport. The end result may be that diuretic agents of this class alter cellular membrane permeability of the neuroepithelial tissues, which in turn allows the aminoglycoside to penetrate into the cell and produce cellular death.

VIII. Clinical Overview

Reliance on reviews of the literature for comparisons of the clinical ototoxicity of the various aminoglycosides has obvious limitations. Differences in clinical protocols, methodologies, patient populations, and dosage regimens make comparisons difficult. In general, patients who are candidates for aminoglycoside therapy have multisystem disease and may be too ill to cooperate in the evaluation of their cochlear or vestibular functions. It would be difficult to find a clinical setting where a side-by-side comparison of several aminoglycosides could be studied under the same protocol. Additionally, some of the older aminoglycosides have fallen into a pattern of usage and application that would obviate comparison with the newer ones. Neomycin, for example, is generally limited to topical use, since ototoxicity has been associated with every other mode of application, including parenteral, oral, intraperitoneal, wound irrigation, and on burned skin. Streptomycin is, for the most part, restricted to the therapy of tuberculosis, on an intermittent dosage schedule, or of certain other rare diseases. The aminoglycosides now in common use for serious gram-negative bacterial infections are tobramycin, gentamicin, kanamycin, and amikacin. Although the literature attests to the ototoxicity of each of the aminoglycosides (Table 3), or in some instances to a comparison of two, there are no retrospective or prospective studies of all four (i.e., amikacin, gentamicin, kanamycin, and tobramycin) done by the same investigator in the same hospital setting under a single protocol. Therefore, for all the reasons mentioned above, the data presented in Table 3 cannot necessarily be directly compared.

Evaluations of ototoxicity in children are not nearly as abundant as reports for adults treated with aminoglycosides. One probable reason is the difficulty of performing the tests in children. Those reports that are available, however, indicate that children may be less susceptible to ototoxicity than are adults.

A. Streptomycin Sulfate USP

Description

Streptomycin sulfate USP is the sulfated salt of a kind of streptomycin or a mixture of two or more such salts [according to U.S. Food and Drug Administration (FDA) certification and *United States Pharmacopeia* (USP) standards].

Clinical Experience

Introduced in 1947, streptomycin still has a place in the treatment of serious tuberculosis and of certain other less common diseases, such as plague and severe brucellosis;

Table 3 Reports in the Literature on Ototoxicity (Auditory or Vestibular) Attributed to Various Aminoglycosides

Drug (reference)	Number of patients or treatment courses	Number of auditory effects	Number of vestibular effects
Amikacin [87]	36	6 (Tinnitus, 2; audiometric loss, 2; both, 2)	0
Amikacin [123]	23	2 (Audiometric loss, 1; midfrequency loss, 1)	0
Amikacin [14]	67	9 (Tinnitus, 2; symptomatic hearing, loss, 1; high-frequency loss, 6)	0
Amikacin [79]	1548	79 (High-frequency loss, 71; conversational loss, 8)	10 (Vestibular alterations)
Gentamicin [5]	1484	7	37[a]
Gentamicin [92]	34	0	6 (Vestibular symptoms, 2; abnormal electronystagmograms, 3; dizziness, 1)
Gentamicin [16]	45	0	12 (Abnormal caloric tests, 7; vertigo, 1; unsteadiness, 4)
Kanamycin [40]	106	22 (Audiometric loss, 13; conversational loss, 4; tinnitus, 5)	5 (Vertigo and/or tinnitus)
Neomycin [58]	63	5	1[b]
Tobramycin [9]	3506	14 (Audiometric loss, 4; tinnitus, 5; hearing decrease, 5)	18[c] (Dizziness, 8; nystagmus, 7; ataxia, 1; vestibular disturbances, 2)

[a] Eight of these patients also have auditory effects.
[b] This patient also had auditory effects.
[c] In 21 patients, of whom 5 also had auditory effects.

Source: Adapted from [91]; by permission of The University of Chicago Press. Copyright 1976, The University of Chicago.

given orally to cholera carriers, it rarely is used in other gram-negative infections in which preferred agents have failed.

Incidence of Ototoxicity

As is the case with most other aminoglycosides, streptomycin toxicity seems to be more selective for one portion of the eighth nerve than for the other; it is more toxic to the vestibular portion. Two grams daily for 60–120 days produces some degree of vestibular disturbance in 75% of patients; 1 g daily reduces this incidence to about 25% [44]. Permanent impairment has been reported, but compensatory mechanisms, such as visual cues and deep proprioceptive sensation, frequently permit satisfactory ambulation and orientation, especially in younger patients. Vestibular disturbances have been recognized in the offspring of tuberculous mothers treated with streptomycin during pregnancy. With the introduction of newer and safer antituberculosis drugs, this risk should no longer be encountered. Presumably, "safe" doses of streptomycin for adults have been proposed to be a maximum of 16–19 mg/kg per day up to 1 g/day [47,122].

Clinical deafness has been reported rarely following streptomycin therapy, but various degrees of hearing impairment, especially in the high-tone range and later in the conversational range, have occurred. A measurable decrease in hearing has been reported in 4–15% of patients treated for more than 1 week [44].

B. Neomycin Sulfate USP

Description

Neomycin sulfate USP is a mixture of the sulfate salt of neomycin B and neomycin C. The ratio of B to C varies widely (according to FDA certification and USP standards).

Clinical Experience

Neomycin was first used experimentally in the early 1950s for the treatment of tuberculosis. Although there were clinical successes, virtually every patient who was saved from tuberculosis became deaf [134]. Thus, although neomycin is structurally similar to streptomycin, early in clinical application it was known to be ototoxic, but unlike streptomycin the toxic effect was on the auditory portion of the eighth cranial nerve.

Once ototoxicity was recognized as a significant hazard with parenteral neomycin, other applications were sought. The drug has been used for gut sterilization preoperatively, in wound irrigation, for instillation into body cavities, as an aerosol for inhalation, and as topical treatment for burns. Neomycin has caused ototoxicity as a result of systemic absorption in all these applications, particularly when it was used topically in burn patients.

Neomycin sulfate is commercially available in tablet form or as a sterile powder. The tablets (usually 0.5 g) and at least one solution (125 mg/5 ml) are commercially available for oral use [26]. Sterile powder can be diluted in Sodium Chloride Injection USP for intramuscular use only in hospitalized patients with urinary tract infections due to organisms not susceptible to less toxic antibiotics. A variety of creams and ointments containing neomycin, alone or in combination with other antibiotics or steroids, is available.

Orally, neomycin has been used for suppression of bowel flora (preoperatively or in hepatic coma) and for diarrhea due to enteropathogenic *Escherichia coli*.

Incidence of Ototoxicity

Reports of hearing impairment due to parenteral use of neomycin are now becoming rare, but one of the more recent was a retrospective study from Czechoslovakia in 1972 in which 9 of 26 patients receiving total doses of less than 5 g developed nearly complete deafness [122]. Carr et al. [20] in 1950 were the first published reference on ototoxicity of neomycin, but between 1950 and 1972, numerous reports have appeared. Doses of 0.5 to 2.0 g/day have resulted in partial to total deafness. The hearing loss may appear after cessation of even short-term therapy and may progress to complete deafness. Waisbren [133] in 1956 estimated the incidence of deafness at about 10%.

The serum concentration of neomycin that may be regarded as producing ototoxicity has not been defined. Certain patients with an incompetent intestinal mucosa who receive large oral doses may be at risk because of greater than usual absorption.

It is evident that without appropriate precautionary measures, the risk of ototoxicity to neomycin is considerable.

C. Kanamycin Sulfate USP

Description

Kanamycin sulfate is a mixture of the sulfated salts of kanamycin A and kanamycin B, with no more than 5% of the latter (according to FDA certification standards).

Clinical Experience

Kanamycin today is usually recommended in the therapy of serious infections due to gram-negative organisms not susceptible to antibiotics with less toxicity. Although used extensively in early clinical application for long-term therapy of tuberculosis, the drug no longer is indicated for long-term administration because of its ototoxic liability [71]. Kanamycin still finds favor among many pediatricians, primarily because of the low incidence of reported ototoxicity in children. It is active in vitro against *Staphylococcus aureus* and many gram-negative organisms (with the exception of *Pseudomonas aeruginosa*) however, it is not the drug of choice for *S. aureus* infections and is not recommended for use in *Pseudomonas* infections [71].

Incidence of Ototoxicity

The ototoxic effect of kanamycin, like neomycin, is almost exclusively on the auditory branch of the eighth nerve, but unlike neomycin, the progression of hearing impairment usually subsides upon cessation of therapy [125]. Kreis [78] in 1966 cited 21 studies of kanamycin ototoxicity in which the incidence averages approximately 30%, but the author appropriately pointed out: "Such a percentage has no precise meaning, because it depends on the type of impairment taken into account, the nature of the therapy and the type of case." Kreis also postulated that toxicity is related not so much to the size of the dose as to the length of time kanamycin blood levels remain elevated.

Another study reported that ototoxicity occurred in 22 of 106 patients receiving kanamycin and was directly related to total dose, daily dose, and duration of therapy [40]. Other reports are cited in this article which indicate an incidence of ototoxicity in 25 of 75 patients who received 1 g/day, 17 of 21 who received 0.5–1.5 g/day for 3 months or more, 2 of 12 who received 3 g/week, and 11 of 110 who received up to 6 g/week for 2 months or more. The author recommended an upper limit of 15 mg/kg per day.

In 72 tuberculosis patients treated with 1 g of kanamycin for periods varying from 30 to over 180 days, ototoxicity occurred in 40 (57%) [100].

In a panel discussion following a symposium on kanamycin, one speaker concluded, on the basis of clinical observations in over 1200 infants and children aged 1 day to 15 years, that children probably tolerate kanamycin better than do adults [141]. The difficulty of evaluating young infants was recognized, but toxicity appeared to have occurred in only two infants who were receiving intensive therapy for miliary tuberculosis.

D. Gentamicin Sulfate USP

Description

Gentamicin sulfate is a mixture of three sulfated salts from the gentamicin C complex with the following composition limits (according to FDA certification standards): gentamicin C, 25–50%; gentamicin C_{1a}, 15–40%; and gentamicin C_2, 20–50%.

Clinical Experience

Gentamicin, first introduced in the United States in 1969 for intramuscular use, quickly replaced the earlier aminoglycosides for systemic therapy because of its broader antibacterial spectrum against gram-negative organisms and its apparently lower potential for ototoxicity and nephrotoxicity. Ten years of clinical experience have resulted in awareness of the potential of gentamicin for nephrotoxicity and ototoxicity, and numerous reviews and reports of these side effects have appeared in the literature. It was recognized early that gentamicin, although active in vitro against *Mycobacterium tuberculosis,* is much less effective than streptomycin against experimental tuberculosis in mice. Gentamicin has not been used clinically for treating human tuberculosis in regimens like those for streptomycin and neomycin [104].

The primary clinical application of gentamicin is the treatment of serious infections due to susceptible gram-negative organisms. It is frequently used concurrently with carbenicillin or a cephalosporin to provide broad-spectrum antibacterial coverage in patients with life-threatening illness and sometimes with a semisynthetic penicillin for serious staphyloccocal infections. Although gentamicin is given primarily by the parenteral route, various topical preparations are available for dermatologic and ophthalmic use.

Incidence of Ototoxicity

Some of the earliest clinical investigations with gentamicin documented ototoxicity [65]. Additional clinical studies have demonstrated that vestibular toxicity is more frequent than cochlear toxicity [59,64]. Ototoxicity usually occurs in 2–3% of patients, but in selected high-risk groups this incidence may increase to 10–20% [124]. One author, reviewing clinical trial case reports during the investigational studies of gentamicin, found possible drug-related ototoxic effects in 88 of 3560 patients (2.4%) [92].

Several follow-up studies of neonates or infants treated with gentamicin have been reported [29,30,84]. In one series, 1 of 35 children had severe impairment of hearing. In another report, 1 of 67 cases demonstrated ototoxicity. Consistent with the kanamycin experience, infants and children appear to be more resistant to gentamicin ototoxicity than are adults.

Topical application of gentamicin on large burn areas has resulted in ototoxicity [13].

E. Tobramycin Sulfate USP

Description

Tobramycin sulfate is a monocomponent sulfated salt of Factor VI of the nebramycin complex (according to FDA certification and USP standards).

Clinical Experience

Introduced in the United States in 1974, tobramycin is structurally similar to the kanamycin family, especially kanamycin B, and the gentamicin C complex, especially gentamicin C_{1a}. Its in vitro activity is similar to that of gentamicin, but it is more active against *P. aeruginosa*.

Tobramycin has been studied and used clinically in much the same way as gentamicin, but no topical dosage forms are commercially available.

Incidence of Ototoxicity

Ototoxic effects of tobramycin are about evenly divided between those in the cochlear and those in the vestibular apparatus. In a review of multiple clinical studies consisting of 3506 clinical trial patients, 23 cases (0.6%) of ototoxicity were felt by the investigators to be attributable to tobramycin and 13 (0.4%) were considered of doubtful relationship [91]. A further analysis of the data led to the conclusion that the overall incidence of drug-related ototoxicity was 0.6% (21 patients), with 7 of the 21 having auditory effects only, 9 with vestibular effects only, and 5 with both auditory and vestibular effects.

To date, no published large-scale prospective controlled studies have directly compared tobramycin with gentamicin to evaluate the incidence of ototoxicity, but it would appear that tobramycin is at least as safe in this regard as gentamicin, the principal difference being in the type of involvement—vestibular and cochlear effects are about equal for tobramycin, whereas vestibular effects are predominant over cochlear for gentamicin. Based on the clinical literature, it appears that gentamicin is more vestibulotoxic than is tobramycin. One author estimated the incidence of ototoxicity of tobramycin, gentamicin, and amikacin to be about 1%, 2%, and 3%, respectively [63]. It would be expected that when otologic function tests are done frequently, these figures will be somewhat higher, especially in seriously ill patients with clinical "risk factors."

F. Amikacin Sulfate

Description

Amikacin sulfate is a monocomponent sulfated salt of amikacin (according to FDA certification).

Clinical Experience

Amikacin is a semisynthetic derivative of kanamycin and exhibits the same pharmacologic properties. Its antimicrobial spectrum, however, is broader and includes *P. aeruginosa.* In animal studies, on an equal-weight basis, the ototoxic and nephrotoxic effects are less than those for gentamicin and tobramycin. However, from a clinical viewpoint this lesser toxicity potential of amikacin is probably offset by the fact that recommended doses are four to five times those for gentamicin and tobramycin. The clinical indications for amikacin are similar to those for gentamicin and tobramycin.

Incidence of Ototoxicity

One author found ototoxicity (high-tone hearing loss) in 13 of 55 (24%) treatment courses of amikacin [14]. Audiograms returned to normal in three patients; in one patient the toxicity developed after therapy was stopped.

In summary review from multiple clinical trials consisting of 1548 patients, the investigators reported high-tone hearing loss in 71 patients (4.59%) and conversational hearing loss in 8 patients (0.52%) [79]. Of 328 pre- and post-treatment audiograms examined, 11 (3.35%) demonstrated a 15 dB loss after therapy. Vestibular dysfunction occurred in 10 of the 1548 patients (0.52%).

In another study of 49 treatment courses of amikacin in neutropenic patients, hearing loss was found in four patients and vestibular dysfunction in one [129].

A review article on aminoglycosides reported that in controlled studies comparing amikacin and gentamicin, ototoxicities of the two drugs were similar [4]. In an earlier study, Smith et al. [113] reported that the incidence of gentamicin ototoxicity (6% in 34 patients) did not differ significantly from that of amikacin (10% of 30 patients).

Apparently, the ototoxicity of amikacin is manifested to a greater extent as cochlear rather than vestibular damage.

G. Newer Aminoglycosides

Sisomicin and netilmicin are two new aminoglycoside drugs currently with clinical investigation status. Both are monocomponent aminoglycosides that closely resemble gentamicin C_{1a}, a component of the gentamicin complex (C_1, C_{1a}, and C_2). Animal studies have shown both of these agents to be ototoxic and nephrotoxic. Sisomicin appears to be slightly more vestibulotoxic than does gentamicin.

Vestibular or auditory toxicity has been observed in patients treated with sisomicin [39,75,83,95]. The incidence of ototoxicity appears to be similar to that of gentamicin [75,83].

Little is known at this time about the extent of ototoxicity with netilmicin in patients. It has been shown to be ototoxic [94]. When the dosage used in clinical trials (4–6 mg/kg per day) is compared to that of gentamicin (3–5 mg/kg per day), the two drugs will probably prove to be similar in regard to eighth-nerve impairment.

IX. Clinical Risk Factors

Clinically, ototoxicity manifests itself with varying degrees of frequency among different patient groups. It is associated not only with the structural characteristics of the antibiotic drug but with various other clinical "risk factors" as well.

The following clinical risk factors are commonly reported as predisposing or possibly predisposing to aminoglycoside ototoxicity:

1. Impaired renal function [7,32,64,91,92,124,137]
2. Intrinsic ototoxic potential of the drug [15,93]
3. Concomitant administration of other agents that have ototoxic potential (e.g., kidney "loop" diuretics [48,110] or other aminoglycosides)
4. The route of administration (e.g., intraventricular administration into the central nervous system) [12]
5. Total amount of drug given during the treatment period (i.e., total grams [9] or total milligrams per kilogram of body weight [1]
6. Total daily dose (mg/kg) [9,79]
7. Duration of therapy [1,9,78,120]
8. "Hereditary factors" [68,126]
9. Prior exposure to aminoglycosides [43,78,79,91]
10. Exposure to intense noise [25,28,66,77,119]
11. Patient's age (most reports indicate age is not significant) [5,9,27,40,50,79,92]

Renal function is the foremost risk factor cited. Virtually all studies in patients with decreased renal function suggest a greater frequency of ototoxicity than in those with normal renal function [7,32,64,91,92,124,137]. The risk can be reduced or eliminated by maintaining serum aminoglycoside levels within a predetermined range [112].

A common denominator for effects on both renal function and eighth-nerve function seems to be the persistence of these drugs in the body [51]. The retention and high concentrations of aminoglycoside antibiotics in the perilymph probably best explain their tendency to produce ototoxicity.

Findings are conflicting on the contribution of such factors as peak or trough serum antibiotic levels and long-term treatment. Federspil [36] showed that there is no critical serum level for gentamicin before it can penetrate the inner ear; he concluded that the rapid rise in serum levels may not be as harmful as has been postulated. Furthermore, long-term therapy produces antibiotic levels in the inner ear that are no different from those found after a single injection. Thus, although long-term therapy may be associated with an increasing incidence of ototoxicity, it apparently does not greatly potentiate the effect.

Bendush and Weber [9] studied the interaction of dose and length of therapy for 3474 patients treated with tobramycin (Table 4). When the data on patients who developed evidence of nephrotoxicity and/or ototoxicity were analyzed, it was concluded that increased daily dosage (>3.0 mg/kg per day) or prolonged therapy (>10 days) did not significantly increase the risk of toxicity. However, when both "increased dose" and "prolonged therapy" were involved, the incidence of toxicity significantly increased approximately sixfold (P < 0.01).

Comparative animal studies using equal and relatively large doses have shown ototoxicity to occur in the following order: Neomycin > gentamicin > tobramycin > kanamycin. Amikacin is probably similar to kanamycin. Because usual therapeutic doses of amikacin and kanamycin in humans are four to five times greater than those for gentamicin or tobramycin, extrapolation of comparative animal data to humans is clouded. Consequently, it seems likely that amikacin and kanamycin will prove to be no safer than gentamicin or tobramycin. In fact, because of the larger dose requirements, they will probably prove

Table 4   Relationship of Daily Dose and Duration of Tobramycin Therapy to Reports
of Renal and/or Ototoxic Effects[a]

| Duration of therapy | Dose (mg/kg per day) | |
(days)	≤3.0	>3.0
≤10	7/1057 (0.7)[b]	14/1136 (1.2)
>10	10/535 (1.9)	33/746 (4.4)

[a]64 of 3474 patients who experienced drug-related renal and/or ototoxic effects.
[b]Data are given as fractions (percentages) of patients who had adverse reactions.

Source: Reprinted from [9]; by permission of The University of Chicago Press.
Copyright 1976, The University of Chicago.

to be slightly more ototoxic. Comparison of various published clinical studies for gentamicin or tobramycin indicates that the two appear to be approximately equal in terms of
cochleotoxicity and that gentamicin is more vestibulotoxic.

The clinician should choose the agent with the lowest potential for toxicity, especially
for patients who have risk factors present. Furthermore, if hearing or vestibular dysfunction is present, whether due to prior aminoglycosides or not, it is advisable to choose an
agent that is least likely to affect the impaired system.

The concomitant administration of ethacrynic acid or furosemide will potentiate the
ototoxicity of aminoglycosides [48,67,82,85,96,108,110]. This effect is particularly
prominent in patients with uremia. Caution should also be used if neomycin is given by
inhalation or peritoneal lavage to a patient already receiving a parenteral aminoglycoside,
since the two aminoglycosides may be additive in terms of ototoxicity.

If the aminoglycoside is given by routes other than intramuscular or intravenous, the
clinician must continue to observe the patient carefully. Drugs given directly into the
cerebrospinal fluid may find their way into the fluids of the inner ear [12]. The same
is true if the aminoglycoside is instilled into an ear canal with a perforated tympanic
membrane [50]. In fact, such topical therapy may be hazardous in patients with otitis
media, although parenteral therapy is probably more important in terms of drug accumulation in the middle and inner ear in response to altered permeability of the tissue due to
the inflammatory response [37].

If the patient is pregnant, the aminoglycoside may cross the placenta and produce
ototoxicity in the offspring. Kern [73] reported a case of profound hearing loss in an
infant whose mother had received dihydrostreptomycin; however, other investigators
were unable to confirm this finding [23,130].

Familial occurrence of aminoglycoside-induced hearing loss has been reported [68,
126]. Tsuiki and Murai [126] concluded that a hereditary susceptibility of the cochlea
must be inherited as a dominant trait and is triggered somehow by the aminoglycoside
to produce hearing impairment.

Persons exposed to intense noise may experience auditory impairment. Patients with
such a history are probably more susceptible to further loss of function when they receive an aminoglycoside, especially if exposed to noise at the same time. The combined
effect of noise and neomycin has been investigated in the cochlea of guinea pigs [25,28,

66,77]. These studies have shown neuroepithelial hair cell damage and reductions of endocochlear potential [28,77,119]. The implications of these findings should be kept in mind when patients receiving aminoglycosides are exposed to intense noise in the hospital (e.g., radios, television, dialysis machinery, or premature infants in incubators) or when they are exposed to environmental noises as aminoglycoside-treated outpatients.

Some reports indicate that elderly patients are more susceptible than young patients to aminoglycoside ototoxicity [5]. This is probably due to reduced renal function or underlying presbycusis. Other reports have indicated that age as such is probably not significant [9,40,79,92], except that children seem to be less susceptible [27]. Even premature infants appear to tolerate these agents well [27], although careful renal monitoring is in order since they excrete aminoglycoside slowly because of renal immaturity.

X. Clinical Manifestations and Recommendations

Ototoxicity associated with currently popular aminoglycosides is relatively infrequent today. This is probably due to two principal factors:

1. Newer agents have less potential for causing ototoxicity.
2. Clinicians have a greater awareness of this side effect and therefore have learned how to prevent it by careful monitoring of kidney function, serum antibiotic levels, and changes in eighth-nerve function.

When selecting an aminoglycoside the clinician must weigh the known ototoxic characteristics of the drug, and its interaction with apparent clinical rick factors, against the potential benefits.

A. Vestibular Toxicity

The initial symptoms of vestibular toxicity usually go unrecognized. They may consist of nausea, vomiting, cold sweating, and vertigo or dizziness in the upright position. The onset of the clinical manifestations of toxic labyrinthitis is usually gradual. The extent and severity depends mainly on the amount of drug given and the length of therapy in relation to underlying clinical risk factors.

As vestibular toxicity progresses, the vertigo may become disturbing; infrequently, the patient experiences a sensation of rotating within his environment or of the environment rotating around him. If in bed, he may hold on to the sides of the bed. If he tries to walk, the gait is unsteady and he may reel and fall. On physical examination the patient may have nystagmus and past-pointing, although such findings are rare. The ataxia and related clinical features are a reflection of disturbed equilibrium; therefore, muscular incoordination as seen in cerebellar ataxia is not a feature of advanced toxic labyrinthitis.

Equilibrium involves an interaction of the neuromuscular-joint-tendon system with the optic pathways and the vestibular apparatus. Patients with impaired vestibular function can adapt by using visual cues; therefore, they may only have serious trouble with locomotion when surrounded by darkness.

Impaired vestibular function may be documented by use of caloric stimulation or electronystagmography (ENG). Both methods depend upon the occurrence of nystagmus.

B.   Auditory Toxicity

Tinnitus or a sensation of fullness or aching in the ears may herald the onset of auditory toxicity. An audiogram will provide objective evidence of hearing impairment. Significant impairment may occur in the absence of symptomatology. Also, the patient's condition or underlying disease state may mask these symptoms.

The clinician should take careful note of these symptoms. It is useful to use the "watch-test" daily on rounds to check for hearing. Obviously, this is a gross test that will not detect early toxicity in the high-tone range.

If damage to the auditory neuroepithelium occurs, it is usually early in the course of therapy, at about 3–5 days. If therapy is stopped, the symptoms nearly always subside and clinical deafness may not occur; permanent changes persist on the audiogram, although some functional improvement may occur.

So-called latent (delayed) or progressive ototoxicity has been reported up to several weeks following therapy with dihydrostreptomycin [32] and kanamycin [6,49]. It is difficult to postulate a mechanism to explain such a finding. This kind of ototoxicity has apparently not been reported in association with other aminoglycosides. In fact, Lehmann and co-investigators [80] followed 18 patients for 9 months after tobramycin therapy and reported that none of them experienced any changes in pure-tone audiograms.

Improvement in audiogram function has been reported after discontinuation of kanamycin [41] and tobramycin [91]. It is subtle and reflects increasing function following removal of the toxic agent. For the most part the impairment is irreversible and permanent, causing partial or complete deafness if the conversation tones are affected.

C.   Auditory and Vestibular Testing

It is easy to recommend that all patients receiving aminoglycosides have routine testing of auditory and vestibular function before, during, and after therapy with aminoglycosides. However, reliable results will not be attainable in some seriously ill patients or those with a decreased ability to respond. Also, there are inherent limitations in the sensitivity and reproducibility of the tests, especially ENG.

Patients with clinical risk factors should, whenever possible, have auditory and vestibular tests early in the treatment period and following therapy. This is especially true if the patient has impaired renal function or an underlying condition that requires repeated courses of aminoglycoside therapy.

The decision as to the need for audiograms and ENGs, and the timing of the tests, is dictated by the patient's history, physical examination, clinical course during therapy, and by the known ototoxic features of the aminoglycoside used.

D.   Guidelines for Use of Aminoglycosides

Whenever treatment with an aminoglycoside is initiated, the patient or the patient's relatives should be informed of the potential risk of ototoxicity.

The following recommendations should prove useful in preventing or limiting ototoxicity due to aminoglycosides.

1. Obtain a thorough history and examine the patient with attention to clinical risk factors.
2. Assess auditory and vestibular function (clinical or laboratory) before, during, and following therapy, especially in high-risk patients.
3. Once aminoglycoside therapy has been initiated, observe the patient for signs or symptoms of ototoxicity. If repeated courses of aminoglycoside therapy are likely (e.g., in patients with cystic fibrosis), then eighth-nerve function should be measured 1–2 weeks after therapy when the illness will have the least impact on the accuracy of the tests.
4. Monitor renal function; use aminoglycosides with caution in premature and newborn infants because of renal immaturity.
5. Follow recommended dosage; monitor serum concentrations of aminoglycoside in high-risk patients for evidence of elevated trough or increasing peak levels.
6. Reduce or discontinue aminoglycosides if the patient develops evidence of renal or eighth-nerve dysfunction.
7. Maintain hydration and urinary output; hypotension may impair renal function and thus increase the risk of ototoxicity.
8. Avoid prolonged aminoglycoside therapy; otherwise, maintain the lowest possible dosage.
9. Avoid concomitant and sequential use of other ototoxic agents.
10. Prevent exposure to intense noise during and immediately following aminoglycoside administration.
11. The following circumstances have been associated with ototoxicity as reported in humans or shown in laboratory animals.

    a. Local wound irrigation
    b. Inhalation therapy
    c. Intraperitoneal lavage
    d. Oral ingestion
    e. Topical use in burn patients
    f. Direct instillation into ear canal with perforated tympanic membrane
    g. Parenteral administration in patients with otitis media
    h. Intrathecal or intraventricular administration

Finally, the intrinsic ototoxic potential of the aminoglycoside should be considered. Indeed, dihydrostreptomycin is no longer available because, when compared with streptomycin, it is associated with more severe hearing impairment and delayed auditory toxicity. Neomycin is no longer routinely given by the parenteral route because safer agents are available. Compared to older aminoglycosides, the newer ones (tobramycin, gentamicin, and amikacin) have distinct clinical advantages (i.e., broader spectrum of in vitro antimicrobial activity, less tendency to induce plasma-mediated bacterial resistance, or better tolerance with regard to ototoxicity). In conclusion, the newer aminoglycosides have similar ototoxic potential in humans; however, gentamicin is apparently more vestibulotoxic than tobramycin or amikacin, and amikacin appears to be slightly more cochleotoxic than either tobramycin or gentamicin.

References

1. Acar, J.R., Guibert, J.M., Bourquelot, P., and Brisset, J.M.: Infections urinaires à bacilles gram négatif multi-résistants. *Nouv. Presse Med.* 27(Suppl. 17):69–72, 1974.
2. Akiyoshi, M., Sato, K., Nakada, H., Nara, T., Tajima, T., Sasaki, K., and Ogawa, M.: Ototoxicity of tobramycin in guinea pigs. *Chemotherapy* 23:1542–1543, 1975.
3. Amikin. *Physicians' Desk Reference* (33rd ed.). Oradell, N.J.: Medical Economics Co., 1979, pp. 694–696.
4. Appel, G.B., and Neu, H.C.: Gentamicin in 1978. *Ann. Intern. Med.* 89:528–538, 1978.
5. Arcieri, G.M., Falco, F.G., Smith, H.M., and Hobson, L.B.: Clinical research experience with gentamicin. Incidence of adverse reactions. *Med. J. Aust.* 1(Suppl.):30, 1970.
6. Ballantyne, J.: Ototoxicity: a clinical review. *Audiology* 12:325, 1973.
7. Banck, G., Belfrage, S., Juhlin, I., Nordstro, L., Tjernstr, O., and Toremalm, N.G.: Retrospective study of the ototoxicity of gentamicin. *Acta Pathol. Microbiol. Scand.* 24(Suppl.):54–57, 1973.
8. Beard, N.S., Armentrout, S.A., and Weisberger, A.S.: Inhibition of mammalian protein synthesis by antibiotics. *Pharmacol. Rev.* 21:213–245, 1969.
9. Bendush, C.L., and Weber, R.: Clinical results with tobramycin sulfate: a summary of worldwide experience. *J. Infect. Dis.* 134(Suppl.):S219–S234, 1976.
10. Benitez, J.T., Schuknecht, H.F., and Brandenburg, J.H.: Pathologic changes in human ear after kanamycin. *Arch. Otolaryngol.* 75:192–197, 1962.
11. Benveniste, R., and Davies, J.: Structure-activity relationships among the aminoglycoside antibiotics: role of hydroxyl and amino groups. *Antimicrob. Agents Chemother.* 4:402–409, 1973.
12. Berg, K.: The toxic effect of streptomycin on the vestibular and cochlear apparatus. *Acta Otolaryngol. (Stockh.)* 97(Suppl.):1–77, 1951.
13. Bernard, P.A.: Tolérance cochléovestibulaire du sulfate de gentamicin chez l'homme. *Gaz. Med. Fr.* 81:85, 1974.
14. Black, R.E., Lau, W.K., Weinstein, R.J., Young, L.S., and Hewitt, W.L.: Ototoxicity of amikacin. *Antimicrob. Agents Chemother.* 9(6):956–961, 1976.
15. Brazil, O.V., Corrado, A.P., and Berti, F.A.: Neuromuscular blockage produced by streptomycin and some of its degregration products. In *Curare and Curare-like Agents,* edited by Bovet, D., Bovet-Nitti, F., and Marini-Bettolo, G. Amsterdam, Elsevier, 1959, p. 415.
16. Brogard, J.M., Conraux, C., Collard, M., Jahn, H., and Lavillaureix, J.: Étude de la tolérance cochléovestibulaire de la tobramycine chez l'homme: étude préliminaire. *Nouv. Presse Med.* 27(Suppl. 17):75–79, 1974.
17. Brown, J.A., and Hinshaw, H.C.: Toxic reaction of streptomycin on the eighth nerve apparatus. *Proc. Staff Meet., Mayo Clin.* 21:347–352, 1946.
18. Brummett, R.E., Himes, D., Saine, B., and Vernon, J.A.: A comparative study of the ototoxicity of tobramycin and gentamicin. *Arch. Otolaryngol.* 96:505–512, 1972.
19. Brummett, R.E., Meikle, M.M., and Vernon, J.A.: Ototoxicity of tobramycin in guinea pigs. *Arch. Otolaryngol.* 94:59–63, 1971.
20. Carr, D.T., Brown, H.A., and Psuetz, K.H.: Occurrence of deafness in neomycin therapy. *JAMA* 144:65, 1950.
21. Causse, R.: Action toxique vestibulaire et cochléaire de la streptomycine au point de vue expérimental. *Ann. Otolaryngol. (Paris)* 66:518–538, 1949.
22. Causse, R., Gondet, I., and Vallancien, B.: Action vestibulaire de la streptomycine chez la souris. *C. R. Soc. Biol. (Paris)* 142:747–749, 1948.

23. Conway, N., and Birt, B.D.: Streptomycin in pregnancy: effect on the foetal ear. *Br. Med. J.* 2:260–263, 1965.

24. Corrado, A.P.: Tentativa de correlação estrutura quimica e atividade tóxica de antibióticos oligossacarídicos do grupo da estreptomicina. *Acta Physiol. Lat. Am.* 20:295, 1970.

25. Dayal, V.S., Kokshanian, A., and Mitchell, D.P.: Combined effects of noise and kanamycin, *Ann. Otol. (St. Louis)* 80:897–902, 1971.

26. Dukes, M.N.G., ed.: *Meyler's Side Effects of Drugs,* Vol. 8 (Manten, A.: Antibiotic drugs). New York, American Elsevier, 1975, p. 624.

27. Eichenwald, H.F.: Some observations on dosage and toxicity of kanamycin in premature and full-term infants. *Ann. N.Y. Acad. Sci.* 132:984–991, 1966.

28. Eldredge, D.H., and Cowell, W.P.: A laboratory method in the study of acoustic trauma. *Laryngoscope* 68:465, 1958.

29. Elfving, J., Pettay, O., and Raivio, M.: A follow-up study on the cochlear, vestibular and renal function in children treated with gentamicin in the newborn period. *Chemotherapy (Basel)* 18:141, 1973.

30. Elfving, J., and Pettay, O.: Gentamicin in the treatment of sepsis in newborn infants. *Acta Pathol. Microbiol. Scand.* 81(Suppl. 241):124, 1973.

31. Engstrom, H. Ades, H.W., and Anderson, A.: *Structural Pattern of the Organ of Corti.* Stockholm, Almqvist & Wiksell, 1966.

32. Erlanson, P., and Lundgren, A.: Ototoxic side effects following treatment with streptomycin, dihydrostreptomycin and kanamycin. *Acta Med. Scand.* 176:147–163, 1964.

33. Escher, F., and Rupp, F.: Die Schutzfunktion des Vitamin A bei der Streptomycin-Intoxikation. *Acta Otolaryngol. (Stockh.)* 43:311–321, 1953.

34. Farkashidy, J., Black, R.G., and Briant, T.D.R.: Effect of kanamycin on the internal ear: an electrophysiological and electron microscopic study. *Laryngoscope* 73:713–727, 1963.

35. Federspil, P.E.: Experimental and clinical studies on the ototoxicity of tobramycin. In *Tobramycin:* Selected Proceedings from the 8th International Congress of Chemotherapy, Athens, September 8–15, 1973. Amsterdam, Excerpta Medica, 1974, pp. 38–43.

36. Federspil, P.: Morphologische Untersuchungen zur Ototoxizität von Gentamycin und Tobramycin. *Arzneim. Forsch.* 23:1739–1745, 1973.

37. Federspil, P : Ototoxizität der Aminoglykosid-Antibiotica und Otitis media. *Arch. Rhino. Laryngol.* 207:487–488, 1974.

38. Federspil, P., Schatzle, W., and Tiesler, E.: Pharmacokinetics and ototoxicity of gentamicin, tobramycin, and amikacin. *J. Infect. Dis.* 134(Suppl.):S200–S205, 1976.

39. Feld, R., Valdivieso, M., Bodey, G.P., and Rodriguez, V.: A comparative trial of sisomicin therapy by intermittent versus continuous infusion. *Am. J. Med. Sci.* 274:197–188, 1977.

40. Finegold, S.M.: Toxicity of kanamycin in adults. *Ann. N.Y. Acad. Sci.* 132:942–956, 1966.

41. Finegold, S.M., Winfield, M.E., Aronsohn, R.B., Hewitt, W.L., and Guze, L.B.: Clinical experience with kanamycin. *Ann. N.Y. Acad. Sci.* 76:319, 1958.

42. Gailiunas, P., Dominguez-Moreno, M., Lazarus, J.M., Lowrie, E.G., Gottlieb, M.N., and Merrill, J.P.: Vestibular toxicity of gentamicin. *Arch. Intern. Med.* 138:1621–1624, 1978.

43. Geddes, A.M., Goodall, J.A.D., Speirs, C.F., and Williams, J.D.: Clinical and pharmacological studies with tobramycin. In *Tobramycin:* Selected Proceedings from the 8th International Congress of Chemotherapy, Athens, September 8–15, 1973. Amsterdam, Excerpta Medica, 1974, pp. 51–54.

44. Goodman, L.S., and Gilman, A., eds.: *The Pharmacological Basis of Therapeutics* (4th ed.). 1970, p. 1247.

45. Gordon, E.E., and DeHartog, M.: The relationship between cell membrane potassium ion transport and glycolysis: the effect of ethacrynic acid. *J. Gen. Physiol.* 54: 650-663, 1969.

46. Graf, K.: Histologische Veranderungen des Innenohres nach Behandlung der Meningitis Tuberculosa mit Streptomycin. *Acta Otolaryngol. (Stockh.)* 39:121-131, 1951.

47. Graf, K.: Streptomycindosierung und neurotoxische Nebenwirkungen. *Schweiz. Med. Wochenschr.* 79:793, 1949.

48. Hanzelik, E., and Peppercorn, M.: Deafness after ethacrynic acid. *Lancet* 1:416, 1969.

49. Harrison, W.H.: Ototoxicity of dihydrostreptomycin. *Q. Bull. Northwest. Univ. Med. Sch.* 28:271, 1954.

50. Hawkins, J.E.: Comparative otopathology: aging, noise, and ototoxic drugs. *Adv. Otorhinolaryngol.* 20:125, 1973.

51. Hawkins, J.E.: Drug ototoxicity. In *Handbook of Sensory Physiology, V: Auditory System, Part 3: Clinical and Special Topics,* edited by Keidel, W.D., and Neff, W.D. Berlin, Springer-Verlag, 1976, p. 707.

52. Hawkins, J.E., and Lurie, M.H.: The ototoxicity of streptomycin. *Ann. Otol. (St. Louis)* 61:789-806, 1952.

53. Hawkins, J.E., and Lurie, M.H.: The ototoxicity of dihydrostreptomycin and neomycin in the cat. *Ann. Oto. Rhinol. Laryngol.* 62:1128-1148, 1953.

54. Hawkins, J.E., and O'Shanny, W.J.: Functional analysis of the chronic neurotoxic action of streptomycin. *Fed. Proc.* 7:225, 1948.

55. Hawkins, J.E., Johnsson, L.G., Stebbins, W.C., and Moody, D.B.: Antibiotics and the inner ear. *Trans. Am. Acad. Ophthalmol. Otolaryngol.* 63:206-218, 1959.

56. Hawkins, J.E., Wolcott, H., and O'Shanny, W.J.: Ototoxic effects of streptomycin and dihydrostreptomycin pantothenates in the cat. *Antibiot. Annu.* 1956-1957:554-563, 1957.

57. Hawkins, J.E., Johnsson, L., and Aran, J.: Comparative tests of gentamicin ototoxicity. *J. Infect. Dis.* 119:417-426, 1969.

58. Hewitt, W.L.: Reflections on the clinical pharmacology of gentamicin. *Acta Pathol. Microbiol. Scand. [B]* 81(Suppl. 241):151-156, 1973.

59. Hewitt, W.L.: Gentamicin: toxicity in perspective. *Postgrad. Med. J.* 50(Suppl. 7): 55-59, 1974.

60. Hinshaw, H.C., Feldman, W.H., Carr, D.T., and Brown, H.: The clinical administration of dihydrostreptomycin in tuberculosis: a preliminary report. *Am. Rev. Tuberc.* 58:525-530, 1948.

61. Hobson, L.D., Tompsett, R., Muschenheim, C., and McDermott, W.: A laboratory and clinical investigation of dihydrostreptomycin. *Am. Rev. Tuberc.* 58:501-524, 1948.

62. Iinuma, T., Mizukoshi, O., and Daly, J.F.: Possible effects of various ototoxic drugs upon the ATP-hydrolyzing system in the stria vascularis and spiral ligament of the guinea pig. *Laryngoscope* 77:159-170, 1969.

63. Jackson, G.G.: Present status of aminoglycoside antibiotics and their safe, effective use. *Clin. Ther.* 1(3):200-215, 1977.

64. Jackson, G.G., and Ancieri, G.: Ototoxity of gentamicin in man: a survey and controlled analysis of clinical experience in the U.S. *J. Infect. Dis.* 124(Suppl.): 130-137, 1971.

65. Jao, R.L., and Jackson, G.G.: Gentamicin sulfate, new antibiotic against gram-negative bacilli. *JAMA* 189:817-822, 1964.

66. Jauhiainen, T., and Jauhiainin, M.: Combined effects of noise and neomycin on the cochlea. *Acta Otolaryngol. (Stockh.)* 73:387, 1972.

67. Johnson, A.H., and Hamilton, C.H.: Kanamycin ototoxicity—possible potentiation by other drugs. *South. Med. J.* 63:511-513, 1970.

68. Johnsonbaugh, R.E., Drexler, H.G., Light, I.J., and Sutherland, J.M.: Familial occurrence of drug-induced hearing loss. *Am. J. Dis. Child.* 127:245, 1974.

69. Johnsson, L.G., and Hawkins, J.E.: Strial atrophy in clinical and experimental deafness. *Laryngoscope* 82:1105-1125, 1972.

70. Judah, J.D., and Ahmed, K.: The biochemistry of sodium transport. *Biol. Rev.* 39:160-193, 1964.

71. Kantrex Injection. *Physicians' Desk Reference* (34th ed.). Oradell, N.J.: Medical Economics Co., 1979, pp. 704-705.

72. Keller, H., Krupe, W., Sous, H., and Muckter, H.: Further investigations with a view to enhancing the tolerance of streptomycin, viomycin, and neomycin by means of pantothenic acid. *Antibiot. Annu.* 1956-1957:549-553, 1957.

73. Kern, G.: Zur Frage der intrauterinen Streptomycinschadigung. *Schweiz. Med. Wochenschr.* 92:77-79, 1962.

74. Klahr, S., Bourgoig, J.J., Yates, J., and Bricker, N.: Inhibition of glycolysis by ethacrynic acid and furosemide. *Fed. Proc.* 30:608, 1971.

75. Klastersky, J., Hensgens, C., Gerard, M., and Daneau, D.: Comparison of sisomicin and gentamicin in bacteriuric patients with underlying diseases of the urinary tract. *Antimicrob. Agents Chemother.* 7:742-474, 1975.

76. Kluyskens, P.: La Mercaptoethylamine et l'intoxication à la dihydrostreptomycine. *C. R. Soc. Biol. (Paris)* 147:733-735, 1953.

77. Kohonen, A.: Effect of some ototoxic drugs upon the pattern and innervation of cochlear sensory cells in the guinea pig. *Acta Otolaryngol. (Stockh.)* 208(Suppl.): 1-70, 1975.

78. Kreis, B.: Kanamycin toxicity in adults. *Ann. N.Y. Acad. Sci.* 132:957-967, 1966.

79. Lane, A.Z., Wright, G.E., and Blair, D.C.: Ototoxicity and nephrotoxicity of amikacin: an overview of Phase II and Phase III experience in the United States. *Am. J. Med.* 105-111 (June), 1977.

80. Lehmann, W., Hausler, R., and Waldvogel, F.A.: A clinical study on the ototoxic effects of tobramycin. *Arch. Otorhinolaryngol.* 212:203-211, 1976.

81. Line, D.H., Poole, G.W., and Waterworth, P.M.: Serum streptomycin levels and dizziness. *Tubercle (Lond.)* 51:76, 1970.

82. Maher, J.F., and Schreiner, G.E.: Studies on ethacrynic acid in patients with refractory edema. *Ann. Intern. Med.* 62:15-19, 1965.

83. Maki, D.G., Agger, W.A., and Craig, W.A.: Comparative clinical study of sisomicin and gentamicin. *Clin. Res.* 26.156A, 1978 (Abstr.).

84. Mathieu, H., Narcy, P., and Le Poncin Charachon, D.: Tolérance cochléaire de la gentamicin chez l'enfant. *Gaz. Med. Fr.* 81:77, 1974.

85. Mathog, R.H., and Klein, W.J.: Ototoxicity of ethacrynic acid and aminoglycoside antibiotics in uremia. *N. Engl. J. Med.* 280(22):1223-1224, 1969.

86. McGee, T.M., Webster, J., and Williams, M.: Histological and functional changes in the ears of cats after subcutaneous administration of gentamicin. *J. Infect. Dis.* 119: 432-439, 1969.

87. Meyer, R.D., Lewis, R.P., Carmalt, E.D., and Rinegold, S.M.: Amikacin therapy for serious gram-negative bacillary infections. *Ann. Intern. Med.* 83:790-800, 1975.

88. Mihaly, J.P., Thompson, E.A., Billow, B.W., Gittens, S.A., and Simmons, D.M.: The pantothenate of streptomycin in the treatment of pulmonary tuberculosis. *Antibiot. Annu.* 1957-1958:602-608, 1958.

89. Mora, P.T., Young, B.G., and Shear, M.J.: Reduction of toxicity of cationic macromolecules by complexing with anionic derivatives of synthetic polyglucoses. *Nature* 184:431-432, 1959.

90. Muraveiskaya, V.S.: Determination of streptomycin, dihydrostreptomycin, colimycin, and monomycin in labyrinth fluid of guinea pigs. *Antibiotiki* 10:245, 1965.

91. Neu, H.C., and Bendush, C.L.: Survey of tobramycin ototoxicity. *J. Infect. Dis.* 134(Suppl.):S206-S218, 1976.

92. Nordstrom, L., Banck, G., Belfrage, S., Juhlin, I., Tjernstrom, O., and Toremalm, N.G.: Prospective study of the ototoxicity of gentamicin. *Acta Pathol. Microbiol. Scand. [B]* 81(Suppl. 241):58-61, 1973.

93. Owada, K.: Experimental studies on the toxicity of kanamycin, its hydrolyzed products and neomycin. *Chemotherapia* 5:277, 1962.

94. Panwalker, A.P., Malow, J.B., Zimelis, V.M. and Jackson, G.G.: Netilmicin: clinical efficacy, tolerance, and toxicity. *Antimicrob. Agents Chemother.* 13:170-176, 1978.

95. Parry, M.F., Neu, H.C., Garvey, G.J., Ortiz-Neu, C., and Appel, G.B.: Sisomicin therapy of serious gram-negative infections. *Infection* 4(Suppl. 4):439-445, 1976.

96. Pillay, V.K.G., Schawartz, F.D., Aimi, K., and Kark, R.M.: Transient and permanent deafness following treatment with ethacrynic acid in renal failure. *Lancet* 1:77-79, 1969.

97. Poutsiaka, J.W., Thomas, B.G.H., Lingar, C.R., and Hobson, L.B.: Laboratory studies on distrycin, a 1:1 mixture of streptomycin and dihydrostreptomycin. Transactions of the 12th Conference on the Chemotherapy of Tuberculosis, Veterans Administration, Washington, D.C., February 1953, pp. 291-293.

98. Quick, C.A., and Duvall, A.J.: Early changes in the cochlear duct from ethacrynic acid: an electronmicroscopic evaluation. *Laryngoscope* 80:954-965, 1970.

99. Reddy, J.B., and Igaraski, M.: Changes produced by kanamycin. *Arch. Otolaryngol.* 76:146-150, 1962.

100. Rempt, E.: Gehorschaden bei Kanamycinlangzeittherapie. *Z. Laryngol. Rhinol.* 49:504, 1970.

101. Rossi, G., and Olivieri, A.: The salts derived from the combination of streptomycin and neomycin with glucuronic acid. *AMA Arch. Otolaryngol.* 69:188-196, 1959.

102. Ruedi, L.: Some animal experimental findings on the functions of the inner ear. *Ann. Otol. (St. Louis)* 60:993-1023, 1951.

103. Ruedi, L., Furrer, W., Escher, F., and Luthy, F.: Toxische Wirkungen des Strepto-mycins. *Acta Otolaryngol. (Stockh.)* 78(Suppl.):66-77, 1949.

104. Sanders, W.E., et al.: Activity of gentamicin against mycobacteria in vitro and against *mycobacterium tuberculosis* in mice. *J. Infect. Dis.* 124(Suppl.):33, 1971.

105. Schacht, J.: Biochemistry of neomycin ototoxicity. *J. Acoust. Soc. Am.* 59:940-944, 1976.

106. Schacht, J.: Interaction of neomycin with phosphoinositide metabolism in guinea pig inner ear and brain tissues. *Ann. Otol. (St. Louis)* 83:613-618, 1974.

107. Schatz, A., Bugie, E., and Waksman, S.A.: Streptomycin, a substance exhibiting antibiotic activity against gram-positive and gram-negative bacteria. *Proc. Soc. Exp. Biol. Med.* 55:66, 1944.

108. Schneider, W.J., and Becker, L.E.: Acute transient hearing loss after ethacrynic acid therapy. *Arch. Intern. Med.* 117:715-717, 1966.

109. Schuknecht, H.F.: Ablation therapy in the management of Menière's disease. *Laryngoscope* 66:859-871, 1950.

110. Schwartz, G.H., David, D.S., Riggio, R.R., Stenzel, K.H., and Rubin, A.L.: Oto-toxicity induced by furosemide. *N. Engl. J. Med.* 282:1413-1414, 1970.

111. Smith, C.A., Lowry, O.H., and Wu, M.L.: The electrolytes of the labyrinthine fluids. *Laryngoscope* 64:141-153, 1954.

112. Smith, C.R., Lipsky, J.J., and Lietman, P.S.: Relationship between aminoglycoside-induced nephrotoxicity and auditory toxicity. *Antimicrob. Agents Chemother.* 15:780-782, 1979.

113. Smith, C.R., Baughman, K.L., Edwards, C.Q., Rodgers, J.F., and Lietman, P.S.: Controlled comparison of amikacin and gentamicin. *N. Engl. J. Med.* 296:349-353, 1977.

114. Sparwald, E., Lange, G., and Leupe, M.: Veranderungen an den dunklen Zellen der Crista ampullaris nach Streptomycin-Medikation beim Meerschweinchen. *Arch. Klin. Exp. Ohren-Nasen-Kehlkopfheilkd.* 199:587–590, 1971.

115. Sparwald, W., Merck, W., and Leupe, M.: Über den zeitlichen Zusammenhang zwischen Schädigung der Sinneszellen und der dunklen Zellen der Crista ampullaris des Meerschweinchen-labyrinthes bei Streptomycininotoxikation. *Arch. Klin. Exp. Ohren-Nasen-Kehlkopfheilkd.* 202:687–689, 1972.

116. Spoendlin, H.: Zur Ototoxizität des Streptomyzins. *Pract. Otorhinolaryngol.* 28: 305–322, 1966.

117. Stebbins, W.C., Miller, J.M., Johnsson, L.G., and Hawkins, J.E.: Ototoxic hearing loss and cochlear pathology in the monkey. *Ann. Otol. (St. Louis)* 78:598–602, 1969.

118. Stevenson, L.D., Alvord, E.C., and Correll, J.W.: Degeneration and necrosis of neurons in eighth cranial nuclei caused by streptomycin. *Am. J. Pathol.* 23:875–876, 1947.

119. Stockwell, C.W., Ades, H.W., and Engstrom, H.: Pattern of hair cell damage after intense auditory stimulation. *Ann. Otol. (St. Louis)* 78:1144, 1969.

120. Stone, H.H., Kolb, L.D., Geheber, C.E., and Currie, C.A.: Treatment of gram-negative infections in surgical patients with a new aminoglycoside–tobramycin. In *Tobramycin:* Selected Proceedings from the 8th International Congress of Chemotherapy, Athens, September 8-15, 1973. Amsterdam, Excerpta Medica, 1974, pp. 102–109.

121. Stupp, H., Kupper, K., Lagler, F., Sous, H., and Quante, M.: Inner ear concentrations and ototoxicity of different antibiotics in local and systemic application. *Audiology* 12:350–363, 1973.

122. Supacek, I.: Toxic hearing loss due to neomycin. A study of 26 cases. *Audiology* 11(Suppl.):29, 1972.

123. Tally, F.P., Louie, T.J., Weinstein, W.M., Bartlett, J.G., and Gorbach, S.L.: Amikacin therapy for severe gram-negative sepsis: emphasis on infections with gentamicin-resistant organisms. *Ann. Intern. Med.* 83:484–488, 1975.

124. Tjerstrom, O., Banck, G., Belfrange, S., Juhlin, I., Nordstrom, L., and Toremalm, N.G.: The ototoxicity of gentamicin. *Acta Pathol. Microbiol. Scand.* 81(Suppl. 241):73–78, 1973.

125. Tompsett, R.: Relation of dosage to streptomycin toxicity. *Ann. Otol. (St. Louis)* 57:181, 1948.

126. Tsuiki, T., and Murai, S.: Familial incidence of streptomycin hearing loss and hereditary weakness of the cochlea. *Audiology* 10:315–322, 1971.

127. Tyberghein, J.: Influence of some streptomyces antibiotics on the cochlear microphonics in the guinea pig. *Acta Otolaryngol. (Stockh.)* 171(Suppl.):1–56, 1962.

128. Umbreit, W.W.: The action of streptomycin: V. A new metabolic intermediate. *J. Bacteriol.* 66:74–81, 1953.

129. Valdivieso, M., Feld, R., Rodriguez, V., and Bodey, G.P.: Amikacin therapy of infections in neutropenic patients. *Am. J. Med. Sci.* 270:453, 1975.

130. Varpela, E., and Kietalahti, J.: Streptomycin medication during pregnancy and the child's hearing. *Ann. Paediatr. Feen.* 11:38–45, 1965.

131. Vazquez, D., and Monro, R.E.: Effects of some inhibitors of protein synthesis on the binding of aminoacyl t-RNA to ribosomal subunits. *Biochim. Biophys. Acta* 142:155–173, 1967.

132. Voldrich, L.: The kinetics of streptomycin, kanamycin, and neomycin in the inner ear. *Acta Otolaryngol. (Stockh.)* 60:243, 1965.

133. Waisbren, B.A.: Neomycin. *Practitioner* 176:39–46, 1956.

134. Waksman, S.A., ed.: *Neomycin: Its Nature and Practical Application.* Baltimore, Md., Williams & Wilkins, 1958 (published for The Institute of Microbiology).

135. Welles, J.S., Emmerson, J.L., Gibson, W.R., Nicklander, R., Owen, N.V., and Anderson, R.C.: Preclinical toxicology studies with tobramycin. *Toxicol. Appl. Pharmacol.* 25:398–409, 1973.

136. Wersall, J., and Hawkins, J.E., Jr.: The vestibular sensory epithelia in the cat labyrinth and their reactions in chronic streptomycin intoxication. *Acta Otolaryngol. (Stockh.)* 54:1–22, 1962.

137. Wersall, J., Lundquist, P.G., and Bjorkroth, B.: The ototoxicity of gentamicin. *J. Infect. Dis.* 119:410–416, 1969.

138. West, B.A., Brummett, R.E., and Himes, D.L.: Interaction of kanamycin and ethacrynic acid. *Arch. Otolaryngol.* 98:32–37, 1973.

139. Wilson, P., and Ramsden, R.T.: Immediate effects of tobramycin on human cochlea and correlation with serum tobramycin levels. *Br. Med. J.* 1:259–261, 1977.

140. Ylikoski, J.: Degeneration of neural elements in the cochlea of the guinea pig after damage to the organ of Corti by ototoxic antibiotics. *Acta Otolaryngol. (Stockh.)* 326(Suppl.):23–41, 1974.

141. Yow, Martha, mod.: Panel discussion: Kanamycin in pediatric practice with special reference to observations on ototoxicity. *Ann. N.Y. Acad. Sci.* 132:1037–1044, 1979.

142. Antibiotic concentration levels in perilymph studies. *JAMA* 193:29, 1965.

CLINICAL USE

20

AMINOGLYCOSIDE DOSING REGIMENS: ONCE DAILY, THRICE, OR CONTINUOUS ADMINISTRATION?

W. LEIGH THOMPSON   Case Western Reserve
University, Cleveland, Ohio

Aminoglycoside efficacy and toxicity are related, in general, to daily dose. Small doses are neither effective nor toxic; larger doses may be effective and the largest doses are certainly toxic [8,35,37]. However, is efficacy or toxicity more closely related to average aminoglycoside concentration, or maximum concentrations, or minimum ones? There are three possible ways of adjusting aminoglycoside doses for slowed elimination in renal failure. Some authorities recommend changing the dosing interval, but not the dose, keeping constant the maximum and minimum concentrations of aminoglycosides [11,18, 23,24,29,31]. Others recommend dosage adjustment at the same interval [4,5,16], with less oscillation about the same average concentration. Still others hedge, recommending both or either adjustment [6,9,12,15,21,22,30,33]. If efficacy, or toxicity, were clearly related to maximum or minimum concentrations, there would be little disagreement in regimens. In most animal and clinical studies of aminoglycoside efficacy and toxicity, only a single regimen has been employed, and data on the influence of concentration oscillations are sparse. Before considering existing data, let us review the principles to consider in relating dosage regimens to drug concentrations and effects.

I. Aminoglycoside Kinetics

Will the average plasma concentration of aminoglycosides be proportional to the total daily dose? There are two factors that determine this relationship: distribution and elimination of the drug. Most of aminoglycosides in the body are present in extracellular fluids at equal concentrations throughout and there is little association of aminoglycosides with plasma proteins. Thus, the plasma water concentration of aminoglycosides is proportional to the total body content of these drugs, and that proportionality constant is the distribution coefficient (also called distribution "volume" because its units are more often volume than mass). The relation of aminoglycoside content (or plasma water concentration) to dose is also determined by elimination, which for these drugs is predominantly by glomerular filtration. This passive process "clears" a constant amount of plasma water of aminoglycoside, whether the drug concentration is small or large. This means that the fraction of drug in the body eliminated each minute is constant and that the mass of drug eliminated is directly proportional to its concentration in plasma water. Thus, to a large extent the concentration of aminoglycoside in plasma water will be directly proportional to the dose, and that proportionality constant is the plasma (water) clearance. If clearance is 120 ml/min and aminoglycoside concentration is 2 mg/L, drug elimination will be 240 μg/min or 346 mg/day. To maintain an average plasma water concentration of 2 mg/L will require giving 346 mg/day of aminoglycoside. For an average concentration twice as great, give twice as much drug. To determine the total daily dose required, measure or estimate the clearance and multiply by the average concentration desired. The average aminoglycoside concentration observed will be as accurate as your estimate of aminoglycoside clearance.

II. Tissue Accumulation

There are two exceptions to this proportionality of aminoglycoside concentration and dose. Although most of aminoglycoside is eliminated by glomerular filtration, there is active uptake into the proximal tubular epithelial cells of the kidney of a small fraction of the daily dose. This fraction, however, is greatly concentrated within the lysosomes. Similar accumulation, to lesser tissue concentrations, also occurs in other tissues. When dogs are given 450 mg/kg of gentamicin or tobramycin over 10 days, the total accumulation in soft tissues at the end of treatment is 6% of the total dose [28]. With current techniques it is not possible to measure with accuracy a 6% difference in plasma water concentration or in clearance, and thus the tissue accumulation might be neglected in determining disposition of the bulk of drug injected. Tissue stores of aminoglycosides are of importance in two instances. First, this fraction of aminoglycoside is excreted slowly. In patients it is excreted with a first-order rate constant of 0.006 hr^{-1} [32], and this is 50 times slower than the elimination by normal glomerular filtration (0.3 hr^{-1}). Thus, small concentrations of aminoglycosides will be found in plasma and urine for weeks after the last dose, because of gradual release from tissue stores. The slow elimination of aminoglycoside from the proximal tubular cells may well reflect cell turnover rather than active exocytosis, for the rates are similar. The second importance of the tissue storage is in regard to minimum aminoglycoside concentrations in plasma during long dosing intervals. If tissue stores were neglected, one might predict that once-daily injection of 240 mg of aminoglycoside would cause plasma concentrations in a subject with clearance of 120 ml/min and distribution coefficient of 18 L to be: maximum 13.3 mg/L, average 1.39 mg/L, and minimum 0.001 mg/L. The actual minimum concentrations in plasma water will be small, perhaps up to 0.5 mg/L, and will depend upon the duration of treatment, the extent of tissue stores, the rate of liberation from tissue stores, and the rate of aminoglycoside excretion. Although the highest tissue concentrations of aminoglycosides are observed in kidney cortex, and this store may be eliminated directly into urine, the greatest mass of aminoglycosides is in other tissues, and this mass must pass through blood into urine [28].

III. Distribution of Aminoglycosides

The second exception to the proportionality of aminoglycoside concentration and dose occurs just after injection, before distribution is complete. Aminoglycosides usually distribute in the body so that the amount in the entire plasma volume is about one-sixth of the total mass of aminoglycosides in the body. If the aminoglycoside is injected instantaneously into plasma, the bolus may mix with only a fraction of the plasma volume during the first circulation, with very high concentrations momentarily. If the aminoglycoside is injected over 2 or 3 minutes, so that it mixes fully with the plasma volume, much of the dose may be restricted to plasma volume at the end of injection, with concentrations about six times greater than those achieved a few minutes later when the aminoglycoside has distributed fully. During this initial phase of distribution, relatively high concentrations of drug in plasma lead to elimination of greater amounts of drug than when distribution is complete. The magnitude of this distributive phase has been examined in dogs [28]. When doses of 45 mg/kg were injected in dogs in 1 min, distribution had a half-time of 4 min [28]. Neglecting this initial phase, which was 99% completed within 30 min, and analyzing only plasma concentrations measured after distribution led to an underestimation of total plasma clearance of 12.4% and an underestimation of the actual average plasma concentration of 11%.

IV. Maximum and Minimum Concentrations

With these general principles we can relate the plasma concentrations of aminoglycosides to the dose and examine how the concentrations in plasma relate to effects. We must in this case consider not only the concentrations of aminoglycosides in plasma, but at sites of action. If aminoglycoside transport to an active site has a *threshold*, the effects will be greater with wide oscillations in concentration than with a constant plasma concentration below the threshold. A threshold may occur when there is a constant force opposing drug entry or action, as with the bulk transport of fluid and drug out of the cerebrospinal fluid or when the mechanism of entry or action is activated only by a high concentration of drug. Active uptake of aminoglycosides by some bacteria may be "inducible" in this way. If antimicrobial action is described by static in vitro systems used to determine "minimum inhibitory concentrations," then such concentrations may be regarded as threshold levels below which activity is not demonstrated. If, on the other hand, the aminoglycoside action is reversible, and "escape" occurs during periods of small concentrations, the effects may be greater when concentrations are maintained constant and "escape" is prevented. If aminoglycoside action has a *maximum*, with no additional effects of supramaximal concentrations, then overall effects may be blunted by momentary excursions to high concentrations. These possibilities are further complicated by three additional factors. First, the mechanisms of toxicity of aminoglycosides in the kidney, in the cochlea, to vestibular function, and at myoneural junctions are not defined, and therefore concentration–response relations are impossible to predict. Second, the kinetics of bacterial replication must be considered. If all bacteria are killed with a single brief exposure to an effective concentration, then only a single dose of aminoglycoside is required. If aminoglycosides delay regrowth and alter regrowth of the survivors, other regimens may be optimal. Finally, the sites of action must be considered. A regimen with optimal antimicrobial action in bloodstream infection may be an inadequate regimen for the same organism sequestered in an abscess, in spinal fluid, or in an endocardial vegetation.

The relationship of the time course of aminoglycoside concentrations to effects has not been well established. Some investigators incriminate momentary maximum concentrations for toxicity [25] and others incriminate excessive minimum plasma concentrations [7,13,14,20,25]. But most animal and clinical studies have employed the same dosage regimen, so that maximum, average, and minimum concentrations increase to almost the same extent when clearances are reduced. For example, in a patient with a distribution coefficient of 18 L and a clearance of 120 ml/min, a dose of 80 mg every 8 hr injected rapidly will cause plasma concentrations to oscillate from 4.6 to 0.2 mg/L with an average of 1.4 mg/L. If clearance is reduced to 40 ml/min, these concentrations will become, at equilibrium, 6.8 to 2.3 mg/L with an average of 4.2 mg/L, and each concentration has increased by about 2.5 mg/L.

Only a few studies have examined directly the influence of dosage regimen, at the same total dose, on toxicity. In 18 rabbits given gentamicin 7.5 mg/kg per day for up to 4 weeks, once daily injections were less nephrotoxic than the same dose given as three injections daily [10]. However, patients given gentamicin 160 mg once daily intramuscularly had more reduction in creatinine clearance than those given 60 or 80 mg three times daily for 8–12 days [19]. Continuous intravenous infusion of aminoglycosides in seriously ill neutropenic patients has been thought to be less nephrotoxic than when comparable doses were given as intravenous injections every 6 hr [17,36]. However, in these studies the patients had reduced clearances, there was little oscillation in concentration achieved at 6 hr intervals, and in many cases the "continuous" infusions were interrupted by technical problems [17,36].

V. Nephrotoxicity and Dosage Regimen

To examine nephrotoxicity as related to plasma aminoglycoside concentrations, we gave dogs 45 mg/kg per day of gentamicin or tobramycin for 10 days by constant intravenous infusion (with a monitored infusion pump) or once-daily rapid intravenous injection. The average plasma concentrations during the first day of therapy were 9.3 mg/L in all 40 dogs. Maximum plasma aminoglycoside concentrations were more than 200 mg/L in dogs given once-daily injections, and concentrations were constant throughout the dosing interval in those given continuous infusions [28]. After 10 days of treatment, glomerular filtration rates had decreased 48% in the entire group of 40 dogs, but the decrement was twice as great in the dogs treated with constant intravenous infusions (P = 0.001). At the time of sacrifice, at the end of treatment, the tissue concentrations of aminoglycosides did *not* correlate with the extent of reduction in glomerular filtration, even with regard to renal cortical aminoglycoside accumulation. Tissue concentrations were equal in dogs given once-daily injections and in those given constant intravenous infusions [28].

VI. Efficacy and Dosage Regimen

The relation of dosage regimen to efficacy has been examined infrequently. In urinary tract infections with susceptible organisms, once-daily aminoglycoside injections are equally effective with the same dose given three times daily [19]. Very infrequent dosing of aminoglycosides may permit rapid regrowth of uninhibited surviving organisms during periods of reduced drug concentration, and for this reason frequent dosing has been suggested [34]. However, the effects of aminoglycosides on regrowth have not been defined. Even the penetration of aminoglycosides into sites of infection may be related to concentration oscillation. Penicillin, for example, reaches greater average concentrations in fibrin clots when given intermittently than when the same doses are given continuously to animals [1-3].

Efficacy was studied directly in rabbits infected with a serum-resistant strain of *Pseudomonas aeruginosa* that was inhibited in vitro by tobramycin in concentrations of 0.5 mg/L [26]. Left-heart endocarditis was induced with the indwelling arterial catheter technique, and infected rabbits were treated with tobramycin 22.5 mg/kg per day for 7 days. This dose established initial average plasma concentrations of 5 mg/L in both treatment groups. Rabbits treated with once-daily intravenous injections had survival rates, rates of sterilization of blood cultures, and bacterial densities on endocardial vegetations that were significantly better than those of rabbits that were not treated with antibiotics. Rabbits with constant intravenous infusions of tobramycin have responses that were comparable to those in rabbits treated with the same dose injected once daily [26].

One clinical trial of once daily or continuous intravenous infusion of tobramycin has been reported [37]. In 24 patients with cystic fibrosis, tobramycin was given in doses to maintain average plasma concentrations at 4 mg/L. Doses ranged from 5.7 to 22.6 mg/kg per day, with an average of 10.5 ± SD 3.8 mg/kg per day. In patients given once-daily injections, maximum plasma concentrations ranged from 21 to 80 mg/L. After 10 days of treatment the clinical response to tobramycin was equivalent in the two groups, and no ototoxicity or nephrotoxicity was observed in any subject [37].

It is not certain at present whether infrequent dosing or continuous infusion of aminoglycosides will prove more efficacious or less toxic for all or some uses. It does seem clear that intermittent maximal plasma concentrations that are quite great may not be associated with enhanced toxicity. How aminoglycoside concentration oscillations may interact with the time course of concentrations of other antimicrobial agents given simultaneously, to patients with varying rates of elimination, with infections at various sites, caused by organisms of different responses to drugs, remains a challenging area of investigation. For the present it is clear that we must consider not the static model of the in vitro culture or the single-application dose-response curve, but the dynamic interaction of the third dimension: time. If the effects of time on efficacy differs from the effects on toxicity, then the clinical utility of aminoglycosides may be dependent on the interval between doses as well as on the total daily dose.

References

1. Barza, M., Brusch, J., Bergeron, M.G., and Weinstein, L.: Comparison of levels of antibiotic in peripheral loci during continuous and intermittent infusion. In *13th Inter-*

science Conference on Antimicrobial Agents and Chemotherapy. Washington, D.C., American Society for Microbiology, 1973, Abstr. 12.

2. Barza, M., Brusch, J., Bergeron, M.G., and Weinstein, L.: Penetration of antibiotics into fibrin loci in vivo: III. Intermittent vs. continuous infusion and the effect of probenecid. *J. Infect. Dis.* 129:73–78, 1974.

3. Barza, M., and Weinstein, L.: Penetration of antibiotics into fibrin loci in vivo: I. Comparison of penetration and ampitillin into fibrin clots, abscesses, and "interstitial fluid." *J. Infect. Dis.* 129:59–65, 1974.

4. Benner, E.J., Kranbold, J., and Bush, W.: Tobramycin: in vitro activity, dosage nomogram derived from renal handling and clinical efficacy. In *Tobramycin:* Selected proceedings from the 8th International Congress of Chemotherapy, Athens, September 8–15, 1973. Amsterdam, Excerpta Medica, 1974, pp. 16–23.

5. Chan, R.A., Benner, E.J., and Hoeprich, P.D.: Gentamicin therapy in renal failure: a nomogram for dosage. *Ann. Intern. Med.* 76:773–778, 1972.

6. Cutler, R.E., Gyselynck, A.M., Fleet, R., and Forrey, A.W.: Correlations of serum creatinine concentrations and gentamicin half-life. *JAMA* 219:1037–1041, 1972.

7. Dahlgren, J.H., Anderson, E.T., and Hewitt, W.L.: Gentamicin blood levels: a guide to nephrotoxicity. *Antimicrob. Agents Chemother.* 8:58–62, 1975.

8. De Rosa, F., Buoncristiani, U., Capitanucci, P., and Frongillo, R.F.: Tobramycin: toxicological and pharmacological studies in animals and pharmacokinetic research in patients with varying degrees of renal impairment. *J. Int. Med. Res.* 2:100–114, 1974.

9. Dettli, L.C.: Drug dosage in patients with renal disease. *Clin. Pharmacol. Ther.* 16:274–280, 1974.

10. Frame, P.T., Phair, J.P., Watanakunakorn, C., and Bannister, T.W.P.: Pharmacologic factors associated with gentamicin nephrotoxicity in rabbits. *J. Infect. Dis.* 135: 952–956, 1977.

11. Gingell, J.C., and Waterworth, P.M.: Dose of gentamicin in patients with normal renal function and renal impairment. *Br. Med. J.* 2:19–22, 1968.

12. Giusti, D.L., and Hayton, W.L.: Dosage regimen adjustments in renal impairment. *Drug Intell. Clin. Pharm.* 7:382, 1973.

13. Goodman, E.L., Van Gelder, J., Dolmes, R., Hull, A.R., and Sanford, J.P.: Prospective comparative study of variable dosage and variable frequency regimens for administration of gentamicin. *Antimicrob. Agents Chemother.* 8:434–438, 1975.

14. Hewitt, W.L.: Gentamicin: toxicity in perspective. *Postgrad. Med. J.* 50(Suppl. 7): 55–61, 1974.

15. Hull, J.H., and Sarubbi, R.A., Jr.: Gentamicin serum concentrations: pharmacokinetic predictions. *Ann. Intern. Med.* 85:183–189, 1976.

16. Knight, R., Buell, J., Kalaba, R., Rockwell, R., and Jelliffe, R.W.: Computer-assisted kanamycin dose programs. *Clin. Res.* 18:137, 1970.

17. Korner, B.: Gentamicin therapy administered by intermittent intravenous injections. *Acta Pathol. Microbiol. Scand. [B]* 81 (Suppl. 241):15–22, 1973.

18. Kunin, C.M.: A guide to use of antibiotics in patients with renal disease. A table of recommended doses and factors governing serum levels. *Ann. Intern. Med.* 67:151–158, 1967.

19. Labovitz, E., Levison, M.E., and Kay, D.: Single-dose daily gentamicin therapy in urinary tract infection. *Antimicrob. Agents Chemother.* 6:465–470, 1974.

20. Line, G.H., Poole, G.W., and Waterworth, P.M.: Serum streptomycin levels and dizziness. *Tubercle* 51:76–81, 1970.

21. Mawer, G.E.: Nomograms for antibiotic dosage in adults. *Adverse Drug React. Bull.,* No. 52:176–179, 1975.

22. Mawer, G.E., Ahmad, R., Dobbs, S.M., McGough, J.G., Lucas, S.B., and Tooth, J.A.: Prescribing aids for gentamicin. *Br. J. Clin. Pharmacol.* 1:45–50, 1974.

23. McHenry, M.C., Gavan, T.O. Gifford, R.W., Jr., Geurkink, N.A., Van-Ommen, R.A., Town, M.A., and Wagner, J.G.: Gentamicin dosage for renal insufficiency. Adjustments based on endogenous creatinine clearance and serum creatinine concentrations. *Ann. Intern. Med.* 74:192–197, 1971.

24. McHenry, M.C., Vidt, D.G., Gavan, T.L., Van-Ommen, R.A., and Wagner, J.G.: Tobramycin in renal failure. In *Tobramycin:* Selected Proceedings from the 8th International Congress of Chemotherapy, Athens, September 8–15, 1973. Amsterdam, Excerpta Medica, 1964, pp. 67–71.

25. Noone, P., Parsons, T.M.C., Pattison, J.R., Slack, R.C.B., Garfield-Davies, D., and Hughes, K.: Experience in monitoring gentamicin therapy druing treatment of serious gram-negative sepsis. *Br. Med. J.* 1:477–481, 1974.

26. Powell, S.H., Groden, D.L., and Thompson, W.L.: Aminoglycoside nephrotoxicity in dogs: Dependence on drug, dose and regimen. *Scand. J. Infect. Dis.* 23:91–95, 1980.

27. Powell, S.H., Stern, R.C., and Thompson, W.L.: Safety of once daily therapy with high-dose tobramycin. *Pediatr. Res.* 13:466 (Abstr.), 1979.

28. Reiner, N.E., Bloxham, D.D., and Thompson, W.L.: Nephrotoxicity of gentamicin and tobramycin given once daily or continuously in dogs. *J. Antimicrob. Chemother.* 4(Suppl. A):85–101, 1978.

29. Riff, L.J., and Jackson, G.G.: Pharmacology of gentamicin in man. *J. Infect. Dis.* 124(Suppl.)S98–S105, 1971.

30. Sawchuk, R.J., and Zaske, D.E.: Pharmacokinetics of dosing regimens which utilize multiple intravenous infusions: gentamicin in burn patients. *J. Pharmacokinet. Biopharm.* 4:183–195, 1975.

31. Sawchuk, R.J., Zaske, D.E., Cipolle, R.J., Wargin, W.A., and Strate, R.G.: Kinetic model for gentamicin dosing with the use of individual patient parameters. *Clin. Pharmacol. Ther.* 21:362–369, 1976.

32. Schentag, J.J., Jusko, W.J., Plant, M.E., Cumbo, J.J., Vance, J.W., and Abrutyn, E.: Tissue persistence of gentamicin in man. *JAMA* 238:327–329, 1977.

33. Schumacher, G.E.: Practical pharmacokinetic techniques for drug consultation and evaluation. II: A perspective on the renal impaired patient. *Am. J. Hosp. Pharm.* 30:824–830, 1973.

34. Setia, U., and Gross, P.A.: Administration of tobramycin and gentamicin by the intravenous route every 6 hr in patients with a normal renal function. *J. Infect. Dis.* 134(Suppl.):S142–S145, 1976.

35. Valdivieso, M., Feld, R., Rodriguez, V., and Bodey, G.P.: Amikacin therapy of infections in neutropenic patients. *Am. J. Med. Sci.* 270:453–463, 1975.

36. Valdivieso, M., Horikoshi, N., Rodriguez, V., and Bodey, G.P.: Therapeutic trials with tobramycin. *Am. J. Med. Sci.* 268:149–156, 1973.

37. Welles, J.S., Emmerson, J.L., Gibson, W.R., Nickander, R., Owen, N.V., and Anderson, R.C.: Preclinical toxicology studies with tobramycin. *Toxicol. Appl. Pharmacol.* 25:398–409, 1973.

38. Westernfelder, S.R., Welling, B., and Madsen, P.O.: Efficacy and pharmacokinetics of tobramycin in patients with chronic urinary tract infection and various degrees of renal impairment. *Infection* 2:76–79, 1974.

21

COMPARATIVE CLINICAL TRIALS OF AMINOGLYCOSIDES

CRAIG R. SMITH and PAUL S. LIETMAN   The
Johns Hopkins University School of Medicine,
Baltimore, Maryland

Aminoglycoside antibiotics have played an important role in the evolution of comparative drug trials. The first randomized controlled clinical trial was published in 1948, comparing streptomycin and placebo in the treatment of pulmonary tuberculosis [19]. Since that report the scientific basis of clinical trial methodology has progressed as means to eliminate bias have been developed, such as double-blind evaluation and prognostic stratification. The first comparative trial of aminoglycosides in patients with bacterial infection was reported in 1969, 6 years after the introduction of gentamicin [18]. Since that report, 23 other trials comparing aminoglycosides have appeared in the literature. The purpose of this chapter is to review these reports to determine the extent of our knowledge about the relative effectiveness and toxicity of aminoglycoside antibiotics.

I. Methods

The English language literature was reviewed through December 1979 and all published clinical trials comparing aminoglycoside antibiotics in patients with bacterial infections were included in the analysis. Reviews, editorials, abstracts, extended abstracts, and letters to the editor were excluded because they did not provide sufficient information for critical examination.

Each paper was reviewed to determine the year of publication, the methods used, and the results obtained. The methods were examined to determine the study design used, the drugs administered, the patients studied (including those who were excluded), and the outcome criteria and statistical analysis used. The study design was analyzed to find out if it was prospective, controlled, and/or double-blind. Drug administration was reviewed to determine which aminoglycosides were given, how they were given (dose, route, frequency, and duration), how patients were assigned to receive them, and whether concurrent anti-biotics were given. The outcome criteria were examined to determine the method and frequency of data collection, and the definitions of effectiveness, nephrotoxicity, and ototoxicity. The statistical analysis was reviewed to find out the methods used and if beta error was calculated in studies reporting no difference between aminoglycosides. The results of the clinical trials were then analyzed to determine the number of patients enrolled; the number in each group evaluated for efficacy, nephrotoxicity, and oto-toxicity; and the incidence of each outcome. In addition, the number of infections treated and the incidence of urinary tract infection, bacteremia, and *Escherichia coli* infection were also noted. Results were examined to determine if the authors statified results by prognosis. Finally, the authors' conclusions about the relative effectiveness and toxicity fo aminoglycosides were reviewed.

II. General Review

Since 1969, 24 comparative aminoglycoside trials have been reported from 16 investiga-tive groups; four have been reported by Madsen et al. [15-17,20] from Wisconsin, and three by Smith et al. [23-25] from Johns Hopkins. The rate at which these reports have appeared in the literature has been coincident with the rate of development of new aminglycosides and the interest these new agents have evoked. From 1969 to 1976, only eight reports were published. In 1977 and 1978, six reports appeared each year, re-flecting interest in two new aminoglycosides, tobramycin and amikacin. During 1979 only three reports appeared, suggesting that interest in comparing new aminoglycosides in clinical trials may have waned.

The populations studied in these comparative clinical trials have differed. Eight have been conducted in patients with complicated urinary tract infections, five in patients with malignancy and/or neutropenia, and three have been reported in other special populations (patients in a surgical intensive care unit, patients on a urology ward, and patients with cystic fibrosis). Only nine have been reported in general medical populations and none have been done in patients with only gram-negative pneumonia or bacteremia, two groups with a poor prognosis.

All the studies have been prospective and controlled, with the standard aminoglyco-side used in the institution usually serving as the control group. In 20 studies, patients were assigned randomly to treatment groups. However, only 11 papers stated the method used and in only 9 (36%) were patients assigned by a table of random numbers. Only three studies (12%) have been double-blind (in another two the investigators, but not the physicians treating the patients, were blind). The failure of most studies to conduct their evaluation using an unbiased technique lessens the confidence that can be placed in their conclusions.

There are six aminoglycosides that have been studied in these trials. Gentamicin has been studied in 23, tobramycin in 11, amikacin in 8, kanamycin and sisomicin in 3, and netilmicin in 2. There are 50 groups because two trials studied three groups each. Genta-micin has been the control group most often (19 studies), and tobramycin (10 studies) and amikacin (7 studies) have most often been the experimental group. Other antibiotics were given in 16 studies; cephalosporins in 5, carbenicillin or ticarcillin in 4, penicillin or methicillin or nafcillin in 4, and polymixin B in 3.

The dose of gentamicin or tobramycin was reported in 22 studies. Seven trials used 1 mg/kg or less, seven used 1-2 mg/kg, and five gave 2 mg/kg or more. Three studies dosed patients on the basis of body surface area. Doses were given intravenously in 9, intramuscularly in 7, and by either route in 6; 2 did not report the route used. The frequency of administration was 8 hourly in 14, 6 hourly in 3, 12 hourly in 2, and continuously in 1; 4 did not report frequency of administration. The mean duration of treatment was stated in 21 studies; 12 treated patients for an average of 7-9 days, 6 treated patients for 10 or more days, and 3 treated patients for less than 7 days. Aminoglycoside levels were measured in 18 trials (75%), but dosage was altered in patients with renal dysfunction in only 11 studies (44%).

Efficacy was evaluated in 19 (78%) trials, while nephrotoxicity was evaluated in 23 (96%), auditory toxicity in 11 (44%), and vestibular toxicity in only 2 (8%). The average number of patients enrolled in these studies was 106 (range 30-281); 50 or fewer patients were enrolled in 5, 51-150 in 13, and more than 150 in only 6.

The most striking finding of this general review is the failure of these studies to use appropriate methodology to minimize the possibility that bias may have influenced their results. Conducting prospective controlled trials is not adequate to exclude bias. Random allocation, using a table of random numbers, is necessary to increase the probability that recognized or unrecognized risk factors will be equally distributed between two drug treatment groups. Double-blind evaluation is necessary to prevent conscious or unconscious bias from affecting the results. In addition, the heterogeneity of the populations studied, the treatment regimens used, and the evaluations performed make comparison of these studies difficult and necessitates careful examination to assure that appropriate conclusions are drawn.

III. Effectiveness

Evaluating the effectiveness of treatment is often difficult. It is important to clearly identify the population being studied, as well as those who are not being studied, in order to determine to whom the results can be generalized. Random drug allocation and double-blind evaluation are necessary to minimize bias. To determine the validity of the results, it is also necessary to define the diseases being treated and the outcome criteria for effectiveness. Evaluation of the effectiveness of aminoglycosides is complicated by a need to identify other antibiotics being given and to stratify treatment groups by prognosis. The latter consideration is of particular importance in evaluating antibiotics in the treatment of gram-negative infections. Several factors have been shown to influence the outcome of treatment, including severity of underlying disease, bacteremia, shock, neutrophil count, type of infection, and the species and susceptibility of the bacteria causing the infection. As a result, it is often difficult to study sufficient numbers of patients with a similar prognosis in order to compare results in the most rigorous fashion. In addition, it is not possible to treat infections due to organisms that are resistant to one or the other of the aminoglycosides being studied, therefore eliminating the major in vitro difference between drugs. Only differences in pharmacokinetics or potency could then lead to increased effectiveness, and it is not likely that these differences will have a large influence on outcome.

There have been 19 reports comparing the effectiveness of aminoglycosides (Table 1). Four have been conducted in general medical populations, 8 in patients with complicated urinary tract infections, 5 in patients with malignancy and/or neutropenia, and 2 in populations with other characteristics (patients on a urology ward and patients with cystic fibrosis). Definition of the population excluded from each trial has been reported in only 6 of these studies.

The aminoglycosides compared in these trials are outlined in Table 1. Gentamicin, tobramycin, and amikacin have been most often studied. Other antibiotics were given in 10 studies (50%), and in 8 these other antibiotics were effective against gram-negative organisms (Table 2). In these 8 studies, therefore, it is difficult to know if the aminoglycoside or the other antibiotic or both were effective. All but one of these trials were randomized, but in only 8 (40%) was the method of randomization stated to be by a table of random numbers or its equivalent. Two were double-blind and two were single-blind. Only 9 studies defined their criteria for identifying an infection and 7 of these were

Table 1 Patient Populations and Aminoglycosides Compared in Clinical Trials of the Effectiveness of Aminoglycosides

Study (reference)	Patient population[a]	Exclusions stated	Aminoglycosides compared[b]	
			Experimental group	Control group
18	GM	No	Gent	Kana
1	Other	No	Gent	Kana
6	GM	No	Gent	Kana
10	CA/Neut	No	Tob	Gent
20	UTI	No	Gent C_1	Gent
11	UTI	No	Sis	Gent
15	UTI	No	Tob	Gent
26	UTI	No	Tob	Gent
3	CA/Neut	No	Amik	Tob
23	GM	Yes	Amik	Gent
12	CA/Neut	No	Amik	Gent
5	UTI	Yes	Amik	Gent
16	UTI	No	Sis	Gent
4	UTI	No	Tob	Gent
21	Other	Yes	Tob	Gent
17	UTI	No	Net	Amik
9	CA/Neut	Yes	Sis Amik	Gent
14	CA/Neut	Yes	Net Amik	Gent
24	GM	Yes	Tob	Gent

[a]GM, general medical patients; CA/Neut, patients with malignancy and/or neutropenia; UTI, patients with urinary tract infections; Other, patients on a urology ward or patients with cystic fibrosis.

[b]Gent, gentamicin; Kana, kanamycin; Tob, tobramycin; Gent C_1, gentamicin C_1; Sis, sisomicin; Amik, amikacin; Net, netilmicin.

urinary tract infection (UTI) studies using colony counts. In addition, only 10 studies defined their criteria for cure, response, and treatment failure in objective terms. Once again, 7 of these were UTI studies. The other studies used subjective criteria ("improved," "cleared," etc.) or did not state their criteria.

Excluding the studies done exclusively in patients with urinary tract infections, there were an average of 5.5 different types of infections treated in each study. However, 43% of these infections (range 0–89%) were UTIs. Only 5 (46%) of the non-UTI studies had less than 50% UTIs. Only one study treated a large number of patients with pulmonary infections, and 81% of these patients had cystic fibrosis [21]. The mean number of organisms causing infection was 7.1 per study. *E. coli* caused 52% of the infections (range 28–90%) in the UTI studies and 29% (range 5–59%) of the other infections.

The incidence of effectiveness is shown in Table 3. In the experimental group, gentamicin has been effective on the average in 74% of patients, tobramycin in 76%, amikacin in 72%, netilmicin in 82%, and sisomicin in 77%. In the control groups, gentamicin has been effective on the average in 72% of patients, kanamycin in 66%, tobramycin in 57%, and amikacin in 57%. The latter two incidence rates were obtained from one study each

Table 2   Methods of Evaluation and Criteria Used in Clinical Trials Comparing the Effectiveness of Aminoglycosides

Study (reference)	Other antibiotics[a]	Double-blind evaluation	Criteria for infection stated	Outcome criteria stated[b]
18	1, 2, 3	No	No	No
1	1	No	No	No
6	1	No	No	No
10	None	No	No	SUBJ
20	None	No	Yes	+OBJ
11	None	No[c]	Yes	+OBJ
15	None	No	Yes	+OBJ
26	None	No	Yes	+OBJ
3	None	No	No	SUBJ
23	4, 5	Yes	Yes	OBJ
12	6, 7	No	No	SUBJ
5	None	No	No	OBJ
16	None	No	Yes	+OBJ
4	8	No	No	No
21	6, 9	No	No	OBJ
17	None	No	Yes	OBJ
9	6	No	No	SUBJ
14	9	No	No	SUBJ
24	4, 10	Yes	Yes	OBJ

[a] 1, polymixin B; 2, colistin; 3, cephaloridine; 4, clindamycin; 5, penicillin G; 6, carbenicillin; 7, cefazolin; 8, cefamandole; 9, ticarcillin; 10, methicillin/nafcillin.
[b] SUBJ, subjective criteria; OBJ, objective criteria.
[c] Single-blind evaluation (see the text).

[3,17]. The mean response rate in the control groups is 70%. The response rates have varied, depending on the populations studied. Patients with urinary tract infections have had a favorable outcome on the average in 71% of cases, patients with malignancy or neutropenia have responded in 69% of cases, patients with septicemia have responded in 78% of cases, and general medical populations have had a favorable outcome in 73% of cases. The average sample size of the control group is 35, with a mean response rate of 70%. The minimum improvement in response rate that could be detected under these conditions is 29% using a one-tailed test and with α less than 0.05 and β less than 0.2. Therefore, the response rate would have had to be 99% in the experimental group before the difference would have been statistically significant. Given the severity of the underlying diseases in these populations and the multiplicity of the infections being treated, this outcome is highly unlikely. If prognostic stratification were applied to these studies, the results would become meaningless. Having made these observations, it is not hard to

Table 3   Aminoglycosides Compared, Incidence of Effectiveness, and Authors' Conclusions

Study (reference)	Aminoglycosides compared[a]		Incidence of effectiveness (%)		Authors' conclusions[b]
	Experimental group	Control group	Experimental group	Control group	
18	Gent	Kana	81	74	Difference
1	Gent	Kana	90	68	No difference
6	Gent	Kana	73	58	No difference
10	Tob	Gent	45	45	No difference
20	Gent C_1	Gent	53	60	No difference
11	Sis	Gent	88	62	No difference
15	Tob	Gent	64	63	No difference
26	Tob	Gent	80	80	No difference
3	Amik	Tob	55	57	No difference
23	Amik	Gent	77	78	No difference
12	Amik	Gent	72	83	No difference
5	Amik	Gent	68	60	No difference
16	Sis	Gent	70	63	No difference
4	Tob	Gent	100	100	No difference
21	Tob	Gent	93	71	Difference
17	Net	Amik	69	57	No difference
9	Sis	Gent	72	68	No difference
	Amik		70		
14	Net	Gent	95	97	No difference
	Amik		91		
24	Tob	Gent	80	78	No difference

[a]Gent, gentamicin; Kana, kanamycin; Tob, tobramycin; Gent C_1, gentamicin C_1; Amik, amikacin; Sis, sisomicin; Net, netilmicin.

[b]Study 18 stated that gentamicin plus cephaloridine was the best combination tested, but the differences were not statistically significant. Study 21 found that ticarcillin/tobramycin was better than carbenicillin/gentamicin ($P < 0.05$).

understand why every study except one has failed to show a difference in efficacy between aminoglycosides. The one study that found a valid improvement in the experimental group compared ticarcillin-tobramycin and carbenicillin-gentamicin in patients with pulmonary infections, most of whom had cystic fibrosis and *Pseudomonas* infection [21]. In this study it is not possible to know if the difference was due to the ticarcillin or the tobramycin or both and does not imply an intrinsic difference between tobramycin and gentamicin.

In summary, we do not know much about the relative effectiveness of aminoglycosides. The heterogeneity of the populations treated and the small numbers of patients studied make the results of the current trials of little value. If increments in effectiveness of even as large as 20% are of interest, sample sizes of greater than 100 patients in each group are necessary to demonstrate this difference. The imposition of prognostic stratification within these groups would entail yet an even larger sample size. Future investigators should give serious consideration to these problems before initiating studies similar to those that have been reported. Because of the practical problems of enrolling patients in studies and the need for large numbers of patients, it seems that only studies taking many years to complete or multicenter trials would likely be of any utility. Patients with a poor prognosis, such as those with malignancy or neutropenia, gram-negative pneumonia, or gram-negative bacteremia seem most appropriate for the latter approach.

IV. Nephrotoxicity

Nephrotoxicity may be an important factor limiting the clinical utility of aminoglycosides. Most of the comparative trials we reviewed attempted to establish the relative nephrotoxicity of these agents (Table 4). Several considerations are important in comparing the results of these studies. The definition of nephrotoxicity must be stated, the criteria used should be quantitative, and the definition should be of clinical importance. Although aminoglycosides specifically affect proximal tubular cells, measures of proximal tubular damage do not measure functional changes that are usually of clinical importance. In addition, these measures are sensitive to many other insults. For example, aspirin may cause urinary N-acetyl-β-D-glucosaminidase to increase. Since determination of creatinine clearance, a commonly used measure of the glomerular filtration rate (GFR), is difficult on most medical wards, the serum creatinine is the most useful measure of GFR. The relationship between serum creatinine and GFR is not linear. Therefore, at high GFRs (greater than 80 cm^2/min), a small change in creatinine may indicate a large change in GFR. At higher creatinines, the same change in GFR will be associated with a higher rise in creatinine. Definitions of nephrotoxicity should take this relationship into account, as well as the effects of age. Aged patients have a lower serum creatinine for a given GFR than do younger patients. Therefore, in aged patients a given change in GFR will be associated with a larger increment in creatinine. Another factor which must be taken into account is that many of the patients treated are seriously ill and may have other causes of acute renal failure. Because it is often not possible to determine if the aminoglycoside or the other factors are responsible for the renal failure, these patients should be excluded from the evaluation. This determination is often judgmental and requires methodology to prevent conscious or unconscious bias from influencing the decision. Ideally, studies should be double-blind in order to circumvent this difficult problem.

Several other factors may influence the outcome, such as drug dose, duration of treatment, or the concurrent use of cephalosporins. The frequency of monitoring is also an important factor influencing the frequency of toxicity, since toxicity can be recognized only if it is looked for. Those studies that measure renal function irregularly and infrequently will likely show a lower incidence. In addition, since changes in creatinine may not develop until 7-9 days after therapy is discontinued, measurements should be made during this interval. Finally, serum levels should be reduced in patients with renal dysfunction. Ideally, serum levels should be measured frequently and drug dose altered to maintain levels within a predefined range.

All but one of the comparative trials we reviewed assessed nephrotoxicity (Table 4). A quantitative definition was used in 12 reports, a parametric analysis (comparison of means) in 3, a qualitative definition in 3 ["rise" or "increase" in creatinine or blood urea nitrogen (BUN)], and no definition was reported in 5. In the 12 reports using a quantitative measure, there were 10 different definitions of nephrotoxicity. One definition was used 3 times; all 3 reports were from the same institution [23-25]. The serum

Table 4   Aminoglycosides Compared and Definition Used in Clinical Trials Evaluating Nephrotoxicity

Study (reference)	Aminoglycosides compared		Definition[a]	Creatinine clearance measured	Concomitant insults excluded
	Experimental group	Control group			
18	Gent	Kana	3	No	Yes
1	Gent	Kana	3	No	No
10	Tob	Gent	3	No	No
20	Gent C_1	Gent	3	No	No
11	Sis	Gent	2	No	No
15	Tob	Gent	3	No	No
26	Tob	Gent	2	No	No
3	Amik	Tob	1	No	No
23	Amik	Gent	1	No	Yes
12	Amik	Gent	1	No	No
5	Amik	Gent	1	Yes	No
16	Sis	Gent	4	No	No
4	Tob	Gent	4	Yes	Yes
21	Tob	Gent	2	No	No
25	Tob	Gent	1	No	Yes
17	Net	Amik	4	Yes	No
9	Amik	Gent	1	No	No
	Sis			No	
13	Amik	Gent	1	No	No
2	Tob	Gent	1	Yes	No
7,8	Tob	Gent	1	Yes	No
14	Net	Gent	1	No	Yes
	Amik			No	
22	Tob	Gent	1	Yes	Yes
24	Tob	Gent	1	No	Yes

[a] 1, quantitative; 2, qualitative; 3, not stated; 4, parametric analysis.

creatinine was used to define nephrotoxicity in all studies. Six studies also used creatinine clearance, one used cylinduria [21], and one used urinary cast counts, urinary B_2-microglobulin excretion, and a prediction of renal tissue aminoglycoside levels [22]. The frequency of monitoring therapy was mentioned in 20 studies. The frequency was twice weekly or greater in only 12 studies (60%); 3 studies only measured renal function before and after, and 5 measured before, during, and after therapy. Other causes of acute renal failure were differentiated from aminoglycoside nephrotoxicity in only 8 studies (35%).

The incidence of nephrotoxicity has varied from 0 to 77.8% (Table 5). The mean incidence is 15%. Gentamicin has caused nephrotoxicity on the average in 17% of patients, kanamycin in 16%, sisomicin in 16%, tobramycin in 13%, amikacin in 9%, and netilmicin in 2%. Because of differences in the definition of nephrotoxicity and study methodology, however, these rates cannot be used to rank order the relative toxicity of these agents.

Table 5   Frequency of Nephrotoxicity and Authors' Conclusions

Study (reference)	Aminoglycoside compared[a]		Nephrotoxicity frequency (%)[b]		Authors' conclusions
	Experimental group	Control group	Experimental group	Control group	
18	Gent	Kana	6	21	No difference
1	Gent	Kana	2	12	No difference
10	Tob	Gent	3	7	No difference
20	Gent C_1	Gent	NA	NA	Gent C_1 > Gent
11	Sis	Gent	10	13	No difference
15	Tob	Gent	NA	NA	No difference
26	Tob	Gent	5	18	Gent > Tob
3	Amik	Tob	20	22	No difference
23	Amik	Gent	8	11	No difference
12	Amik	Gent	8	9	No difference
5	Amik	Gent	13	13	No difference
16	Sis	Gent	NA	NA	No difference
4	Tob	Gent	0	0	No difference
21	Tob	Gent	12	11	No difference
25	Tob	Gent	13	21	No difference
17	Net	Amik	NA	NA	No difference
9	Amik	Gent	8	15	A < G + S
	Sis		22		No difference
13	Amik	Gent	0	15	No difference
2	Tob	Gent	9	29	Gent > Tob
7	Tob	Gent	42	78	Gent > Tob
14	Net	Gent	2	2	No difference
	Amik		6		No difference
22	Tob	Gent	10	23	Gent > Tob
24	Tob	Gent	12	26	Gent > Tob

[a]1, quantitative; 2, qualitative; 3, not stated; 4, parametric analysis.
[b]NA, not applicable, parametric analysis.

The conclusions of these studies must be examined individually to determine the extent of our knowledge about the relative toxicity of these agents. Eighteen trials found no difference between the aminoglycosides studied. This finding is not surprising because the average size of the experimental and control groups in these studies is 48. In a study of this size, using a one-tailed test, finding an incidence of nephrotoxicity of 17% in the control group with α less than 0.05 and β less than 0.2, the experimental group would have to have an incidence of nephrotoxicity of less than 5% for the difference to be significant. Positive results were obtained in some studies, however. One study concluded that gentamicin C_1 was less nephrotoxic than gentamicin [20], and one reported that amikacin was less nephrotoxic than gentamicin or sisomicin [9]. The findings of these two studies, however, are not statistically significant (α greater than 0.05). Five studies have found that tobramycin causes nephrotoxicity less frequently than gentamicin, but four of these studies are not randomized and/or double-blind. However, one study is a prospective, controlled, randomized, double-blind trial, providing strong evidence supporting less frequent nephrotoxicity with tobramycin [24]. None of these studies have found that the severity of the nephrotoxicity caused by gentamicin and tobramycin differs.

Based on this analysis, it is apparent that our current knowledge of the relative nephrotoxicity of aminoglycosides is limited by the failure of most studies to (1) use a standard dosage and duration of treatment, (2) evaluate toxicity in an unbiased manner, (3) define nephrotoxicity uniformly, (4) monitor renal function frequently during and after therapy, (5) differentiate patients with other causes of acute renal failure from those with aminoglycoside induced nephrotoxicity, and (6) enroll sufficient number of patients to draw meaningful conclusions. Despite these differences, there is considerable evidence, some of which was gathered in well-designed trials, that tobramycin causes nephrotoxicity less frequently than gentamicin. Future work in this area should be directed at developing a uniform definition of nephrotoxicity and performing studies that adhere to methodological standards to ensure that the results are meaningful.

V. Ototoxicity

Ototoxicity is difficult to detect in patients who are seriously ill. Clinical examination of auditory and vestibular function can detect only large changes and is poorly standardized. Bedside audiometry may improve the sensitivity and reproducibility of the testing of auditory function. However, patients must be alert and able to cooperate with the examiner in order to perform the test. Vestibular function testing using electronystagmography can only be done in a specialized laboratory, seriously impeding the use of this method in clinical trials of aminoglycosides. As a result of the difficulties in detection, only 6 studies have attempted to compare the auditory toxicity of aminoglycosides using audiometry, although 11 have mentioned auditory toxicity. Only 2 studies have monitored vestibular function with electronystagmography.

The definition of auditory toxicity has been either (1) $\geqslant$20 dB loss (2 studies) [2,14], (2) $\geqslant$15 dB loss at any frequency bilaterally or unilaterally (3 studies) [12,23,24], or (3) $\geqslant$15 dB loss at two or more frequencies (1 study) [13]. Vestibular toxicity has been defined as either a 33% or greater or a 50% or greater decrease in maximum slow-phase eye speed [2,12]. The frequency of monitoring has been before, during, and after in 3 studies, every 2-3 weeks in 1, and weekly in 2. Concomitant audiologic insults have been differentiated from aminoglycoside ototoxicity in only 2 studies [13,24].

The mean frequency of ototoxicity has been 7%. The mean frequency with gentamicin has been 8%, amikacin 7%, tobramycin 6%, sisomicin 5%, and netilmicin 3%. None of the studies comparing the ototoxicity of aminoglycosides has found a difference between them. However, the mean number of patients studied in the experimental and control groups is only 35. With an average rate of 8% in the control group, it would be impossible to show a statistically significant difference between the groups, even if the experimental group did not experience toxicity. If the experimental group was more ototoxic, these studies can only exclude differences of approximately 25%. These findings indicate that we know almost nothing about the relative ototoxicity of aminoglycosides. Even differences of two- or threefold cannot be excluded by most current studies. Larger, well-designed studies, frequently monitoring auditory and vestibular function, will be needed to detect differences that may be clinically meaningful.

VI. Conclusions

This analysis has revealed several striking findings. First, the heterogeneity of these studies makes comparisons between them difficult. Second, many of the studies comparing aminoglycosides have failed to adhere to methodological standards for comparative drug trials. This deficiency detracts from the confidence that can be placed in the conclusions of these studies. Third, most trials have studied small numbers of patients, probably explaining their failure to demonstrate differences between aminoglycosides. However, clinically meaningful differences may exist that could not have been detected with the frequency of monitoring and the numbers of patients studied. Finally, the only difference between aminoglycosides that has been demonstrated consistently and in well-designed studies is that tobramycin causes nephrotoxicity less frequently than does gentamicin.

References

1. Cox, C.E., and Harrison, L.H.: Comparison of gentamicin and polymyxin B-kanamycin in therapy of bacteremia due to gram-negative bacilli. *J. Infect. Dis.* 124(Suppl.):S156–S163, 1971.
2. Fee, W.E., Jr., Vierra, V., and Lathrop, G.R.: Clinical evaluation of aminoglycoside toxicity: tobramycin versus gentamicin, a preliminary report. *J. Antimicrob. Chemother.* 4(Suppl.):31–36, 1978.
3. Feld, R., Valdivieso, M., Bodey, G.P., and Rodriguez, V.: Comparison of amikacin and tobramycin in the treatment of infection in patients with cancer. *J. Infect. Dis.* 135:61–66, 1977.
4. Gentry, L.O.: Efficacy and safety of cefamandole plus either gentamicin or tobramycin in therapy of severe gram-negative bacterial infections. *J. Infect. Dis.* 137(Suppl.): S144–S149, 1978.
5. Gilbert, D.N., Eubanks, N., and Jackson, J.: Comparison of amikacin and gentamicin in the treatment of urinary tract infections. *Am. J. Med.* 62:924–929, 1977.
6. Holloway, W.J., and Taylor, W.A.: Gentamicin and kanamycin in the treatment of gram-negative sepsis: a comparative study. *J. Infect. Dis.* 124(Suppl.):S180–S184, 1971.
7. Kahlmeter, G., Hallberg, T., and Kamme, C.: Gentamicin and tobramycin in patients with various infections—concentrations in serum and urinary excretion. *J. Antimicrob. Chemother.* 4(Suppl.):37–45, 1978.
8. Kahlmeter, G., Hallberg, T., and Kamme, C.: Gentamicin and tobramycin in patients

with various infections—nephrotoxicity. *J. Antimicrob. Chemother.* 4(Suppl.): 47–52, 1978.

9. Keating, M.J., Bodey, G.P., Valdivieso, M., and Rodriguez, V.: A randomized comparative trial of three aminoglycosides—comparisons of continuous infusions of gentamicin, amikacin and sisomicin combined with carbenicillin in the treatment of infections in neutropenic patients with malignancies. *Medicine (Baltimore)* 58:159–170, 1979.

10. Klastersky, J., Hensgens, C., Henri, A., and Daneau, D.: Comparative clinical study of tobramycin and gentamicin. *Antimicrob. Agents Chemother.* 5:133–138, 1974.

11. Klastersky, J., Hensgens, C., Gerard, M., and Daneau, D.: Comparison of sisomicin and gentamicin in bacteriuric patients with underlying diseases of the urinary tract. *Antimicrob. Agents Chemother.* 7:742–747, 1975.

12. Lau, W.K., Young, L.S., Black, R.E., Winston, D.J., Linne, S.R., Weinstein, R.J., and Hewitt, W.L.: Comparative efficacy and toxicity of amikacin/carbenicillin versus gentamicin/carbenicillin leukopenic patients. *Am. J. Med.* 62:959–966, 1977.

13. Lerner, S.A., Seligsohn, R., and Matz, G.J.: Comparative clinical studies of ototoxicity and nephrotoxicity of amikacin and gentamicin. *Am. J. Med.* 62:919–923, 1977.

14. Love, L.J., Schimpff, S.C., Hahn, D.M., Young, V.M., Standiford, H.C., Bender, J.F., Fortner, C.L., and Wiernik, P.H.: Randomized trial of empiric antibiotic therapy with ticarcillin in combination with gentamicin, amikacin or netilmicin in febrile patients with granulocytopenia and cancer. *Am. J. Med.* 66:603–610, 1979.

15. Madsen, P.O., Kjaer, T.B., and Mosegaard, A.: Comparison of tobramycin and gentamicin in the treatment of complicated urinary tract infections. *J. Infect. Dis.* 134 (Suppl.):S150–S152, 1976.

16. Madsen, P.O., Kjaer, T.B., and Mosegaard, A.: Treatment of complicated urinary tract infections. Comparative study of sisomicin and gentamicin. *Urology* 9:635–638, 1977.

17. Maigaard, S., Frimodt-Moller, N., and Madsen, P.O.: Comparison of netilmicin and amikacin in treatment of complicated urinary tract infections. *Antimicrob. Agents Chemother.* 14:544–548, 1978.

18. Martin, C.M., Cuomo, A.J., Geraghty, M.J., Zager, J.R., and Mandes, T.C.: Gram-negative rod bacteremia. *J. Infect. Dis.* 119:506–517, 1969.

19. Medical Research Council: Streptomycin treatment of pulmonary tuberculosis. *Br. Med. J.* 2:769–782, 1948.

20. Mosegaard, A., Welling, P.G., and Madsen, P.O.: Gentamicin and gentamicin C_1 in the treatment of complicated urinary tract infections: comparative study of efficacy, tolerance, and pharmacokinetics. *Antimicrob. Agents Chemother.* 7:328–332, 1975.

21. Parry, M.F., and Neu, H.C.: A comparative study of ticarcillin plus tobramycin versus carbenicillin plus gentamicin for the treatment of serious infections due to gram-negative bacilli. *Am. J. Med.* 65:961–966, 1978.

22. Schentag, J.J., Plaut, M.E., Cerra, F.B., Wels, P.B., Walczak, P., and Buckley, R.J.: Aminoglycoside nephrotoxicity in critically ill surgical patients. *J. Surg. Res.* 26: 270–279, 1979.

23. Smith, C.R., Baughman, K.L., Edwards, C.Q., Rogers, J.F., and Lietman, P.S.: Controlled comparison of amikacin and gentamicin. *N. Engl. J. Med.* 296:349–353, 1977.

24. Smith, C.R., Lipsky, J.J., Laskin, O.L., Hellman, D.B., Mellits, E.D., Longstreth, J., and Lietman, P.S.: Double-blind comparison of the nephrotoxicity and auditory toxicity of gentamicin and tobramycin. *N. Engl. J. Med.,* 302:1106–1109, 1980.

25. Wade, J.C., Smith, C.R., Petty, B.G., Lipsky, J.J., Conrad, G., Ellner, J., and Liet-

man, P.S.: Cephalothin plus an aminoglycoside is more nephrotoxic than methicillin plus an aminoglycoside. *Lancet* 2:604–606, 1978.

26. Walker, B.D., and Gentry, L.O.: A randomized, comparative study of tobramycin and gentamicin in treatment of acute urinary tract infections. *J. Infect. Dis.* 134 (Suppl.):S146–S149, 1976.

22

ENDOTRACHEAL ADMINISTRATION OF AMINOGLYCOSIDES

JEAN KLASTERSKY Institute Jules Bordet, Brussels, Belgium

JEAN-PIERRE THYS Saint-Pierre University Hospital, Brussels, Belgium

Despite recent advances in respiratory intensive care and in antibiotic therapy, the prognosis of gram-negative pneumonias remains unchanged and rather poor; the mortality reaches approximately 50% in the general population [12,40], and some studies of cancer patients report results that are even more pessimistic [29,39].

Several factors could explain the severity of this disease. Underlying illnesses and predisposing factors are the rule with gram-negative pneumonias. In some respects, gram-negative bronchopneumonias can be considered as an host-dependent disease occurring only under conditions that decrease the natural capacity of the host to resist bacterial infections. They can be considered most often as opportunistic infections. The problem of the in vitro resistance of the offending organism to the antibiotics that are usually given, aminoglycosides, has not been, until now, an adequate explanation for the poor clinical results observed with gram-negative bronchopneumonias. On the other hand, pharmacologic characteristics of the penetration of these antibiotics within the sputum might be involved in the clinical results observed so far. There is some correlation between the antimicrobial activity of the sputum and the clinical outcome of bronchopulmonary infections [25,41]. Thus, it is important to realize that several factors in bronchial secretions can reduce the antimicrobial activity of aminoglycosides. The penetration of these drugs through the blood–bronchus barrier is relatively poor, especially if the low therapeutic index of these antibiotics is considered. Moreover, there is some inactivation of aminoglycosides by purulent secretions and thus a loss of antibiotic activity in the site of infection. Thus, both relatively low levels of aminoglycosides in bronchial secretions and inactivation can interfere with the possibility of achieving, with systemic aminoglycosides, an adequate antimicrobial activity within the bronchial secretions, despite adequate blood levels of these antibiotics. To confirm this hypothesis, we have conducted a series of studies which indicate that when aminoglycosides are given systemically to patients with bronchopulmonary infections, adequate concentrations of antibiotics in blood do not result in adequate concentrations of antibiotics in the bronchial lumen. The next logical step appeared to be the direct administration—by endotracheal instillation or by aerosol—of the antibiotic to the site of the infection in order to achieve adequate levels within the bronchial tree without major systemic adverse effects. This was investigated in another series of experiments. Finally, endotracheal administration of aminoglycosides were used for both the prevention and the treatment of gram-negative bronchopulmonary infections in severely ill patients. We have treated mostly neurosurgical patients who were tracheotomized or intubated. In those patients, because they are unconscious and uncooperative with physiotherapists, the incidence of bronchopulmonary infections is very high and deserves special attention as far as antimicrobial therapy is concerned.

These various aspects of our work, as well as the pertinent literature, are reviewed here in an attempt to delineate the role of endotracheally administered aminoglycosides as a therapy for bronchopulmonary gram-negative infections.

I. Penetration of Aminoglycosides into Bronchial Secretions

Several studies of the penetration of antibiotics into the bronchial secretions have been recently summarized [20]. These investigations had been frequently performed according to different protocols on small number of patients with various bronchopulmonary conditions. Moreover, the timing of the collection of bronchial secretions specimens, as well as the techniques for assaying antibiotics in this fluid, were numerous. It is therefore not surprising that information about the antimicrobial drug concentrations in bronchial secretions is sometimes conflicting. Nevertheless, it can be estimated that the ratios between the antibiotic levels in bronchial secretions and in serum are approximately 3–6% for ampicillin and amoxycillin, 10–20% for carbenicillin and about 10–30% for tetracycline; the bronchial penetration of cephalosporins seems to be very low.

The penetration of the aminoglycosides through the blood–bronchus barrier varies roughly from 10 to 40% (Table 1). Pennington and Reynolds [28] developed an animal model for the study of the penetration of antibiotics into bronchial secretions after intravenous injection. These experiments were performed in uninfected dogs whose bronchial secretions had been stimulated by pilocarpine. After the intravenous injection of gentamicin (1.7 mg/kg), the peak level in bronchial secretions, obtained at 15 min, was 5.1 μg/ml, approximately 26% of the peak concentration in serum (Table 1). Two hours after the injection, bronchial levels were about 3.3 μg/ml and 52% of serum levels. Clearance of antibiotic from the bronchial secretions was slower than from the serum.

In another study with the same experimental model, these authors administered gentamicin parenterally at the same dosage (1.7 mg/kg) by three methods: the rapid intravenous injection and the intramuscular injection in one dose or in two divided doses [31]. With these two modes of intramuscular administration, the peaks serum levels were observed

Table 1   Penetration of Various Aminoglycosides into Bronchial Secretions After IM or IV Administration

Authors (reference)	Antibiotic	Dosage (mg/kg)	Mode of administration (IM, IV)	Peak levels (μg/ml) Serum (S)	Peak levels (μg/ml) Bronchial secretions (BS)	Ratio BS/S × 100 (%)
Pennington and Reynolds [28][a]	Gentamicin	1.7	IV	19.1	5.1	26
Odio et al. [26]	Gentamicin	2	IV	6.8	1.8	26.5
Marks et al. [24]	Gentamicin	1–1.25	IV	5.7	1.3	22.8
Wong et al.	Gentamicin	1.7	IV	–	–	40
Pennington and Reynolds [30][a]	Tobramycin	1.7	IV	11.3	1.15	10.2
McCrae et al. [23]	Tobramycin	1.7–4	IV	5.7–8.7	0.4–0.7	4.6–12
Thys and Klastersky [46]	Netilmicin	2.5	IV	12.4	1.1	8.9
Bergogne-Berezin et al. [3]	Amikacin	500 mg[b]	IM	33	3.6	10.9
Thys and Klastersky [42]	Amikacin	3.75 7.5	IM IM	9.7 21.8	3.5 5.4	36.1 24.8

[a] Studies in dogs.
[b] Total dose.

at 1 or 2 hr, later than in the preceding study. This observation explains why the ratios between the peaks bronchial and serum concentrations in that study have been found to be higher (about 40%). In tracheostomized patients, Odio et al. [26] administered gentamicin (2.0 mg/kg) intramuscularly: peak concentrations in the bronchial secretions and the serum—observed at the same time 1 hr after the injection—were 1.8 and 6.8 μg/ml, respectively, with a ratio between the peaks of 26.5%. With the same drugs in patients with cystic fibrosis, treated with a daily dose of 1–2 mg/kg of gentamicin, Saggers and Lawson [37] found a ratio between the levels in bronchial secretion and serum of 5–20%. After the intravenous administration of gentamicin (4–5 mg/kg, in four divided doses) to three patients with cystic fibrosis, Marks et al. [24] observed a ratio between the peak concentrations in sputum and blood of approximately 20%. Wong et al. [48] observed higher bronchial levels of gentamicin in patients with respiratory tract infections which corresponded to 40% of serum levels. In their experimental model employing tobramycin, Pennington and Reynolds [30] observed that 5 min after the intravenous injection of tobramycin (1.7 mg/kg), the peak serum level was 11.3 μg/ml; in bronchial secretions, the peak tobramycin level occurred somewhat later (2 hr) and was lower (1.15 μg/ml) than that observed with gentamicin under similar experimental conditions.

The ratio between the peak concentration in bronchial secretion and in serum has been found at 10.2%. In patients with cystic fibrosis treated with tobramycin, this ratio has been found between 4.6 and 12%, after one intravenous injection of 1.7-4 mg/kg [23].

More recently, we administered netilmicin (7.5 mg/kg) either by intermittent injections or by continuous infusion to 10 tracheostomized patients in a crossover fashion [46]. After the bolus injection, the mean peak serum levels in 10 patients was 12.4 μg/ml at 10 min; 8 hr later, the mean serum levels was 1.0 μg/ml. During the continuous infusion, the mean serum level remained stable at 3.3 μg/ml. After the bolus injection, the mean peak concentration of netilmicin in the bronchial secretions was obtained at 30 min and reached 1.1 μg/ml; 8 hr later, the mean bronchial secretion concentration was 0.4 μg/ml. The mean bronchial level during the continuous infusion experiment was 0.6 μg/ml. With the bolus injection, the ratio between the peak concentrations in bronchial secretions and serum was 8.9%; the ratio between the antibiotic levels in the bronchial secretions and in blood was 18.2% during the continuous infusion. During the first 2 hr of the bolus and continuous intravenous administration, the areas under the curves, in the serum and in the sputum, were considerably greater with the bolus injections than with the continuous infusion (206% and 150%, respectively; $P < 0.001$). However, the areas under the curves calculated for the total 8 hr duration of the experiment were similar with the two modes of administration, in the serum and in the sputum. Moreover, the penetration of netilmicin (ratio between the areas under the curves in bronchial secretions and in serum x 100) across the blood-bronchus barrier was found identical (19.3%) with the two modes of administration. Bronchial levels of amikacin after systemic administration have also been evaluated. Bergogne-Berezin et al. [3] observed, after the intramuscular injection of 500 mg in tracheostomized patients, peak levels in serum and bronchial secretions at 33 and 3.6 μg/ml, respectively, thus an approximate ratio of 11%.

We determined the levels of amikacin after intramuscular administration at two different dosages (3.75 and 7.5 mg/kg, twice daily) in serum and in bronchial secretions in nine tracheostomized patients who were studied in a crossover fashion on two consecutive days [42]. With the low dose, the mean peak serum level was 9.7 μg/ml; 12 hr after the injection, it was 1.7 μg/ml. With the high dose, the mean peak serum level was 21.8 μg/ml and 12 hr later, the mean serum level was 3.5 μg/ml. In bronchial secretions, the mean serum level was 3.5 μg/ml. In bronchial secretions, the mean peak level after the low dose was 3.5 μg/ml; 12 hr later, this level was 1.0 μg/ml; with the higher dose, the mean peak levels was 5.4 μg/ml, and after 12 hr, 1.8 μg/ml. The ratios between the peak concentration in bronchial secretions and in serum were found to be 36.1% and 24.8% for the low and the high dose, respectively. The ratios between the areas under the curves obtained in bronchial secretions and in serum were found to be significantly different ($P < 0.05$): 48.1% for the low dose of amikacin and 38.4% for the higher dose.

To summarize these studies of the penetration of aminoglycosides into bronchial secretions, it can be said that in most studies with gentamicin, the ratios between the peak concentration of this antibiotic in bronchial secretions and in serum are approximately 20-25%. Fewer studies with tobramycin suggest that this ratio might be lower with that drug (5-10%) and that the penetration of amikacin through the blood-bronchus barrier is in the range 10-36%. At this point, one can conclude that, although the penetration of the aminoglycosides through the blood-bronchus barrier is surely not negligible, the low therapeutic index of these drugs results in absolute levels of these antibiotics in the bronchial secretions which are low and often inadequate to inhibit most strains of Enterobacteriacae isolated from clinical material. It must be noted in addition that the ratios between the levels in bronchial secretions and in serum were calculated here on the basis of peak concentrations. After its peak, the concentration of aminoglycoside decreases more rapidly in blood than in bronchial secretions. Thus, the ratio between these concentrations increases; however, it should be stressed that the blood levels are usually below any therapeutically significant concentration 2 hr after the injection and thus that even a relatively high proportion of penetration into the sputum is clinically meaningless.

II. Interaction Between Aminoglycosides and Bronchial Secretions

It has been demonstrated for ampicillin [25] and cephalexin [13] that severe bronchial inflammation increases the penetration of antibiotics into bronchial secretions; when the inflammation subsides, the penetration of these antibiotics into the bronchial lumen is reduced. To what extent bronchial inflammation can enhance the penetration of aminoglycosides through the blood-bronchus barrier is not known. In Wieser's series [47], inflammation of the respiratory tract with production of purulent sputum does not appreciably increase levels of gentamicin. Similarly, in our hands, a correlation between the degree of inflammation of the bronchial mucosa and the level of netilmin in the bronchial secretions has not been very satisfactory [46]. Moreover, it appears likely, as discussed next, that inflammation might be an adverse factor rather than a favorable one for the antimicrobial activity of aminoglycosides within the bronchial secretions, since gentamicin can be inactivated by purulent material [6]. When netilmicin is added to whole sputum and to its supernatant obtained after centrifugation, the antimicrobial activity is reduced in the whole sputum as compared to the supernatant [45]. Variable amounts of netilmicin can be recovered from the whole sputum by successive washings with water, suggesting a partially reversible binding of the antibiotic to the purulent material of the sputum rather than an inactivation. The decrease of antibiotic activity in whole sputum can be roughly correlated to the degree of purulence; very purulent sputum (50-100 pus cells per high-power field) inactivates 36-85% of netilmicin, whereas there is no decrease of activity of the aminoglycoside in mucoid sputum (1-12 pus cells). To test this observation further, various numbers of leucocytes (from 10,000 to 650,000 per mm^3) were mixed with aliquots of a pool of supernatant of sputum, to which netilmicin (10 μg/ml) was then added; a linear relationship was observed between the number of leucocytes and the degree of inactivation of netilmicin [44]. Similar data have been observed with amikacin [44].

In sputum from patients with cystic fibrosis, the activity of neomycin and, to a lesser degree, the activity of gentamicin and tobramycin, was found to be decreased against strains of *Pseudomonas aeruginosa;* the minimal bactericidal activities of these antibiotics in broth were reduced by a factor of 2-16 when tested in sputum [9]. It has been shown that neomycin and gentamicin bind strongly to ribosomes and ribosomal ribonucleic acid [7]; moreover, complex formation of DNA with neomycin with consequent partial inactivation of the antibiotic has been shown to occur in the bronchial secretions of patients with cystic fibrosis [33]. In addition, some binding of gentamicin to glycoproteins of the bronchial mucus has been observed as well [38].

To conclude, it appears that a partial inactivation of aminoglycosides by purulent bronchial secretions occurs and might be an important factor which, in addition to the relatively poor penetration of these antibiotics through the blood–bronchus barrier, contributes to the disappointing clinical results in the therapy of severe gram-negative bacillary bronchopulmonary infections with aminoglycosides alone. Therefore, the endotracheal administration of aminoglycosides has been attempted in order to achieve adequate antibiotic levels at the site of the infection.

III. Pharmacology of Endotracheally Administered Aminoglycosides

Local antibiotic therapy of respiratory tract infection may be performed by two methods: either by direct instillation into the tracheobronchial tree via a catheter introduced into the trachea through a tracheostomy or an endotracheal tube, or by aerosol either via the upper airways or through some respiratory equipment.

The relative merits of these two modalities of topical administration are not entirely known. It has been suspected that the distribution of particles aerosolized into the tracheobronchial tree is not homogeneous, resulting in large amounts of aerosolized material being collected in the midline pulmonary structures and very little reaching the peripheral lung parenchyma [27]. Animal studies comparing the pulmonary distribution of particles given by endotracheal instillation or by aerosol inhalation have suggested that the latter mode of administration results in a more widespread distribution of the aerosolized material [5]. However, Ramirez and O'Neill [34], using pulmonary scans after endotracheal instillation of radioactive material, has shown an excellent distribution within the bronchial tree of the material directly injected into the trachea. Direct endotracheal instillation appears to be a convenient technique for endotracheal administration of antibiotics, especially in patients who are not cooperative enough for effective aerosol therapy. For such a therapy, the patient must have a tracheostomy or an endotracheal tube; conceivably, endotracheal injections could be performed via a transtracheal catheter. In a crossover study comparing the intratracheal instillation with the aerosol, 40 mg of gentamicin was administered to 14 children with a tracheostomy or an endotracheal intubation [2]. The bronchial levels of gentamicin, obtained 1 hr after the administration, was found to be 690 and 55.2 μg/ml for direct intratracheal administration and aerosolization, respectively. The corresponding mean blood serum levels were 0.79 and 0.13 μg/ml. A similar comparison between the levels of gentamicin in bronchial secretions and in blood with the two techniques of intrabronchial therapy has been made by Odio [26]. The bronchial levels 1 hr after the endotracheal instillation of 2 mg/kg of gentamicin was 480 μg/ml, and such high levels were found to persist for several hours. Four and 6 hr after the instillation, the bronchial levels were still 43.4 and 13.8 μg/ml, respectively. With the aerosol technique, the mean bronchial level obtained was 22.2 μg/ml. The mean peak blood level observed after intratracheal administration was 1.0 μg/ml; the corresponding value for aerosol technique was 0.2 μg/ml. As far as the aerosolization of gentamicin is concerned, the levels of gentamicin found by Regula et al. [35] within the bronchial secretions and in blood are similar to those observed by Odio [26].

Thus, it can be concluded that both direct endotracheal instillation and aerosolization of aminoglycosides into the bronchial tree results in high and sustained levels of these antibiotics in the bronchial secretions. However, the antibiotic levels resulting from aerosol are much lower than those observed after direct instillation. Nevertheless, these concentrations are adequate to inhibit most strains of Enterobacteriacae or of *Pseudomonas* and are probably sufficient to overcome, in part at least, the inactivation of aminoglycoside by purulent bronchial secretions, which could be quantitatively important since 1 ml of pus could inactivate 700 μg of gentamicin [6].

We have studied the levels of sisomicin in the bronchial secretions and blood after the endotracheal instillation of this drug at two different dosages, 0.35 and 0.70 mg/kg [43]. We demonstrated that the bronchial levels of sisomicin were very high after endotracheal injection. With the dosage of 0.35 mg/kg, the mean peak level after 30 min was 1080 μg/ml; after 4 and 8 hr, the levels were 92 and 22 μg/ml, respectively (Figure 1). With the higher dosage (0.70 mg/kg), the mean peak was 1309 μg/ml; 4 and 8 hr after the instillation, the bronchial levels were 124 and 51 μg/ml, respectively. Simultaneously, the mean peak serum levels of sisomicin did not exceed 0.16 μg/ml with the low dose, and the corresponding value for the higher dosage was 0.41 μg/ml.

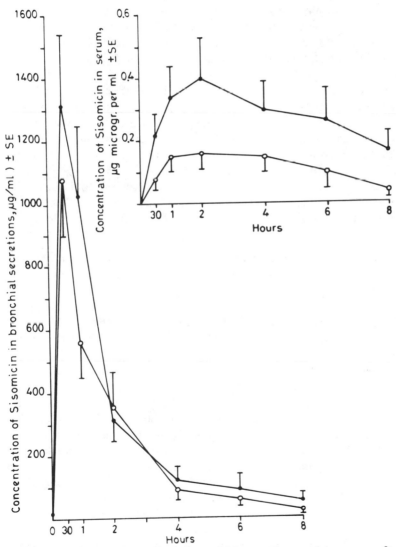

Figure 1   Levels of sisomicin in bronchial secretions and in serum after endotracheal injection of 0.35 (o) and 0.70 (●) mg/kg.

The levels of gentamicin or sisomicin in bronchial secretions were also measured in several studies in which the clinical effectiveness of endotracheally administered antibiotics was studied [15,16,18]. These studies confirm that very high levels of these antibiotics can be achieved in the bronchial secretions and can persist there for several hours after the administration.

Local antibiotic therapy to the tracheobronchial tree may produce bronchial irritation with bronchospasm and/or cough. This side effect was reported with polymyxin [10]; on the other hand, kanamycin, gentamicin, and sisomicin are usually well tolerated [1,4, 10,14,19,22]. However, serious reactions in some asthmatic subjects have been observed after inhalation of gentamicin [8].

Another possible limitation to the use of endotracheal administration of aminoglycosides is the fear of potential toxicity resulting from systemic absorption of the locally applied aminoglycoside, especially if systemic aminoglycosides are given at the same time. In fact, the blood levels observed after the endotracheal instillation of aminoglycosides in all the studies discussed above are extremely low [2,26,43]. However, higher serum levels have been noted after endotracheal instillation of gentamicin in patients with impaired renal function [19].

Therefore, it might be indicated to reduce the systemic dosage of gentamicin and to monitor the serum levels in order to avoid toxicity when systemic gentamicin is combined with tracheal instillation, in this type of patient [19,32]. On the other hand, very small amounts of aerosolized aminoglycosides are absorbed from the tracheobronchial [1,2,4, 14,19,22,26]; with this kind of local therapy systemic side effects are extremely unlikely and, as a matter of fact, have not yet been reported.

In conclusion, both instillation and aerosolization of aminoglycosides into the bronchial tract result in high levels of these antibiotics within the bronchial secretions. This form of administration is well tolerated locally and does not result in serious systemic side effects.

IV. Clinical Studies with Endotracheally Administered Gentamicin to Prevent Infections of the Respiratory Tract in Patients with Tracheostomy

Preliminary studies have suggested that endotracheally administered antibiotics might be effective in reducing the colonization of the respiratory tract by potentially pathogenic microorganisms. Lepper et al. [21] showed that polymyxin B given by aerosol inhalation prevented infections with *P. Aeruginosa* in tracheostomized patients with poliomyelitis. Later, Rose et al. [36] evaluated sodium colistimethate aerosol in 20 adult male patients carrying gram-negative bacilli in their respiratory tract; in 18 of 20 patients, eradication or marked reduction of these colonizing microorganisms could be obtained. Unfortunately, no attempt was made in that study to draw conclusions about a possible effect of the aerosolized drug on the clinical course. Therefore, a controlled study was undertaken to investigate the effectiveness of endotracheally administered gentamicin for the prevention of severe infections of the respiratory tract in debilitated subjects with tracheostomy [16]. The daily dose of gentamicin, 80 mg given three times daily, was slowly injected through a plastic catheter placed deeply in the trachea. A total of 85 patients were studied; 43 received gentamicin and 42 were treated with a a placebo. The prophylactic administration of endotracheal gentamicin to these hospitalized patients with tracheostomy resulted in a significant decrease of the frequency of purulent tracheal aspirates and of bacteriologically positive tracheal aspirates. Eleven of the 43 gentamicin-treated patients never had a positive culture of their tracheal aspirate, but this occurred in only 1 of 42 patients who received the placebo. The number of bacteriologically proven pulmonary infections, which required therapy with systemic antibiotics was significantly higher ($P < 0.01$) in the placebo-treated group. Among the gentamicin-treated patients, 2 (8.6%) of 23 deaths were caused by pulmonary infection, whereas in the placebo group, 4 (25%) of 16 deaths were caused by respiratory infection. No striking differences in the susceptibility to gentamicin between the gram-negative microorganisms isolated from gentamicin-treated and placebo-treated patients could be found. However, the microorganisms that were isolated from patients who received gentamicin were slightly more resistant to gentamicin than were those that were isolated from the control group.

That study clearly suggests that endotracheally administered gentamicin can prevent the colonization of the respiratory tract by gram-negative bacilli and can reduce the frequency of serious gram-negative tracheobronchial infections. However, the use of endotracheally administered gentamicin for the prevention of tracheobronchial infections in patients with tracheostomy might represent a two-edged sword, some selection of gentamicin-resistant microorganisms occurred in the patients who received endotracheal prophylactic therapy. Particularly, colonization by gentamicin-resistant *Providence* strains has been observed and led occasionally to serious infection.

The potential hazard of the selection of antibiotic-resistant strains should stimulate attempts to use, whenever possible, drugs with a less unique antimicrobial activity than gentamicin. Therefore, we have attempted to prevent infection of the tracheobronchial tree in tracheostomized patients with a combination of aminosidin, a "kanamycin-like" aminoglycoside, and polymyxin B, in order to limit the emergence of gentamicin-resistant strains in our hospital [17]. A group of patients treated endotracheally with the combination aminosidin/polymyxin B has been compared to another group receiving endotracheal gentamicin. The dosage of gentamicin was 80 mg and the combination consisted of 250 mg of aminosidin plus 50 mg of polymyxin given every 8 hr. The frequency with which positive cultures of the bronchial secretions were obtained in gentamicin-treated and aminosidin/polymyxin-treated patients was similar to the figures obtained in the gentamicin-treated patients mentioned previously. Systemic antibiotics were given with similar frequency and for similar periods of time in both groups. The number of bacteriologically proven pulmonary infections was identical in both groups. All the infections in the gentamicin-treated group were caused by gentamicin-resistant strains [minimum inhibitory concentration (MIC) $\geqslant 6$ μg/ml]. Resistance to aminosidin or polymyxin B was not always present in the strains that were responsible for the infections in the aminosidin-polymyxin group. Death which could be attributed to infection occurred in two patients in each series. Both regimens were similarly effective in preventing colonization of bronchial secretions by potential pathogens and were associated with a similar frequency of infectious episodes. Thus, the favorable response obtained with the aminosidin/polymyxin combination suggests that it might be used in the prevention of serious tracheobronchial infections in tracheostomized patients. No major adverse effects from the endotracheal instillation of aminosidin/polymyxin combination were observed in that study.

V. Clinical Studies with Endotracheal Gentamicin for Therapy of Bronchopulmonary Infections in Tracheostomized Patients

Since endotracheally administered antibiotics (i.e., gentamicin) appear to be effective in preventing serious gram-negative infections in tracheostomized patients, a logical further step in this investigation was to study whether endotracheally administered antibiotics would be effective as a therapy of gram-negative infections of bronchopulmonary infections. Results obtained by Bilodeau et al. [4] in 200 patients with bronchopulmonary suppurative diseases who received kanamycin aerosol therapy under intermittent positive pressure were encouraging; the method was well tolerated, and the clinical improvement was rapid and often lifesaving. The favorable pharmacokinetic characteristics of endotracheally administered gentamicin, as discussed here, suggested a potential benefit in bronchopulmonary infections of that type of therapy.

Therefore, a study was performed by us to compare the effectiveness of intratracheally and intramuscularly administered gentamicin in severely ill, tracheostomized patients, who presented a serious tracheobronchial or pulmonary infection caused by gram-negative rods [15]. Fifteen tracheostomized patients with a serious underlying illness (disseminated malignant disease or intracranial surgery) were studied in a prospective randomized trial. Patients previously randomized were treated by intramuscular or endotracheal gentamicin (240 mg daily). Two of eight patients were cured by the intramuscular therapy, but all seven patients who received gentamicin endotracheally were cured. The difference between these two groups was statistically significant (P < 0.01). The more favorable results obtained by endotracheal administration correlated with the high levels of gentamicin that were detected within the bronchial secretions. Such high levels could not be obtained by intramuscular injections. It was decided to limit this comparative study to a small number of patients because the data strongly suggested a superiority for intratracheal administration of gentamicin over the intramuscular route.

However, there was no definitive evidence that patients such as those studied in that investigation should always receive antibiotics via the endotracheal route. Physiotherapy and careful bronchial toilet might be sufficient to achieve cure in the patients who have infections confined only to the bronchial tree. Since endotracheal administration of gentamicin might result in a better bronchial toilet and improve the patient just by that mechanism, it seemed advisable to compare endobronchial administration of both gentamicin and saline.

To assess the value of endotracheal aminoglycosides in the treatment of gram-negative bronchopneumonia, endotracheal sisomicin or a placebo was administered respectively to 2 groups of 18 and 20 unconscious and tracheostomized or intubated patients with bronchopulmonary infection [18]. Gram-negative bronchopneumonia was defined as clinical and radiological lung involvement with fever and grossly purulent sputum with gram-negative rods on smears, which could be cultured as predominant organisms. Endotracheal sisomicin (25 mg, three times daily) or an endotracheal placebo were administered; in addition, all the patients received systemically a combination of sisomicin (225 mg) and carbenicillin (30 g) daily. Therapy was considered a success if the clinical and radiological signs of bronchopulmonary infection improved and if the patient survived the infectious episode or died after the completion of therapy from causes unrelated to infection. When the infection persisted clinically and required a change in antimicrobial therapy or when the patient died during the therapy with the trial antibiotics, primarily as a result of the infection, the therapy was considered a failure. It must be stressed that our trial of endotracheal antimicrobial therapy was performed under demanding conditions. All patients received in addition to endotracheal therapy a drug or a combination of drugs that were active against the offending pathogens. This systemic therapy was designed to suppress any eventual marginal effect of the endotracheal therapy. A placebo was used to control a possible effect of the endotracheal instillation on bronchial toilet.

Table 2   Clinical and Bacteriological Results Following Endotracheal Treatment with Sisomicin or a Placebo

	Placebo		Sisomicin	
	Total number of patients	Infections caused by *P. aeruginosa*	Total number of patients	Infections caused by *P. aeruginosa*
Number of patients	20	12	18	12
Favorable responses	9 (45%)	6 (50%)	14 (79%)	9 (75%)
Bacteriological results				
No changes	6	5	10	5
Pathogens decreased	10	5	8	7
Pathogens increased	4	2	0	0
Colonization	5	3	7	3
Superinfection	3	1	4	1
Deaths				
As a result of infection	4	1	0	0
As a result of underlying disease	4	1	5	4

A favorable clinical response was obtained in 14 (75%) patients who were treated with sisomicin and in 9 (45%) patients received the placebo ($P < 0.05$), as indicated in Table 2. Endotracheal therapy with sisomicin was well tolerated and resulted in high levels of sisomicin and in elevated bactericidal activity within the bronchial secretions, as shown in Table 3. This last study indicates that endotracheal therapy with sisomicin, in addition to systemic therapy with carbenicillin and sisomicin, is beneficial to patients with severe bronchopulmonary infections caused by gram-negative bacilli.

Table 3   Levels of Sisomicin and Bactericidal Activities in Bronchial Secretions and in Serum Following Endotracheal Treatment with Sisomicin or a Placebo

	Placebo		Sisomicin	
	Bronchial secretions	Serum	Bronchial secretions	Serum
Level of sisomicin (μg/ml)				
Mean	0.45	1.04	1051.0	1.17
Standard error	±0.11	±0.25	±271.5	±0.28
Bactericidal activity (geometric mean)				
Without penicillinase	0	1/8	1/256	1/8
With penicillinase	0	1/2	1/256	1/2

VI. Conclusions

The poor results of therapy with aminoglycosides of gram-negative bronchopneumonia may be due, at least in part, to inadequate antimicrobial activity at the site of infection. The low therapeutic index of the aminoglycosides limits their dosage when systemic therapy is used; relatively poor penetration into the bronchial secretions and inactivation by leucocytes within the bronchial secretions are possibly responsible for a reduced antimicrobial activity at the site of infection. Various data indicate that the administration of gentamicin or sisomicin, by direct injection or by aerosolization, to tracheostomized patients results in high and sustained concentrations of the drugs in the bronchial secretions. Moreover, it has been shown that endotracheally administered gentamicin can prevent the colonization of the respiratory tract by gram-negative bacilli and can reduce accordingly the frequency of serious bronchopulmonary infections. The efficacy of endotracheal sisomicin in the treatment of gram-negative bronchopneumonias has been well demonstrated; a favorable clinical response was observed in 75% of the patients treated with sisomicin but in only 45% of the patients receiving a placebo. It is thus suggested that endotracheally administered aminoglycoside should be used as an important adjunct to systemic antibiotic in the management of such infections. Since the serum levels of gentamicin or sisomicin after an endotracheal injection are low, it seems wise to use both endotracheal and systemic administrations in cases where tracheobronchial and parenchymal infections are present together.

The use of endotracheally administered aminoglycosides for the prevention and therapy of bronchopulmonary infections caused by gram-negative bacilli in patients with tracheostomy might, however, represent a two-edged sword. It may favor the selection, within the hospital environment, of resistant microorganisms. This has been shown clearly in our study, in which gentamicin was used to prevent bronchopulmonary infections in tracheostomized patients. A similar conclusion was reached by Feeley et al. [11], who used aerosols with polymyxin in tracheostomized patients; they showed that the polymyxin aerosol can be effective in preventing pneumonia caused by sensitive organisms, especially *P. aeruginosa*. When administered for prolonged periods, however, it was found to be a dangerous form of therapy because of the selection of polymyxin-resistant microorganisms which became serious pathogens. In an attempt to avoid the emergence of resistant bacilli, some forms of alternating-cycle method of endotracheal administration of antibiotics may thus be necessary to take advantage of the efficacy of this form of administration to prevent gram-negative bronchopneumonia in susceptible patients. Adequate isolation of hospitalized patients who are treated for prolonged periods with endotracheal antibiotics such as gentamicin should also be considered in order to avoid the dissemination throughout the hospital of resistant strains which are more likely to appear during this form of therapy. In addition, careful and frequent bacteriologic monitoring of the patients undergoing topical antimicrobial therapy may help to detect early bacterial colonization by resistant bacteria.

The problem of emergence of antibiotic-resistant organisms as a consequence of endotracheal administration of aminoglycosides is important but should not be overemphasized. In our recent therapeutic study, we observe only one instance of colonization by a sisomicin-resistant organism among the patients who received sisomicin endotracheally. This relatively low rate of emergence of antibiotic-resistant strains in our therapeutic study might be explained by the fact that endotracheal therapy was given to only 7% of patients who were admitted to the intensive care neurosurgical unit over the 3 years of the study. Therefore, we suggest that endotracheal therapy should be reserved only for patients with documented gram-negative bacillary bronchopneumonias.

The local tolerance to endotracheally administered aminoglycosides is excellent. As far as systemic toxicity is concerned, the increment of the serum levels of gentamicin or sisomicin due to the topical administration might be considered as negligible in patients with normal renal function. However, in severe renal failure, the aminoglycosides might accumulate and thus be responsible for excessive blood levels and toxicity. In such cases, regular monitoring of the antibiotic blood levels is mandatory.

In summary, our study strongly suggests that endotracheal therapy with aminoglycosides should be a part of the therapy of severe gram-negative bronchopneumonias. When one takes into account the large number of patients at risk of developing severe gram-negative bronchopneumonias in the modern hospital environment, the improvement in clinical response observed here appears to be important enough to be considered. In addition, endotracheal administration of aminoglycosides can prevent serious gram-negative bronchopneumonias in tracheostomized or intubated patients. However, prolonged endotracheal administration of antibiotics could favor the emergence of antibiotic-resistant strains. This potential hazard suggests the need to restrict prophylaxis with endotracheal antibiotics to high-risk patients and to perform this type of management under conditions likely to minimize the development and spread of resistant strains.

References

1. Ayres, S.M., Griesbach, J., and Giannelli, S., Jr.: A study of bronchial irritation and systemic absorption of aerosolized kanamycin. *Curr. Ther. Res.* 14:153–157, 1972.
2. Baran, D., Dachy, A., and Klastersky, J.: Concentration of gentamicin in bronchial secretions of children with cystic fibrosis or tracheostomy (comparison between the intramuscular route, the endotracheal instillation and aerosolization). *Int. J. Clin. Pharmacol.* 12:336–341, 1975.
3. Bergogne-Berezin, E., Even, P., and Berthelot, G.: Pharmacocinétique de l'amikacine dans les sécrétions bronchiques. *Rev. Fr. Mal. Resp.* 6:385–392, 1978.
4. Bilodeau, M., Roy, J.C., and Giroux, M.: Studies of absorption of kanamycin by aerosolization. *Ann. N.Y. Acad. Sci.* 132:870–878, 1966.
5. Brain, J.D., Knudson, D.E., Sorokin, S.P., and Davis, M.A.: Pulmonary distribution of particles given by intratracheal instillation or by aerosol inhalation. *Environ. Res.* 11:13–33, 1976.
6. Bryant, R.E., and Hammond, D.: Interaction of purulent material with antibiotics used to treat *Pseudomonas* infections. *Antimicrob. Agents Chemother.* 6:702–707, 1974.
7. Dahlberg, A.E., Horodyski, F., and Keller, P.: Interaction of neomycin with ribosomes and ribosomal ribonucleic acid. *Antimicrob. Agents Chemother.* 13:331–339, 1978.
8. Dally, M.B., Kurrle, S., and Breslin, A.B.X.: Ventilatory effects of aerosol gentamicin. *Thorax* 33:54–56, 1978.
9. Davis, S.D., and Bruns, W.T.: Effects of sputum from patients with cystic fibrosis on the activity in vitro of 5 antimicrobial drugs on *Pseudomonas aeruginosa*. *Am. Rev. Resp. Dis.* 117:176–178, 1978.
10. Dickie, K.J., and de Groot, W.J.: Ventilatory effects of aerosolized kanamycin and polymyxin. *Chest* 63:694–697, 1973.
11. Feeley, T.W., Du Moulin, G.C., Hedley-Whyte, J., Bushnell, L.S., Gilbert, J.P., and Feingold, D.S.: Aerosol polymyxin and pneumonia in seriously ill patients. *N. Engl. J. Med.* 293:471–475, 1975.

12. Graybill, J.R., Marshall, L.W., Charache, P., Wallace, C.K., and Melvin, V.B.: Nosocomial pneumonia. A continuing major problem. *Am. Rev. Resp. Dis.* 108:1130–1140, 1973.

13. Halprin, G.M., and McHahon, S.M.: Cephalexin concentrations in sputum during acute respiratory infections. *Antimicrob. Agents Chemother.* 3:703–707, 1973.

14. Hellström, P.E., Gruenwaldt, G., and Scheer, M.: Serumkonzentrationen nach Aerosol-Inhalation und intravenöser Injektion von Sisomicin: Vorläufiger Bericht. *Infection* 4:412–424, 1976.

15. Klastersky, J., Geuning, C., Mouawad, E., and Daneau, D.: Endotracheal gentamicin in bronchial infections in patients with tracheostomy. *Chest* 61:117–120, 1972.

16. Klastersky, J., Huysmans, E., Weerts, D., Hensgens, C., and Daneau, D.: Endotracheally administered gentamicin for the prevention of infections of the respiratory tract in patients with tracheostomy: a double-blind study. *Chest* 65:650–654, 1974.

17. Klastersky, J., Hensgens, C., Noterman, J., Mouawad, E., and Meunier-Carpentier, F.: Endotracheal antibiotics for the prevention of tracheobronchial infections in tracheostomized unconscious patients. A comparative study of gentamicin and aminosidin-polymyxin B combination. *Chest* 68:302–306, 1975.

18. Klastersky, J., Carpentier-Meunier, F., Kahan-Coppens, L., and Thys, J.P.: Endotracheally administered antibiotics for gram negative bronchopneumonia. *Chest* 75:586–591, 1979.

19. Lake, K.B., Van Dyke, J.J., and Rumsfeld, J.A.: Combined topical pulmonary and systemic gentamicin: the question of safety. *Chest* 68:62–64, 1975.

20. Lambert, H.P.: Clinical significance of tissue penetration of antibiotics in the respiratory tract. *Scand. J. Infect. Dis.* 14(Suppl.):262–266, 1978.

21. Lepper, M.H., Kofman, S., Blatt, N., Dowling, H.F., and Jackson, G.G.: Effect of eight antibiotics used singly and in combination on the tracheal flora following tracheotomy in poliomyelitis. *Antibiot. Chemother.* 4:829–832, 1954.

22. Lifschitz, M.I., and Denning, C.R.: Safety of kanamycin aerosol. *Clin. Pharmacol. Ther.* 12:91–95, 1971.

23. McCrae, W.M., Raeburn, J.A., and Hanson, E.J.: Tobramycin therapy of infections due to *Pseudomonas aeruginosa* in patients with cystic fibrosis: effect of dosage and concentration of antibiotic in sputum. *J. Infect. Dis.* 134(Suppl.):S191–S193, 1976.

24. Marks, M.I., Prentice, R., Swarson, R., Cotton, E.K., and Eickhoff, T.C.: Carbenicillin and gentamicin: pharmacologic studies in patients with cystic fibrosis and *Pseudomonas* pulmonary infections. *J. Pediatr.* 79:822–828, 1971.

25. May, J.R., and Delves, D.M.: Treatment of chronic bronchitis with ampicillin: some pharmacological observations. *Lancet* 1:929–933, 1965.

26. Odio, W., Van Laer, E., and Klastersky, J.: Concentrations of gentamicin in bronchial secretions after intramuscular and endotracheal administration. *J. Clin. Pharmacol.* 15:518–524, 1975.

27. Patterson, C.D., and Kamp, G.H.: Retention of liquid aerosols in the lung. *Am. Rev. Resp. Dis.* 95:443–446, 1967.

28. Pennington, J.E., and Reynolds, H.Y.: Concentrations of gentamicin and carbenicillin in bronchial secretions. *J. Infect. Dis.* 128:63–68, 1973.

29. Pennington, J.E., Reynolds, H.Y., and Carbone, P.P.: *Pseudomonas* pneumonia: a retrospective study of 36 cases. *Am. J. Med.* 55:155–160, 1973.

30. Pennington, J.E., and Reynolds, H.Y.: Tobramycin in bronchial secretions. *Antimicrob. Agents Chemother.* 4:299–301, 1973.

31. Pennington, J.E., and Reynolds, H.Y.: Pharmacokinetics of gentamicin sulfate in bronchial secretions. *J. Infect. Dis.* 131:158–162, 1975.

32. Peromet, M., Schoutens, E., and Yourassowski, E.: Endotracheal administration of

gentamicin: relationship between obtained serum levels and creatininemia. *Chemotherapy* 19:211–214, 1973.

33. Potter, J.L., Matthews, L.W., Spector, S., and Lemm, J.: Complex formation between basic antibiotics and deoxyribonucleic acid in human pulmonary secretions. *Pediatrics* 36:714–720, 1965.

34. Ramirez, J.R., and O'Neill, E.F.: Endobronchial polymyxin B: experimental observations in chronic bronchitis. *Chest* 58:352–357, 1970.

35. Regula, H., Wieser, O., Naumann, P., Wundt, W., and Kleinhuber, U.: Pharmakokinetische Untersuchungen über Sputum-, Serum- und Urinkonzentration von Gentamycin nach Aerosol-Inhalation. *Int. J. Clin. Pharmacol.* 7:95–100, 1973.

36. Rose, H.D., Pendharker, M.B., Snider, G.L., and Kory, R.C.: Evaluation of sodium colistimethate aerosol in gram negative infections of the respiratory tract. *J. Clin. Pharmacol.* 10:274–281, 1970.

37. Saggers, B.A., and Lawson, D.: In vivo penetration of antibiotics into sputum in cystic fibrosis. *Arch. Dis. Child.* 43:404–409, 1968.

38. Saggers, B.A., and Lawson, D.: The differential attachment of antibiotics to glycoprotein and blood lymphocytes. *J. Clin. Pathol.* 23:266–268, 1970.

39. Sickles, E.A., Young, V.M., Greene, W.H., and Wiernik, P.H.: Pneumonia in acute leukemia. *Ann. Intern. Med.* 79:528–534, 1973.

40. Stevens, R.M., Teres, D., Skillman, J.J., and Feingold, D.S.: Pneumonia in an intensive care unit: a 30-month experience. *Arch. Intern. Med.* 134:106–111, 1974.

41. Stewart, S.M., Anderson, I.M.E., Jones, G.R., and Calder, M.A.: Amoxycillin levels in sputum, serum and saliva. *Thorax* 29:110–114, 1974.

42. Thys, J.P., Klastersky, J.: Concentrations of amikacin in serum and in bronchial secretions after intramuscular administration. Abstr. 389 18th International Conference on Antimicrobial Agents and Chemotherapy, Atlanta, Ga., 1978.

43. Thys, J.P., and Klastersky, J.: Concentrations of sisomicin in serum and in bronchial secretions after intratracheal administration. *Curr. Chemother.* 2:920–921, 1978.

44. Thys, J.P., Husson, M., and Klastersky, J.: Inactivation of netilmicin and amikacin by leucocytes. In press.

45. Thys, J.P., Husson, M., and Klastersky, J.: Assay of netilmicin in sputum. In press.

46. Thys, J.P., and Klastersky, J.: Concentrations of netilmicin in bronchial secretions and in serum during intermittent versus continuous infusion. A cross-over study in humans. *J. Infect. Dis.* 120:634, 1979.

47. Wieser, O., Regula, H., and Wundt, W.: Die Ausscheidung von Gentamicin über den Bronchialbaum. *Dsch. Med. Wochenschr.* 96:870–872, 1971.

48. Wong, G.A., Pierce, T.H., Goldstein, E., and Hoeprich, P.D.: Penetration of antimicrobial agents into bronchial secretions. *Am. J. Med.* 59:219–223, 1975.

23

AMINOGLYCOSIDES IN PEDIATRICS

JANE D. SIEGEL and GEORGE H. McCRACKEN, Jr.
The University of Texas Health Sciences Center at
Dallas, Dallas, Texas

I. Unique Characteristics of Pediatric Pharmacology

For the past two decades, the aminoglycosides have been used as first-line drugs for treatment of serious infections in five specialized areas of pediatrics: (1) newborn nursery, (2) cyctic fibrosis, (3) oncology, (4) surgery, and (5) burns. In the early years of aminoglycoside usage, differences in the metabolism and excretion of these drugs in infants, especially neonates, as compared to adults were not appreciated. Excessive dosages of aminoglycosides caused irreversible ototoxicity, whereas inadequate dosages resulted in subtherapeutic body fluid concentrations that were associated with delayed bacteriologic cure. As pediatric pharmacology has gained recognition as a specialized field, the pharmacokinetics of antibiotics have been defined for each age group, and safe and effective dosage schedules have been established.

Because the volume of distribution of aminoglycosides approximates the extracellular

527

fluid space and these drugs do not penetrate well into cells, it has been assumed that they distribute primarily in this space [101]. The variation of the extracellular fluid volume with age best explains the age-related dose response to aminoglycosides. The extracellular fluid space is approximately 40–44% of body weight in the newborn period, 26% from 6 months to 5 years, and 18% in older children and adults [39]. When expressed as a fraction of body surface area, the volume of extracellular fluid decreases from 7 L/m^2 to 6 L/m^2 in the first 3 months of life and remains remarkably constant throughout the remainder of infancy and childhood. Consequently, several authors have suggested that aminoglycoside dosages be based on surface area rather than body weight [66,102,113]. Body surface area dosage compensates for all but the extreme variations in body habitus. For example, severely cachectic patients achieve lower peak serum levels [102], whereas obese patients have higher concentrations of drug in serum [99,102] than anticipated because of the relatively larger and smaller volumes of drug distribution, respectively.

It is fortunate that the most unpredictable group of pediatric patients, newborn infants, is the best studied. Absorption, distribution, metabolism, and excretion of drugs are constantly changing during the transition from intrauterine to extrauterine life. Maturity, reflected by gestational and chronological age, is a major determinant of the fate of drugs administered to the newborn. Therefore, data must be collected separately for each of these groups. In addition, transplacental transport, renal immaturity, protein binding, and excretion into breast milk must be considered for a complete understanding of the pharmacokinetics of aminoglycosides during the neonatal period.

The aminoglycosides are used in pregnancy most frequently to treat amnionitis and intrauterine bacterial infections that occur during the last trimester. The fetus is subject to the effects of any maternally administered drugs that are able to traverse the placenta either passively by simple diffusion or by active transport after temporary combination with membrane constituents. The factors that enhance passage of drugs across the placenta are lipid solubility; existence in the un-ionized, unbound state; molecular weight of less than 1000; increased placental blood flow, placental maturation (drug transfer is slowest in midgestation and most rapid in the first and third trimesters); and a more rapid rate of maternal administration. The placenta possesses four mechanisms of drug metabolism: oxidation, reduction, hydrolysis, and conjugation; but these activities are so much lower than those of either the maternal or fetal liver when compared on a unit per weight basis that the placenta cannot be considered an important site for xenobiotic drug metabolism.

Once the drug has crossed the placental barrier, fetal distribution is influenced by the permeability of the membranes surrounding specific organs or body components. Most drugs have a ubiquitous distribution pattern in the fetus. Drugs with a high degree of lipid solubility, however, are concentrated in organs with increased lipid fractions (liver, ovary, and adrenal glands). Although the blood–brain barrier is more permeable to many drugs early in development, the affinity of the brain for lipophilic drugs is decreased due to its low myelin and high water content [80].

Transplacental transport of antibiotics has been studied extensively. However, there are relatively few studies of the aminoglycosides [8,9,25,42,44,62,118]. Representative studies of drug levels in the fetus and in amniotic fluid after maternal administration are listed in Table 1. These doses are generally lower than those currently recommended. Maternal serum values are less than those reported in nonpregnant adults because of the increased plasma volume and glomerular filtration rate that occurs in the first 5 months

Table 1   Transplacental Transport of Aminoglycosides

| Drug | Dose/route | Peak serum concentration (μg/ml) | | Concentration in μg/ml (hr after dose) | | |
		Maternal (hr after dose)	Cord (hr after dose)	Amniotic fluid	Fetal kidney	Fetal urine
Amikacin [8]	7.5 mg/kg IM 1st and 2nd trimesters	18.0 (1)	3.4 (2)	1.04 ($\geqslant$3.5)	22 (12)	24 (3)
Gentamicin [8]	40 mg IM prior to delivery	3.65 (0.5)	1.25 (1–2)	—	—	—
Kanamycin [42]	500 mg IM prior to delivery	18.6 (1)	9.0 (1.75)	5.5 (6)	—	—
Streptomycin [4]	100,000 units sub Q prior to delivery	7.6 units/ml (1.5)[a]	5.9 units/cm³ (1.5)[a]	—	—	—
Tobramycin [9]	2 mg/kg IM 1st and 2nd trimesters	4.0 (1)	0.6 (2.5)	0.25 ($\geqslant$4)	7.2 (34)	3.4 (25)

[a]Not a peak value; this was the only time that samples were obtained.

of pregnancy. These drugs traverse the placenta and achieve levels in cord serum that vary from 15 to 70% of the concentrations in maternal serum. Higher concentrations of amino-glycosides are present in fetal kidney and urine, and plateau after 3–6 hr at a low, often subtherapeutic, level in amniotic fluid. The only documented cases of toxicity in infants and children resulting from antepartum aminoglycoside therapy have been ototoxicity after prolonged administration of streptomycin and dihydrostreptomycin for treatment of tuberculosis [17,93,112].

After birth, the neonate must rely on his own enzyme systems and kidneys for bio-transformation and excretion of drugs. The immaturity of hepatic enzymes has little, if any, influence on the fate of aminoglycosides because these agents are excreted primarily in unaltered form by glomerular filtration. An understanding of neonatal kidney function is essential for predicting drug levels in the serum. Prior to birth, the kidney does not function as an organ of elimination. Induction of new nephrons is completed by week 34 of gestation, but the glomerulus in a full-term neonate is only one-third of its eventual adult diameter. These observations may account for Arant's [1] findings that creatinine clearance did not increase significantly from 28 to 34 weeks gestation, but did increase from 0.45 ml/min in a group of infants <34 weeks gestation to 1.01 ml/min at 34–37 weeks and to 2.24 ml/min in the group ⩾38 weeks of gestational age. Studies conducted in the first week of life have demonstrated that the renal plasma flow and consequently glomerular filtration rate are small at birth but increase substantially by 3 days of age [15,104]. As the kidney matures, creatinine clearance rates continue to increase to a level of 113 ml/min per 1.73 m² achieved by 3 years of age. Any factor that decreases renal plasma flow will delay drug elimination. This mechanism may explain the prolonged serum half-life of aminoglycosides observed in hypoxemic infants [81]. Immaturity of the glomerulus and decreased renal plasma flow are also responsible for apparent decreases in tubular function.

Serum creatinine is the most convenient index of renal function in the newborn infant. At birth, creatinine levels reflect maternal renal function, but decrease rapidly over the first 24–48 hr of life. Stonestreet and Oh [109] have demonstrated an inverse relationship between plasma creatinine and postnatal age for the first month of life in infants from 26–36 weeks of gestation. The plasma creatinine ranged from 0.8 to 1.8 mg/dl in the first 10 days of life compared to values that were consistently <1.0 mg/dl beyond 1 month of age. A similar temporal relationship exists for term babies, but the creatinine values remain ⩽1 mg/dl after 2 days of age [4]. These rapidly changing renal processes necessitate pharmacokinetic studies in different gestational and chronological age categories.

The affinity of antibiotics for albumin has clinical relevance with regard to tissue distribution and displacement of bilirubin from albumin binding sites. The latter has been associated with bilirubin encephalopathy in infants treated prophylactically with sulfa-soxazole [105]. Although early studies suggested that gentamicin interfered with the binding of bilirubin to albumin, further work has conclusively disproved this finding [85]. The aminoglycosides have little, if any, binding to serum proteins and have not been incriminated as a cause of kernicterus in infants treated with these agents.

The final consideration is of excretion into human milk. There are no documented cases of aminoglycoside toxicity in a suckling neonate of a mother who is receiving aminoglycoside therapy. This is most likely due to the low total amount of drug ingested and the poor absorption of aminoglycosides from the gastrointestinal tract. Because these drugs are excreted into human milk primarily by passive diffusion, the same factors influencing transplacental passage are operative. There are few studies of the excretion of antibiotics into human milk, and only two [68,84] are specifically concerned with aminoglycosides: kanamycin and dihydrostreptomycin. A 1-g kanamycin dose administered intramuscularly to mothers gave an average milk concentration of 18.4 μg/ml 1 hr after injection as compared to 0.3-1.3 μg/ml after a similar dose of dihydrostreptomycin. Simultaneous maternal serum and milk serum ratios were 55.2 μg/ml, 0.3 for kanamycin, and 14-18 μg/ml, 0.02-0.1 for dihydrostreptomycin.

II. Methods of Pharmacologic Evaluation of Aminoglycosides in Infants and Children

The initial step in evaluating an antibiotic for potential use in the neonate is the determination of in vitro susceptibilities against the most frequently encountered bacterial pathogens in this age group. For a new aminoglycoside to be clinically useful in the newborn nursery, it should be active against most *Escherichia coli* and *Klebsiella-Enterobacter* and *Pseudomonas* species. Emergence of strains resistant to the aminoglycosides currently in use makes continuing surveillance mandatory. Cross-resistance with presently available drugs provides no advantage to the newer agents. As illustrated in Table 2, semiannual surveillance of eight neonatal nurseries in North America over the past 6 years has shown intermittent resistance of coliforms to ampicillin and kanamycin. By comparison, an increasing number of *E. coli* and *Klebsiella-Enterobacter* strains isolated from patients enrolled in the Neonatal Meningitis Cooperative Study have become resistant to ampicillin and, to a lesser degree, kanamycin. Selective pressure exerted by both parenterally [12,19,31,34,38] and orally [12,29] administered aminoglycosides has been sufficiently documented to warrant the recommendation of alternating usage of kanamycin and gentamicin within the neonatal nursery at 2-3 year intervals. Sporadic outbreaks of nosocomial infection secondary to multiply resistant strains of coliforms gives the impetus for development and testing of new, broader spectrum aminoglycosides. A similar approach is applicable to older children with cystic fibrosis, malignancies, and postoperative infections.

Basic pharmacokinetic data must be obtained by study of sick neonates and children because of the ethical injunctions against testing drugs in healthy pediatric patients. The antimicrobial agent to be tested must first have completed Phase I studies in adults (administration of drugs to healthy adult volunteers in order to obtain data regarding absorption, metabolism, excretion, and acute safety). These data serve as a guideline for estimating initial dosages in neonates at the time that pharmacokinetic studies are undertaken.

Once the primary physician decides to treat an infant or child with antimicrobial agents, the investigator may substitute a single dose of the test drug which obviates exposure to a prolonged period of either possible ineffective therapy or toxic concentrations. Multiple serum samples are obtained by heelstick and are assayed by a micromethod technique requiring only 0.02 ml specimens. Urine is collected in 4 hr fractions for the 12- or 24-hr test period in order to determine the rate and amount of antibiotic excretion.

Traditionally, serum concentration-time curves have been constructed and analyzed by a one-compartment model. Recent studies have demonstrated that the initial distribution phase after the first dose and the final tissue washout phase after discontinuation of drug follow a biphasic pattern in neonates [103], older children [33], and adults [98] and are most accurately fit to a two-compartment pharmacokinetic model. The clinical

Table 2 Aminoglycoside Susceptibilities of Selected Enterobacteriaceae Isolated from Eight North American Nurseries 1973–1978 and from Neonatal Meningitis Cooperative Study Patients (1971–1978)

	Ampicillin, ≤10 μg/ml	Kanamycin, ≤15 μg/ml	Amikacin, ≤15 μg/ml	Gentamicin, ≤5 μg/ml	Netilmicin, ≤5 μg/ml
Escherichia coli					
Nurseries	1034/1221[a] (85)[b]	1170/1221 (96)	1210/1221 (99)	1163/1221 (95)	1208/1221 (99)
Meningitis	71/87 (82)	76/86 (88)	83/85 (98)	78/80 (98)	22/23 (96)
Klebsiella-Enterobacter					
Nurseries	82/534 (15)	495/534 (93)	529/534 (99)	517/534 (97)	528/534 (99)
Meningitis	15/31 (48)	28/29 (97)	32/32 (100)	28/30 (93)	22/24 (92)

[a]Total number of organisms susceptible/total number of organisms tested.
[b]Cumulative percentage of susceptible organisms.

application of this model may be important for two reasons: (1) more accurate prediction of serum drug levels [33]; and (2) the persistence of subinhibitory concentrations of aminoglycosides in the urine for days after the antibiotic is discontinued may exert a selective pressure for development of resistant strains of Enterobacteriaceae residing in the environment of neonatal intensive care units.

Serum half-life, area under the curve, volume of distribution, and plasma clearance values are analyzed in relation to birthweight and gestational and chronologic ages. The relationships presented in Figures 1 and 2 are representative of those observed in neonates after intramuscular administration of the aminoglycosides. Such values are utilized to formulate preliminary dosage schedules and intervals of administration for infants and young children. These regimens are then tested in large numbers of pediatric patients, who require many days of therapy. Serum concentrations are monitored to determine if accumulation occurs after multiple doses. Although most studies on newborn infants have found no evidence of aminoglycoside accumulation in serum, there are two recent reports of higher peak and trough serum values after multiple doses of gentamicin [18] and netilmicin [103] in low-birthweight babies (mean birthweight 1400 g). Hypoxemia was documented in a number of the infants who received netilmicin [103]. Additional studies are required in low-birthweight, premature infants in order to define the pharmacokinetics of aminoglycosides in this special group of neonates.

Most studies evaluate the pharmacology of the aminoglycosides after intramuscular

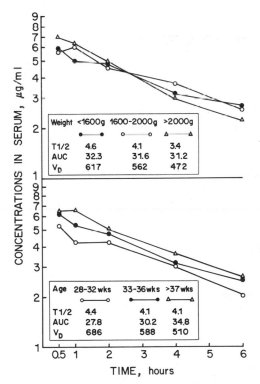

Figure 1   Serum concentration–time curves and pharmacokinetic values of netilmicin according to birth weight (top) and gestational age (bottom). Dosage was 3 mg/kg. (Reprinted from [103].)

administration. With the improved survival of infants ≤2000 g at birth in whom muscle mass is markedly decreased and the circulatory system is unstable, the intravenous route is often preferred. Bioavailability data [73,103,111] for several aminoglycosides after intravenous and intramuscular administration to neonates are presented in Table 3. There is no significant difference for these two routes of administration when 20-min constant intravenous infusions of kanamycin, gentamicin, netilmicin, or amikacin are used.

There have been few pharmacokinetic studies of orally administered aminoglycosides. An early report on adults [11] indicated minimal systemic absorption by recovering only 1–2% of a single orally administered dose of gentamicin from the urine. Nunnery and Riley [83] administered gentamicin orally to a group of infants under 1 year of age who had shigella dysentery. More than 10% of the drug was excreted in the urine during the acute phase of illness, compared to only 2% after acute inflammation had apparently subsided. Increased absorption when the intestinal mucosal lining is inflamed or impaired has been suggested in two studies [13,64] of infants with gastroenteritis who were treated orally with neomycin. Serum drug levels in infants who received both orally (10–15 mg/kg per day) and parenterally (5–7.5 mg/kg per day) administered gentamicin for the therapy of necrotizing enterocolitis ≥ 10 μg/ml have been reported.

Assessment of renal toxicity in neonates and children does not differ from that in adults. The serum creatinine is evaluated according to the normal values for age and urinalysis is followed for development of tubular dysfunction. Ototoxicity is difficult to evaluate because infants and young children are unable to provide sufficient cooperation

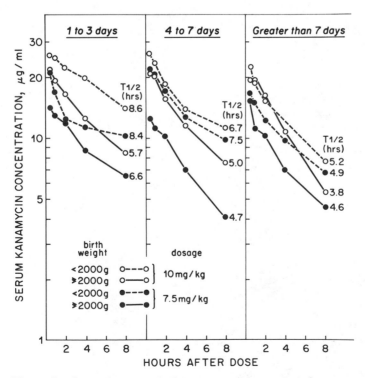

Figure 2   Serum kanamycin dose-response curves in infants grouped according to birth weight and chronologic ages. The dose of kanamycin was 7.5 or 10 mg/kg administered intramuscularly every 12 hr. (Reprinted from [49].)

Table 3 Pharmacokinetics of Kanamycin, Gentamicin, Netilmicin, and Amikacin After Intramuscular and Intravenous Administration to Newborn Infants

Drug/dosage	Median birthweight route Median gestation age	Predose	Serum concentration (µg/ml)						T½ (hr)	AUC (µg/ml per hr)
			0[a]	1/2	1	2	4	6		
Amikacin [111]	IV (6)	2.1	–	6.5	7.9	6.5	2.9	–	–	–
	IM (3)	2.0	–	6.6	8.1	6.5	2.7	–	–	–
Kanamycin, 7.5 mg/kg [73]	2105 g/IV (12)[b] 36 weeks	6.8 ± 0.65[c] (3.9–10.8)	20.5 ± 1.23 (13–29.4)	21.4 ± 1.35 (15.8–27)	18.4 ± 0.74 (14–22.8)	15.9 ± 0.73 (12.4–19)	12.2 ± 0.63 (9.6–16.5)	–	4.5	97
	IM (12)	6.8 ± 0.41 (3.6–8.8)	–	21.8 ± 0.92 (16.5–28)	20.8 ± 0.65 (17.8–26)	17.3 ± 0.72 (12.2–18.8)	12.6 ± 0.77 (11.5–15.1)	–	4.3	103
Gentamicin, 2.5 mg/kg [73]	2105 g/IV (9) 36 weeks	1.5 ± 0.27 (0.7–3.4)	4.2 ± 0.84 (2.2–9.4)	5.2 ± 0.7 (2.6–7.9)	5.0 ± 0.49 (2.8–7.5)	4.0 ± 0.36 (2.8–5.5)	3.0 ± 0.42 (2.0–6.2)	–	4.3	32
	IM (9)	1.8 ± 2.5 (0.8–0.28)	–	5.6 ± 0.32 (4.5–7.4)	6.3 ± 0.45 (4.0–7.5)	4.8 ± 0.35 (3.6–6.8)	3.5 ± 0.43 (2.0–5.4)	–	3.7	28
Netilmicin, 4 mg/kg [103]	2248 g/IV (10) 36 weeks	1.8 ± 0.29 (0.5–2.8)	7.6 ± 0.75 (4.9–12.4)	8.5 ± 0.73 (5.0–12.8)	7.2 ± 0.65 (5.2–11.6)	6.4 ± 0.77 (3.6–10.3)	3.9 ± 0.39 (1.9–5.8)	3.0 ± 0.37 (1.4–4.2)	3.7	36
	IM (10)	2.3 ± 0.34 (1.1–4.2)	–	8.9 ± 0.47 (6.4–11.2)	7.5 ± 0.45 (4.8–10.0)	6.6 ± 0.52 (4.4–9.2)	4.0 ± 0.40 (2.0–5.8)	3.1 ± 0.34 (1.5–4.8)	3.6	36

[a] Specimen obtained at completion of 20 min infusion.
[b] Number of patients.
[c] Mean (± SEM): range of values shown underneath.

535

for routine audiometric testing. Auditory-evoked brain stem responses have been described recently [108] as an effective means to identify cochlear dysfunction. This procedure is particularly valuable during the neonatal period because it is essentially unaffected by the baby's state of arousal. Although abnormalities associated with drug ototoxicity may not always be clearly distinguished from those caused by birth asphyxia, prematurity, or the disease for which the drug is prescribed, controlled studies in progress in our nurseries suggest that measurement of auditory-evoked brain stem responses may be useful for identifying drug toxicity in the neonate and young infant.

III. Pharmacology of Each Aminoglycoside

Each of the aminoglycosides has been more thoroughly evaluated in neonates than in older infants and children. As drugs with a broader spectrum of activity have become available, studies in patients with cystic fibrosis and malignancies have broadened our knowledge of pharmacokinetics in older children. Data for each drug will be presented by age group. A composite of recommended dosage schedules and pharmacokinetic data for each drug is illustrated in Tables 4 and 5.

A. Neomycin

Neomycin is best known to the pediatrician for its oral use in suppression of the ammonia-producing gastrointestinal microflora in patients with hepatic failure and for treatment of diarrhea caused by enteropathogenic strains of E. coli in young infants. Efficacy of this regimen in the latter disease has been questioned. Studies from our institution [82] have demonstrated a more rapid bacteriologic and clinical response in infants treated for 3 to 5 days with orally administered neomycin compared to infants given placebo. Prompt eradication is especially important for control of nursery outbreaks of diarrhea caused by one of the enteropathogenic strains of E. coli.

Following intramuscular administration of 7 mg/kg of neomycin sulfate to premature infants, a mean peak serum level of approximately 12 μg/ml occurred 2 hr after the dose was given [3]. The serum half-life was inversely related to postnatal age: 5-6 hr in infants 4-10 days old and 3-4 hr in infants 13-21 days old. At present, there is no indication for parenterally administered neomycin in neonates.

B. Streptomycin

Owing to the increased number of resistant strains of Enterobacteriaceae, streptomycin is no longer a useful drug for therapy of infections caused by these organisms in infants and children. The only indication for the use of streptomycin as a single drug is in the therapy of tularemia. Streptomycin combined with tetracycline is indicated for treatment of Yersinia pestis (plague) and Pseudomonas mallei (glanders) infections. In combination with isoniazid and para-aminosalicylic acid (PAS), rifampin, or ethambutal, streptomycin is effective for therapy of disseminated (including central nervous system disease) or extensive local tuberculosis. When administered intramuscularly, this drug diffuses readily into peritoneal and pleural spaces but is variable in its degree of meningeal penetration.

Table 4 Comparative Pharmacokinetics of Aminoglycosides in Neonates

Drug	Dose (mg/kg)	Birthweight (g)	Age groups (days)	Peak serum concentration (μg/ml)	Serum half-life (hr)	Volume of distribution (ml/kg)	Plasma clearance (ml/min per 1.73 M^2)
Amikacin	7.5	≤2000	≤7	17.0	6.5	565	22.0
			>7	18.9	5.5	502	24.6
		>2000	≤7	18–20	5.0–6.5	510–570	27.0–30.0
			>7	17.4	4.9	594	36.4
Gentamicin	1.5[a]	≤2000	≤7	1.5–2.2	10.5–14	690–780	12–16.5
		>2000		2.5–2.7	4.5–5.5	520–610	30–34
Kanamycin	10.0[b]	All infants >7		3.0	3.2	488	56.2
		≤2000	≤7	26.0	6.5–8.5	520–715	23.6
			>7	22.3	5.2	547	25.4
		>2000	≤7	21.2	5.4	460–620	29.3–36.3
			>7	19.6	3.8	497	45.5
Netilmicin	3.0	≤2000	<7	6.0	4.7	609	30.8
			≥7	5.6	4.1	599	34.1
		>2000	<7	6.9	3.4	472	38.8
Tobramycin	2.0	≤2500	<7	4.9–5.6	8.6	378–431	11.0
			≥7	5.0–5.4	6.0–9.8	339–385	8.6–14.3
		>2500	<7	4.9	5.1	390	25.3
			≥7	4.5	4.0	435	35.9

[a]Recommended dose 2.5 mg/kg for all neonates.
[b]Recommended dose 7.5 mg/kg for neonates ≤2000 g at birth and ≤7 days old.

537

Table 5   Comparative Pharmacokinetic Properties of Aminoglycosides in Infants
and Children

Drug	Dose (mg/kg)	Age (yr)	Peak serum concentration (μg/ml)	Serum half-life (hr)	Volume of distribution (ml/mg)	Plasma clearance (ml/mg per 1.73 m^2)
Amikacin	10-20[a] (420 mg/m^2)	4-16	28.7	1.6	320	120
Gentamicin	2.5	0.5-5	4.4	–	422[b]	–
	2.4[c]	5-10	5.3	–	288	–
	2.0	10-15	6.0	–	266	–
Kanamycin	10	0.1-0.5	18.0	2.0	506	133
		1-12	18.4	1.5	422	114
Tobramycin	1.3 (40 mg/m^2)	2-6	3.9	2.0	490	124
		12-18	4.2	1.4	400	195

[a] Recommended dose 300 mg/m^2 or 7.5-10.0 mg/kg.
[b] Volumes of gentamicin distribution calculated after 1 mg/kg doses.
[c] Recommended dose 60 mg/m^2 for all ages or 2.0 mg/kg 5-10 years; 1.5 mg/kg 10-15 years.

Demonstration of in vitro synergism of streptomycin and a penicillin against *Strepto-coccus viridans,* enterococcus, and *Streptococcus bovis* has led to use of combined therapy for bacterial endocarditis caused by these organisms. Clinical data to support the superiority of combination therapy exists for endocarditis caused by viridans streptococcus [117] and enterococcus [53].

There are few studies of the pharmacokinetics of streptomycin in newborn and young infants. Peak serum concentrations of 17-42 μg/ml (mean 29 μg/ml) were observed 2 hr after a 10 mg/kg dose was given to premature infants [2]. In full-term infants, serum concentrations of 18-25 μg/ml were noted 0.5 hr after a 6.6 mg/kg dose was given [51]. The serum half-life in premature infants was 7 hr. Premature infants 1-3 days of age excreted 30% of the administered dose within 12 hr compared to 70% excretion by older children.

C. Kanamycin

Because of rapidly increasing resistance to streptomycin, kanamycin was readily accepted for use in neonates before adequate pharmacokinetic studies had been completed. This is apparent in a study of the ototoxic effects reported by Yow and co-workers [121] in 1962, where dosages of kanamycin given to 30 patients who ranged in age from less than 1 month to 14 years varied from 10 mg/kg per day to 100 mg/kg per day. Following the studies of Axline and Simon [2] and Eichenwald [30], a dosage schedule of 7.5 mg/kg given every 12 hr to all neonates, regardless of birthweight and chronological age, was adopted. This regimen was also used in infants and children. Dosages of 6.3-8.5 mg/kg given to premature infants produced mean serum concentrations of 17.5 μg/ml at 1 hr and 6 μg/ml at 12 hr after the dose [106]. Serum half-life values were inversely correlated with postnatal age. For example, average half-life values of 8.9 hr in preterm infants less than 48 hr of age and of 6 hr in infants 5–22 days old have been reported [106].

Reappraisal of kanamycin pharmacokinetics in 1975 [49] revealed an unexpectedly low peak serum concentration in the higher-birthweight babies who received 7.5 mg/kg dosages. When 7.5 and 10.0 mg/kg dosages were compared, a disproportionately large difference occurred in term infants during the first week of life; for an increase in dose of only 33%, the mean peak serum concentrations after 10 mg/kg were approximately 60% greater than measured after 7.5 mg/kg doses in infants less than 1 week of age. Such disparities were not observed in neonates older than 1 week of age. One possible explanation for this phenomenon is saturation of tissue binding sites by the smaller dose. Thus, we have suggested that the 10 mg/kg dose be used for all newborn infants except those less than 2000 g and less than 1 week of age. In the latter infants, 7.5 mg/kg doses administered every 12 hr are satisfactroy.

Serum kanamycin half-life values correlated inversely with gestational and chronologic ages and varied from 3 to 4 hours in term infants older than 7 days of age to 8.5 hr for infants 1–3 days old with gestational ages of 30–33 weeks [49]. The volume of kanamycin distribution showed a similar inverse relationship to birthweight and chronologic age, with values ranging from approximately 500 to 800 ml/kg. By contrast, plasma clearance was directly related to postnatal age and birthweight: clearances of 15.0 ml/min per 1.73 m^2 were observed in the first several days of life, compared to 45 ml/min per 1.73 m^2 by 7 days of life.

Mean cerebrospinal fluid (CSF)/serum ratios (penetration) for kanamycin ranged from 10 to 40%, the higher values occurring in patients with inflamed meninges, as indicated by the CSF cell count and protein content [30,49,76]. Several studies [14,30,49,76] have documented subtherapeutic levels of kanamycin in the CSF. The quantity of drug traversing the blood–brain barrier appears to be directly proportional to the administered parenteral dose. For example, kanamycin was undetectable in 9 CSF samples obtained 1–3 hr after a slow intravenous infusion of a 5 mg/kg dose administered on the second day of therapy for bacterial meningitis [14]. Simultaneous mean serum concentrations ranged from 5.5 to 7.8 μg/ml. Following an intramuscular dose of 7.5 mg/kg, the peak concentration in cerebrospinal fluid occurred 3–6 hr later and averaged 5.6 μg/ml (range 3–12 μg/ml) or 43% of the simultaneous serum concentration [49]. It is the low toxic/ therapeutic ratio of all the aminoglycosides that prohibits systemic administration of a quantity of drug sufficient to consistently achieve bactericidal levels in the CSF.

Oral administration of kanamycin or gentamicin has been recommended for the treatment [5] and prevention [29,43] of necrotizing enterocolitis in high-risk newborn infants. Bell et al. [5] presented data suggesting that suppression of the gastrointestinal microflora may prevent necrosis and perforation of the ischemic bowel. Although the desired effect on the microflora has been demonstrated in two other studies [6,12], this mode of therapy has not been evaluated in a controlled fashion in a large number of infants. On the other hand, three groups [12,29,43] have evaluated the efficacy of orally administered aminoglycosides in preventing necrotizing enterocolitis in high-risk infants. These studies demonstrated substantial reduction of disease rates in treated compared to untreated infants. However, the significant increase in the prevalence of infants with kanamycin-resistant, gram-negative enteric flora [12] is a definite contraindication to the widespread use of these drugs administered orally for prophylaxis. When combined with parenteral drugs for therapy of necrotizing enterocolitis in selected patients, the recommended dosage of kanamycin is 30 mg/kg per day given in two to four doses through the nasogastric tube, which is then clamped for 1 hr [72]. When other aminoglycosides are used, the total daily dose is approximately twice that administered parenterally.

Kanamycin pharmacokinetics have been evaluated in children beyond the neonatal period [7,45,46]. These studies are consistent in the finding that the 10 mg/kg dosage administered every 8 hr in infants and children older than 1 month of age is necessary to produce the desired therapeutic serum levels of 20-30 μg/ml. The importance of achieving such concentrations was illustrated by the finding that only 2 of 23 sera containing ≤20 μg/ml of kanamycin produced bactericidal titers of 1:2-1:4 against coliform bacilli, whereas the 3 sera containing ≥21 μg/ml had bactericidal titers of 1:2 against two test organisms with minimum inhibitory concentration values (MICs) of 5 μg/ml each [45]. A 10 day course of therapy with the 30 mg/kg daily regimen will result in a total dosage of 300 mg/kg, which is believed to be within the safe range with regard to ototoxicity [121].

D. Gentamicin

When introduced into clinical pediatrics in the late 1960s, gentamicin offered two advantages over kanamycin: activity against resistant Enterobactericaceae which were emerging rapidly in many nurseries across the country, and efficacy against *Pseudomonas aeruginosa* infections of patients with cystic fibrosis and with burns. Studies in neonates have delineated the pharmacokinetics of gentamicin in relationship to birthweight, gestational age, and chronologic age [14,65,69,73,74,78,86,122]. Mean peak serum concentrations of 3.5-6.0 μg/ml have been found in neonates within 1 hr after a 2.5 mg/kg dose of gentamicin and were unrelated to the rate of creatinine clearance. The mean serum values 12 hr after this dose were 0.5-1.0 μg/ml. Although most studies have shown no evidence of accumulation of gentamicin during a 5-7 day course of therapy, a recent report of gentamicin pharmacokinetics in low-birthweight (1365 g), preterm (gestational age 30.5 weeks) infants suggests that accumulation may occur in some premature infants [18].

The concentration of gentamicin in urine was greater during the first 4 hr after the dose [74]. The urinary concentrations were variable and in the range 2-135 μg/ml. Urinary excretion, expressed as percent of dose given, correlated directly with postnatal age and rates of creatinine clearance, but was independent of birthweight and dosage. Approximately 10% of the dose given to infants 0-3 days old was excreted within 12 hr compared to 40% excreted during the same period by infants 5-40 days of age.

The serum half-life of gentamicin correlated inversely with the rate of creatinine clearance, birthweight, and postnatal age [69,74]. A mean half-life value of approximately 14 hr has been seen in infants with birthweights of 800-1500 g and who were less than 1 week of age compared to 4.5 hr in infants who were more than 2500 g at birth. The half-life of gentamicin was approximately 3 hr in most neonates greater than 2 weeks of age, regardless of body weight. The volume of gentamicin distribution was largest in low-birthweight infants and decreased with increasing birthweight and postnatal age. The peak serum concentration of drug correlated inversely with the volume of distribution.

Data gathered by the Neonatal Meningitis Cooperative Study Group indicated that lumbar intrathecal or intraventricular administration of gentamicin were not efficacious in the therapy of neonatal meningitis caused by gram-negative enteric organisms [70,71]. The CSF concentrations after intramuscular, lumbar intrathecal, and intraventricular administration of gentamicin are presented in Table 6. Concentrations of gentamicin in CSF of infants with meningitis were 0.3–3.7 μg/ml (mean 1.6 μg/ml) 1–6 hr after 2.5 mg/kg doses were administered systemically. Gentamicin concentrations in CSF obtained 18–24 hr after a 1 mg dose into the lumbar intrathecal space and 1–6 hr after intramuscular administration were 0.5–3.4 μg/ml (mean, 1.6 μg/ml) [70]. Cerebrospinal fluid specimens taken 2–4 hr after the 1 mg lumbar intrathecal dose contained concentrations of approximately 18–40 μg/ml. Instillation of 2.5 mg gentamicin directly into the ventricles resulted in ventricular fluid concentrations of 10–130 μg/ml (mean, 48 μg/ml) 1–6 hr after the dose compared to 0.1–3.0 μg/ml (mean, 1.1 μg/ml) after systemic therapy only. The higher values were seen in infants who had reduced flow of fluid out of the ventricles, resulting in accumulation of gentamicin after two or three doses. Concentrations of 100 μg/ml or greater may be observed in such cases. Thus, despite gentamicin ventricular and lumbar fluid levels that greatly exceed the MIC values for most coliform organisms, the mortality rate was not improved. Indeed, the case-fatality rate was approximately threefold greater in infants with ventriculitis who were treated with systemic plus intraventricular gentamicin compared to those with ventriculitis who received systemic therapy only [71]. The explanation for the increased mortality in the intraventricular treatment group is unknown, but may relate to the procedure itself (ventricular taps) or to the high ventricular fluid concentrations of gentamicin, or both.

Gentamicin is the first aminoglycoside to have been studied extensively in children beyond the neonatal period [28,33,37,41,56,88,100–102]. An age-dependent dose response to gentamicin has been demonstrated [101] and is best explained by the age-related decrease in the extracellular fluid space and, consequently, the volume of gentamicin distribution. To achieve equivalent serum concentrations in the different age groups of pediatrics, the dosage must be adjusted appropriately. The recommended schedule calculated on the basis of body weight is 2.5 mg/kg for ages 0.5–5 years, 2.0 mg/kg for ages 5–10 years, and 1.5 mg/kg for individuals greater than 10 years of age. Use of a single body surface area dosage of 60 mg/m² will produce similar peak serum gentamicin concentrations in all age groups [102].

Improved predictability of serum gentamicin concentrations in children has been accomplished by using the two-compartment model of pharmacokinetics [33]. In children as well as in adults, gentamicin serum concentrations decline in a biphasic manner, reflecting

Table 6   Cerebrospinal Fluid Concentrations of Gentamicin in Infants With Meningitis Treated With Systemic Gentamicin With or Without Lumbar Intrathecal or Intraventricular Gentamicin Administration

| | Hours after gentamicin administration | | | | |
| | Lumbar intrathecal (1 mg) | | | Intraventricular (2.5 mg) | |
	0[a]	2–4	18–24	1–6	16–24
Lumbar CSF	1.6 (0.3–3.7)	30[b] (18.4–>40)	1.6 (0.5–3.4)	32 (8–85)	3.2 (1.8–4.2)
Ventricular fluid	1.1 (0.1–3.0	—	—	48 (10–130)	8.1 (1–24)

[a]Systemic gentamicin 2.5 mg/kg therapy only.
[b]Four samples only.

an initial distribution phase and a final washout (elimination) phase. Although the one-compartment model gives a reasonably accurate estimate of drug levels during the steady state, it does not account for extravascular distribution and therefore, if used to predict a dose response, it consistently underestimates serum levels after multiple doses. The two-compartment model has been utilized in adults [97] to demonstrate tissue binding of drug after the first dose and may identify those patients who experience abnormal tissue accumulation before detectable changes in renal function occur. Further study is necessary to establish the applicability of this model to pediatric pharmacokinetic studies.

E. Tobramycin

Tobramycin is not used as a first-line drug in the neonatal nursery because of the limited clinical experience [60,61] and lack of long-term follow-up studies of neonates for ototoxicity. Although tobramycin has increased in vitro activity against *P. aeruginosa* as compared to gentamicin, there are no clinical studies to indicate enhanced efficacy in the treatment of *Pseudomonas* infections. After a 2.0 mg/kg dose mean peak serum concentrations of 4–6 μg/ml have been observed at 0.5–1 hr. The concentrations at 12 hr after the dose were 1.5–2 μg/ml in infants $\leqslant$2000 g at birth and 0.5–1 μg/ml in higher-birthweight babies [61]. The drug concentrations in urine varied from 25 to 32 μg/ml. Excretion in urine, expressed as percentage of the dose, correlated directly with postnatal age. Average excretion values were 15–25% of the administered dose during the first week of life and 25–40% in older neonates.

The half-life of tobramycin in serum correlated inversely with birthweight and postnatal age [61]. Infants less than 1500 g at birth had more variable half-life values (range 9–17 hr, mean 8.7 hr) than did older neonates who were 2500 g or greater at birth (range, 3–4.5 hr, mean 3.9 hr). The mean volume of tobramycin distribution in neonates was in the range 339–435 ml/kg and was unrelated to postnatal age. Plasma clearance values increased during the postnatal period from 9 to 11 ml/min per 1.73 m^2 in infants less than 1500 g at birth to 25–35 ml/min per 1.73 m^2 in the larger, older babies. No accumulation of drug in serum was observed at completion of therapy.

Studies in children beyond the neonatal period have been performed in patients with malignancies and cystic fibrosis [47,75,87]. Evaluation of tobramycin in children with malignancies has led to a suggested increase in dosage from 3–5 mg/kg per day given at 8 hr intervals to 8–10 mg/kg (240–300 mg/m^2) per day given at 4–6 hr intervals. Utilizing this schedule, the following pharmacokinetic values were noted: mean peak serum concentration after a 60 min infusion of 2.0 mg/kg, 4.2 μg/ml; serum half-life, 1.6 hr; plasma clearance rate, 164 ml/min per 1.73 m^2 and volume of distribution, 420 ml/kg [47]. Although other studies [59,75] support the increase in dosage, additional studies are necessary to determine the optimal interval of administration. Most children who have received lower doses (<2 mg/kg) achieved peak serum levels $\leqslant$3 μg/ml and all have had undetectable antimicrobial activity in predose samples.

Tobramycin levels in sputum are less than 10% of simultaneous serum concentration and do not appear to relate to eradication of *P. aeruginosa* from sputum cultures [87]. Sputum concentrations of tobramycin which exceed the MIC values of the infecting organism do not ensure a bacteriologic or clinical response [75,87]. It is possible that antimicrobial therapy in patients with cystic fibrosis may exert its beneficial effect by decreasing the colony count [20,87,115] of *Pseudomonas* in respiratory secretions and in adjacent pulmonary tissue. This may not be reflected in studies of expectorated sputum. In a double-blind controlled trial of tobramycin versus placebo for acute pulmonary exacerbations in patients with cystic fibrosis, Wientzen and co-workers [115] found a statistically significant difference in the clinical and bacteriologic responses between the two groups. Although *Pseudomonas* or *Staphylococcus aureus* were not eradicated from the sputum of any of these patients, a larger number of patients in the tobramycin-treated group demonstrated a decrease of one log unit or greater in colony counts of at least one organism, whereas a greater number of patients in the placebo group showed an increase in colony counts. This observation correlated with the improved clinical response of the tobramycin-treated group of patients. In another study, tobramycin levels in muscle and fat were found to be 17% and 24%, respectively, of simultaneous serum levels when measured 1½ hours after intramuscular administration [23]. At 1 and 2 hr after the dose, concentrations of tobramycin in spleen were shown to be less than those in muscle. The interpretation of the investigator was that the blood supply did not appear to be a major factor in determining tissue levels of tobramycin. Additional information is required before this conclusion can be accepted. Nevertheless, the paucity of published data on the use of tobramycin in children precludes its routine use and relegates its position to a limited-purpose aminoglycoside that should be considered for therapy of infections caused by organisms that are resistant to kanamycin and gentamicin.

F. Amikacin

Amikacin is the most recently licensed aminoglycoside and thus the one for which we have the least clinical experience in children. Similar to tobramycin, this drug has been reserved for therapy of infections caused by multiply-resistant Enterobacteriaceae. The pharmacokinetic properties of amikacin are similar to those of kanamycin in newborn infants [50]. Serum concentration–time curves after a 7.5 mg/kg dose were comparable for these two aminoglycosides, with the exception of lower peak serum kanamycin levels in the older, higher-birthweight babies. Mean peak serum concentrations of 15–20 μg/ml occurred 0.5–1 hr after 7.5 mg/kg doses of amikacin and were independent of birthweight and chronologic age [48,50]. The average trough concentrations 12 hr after the dose were 3–6 μg/ml. Urinary amikacin concentrations ranged from 50 to 650 μg/ml and the average urinary excretion of drug in 12 hr after administration was 30–50% of the dose.

There are few reports of CSF concentrations of amikacin [50,111,120]. Penetration appears to be erratic. In the presence of uninflamed meninges in 1-day-old infants, concentrations in serum ranged from 15 to 29 μg/ml, while those in CSF were ≤0.2–2.7 μg/ml when measured 1–4 hr after a single 10 mg/kg dose administered by slow intravenous infusion [120]. The highest concentration reported has been 9.2 μg/ml after a 7.5 mg/kg intramuscular dose in an infant with meningitis [50]. Amikacin concentrations in ventricular fluid 12 hr after 1 or 2 mg intraventricular doses and 2–8 hr after intramuscular doses were 4.5–11.6 μg/ml (mean, 7.3 μg/ml).

Serum half-life values of amikacin in newborns were inversely correlated with gestational and chronologic ages. Values of 7–8 hr were noted in low-birthweight infants 1–3 days old and of 4–5 hr in full-term infants who were greater than 1 week of age. The volume of amikacin distribution remained constant at 500–600 ml/kg throughout the first month of life. Plasma clearance rates were usually within the narrow range 20–36 ml/min per 1.73 m^2 during the neonatal period.

The dosage schedule has not been clearly established for neonates. Although a 10 mg/kg "loading" dose was recommended initially, more recent experience indicates that this is not necessary to achieve therapeutic levels [48,50]. Doses of 7.5 mg/kg given intramuscularly every 12 hr appear to be effective in the first week of life. Based on the pharmacokinetics of other aminoglycosides, infants greater than 2000 g at birth and older than 7 days of age should receive amikacin at 8 hr intervals. However, additional experience is necessary before this dosage schedule can be recommended.

Similar to gentamicin and tobramycin, most studies of amikacin in older children have been performed in patients with either cystic fibrosis or malignancies [16,65,113,120]. Beyond the neonatal period, dosages based on body surface area rather than body weight appear to produce more predictable serum concentrations independent of age [66,113]. After a 420 mg/m^2 dose was administered by rapid intravenous infusion to children older than 4 years of age, an average serum concentration at 1–1¼ hr of 28.7 μg/ml was observed [113]. Because many of the individual values exceeded the theoretically toxic concentration of 30–35 μg/ml, 300 mg/m^2 may be a more suitable dosage [66]. The total clearance of amikacin was 120 ml/min per 1.73 m^2 [113]. When the total clearance was compared to renal clearance in each patient, an average of 82% of the drug was cleared by the kidneys. By 3.75 hr after administration of amikacin, only 17 ± 6% of the injected dose remained available to the body. Total clearance appeared to be directly correlated with age [16,113]. Once the phase of rapid distribution was completed by 45–75 min, plasma concentrations decreased monoexponentially and reached an equilibration phase, during which the serum half-life was 1.6 hr and the volume of distribution was 32% of body weight; both values were independent of age. This biphasic decline was most accurately analyzed by the two-compartment pharmacokinetic model [113].

Daschner [24] has evaluated amikacin concentrations in several body tissues. At 1.5 hr after a single intramuscular dose of 7.5 mg/kg, the mean peak serum concentration of amikacin was 14.9 μg/ml (range, 7–19.3 μg/ml). Serum levels declined to 0.18 μg/ml 13 hr after administration. Concentrations in muscle and fat peaked at 1.5 hr. The decline of these tissue concentrations was proportionately slower than those in serum, so that the levels in fat and muscle were 77% of that in serum at 12 hr compared with 14% at 1.5 hr. These data suggest that repeated doses may result in accumulation of drug in muscle and fat. The superior blood supply of muscle did not provide significantly greater drug levels in this tissue than in fat. It is possible that these observations may explain in part the delay in clearance of bacteria from parenchymal infections despite the presence of seemingly adequate serum concentrations, a phenomenon that we have observed in two neonates with necrotizing enterocolitis and peritonitis.

It is difficult to evaluate efficacy of amikacin in children because of the severity of the underlying disease in most treated patients. Those with cystic fibrosis appear to improve clinically after therapy with amikacin [120] as they do with gentamicin and tobramycin, despite the persistence of positive sputum cultures. Although amikacin appears to be efficacious for treatment of urinary tract infections in doses ranging from 5 to 15 mg/kg per day [63,79], this drug should be reserved for infections caused by organisms resistant to other aminoglycosides.

G. Sisomicin and Netilmicin

These two drugs are structurally related to gentamicin and are the two newest aminoglycosides which, at this time, are not licensed for use in the United States. The spectrum of

activity of these drugs is similar to that of gentamicin with a few isolated differences in
in vitro susceptibilities. Netilmicin has the added advantage that an increasing number of
Enterobacteriaceae possess adenylating enzymes mediating resistance to all the other
aminoglycosides except netilmicin and amikacin [58]. In a rabbit model of *E. coli* meningitis, sisomicin and netilmicin appeared to have a more rapid bactericidal effect than did
tobramycin, amikacin, and gentamicin, as judged by the degree of reduction in bacterial
colony counts in cerebrospinal fluid [96,110]. An unexplained finding in these studies
was that netilmicin [96] and sisomicin [110] had a bactericidal effect at concentrations
approximately equal to the mean blood concentration (MBC) for the organisms compared
to concentrations 2-5 times the MBC required with the other aminoglycosides. Finally,
animal studies [57,67,77] indicate that netilmicin may have less ototoxicity and nephrotoxicity than does gentamicin.

Although both netilmicin and sisomicin have been studied extensively in adults, the
few studies in neonates have been performed with netilmicin only [94,95,103]. Mean
peak serum concentrations varied from approximately 5.5 to 7.0 μg/ml in all neonates
tested at ½ hr after intramuscular administration of 3 mg/kg [103]. Average serum values
of 1.0-1.5 μg/ml were observed 12 hr after the dose. The average concentrations of netilmicin in urine were 46 μg/ml and 29 μg/ml for the first and second 3 hr study periods,
respectively, after a 3 mg/kg dose. These values increased to 69 μg/ml and 103 μg/ml after
a 4 mg/kg dose.

The serum half-life of netilmicin was inversely related to birthweight, gestational age
(see Figure 1), and chronologic age. The mean half-life varied from 4.7 hr in infants equal
to or less than 2000 g at birth in the first week of life to 3.8 hr in infants greater than
2000 g at birth and 7 days of age and older. Volume of netilmicin distribution showed a
similar relationship and ranged from 696 ml/kg in infants 28-32 weeks of gestation to
510 ml/kg in infants greater than 37 weeks of gestational age. The serum concentration-
time curves and area-under-the-curve values after 3 mg/kg doses of netilmicin or gentamicin were similar. The mean peak serum concentration and half-life of gentamicin, 6.9
μg/ml and 4.0 hr, respectively, were slightly greater than the corresponding values for
netilmicin, 6.3 μg/ml and 3.5 hr, respectively, when these drugs were administered to the
same babies for comparative purposes. More variability in the range of serum concentrations has been observed with gentamicin than with netilmicin in studies performed in
adults as well [103].

The washout phase of up to 11 days after discontinuation of netilmicin has been most
accurately described in neonates by the two-compartment model of pharmacokinetics
[103]. Netilmicin activity in the serum and urine remained detectable for up to 11 and
14 days, respectively, after the antibiotic was stopped. The serum concentration curve
declined in a biphasic manner (Figure 3). The average terminal half-life of 62.4 hr is within
the range determined for gentamicin in adults [98]. However, in spite of this long terminal
half-life and tissue accumulation, dosing intervals of two to three times the half-life appear
to be appropriate for most neonates.

Similar to gentamicin [18], netilmicin appears to accumulate in low-birthweight,
premature infants [103]. After 4 mg/kg doses of netilmicin given every 12 hr for an
average of 6.4 days, the mean trough value in premature infants (mean weight, 1415 g;
mean gestational age, 32 weeks) increased from 2.2 μg/ml on the second day of therapy
to 5.6 μg/ml on the final day. This same group of infants did not show the expected decrease
in serum creatinine observed in term infants during the first 2 weeks of life and they required mechanical ventilation. Thus, hypoxemia superimposed on immature renal function

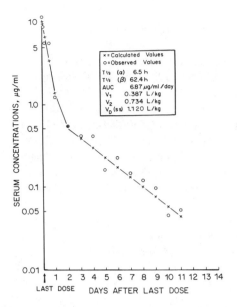

Figure 3 Serum concentrations of netilmicin measured for 11 days after last 4 mg/kg dose in 12 neonates. (Reprinted from Ref. 103.)

is a possible explanation of drug accumulation in this special group of babies.

On the basis of these pharmacokinetic studies, a dose schedule of 3.0 mg/kg administered every 12 hr should produce serum levels that are within the therapeutic range. It is possible that 4 mg/kg doses on an 8 hr schedule should be given to infants greater than 2000 g and older than 1 week of age, but additional studies of long-term toxicity must be completed first.

IV. Toxicity of the Aminoglycosides

As in adults, the major adverse effects of aminoglycosides involve impairment of vestibular, cochlear, and renal function and, rarely, neuromuscular blockade. Although one study [63] has reported transient elevation of SGOT up to 279 units (normal $\leqslant 48$ units) in the absence of hepatomegaly or other evidence of abnormal liver function, hepatotoxicity and hematologic toxicity are generally not associated with aminoglycoside therapy. Acute toxic reactions are unusual in the neonate and child.

Transient cylinduria and proteinuria may occur after prolonged administration of any of the aminoglycosides. However, significant elevations in blood urea nitrogen (BUN) and creatinine values are rarely observed. The applicability of maintaining peak serum gentamicin concentrations less than 10 μg/ml and trough values less than 2 μg/ml to prevent nephrotoxicity in adults [22] has not been systematically assessed in infants and children.

Neomycin [10,13,64], streptomycin [91,93], and kanamycin [40,121] have been implicated as a cause of sensorineural hearing losses in infants and children and gentamicin [32] has been associated with vestibular impairment. It is, however, difficult to incriminate the aminoglycoside as the single causative agent of hearing loss in most studies because of other high-risk factors present in these patients. For example, asphyxia, hyperbilirubinemia, or incubator exposure have been associated independently with ototoxicity [116].

Although animal studies have demonstrated a synergistic effect of noise combined with neomycin [52] or kanamycin [26] on development of ototoxicity, this has not been substantiated in the human neonate exposed to both incubator noise and kanamycin [116]. A familial predisposition toward cochlear damage has been observed following therapy with streptomycin [54,93], but not with the other aminoglycosides.

Johnsonbaugh et al. [55] performed follow-up studies on 98 infants who received short courses of streptomycin sulfate during the neonatal period. The mean total dosage was 118 mg/kg given for an average of 3.2 days, or 37 mg/kg per day. Hearing loss ascribable to the drug was not observed in these children when they were evaluated at 7½–8 years of age.

Eichenwald [30] demonstrated no significant differences in the frequency or type of hearing loss in long-term follow-up studies of newborn infants treated with either kanamycin or streptomycin when compared with untreated controls. In infants and children, kanamycin ototoxicity appears to be related primarily to total dosages. High-frequency sensorineural hearing loss in infants with normal renal function appears to be more likely if the total dosage exceeds 500 mg/kg per day [121]. Total daily dosage may not be as significant in children as in adults [35]. Additional information is necessary to define the critical factors associated with aminoglycoside ototoxicity in neonates and young infants.

One long-term study of infants who received gentamicin during the neonatal period found vestibular dysfunction in 2 of 28 infants evaluated at 2–3¾ years of age [32]. Because of the complicated medical histories and other neurological dysfunctions in these two patients, a direct cause-and-effect relationship could not be concluded with certainty. Data from the first Neonatal Meningitis Cooperative Study [70] indicated that only 1 (1.3%) of 79 infants who received a minimum of 5–7.5 mg/kg per day of gentamicin for 3 weeks or greater developed profound deafness that may have been drug-related.

In a prospective evaluation of long-term toxicity of gentamicin and kanamycin [36] 86 infants who received one of these two drugs during the neonatal period underwent yearly audiometric, vestibular, and psychometric evaluations for 4 years. Neither gentamicin nor kanamycin could be incriminated as the sole agent responsible for sensorineural hearing impairment. Because the infants in this study received doses lower than those currently recommended (5–6 mg/kg per day vs. 5–7.5 mg/kg per day for gentamicin and 15 mg/kg per day vs. 15–30 mg/kg per day for kanamycin), continued evaluation is necessary for these as well as the newer aminoglycosides.

There are no long-term toxicity data in children for tobramycin, amikacin, and netilmicin. Acute toxicity in this age group is exceedingly rare. One report [66] of amikacin therapy in 18 patients with cystic fibrosis documented a rise in serum creatinine from 0.8 mg/dl to 4.5 mg/dl in one patient after 6 days of therapy. This individual also had cardiovascular collapse. Serum amikacin levels were not elevated. In addition, two patients had unilateral hearing impairment without alteration of renal function.

Neuromuscular blockade caused by aminoglycosides has been associated most frequently with the instillation of neomycin directly onto serosal surfaces during surgery [114]. However, streptomycin [90], kanamycin [92], and gentamicin [114] administered by the intramuscular and intravenous routes have been implicated in children as well as adults. Intravenous bolus doses may be more likely to cause this unusual adverse effect. Its rare occurrence in recent years is probably due to the infrequent use of neomycin, the aminoglycoside with the most potent neuromuscular blocking action in animals [89]. The very young postoperative surgical infant with a highly variable fluid volume and renal function is at highest risk. Diagnosis is made by nerve conduction studies which reveal a progressive fatigue and post-tetanic facilitation characteristic of a nondepolarizing, curare-like neuromuscular block. Reversal is achieved by neostigmine and/or calcium. Prophylactic treatment with calcium is not indicated because this cation may interfere with the antimicrobial activity of aminoglycosides against certain organisms [35].

V. Aminoglycosides in Renal Failure

Because renal excretion accounts for the elimination of approximately 80% of an aminoglycoside dose, the fate of these drugs is severely affected by reduction in renal function. During the first month of life creatinine clearance is markedly reduced and does not reach adult levels until 3 years of age [15]. The serum half-life of the aminoglycosides is inversely related to rates of creatinine clearance and therefore is prolonged in infancy. Further reduction in renal function for any reason may result in accumulation of drug in serum unless the dosage schedule is appropriately altered. In adults with renal failure, the creatinine clearance [27] or serum creatinine [21,27] values have been utilized to predict serum half-life and to formulate appropriate dosing intervals. A widely used guide for formulating dosage schedules is the Dettli nomogram [27]. This nomogram utilizes the relationship between specific drug elimination rate constants and creatinine clearance or serum creatinine to estimate the serum half-life. This method has not, however, been evaluated systematically in the pediatric age group.

A nomogram based on studies in 15 children ages 8–17 years has been constructed and shows a linear relationship between serum half-life and creatinine clearance [119]. Yashioka et al. [119] recommend the usual dose for this age group of 1 mg/kg to be repeated at intervals that are three times the serum half-life value estimated from the nomogram. Because of the difficulty in obtaining accurate creatinine clearance values in neonates and young infants, this nomogram is not always applicable.

Preliminary data from our laboratory [107] indicate that the estimated serum half-life of gentamicin at three to four times the serum creatinine is applicable to neonates and infants. Pharmacokinetic data have been derived from 17 patients in chronic renal failure. These patients ranged in age from <1 week to 12 years, with a median age of 5 months; six children were less than 1 month old. Serum half-life values from these patients were plotted against serum creatinine values. The half-life was approximated more closely by 3.6 times the serum creatinine than by the Dettli nomogram when the serum creatinine was 4 mg/dl or less. Higher serum creatinine values result in increased variation.

Once the serum half-life has been estimated by one of the methods described above, the loading dose may be repeated at intervals twice that of the half-life. Whenever possible, peak and trough serum samples should be assayed and serum half-life times calculated for the individual patient. If drug assays are not available to the clinician, an estimate of the antibiotic concentration in the serum may be derived from determining serum inhibitory titers against the patient's infecting organism or against a representative stock laboratory strain. As a rule, the concentration of aminoglycoside is considered within the effective and nontoxic range with titers against susceptible pathogens are 1:2 to 1:8.

Preliminary data from six pediatric patients in acute renal failure indicate no relation between serum creatinine and serum half-life values.

VI. Conclusions

When used appropriately, aminoglycosides are safe and efficacious for the therapy of serious pediatric infections caused by sensitive strains of Enterobacteriaceae. The rapidly changing physiologic processes of neonates and young children must be taken into account when prescribing these agents to young infants. Birthweight, chronologic age, and renal function are the three major determinants of drug absorption, metabolism, and excretion in the neonate. Although dosage recommendations have been made for all pediatric patients on a weight basis, recent evidence suggests that beyond the newborn period, body surface area is a more practical and accurate basis for prescribing aminoglycoside dosage because it eliminates the necessity for age-related adjustments. Predictability of drug levels in serum is increased further by application of the two-compartment model of pharmacokinetics.

Neomycin and streptomycin are drugs of choice only in a few specific infections. Although kanamycin, gentamicin, tobramycin, and amikacin have broad clinical application, the susceptibility patterns of Enterobacteriaceae in each hospital must guide the pediatrician's prescribing habits. Amikacin, tobramycin, netilmicin, and sisomicin should be reserved for organisms resistant to kanamycin and gentamicin. Efficacy and toxicity studies must be continued with the newer aminoglycosides to confirm preliminary findings in small numbers of patients.

Infants and children in acute renal failure require modified dosage schedules based on the calculation of serum half-life for each patient individually. The reliability of nomograms established in studies of adults in chronic renal failure have not been confirmed in infants and children. Therefore, measurement of drug levels and calculation of serum half-life is required for each patient.

References

1. Arant, B.S.: Developmental patterns of renal functional maturation compared in the human neonate. *J. Pediatr.* 92:705–712, 1978.
2. Axline, S.G., and Simon, H.J.: Clinical pharmacology of antimicrobials in premature infants: I. Kanamycin, streptomycin and neomycin. *Antimicrob. Agents Chemother.,* 135–141, 1964.
3. Axline, S.G., Yaffe, S.J., and Simon, H.J.: Clinical pharmacology of antimicrobials in premature infants. II. Ampicillin, methicillin, oxacillin, neomycin and colistin. *Pediatrics* 39:97–107, 1967.
4. Barratt, T.M., and Chantler, C.: Clinical assessment of renal function. In *Pediatric Nephrology,* edited by Rubin, M.I. Baltimore, Md., Williams & Wilkins, 1976, p. 62.
5. Bell, M.J., Kosloske, A.M., Benton, C., and Martin, L.W.: Neonatal necrotizing enterocolitis: prevention of perforation. *J. Pediatr. Surg.* 8:601–605, 1973.
6. Bell, M.J., Shackleford, P.G., Feigin, R.D., Ternberg, J.L., and Brotherton, T.: Alterations in gastrointestinal microflora during antimicrobial therapy for necrotizing enterocolitis. *Pediatrics* 63:425–428, 1979.
7. Berger, S.H., and Wehrle, P.F.: Kanamycin serum levels in infants and children. *Ann. N.Y. Acad. Sci.* 76:136–139, 1958.
8. Bernard, B., Abate, M., Threlen, P.F., Attar, H., Ballard, C.A., and Wehrle, P.F.: Maternal-fetal pharmacological activity of amikacin. *J. Infect. Dis.* 135:925–932, 1977.
9. Bernard, B., Garcia-Cazares, S.J., Ballard, C.A., Thrupp, L., Mathies, A.W., and Wehrle,

P.F.: Tobramycin: maternal-fetal pharmacology. *Antimicrob. Agents Chemother.* 11:688–694, 1977.

10. Beukelaer, M.M., Travis, L.B., Dodge, W.F., and Guerra, F.A.: Deafness and acute tubular necrosis following parenteral administration of neomycin. *Am. J. Dis. Child.* 121:250–252, 1971.

11. Black, J., Calesnick, B., Williams, D., and Weinstein, M.J.: Pharmacology of gentamicin, a new broad-spectrum antibiotic. *Antimicrob. Agents Chemother.,* 138–147, 1963.

12. Boyle, R., Nelson, J.S., Stonestreet, B.J., Peter, G., and Oh, W.: Alterations in stool flora resulting from oral kanamycin prophylaxis of necrotizing enterocolitis. *J. Pediatr.* 93:857–861, 1978.

13. Buetow, K.C., Cheung, C.-S., and Finberg, R.: Prolonged use of neomycin in premature infants. *Am. J. Dis. Child.* 104:76–81, 1962.

14. Chang, M.J., Escobedo, M., Anderson, D.C., Hillman, L., and Feigin, R.D.: Kanamycin and gentamicin treatment of neonatal sepsis and meningitis. *Pediatrics* 56:695–699, 1975.

15. Chantler, C.: Newborn disorders. In *Clinical Pediatric Nephrology,* edited by Lieberman, E., Philadelphia, J.B. Lippincott, 1976.

16. Cleary, T., Pickering, L.K., Kramer, W., Kohl, S., and Van Eys, J.: Pharmacokinetics of intravenously administered amikacin in pediatric patients. Abstract, 18th Interscience Conference on Antimicrobial Agents and Chemotherapy, October 1–4, 1978, Atlanta, Ga.

17. Conway, N.: Streptomycin in pregnancy: effect on the foetal ear. *Br. Med. J.* 2:260–263, 1965.

18. Coyer, W.F., Wesbey, G.E., Cech, K.L., and Robinson, J.A.: Intravenous gentamicin pharmacokinetics in the small preterm infant. *Pediatr. Res.* 12:Abstr. 240, 1978.

19. Crosson, F.J., and Moxon, E.R.: Factors influencing kanamycin resistance in gram negative enteric neonatal sepsis. *Pediatrics* 61:488–489, 1978.

20. Crozier, D.N., and Khan, S.R.: Tobramycin in treatment of infections due to *Pseudomonas aeruginosa* in patients with cystic fibrosis. *J. Infect. Dis.* 134(Suppl.):S187–S190, 1976.

21. Cutler, R.E., Gyselynck, A., Fleet, P., and Forrey, A.W.: Correlation of serum creatinine concentration and gentamicin half-life. *JAMA* 219:1037–1041, 1972.

22. Dahlgren, J.G., Anderson, E.T., and Hewitt, W.L.: Gentamicin blood levels: a guide to nephrotoxicity. *Antimicrob. Agents Chemother.* 8:58–62, 1975.

23. Daschner, F.: Tobramycin serum levels and tissue content in children. *Chemotherapy* 23:293–298, 1977.

24. Daschner, F., Reiss, E., and Engert, J.: Distribution of amikacin in serum, muscle, and fat in children after a single intramuscular injection. *Antimicrob. Agents Chemother.* 11:1081–1083, 1977.

25. Daubenfeld, O., Modde, H., and Hirsch, H.A.: Transfer of gentamicin to the foetus and the amniotic fluid during a steady state in the mother. *Arch. Gynaekol.* 217:233–240, 1974.

26. Dayal, V.S., Kokshanian, A., and Mitchell, D.P.: Combined effects of noise and kanamycin. *Ann. Otol.* 80:897–902, 1971.

27. Dettli, L.: Individualization of drug dosage in patients with renal disease. *Med. Clin. North Am.* 58:977–985, 1974.

28. Echeverria, P., Siber, G.R., Paisley, J., et al.: Age-dependent dose response to gentamicin. *J. Pediatr.* 87:805–808, 1975.

29. Egan, E.A., Mantilla, G., Nelson, R.M., and Eitzman, D.V.: A prospective controlled trial of oral kanamycin in the prevention of neonatal necrotizing enterocolitis. *J. Pediatr.* 89:467–470, 1976.

30. Eichenwald, H.F.: Some observations on dosage and toxicity of kanamycin in premature and fullterm infants. *Ann. N.Y. Acad. Sci.* 132:984–991, 1966.

31. Eidelman, A.J., and Reynolds, J.: Gentamicin-resistant *Klebsiella* infections in a neonatal intensive care unit. *Am. J. Dis. Child.* 132:421–422, 1978.

32. Elfing, J., Pettay, O., and Raivio, M.: A follow-up study on the cochlear, vestibular and renal function in children treated with gentamicin in the newborn period. *Chemotherapy* 18:141–153, 1973.

33. Evans, W.E., Feldman, S., Barker, L.F., Osse, M., and Chaudhary, S.: Use of gentamicin serum levels to individualize therapy in children. *J. Pediatr.* 93:133–137, 1978.

34. Faden, H., Neter, E., McLaughlin, S., and Giacoia, G.: Gentamicin-resistant *Staphylococcus aureus:* emergence in an intensive care nursery. *JAMA* 241:143–145, 1979.

35. Finegold, S.M.: Toxicity of kanamycin in adults. *Ann. N.Y. Acad. Sci.* 132:942–956, 1966.

36. Finitzo-Hieber, T., McCracken, G.H., Roeser, R.J., Allen, D.A., Chrane, D.F., and Morrow, J.: Ototoxicity in neonates treated with gentamicin and kanamycin: results of a four-year controlled follow-up study. *Pediatrics* 63:443–450, 1979.

37. Finkelstein, E., and Hall, K.: Aminoglycoside clearance in patients with cystic fibrosis (letter). *J. Pediatr.* 94:163–164, 1979.

38. Franco, J.A., Eitzman, D.V., and Baer, H.: Antibiotic usage and microbial resistance in an intensive care nursery. *Am. J. Dis. Child.* 126:318–321, 1973.

39. Friis-Hansen, B.: Body water compartments in children: changes during growth and related changes in body composition. *Pediatrics* 28:169–181, 1961.

40. Frost, J.O., Hawkins, J.E., and Daly, J.F.: Kanamycin: II. Ototoxicity. *Am. Rev. Resp. Dis.* 82:23–30, 1960.

41. Glew, R.H., Moellering, R.C., and Burke, J.F.: Gentamicin dosage in children with extensive burns. *J. Trauma* 16:819–823, 1976.

42. Good, R.G., and Johnson, G.H.: The placental transfer of kanamycin during late pregnancy. *Obstet. Gynecol.* 38:60–62, 1971.

43. Grylack, L.F., and Scanlon, J.W.: Oral gentamicin therapy in the prevention of neonatal necrotizing enterocolitis. *Am. J. Dis. Child.* 132:1192–1194, 1978.

44. Heilman, D.H., Heilman, F.R., Hinshaw, H.C., Nichols, D.R., and Henell, W.E.: Streptomycin: absorption, diffusion, excretion and toxicity. *Am. J. Med. Sci.* 210:576–584, 1945.

45. Hieber, J.P., and Nelson, J.D.: Reevaluation of kanamycin dosage in infants and children. *Antimicrob. Agents Chemother.* 9:899–902, 1976.

46. High, R.H., Sarria, A., and Huang, N.N.: Kanamycin in the treatment of infections in infants and children. *Ann. N.Y. Acad. Sci.* 76:289, 1958.

47. Hoecker, J.L., Pickering, L.K., Swaney, J., Kramer, W.G., Van Eys, J., Feldman, S., and Kohl, S.: Clinical pharmacology of tobramycin in children. *J. Infect. Dis.* 137:592–596, 1978.

48. Howard, J.B., and McCracken, G.H.: Pharmacological evaluation of amikacin in neonates. *Antimicrob. Agents Chemother.* 8:86–90, 1975.

49. Howard, J.B., and McCracken, G.H.: Reappraisal of kanamycin usage in neonates. *J. Pediatr.* 86:949–956, 1975.

50. Howard, J.B., McCracken, G.H., Trujillo, H., and Mohs, E.: Amikacin in newborn infants: comparative pharmacology with kanamycin and clinical efficacy in 45 neonates with bacterial diseases. *Antimicrob. Agents Chemother.* 10:205–210, 1976.

51. Hunt, A.D., and Fell, B.: Streptomycin intramuscular dosage per unit body weight correlated with serum levels in infants and children. *Pediatrics* 4:163–169, 1949.

52. Jauhiainen, T., Kohonen, A., and Jauhiainen, M.: Combined effect of noise and neomycin on the cochlea. *Acta Otolaryngol. (Stockh.)* 73:387–390, 1972.

53. Jawetz, E., and Sonne, M.: Penicillin-streptomycin treatment of enterococcal endocarditis. *N. Engl. J. Med.* 274:710–715, 1966.

54. Johnsonbaugh, R.E., Drexler, H.G., Light, I.J., and Sutherland, J.M.: Familial occurrence of drug-induced hearing loss. *Am. J. Dis. Child.* 127:245–247, 1974.

55. Johnsonbaugh, R.E., Drexler, H.G., Sutherland, J.M., et al.: Audiometric study of streptomycin-treated infants. *Am. J. Dis. Child.* 112:43–45, 1966.

56. Jusko, W.J., Baliah, G., Kim, K.H., Gerbracht, L.M., and Yaffe, S.J.: Pharmacokinetics of gentamicin during peritoneal dialysis in children. *Kidney Int.* 9:430–438, 1976.

57. Jzaraski, M., Levy, J.K., and Jerger, J.: Comparative toxicity of netilmicin and gentamicin in squirrel monkeys *(Saimiri seiureus)*. *J. Infect. Dis.* 137:476–480, 1978.

58. Kabins, S.A., Nathan, C., and Cohen, S.: In vitro comparison of netilmicin, a semisynthetic derivative of sisomicin and four other aminoglycoside antibiotics. *Antimicrob. Agents Chemother.* 10:139–145, 1976.

59. Kafezis, D.A., Sinoniotis, C.A., Kitsiou-Tzeli, S., Papadatos, C.J.: Tobramycin dosage in infants and children [letter]. *Lancet* 2:1264, 1978.

60. Kannan, M.M., Dalton, H.P., and Escobar, M.R.: Tobramycin in the neonatal period. *Va. Med. Month.* 100:1030–1034, 1973.

61. Kaplan, J.M., McCracken, G.H., Thomas, M.L., Horton, L.J., and Davis, N.: Clinical pharmacology of tobramycin in newborns. *Am. J. Dis. Child.* 125:656–660, 1973.

62. Kauffman, R.E., Morris, J.A., and Azarnoff, D.L.: Placental transfer and fetal urinary excretion of gentamicin during constant rate maternal infusion. *Pediatr. Res.* 9: 104–107, 1975.

63. Khan, A.J., Evans, H.E., Jhaveri, R., Chang, C.J., and Hochstein, L.: Amikacin pharmacokinetics in the therapy of childhood urinary tract infection. *Pediatrics* 58: 873–876, 1976.

64. King, J.F.: Severe deafness in an infant following oral administration of neomycin. *J. Med. Assoc. Ga.* 51:530–531, 1962.

65. Klein, J.O., Herschel, M., Theraken, R.M., et al.: Gentamicin in serious neonatal infections: absorption, excretion and clinical results in 25 cases. *J. Infect. Dis.* 124(Suppl.):S224–S231, 1971.

66. Lau, W.K., Young, L.S., Osher, A.B., and Dooley, R.R.: Amikacin therapy of exacerbations of *Pseudomonas aeruginosa* infections in patients with cystic fibrosis. *Pediatrics* 60:372–377, 1977.

67. Luft, F.C., Yum, M.N., Kleit, S.A.: Comparative nephrotoxicities of netilmicin and gentamicin in rats. *Antimicrob. Agents Chemother.* 10:845–849, 1976.

68. McCracken, G.H.: Clinical pharmacology of antibacterial agents. In *Infectious Diseases of the Fetus and Newborn Infant,* edited by Remington, J.S., and Klein, J.O. Philadelphia, W.B. Saunders, 1976, p. 1029.

69. McCracken, G.H., Chrane, D.F., and Thomas, M.L.: Pharmacologic evaluation of gentamicin in newborn infants. *J. Infect. Dis.* 124(Suppl.):S214–S223, 1971.

70. McCracken, G.H., and Mize, S.G.: A controlled study of intrathecal antibiotic therapy in gram negative enteric meningitis of infancy. Report of the Neonatal Meningitis Cooperative Study Group. *J. Pediatr.* 89:66–72, 1976.

71. McCracken, G.H., Mize, S.G., et al.: Intraventricular therapy of neonatal meningitis caused by gram negative enteric bacilli. *Pediatr. Res.* 13:Abstr., 1979.

72. McCracken, G.H., and Nelson, J.D.: *Antimicrobial Therapy for Newborns.* New York, Grune & Stratton, 1977, p. 131.

73. McCracken, G.H., Threlkeld, N., and Thomas, M.L.: Intravenous administration of kanamycin and gentamicin in newborn infants. *Pediatrics* 60:463–466, 1977.

74. McCracken, G.H., West, N.R., and Horton, L.J.: Urinary excretion of gentamicin in the neonatal period. *J. Infect. Dis.* 123:257–262, 1971.

75. McCrae, W.M., Raeburn, J.A., and Hanson, E.J.: Tobramycin therapy of infections due to *Pseudomonas aeruginosa* in patients with cystic fibrosis: effect of dosage

and concentration of antibiotics in sputum. *J. Infect. Dis.* 134(Suppl.):S191–S193, 1976.

76. McDonald, L.L., and St. Geme, J.W.: Cerebrospinal fluid diffusion of kanamycin in newborn infants. *Antimicrob. Agents Chemother.* 2:41–44, 1972.

77. Miller, G.H., Arceiri, G., Weinstein, M.J., and Waitz, J.A.: Biological activity of netilmicin, a broad-spectrum semisynthetic aminoglycoside antibiotic. *Antimicrob. Agents Chemother.* 10:827–836, 1976.

78. Milner, R.D.G., Ross, J., Froud, D.J.R., and Davis, J.A.: Clinical pharmacology of gentamicin in the newborn infant. *Arch. Dis. Child.* 47:927–932, 1972.

79. Mimica, J., and Toporovsky, J.: Summary of a clinical trial of amikacin in the treatment of pyelonephritis in children. *J. Infect. Dis.* 134(Suppl.):S420, 1976.

80. Mirkin, B.L.: Perinatal pharmacology: placental transfer, fetal localization and neonatal disposition of drugs. *Anesthesiology* 43:156–170, 1975.

81. Myers, M.G., Roberts, R.F., and Mirhij, N.J.: Effects of gestational age, birthweight, and hypoxemia on pharmacokinetics of amikacin in serum of infants. *Antimicrob. Agents Chemother.* 11:1027–1032, 1977.

82. Nelson, J.D.: Duration of neomycin therapy for enteropathogenic *Escherichia coli* diarrheal disease: a comparative study of 113 cases. *Pediatrics* 48:248–258, 1971.

83. Nunnery, A.W., and Riley, H.D.: Gentamicin: pharmacologic observations in newborns and infants. *J. Infect. Dis.* 119:402–405, 1969.

84. O'Brien, J.E.: Excretion of drugs in human milk. *Am. J. Hosp. Pharm.* 31:844–854, 1974.

85. Odell, G.B., Cukier, J.D., and Maglalong, A.C.: Commentary. *J. Pediatr.* 86:614, 1975.

86. Paisley, J.W., Smith, A.L., and Smith, D.H.: Gentamicin in newborn infants: comparison of intramuscular and intravenous administration. *Am. J. Dis. Child.* 126:473–477, 1973.

87. Parry, M.F., and Neu, H.C.: Tobramycin and ticarcillin therapy for exacerbations of pulmonary disease in patients with cystic fibrosis. *J. Infect. Dis.* 134(Suppl.): S194–S197, 1976.

88. Parry, M.F., Neu, H.C., Merlino, M., Gaerlan, P.F., Ores, C.N., and Denning, C.R.: Treatment of pulmonary infections in patients with cystic fibrosis: a comparative study of ticarcillin and gentamicin. *J. Pediatr.* 90:144–148, 1977.

89. Pindell, M.H.: The pharmacology of kanamycin. *Ann. N.Y. Acad. Sci.* 132:805–810, 1966.

90. Pittinger, C.B., Eryasa, Y., and Adamson, R.: Antibiotic-induced paralysis. *Anesth. Analg.* 49:487–501, 1970.

91. Prazic, M., Salaj, B., and Subotic, R.: Familial sensitivity to streptomycin. *J. Laryngol. Otol.* 78:1037–1043, 1964.

92. Ream, C.R.: Respiratory and cardiac arrest after intravenous administration of kanamycin with reversal of toxic effects by neostigmine. *Ann. Intern. Med.* 59: 384–387, 1963.

93. Robinson, G.C., and Cambon, K.G.: Hearing loss in infants of tuberculous mothers treated with streptomycin during pregnancy. *N. Engl. J. Med.* 271:949–951, 1964.

94. Rubio, T., Wirth, F., Wellman, L., and Simmons, B.: Pharmacokinetics of netilmicin in newborn infants. Abstract, 18th Interscience Conference on Antimicrobial Agents and Chemotherapy, October 1–4, 1978, Atlanta, Ga.

95. Schauf, V., Chindasilpa, V., Hamilton, L., and Riff, L.: Netilmicin pharmacology in pediatric patients. *Pediatr. Res.* 12:Abstr. 267, 1978.

96. Scheld, C.M., Brown, R.S., and Sande, M.A.: Comparison of netilmicin and gentamicin in the therapy of experimental *Escherichia coli* meningitis. *Antimicrob. Agents Chemother* 13:899–904, 1978.

97. Schentag, J.J., Cumbo, T.J., Jusko, W.J., and Plaut, M.E.: Gentamicin tissue accumulation and nephrotoxic reactions. *JAMA* 240:2067–2072, 1978.

98. Schentag, J.J., Jusko, W.J., Plaut, M.E., Cumbo, T.J., Vance, J.W., Abrutyn, E.: Tissue persistence of gentamicin in man. *JAMA* 238:327–329, 1977.

99. Schwartz, S.N., Pazin, G.J., Lyon, J.A., Ho, M., and Pasculle, A.W.: A controlled investigation of the pharmacokinetics of gentamicin and tobramycin in obese subjects. *J. Infect. Dis.* 138:499–505, 1978.

100. Shwachman, H., Khaw, K.-J., and Kowalski, S.M.: The management of cystic fibrosis. *Clin. Pediatr.* 14:1115–1118, 1975.

101. Siber, G.R., Escheverria, P., Smith, A.L., Paisley, J.W., and Smith, D.H.: Pharmacokinetics of gentamicin in children and adults. *J. Infect. Dis.* 132:637–651, 1975.

102. Siber, G.R., Smith, A.L., and Levin, M.J.: Predictability of peak serum gentamicin concentration with dosage based on body surface area. *J. Pediatr.* 94:135–138, 1979.

103. Siegel, J.D., McCracken, G.H., Thomas, M.L., and Threlkeld, N.: Pharmacokinetic properties of netilmicin in newborn infants. *Antimicrob. Agents Chemother.* 15: 246–253, 1979.

104. Siegel, S.R., and Oh, W.: Renal function as a marker of human fetal maturation. *Acta Paediatr. Scand.* 65:481–485, 1976.

105. Silverman, W.A., Anderson, D.H., Blanc, W.A., et al.: A difference in morbidity rate and incidence of kernicterus among premature infants alloted to two prophylactic antibacterial regimens. *Pediatrics* 18:614–624, 1956.

106. Simon, H.J., and Axline, S.G.: Clinical pharmacology of kanamycin in premature infants. *Ann. N.Y. Acad. Sci.* 132:1020–1025, 1966.

107. Sirinavin, S., McCracken, G.H., and Nelson, J.D.: Determining gentamicin dosage in infants and children with renal failure. *J. Pediatr.* 96:331–334, 1980.

108. Starr, A., Amlie, R.N., Martin, W.H., and Sanders, S.: Development of auditory function in newborn infants revealed by auditory brainstem potentials. *Pediatrics* 60:831–839, 1977.

109. Stonestreet, B.S., and Oh, W.: Plasma creatinine levels in low birthweight infants during the first three months of life. *Pediatrics* 61:788–789, 1978.

110. Strausbaugh, L.J., Mandalerio, C.D., and Sande, M.A.: Comparison of four aminoglycoside antibiotics in the therapy of experimental *E. coli* meningitis. *J. Lab. Clin. Med.* 89:692–701, 1977.

111. Trujillo, H., Manotas, R., Londono, R., Ramirez, J.I., Uribe, A., deVidal, E.L., and Agudelo, N.: Clinical and laboratory studies with amikacin in newborns, infants and children. *J. Infect. Dis.* 134(Suppl.):S406–S411, 1976.

112. Varpela, E., Hietalahti, J., and Aro, M.J.T.: Streptomycin and dihydrostreptomycin medication during pregnancy and their effect on the child's inner ear. *Scand. J. Resp. Dis.* 50:101–109, 1969.

113. Vogelstein, B., Kowarski, A.A., and Lietman, P.S.: The pharmacokinetics of amikacin in children. *J. Pediatr.* 91:333–339, 1977.

114. Warner, W.A., and Sanders, E.: Neuromuscular blockade associated with gentamicin therapy. *JAMA* 215:1153–1154, 1971.

115. Wientzen, R.L., Nelson, J.D., McCracken, G.H., Kramer, R., and Prestidge, C.: Double-blind controlled trial of tobramycin or placebo therapy for acute pulmonary exacerbations of cystic fibrosis. Abstract, 18th Interscience Conference on Antimicrobial Agents and Chemotherapy, October 1–4, 1978, Atlanta, Ga.

116. Winkel, S., Bonding, P., Larsen, P.K., and Roosen, J.: Possible effects of kanamycin and incubation in newborn children with low birthweight. *Acta Paediatr. Scand.* 67:709–715, 1978.

117. Wolfe, J.C., and Johnson, W.D.: Penicillin-sensitive streptococcal endocarditis. *Ann. Intern. Med.* 81:178–181, 1974.

118. Yashioka, H., Monma, T., and Matsuda, S.: Placental transfer of gentamicin. *J. Pediatr.* 80:121–123, 1972.
119. Yashioka, H., Takimato, M., Matsuda, I., and Hattori, S.: Dosage schedule of gentamicin for chronic renal insufficiency in children. *Arch. Dis. Child.* 53:334–337, 1978.
120. Yow, M.D.: An overview of pediatric experience with amikacin. *Am. J. Med.* 62:954–958, 1977.
121. Yow, M.D., Tengg, N.E., Bangs, J., Bangs, T., and Stephenson, W.: The ototoxic effects of kanamycin sulfate in infants and children. *J. Pediatr.* 60:230–242, 1962.
122. Zoumboulakis, D., Anagnostakis, D., Arseni, A., Nicolopoulos, D., and Matsaniotis, N.: Gentamicin in the treatment of purulent meningitis in neonates and infants. *Acta Paediatr. Scand.* 62:55–58, 1973.

24

AMINOGLYCOSIDE USE IN THE COMPROMISED HOST

GERALD P. BODEY, Sr.   University of Texas System
Cancer Center, M. D. Anderson Hospital and
Tumor Institute, Houston, Texas

The management of infection in the compromised host continues to be difficult despite the great progress achieved in recent years. Many disease processes and therapeutic measures compromise host defense mechanisms against infection. The infectious complications associated with various types of patients, such as marrow transplant recipients, patients with hematological diseases, and patients receiving cancer chemotherapy, have been extensively reviewed [3,31,33]. The compromised host is susceptible to infections caused by a wide variety of bacterial and nonbacterial pathogens. However, the majority are gram-negative bacillary infections, most often due to *Escherichia coli, Klebsiella pneumoniae,* and *Pseudomonas aeruginosa.*

Extensive experience has been gained with the aminoglycoside antibiotics because of their broad spectrum of activity against gram-negative bacilli. Generally, aminoglycosides have been used in combination with one or more other antibiotics as initial therapy for presumed or proven infection in the compromised host. Most studies evaluating the efficacy of antibiotic regimens exclusively in compromised hosts have been limited to patients with blood dyscrasias and patients undergoing cancer chemotherapy. The major deficiency in host defense mechanisms in these patients has been neutropenia, which appears to be one of the most important defects influencing both the frequency of infectious complications and their response to therapy [4]. Because of the wealth of information available in patients with cancer, this review will concentrate on studies in this group of patients.

I. Aminoglycosides as Single-Agent Therapy

Prior to the introduction of gentamicin, few studies were conducted to evaluate the efficacy of antibiotics in the compromised host. Since carbenicillin became available, most studies in these patients have been with combinations of antibiotics. Hence, data evaluating aminoglycosides alone for the treatment of infections in compromised hosts are rather limited. Early studies with gentamicin and tobramycin used conventional dosage schedules. More recently, continuous infusion schedules of aminoglycosides have been advocated for neutropenic patients who often fail to respond to appropriate antibiotic therapy.

A. Intermittent Schedules

Gentamicin

Gentamicin represented a potential important advance in antibiotic therapy because its spectrum of activity included most Enterobacteriaceae and *Pseudomonas aeruginosa*. At the time of its introduction, the frequency of *Pseudomonas* infections was increasing, especially among neutropenic patients. This organism caused 35% of infectious deaths among patients with acute leukemia [18]. Although the polymyxin antibiotics had antipseudomonal activity in vitro, few neutropenic patients who were treated with these antibiotics for serious *Pseudomonas* infections survived [43].

Bodey et al. evaluated gentamicin alone during 122 episodes of infection in 101 cancer patients [7]. About half of the patients received it as initial therapy for presumed gram-negative bacillary infections. The remaining patients were given gentamicin after they had failed to respond to other antibiotic regimens. Fifty-two percent of the 99 gram-negative bacillary infections were cured (Table 1). Most of these infections were episodes of

Table 1   Response of Gram-Negative Bacillary Infections to Aminoglycoside Antibiotics

Antibiotic	Gentamicin		Tobramycin		Amikacin		Netilmicin	
	No.	% Cures	No.	% Cures	No.	% Cures	No.	% Cures
Klebsiella-Enterobacter	20	60	18	61	26	81	14	50
Serratia marcescens	22	45	2	50	—	—	3	0
Pseudomonas aeruginosa	22	45	10	40	10	80	12	42
Escherichia coli	8	75	11	36	26	81	16	100
Proteus spp.	3	100	3	100	—	—	—	—
Other gram-negative bacilli	3	67	2	50	7	57	8	50
Multiple organisms	21	38	13	77	22	41	10	50
Total gram-negative bacilli	99	52	59	58	91	63	63	59

pneumonia, septicemia, and cellulitis, and the predominant infecting organisms were *Klebsiella pneumoniae, Serratia marcescens,* and *P. aeruginosa.*

These patients' response to gentamicin therapy was related to their initial neutrophil count and to the change in neutrophil counts during their infection (Table 2). Although 79% of patients with an adequate neutrophil count (>1000 per mm^3) initially were cured, only 23% of patients with severe neutropenia (<100 per mm^3) initially were cured ($P < 0.001$). Among those patients whose neutrophil count increased during their infection, 74% were cured, whereas among those patients whose neutrophil count decreased or remained less than 1000 per mm^3, only 31% were cured ($P < 0.001$). The group with the poorest prognosis were those patients whose neutrophil count remained below 100 per mm^3. Gentamicin cured only 19% of the gram-negative bacillary infections in these patients.

The effect of changes in neutrophil count on response to gentamicin therapy is illustrated by the clinical course of a 49-year-old woman who was severely neutropenic following therapy for acute myelogenous leukemia (Figure 1). She initially received carbenicillin and cephalothin when she became febrile without benefit. Gentamicin was substituted for carbenicillin when *K. pneumoniae* was cultured from two blood specimens. Gentamicin was discontinued after 12 days because daily blood cultures were negative during this time, although the patient remained febrile. No circulating neutrophils were detected in her blood while she received gentamicin. *K. pneumoniae* was cultured again from two blood specimens while she was no longer receiving gentamicin. Because all the isolates of *K. pneumoniae* had been sensitive to gentamicin in vitro, this drug was reinstituted. Her temperature defervesced promptly and she recovered from her infection. However, at this time, her leukemia remitted and her neutrophil count recovered.

Table 2   Effect of Neutrophil Count on Response to Aminoglycosides in Gram-Negative Bacillary Infections[a]

Antibiotic: Schedule:	Gentamicin Intermittent		Tobramycin Intermittent		Amikacin[b] Continuous		Netilmicin[b] Continuous	
	No.	% Cures	No.	% Cures	No.	% Cures	No.	% Cures
Total	63	52	59	58	70	74	53	62
Neutrophil count/mm^3								
<100	22	23	21	24	13	69	11	73
101–1000	17	53	10	70	14	79	8	50
>1000	24	79	28	79	43	74	34	59
Change in neutrophil count during therapy								
Decreased	32	31	23	39	—	—	15	47
Increased	31	74	36	69	—	—	38	66
<100, increased	6	33	14	29	7	100	6	83
<100, unchanged	16	19	7	14	6	33	5	60

[a]Data on neutrophil counts were not available from all patients.
[b]Patients whose neutrophil counts were >1000 per mm^3 received these antibiotics on an intermittent schedule.

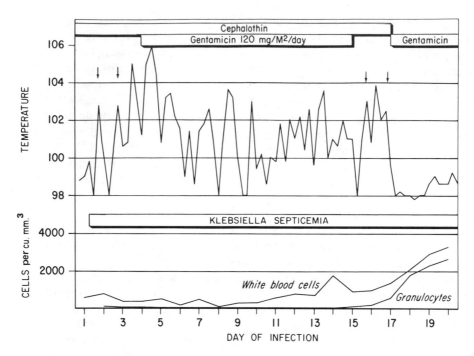

Figure 1   Case of a neutropenic patient with *Klebsiella* septicemia who failed to respond to gentamicin therapy. A second course of gentamicin therapy was successful because it was associated with return of her neutrophil count to normal levels. The arrows indicate when *K. pneumoniae* was cultured from her blood. [Reprinted from Bodey et al., *J. Infect. Dis.* 124(Suppl.):S174–S179, 1971; by permission of The University of Chicago Press. Copyright 1971, The University of Chicago.]

Gentamicin was used to treat gram-negative bacillary infections in 33 cancer patients, most of whom had acute leukemia, with similar results [16,46]. However, the majority of these patients received other antibiotics in combination with gentamicin. Most of the infections were caused by *Escherichia coli*, *Klebsiella* spp., and *P. aeruginosa*. Only 18 (55%) of these patients were cured of their infections. Although the effect of neutrophil count was not considered, these investigators noted that only those patients who achieved a remission of their leukemia or "control of the underlying disease" were cured of their infection. One of these investigators concluded with respect to *Pseudomonas* infections in leukemia patients that "the critical factor for success of therapy in our experience has been induction of bone marrow remission, with repopulation of the peripheral blood with normal polymorphonuclear leukocytes. We have yet to document a clinical and bacteriologic cure of Pseudomonas bacteremia in a patient with unremitting leukemia."

Jackson and Riff examined the results of gentamicin alone in 21 patients with *Pseudomonas* bactermia [21]. None of 7 patients with "rapidly fatal disease" responded to gentamicin, whereas 6 of 14 patients with less serious diseases responded. Five of the former 7 patients had persistent bacteremia regardless of peak serum concentrations varying from less than 1 to more than 4 μg/ml of gentamicin. This was not true for the remaining patients. Bacteremia persisted in 4 of 5 patients whose peak serum concentration was less than 2 μg/ml but in none of 6 patients whose peak serum concentration was at least 4 μg/ml. Among the 14 patients without rapidly fatal disease, none of 3 whose serum concentration of gentamicin was less than 1 μg/ml survived, 2 of 6 whose serum concentration was 1 to 2 μg/ml survived, and 4 of 5 patients whose serum concentration exceeded 4 μg/ml survived.

Tobramycin

Tobramycin was considered to be of potentially greater value for the neutropenic patient because it was more active than gentamicin in vitro against strains of *P. aeruginosa* [38]. In retrospect, the doses of gentamicin used in initial studies had been conservative; therefore, many investigators utilized higher doses of tobramycin in their studies. Valdivieso et al. treated 82 infections in cancer patients with tobramycin [41]. The majority of these patients received tobramycin after failing to respond to other antibiotic regimens. A dose of 50 mg/m^2 was administered intravenously over a 2 hr period every 6 hr. The cure rate was 58% for the 59 gram-negative bacillary infections (Table 1). The majority of infections were episodes of pneumonia and septicemia, and the most frequent etiologic organisms were *K. pneumoniae, P. aeruginosa,* and *E. coli.* Despite the superiority of tobramycin over gentamicin against *P. aeruginosa* in vitro, this has not resulted in substantially greater efficacy against *Pseudomonas* infections in the compromised host.

As with gentamicin, response to tobramycin therapy was related to the patients' neutrophil counts (Table 2). The cure rate among patients with an adequate neutrophil count initially was 79%, whereas the cure rate among patients with severe neutropenia was only 24% (P < 0.001) Those patients whose neutrophil count increased during their infection had a higher cure rate than those patients whose neutrophil count decreased or remained inadequate (69% vs. 39%, P = 0.04). Only 1 of 7 patients (14%) who remained severely neutropenic during his infection was cured with tobramycin. Response to tobramycin was related to its in vitro activity against some of the gram-negative bacilli causing the infections. Fourteen of 28 (50%) infections caused by gram-negative bacilli that were inhibited in vitro by 0.39 μg/ml or less of tobramycin were cured, whereas only 2 of 8 (25%) infections caused by gram-negative bacilli that were inhibited by 0.78 to 12.5 μg/ml were cured. After at least 3 days of therapy, serum was collected from 17 patients immediately after an infusion of tobramycin. Five of the 7 patients who had a serum concentration greater than 4.0 μg/ml (5.1 to 10.5 μg/ml) were cured, whereas only 2 of 8 patients who had a lower serum concentration (2.2 to 3.6 μg/ml) were cured.

B. Continuous Therapy

Since aminoglycoside therapy of infections in neutropenic patients proved to be suboptimal, different schedules were explored in an attempt to improve their efficacy. The failure of infections in neutropenic patients was not solely related to the in vitro sensitivity of the infecting organisms, the peak serum concentrations of antibiotic, or to the dose of drug administered. Some patients failed to respond despite adequate peak serum concentrations and infecting organisms that were highly susceptible to the aminoglycosides in vitro.

Data became available which suggested that better results might be obtained in the neutropenic patient by maintaining a constant serum concentration. Rolinson exposed carbenicillin-sensitive strains of *P. aeruginosa* to this antibiotic for several hours, after which the residual antibiotic was destroyed by penicillinase [35]. The surviving organisms began to proliferate after only a brief lag phase. Similar effects have been obtained when cephalothin-sensitive strains of *E. coli, K. pneumoniae,* and *Proteus mirabilis* have been exposed to cephalothin [8]. These data indicate that unlike gram-positive cocci, gram-negative bacilli recover rapidly from the effects of these antibiotics. Whether aminoglycosides produce the same effects has not been clearly demonstrated because of difficulties in destroying these antibiotics once they have been introduced into culture media.

Different schedules of carbenicillin were used to treat experimental *Pseudomonas* infections in mice [35]. A high-dose intermittent schedule failed to eradicate these infections. When the serum concentration was high, the number of organisms at the site of infection decreased, but several hours later when the serum concentration had decreased substantially, the number of organisms increased to the pretreatment level. However, when a more frequent dosage schedule of carbenicillin was utilized, the antibiotic effectively destroyed the organisms and eradicated the infection.

The neutropenic patient is unable to mount an adequate inflammatory response at a site of infection and thus is primarily dependent upon antibiotic therapy for control of infection. Pharmacological studies indicate that when aminoglycosides are administered by conventional 6–8 hr schedules there can be intervals between doses when the serum concentration falls below the concentration required to inhibit the infecting organism in vitro. For example, serum concentrations were determined in patients who had received tobramycin at a dose of 50 mg/m^2 over 2 hr every 6 hr [19]. On day 3 of therapy, the peak serum concentration following a dose was $5.6 \ \mu\text{g/ml}$, but at 4 hr the serum concentration was $2.6 \ \mu\text{g/ml}$ and at 6 hr it was only $1.6 \ \mu\text{g/ml}$. To obtain maximum destruction of infecting organisms, it is likely that the serum concentration must be considerably higher than the minimum inhibitory concentration (MIC). It is possible that, in neutropenic patients receiving intermittent schedules of aminoglycoside therapy, the infecting organisms are able to proliferate again during intervals when the concentration of antibiotic is inadequate, as was demonstrated with carbenicillin in the animal model. Hence, continuous infusion schedules have been investigated with the newer aminoglycosides in these patients.

Pharmacology studies were instituted with gentamicin, tobramycin, and sisomicin in neutropenic patients receiving these drugs as therapy for presumed or proven infections [5]. The dosage schedules and average serum concentrations after the first 6 hr are shown in Table 3. The daily dose of drug was adjusted to try to maintain an optimum

Table 3   Schedule for Continuous Infusions of Aminoglycoside Antibiotics

Antibiotic	Loading dose (mg/m^2)	Daily dose (mg/m^2)	Maximum dose administered (mg/m^2)	Average serum concentration $(\mu\text{g/ml})$[a]	Estimated optimum serum concentration $(\mu\text{g/ml})$
Gentamicin	60	300	360	4.4 ± 0.3	4–6
Tobramycin	60	300	360	2.2 ± 0.3	4–6
Sisomicin	40	160	200	2.6 ± 0.2	3–4
Amikacin	150	800	800	8.3 ± 1.6	12–16
Netilmicin	60	240	240	2.9 ± 0.5	?

[a] Average ± SE of the mean 6 hr after onset of therapy.

serum concentration. There was considerable variability in the serum concentration from patient to patient despite the fact that the dosage administered was calculated from the patients' body surface area (Table 4). Furthermore, the serum concentration varied from day to day in the same patient despite the use of infusion pumps and adjustments in the dosage to try to maintain a constant serum concentration. Interestingly, the mean serum concentrations of tobramycin were consistently lower than those of gentamicin even though the patients received a similar dosage schedule. Nephrotoxicity occurred in 3 of 33 patients studied (2 gentamicin, 1 tobramycin). A patient who received gentamicin had progressive renal failure for 6 days after the drug was discontinued at which time he died of his infection.

Amikacin

The introduction of amikacin presented an opportunity to assess the efficacy of the continuous infusion schedule in neutropenic patients [9]. Neutropenic patients with presumed gram-negative bacillary infection received the drug according to the schedule in Table 3. Cancer patients with adequate neutrophil counts received a dosage schedule of 150 mg/m^2 administered intravenously over 30 min every 6 hr. Approximately one-half of the patients entered on this study had failed to respond to prior antibiotic regimens which did not include an aminoglycoside. The majority of 134 infections were episodes of pneumonia, urinary tract infection, septicemia, and cellulitis. Most of the 91 gram-negative bacillary infections were caused by *E. coli, K. pneumoniae,* and *P. aeruginosa* (Table 1). The overall cure rate for gram-negative bacillary infections was 63%. The cure rates for patients with neutropenia receiving continuous infusion therapy and for patients with adequate neutrophil counts receiving intermittent therapy were similar (Table 2). All 7 patients who had an initial neutrophil count of less than 100 per mm^3 that increased during their infection were cured with amikacin. However, only 2 of the 6 patients with persistent severe neutropenia were cured with the continuous infusion schedule.

Pharmacology studies were conducted after patients had received 3-4 days of amikacin therapy by either schedule [10]. The average peak serum concentration with the intermittent schedule was 18.3 μg/ml and the average serum concentration was maintained at about 2.5 μg/ml. The serum concentration was maintained above 8.0 μg/ml in most patients with the continuous infusion schedule. Response to therapy in neutropenic patients who received amikacin by continuous infusion was related to the serum concentration. The cure rate was 83% for the patients in whom the serum concentration was maintained above 10 μg/ml, but it was 63% for the patients in whom the serum concentration was maintained below 10 μg/ml. In vitro sensitivities were available for 25 of the 27 gram-negative bacilli causing infection in patients who received the continuous infusion schedule. Six of the 12 (50%) infections caused by organisms with MIC values of greater

Table 4 Serum Concentrations During Continuous Infusions of Aminoglycoside Antibiotics

	Gentamicin	Tobramycin	Sisomicin	Netilmicin
Range, day 1	2.6–6.2	1.0–3.7	1.3–3.6	0.8–4.1
Range, day 7	1.6–5.6	2.6–4.5	1.1–3.5	1.6–6.2
Widest range in the same patient	1.6–8.2	1.7–5.4	1.7–7.2	1.9–4.8

than 0.39 µg/ml were cured, whereas 11 of 13 (85%) infections caused by organisms with MIC values of 0.39 µg/ml or less were cured. Hence, response to amikacin was related to the in vitro sensitivities of the infecting organisms.

Interestingly, the frequency of nephrotoxicity was lower among those patients who received amikacin by continuous infusion. Considering only patients with normal renal function initially, 21% of those receiving amikacin by the intermittent schedule developed azotemia (blood urea nitrogen >20 mg/dl or serum creatinine >1.5 mg/dl) compared to 6% of those who received amikacin by continuous infusion (P = 0.008). Serial serum concentrations were determined during 77 episodes when patients received amikacin by continuous infusion. Studies were conducted in patients with documented infection and in patients with a fever of undetermined origin. The frequency of nephrotoxicity was related to the serum concentrations of amikacin which was maintained in these patients. Six of 15 (40%) patients in whom the serum concentration was maintained above 15 µg/ml developed azotemia, whereas only 4 of 62 (6%) patients in whom the serum concentration was maintained below 15 µg/ml developed azotemia (P = 0.03). Similarly, audiotoxicity occurred in 3 (20%) of the former patients but in only 2 (3%) of the latter patients.

A major advantage of amikacin is its activity against gram-negative bacilli that are resistant to other aminoglycoside antibiotics. Some investigators have suggested that for this reason it should be selected as the aminoglycoside used in initial regimens for presumed infection in the compromised host. Only a limited experience has been obtained with the use of amikacin in compromised hosts infected by such resistant organisms. Twenty-four infections in cancer patients caused by gentamicin-resistant gram-negative bacilli were treated with amikacin [39]. Most of the infecting organisms were *P. aeruginosa* and the *Klebsiella-Enterobacter-Serratia* group. Most of the infections were septicemias and urinary tract infections. The overall cure rate was 54%. The cure rate was similar among neutropenic patients who received amikacin by continuous infusion (50%) and among patients with adequate neutrophil counts who received it by an intermittent schedule (58%). Five of the 11 (45%) episodes of septicemia were cured, but all 6 episodes that failed to respond occurred in neutropenic patients. The MIC of amikacin against 86% of the infecting organisms was 6.25 µg/ml or less, a concentration that can be readily achieved with intermittent and continuous infusion schedules.

Sisomicin

The effect of schedule of aminoglycoside on response in neutropenic patients was explored further in a prospective randomized trial of sisomicin [14]. One hundred and two patients were randomized to receive sisomicin by intermittent or continuous infusion during 121 evaluable febrile episodes. The dose of sisomicin by intermittent infusion was 30 mg/m² given over 30 min every 6 hr. Patients allocated to the continuous infusion schedule received a loading dose of 30 mg/m² over 30 min followed by 120 mg/m² daily. The dose of sisomicin was reduced if serum concentrations were above 4 µg/ml or the patients developed azotemia. However, no patients received more than 120 mg/m² daily. All of these patients had received prior antibiotic therapy (carbenicillin plus a cephalosporin) and had failed to respond after 48–72 hr.

The results of this study are presented in Table 5. There were 73 episodes of infection, which included 32 pneumonias, 24 septicemias, and 12 soft tissue infections. The majority of infections were caused by *K. pneumoniae* and *P. aeruginosa*. The cure rates for all identified infections were 46% with the intermittent schedule and 61% with the continuous

Table 5 Sisomicin in Neutropenic Patients: Continuous vs. Intermittent Therapy

	Intermittent		Continuous	
	Episodes	% Cures	Episodes	% Cures
Documented infections	35	46	38	61
Gram-negative bacillary infections	25	44	31	61
Klebsiella pneumoniae	9	67	10	80
Pseudomonas aeruginosa	3	0	5	40
Escherichia coli	3	67	0	—
Others	3	0	6	50
Multiple organisms	7	43	10	60
Septicemia	12	33	12	42
Pneumonia	18	60	14	50
Severe neutropenia	15	40	21	50

schedule. Considering only those infections caused by gram-negative bacilli, the cure rates were 44% and 61%, respectively. Among patients with severe neutropenia, the cure rate was 40% with the intermittent schedule and 50% with the continuous schedule. Although these results suggest that the continuous infusion schedule is more effective in neutropenic patients, none of these differences were statistically significant. In this randomized trial, among those patients with normal renal function initially, azotemia occurred in 13% of the patients receiving the continuous infusion schedule and in 21% of the patients receiving the intermittent schedule.

The efficacy of sisomicin was related to the in vitro sensitivity of the infecting organisms. Twenty of the 29 (69%) infections caused by organisms that had a MIC of less than 0.2 μg/ml were cured, whereas only 7 of the 22 (32%) infections caused by organisms that had a minimum inhibitory concentration of 0.2 μg/ml or greater were cured (P = 0.02). Similar results were obtained in patients who received either the intermittent or continuous infusion schedule. In this study, there was no correlation between response and serum concentration of sisomicin or serum bactericidal activity among patients who were treated with the continuous infusion schedule, although only a few patients were studied.

Netilmicin

Netilmicin is a new semisynthetic aminoglycoside, derived from sisomicin, that is more active than other aminoglycosides in vitro against *E. coli* and *K. pneumoniae*, including strains resistant to gentamicin [22]. In animal studies, it also has less audiotoxicity and nephrotoxicity. Because of these potential advantages, it has been evaluated in cancer patients, using an intermittent schedule in patients with adequate neutrophil counts and a continuous infusion schedule in neutropenic patients (Table 3). The intermittent dosage schedule was 60 mg/m^2 administered intravenously over 30 min every 6 hr. Pharmacology studies indicated that the continuous infusion maintained the serum concentration between 2 and 3 μg/ml [45]. Again, there were substantial differences in the serum concentrations in patients receiving the same dose by continuous infusion. Also, the serum concentration fluctuated considerably from day to day in the same patient (Table 4).

One hundred episodes of documented infection in 92 cancer patients were evaluable for response [44]. Thirty-four of these episodes occurred in neutropenic patients. Over 60% of the patients received netilmicin as initial therapy. The most common types of infection were pneumonias, urinary tract infections, septicemias, and soft tissue infections. Sixty-three of the 100 identified infections were caused by gram-negative bacilli, and the cure rate for these infections was 59% (Table 1). Considering only infections caused by single gram-negative bacilli, the cure rate was 59% for patients with adequate neutrophil counts who received netilmicin by the intermittent schedule (Table 2). The cure rate for patients with severe neutropenia who received netilmicin by the continuous infusion schedule was 73%. This cure rate among patients with severe neutropenia was impressive.

As was true with the other aminoglycosides, response to netilmicin was related to the in vitro sensitivity of the infecting organisms. Eight of the 9 infections caused by organisms that had an MIC value of 0.012 to 0.025 μg/ml were cured compared to 3 of 9 infections caused by organisms that had a minimum inhibitory concentration of 0.78–1.56 μg/ml. There was no correlation between the serum concentration and cure rate for patients receiving netilmicin by continuous infusion. The ratio between the serum concentration and minimum inhibitory concentration of the infecting organism was determined in 15 neutropenic patients who received continuous infusion therapy. Eight of 10 patients whose ratio was greater than 3 were cured, whereas only 1 of 5 patients were cured whose ratio was less than 3 (P = 0.09).

The importance of dosage schedule can be appreciated more fully if the results with gentamicin and tobramycin are compared to those with amikacin and netilmicin (Table 2). The studies are not entirely comparable, since the relative distribution of sites of infection and causative organisms are not identical, a problem that also exists with most randomized comparative trials, however. Among patients with an initial neutrophil count of less than 100 per mm^3, the cure rate was 23% for the 43 patients who received intermittent gentamicin or tobramycin, compared to 71% for the 24 patients who received continuous amikacin or netilmicin (P < 0.001). Considering only those patients whose neutrophil count increased from less than 100 per mm^3 to greater than 100 per mm^3 during their infection, the cure rates were 30% and 92%, respectively (P = 0.002). The results were suboptimal with both schedules among patients with persistent severe neutropenia, although they were somewhat better with the continuous schedule (17% vs. 45%, P = 0.19).

II.   Aminoglycosides in Combination with Other Agents

The majority of therapeutic studies with aminoglycoside antibiotics in cancer patients have used them in combination with other antibiotics, primarily semisynthetic penicillins and cephalosporins. Several reasons exist for this practice. Although the aminoglycosides provide a broad spectrum of antibacterial activity, their spectrum does not include all the pathogens potentially infecting these patients. When used alone, the aminoglycosides have been suboptimal in many neutropenic patients. Recent evidence indicates that synergistic combinations of antibiotics appear to provide optimum efficacy, especially in neutropenic patients.

Klastersky has emphasized the importance of synergism in antibiotic therapy. In an early study, he treated patients with cephalothin plus gentamicin or ampicillin plus gentamicin [26]. Sixteen of the 28 patients (58%) responded to the former combination, and 18 of the 27 patients (67%) responded to the latter combination. The antibiotics used were tested in vitro by the agar dilution checkerboard method for synergistic activity against the infecting organisms. Nineteen of 24 (80%) infections were cured when the antibiotic combination used was found to be synergistic, but only 15 of 31 (49%) infections were cured when the antibiotic combination was not synergistic. Synergistic combinations resulted in higher inhibitory activity of the serum than did nonsynergistic combinations.

In a subsequent study [24], he evaluated combinations of gentamicin plus cephalothin (18 patients), gentamicin plus ampicillin (23 patients), and carbenicillin plus gentamicin (57 patients). Nineteen of the 98 patients had leukopenia. Forty of the 51 infections (78%) treated with a synergistic combination were cured, whereas only 25 of the 47 infections (53%) treated with a nonsynergistic combination were cured ($P < 0.01$). The geometric mean for the peak bactericidal activity of diluted serum from patients treated with synergistic combinations was 1:16, whereas that in the remaining patients was 1:4.

Anderson et al. reviewed their experience in 173 infections treated with antibiotic combinations to which the infecting organisms were sensitive [1]. Nearly all these patients received an aminoglycoside plus a semisynthetic penicillin or cephalosporin. The response rate was 80% for the 83 infections treated with a synergistic combination but was only 64% for the 90 infections treated with a nonsynergistic combination ($P < 0.05$). Among patients with "rapidly fatal disease," 73% of the 26 who received a synergistic combination responded compared to only 43% of the 28 who received a nonsynergistic combination ($P < 0.05$). Among patients with "ultimately fatal" and "nonfatal" underlying diseases, this difference was less impressive (82% vs. 70%). The importance of synergistic antibiotic combinations was most important in patients with neutropenia (<2000 per mm^3). Among these patients, 79% of 29 patients who received a synergistic combination responded compared to only 33% of 24 patients who received a nonsynergistic combination ($P < 0.001$).

A variety of antibiotic combinations have been evaluated for the treatment of infections in cancer patients, especially in neutropenic patients. Many of these studies have been designed as prospective randomized trials. Table 6 summarizes the results of several of the larger noncomparative studies in which gentamicin, tobramycin, or amikacin were combined with carbenicillin or a cephalosporin. Two studies have used a three-drug regimen. Although other studies have been conducted in compromised hosts, many of them have not been presented in a fashion to permit easy comparison.

The first reported study of the combination of gentamicin plus carbenicillin for the treatment of infections in the compromised host was reported by Rodriguez et al. [34]. Thirty-two infections in cancer patients, most of whom had neutropenia, were treated with this combination (Table 6). The majority of infections were episodes of septicemia and pneumonia, and the predominant organisms were the *Klebsiella-Enterobacter-Serratia* group. The overall cure rate and the cure rate for gram-negative bacillary infections were both 50%. All of the *Pseudomonas* and *Proteus* infections responded, whereas only about half of the other gram-negative bacillary infections responded. The majority of failures occurred among severely neutropenic patients with *Klebsiella-Enterobacter-Serratia* infections. Lack of response could not be correlated with the in vitro sensitivity of the infecting organisms or the serum concentrations of the antibiotics in these patients.

Schimpff et al. used the combination of gentamicin plus carbenicillin for initial therapy of presumptive infection in cancer patients with neutropenia [36]. The "improvement

Table 6 Combination Regimens Used to Treat Infections in Compromised Hosts

Regimen[a]	Infections No.	% Cures	GNB infections No.	% Cures	Septicemias No.	% Cures
Gent + carb	32	50	24	50	15	47
Gent + carb	60	55	41	54	26	50
Tobra + carb	125	70	62	69	8	63
Amik + ceph	47	83	14	57	17	65
Gent + carb + ceph	33	82	27	76	18	?
Amik + carb + cef	39	67	19	63	14	64

[a]Gent, gentamicin; carb, carbenicillin; ceph, cephalothin; cef, cefazolin; tobra, tobramycin; amik, amikacin.

rate" among the 60 patients who were found to have evidence of infection was 55% (Table 6). Fourteen of 21 (67%) patients with *Pseudomonas* infections responded to the combination. However, only 8 of 20 patients with infections caused by other gram-negative bacilli (primarily *E. coli* and *Klebsiella* sp.) and 3 of 6 patients with infections caused by gram-positive cocci responded. They concluded that "improvement occurred regardless of tumor type or the initial granulocyte count; improvement usually occurred before the granulocyte count rose to greater than 1000 per cubic millimeter. However, persistent response did depend on a return to normal hematopoietic status. . . ." This lack of correlation between neutrophil count and response was due to the effect of carbenicillin, since the response of infections caused by organisms resistant to carbenicillin was substantially lower among these neutropenic patients.

Tobramycin has been used in combination with several other antibiotics for the treatment of infections in neutropenic patients. Issell et al. studied the combination of tobramycin plus carbenicillin as initial therapy during 125 documented infections, of which 80 occurred in neutropenic patients [20]. Tobramycin was administered by the continuous infusion schedule shown in Table 3. The majority of infections were pneumonias, septicemias, and soft tissue infections. The predominant etiologic agents were *K. pneumoniae* and *P. aeruginosa*. Seventy percent of these infections were cured, including 69% of the gram-negative bacillary infections (Table 6). Nine (82%) of the 11 *Pseudomonas* infections and 17 (63%) of the 27 *Klebsiella* infections were cured with this combination. Eighty percent of the 20 gram-positive coccal infections also responded. Interestingly, 55% of 31 gram-negative bacillary pneumonias were cured. Overall, there was no correlation between response rate and initial neutrophil count. However, among patients with neutropenia, the cure rate was 76% for the patients whose neutrophil count increased compared to 60% for the patients whose neutrophil count remained less than 1000 per mm^3 (P = 0.005). Among the 26 patients with persistent severe neutropenia, the cure rate was 58%. The suboptimal effect of aminoglycosides in neutropenic patients appeared to be overcome in this study by the continuous infusion schedule of tobramycin. The cure rate was 76% for 41 infections caused by organisms sensitive to both antibiotics, 67% for 33 infections caused by organisms sensitive to tobramycin only, and 67% for 9 infections caused by organisms sensitive to carbenicillin only.

Infections in neutropenic patients		Change in PMN count: increase		Change in PMN count: no increase		Reference
No.	% Cures	No.	% Cures	No.	% Cures	
24	54	—	—	—	—	34
—	—	—	—	—	—	36
80	69	38	76	42	60	21
—	—	9	100	8	75	17
26	81	14	100	19	68	2
39	67	22	73	17	59	27

Tobramycin plus clindamycin was used to treat 31 episodes of infection in leukemic patients and 24 apparently were cured [13]. An additional 5 patients responded when co-trimoxazole was added. Although clindamycin was included for coverage against anaerobic infections, these occurred infrequently in this study. It is likely that most, if not all, of these patients had neutropenia, although this is not stated in the report.

Two studies have reported results with the combination of tobramycin plus cephalo-thin as initial therapy for presumed infection in neutropenic cancer patients. In one study, 20 patients were entered on study and 14 (70%) responded to therapy. However, there was no documented infection in 12 of the 20 episodes [11]. In the second study, 29 episodes of infection in 15 patients with acute leukemia were treated [42]. During 3 episodes, cephalothin was not used and during an additional 10 episodes carbenicillin was substituted for cephalothin because of failure to respond after 2-4 days of therapy. There were 21 "satisfactory responses," although apparently only 11 responded to tobra-mycin plus cephalothin. The majority of infections were septicemias and respiratory tract infections and most of the infecting organisms were gram-negative bacilli.

Amikacin plus cephalothin has been used as an empiric regimen for the treatment of 47 documented infections in neutropenic cancer patients [17]. The majority of infections were caused by gram-negative bacilli or multiple organisms and E. coli was the predomi-nant organism. The overall response rate was 83% in these documented infections; how-ever, only 8 of the 14 (57%) infections caused by single gram-negative bacilli responded (Table 6). Five of the 6 failures occurred in patients with persistent neutropenia. Indeed, 6 of 8 patients with bacteremia whose neutrophil count failed to increase did not respond to this regimen. Nephrotoxicity, presumably due to the antibiotic regimen, was observed in only 7% of episodes.

Two studies have reported results with three drug combinations in neutropenic cancer patients. Gentamicin, cephalothin, and carbenicillin was used to treat 33 documented bacterial infections in 23 patients [2]. Gram-negative bacilli were the etiologic agents in 27 of these episodes, predominantly Klebsiella spp. and E. coli. The majority of infections were pneumonias and septicemias. The overall cure rate was 82%, and the regimen ap-peared to be equally effective regardless of the site of infection or infecting organism. The response rate was similar for patients with neutrophil counts less than and greater than

500 per mm^3 (88% vs. 86%). Patients whose neutrophil count increased to greater than 500 per mm^3 had a response rate of 100%, whereas patients whose neutrophil count remained less than 500 per mm^3 had a response rate of 79%. Nephrotoxicity was observed during 25% of these episodes, but in about half of the patients it probably was not due to the antibiotics.

Klastersky et al. evaluated the combination of carbenicillin, cefazolin, and amikacin during 39 episodes of infection in cancer patients with neutropenia [27]. A favorable clinical response was obtained in 67% of these patients, including 12 (63%) of the 19 patients with gram-negative bacillary infections (Table 6). A response rate of 67% was obtained in patients with persistent severe neutropenia and in patients whose neutrophil count was greater than 100/mm^3. The neutrophil count increased in 22 patients during their infection and 16 (73%) responded to therapy. In the remaining 17 patients, the neutrophil count decreased or remained unchanged and 10 (59%) of these patients responded to therapy. Only 2 of 7 patients with pulmonary infections responded. Response to this combination was related to the in vitro activity of the antibiotics against the infecting organisms. Eight (67%) of 12 infections caused by organisms sensitive to all 3 antibiotics responded, whereas only 4 (36%) of 11 infections caused by organisms sensitive to 1 or 2 antibiotics responded. None of the 3 infections caused by organisms sensitive only to amikacin responded. Nephrotoxicity was not a problem with this combination since it occurred in only 4% of the patients.

It is difficult to derive firm conclusions from these studies because of the diversity of patients, sites of infection, infecting organisms, and antibiotic regimens. Most of these studies included only a few patients with gram-negative bacillary infections. Three antibiotic regimens may be superior to two antibiotic regimens, but the combination of carbenicillin plus tobramycin administered by continuous infusion was nearly as effective as the three-antibiotic regimens. Although the combination of gentamicin plus carbenicillin and cephalothin produced the best results by every parameter measured, it was associated with a rather high frequency of nephrotoxicity. Further studies of this combination are needed to confirm these results in a larger group of patients.

III. Comparative Studies of Antibiotic Combinations

A variety of prospective randomized trials of antibiotic combinations for initial therapy of presumed infection have been conducted in the compromised host. Most of the studies that have included an aminoglycoside in at least one arm are summarized in Tables 7 to 9. These studies can be divided into three categories: comparisons of regimens with and without an aminoglycoside, comparisons of regimens including the same aminoglycoside, and comparisons of regimens using different aminoglycosides. These studies are not entirely comparable because of the different eligibility criteria, antibiotic schedules, and types of patients entered.

A. Comparisons of Regimens With and Without an Aminoglycoside

The first prospective randomized trial of antibiotic combinations for the treatment of infections in cancer patients with neutropenia compared carbenicillin plus cephalothin (carb-ceph) to carbenicillin plus kanamycin (carb-kan) [32]. Since optimal results had not been obtained when aminoglycoside antibiotics were administered by intermittent infusion schedules, it was anticipated that the combination of carb-ceph would be more effective despite the broader antibacterial coverage provided by carb-kan. There were 98 evaluable infections entered on this study. Sixty percent of those treated with carb-ceph and 58% of those treated with carb-kan were cured (Table 7). The majority of infections were episodes of septicemia and pneumonia and the predominant organisms were *P. aeruginosa* and *Klebsiella* sp. Although carb-kan was less effective than carb-ceph against gram-negative bacillary infections, the difference was not statistically significant (48% vs. 63%). Fourteen of the 16 (88%) *Pseudomonas* infections were cured, evidence of response to carbenicillin, since neither of the other antibiotics had activity against these organisms. Excluding the *Pseudomonas* infections, the cure rates for the remaining gram-negative bacillary infections were 44% for carb-ceph and 25% for carb-kan. The effect of neutrophil count on response was determined for all infections except those caused by *P. aeruginosa*. The latter were excluded because they were responsive only to carbenicillin, which has been found to be effective even in severely neutropenic patients. Among these remaining infections, 67% of the 21 occurring in patients with adequate neutrophil counts responded compared to only 29% of the 24 occurring in patients with severe neutropenia. The cure rates among patients with severe neutropenia were 15% for carb-ceph and 45% for carb-kan. This study demonstrated that the two regimens were equally effective and both were suboptimal in neutropenic patients.

A prospective randomized trial of carbenicillin alone, gentamicin alone, or carbenicillin plus gentamicin was conducted in cancer patients with gram-negative bacillary infections [25]. Very few of these patients were neutropenic during their infection. The cure rate was highest among those patients who received the combination regimen, 83% vs. 57% for gentamicin alone and 50% for carbenicillin alone (Table 7). Considering only those infections caused by organisms that were sensitive in vitro to the antibiotics used, 65% of the 17 infections treated with carbenicillin responded, 57% of the 21 infections treated with gentamicin responded, and 91% of the 21 infections treated with the combination responded. The difference between the combination and gentamicin alone was the only significant difference (P < 0.02). Among patients with *Pseudomonas* infections, 9 of 10 responded to the combination, 4 of 9 responded to gentamicin, and 6 of 12 responded to carbenicillin. For those patients receiving carbenicillin plus gentamicin, the combination was synergistic against 68% of the organisms causing infections. The serum bactericidal activity was considerably higher in patients receiving the combination than in patients receiving either drug alone. The response of *Pseudomonas* infections to carbenicillin alone was lower in this study than in other studies, which may have been due to the every 8 hr schedule of drug administration. Also, the dose of gentamicin used was 6 to 8.5 mg/kg, which is higher than has been customarily used, but toxicity was observed infrequently despite this higher dose.

Klastersky et al. compared carbenicillin plus cephalothin (carb-ceph) to carbenicillin plus cephalothin plus gentamicin (carb-ceph-gent) as initial therapy for presumptive gram-negative bacillary infections in cancer patients [28]. Only 14 of these 81 patients had leukopenia. A favorable result was obtained in 80% of the patients who received carb-ceph and in 76% of the patients who received carb-ceph-gent for their infections (Table 7). The majority of patients in both groups had septicemia and the cure rates were 80% and 79%, respectively. The results in respiratory infections were less impressive, which has been true in most studies. A wide variety of gram-negative bacilli was responsible for these infections, and although there were differences in response rates between the two regimens for some bacteria, none were statistically significant, because of small numbers

Table 7 Comparisons of Antibiotic Regimens With and Without an Aminoglycoside

Regimen	GNB infections		Pseudomonas-Proteus	
	No.	% Cures	No.	% Cures
1. Carb + ceph vs. carb + kan				
Carb + ceph	16	63	7	86
Carb + kan	25	48	9	89
2. Carb vs. gent vs. carb + gent				
Carb	22	50	17	53
Gent	23	57	12	58
Carb + gent	23	83	15	87
3. Carb + ceph vs. carb + ceph + gent				
Carb + ceph	40	80	20	75
Carb + ceph + gent	41	76	20	75
4. Carb (ceph) vs. carb (ceph) + gent				
Carb	—	—	12	83
Carb + gent	—	—	14	93
Ceph	11	64	0	—
Ceph + gent	12	75	0	—
5. Carb + ceph vs. carb + gent vs. ceph + gent				
Carb + ceph	50	50	8	50
Carb + gent	83	63	16	63
Ceph + gent	64	52	13	38

[a] Defined in various studies as <1000 PMN/mm^3 or <500 PMN/mm^3 or not specifically defined.

in each group. Failure to respond was not related to suboptimal antibacterial activity of the patients' sera. Renal functional impairment occurred in one patient who received carb-ceph, but it occurred in 7 patients who received carb-ceph-gent. Hence, in this group of patients, the addition of gentamicin did not improve the efficacy of carb-ceph but did appear to increase the frequency of nephrotoxicity.

An attempt was made to ascertain whether a combination regimen was more effective than a single antibiotic against susceptible organisms causing infection in neutropenic patients [6]. All patients were treated with carbenicillin plus cephalothin when they became febrile. When the infecting organism was identified but no later than 48 hr after onset of fever, the patients were entered on study. Patients with *Pseudomonas* and *Proteus* infections were randomly allocated to receive carbenicillin alone or carbenicillin plus gentamicin (carb-gent). Patients with infections caused by other gram-negative bacilli that were susceptible to cephalothin in vitro were randomly allocated to receive cephalothin or cephalothin plus gentamicin (ceph-gent). The cure rates for *Pseudomonas* and *Proteus* infections were 83% with carbenicillin alone and 93% with carb-gent (Table 7). The cure rates for other gram-negative bacillary infections were 64% with cephalothin alone and 75% with ceph-gent. When the two groups are combined, 74% of the patients

Klebsiella-Enterobacter-Serratia		Septicemia		Respiratory		Neutropenia[a]		Reference
No.	% Cures	No.	% Cures	No.	% Cures	No.	% Cures	
								32
6	33	10	60	23	52	33	52	
13	31	15	53	25	52	28	46	
								25
0	–	4	50	9	44	5	20	
7	71	3	33	12	58	4	25	
1	100	5	80	10	80	3	33	
								28
10	90	20	80	9	66	9	55	
8	62	19	79	13	54	5	60	
								6
0	–	5	60	–	–	7	57	
0	–	9	88	–	–	10	90	
3	67	7	57	–	–	9	67	
8	63	7	71	–	–	9	67	
								37
15	60	–	–	–	–	135	72	
22	55	–	–	–	–	156	71	
17	35	–	–	–	–	162	68	

receiving a single antibiotic were cured compared to 81% of the patients receiving a combination. Seven of 12 (58%) episodes of septicemia were cured with a single antibiotic and 13 of 16 (81%) were cured with a combination. Among the patients who had severe neutropenia at the onset of their infection, 6 of 11 (55%) who received a single antibiotic were cured, whereas 10 of 14 (71%) who received a combination were cured. The patients' neutrophil counts either decreased or remained the same during 32 infections. Twelve of 17 (71%) who received a single antibiotic were cured, whereas 13 of 15 (87%) who received a combination were cured. Although none of these differences were statistically significant, the results suggest that combinations may be preferable to single antibiotics, especially in patients with the poorest prognosis. These results are basically in agreement with those of Klastersky et al. [25].

The European Organization for Research on Treatment of Cancer conducted a prospective randomized study of carbenicillin (or ticarcillin) plus cephalothin (carb-ceph), carbenicillin (or ticarcillin) plus gentamicin (carb-gent), and cephalothin plus gentamicin (ceph-gent) for the treatment of presumed infection in cancer patients with neutropenia [37]. There were 625 febrile episodes, of which 396 were due to infection. Considering only those infections in which the etiologic agents were identified, 68% were caused by *E. coli, Staphylococcus aureus, Klebsiella* spp., and *P. aeruginosa*. The majority of infec-

Table 8 Comparisons of Regimens Including the Same Aminoglycoside Antibiotics

Regimen	Total infections		GNB infections		Pseudomonas-Proteus	
	No.	% Cures	No.	% Cures	No.	% Cures
Gent + chlora vs. gent + ceph						
Gent + chlora	28	64	14	57	0	—
Gent + ceph	27	78	14	93	0	—
Amik + pen vs. amik + carb						
Amik + pen	—	—	54	56	27	48
Amik + carb	—	—	63	64	22	59

tions were septicemias and infections of the respiratory tract, oral cavity, and soft tissues. Only 22 of the 305 organisms tested in vitro were resistant to both antibiotics used in the therapeutic regimens. Carb-ceph had a disproportionately large share of these resistant organisms. There were no significant differences in the efficacy of the regimens tested and no differences were observed when carbenicillin was substituted for ticarcillin. The overall cure rates for documented infections were 72% with carb-ceph, 71% with carb-gent, and 68% with ceph-gent (Table 7).

These authors also found that the initial neutrophil count and subsequent change were

Table 9 Comparisons of Regimens Using Different Aminoglycoside Antibiotics

Regimen	Total infection		GNB infections		Pseudomonas-Proteus	
	No.	% Cures	No.	% Cures	No.	% Cures
1. Amik vs. tobra						
Amik	60	55	29	69	7	57
Tobra	51	57	18	67	5	40
2. Carb + amik vs. carb + gent						
Carb + amik	50	72	—	—	6[a]	67
Carb + gent	47	83	—	—	5[a]	20
3. Carb + gent vs. carb + amik vs. carb + siso						
Carb + gent	115	67	55	67	9	78
Carb + amik	102	68	44	64	11	82
Carb + siso	93	67	43	74	12	83

[a] Septicemia only.
[b] GNB septicemia only.

Klebsiella-Enterobacter-Serratia		Septicemia		Respiratory		Neutropenia		Reference
No.	% Cures	No.	% Cures	No.	% Cures	No.	% Cures	
								40
4	50	8	50	12	67	25	60	
8	100	8	88	14	64	22	82	
								29
17	65	25	52	31	45	0	—	
17	59	27	70	19	42	0	—	

the most important prognostic factors. Patients with persistent severe neutropenia had a cure rate of only 44%. Patients with initial severe neutropenia whose neutrophil count subsequently increased during their infection had a cure rate of 87%. Carb-gent was the most effective regimen among patients who had an initial neutrophil count of 500 per mm^3 or less which did not increase during their infection. The cure rates were 62% for carb-gent, 39% for carb-ceph, and 44% for ceph-gent (P = 0.08). Only 22% of patients whose neutrophil count did not increase by at least 100 per mm^3 were cured, whereas 88% of patients whose neutrophil count increased by greater than 100 per mm^3 were cured.

Klebsiella-Enterobacter-Serratia		Septicemia		Respiratory		Neutropenia		Reference
No.	% Cures	No.	% Cures	No.	% Cures	No.	% Cures	
								15
7	71	11	64	24	33	0	—	
8	75	6	50	32	50	0	—	
								30
9[a]	44	23[b]	65	—	—	—	—	
6[a]	67	17[b]	59	—	—	—	—	
								23
32	59	27	44	44	45	107	67	
20	55	29	72	40	45	86	69	
22	64	24	75	30	50	89	67	

The carb-ceph regimen was least effective against *E. coli* infections. Only 44% of 18 infections responded to this regimen compared to 73% of 30 infections that responded to carb-gent and 67% of 27 infections that responded to ceph-gent ($P < 0.05$). This difference was due primarily to the fact that some strains of *E. coli* treated with carb-ceph were resistant to both antibiotics. Fewer *Klebsiella* infections responded to ceph-gent (35%) than to the other two regimens (60% and 55%). Ceph-gent was less effective against *Pseudomonas* infections (38%) than the other two regimens (carb-ceph 50%, carb-gent 63%).

Renal dysfunction occurred in 16% of patients who received ceph-gent, in 6% of patients who received carb-ceph, and in 3% of patients who received carb-gent ($P < 0.05$). Although most differences were marginal, the combination of carb-gent appeared to be the most effective and least toxic regimen in this cooperative group study.

B. Comparisons of Regimens Including the Same Aminoglycoside Antibiotics

A comparative study of gentamicin plus cephalothin (gent-ceph) versus gentamicin plus chloramphenicol (gent-chlora) was conducted during 55 episodes of infection in cancer patients with neutropenia [40]. All patients were treated initially with carbenicillin plus cephalothin but were entered on this study after 48 hr if they were found to be infected with an organism resistant to cephalothin in vitro (excluding *Pseudomonas* and *Proteus* infections) or if they had an infection but no etiologic agent could be identified. The major objective of this study was to determine whether a bacteriostatic agent (chloramphenicol) would be more effective in combination with gentamicin against infections caused by cephalothin-resistant bacteria than cephalothin in combination with gentamicin even though the organisms causing infection were resistant to cephalothin. It was anticipated that, although cephalothin was a bactericidal antibiotic, it would be of little benefit in these circumstances. Ancillary objectives were to determine whether cephalothin potentiated the nephrotoxicity of gentamicin and whether chloramphenicol prevented recovery from drug-induced neutropenia. Half of the patients with infections caused by cephalothin-resistant organisms were randomized to each regimen (Table 8). Fourteen of the 27 patients with infections in which the etiologic agent could not be determined were randomized to gent-chlora. The overall cure rates were 78% for gent-ceph and 64% for gent-chlora ($P = 0.25$). The majority of infections were pneumonias and septicemias. All the identified organisms causing infection were gram-negative bacilli, of which the most common was *E. coli*. Among the infections caused by cephalothin-resistant organisms, the cure rates were 93% for gent-ceph and 57% for gent-chlora ($P = 0.07$). Among the infections in which the etiologic agent could not be determined, the cure rates were 62% for gent-ceph and 71% for gent-chlora.

Six of 8 (75%) patients with severe neutropenia responded to gent-ceph, whereas only 4 of 12 (33%) similar patients responded to gent-chlora ($P = 0.07$). The response rates were not significantly different for gent-ceph and gent-chlora among patients whose neutrophil counts decreased during their infection (73% vs. 62%). Twenty-nine organisms (including multiple organisms from the same infection) were tested in vitro and response to therapy was related to their in vitro sensitivities. Among the patients infected with these organisms, 14 of 15 (93%) responded to gent-ceph, whereas only 8 of 14 (57%) responded to gent-chlora, even though all the organisms were resistant to cephalothin in vitro ($P = 0.03$). Twelve organisms proved to be resistant to chloramphenicol in vitro.

Among the patients infected with these organisms, all 7 responded to gent-ceph but only 1 of 5 responded to gent-chlora ($P < 0.01$). Azotemia occurred in 7% of patients who received gent-ceph and in 11% of patients who received gent-chlora. Unusual prolongation of myelosuppressive toxicity from antitumor agnets was not observed in patients receiving gent-chlora. Gent-chlora was not an effective regimen for the treatment of infections caused by cephalothin-resistant organisms. However, this regimen was more effective than previous studies of gentamicin alone in neutropenic patients, suggesting that the antagonism between these two antibiotics observed in vitro probably was not of consequence clinically [12]. Ceph-gent appeared to be more effective than previous studies of gentamicin alone in these neutropenic patients. This may be due to synergistic activity of the combination.

Klastersky et al. compared amikacin plus penicillin G (amik-pen) to amikacin plus carbenicillin (amik-carb) as initial therapy in cancer patients [29]. Neutropenic patients were excluded from this study. Of the 250 patients entered on this study, 117 had gram-negative bacillary infections. Bacteremia was documented during 44% of these infections. Considering only the gram-negative bacillary infections, responses were observed in 56% of patients receiving amik-pen and 64% of patients receiving amik-carb (Table 8). However, patients who developed superinfections were considered failures, although their initial infection may have responded. Among patients with bacteremia, the response rates were 52% and 70%, respectively. These regimens were least effective in patients with respiratory infections, curing less than half of the infections. Patients with infections caused by *E. coli* and *P. aeruginosa* had a better prognosis if treated with amik-carb than if treated with amik-pen. The cure rates for *E. coli* infections were 70% and 56%, respectively and for *Pseudomonas* infections were 67% and 46%, respectively. However, neither of these differences were statistically significant. Infections caused by other gram-negative bacilli responded equally well to both regimens (61% vs. 57%). An important factor influencing the outcome was whether the combination used demonstrated synergistic activity in vitro against the infecting organism. Hence, when synergism was demonstrated, amik-pen cured 70% of infections and amik-carb cured 62% infections. When synergism was absent, amik-pen cured 42% of infections and amik-carb cured 55% of infections. The overall difference in results between synergistic and nonsynergistic therapy was statistically significant (66% vs. 48%, $P < 0.05$). Synergism was demonstrated more often with amik-carb (64%) than with amik-pen (53%). Patients whose serum antibacterial activity was at least 1:8 responded more frequently than patients whose serum antibacterial activity was less than 1:8 (68% vs. 48%, $P = 0.05$). The authors suggest that if the patient's serum antimicrobial activity is $<1:8$ when an aminoglycoside is used alone for therapy of serious infections in nonneutropenic patients, a second drug should be added to obtain optimum results.

C. Comparisons of Regimens Using Different Aminoglycoside Antibiotics

A randomized study of tobramycin versus amikacin as initial therapy for presumed gram-negative bacillary infections was conducted in cancer patients, none of whom had neutropenia [15]. Patients received tobramycin during 51 documented infections and amikacin during 60 documented infections. The cure rate was 57% with tobramycin and 55% with amikacin (Table 9). The most frequent infections were pneumonias, urinary tract infections, and septicemias. Forty-seven infections were caused by gram-negative bacilli. Tobramycin cured 67% of gram-negative bacillary infections and amikacin cured 69%. All the gram-negative bacilli were sensitive to the antibiotic that the patient received. Cure rates with either aminoglycoside were not higher for those infections caused by the most sensitive organisms. Considering only those patients with normal renal function initially, the frequency of azotemia was 21% with tobramycin and 15% with amikacin. Hence, in this study, these two aminoglycosides were equal in efficacy and toxicity.

Lau et al. randomized "leukopenic patients" (<2000 neutrophils per mm^3) to receive carbenicillin plus amikacin (carb-amik) or carbenicillin plus gentamicin (carb-gent) during 273 evaluable febrile episodes [30]. An infecting organism was cultured during 97 episodes. Responses were observed in 83% treated with carb-gent and in 72% treated with carb-amik (Table 9). The majority of gram-negative bacillary infections were caused by *K. pneumoniae, E. coli,* and *P. aeruginosa.* No differences were observed between the two regimens in the treatment of these infections. The antibiotic combinations were tested against the infecting organisms in vitro for synergistic activity. The mortality rate was 18% among patients who received a regimen with synergistic activity, whereas it was 44% among patients who received a regimen without synergistic activity. The mortality rate was 71% among patients whose infecting organism was inhibited in vitro by only one of the antibiotics in the combination.

Ototoxicity was documented in 20% of patients receiving carb-amik and in 14% of patients receiving carb-gent. Nephrotoxicity occurred in 8% and 8.5%, respectively. The frequency of nephrotoxicity correlated with the serum concentrations of the aminoglycosides. Among patients who received carb-amik, 32% of those patients whose serum concentration exceeded 10 μg/ml developed nephrotoxicity, whereas only 2% of the remaining patients developed nephrotoxicity. Among patients who received carb-gent, 24% of patients whose "trough" serum concentration exceeded 2 μg/ml developed nephrotoxicity, whereas only 5% of the remaining patients developed this complication. These authors concluded that, although both regimens were equally effective, amikacin should be used routinely in situations where there is a high probability of infections being caused by gentamicin-resistant gram-negative bacilli.

The combinations of carbenicillin plus gentamicin (carb-gent), sisomicin (carb-siso), and amikacin (carb-amik) were studied as initial therapy for presumed infections in cancer patients [23]. All patients either were neutropenic or were expected to become neutropenic as a consequence of their cancer chemotherapy. The aminoglycosides were administered by continuous infusion schedules. There were 520 evaluable febrile episodes in 281 patients, but infection was documented during 310 episodes. The majority of infections were pneumonias and septicemias. Aerobic gram-negative bacilli caused 65% of those infections in which the etiologic agent could be identified. The response rates for documented infections were 67% for carb-gent, 68% for carb-amik, and 67% for carb-siso (Table 9). Relapses after therapy was discontinued occurred significantly more frequently with carb-siso (P <0.05). The response of septicemias was significantly lower for carb-gent (44%) than for carb-siso (75%) and carb-amik (72%) (P = 0.06, P = 0.03). Only 45%–50% of pneumonias responded to these regimens. Considering only those pneumonias caused by gram-negative bacilli, the cure rates were 29% for carb-gent, 35% for carb-amik, and 57% for carb-siso. There were no significant differences among the various regimens related to the various gram-negative bacilli causing infections. However, regardless of the regimen, *Pseudomonas* infections responded substantially better than did *Klebsiella* infections (85% vs. 55%, P = 0.01). The response rate for the 27 *E. coli* infections was 78%. Even among the 31 infections caused by multiple organisms, the response rate was 77%.

Half of the documented infections occurred in patients with severe neutropenia and they responded nearly as well as patients with higher initial neutrophil counts (63% vs. 72%). Patients whose neutrophil count increased during their infection had significantly higher response rates than did those whose neutrophil count remained stable or decreased (85% vs. 59%, P = 0.001). This difference was observed with all three regimens and all three were equally effective in these patients. The response rate was 83% among infections caused by organisms that were sensitive to both antibiotics in vitro but was only 66% among infections caused by organisms that were resistant to carbenicillin (P = 0.03). The response rates were examined related to neutrophil count and in vitro susceptibility of the infecting organism to carbenicillin. Combining all three regimens, the response rate for infections caused by carbenicillin-sensitive organisms was 100% among the 23 patients whose neutrophil count increased and 75% among the 41 patients whose neutrophil count did not increase (P = 0.05). The response rate for infections caused by carbenicillin-resistant organisms was 80% among the 30 patients whose neutrophil count did not increase (P = 0.05).

The majority of aerobic gram-negative bacilli were tested in vitro against the aminoglycoside that was used to treat the infection. All the organisms causing infections that were treated with sisomicin or amikacin were sensitive to these antibiotics, but 16% of the 32 organisms causing infections that were treated with gentamicin were resistant to 6.25 μg/ml or greater of this antibiotic. The response rates for each antibiotic combination were related to the in vitro sensitivities of the gram-negative bacilli causing infection. Combining all three regimens, the response rate was 78% for infections caused by organisms inhibited by 0.78 μg/ml or less of the aminoglycoside, but the response rate was only 43% for infections caused by more resistant organisms. Response was also related to the ratio between the serum concentration of the aminoglycoside and the concentration required to inhibit the infecting organism in vitro. If this ratio was greater than 10, the cure rate was 85%; if this ratio was 4–10, the cure rate was 67%; but if this ratio was less than 4, the cure rate was only 42% (P = 0.02).

Azotemia occurred in 7% of patients treated with carb-amik, 10% of patients treated with carb-gent, and 14% of patients treated with carb-siso. Azotemia occurred significantly more often among patients who had documented infections than among patients with fevers of unknown origin. Considering only patients with documented infections, carb-siso caused azotemia significantly more often than carb-amik (22% vs. 8%, P < 0.01). Considering both response to therapy and frequency of toxicity, the optimum serum concentration for amikacin is 12–16 μg/ml, for sisomicin is 2–4 μg/ml, and for gentamicin is 4–6 μg/ml. This study indicates that, for routine use, all three regimens are equally effective, although carb-siso causes more nephrotoxicity.

D. Summary of Combination Regimens

The majority of studies have failed to demonstrate any clear-cut superiority for any single antibiotic regimen. The results of all the studies cited in this chapter have been combined in Tables 10 and 11. This is not entirely satisfactory because of the many differences among the studies. Considering only *Pseudomonas* and *Proteus* infections, the cure rate was 71% for carbenicillin alone, 74% for carbenicillin plus an aminoglycoside, but only 48% for an aminoglycoside alone. The differences in response between carbenicillin and an aminoglycoside and between carbenicillin plus an aminoglycoside and an aminoglycoside are statistically significant (P = 0.01 and P < 0.001, respectively). Considering infec-

Table 10 Summary of Single vs. Combination Antibiotic Regimens

Regimen	Pseudomonas-Proteus infections		Klebsiella-Enterobacter-Serratia infections	
	No.	% Cures	No.	% Cures
Carbenicillin	73	71	—	—
Aminoglycoside	64	48	—	—
Carbenicillin + aminoglycoside	171	74	—	—
Cephalothin	—	—	34	65
Aminoglycoside	—	—	267	57
Cephalothin + aminoglycoside	—	—	41	58

tions caused by *K. pneumoniae, Enterobacter* spp., and *S. marcescens,* the combination of cephalothin plus an aminoglycoside was not substantially more effective than either cephalothin nor an aminoglycoside alone. Also, an aminoglycoside alone or cephalothin alone were about as effective for the treatment of these infections.

Six combination regimens have been used extensively as initial therapy for presumed infection in cancer patients (Table 11). The cure rates for all documented infections and for gram-negative bacillary infections are remarkably similar for all these regimens. Regimens that included an aminoglycoside were not more effective than carb-ceph, nor was any one regimen containing an aminoglycoside significantly better than any other. However, the combinations of carb-siso and ceph-gent may be associated with a higher frequency of nephrotoxicity.

IV. Conclusions

Although there is not total unanimity of opinion from the studies in compromised hosts, some conclusions are warranted. Aminoglycoside antibiotics are quite effective in most patients with adequate neturophil counts but produce suboptimum results in neutropenic patients when administered by intermittent schedules. The continuous infusion schedule of administration appears to improve the results in neutropenic patients. However, some patients still fail to respond, even though the infecting organism is sensitive to the aminoglycoside in vitro and adequate serum concentrations are maintained. Gentamicin, tobra-

Table 11 Summary of Therapy with Combination Regimens

Regimen	Total infections		GNB infections	
	No.	% Cures	No.	% Cures
Carb + ceph	187	69	106	63
Ceph + gent	189	69	90	60
Carb + gent	410	67	226	63
Carb + amik	152	69	107	64
Carb + siso	93	67	43	74
Carb + tobra	125	70	62	69

mycin, amikacin, and sisomicin are equally effective for routine use. Recent evidence in animals and humans suggests that nephrotoxicity may be less with tobramycin, a factor of potential importance to compromised hosts who experience multiple episodes of infection. Amikacin should be utilized routinely in those institutions where gentamicin-resistant gram-negative bacilli are a frequent cause of infection. Antibiotic combinations provide a broader spectrum of activity against those organisms potentially causing infection in the compromised host. Synergistic combinations may be more effective than are nonsynergistic combinations, and this appears to be of greater importance in the neutropenic patient. This concept is not universally accepted and further studies are desirable. The advantages of continuous infusions of aminoglycosides to neutropenic patients are less obvious when they are administered in combination with other agents. Despite the disadvantages associated with the aminoglycosides, their broad spectrum of activity and synergism with other antibiotics portends a continuing important role for them in the management of infection in the compromised host.

References

1. Anderson, E.T., Young, L.S., and Hewitt, W.L.: Antimicrobial synergism in the therapy of gram-negative rod bacteremia. *Chemotherapy* 24:45–54, 1978.
2. Bloomfield, C.D., and Kennedy, B.J.: Cephalothin, carbenicillin and gentamicin combination therapy for febrile patients with acute non-lymphocytic leukemia. *Cancer* 34:431–437, 1974.
3. Bodey, G.P.: Infections in cancer patients. *Cancer Treat. Rev.* 2:89–128, 1975.
4. Bodey, G.P., Buckley, M., Sathe, Y.S., and Freireich, E.J.: Quantitative relationships between circulating leukocytes and infection in patients with acute leukemia. *Ann. Intern. Med.* 64:328–340, 1966.
5. Bodey, G.P., Chang, H.Y., Rodriguez, V., and Stewart, D.: Feasibility of administering aminoglycoside antibiotics by continuous intravenous infusion. *Antimicrob. Agents Chemother.* 8:328–333, 1975.
6. Bodey, G.P., Feld, R., and Burgess, M.A.: β-Lactam antibiotics alone or in combination with gentamicin for therapy of gram-negative bacillary infections in neutropenic patients. *Am. J. Med. Sci.* 271:179–186, 1976.
7. Bodey, G.P., Middleman, E., Umsawasdi, T., and Rodriguez, V.: Infections in cancer patients—results with gentamicin sulfate therapy. *Cancer* 29:1697–1701, 1972.
8. Bodey, G.P., and Pan, T.: Effect of cephalothin on growth patterns of microorganisms. *J. Antibiot. (Tokyo)* 29:1092–1095, 1976.
9. Bodey, G.P., Rodriguez, V., Valdivieso, M., and Feld, R.: Amikacin for treatment of infections in patients with malignant diseases: administration by continuous intravenous infusion in the presence of neutropenia. *J. Infect. Dis.* 134(Suppl.):S421–S427, 1976.
10. Bodey, G.P., Valdivieso, M., Feld, R., and Rodriguez, V.: Pharmacology of amikacin in humans. *Antimicrob. Agents Chemother.* 5:508–512, 1974.
11. Cohen, J.D., and Miale, T.D.: Tobramycin and cephalothin for treatment of suspected sepsis in neutropenic children with cancer. *J. Infect. Dis.* 134(Suppl.):S175–S177, 1976.
12. Derrington, A.W.: Combinations of antibiotics. *Med. J. Aust.* 2:238–243, 1971.
13. Falk, R.H., Gillett, A.P., and Melikian, V.: Tobramycin and clindamycin in the treatment of febrile leukaemic patients. *J. Antimicrob. Chemother.* 3:317–322, 1977.
14. Feld, R., Valdivieso, M., Bodey, G.P., and Rodriguez, V.: A comparative trial of sisomicin therapy of intermittent versus continuous infusion. *Am. J. Med. Sci.* 274:179–188, 1977.
15. Feld, R., Valdivieso, M., Bodey, G.P., and Rodriguez, V.: Comparison of amikacin

and tobramycin in the treatment of infection in patients with cancer. *J. Infect. Dis.* 135:61–66, 1977.

16. Haghbin, M., Armstrong, A., and Murphy, M.L.: Intravenous gentamicin for infectious complications in children with acute leukemia. *J. Infect. Dis.* 124(Suppl.): S192–S197, 1971.

17. Hahn, D.M., Schimpff, S.C., Young, V.M., Fortner, C.L., Standiford, H.C., and Wiernik, P.H.: Amikacin and cephalothin: empiric regimen for granulocytopenic cancer patients. *Antimicrob. Agents Chemother.* 12:618–624, 1977.

18. Hersh, E.M., Bodey, G.P., Nies, B.A., and Freireich, E.J.: Causes of death in acute leukemia. A ten year study of 414 patients from 1954–1963. *JAMA* 193:105–109, 1965.

19. Horikoshi, N., Valdivieso, M., and Bodey, G.P.: Clinical pharmacology of tobramycin. *Am. Med. Sci.* 266:453–458, 1973.

20. Issell, B.F., Keating, M.J., Valdivieso, M., and Bodey, G.P.: Continuous infusion tobramycin combined with carbenicillin for infections in cancer patients. *Am. J. Med. Sci.*, 277:311–318, 1979.

21. Jackson, G.G., and Riff, L.J.: Pseudomonas bacteremia: pharmacologic and other bases for failure of treatment with gentamicin. *J. Infect. Dis.* 124(Suppl.):S185–S191, 1971.

22. Kabins, S.A., Nathan, C., and Cohen, S.: In vitro comparison of netilmicin, a semisynthetic derivative of sisomicin, and four other aminoglycoside antibiotics. *Antimicrob. Agents Chemother.* 10:139–145, 1976.

23. Keating, M.J., Bodey, G.P., Valdivieso, M., and Rodriguez, V.: A randomized comparative trial of three aminoglycosides—comparison of continuous infusions of gentamicin, amikacin and sisomicin combined with carbenicillin in the treatment of infections in neutropenic patients with malignancies. *Medicine* 58:159, 1979.

24. Klastersky, J., Cappel, R., and Daneau, D.: Clinical significance of in vitro synergism between antibiotics in gram-negative infections. *Antimicrob. Agents Chemother.* 2:470–475, 1972.

25. Klastersky, J., Cappel, R., and Daneau, D.: Therapy with carbenicillin and gentamicin for patients with cancer and severe infections caused by gram-negative rods. *Cancer* 31:331–336, 1973.

26. Klastersky, J., Cappel, R., Swings, G., and Vandenborre, L.: Bacteriological and clinical activity of the ampicillin/gentamicin and cephalothin/gentamicin combinations. *Am. J. Med. Sci.* 262:283–290, 1971.

27. Klastersky, J., Debusscher, L., Weerts-Ruhl, D., and Prevost, J.M.: Carbenicillin, cefazolin and amikacin as an empiric therapy for febrile granulocytopenic cancer patients. *Cancer Treat. Rep.* 61:1433–1439, 1977.

28. Klastersky, J., Henri, A., Hensgens, C., and Daneau, D.: Gram-negative infections in cancer. Study of empiric therapy comparing carbenicillin-cephalothin with and without gentamicin. *JAMA* 227:45–48, 1974.

29. Klastersky, J., Meunier-Carpenter, F., and Prevost, J.M.: Significance of antimicrobial synergism for the outcome of gram negative sepsis. *Am. J. Med. Sci.* 273:157–167, 1977.

30. Lau, W.K., Young, L.S., Black, R.E., Winston, D.J., Linne, S.R., Weinstein, R.J., and Hewitt, W.L.: Comparative efficacy and toxicity of amikacin/carbenicillin versus gentamicin/carbenicillin in leukopenic patients. A randomized prospective trial. (U.S. Amikacin Symposium, November 1976.) *Am. J. Med.* 212–219, 1977.

31. Levine, A.S., Graw, R.G., Jr., and Young, R.C.: Management of infections in patients with leukemia and lymphoma: current concepts and experimental approaches. *Semin. Hematol.* 9:141–179, 1972.

32. Middleman, E.A., Watanabe, A., Kaizer, H., and Bodey, G.P.: Antibiotic combinations

for infections in neutropenic patients. Evaluation of carbenicillin plus either cephalothin or kanamycin. *Cancer* 30:573–579, 1972.

33. Neiman, P.E., Thomas, E.D., Reeves, W.C., Ray, C.G., Sale, G., Lerner, K.G., Buckner, C.D., Clift, R.A., Storb, R., Weiden, P.L., and Fefer, A.: Opportunistic infection and interstitial pneumonia following marrow transplantation for aplastic anemia and hematologic malignancy. *Transplant. Proc.* 8:663–667, 1976.

34. Rodriguez, V., Whitecar, J.P., Jr., and Bodey, G.P.: Therapy of infections with the combination of carbenicillin and gentamicin. In *Proceedings of the 9th Interscience Conference on Antimicrobial Agents and Chemotherapy,* edited by Hobby, G.L. Washington, D.C.: American Society for Microbiology, 1970. *Antimicrob. Agents Chemother.,* 386–390, 1970.

35. Rolinson, G.N.: Plasma concentrations of penicillin in relation to the antibacterial effect. In *Biological Effects of Drugs in Relation to Their Plasma Concentration,* edited by Davies, D.S., and Prichard, B.N.C. New York, Macmillan, 1973, pp. 183–189.

36. Schimpff, S., Satterlee, W., Young, V.M., and Serpick, A.: Empiric therapy with carbenicillin and gentamicin for febrile patients with cancer and granulocytopenia. *N. Engl. J. Med.* 284:1061–1065, 1971.

37. The EORTC International Antimicrobial Therapy Project Group: Three antibiotic regimens in the treatment of infection in febrile granulocytopenic patients with cancer. *J. Infect. Dis.* 137:14–29, 1978.

38. Traub, W.H., and Raymond, E.A.: Evaluation of the in vitro sensitivity of tobramycin as compared with that of gentamicin sulfate. *Appl. Microbiol.* 23:4–7, 1972.

39. Valdivieso, M., and Bodey, G.P.: Amikacin therapy of severe infections produced by gram-negative bacilli resistant to gentamicin. *Am. J. Med. Sci.* 273:177–184, 1977.

40. Valdivieso, M., Bodey, G.P., Burgess, M.A., and Rodriguez, V.: Therapy of infections in neutropenic patients—results with gentamicin in combination with cephalothin or chloramphenicol. *Med. Pediatr. Oncol.* 2:99–108, 1976.

41. Valdivieso, M., Horikoshi, N., Rodriguez, V., and Bodey, G.P.: Therapeutic trials of tobramycin. *Am. J. Med. Sci.* 268:149–156, 1974.

42. Vincent, P.C., Jennis, F., Hilmer, R., and Fabre, S.: Tobramycin and cephalothin sodium in treatment of infected patients with acute leukemia. *J. Infect. Dis.* 134 (Suppl.):S170–S174, 1976.

43. Whitecar, J.P., Jr., Bodey, G.P., and Luna, M.: Pseudomonas bacteremia in cancer patients. *Am. J. Med. Sci.* 260:216–223, 1970.

44. Yap, B.S., and Bodey, G.P.: Netilmicin in the treatment of infections in patients with cancer. *Arch. Int. Med.* 139:1259–1262, 1979.

45. Yap, B.S., Stewart, D., and Bodey, G.P.: Clinical pharmacology of netilmicin. *Antimicrob. Agents Chemother.* 12:717–720, 1977.

46. Young, L.S.: Gentamicin: Clinical use with carbenicillin and in vitro studies with recent isolates of *Pseudomonas aeruginosa. J. Infect. Dis.* 124(Suppl.):S202–S206, 1971.

25

AMINOGLYCOSIDE KINETICS AND CLINICAL USE IN RENAL INSUFFICIENCY

ANDREW WHELTON   The Johns Hopkins University
School of Medicine, Baltimore, Maryland

The increasing interest and awareness of the clinical importance of drug therapy in renal functional impairment is a phenomenon of recent years. This new branch of clinical pharmacology and therapeutics has developed as a direct result of the remarkable advances that have been made in the conservative medical management and dialysis therapy of patients with chronic or acute forms of renal failure. It is of interest that at the present time that some 40,000 individuals in the United States are dependent upon chronic dialysis for survival [51]. This, of course, represents only a very small fraction of the general patient population who manifest some degree of clinically important renal failure. As the clinical volume of patients with renal insufficiency has expanded, so have the medical complications of these patients. One such complication that has received particular physician attention is the recognition that patients with renal failure have an increased susceptibility to any of a whole host of bacterial, fungal, or viral infections. As a result, the use of antibiotic chemotherapeutic agents in patients with some degree of renal impairment is a common day-to-day problem not only for the specialist in renal diseases, but for the entire cross section of practicing physicians and surgeons. It is also safe to project that the issue of drug therapy in renal failure will continue to expand and become more complicated.

Obviously, the focus of attention in this chapter is on the multifactorial considerations involved in the safe and effective use of the aminoglycoside group of antibiotics in individuals with gram-negative bacterial infections complicating acute and chronic forms of renal failure. It is important to point out at this juncture that in the consideration of antibiotic therapy in renal failure the clinician often has a substantial leeway of safety between the therapeutic and toxic serum concentrations of, for example, the penicillins or cephalosporins, but that nothing could be further from the truth when one considers the clinical use of aminoglycosides in renal failure. In the case of the aminoglycosides, there is in essence no appreciable safety buffer zone between the upper reaches of the therapeutic serum concentration range and the development of potentially toxic serum levels. Since the toxicity of the aminoglycosides is characteristically related to the dose and duration of therapy and manifests itself as ototoxicity, nephrotoxicity, and neurotoxicity, it is obvious that the clinician has to contend with extremely serious side effects when prescribing the aminoglycosides, as opposed to transient skin rashes or gastroenteritis that may be seen as the predominant side effects of other classes of antibiotics. Acknowledging the very serious nature of the dose-related toxic side effects of the aminoglycosides and recognizing that the latter antibiotics are dependent upon the kidney for elimination purposes, it is clear that the use of the aminoglycosides in renal failure must be undertaken with particular clinical care and caution.

A logical approach to the safe and effective use of aminoglycosides in renal failure requires a systematic review of all factors entering into clinical therapeutic decisions. The component parts, which make up the clinical data base, allow the development of logical decisions to be summarized and examined sequentially:

1. Identification of the patient's degree of renal functional impairment.
2. Assessment of the biologic serum half-life ($T\frac{1}{2}$) of the aminoglycoside antibiotic in the presence of normal renal function and in the setting of varying degrees of renal insufficiency. Identification of alternative routes of excretion or biotransformation of the aminoglycoside in renal failure, together with recognition of the systemic dose-related toxic side effects of the compound.
3. Quantification of the degree of serum protein binding of the aminoglycoside and evaluation of the influence of peritoneal dialysis or extracorporeal hemodialysis on the serum levels of the compound.
4. Development of reference drug dosage tables and nomograms for use in the setting of varying degrees of renal functional impairment to provide a means of therapeutic guidance in aminoglycoside dosing. In addition, specific comments about the avoidance of dosing rules of thumb for aminoglycosides in renal failure.

I. Renal Functional Assessment

No one single test of renal function will provide information pertinent to all pathophysiologic aspects of renal function [94]. A full evaluation of a patient's renal function would involve the correlation of information derived from the patient's medical history, physical examination, microscopic and chemical urinalysis, renal clearance evaluation, radiographic and isotopic examinations, and renal biopsy when necessary. It is clear that in the instances when reduction or adjustment in drug dosing is necessary as a result of renal functional impairment, the clinician will usually not have available a comprehensive review of kidney function tests as listed above. To obtain all this information simply as a guide to drug therapy would be time consuming, costly, and quite unnecessary. Therefore, a quick, inexpensive, clinical guide to safe aminoglycoside therapy must be utilized. The most convenient and readily available clinical test for this purpose is the determination of serum creatinine and creatinine clearance (glomerular filtration rate). The advantage of creatinine clearance evaluation is that the latter results provide renal functional information of a distinctly quantitative nature, allowing one to follow the therapeutic response to the aminoglycoside management of either renal or extrarenal systemic gram-negative infections. It may be recalled that inulin clearance studies provide the classic and definitive means of measuring glomerular filtration rate (GFR), but the techniques and methodology of measuring inulin clearance are such that they are suitable only for clinical investigative purposes and not for routine quantitative renal functional assessment. Despite some minor discrepancies [1,12,86,94], determination of the creatinine clearance is realistically the best clinical guide to renal functional assessment throughout the spectrum that spans health to severe renal impairment. The creatinine clearance is in large part independent of urine flow rate and hence no special considerations are required in terms of the patient's state of hydration. In the vast majority of stable, chronic renal failure patients, the serum creatinine remains relatively constant throughout a 24-hr period. If changes occur they are within the framework of 10% or less [1,86]. Within this framework, we can estimate the patient's GFR integrated throughout a 24-hr period. These facts make the creatinine clearance eminently suitable for assisting the day-to-day clinical management of patients. Parenthetically, it should be mentioned that in anuric acute or chronic renal failure, the serum creatinine will increase by 0.5–2 mg/dl per day, depending on the muscle bulk and metabolic status of the patient.

In practical terms, calculation of the serum creatinine is generally performed by collecting an accurately timed 24 hr urine collection and a midpoint blood specimen. The creatinine clearance may also be calculated on the basis of a 2-hr urinary collection or a 12-hr urinary collection, provided that the urine collection is accurate. The clearance calculation may be represented by the formula UV/P of Van Slyke, where U is the urinary creatinine concentration (mg/dl), V the urinary volume per unit time, and P the plasma creatinine concentration (mg/dl). A small quantity of creatinine is secreted into the renal tubules, so that creatinine clearance determinations tend to overestimate the GFR. However, the methods routinely used for determination of serum creatinine also measure small amounts of noncreatinine chromogenic material, so that the slight error in serum creatinine determination tends to cancel out the significance of the small fraction of creatinine secreted by the tubule in the UV/P clearance calculation [94]. In the presence of renal disease there are some instances where the creatinine clearance may not be entirely an accurate reflection of the GFR [81], but the relationship between the two is still clinically useful and important.

When creatinine clearance information is not available to the clinician and the serum creatinine is the only information available at the time of making a dose modification decision, it is possible to make an acceptably accurate estimation of creatinine clearance and GFR based on the following formula, which uses the patient's age, weight, and sex in converting serum creatinine into creatinine clearance.

$$\text{Males:} \quad \frac{\text{weight (kg)} \times (140 - \text{age})}{72 \times \text{serum creatinine (mg/dl)}} \quad (1)$$

$$\text{Females:} \quad \frac{\text{weight (kg)} \times (140 - \text{age}) \times 0.85}{72 \times \text{serum creatinine (mg/dl)}} \quad (2)$$

In addition to the equations presented immediately above, creatinine clearance information may also be extrapolated from the serum creatinine by use of a nomogram. Figure 1 presents such a nomogram, adapted from the data of Siersbaek-Nielsen and colleagues [85]. The use of the nomogram is explained in the legend of the figure. The nomogram is to be particularly recommended if the patient is elderly, since the construction of the nomogram takes into account the decreased creatinine production noted in older patients.

Some important correlations between serum creatinine, creatinine clearance (or GFR), and functional nephron mass should be identified, and they are summarized in Figure 2. In the latter figure, it can be seen that the glomerular filtration rate (creatinine clearance) and the functional nephron mass are interchangeable in the ordinate of the graph. When these values are considered in relationship to the serum creatinine, the relationship is hyperbolic in nature, so that a 50% reduction in functional renal tissue or GFR produces a twofold rise in serum creatinine. The practical implication of this relationship is that a greater number of nephrons are destroyed when elevations of serum creatinine are small and unimpressive than in the situation that exists later in the course of renal disease when smaller degrees of nephron damage produce greater and more impressive elevations in serum creatinine. Since the aminoglycosides are almost totally dependent upon renal elimination from the body, and specifically glomerular filtration, it would be possible to insert "aminoglycoside" for "creatinine" in Figure 2.

Some special considerations in regard to the correlations already reviewed for Figure 2 should be identified. The first of these considerations is the clinical instance wherein endogenous creatinine production rates are low, such as in advancing age. In the geriatric population muscle mass usually decreases. Daily creatinine production rates commensurately decrease at a time when glomerular filtration rates are reduced because of the progressive vascular changes of arteriolonephrosclerosis that typically occur in the glomeruli. As can be identified in Figure 1, the reduction in GFR will be matched by a reduction in serum creatinine consequent to its decreased metabolic production. Therefore, the geriatric patient will continue to manifest a "normal" serum creatinine. In these circumstances, a creatinine clearance determination would be necessary to ascertain that despite a normal serum creatinine the patient does in fact have as much as a 30–50% reduction in GFR. The implication for antibiotic use such as the aminoglycoside class of drugs is that serum measurement of the drug would be particularly appropriate and useful since reliance upon serum creatinine alone might lead to erroneously high dosing and stepwise systemic accumulation of the aminoglycosides. The aminoglycosides are particularly hazardous when they accumulate systemically because of their dose-related toxic potential.

In absolute contrast to the geriatric population would be the case of pregnant females in the third trimester of pregnancy. In the latter circumstance the GFR may be increased to 50% or more without a comparative increase in endogenous creatinine generation rates. Hence, as the GFR increases, there may be a slight fall in the serum creatinine level. If

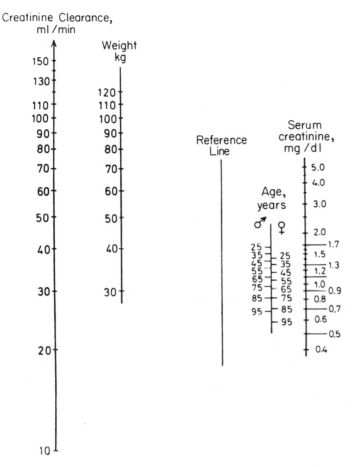

Figure 1 Nomogram that may be used for the estimation of endogenous creatinine clearance from the serum creatinine. To use the nomogram, join the weight to age with a ruler. Keep the ruler at the crossing point of the reference line. Then move the right-hand side of the ruler to the appropriate value of serum creatinine and read the patient's creatinine clearance. Use of the nomogram is particularly recommended in older patients and in obese patients. (Adapted from [85]. Used by permission.)

aminoglycoside therapy is necessary, there will be a tendency to underdose these patients based upon their "normal" serum creatinine in the setting of a significantly increased GFR. Again, appropriate determination of serum aminoglycoside levels would be the most prudent guide to safe and effective therapy.

Other special clinical circumstances are reviewed subsequently in this chapter.

II. Aminoglycoside Serum Half-Life in Normal and Impaired Renal Function

A diagrammatic schema of what happens to any antibiotic chemotherapeutic compound following oral or parenteral administration is represented in Figure 3. In the context of

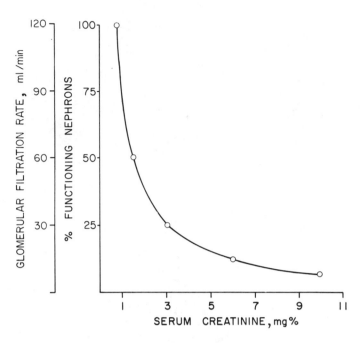

Figure 2   Relationship between the serum creatinine and glomerular filtration rate or functional tubular mass. It should be noted that approximately 75% of renal functional activity is lost before marked elevations in serum creatinine are noted. This relationship also applies for the aminoglycosides, since they are dependent upon glomerular filtration for elimination from the body.

the aminoglycosides, parenteral administration is the only means by which these drugs enter the systemic circulation. The efficacy of an aminoglycoside is related to the local concentration and ease with which it reaches its site of action, which in this instance is the interior of an infecting bacterium. The concentration of the aminoglycoside at the infecting site and the ease with which it reaches the site of action are correlated with the rate of absorption of the aminoglycoside from its parenteral route of action, its distribution through the body compartments, the degree of serum protein and tissue binding, the rate of biotransformation or degradation, and the rate of excretion of the aminoglycoside from the body. As a general rule for the aminoglycoside antibiotics, their therapeutic effectiveness and potential systemic toxicity are directly correlated with the serum concentration of the drug, and thus it allows the clinician easy access to a means of following or predicting therapeutic response to the aminoglycoside and of alerting one to the potential development of systemic toxicity secondary to increasing serum concentrations of these agents.

   The duration of the therapeutic effectiveness of an aminoglycoside is correlated with the rate with which the drug is absorbed from the route of parenteral administration and the rapidity with which it is removed from the systemic circulation. For the aminoglycosides, a healthy individual will demonstrate a relatively constant rate of absorption from the site of intramuscular injection (k_1) and a separate, but similarly relatively constant elimination rate of the drug from the blood (k_2). This relationship is demonstrated in Figure 4. In this instance, an aminoglycoside moves from compartment A into B and then

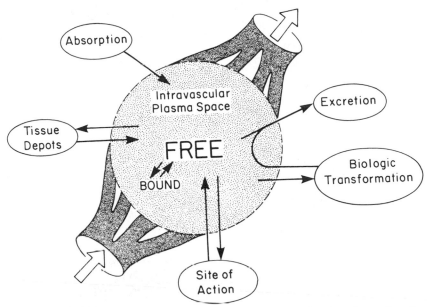

Figure 3   General schema of the fate of an antibiotic following absorption from an oral or parenteral route of administration. For systemic use all aminoglycosides must be administered parenterally.

is eliminated into compartment C. The ascending and descending slopes of the curve represent the constant rates at which the drug moves from compartment A through B into C. In the case of the intramuscular injection of an aminoglycoside, these compartments may be thought as representing, respectively, the interstitial muscle fluid at the site of intramuscular injection, the bloodstream, and then the urine.

From the practical clinical and therapeutic point of view, we are generally concerned with the second rate constant (k_2) presented in Figure 4, since this is directly related to the duration of effective therapeutic action of an aminoglycoside. This elimination rate constant is most conveniently expressed in terms of the time required for a 50% reduction in the measured serum concentration of the aminoglycoside. This time measurement is expressed as the serum $T\frac{1}{2}$ of a drug. Since the aminoglycosides are almost totally dependent upon the kidneys for elimination purposes, a reduction in renal function will obviously result in a prolongation of the serum $T\frac{1}{2}$ (k_2 in Figure 4). In the setting of renal impairment the extension of an aminoglycoside serum $T\frac{1}{2}$ serves as the most useful practical guide to adjustment of antibiotic therapy. As we will see subsequently, the development of dosing nomograms for renal impairment are entirely dependent upon data points derived from assessment of serum $T\frac{1}{2}$ values of aminoglycosides at multiple different levels of deterioration of renal functional activity.

III. Calculation of the Serum $T\frac{1}{2}$ of an Aminoglycoside

From the material already presented, it is obvious that estimation of the serum $T\frac{1}{2}$ of an aminoglycoside in the presence of normal and impaired renal function is of prime importance before recommendations concerning alterations of drug regimens in the presence

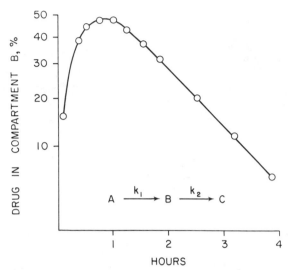

Figure 4 Representation of the serum levels of an aminoglycoside (compartment B) versus time. The upstroke of the curve (k_1) represents the rate of absorption of the drug from compartment A (intramuscular interstitial fluid) and the downstroke (k_2) indicates the rate of removal of the compound from the serum into the urine (compartment C). (Reprinted from [93]. Used by permission.)

of renal disease can be made. Detailed descriptions for the methodology involved in the formulation of the equations used for the calculation of a drug's serum T½ are discussed elsewhere [25,28,93].

Since the method is relatively straightforward, an example of the derivation of the serum T½ for the aminoglycoside gentamicin, in normal and decreased renal function, is presented in Figure 5. Following the intramuscular injection of 80 mg of gentamicin in three individuals with normal, modestly reduced (GFR 20-30 ml/min) and severely impaired (GFR 5 ml/min) renal function, a peak serum level of the compound was reached at 1 hr. Serial determinations of gentamicin serum concentrations were performed thereafter for several hours. A semilogarithmic plot of serum concentration versus time was then made as indicated in Figure 5. Three exponential lines were derived and a slope (k) for each line was developed by regression analysis utilizing the method of least squares [29]. Extrapolation of the plotted straight lines to time zero gives the theoretical peak serum concentration that would have resulted if absorption and distribution throughout the body compartments were instantaneous. The latter concentration (Co) is referred to as the "extrapolated maximum concentration." Since the decrease in the serum level of the drug occurs at a constant rate (K), the concentration (C) of the compound at any time (t) is represented by the equation

$$\log C = \log Co + Kt \tag{3}$$

or

$$C = Co \cdot e^{-Kt} \tag{4}$$

(when logs to the base e are used). Equation (4) may also be written as

$$C/Co = e^{-Kt} \tag{5}$$

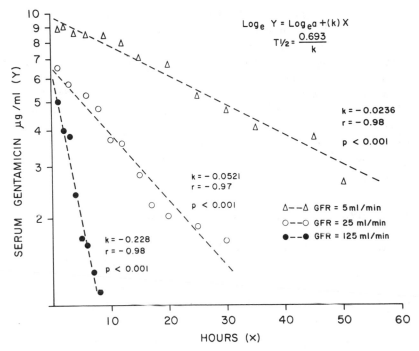

Figure 5 Observed serum gentamicin levels noted in three individuals with different degrees of renal function: (1) A 41 kg adolescent with a 5 ml/min GFR ($\triangle$); (2) a 68 kg adult with a 25 ml/min GFR (o); and (3) a 74 kg adult with a 125 ml/min GFR ($\bullet$). A gentamicin serum T½ for each individual was calculated utilizing the formulas derived in the text. (Reprinted from [93]. Used by permission.)

When 50% reduction or decay in the serum concentration of an aminoglycoside has occurred, C/Co will be equal to 0.5, and the time necessary for this to take place will be the serum T½ of that compound. Therefore, substituting in equation (5), we obtain

$$0.5 = e^{-Kt}$$

or when t equals T½,

$$T^{1/2} = \frac{\log_e 0.5}{K} \tag{6}$$

$$= \frac{0.693}{K}$$

This formulation may now be conveniently utilized to calculate the serum T½ of gentamicin at the three different degrees of renal failure depicted in Figure 5. In this example the exponential line that represents values obtained at a GFR of 125 mg/min has a k value of –0.228. Thus, using equation (6), it indicates that in this circumstance the serum T½ of gentamicin is 0.228/0.693 or 3 hr, whereas at the reduced GFR level of 25 and 5 ml/min, the value is extended to 13 and 29 hr, respectively. This information identifies that at the latter two levels of impaired renal function, approximately fourfold and tenfold reductions in gentamicin dosage or administration schedules will be necessary to prevent toxic accumulation of the compound. It will be obvious from Figure 5 that the serum T½ of an aminoglycoside may in addition be obtained by direct measurement from the exponential plot of a serum drug decay curve. It is again worth noting at this point that, as is depicted for creatinine in Figure 2, significant and potentially toxic increases in the serum level of a compound excreted by the kidney will occur at a time when 75% or more of the functioning renal tissue has been damaged or destroyed. It is in the setting of this degree of impaired renal function that the most careful attention must be given to aminoglycoside dosage and administration schedules.

To scientifically address the issue of how we decide upon the loading dose of an aminoglycoside, we must first measure the volume of distribution of these antibiotics in the body fluids of a typical patient.

A. Volume of Distribution of Aminoglycosides in Body Fluids

The calculation of the apparent volume of distribution (Vc) is quite easy and is based upon the theoretic peak serum concentration of Co of the aminoglycosides as described earlier. The following equation is used:

$$Vc = \frac{\text{Total quantity of aminoglycoside administered}}{Co} \tag{7}$$

In the examples presented in Figure 5, the calculation of the apparent volume of distribution of gentamicin is 13, 12, and 8 L of body fluid for the data representing the three patients with GFRs of 125, 25, and 5 ml/min, respectively. The body weights for each of these individuals were 74, 68, and 41 kg, respectively. These volumes of distribution approximate 20% of the body weight of each of the patients, and this fraction of the body weight closely correlates with the extracellular fluid compartment (ECF) of the individual patients. It is therefore possible to conclude that gentamicin is primarily distributed in the ECF and that, allowing for potential tissue binding, loading doses should be tailored accordingly. As a practical example of the calculation used to identify an aminoglycoside loading dose for renal failure, let us take the patient mentioned above who has a GFR of 5 ml/min and who weighs 41 kg. Twenty percent of that patient's weight would represent an ECF fluid volume of approximately 8 L. If we wish to produce an initial level of 8 mg/ml, then 8 mg/L or 64 mg/8 L will be necessary. It is important to note that if the patient has not received an aminoglycoside previously, a small fraction of the initial drug administration (approximately 20-30%) may not be detected in the serum, since it binds to tissue protein sites that rapidly become saturated. In general, it is preferable to give the loading dose in divided form either 8 or 12 hr apart, since this will avoid the induction of a single high peak of serum drug activity. Good clinical practice dictates that the serum level of the drug be monitored following the initial loading dose of an aminoglycoside in all renal failure patients. Subsequent dosing should be based upon the use of a nomogram (presented later in the chapter) and repeated serum drug monitoring.

It is standard clinical practice to administer the aminoglycosides at repetitive regular intervals so that a therapeutically effective serum level will be maintained at all times. Several authors have derived more complex formulations, which allow one to predict the course of plasma concentration of a drug following repeated administration [15, 25,27,80].

B. Systematic Degradation, Hepatic Elimination, and Alternative Routes of Excretion

The issue of systematic metabolism or biotransformation of the aminoglycosides can be reviewed quickly, since there is no evidence in humans of either systemic metabolism of the aminoglycosides or any form of biotransformation of these drugs.

There is, however, a minimal elimination of the aminoglycosides via the liver in the form of biliary excretion. Studies in humans indicate that when simultaneous determinations of serum and biliary aminoglycoside concentrations were performed that the bile level is generally ½ to ¼ of the value found in the serum [18,71,76,88]. In the detailed kinetic studies of Christopher and colleagues, it was noted that in functionally anephric patients the mean plasma clearance of gentamicin was 2.2 ml/min [18]. Since there is no systemic metabolism of gentamicin, these authors concluded that the latter elimination rate of the drug from the serum compartment in functionally or surgically anephric patients was possibly due to biliary clearance. They point out, however, that for practical therapeutic purposes it would be advisable to assume that no significant hepatic gentamicin clearance takes place, since such an assumption provides a margin of safety in the administration of gentamicin and at the same time avoids the potential risk of overdosage in patients who may lack a significant hepatic route of elimination.

Other than the biliary route of aminoglycosides, no other significant routes of elimination have been described. A comment on the issue of aminoglycoside excretion in patients with extensive burns is appropriate at this point, since there are several studies which indicate that an important acceleration of the serum $T\frac{1}{2}$ of aminoglycoside takes place in the case of drugs that are known to be eliminated by glomerular filtration. One might assume that in such burn patients, increased elimination of the aminoglycoside might occur through the sites of damaged cutaneous tissue; however, no solid data are available to resolve this question. There are detailed measurements of GFR in these patients, and it is of interest that following the immediate resuscitative period, many burn patients demonstrate GFR values 25-100% higher than normal values, perhaps as a result of the iatrogenic volume expansion of these patients or the resultant decrease in plasma oncotic pressure due to low circulating serum protein concentration and resultant increases in transglomerular basement membrane filtration coefficients. The recent studies of Loirat and colleagues [49] have identified that in the burn patients included in their studies, a significant ($P < 0.01$) reduction in the serum $T\frac{1}{2}$ of tobramycin from a controlled (healthy volunteers) value of 1.13 ± 0.14 hr to 0.85 ± 0.19 hr in their patients occurred. Zaske and associates [95,96] have found similar results in burn patients treated with gentamicin or amikacin, and the conclusion from these studies of aminoglycoside pharmacokinetics in burn patients is that increased dosage of aminoglycosides may well be needed when burn patients are being treated with such drugs. These authors realistically recommend that the increased dosage regimens of aminoglycosides should only be given in the setting of repeated monitoring of the serum drug levels.

IV. Serum Protein Binding and the Influence of Dialysis upon Serum Aminoglycoside Levels

A. Serum Protein Binding

In terms of clinical effectiveness, the degree of in vivo activity of an antibiotic chemotherapeutic agent represents the sum of many factors, which include drug potency, degree of absorption, serum protein binding, tissue concentration and binding, serum protein and tissue inactivation of the compound, and the rate of excretion of the compound. One of these factors, which is of great significance and has been the source of much investigation, is that of the serum protein binding of antibiotics. This form of protein binding plays a major role in deciding (1) the quantity of free drug available for antibacterial activity, (2) the degree to which the drug may be eliminated from the body by glomerular filtration, (3) the duration of the compound's serum $T^{1/2}$, and (4) the quantity of drug that is eliminated from the body by extracorporeal hemodialysis or peritoneal dialysis.

Before reviewing the available data on the issue of aminoglycoside serum protein binding, it is appropriate to point out that many conflicting reports on this topic have appeared in the clinical literature since the introduction of gentamicin into clinical practice. It would appear that the methodology utilized in the drug assay systems might have produced an important difference in the reported results. In addition, the serum protein binding of any antibiotic chemotherapeutic agent is influenced by several variables, such as:

1. The serum concentration of the antibiotic in the serum. Lesser degrees of binding exist when the concentration of the drug is increased.
2. The concentration of serum proteins. Less binding is noted when the serum protein concentration falls, such as is seen in chronic renal disease, hepatic disease, or other debilitating illnesses.
3. The pH of the serum and test system.
4. Temperature of the serum and test system.
5. The electrolyte-ionic concentration of the test system.
6. The presence of different drugs competing for the same binding sites.
7. The species of serum albumin used in the test system.
8. The presence of abnormally high concentrations of the free fatty acids palmitate and oleate. These are the two principal components of the mixture of free fatty acids in human serum, which in high concentration cause a decrease in antibiotic serum [30] protein binding.

As already mentioned, controversy still exists concerning the question of gentamicin serum protein binding. The initial widely reported figure of 25–30% binding of gentamicin to serum proteins derives from the classic early observations of Black and colleagues [11], Bulger and associates [16], and Riff and Jackson [76]. A number of important studies since then, including data from our laboratories, have shown that the serum protein binding of gentamicin in humans is not significant and approaches zero [2,28,32, 72]. In essence, for practical clinical purposes gentamicin is not bound to serum proteins. The most recent detailed studies on this issue have been done in a rat model by Pastoriza-Munoz et al. [65]. Using the mean figure for two experimental techniques, they report a 7.5% serum protein binding for gentamicin. Myers and colleagues have also recently addressed this issue [61]. Their detailed investigations indicate that the concentration of the divalent cations magnesium and calcium is of paramount importance in the test system. They demonstrated an inverse relationship between the percent of serum protein binding of gentamicin and the concentration of magnesium and calcium. At physiologic concentrations of calcium and magnesium, they measured a 20% binding result. Heparin was also noted to significantly influence the observed degree of plasma protein binding in this case by increasing the measured figure to 34%. Scholtan and colleagues have also confirmed a similar effect of calcium upon the binding of various aminoglycosides to serum proteins [77,82,83].

In summary, the plasma and serum protein binding results for gentamicin published in the literature are variable with most evidence favoring a binding figure of 20% or less. The listing of factors, as previously presented above, that influence the measured degree of antibiotic binding in any test system can probably account for the divergence noted in the published data referable to the aminoglycosides.

In brief, a review of the clinical literature indicates similarly that in the case of tobramycin, amikacin, and kanamycin, serum protein binding is low or undetectable [8,28, 32,48,52,77,82,90]. However, in the case of streptomycin a 30–35% rate of serum protein binding has been noted by several investigators [14,28,32,83]. Detailed serum protein binding studies have not been published for dibekacin, sisomicin, and netilmicin, but the available pharmacokinetic data indicate that these agents are not significantly bound to human serum protein.

From the renal pharmacologic point of view, the fraction of antibiotic that is unbound to serum proteins defines the amount of free drug available for passage through the glomerular basement membrane as an ultrafiltrate of plasma. The size of the molecule and electrical charge of the compound also influence the passage of an antibiotic through the glomerular basement membrane. The molecular weight of the aminoglycoside varies from 450 to 400 and as such would be expected to readily pass through the glomerular basement membrane. For gentamicin, tobramycin, and the other clinically popular aminoglycosides, we have already noted that in the framework of clinical practice, these antibiotics are cleared from the kidney by glomerular filtration alone without significant modulation by tubular secretion or tubular reabsorption, and hence their clearance resembles that of creatinine. The issue of tubular secretion and/or reabsorption is reviewed in detail in Chapter 8, since this becomes a matter of pivotal importance from the toxicologic rather than the therapeutic point of view. The possible influence of an aminoglycoside upon the fixed negative charge of the glomerular basement membrane is an important issue that has been addressed by Baylis et al. [9].

It is appropriate to parenthetically mention the question of the effect of aminoglycosides upon the albumin binding of bilirubin. It was originally reported that gentamicin in icteric newborns competed with bilirubin for albumin binding [19,42,89]. This has an important implication in the question of the development or exacerbation of kernicterus. Recent pertinent data indicate that there is no significant interaction between gentamicin and bilirubin binding or transport [7,64,92]. In fact, the data of Ballowitz and her colleagues have indicated that it was the stabilizers or solvents used in some of the commercial preparations of aminoglycosides, particularly benzylalcohol, that produced an influence upon bilirubin metabolism in an in vivo rat model [7].

B. Dialysis of Aminoglycosides

In clinical practice the management of patients with both acute and chronic renal failure very frequently entails the use of peritoneal dialysis or extracorporeal hemodialysis utilizing, in the latter instance, one of the many varieties of artificial kidneys now available on the commercial market. The end result of dialysis is the same whether peritoneal dialysis or extracorporeal hemodialysis is utilized; however, the time necessary to produce an equal dialysis end result is quite different with the two modalities of treatment, since peritoneal dialysis usually takes some five to six times longer in achieving the same result as an artificial kidney [13,93]. The creatinine clearance rate produced by most artificial kidneys is of the order of 100–120 ml/min, while during peritoneal dialysis the creatinine clearance across the peritoneal membrane is approximately 15–20 ml/min. These data on the dialysance efficiency of artificial kidneys versus peritoneal dialysis, together with data concerning the serum protein binding and molecular configuration of an antibiotic, act as useful guidelines in predicting the rate of removal of the various antibiotic chemotherapeutic agents during one or other modes of dialysis.

For the aminoglycosides one could predict from the pharmacokinetic data already reviewed in this chapter that the serum level of these drugs would be significantly influenced by extracorporeal hemodialysis or peritoneal dialysis. In clinical practice this is exactly what is found [2,6,18,20,23,24,31,33–35,39,41,43,45,46,50,52–55,60,66–69, 73,78,87,91,93]. The most useful guideline that can be developed from these extensive studies is that approximately a 50% decrease in serum aminoglycoside level will occur during the standard 5–8 hr hemodialysis with good dialyzer blood flow rates (approximately 150–300 ml/min in the adult) or during a 24–36 hr peritoneal dialysis with good mechanical dialysate exchange into and out of the peritoneal cavity (50–70 L of dialysate exchanged in the adult). An alternative way of expressing the rate of dialysance of the aminoglycosides is to measure the serum $T^{1/2}$ of the compound during the interdialysis period of a patient with chronic renal failure (serum $T^{1/2}$, 40–100 hr) and during dialysis (serum $T^{1/2}$, 8–12 hr during extracorporeal hemodialysis). The data relevant to serum $T^{1/2}$ changes of aminoglycosides during peritoneal dialysis are inadequate to develop very strong guideline statements; however, the available data, with exceptions [91], indicate that the decrease in serum $T^{1/2}$ of the aminoglycosides during peritoneal dialysis would be about one-fifth of that noted during extracorporeal hemodialysis.

Clearly, patient variability will be noted. The drug clearance data indicate that supplementary doses will be necessary during dialysis or immediately following dialysis so that therapeutic serum concentrations can be maintained. In the acutely ill septicemic patient undergoing dialysis, it would appear to be clinically appropriate to administer the aminoglycoside intravenously throughout the period of dialysis management to ensure maintenance of a therapeutically effective level for the patient and pathogen in question. Table 1 summarizes the data relevant to dialysance of aminoglycosides and provides a dosage guide to initiation and maintenance of therapy. All therapeutic dosing decisions should be carefully and frequently monitored by the measurement of serum aminoglycoside levels so that safe and effective levels of the drug are achieved in this complicated clinical setting.

C. Nomograms and Reference Dosing Tables as a Guide to Aminoglycoside Therapy in Renal Insufficiency

Since elimination of the aminoglycosides and creatinine is almost entirely dependent upon the renal ultrafiltration of these compounds through the glomeruli, this provides a clinical pharmacokinetic set of circumstances where it becomes possible to develop accurate aminoglycoside dosing nomograms based upon the measurement of concurrent serum creatinine or creatinine clearance. Several such nomograms have been developed based upon the observation of multiple peak and trough level serum drug data points throughout the range of normal to abnormal renal function. An excellent and detailed description of the development of a nomogram for gentamicin usage in renal failure was initially presented in 1972 in the report by Chan et al. [17] and several other dosing

Table 1   Initiation and Maintenance Therapy Guidelines in Patients Undergoing Dialysis Treatment[a]

Treatment drugs	Loading dose (mg/kg body wt)	Interdialysis supplementary dose (mg/kg body wt)[b]	Dialysis drug kinetics [serum $T^{1/2}$ (hr)] [c]
Extracorporeal hemodialysis			
Gentamicin	3-5	3-5	8-12
Tobramycin			
Netilmicin			
Sisomicin			
Dibekacin			
Streptomycin	15	4	8-12
Kanamycin			
Amikacin			
Lividomycin			
Peritoneal dialysis			
Gentamicin	3-5	1	35-60
Tobramycin			
Netilmicin			
Sisomicin			
Dibekacin			
Streptomycin	15	4	35-60
Kanamycin			
Amikacin			
Lividomycin			

[a]Monitor serum levels to ensure accuracy.
[b]Give in two divided doses post dialysis IV or IM.
[c]For acutely ill septicemic patients, drug should be administered during dialysis to maintain effective therapeutic levels.

nomograms for gentamicin and other aminoglycosides have subsequently been published [26,36,40,56,79]. It should also be mentioned that because of the direct correlation between the serum values of creatinine and aminoglycosides at all levels of renal functional activity, it becomes possible to develop a computer-assisted regimen of aminoglycoside dosing at all levels of renal function. This topic is reviewed in detail in Chapter 8 and elsewhere [80]. Several excellent general review texts, which contain drug dosing tables for renal failure, are available in the medical literature and they contain valuable additional reference material [3-5,10,37,38,44,46,57-59,62,63,74,75,84,93].

Figure 6 provides a nomogram as developed for the use of tobramycin. Exactly the same nomogram may be used for gentamicin, netilmicin, sisomicin, and dibekacin because of the similarity in dosing regimens and renal elimination. The nomogram presented in Figure 7 was developed as a guide to the administration of amikacin in renal failure by

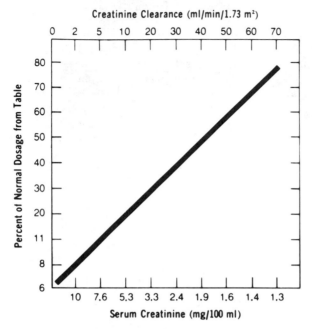

REDUCED DOSAGE NOMOGRAM*

Creatinine Clearance (ml/min/1.73 m²)

*Scales have been adjusted to facilitate
dosage calculations.

TABLE
DOSAGE SCHEDULE GUIDE FOR ADULTS
WITH NORMAL RENAL FUNCTION
(Dosage at Eight-Hour Intervals)

For Patient Weighing		Usual Dose for Serious Infections 1 mg/kg q 8 h (Total, 3 mg/kg/day)		Maximum Dose for Life-Threatening Infections (Reduce as soon as possible) 1.66 mg/kg q 8 h (Total, 5 mg/kg/day)	
kg	lb	mg/dose q 8 h	ml/dose*	mg/dose q 8 h	ml/dose*
120	264	120 mg	3 ml	200 mg	5 ml
115	253	115 mg	2.9 ml	191 mg	4.75 ml
110	242	110 mg	2.75 ml	183 mg	4.5 ml
105	231	105 mg	2.6 ml	175 mg	4.4 ml
100	220	100 mg	2.5 ml	166 mg	4.2 ml
95	209	95 mg	2.4 ml	158 mg	4 ml
90	198	90 mg	2.25 ml	150 mg	3.75 ml
85	187	85 mg	2.1 ml	141 mg	3.5 ml
80	176	80 mg	2 ml	133 mg	3.3 ml
75	165	75 mg	1.9 ml	125 mg	3.1 ml
70	154	70 mg	1.75 ml	116 mg	2.9 ml
65	143	65 mg	1.6 ml	108 mg	2.7 ml
60	132	60 mg	1.5 ml	100 mg	2.5 ml
55	121	55 mg	1.4 ml	91 mg	2.25 ml
50	110	50 mg	1.25 ml	83 mg	2.1 ml
45	99	45 mg	1.1 ml	75 mg	1.9 ml
40	88	40 mg	1 ml	66 mg	1.6 ml

Figure 6 Reduced dosage nomogram developed for use with tobramycin. (Reproduced with permission of Eli Lilly and Company, Indianapolis, Ind.)

AMINOGLYCOSIDE DOSING CHART

1. Select Loading Dose in mg/kg [IDEAL WEIGHT] to provide peak serum levels in range listed below for desired aminoglycoside.

AMINOGLYCOSIDE	USUAL LOADING DOSES	EXPECTED PEAK SERUM LEVELS
Tobramycin Gentamicin	1.5 to 2.0 mg/kg	4 to 10 μg/ml
Amikacin Kanamycin	5.0 to 7.5 mg/kg	15 to 30 μg/ml

2. Select Maintenance Dose (as percentage of chosen loading dose) to continue peak serum levels indicated above according to desired dosing interval and the patient's corrected creatinine clearance.*

	PERCENTAGE OF LOADING DOSE REQUIRED FOR DOSAGE INTERVAL SELECTED			
C(c)cr (ml/min)	half life[†] (hrs)	8 hrs	12 hrs	24 hrs
90	3.1	84%	-	-
80	3.4	80	91%	-
70	3.9	76	88	-
60	4.5	71	84	-
50	5.3	65	79	-
40	6.5	57	72	92%
30	8.4	48	63	86
25	9.9	43	57	81
20	11.9	37	50	75
17	13.6	33	46	70
15	15.1	31	42	67
12	17.9	27	37	61
10*	20.4	24	34	56
7	25.9	19	28	47
5	31.5	16	23	41
2	46.8	11	16	30
0	69.3	8	11	21

*Calculate corrected Creatinine Clearance C(c) cr as:

C(c) cr male - 140-age/serum creatinine
C(c) cr female - 0.85 x C(c) cr male

[†]Alternatively, one half of the chosen loading dose may be given at an interval approximately equal to the estimated half life.

[‡]Dosing for patients with C(c) cr≤10 ml/min should be assisted by measured serum levels.

Figure 7 Reduced dosage nomogram for tobramycin, gentamicin, amikacin, and kanamycin as developed by Sarubbi and Hull. (Reprinted from [79]. Used by permission of the publisher.)

Sarubbi and Hull [79]. As can be seen, the latter nomogram may also be used for genta-micin, tobramycin, and kanamycin dosing. Because of the milligram-for-milligram dosing equivalency, the amikacin data are equally relevant for streptomycin and lividomycin use.

It is unwise to rely entirely upon a nomogram for a full course of aminoglycoside therapy in an individual with renal impairment and, as we have emphasized before, de-termination of peak and trough serum aminoglycoside levels should complement the use of the nomograms identified in either Figure 6 or 7.

We now need to address the issue of the time interval between drug administrations in the patient with significant renal impairment. The issue is simple, but most important, and in summary form asks the question: (1) Do we administer the "usual dose" at an extended time interval; (2) do we administer a reduced dose at the "usual time interval," or (3) do we combine the approach of (1) and (2)? In clinical practice, a combination of the approaches indicated in (1) and (2) is the most useful and reliable, since it reduces the potential for systemic dose-related toxicity and affords a degree of nursing prac-ticality that reduces dosing time administration errors. In fact, this flexibility in the tim-ing of realistic dose intervals is built into the aminoglycoside dosing nomogram presented in Figure 7. A practical illustration of the rationale for combining approaches (1) and (2) can be seen when we clinically correlate serum creatinine $T^{1/2}$ kinetics and aminogly-coside serum $T^{1/2}$ kinetics.

D. Serum Creatinine Correlations as a Guide to Dosing for Aminoglycoside Therapy in Renal Impairment

The similarities in the excretory kinetics of creatinine and the aminoglycosides first led Cutler and Orme to make the therapeutically important observations that if one multi-plied the observed serum creatinine value (in mg/dl) by a factor of 3, it provided an accurate estimate of the serum $T^{1/2}$ of kanamycin in hours [22]. Subsequently, McHenry and colleagues derived a similar conclusion in the case of gentamicin; however, these authors suggested that the serum creatinine (in mg/dl) be multiplied by a factor of 4 so that the resultant number represented the serum $T^{1/2}$ of gentamicin in hours [58]. Similar correlative data have been found for essentially all the aminoglycosides [21,46,47,70]. This is not surprising when one considers that a normal serum creatinine is approximately 1 mg%, and in such a renal circumstance the serum $T^{1/2}$ of all aminoglycosides will be in the range 3-4 hr. As renal function deteriorates, creatinine and aminoglycosides will undergo the same quantitative changes in excretion or retention. Hence, a similar correla-tion is maintained between an increasing level of serum creatinine and the extension of the serum $T^{1/2}$ of an aminoglycoside.

The therapeutic implications of these correlative data are that they provide the clinician with an immediate estimate of the extension of the $T^{1/2}$ of an aminoglycoside at any level of renal functional impairment. For example, if a patient's serum creatinine is measured at 8 mg/dl, one can assume that the serum $T^{1/2}$ of the aminoglycoside selected for therapy, such as tobramycin, will be extended to 24-32 hr. The physician now has two options following the administration of a standard loading dose of the selected aminoglycoside:

1. Administer the usual, normal dosage of the aminoglycoside at an eightfold extension of the usual dosing intervals (i.e., dosing every 8 hr becomes dosing every 64 hr).
2. Administer an eightfold reduction of the normal dose at the usual dosing interval (i.e., an 80 mg dose becomes a 10 mg dose every 8 hr).

There are drawbacks to both of these approaches. In the first instance, the patient would swing from a high, potentially toxic serum concentration of drug immediately post-dosing to a serum level that may well be therapeutically inadequate for several hours before the next dose is administered. Nonetheless, the second method provides a safer and therapeutically more logical approach, since the milligram dosage is drastically reduced when using this second method. It may introduce a nursing or pharmacy dispensing error, and it would therefore be preferable to increase the calculated dose by threefold (i.e., 30 mg) and administer the latter on a 24 hr basis.

The physician should particularly avoid the approach of writing orders for an aminoglycoside dosing time interval which exactly coincides with a multiple of the measured extension of the serum $T^{1/2}$ of the drug. For example, if the drug orders are written for the administration of a specific, reduced dose every 19 or 37 hr, the chances of inducing a nursing or dispensing error are great; hence, use of the standard, clinically accepted time intervals for drug administration, such as every 6, 8, 12, or 24 hr, are preferable.

In using the correlations of serum creatinine and serum $T^{1/2}$ of aminoglycosides as a guide to therapy, it should be remembered that, as mentioned earlier in this chapter, there are vagaries in the accuracy of serum creatinine as a quantitative predictor of renal function impairment. Therefore, the therapeutic guidelines, as developed for renal failure patients by correlating the observed serum creatinine with extension of an aminoglycoside serum $T^{1/2}$, should always be used in association with measurement of the serum concentration of the selected aminoglycoside.

E. Potential Hazards in the Use of a Fixed Rule of Thumb in Aminoglycoside Dosing in Renal Failure

A special circumstance meriting separate comment is the potential hazard of relying upon a hard-and-fast therapeutic guideline for aminoglycoside therapy in renal impairment. In the multiple-reference drug dosing tables that provide therapeutic recommendations in renal failure, many authors will categorize into one group all patients with renal functional impairment of, for example, "GFR 10 ml/min or less." One therapeutic rule of thumb will be cited for each drug used in the latter group of patients. Although this is usually avoided in reference to the aminoglycosides, many clinicians will tend to rely upon one hard-and-fast dosing rule of thumb for all antibiotics used in renal failure, including the aminoglycosides. This is particularly hazardous in the case of the aminoglycosides because they are virtually totally reliant upon the kidney for elimination purposes and, as we have well identified previously, they manifest dose-related, cumulative, systemic toxicity. In the renal failure patients, since the kidneys are already severely damaged, the systemic toxicity of importance would be that of ototoxicity and neurotoxicity.

An example of this dosing issue exemplifies the inherent clinical dangers. Let us follow the course of five individuals with a "GFR 10 ml/min or less" who are placed on a rule-of-thumb aminoglycoside regimen for 10 days with the dosing indicated in Figure 8. The renal function status of each of these patients is noted in the figure, and it can be seen that it varied from a GFR of 10 ml/min to the functionally anephric state. This figure presents the serum gentamicin level as determined each day 1 hr following the administration of the drug and its theoretic distribution through the extracellular fluid compartment. It can be seen that the three individuals, who are either functionally anephric or who possess a GFR of 1 or 2 ml/min, developed systemic gentamicin toxicity within a few days of therapy. On the other hand, the individual with a GFR of 5 ml/min maintained

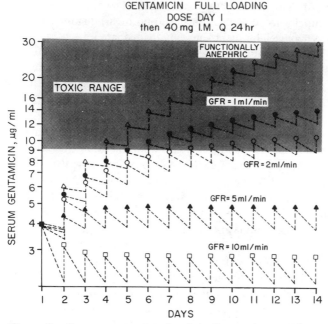

Figure 8 Serum gentamicin levels in five patients with renal impairment of "GFR 10 ml/min or less." The data indicate that following a fixed single-dosing rule of thumb for the latter degree of renal functional impairment is potentially hazardous. Individualization of the dosing regimen for each patient is essential to prevent cumulative dose-related systemic toxicity.

a safe and therapeutically satisfactory serum gentamicin level throughout his period of treatment, whereas the patient with a GFR of 10 ml/min maintained a serum gentamicin level that varied from less than 3 mg/ml 1 hr post-drug administration to less than 1 mg/ml prior to drug administration. In the latter instance, the patient experienced large segments of time in which the serum levels may have been therapeutically inadequate. This is a good example of the danger of relying upon a dogmatic rule of thumb for drug dosing in the case of any antibiotic that is excreted by glomerular filtration and manifests dose-related toxicity as a result of systemic accumulation. The data serve to stress the importance of determining serum levels of aminoglycosides as the best guide to therapy in individuals with severe renal impairment.

References

1. Addis, T., Barrett, E., Poo, L.J., Ureen, H.J., and Lippman, R.W.: The relation between protein consumption and diurnal variations of the creatinine clearance in normal individuals. *J. Clin. Invest.* 30:206–209, 1951.
2. Altmann, G., Bogokowsky, B., Boner, G., and Eliahou, H.E.: Blood levels of ampicillin, kanamycin and gentamicin in the uremic patient. *Isr. J. Med. Sci.* 6:683–690, 1970.
3. Anderson, R.J., Gambertoglio, J.G., and Schrier, R.W.: *Clinical Use of Drugs in Renal Failure.* Springfield, Ill., Charles C Thomas, 1976.

4. Appel, G.G., and Neu, H.C.: Gentamicin in 1978. *Ann. Intern. Med.* 89:528–538, 1978.

5. Appel, G.B., and Neu, H.C.: The nephrotoxicity of antimicrobial agents (three parts). *N. Engl. J. Med.* 296:663–670, 722–728, 784–787, 1977.

6. Atkins, R.C., Mion, C., Despaux, E., Van-Hai, N., Julien, C., and Mion, H.: Peritoneal transfer of kanamycin and its use in peritoneal dialysis. *Kidney Int.* 3:391–396, 1973.

7. Ballowitz, L., Hanefeld, F., Schmid, H., van Eekeren, C., Kretschmer, A., and Wollermann, S.: The influence of various aminoglycoside preparations on bilirubin/albumin binding. *J. Perinat. Med.* 4:168–183, 1976.

8. Barza, M., Samuelson, T., and Weinstein, L.: Penetration of antibiotics into fibrin loci in vivo: II. Comparison of nine antibiotics: effect of dose and degree of protein binding. *J. Infect. Dis.* 129:66–72, 1974.

9. Baylis, C., Rennke, H.R., and Brenner, B.M.: Mechanisms of the defect in glomerular ultrafiltration associated with gentamicin administration. *Kidney Int.* 12:344–353, 1977.

10. Bennett, W.M., Singer, I., and Coggins, C.J.: A guide to drug therapy in renal failure. *JAMA* 230:1544–1553, 1974.

11. Black, J., Calesnick, B., Williams, D., and Weinstein, M.J.: Pharmacology of gentamicin, a new broad-spectrum antibiotic. *Antimicrob. Agents Chemother.* 3:138–137, 1963.

12. Blantz, R.C.: Glomerular filtration. In *Pathophysiology of the kidney,* edited by Kurtzman, N.A. and Martinez-Maldonado, M. Springfield, Ill., Charles C Thomas, 1977, pp. 56–87.

13. Boer, S.T.: Kinetics of peritoneal dialysis. A comparison with the artificial kidney. *Medicine (Baltimore)* 40:243–287, 1961.

14. Boxer, G.E., Jelinek, V.C., and Edison, A.O.: Streptomycin: clearance and binding to protein. *J. Pharmacol. Exp. Ther.* 97:93–104, 1949.

15. Boxer, G.E., Jelinek, V.C., Tompsett, R., DuBois, R., and Edison, A.O.: Streptomycin in the blood. Chemical determinations after single and repeated intramuscular injections. *J. Pharmacol. Exp. Ther.* 92:226–235, 1948.

16. Bulger, R.J., Sidell, S., and Kirby, W.M.M.: Laboratory and clinical studies of gentamicin a new broad-spectrum antibiotic. *Ann. Intern. Med.* 59:593–604, 1963.

17. Chan, R.A., Benner, E.J., and Hoeprich, P.D.: Gentamicin therapy in renal failure: a nomogram for dosage. *Ann. Intern. Med.* 76:773–778, 1972.

18. Christopher, T.G., Korn, D., Blair, A.D., Forrey, A.W., O'Neill, M.A., and Cutler, R.E.: Gentamicin pharmacokinetics during hemodialysis. *Kidney Int.* 6:38–44, 1974.

19. Cukier, J.O., Seungdamrong, S., Odell, J.L., and Odell, G.B.: The displacement of albumin-bound bilirubin by gentamicin. *Pediatr. Res.* 8:339, 1974.

20. Curtis, J.R., McDonald, S.J., and Weston, J.H.: Parental administration of gentamicin in renal failure: patients undergoing intermittent hemodialysis. *Br. Med. J.* 2:537–539, 1967.

21. Cutler, R.E., Gyselynck, A.M., Fleet, R., and Forrey, A.W.: Correlation of serum creatinine concentration and gentamicin half-life. *JAMA* 219:1037–1041, 1972.

22. Cutler, R.E., and Orme, B.M.: Correlation of serum creatinine concentrations and kanamycin half-life. Therapeutic implications. *JAMA* 209:539–542, 1969.

23. Danish, M., Schultz, R., and Jusko, W.: Pharmacokinetics of gentamicin and kanamycin during hemodialysis. *Antimicrob. Agents Chemother.* 6:841–847, 1974.

24. Day, R.E., and White, R.H.: Peritoneal dialysis in children. Review of 8 years experience. *Arch. Dis. Child.* 52:56–61, 1977.

25. Dettli, L.: Dosage of chemotherapeutic agents in patients with renal disease. In *Antibiotics and Hospitals,* edited by Grassi, C., and Ostino, G. New York, Alan R. Liss, 1979, pp. 414–446.

26. Dettli, L.: Drug dosage in patients with renal disease. *Clin. Pharmacol. Ther.* 16: 274–280, 1974.

27. Dettli, L., Spring, P., and Ryter, S.: Multiple dose kinetics and drug dosage in patients with kidney disease. *Acta Pharmacol. Toxicol. Scand.* 3(Suppl.):S211–S224, 1971.

28. Dittert, L.W.: Pharmacokinetics of aminoglycosides. General considerations. (U.S. Amikacin Symposium, November 1976.) *Am. J. Med.* 77–83, 1977.

29. Edwards, A.L.: *Statistical Methods* (2nd ed.). New York, Holt, Rinehart and Winston, 1967, pp. 80–81.

30. Frederickson, D.S., and Gordon, R.S.: Transport of fatty acids. *Physiol. Rev.* 38: 585–630, 1958.

31. Gary, N.E.: Peritoneal clearance and removal of gentamicin. *J. Infect. Dis.* 124 (Suppl.):S96–S97, 1971.

32. Gordon, R.C., Regamey, C., and Kirby, W.M.M.: Serum protein binding of the aminoglycoside antibiotics. *Antimicrob. Agents Chemother.* 2:214–216, 1972.

33. Greenburg, P.A., and Sanford, J.P.: Removal and absorption of antibiotics in patients with renal failure undergoing peritoneal dialysis. *Ann. Intern. Med.* 66:465–470, 1967.

34. Halpren, B.A., Axline, S.G., Coplon, N.S., and Brown, D.M.: Clearance of gentamicin during hemodialysis: comparison of four artificial kidneys. *J. Infect. Dis.* 133: 627–636, 1976.

35. Ho, P.W.L., Pien, F.D., and Kominami, N.: Massive amikacin "overdose." *Ann. Intern. Med.* 91:227–228, 1979.

36. Hull, J.H., and Sarubbi, F.A., Jr.: Gentamicin serum concentrations: pharmacokinetic predictions. *Ann. Intern. Med.* 85:183–189, 1976.

37. Jackson, E.A., and McLeod, D.C.: Pharmacokinetics and dosing of antimicrobial agents in renal impairment: Parts 1 and 2. *Am. J. Hosp. Pharm.* 31:36–52, 137–148, 1974.

38. Jackson, G.G.: Present status of aminoglycoside antibiotics and their safe, effective use. *Clin. Ther.* 1:200–215, 1977.

39. Jaffe, G., Meyers, B.R., and Hirschman, S.Z.: Pharmacokinetics of tobramycin in patients with stable renal impairment, patients undergoing peritoneal dialysis and patients on chronic hemodialysis. *Antimicrob. Agents Chemother.* 5:611–616, 1974.

40. Jelifee, R.W.: Nomogram for kanamycin and gentamicin therapy. Abstracts of the 11th Interscience Conference on Antimicrobial Agents and Chemotherapy, October 19–22, 1971, Atlantic City, N.J., p. 63.

41. Jusko, W.J., Baliah, T., Kim, K.H., Gerbracht, L.M., and Yaffe, S.J.: Pharmacokinetics of gentamicin during peritoneal dialysis in children. *Kidney Int.* 9:430–480, 1976.

42. Kapitulnik, J., Eyal, F., and Simcha, A.J.: Gentamicin and bilirubin binding by plasma. *Lancet* 2:1195, 1972.

43. Krumlovsky, F.A., Emmerman, J., Parker, R.H., Wisgerhof, M., and Del Greco, F.: Dialysis in treatment of neomycin overdosage. *Ann. Intern. Med.* 76:443–446, 1972.

44. Kunin, C.M.: A guide to use of antibiotics in patients with renal disease. A table of recommended doses and factors governing serum levels. *Ann. Intern. Med.* 67:151–158, 1967.

45. Letourneau-Saheb, L., Lapierre, L., Daigneault, R., Prud Homme, M., St.-Louris, G., and Serois, G.: Gentamicin pharmacokinetics during hemodialysis in patients suffering from chronic renal failure. *Int. J. Clin. Pharmacol. Biopharm.* 15:116–120, 1977.

46. Leroy, A., Humbert, G., Oksenkendler, G., and Fillastre, J.-P.: Pharmacokinetics of aminoglycosides in subjects with normal and impaired renal function. *Antibiot. Chemother.* 25:163–180, 1978.

47. Levy, J., and Klastersky, J.: Correlation of serum creatinine concentration and amikacin half-life. *J. Clin. Pharmacol.* 15:705–707, 1975.

48. Lockwood, W.R., and Bower, J.D.: Tobramycin and gentamicin concentrations in the serum of normal and anephric patients. *Antimicrob. Agents Chemother.* 3:125–129, 1973.

49. Loirat, P., Rohan, J., Baillet, A., Beaufils, F., David, R., and Chapman, A.: Increased glomerular filtration rate in patients with major burns and its effect on the pharmacokinetics of tobramycin. *N. Engl. J. Med.* 299:915–919, 1978.

50. Luft, F.C., Brannon, D.R., Stropes, L.L., Costello, R.J., Sloan, R.S., and Maxwell, D.R.: Pharmacokinetics of netilmicin in patients with renal impairment and in patients on dialysis. *Antimicrob. Agents Chemother.* 14:403–407, 1978.

51. Macon, E.J.: Increasing the supply of cadaveric kidneys for transplantation. *Morb. Mortal. Rep., C.D.C.* 28:337–345, 1979.

52. Madhavan, T., Yaremchuk, K., Levin, N., Pohlod, D., Burch, K., Fisher, E., Cox, F., and Quinn, E.L.: Effect of renal failure and dialysis on the serum concentration of the aminoglycoside amikacin. *Antimicrob. Agents Chemother.* 10:464–466, 1976.

53. Mahon, W.A., Exer, J.I., Fenton, S.S., and Cattran, D.C.: Studies on gentamicin clearance during peritoneal and hemodialysis. *Clin. Res.* 21:1018, 1973.

54. Malacoff, R.F., Finkelstein, F.O., and Andriole, V.T.: Effect of peritoneal dialysis on serum levels of tobramycin and clindamycin. *Antimicrob. Agents Chemother.* 8:574–580, 1975.

55. Masur, H., Whelton, P.K., and Whelton, A.: Neomycin toxicity revisited. *Arch. Surg.* 111:822–825, 1976.

56. Mawer, G.E.: Nomograms for antibiotic dosage in adults. *Adverse Drug React. Bull.* 52:176–179, 1975.

57. Mawer, G.E., Ahmad, R., Dobbs, S.M., McGouth, J.H., Lucas, S.B., and Tooth, J.A.: Prescribing aids for gentamicin. *Br. J. Clin. Pharmacol.* 1:45–50, 1974.

58. McHenry, M.C., Gavan, T.L., Gifford, R.W., Jr., Guerkink, N.A., Ban Ommen, R.A., Town, M.A., and Wagner, J.G.: Gentamicin dosages for renal insufficiency: adjustments based on endogenous creatinine clearance and serum creatinine concentration. *Ann. Intern. Med.* 74:192–197, 1971.

59. McHenry, M.C., Wagner, J.G., Hall, P.M., Vidt, D.G., and Gavan, T.L.: Pharmacokinetics of amikacin in patients with impaired renal function. *J. Infect. Dis.* 134 (Suppl.):S343–S348, 1976.

60. Meyer, R.D., Lewis, R.P., and Finegold, S.M.: Amikacin therapy of serious gram-negative bacillary infections in chronic hemodialysis patients. *Chemotherapy* 24: 172–178, 1978.

61. Myers, D.R., DeFehr, J., Bennett, W., Porter, G.A., and Olsen, G.D.: Gentamicin binding to serum and plasma proteins. *Clin. Pharmacol. Ther.* 23:356–360, 1978.

62. Neu, H.C.: Tobramycin: an overview. *J. Infect. Dis.* 134(Suppl.):S3–S19, 1976.

63. Noone, P., Parsons, T.M.C., Pattison, J.R., Slack, R.C.B., Garfield-Davies, D., and Hughes, K.: Experience in monitoring gentamicin therapy during treatment of serious gram-negative sepsis. *Br. Med. J.* 1:477–481, 1974.

64. Odell, G.D., Cukier, J.D., and Maglalang, A.C.: Effects of gentamicin on albumin binding of bilirubin. Commentary. *J. Pediatr.* 86:614, 1975.

65. Pastoriza-Munoz, E., Bowman, R.L., and Kaloyanides, G.: Renal tubular transport of gentamicin in the rat. *Kidney Int.* 16:440–450, 1979.

66. Pechere, J.C., and Dugal, R.: Pharmacokinetics of intravenously administered tobramycin in normal volunteers and in renal-impaired and hemodialyzed patients. *J. Infect. Dis.* 134(Suppl.):S118–S124, 1976.

67. Pechere, J.C., Dugal, R., and Pechere, M.M.: Pharmacokinetics of intravenous amikacin after rapid and slow infusion with special reference to hemodialysis. *Eur. J. Drug. Metab. Pharmacokinet.* 4:49–56, 1979.

68. Pechere, J.C., Dugal, R., and Pechere, M.M.: Pharmacokinetics of netilmicin in renal insufficiency and hemodialysis. *Clin. Pharmacokinet.* 3:395–406, 1978.

69. Pechere, J.C., Pechere, M.M., and Dugal, R.: Clinical pharmacokinetics of sisomicin: dosage schedules in renal-impaired patients. *Antimicrob. Agents Chemother.* 9: 761–765, 1976.

70. Pijck, J., Hallynck, T., Soep, H., Baert, L., Daneels, R., and Boelart, J.: Pharmacokinetics of amikacin in patients with renal insufficiency: relation of half-life and creatinine clearance. *J. Infect. Dis.* 134(Suppl.):S331–S341, 1976.

71. Pitt, H.A., Robert, R.B., and Johnson, W.D.: Gentamicin levels in the human biliary tract. *J. Infect. Dis.* 128:299–302, 1973.

72. Ramirez-Ronda, C.H., Holmes, R.K., and Sanford, J.P.: Effects of divalent cations on binding of aminoglycoside antibiotics to human serum proteins and to bacteria. *Antimicrob. Agents Chemother.* 7:239–245, 1975.

73. Regeur, L., Colding, H., Jensen, H., and Kampmann, J.P.: Pharmacokinetics of amikacin during hemodialysis and peritoneal dialysis. *Antimicrob. Agents Chemother.* 11:214–218, 1977.

74. Reidenberg, M.M.: *Renal Function and Drug Action.* Philadelphia, W.B. Saunders, 1971.

75. Reiner, N.E., Bloxham, D.D., and Thompson, W.L.: Nephrotoxicity of gentamicin and tobramycin given once daily or continuously in dogs. *J. Antimicrob. Chemother.* 4(Suppl. A):85–101, 1978.

76. Riff, L.J., and Jackson, G.G.: Pharmacology of gentamicin in man. *J. Infect. Dis.* 124(Suppl.):S98–S104, 1971.

77. Rosenkranz, H., Scheer, M., and Scholtan, W.: Binding of aminoglycoside antibiotics to human serum proteins: III. Effect of experimental conditions. *Infection* 6:57–64, 1978.

78. Rumpf, K.W., Rieger, J., Doht, B., Ansorg, R., and Scheler, F.: Drug elimination by hemofiltration. *J. Dial.* 1:677–678, 1977.

79. Sarubbi, F.A., and Hull, J.H.: Amikacin serum concentrations: prediction of levels and dosage guidelines. *Ann. Intern. Med.* 89:612–618, 1978.

80. Sawchuck, R.J., and Zaski, D.E.: Pharmacokinetics of dosing regimens which utilize multiple intravenous infusions: gentamicin in burn patients. *J. Pharmacokinet. Biopharm.* 4:183–195, 1976.

81. Schirmeister, J. Von, Willmann, H., Kiefer, H., and Hallauer, W.: Für und wieder die Brauchbarkeit der endogen Kreatininclearance in der Funktionellen Nierendiagnostik. *Dtsch. Med. Wochenschr.* 89:1640–1647, 1964.

82. Scholtan, W., and Rosenkranz, H.: Ionic binding of aminoglycosides to human serum albumin in the absence of divalent cations. IV. Effect of structure, pH and concentration. *Infection* 6:130–136, 1978.

83. Scholtan, W., Rosenkranz, H., and Scheer, M.: Binding of sisomicin and streptomycin to human albumin. *Infection* 6:162–165, 1978.

84. Schreiner, G.E., and Teehan, B.P.: Dialysis of poisons and drugs—annual review. *Am. Soc. Artif. Int. Organs* 15:513–544, 1971.

85. Siersbaek-Nielsen, K., Hansen, J.M., Kampmann, J., and Kristensen, M.: Rapid evaluation of creatinine clearance. *Lancet* 1:1133–1134, 1971.

86. Sirota, J.H., Baldwin, D.S., and Villareal, H.: Diurnal variations of renal function in man. *J. Clin. Invest.* 29:187–192, 1950.

87. Smithivas, T., Hyams, P.J., Matalon, P., Simberkoff, M.S., and Rahal, J.J.: The use of gentamicin in peritoneal dialysis: I. Pharmacologic results. *J. Infect. Dis.* 124 (Suppl.):S77–S83, 1971.

88. Smithivas, T., Hyams, P.J., and Rahal, J.J.: Gentamicin and ampicillin in human bile. *J. Infect. Dis.* 124(Suppl.):S106–S113, 1971.

89. Stern, L.: Drug interactions—Part II. Drugs, the newborn infant, and the binding of bilirubin to albumin. *Pediatrics* 49:916–918, 1972.
90. Ullman, V.W.E.: The protein binding of ticarcillin and tobramycin. *J. Antimicrob. Chemother.* 2:213–214, 1976.
91. Weinstein, A.J., Karchmer, A.W., and Moellering, R.C.: Tobramycin concentrations during peritoneal dialysis. *Antimicrob. Agents Chemother.* 4:432–434, 1973.
92. Wennberg, R.P., and Rasmussen, L.F.: Effects of gentamicin on albumin binding of bilirubin. *J. Pediatr.* 86:611–613, 1975.
93. Whelton, A.: Antibacterial therapy in renal insufficiency. A review. *Antibiot. Chemother.* 18:1–48, 1974.
94. Whelton, A., and Sapir, D.G.: Clinical tests of renal function. In *Practice of Medicine (Tice)*, Vol. 2. New York, Harper & Row, 1976, pp. 1–9.
95. Zaske, D.E., Sawchuck, R.J., Gerding, D.N., and Strate, R.G.: Increased dosage requirements of gentamicin in burn patients. *J. Trauma* 16:824–828, 1976.
96. Zaske, D.E., Sawchuck, R.J., and Strate, R.G.: The necessity of increased doses of amikacin in burn patients. *Surgery* 84:603–608, 1978.

26
CLINICAL USE OF AMINOGLYCOSIDES

HAROLD C. NEU   College of Physicians and Surgeons,
Columbia University, New York, New York

Aminoglycosides have found extensive use in clinical medicine in the three and a half decades since Waksman's group first isolated streptomycin. In many situations the availability of an aminoglycoside antibiotic has meant the difference between survival and death. In other situations their use has been less clearly deliniated, and drugs with a much wider therapeutic ratio (benefit vs. toxicity) may now be considered the preferred agents. This chapter examines the use of the aminoglycosides in different types of infection and attempts to provide some perspective about their use in comparison with other agents.

I. Upper Respiratory Tract Infections

Otitis externa is a superficial infection of the squamous epithelium of the external auditory canal. It is most often the result of excess moisture in the ear and occasionally due to chronically draining middle ear infection. *Pseudomonas aeruginosa* and *Staphylococcus aureus* are the organisms most often found. Although topical aminoglycoside creams and solutions have been used to treat this infection, there is no evidence that gentamicin or tobramycin are more effective than weak acid solutions or antibiotic drops that contain polymyxin B, neomycin, and hydrocortisone.

Malignant external otitis, in contrast to otitis externa, is a severe form of external otitis that is caused by *P. aeruginosa*. There is invasion of the cartilage and bone of the area, producing an osteomyelitis which if untreated will progress to cranial nerve palsies and death. This is a disease primarily of diabetic individuals over 55 years of age. Successful treatment depends upon early diagnosis, aggressive surgical removal of devitalized tissue, and the use of aminoglycosides combined with an antipseudomonas penicillin-carbenicillin, ticarcillin, azlocillin, or piperacillin. The largest number of cases have been treated with gentamicin, but the published results do not demonstrate the superiority of gentamicin, tobramycin, or amikacin. Therapy should be directed at several aspects. It is useful to determine the inhibitory concentration needed for the infecting *Pseudomonas* and to select the agents with the best inhibitory index. Serum levels should be monitored to be certain that serum levels exceed the minimum inhibitory concentration (MIC) of the organism. In view of the age of the patients and the necessity to treat for 6 weeks, tobramycin probably would be the preferred agent. Tobramycin or gentamicin plus ticarcillin or carbenicillin would appear to be preferable to use of aminoglycoside alone, but there is no statistical basis for this statement.

Acute otitis media of childhood, even that of the newborn, should be treated with agents other than aminoglycosides, since *Streptococcus pneumoniae, Haemophilus influenzae,* and less frequently, *Streptococcus pyogenes* are the infecting organisms. Gram-negative Enterobacteriaceae rarely are cultured from middle ear fluid, and are the result of inadequate culture technique in which external canal material contaminates the sample.

Sinusitis is an infection of the paranasal sinuses which is the complication of a preceding viral upper respiratory tract infection. Although gram-negative bacteria have been isolated in 0–24% of acute antral sinusitis, their role in this infection is dubious [39]. In particular, gram-negative bacteria such as *Escherichia coli, Klebsiella,* or *Pseudomonas* may be cultured from the patient who has subacute or chronic sinusitis. There is no evidence that aminoglycosides have contributed to the cure of sinusitis. Epiglottitis is due to *H. influenzae.* The ampicillin-resistant isolates should not be treated with aminoglycosides even though the organisms are susceptible; rather, chloramphenicol would be the antibiotic of choice.

Infections of the oral cavity are due primarily to the endogenous oral flora, which is anaerobic or faculative, although *S. aureus, Haemophilus,* and *Neisseria* can be important in the production of oral disease. There is no evidence to suggest that topical or systemic use of aminoglycosides will alter plaque formation or decrease diseases of the gingiva and peridontium. Similarly, aminoglycosides are not helpful in the extensive necrotizing stomatitis due to anaerobic species, nor in pharyngeal or parapharyngeal space infections, even when gram-negative enteric organisms are the species cultured. Agents of this class have no role in the treatment of streptococcal infection by either the parenteral or topical route. The rare pharyngeal infection in which aminoglycoside therapy (streptomycin) might be indicated would be tuberculosis of the pharynx or ulceration of the oral pharynx in tularemia.

II. Infections After Surgery of the Neck

The one area of the head–neck in which aminoglycosides may have an important role is in therapy of nosocomially acquired gram-negative infections that follow extensive surgery for malignancy of the oral pharynx. *Klebsiella, E. coli,* and particularly *Pseudomonas* have caused serious postoperative wound infection. Selection of the aminoglycoside should be based upon the susceptibility patterns of the organisms infecting the patient, since therapy usually is not initiated until culture results are available. The physician must be certain to separate colonization from infection when considering therapy in this area since most of these patients will be colonized with gram-negative bacteria even though they are not infected. It is probably wise in such infections to always treat with a penicillin as well because of the oral streptococcal species present.

III. Pleuropulmonary Infections

Acute bronchitis primarily is a viral illness. In contrast in chronic bronchitis and acute infectious exacerbations of bronchitis, bacteria seem to be involved. The most frequently cultured organisms are *H. influenzae* and *S. pneumoniae* [38]; *S. aureus* and gram-negative enteric bacilli are infrequent causes of acute infection in chronic bronchitis. Only 5–10% of sputum samples will contain these species. Since there are many questions [99] about the overall benefit of antimicrobial therapy in exacerbations of chronic bronchitis, aminoglycosides used either as a parenteral agent or by aerosol [88] are not appropriate.

Acute pneumonia due to the most common agents—*S. pneumoniae, S. aureus,* and anaerobic mouth flora such as *Bacteroides melaninogenicus, Fusobacteria,* and anaerobic streptococci—should not be treated with aminoglycosides, because these agents have poor to no activity against these organisms.

Against *K. pneumoniae* the aminoglycosides do have a role, as they do with a number of the other gram-negative enteric species that are acquired in the hospital. Kanamycin was used with success to treat *Klebsiella* pneumonitis in the 1960s [11]. However, with the availability of the cephalosporins and the increased resistance of many hospital isolates of *Klebsiella* to kanamycin, it was replaced by gentamicin. Satisfactory results of the therapy of gram-negative pneumonia have been reported by Louria et al. [64] and by Hodges and Saslaw [42]. Bodey et al. [9] reported only a 45% response in the immunocompromised patient to pneumonia due to various gram-negative bacteria which was treated with gentamicin alone. In a 1974 review article, McHenry et al. [74] recommended gentamicin as initial therapy for most life-threatening pneumonias due to gram-negative bacilli. There are data on the use of tobramycin in *Klebsiella* and other gram-negative pneumonias [1,6], with an 80% overall satisfactory response and 24 of 27 patients with *Klebsiella* pneumonia having a favorable response. Trenholme et al. [103] had a good response to *Klebsiella* pneumonia in three patients treated with amikacin. Other investigators [62,116] have reported that amikacin was successful not only in treating pneumonia due to gentamicin-susceptible gram-negative bacilli, but also that due to gentamicin-resistant organisms [62].

The value of gentamicin, tobramycin, or amikacin in the therapy of the less common gram-negative pneumonias is not well established. Although isolated cases of *Proteus, Providencia, Serratia, Acinetobacter,* or *Enterobacter* pneumonia have been cured, there have been an equal number of failures [1,23]. The ready colonization of the tracheal airway with gram-negative microorganisms makes difficult the interpretation of bacteriologic results of therapy of infections in the respiratory tract. Pathogens are rarely eliminated from the sputa of the individual who is most likely to develop gram-negative infections. Thus, clinical and radiologic responses commonly are used to judge the efficacy of therapy; and in the presence of underlying structural lung disease, evaluation is complex.

Pneumonia due to *P. aeruginosa* must be considered separately from other gram-negative pneumonias since mortality in this type of pneumonia is excessively high and with bacteremia is said to approach 100% [45]. Waldvogel et al. [106] treated patients with *Pseudomonas* bronchopneumonia with tobramycin with satisfactory response. Blair and colleagues [8] obtained good response in *Pseudomonas* pneumonia but also gave the patients carbenicillin. Bendush and Weber [6], in a review of the overall experience with tobramycin, showed that only 13 of 25 patients in whom the infection was caused by *Pseudomonas* responded favorably. Pines et al. [92] also did not find gentamicin that useful in the treatment of purulent bronchial infection due to *Pseudomonas.* Klastersky et al. [55] compared the effectiveness of tobramycin and gentamicin in *Pseudomonas* pulmonary infection and could not show a superiority of either compound. Similar results were found by Carmalt et al. [14]. Of note is the fact that only 3 of 6 patients with *P. aeruginosa* pneumonia treated by Trenholme et al. [103] with amikacin were classified as showing a good response. Sisomicin also has been used to treat *Pseudomonas* pneumonia [2,110], but results have been more impressive than treatment with the other agents. Netilmicin also has been effective in treating selected patients with pulmonary disease [49]. Overall review of the treatment of *Pseudomonas* pneumonia does not show that any one agent is most effective. However, with combined therapy (i.e., use of an anti-pseudomonas penicillin plus an aminoglycoside), there are data to suggest that tobramycin plus ticarcillin is more effective than is gentamicin plus carbenicillin [53,86]. At the present time, use of two agents, an aminoglycoside and a β-lactam compound, would be recommended in the treatment of gram-negative enteric pneumonias, particularly those due to *Pseudomonas.* Selection of the aminoglycoside should be based on susceptibility of the isolates and the expected potential for a toxic reaction, since therapy would last for at least 2 weeks.

Pneumonitis in the patient with cystic fibrosis has been treated with all the aminoglycosides, with varying degrees of success. Boxerbaum et al. [10] studied 80 patients with cystic fibrosis who had chronic pulmonary infection due to *Pseudomonas.* They noted neither a dramatic response nor were they able to eradicate the *Pseudomonas* from the sputum. Parry and Neu [85] noted similar results and could not show that ticarcillin plus gentamicin was superior to ticarcillin alone. A number of investigators have studied the values of tobramycin in patients with cystic fibrosis [24,40,73,87]. The results of the studies all showed clinical improvement but a failure to eradicate the *Pseudomonas* from the sputum. Sisomicin has also been considered to be effective [43], but a direct comparative study has not been done.

Comparative studies combining ticarcillin and tobramycin have shown better results than those with the drugs alone [43,85].

Amikacin has been utilized by a number of groups [27,44] to treat exacerbations of pulmonary infection in children with cystic fibrosis. Although the studies show improvement, it is not superior to that seen with other agents. At the present time there are insufficient data to establish the superiority of one aminoglycoside over another except when the infection is due to an isolate that is resistant to one agent but susceptible to another. Combination therapy of an aminoglycoside with ticarcillin seems to be the best program available. Interestingly, toxic reactions of the young individual with cystic fibrosis are rare. Thus, maximal doses of the aminoglycoside must be used. As noted in Chapter 5, the aminoglycosides have a short life in these patients, and levels below the inhibitory level are frequently used.

Other Forms of Pneumonia

Specific pneumonia due to plague or tularemia are still best treated with streptomycin. Lung abscess and empyema are due primarily to anaerobic species. Even when these infections are due to a mixture of aerobic and anaerobic organisms, the place of aminoglycosides is poorly established, since these agents are much less effective under anaerobic conditions. Therapy should be an aminoglycoside combined with penicillin or clindamycin. Drainage of empyema will still be necessary.

Streptomycin has a role to play in the chemotherapy of tuberculosis. It is useful in twice-weekly programs. Amikacin may be effective in those pulmonary infections that are due to the atypical mycobacteria.

IV. Urinary Tract Infections

Aminoglycosides have had their greatest use in the therapy of urinary tract infection. If an organism is susceptible to one of the members of this group of drugs, it will be effective therapy of either upper- or lower-tract infection. Resistant bacteria will not be eradicated, and bacteria will not be eradicated if there are indwelling urethral catheters or major structural abnormalities of the renal system, such as renal stones.

Petersdorf and Turck [89] showed that kanamycin was an effective therapy in young patients who had bacteriuria refractory to other drugs. Ronald et al. [93] showed that single-dose therapy with kanamycin was effective in curing infection localized to the bladder, but not if the infection came from the kidneys.

Numerous studies have documented the value of gentamicin in uncomplicated and complicated urinary tract infections [16,19,21,58,97]. The dosage programs have been variable, but the study of Labovitz et al. [58] indicated that a single daily dose was effective. Note has been made of the increased efficacy of gentamicin for urinary tract infections if the urine is more alkaline [63]. Minuth et al. [77] suggested that a concentrated urine inhibited gentamicin activity, but detailed analysis of sodium and divalent cations did not clarify the role of the components in decreasing activity of gentamicin in urine.

Response to tobramycin therapy of urinary tract infection has been equally good [1,7,109]. However, patients with severe structural defects either relapsed or were reinfected. A number of comparative studies of tobramycin and gentamicin have shown no difference in response to therapy [65,107]. Combined therapy of urinary tract infection using gentamicin or tobramycin with carbenicillin or ticarcillin also has not demonstrated greater efficacy of one agent. As with gentamicin, once-daily therapy of urinary infections with tobramycin is as effective as the twice- or thrice-daily programs [6].

Success in the treatment of urinary infection with amikacin is comparable to that of other drugs [20,36,72]. Amikacin could be utilized as a single daily dose, but amikacin has proved most useful in treating infections of the urine due to bacteria resistant to other aminoglycosides, although some resistance to it also has developed in patients with complicated urinary infections [23].

Sisomicin [54,66,102] and netilmicin [56,84] have also been shown to be effective in curing urinary infections. None of the studies would suggest that the agents are superior to the earlier aminoglycosides. Thus, any aminoglycoside would be a useful compound to use to treat urinary tract infections due to organisms that were not susceptible to less toxic agents. It does seem likely that single-daily-dose therapy is as effective as multiple doses and that there are fewer relapses after therapy with aminoglycosides than after use of other agents such as the β-lactam compounds. But the precise duration of therapy is unknown. A single dose (total daily dose) will cure lower tract (cystitis); whether 7, 10, 14, or 21 days of therapy are needed for pyelonephritis is yet to be established. In view of the nephrotoxicity of the agents, 10–14 day therapy should be the initial form of therapy utilized.

V. Gram-Negative Sepsis

Bacteremia due to gram-negative bacilli constitutes a major problem in medical and surgical services of hospitals throughout the world. Although bacteremia can be a self-limited event, in certain patient populations the mortality has remained excessively high. Early diagnosis of the entity of gram-negative sepsis combined with an early initiation of therapy has improved the survival of patients who previously always died. Aminoglycosides have contributed to the therapy of this serious illness.

All the aminoglycosides have been utilized to treat suspected sepsis. Kanamycin was regarded as the agent of choice in the 1960s before many bacteria developed resistance to the compound and before bacteria such as *P. aeruginosa* and *Serratia* became so important [12]. Although there are hospitals in which *Pseudomonas* sepsis is rare and in which most *E. coli, Klebsiella,* and *Proteus* are susceptible to kanamycin, there would seem to be little use for kanamycin today. In contrast, in many institutions that do not treat patients with hematologic malignancy or transplant patients, both gentamicin or tobramycin inhibit most of the important organisms causing sepsis. Numerous reports [22,70] have documented the efficacy of gentamicin in the treatment of sepsis. However, a number of authors have noted that failure to achieve adequate serum levels undoubtedly has contributed to a continued mortality and to breakthrough bacteremia [2,82]. Thus, close attention to both in vitro inhibitory levels and to the levels achieved in serum is necessary for success. Quite misleading data can be gleaned from some reports in which overall fatality rates are compared and show a 15% fatality with kanamycin and 39% with gentamicin [57]. As the authors themselves have pointed out, the amount of gentamicin used at the institution in the early years may have been suboptimal, since only 3 mg/kg per day was the dose used in the early years of experience with gentamicin.

Tobramycin has been used to treat septicemia due to susceptible organisms such as *E. coli, Klebsiella, Enterobacter, P. aeruginosa,* and *Acinetobacter* [1,6,8,14,41,48,106]. Doses used ranged from 1 to 2 mg/kg every 8 hr in normal renal function. Overall cure rates were comparable to those for other agents. It is clear, however, that bacteremia in markedly neutropenic patients may not be effective even when the organisms are susceptible. This problem is discussed in detail in Chapter 24.

Amikacin would appear to be the agent of choice in hospital settings in which there is a serious problem with resistance to the other aminoglycosides. Meyer et al. [76] and Talley and Gorbach [101] have shown the effectiveness of amikacin in gram-negative bacteremia. The Meyer [76] study showed that 14 of 18 patients were cured and that 6 of these had not previously responded to gentamicin. Talley [100] also reviewed the overall efficacy of amikacin in bacteremia in 152 patients and showed a favorable outcome in 81%, with 73% cured. The failures occurred in patients with mixed aerobic-anaerobic infections and those with pneumonia. The cure rate was 66% for the patients infected with gentamicin-resistant species. Careful analysis of these data again showed that a number of patients who had failed on gentamicin did so due to inadequate dosage. It is clear, however, that amikacin in some situations was successful because of greater ease in adjusting the dosage so that the serum levels stayed above the inhibitory level.

Sisomicin also has proved to be an effective therapy of gram-negative septicemia due to susceptible organisms [37,68,81]. Analysis of cures and failures among the patients treated demonstrates that success is related to maintenance of adequate serum levels. Netilmicin proved to be useful as therapy of bacteremia due to organisms susceptible or resistant to gentamicin [49,56,84,104]. Cure rates are comparable to those achieved with earlier patients who had similar sources for the bacteremia. There are fewer data concerning dibekacin, but it has also been an effective agent to treat bacteria [105].

These studies of the effectiveness of aminoglycosides point out that all the agents are effective when the organisms are susceptible, when the drugs are administered in a correct fashion, and the patient's underlying disease can be held in check. Direct comparative studies of these agents are discussed elsewhere in this volume. But in terms of efficacy, no differences have been noted.

Most studies of the use of aminoglycosides in the compromised patient (see Chapter 24) have demonstrated the value of combination therapy of the aminoglycoside with another drug of the β-lactam class—a penicillin or cephalosporin. The advantage of this approach is that should the diagnosis be incorrect and the patient be septic with an aerobic streptococcus, such as S. pneumoniae or an anaerobic species such as B. fragilis, the outcome will not be dependent on an agent—the aminoglycoside that is ineffective against these bacteria. Studies of combined therapy have tended to yield better results in treatment of sepsis, particularly if the agents act synergistically against the infecting organism [3,5,9]. But not all investigators subscribe to the concept of the use of two agents [57].

VI. Intraabdominal Infections

Peritonitis and secondary abscess formation in the abdomen or pelvis in the majority of patients is due to a mixed facultative group of organisms: E. coli and enterococci and the obligate anaerobes B. fragilis, Peptococcus, Peptostreptococcus, and Clostridium. The aminoglycosides are not active against the obligate anaerobes; hence, other agents must be utilized with the aminoglycosides. Each of the aminoglycosides commercially available has been employed in the treatment of serious abdominal infections. Kanamycin was used both parenterally and as an irrigating solution. The latter form of therapy should be viewed with disfavor except in patients who are undergoing peritoneal dialysis. Combination therapy of gentamicin and clindamycin has been an effective form of therapy in mixed bacterial infection [4,28]. Reports of the use of tobramycin [46] in surgical infections are quite similar to those with gentamicin, as are those with sisomicin, amikacin, and netilmicin. Selection of the agent should be based upon the antimicrobial susceptibilities of the bacteria in the institution and the considerations of toxic reactions discussed elsewhere in this volume.

E. coli and *Klebsiella* are the most important bacteria found in biliary infections, and their susceptibility to semisynthetic penicillins and cephalosporins would make these agents the primary therapy unless a resistant agent was found. This is less likely with the availability of cefoxitin, cefamandole, and the newest agents—moxalactam, cefoperazone, and cefotaxime. There has been question about the ability of aminoglycosides to achieve adequate biliary levels, but more recent studies have demonstrated that adequate levels of amikacin and of the other aminoglycosides can be obtained (see Chapter 5). In fact, some authors recommend use of an aminoglycoside and clindamycin as initial therapy for acute cholangitis. Thus, in a superinfection due to an organism resistant to other antibiotics, the aminoglycosides may be required. It has been our experience, however, that unless the drainage tubes can be removed, therapy will be unsuccessful.

Perirectal lesions are much more common than realized. Aminoglycosides have been advocated in combination with a drug with anaerobic activity such as clindamycin, chloramphenicol, or metronidazole. In the neutropenic individual, perirectal abscess due to *P. aeruginosa* is a life-threatening event, and all agents of this class have been used, with varying degrees of success.

VII. Cardiovascular Infections

Aminoglycosides have become an important component in the therapy of a number of forms of endocarditis. There is still some controversy about the best agent, duration of use, and whether all patients with endocarditis should receive a cell-wall inhibiting agent (penicillin, cephalosporin, or vancomycin) and an aminoglycoside.

Penicillin-susceptible streptococcal endocarditis is that caused by the various strains of the *viridans* group streptococci and nonenterococcal group D streptococci such as *S. bovis.* The in vitro synergism of penicillin and the aminoglycosides has been associated with more rapid eradication of bacteria from animal models of endocarditis [25]. The Cornell–New York Hospital group [114] treated over 200 patients with endocarditis with penicillin 10–200 million units per day for 4 weeks plus streptomycin 0.5 g every 12 hr for the first 2 weeks. Other investigators have been less convinced of the necessity to use combined therapy and have reported excellent results with penicillin alone [31,52,69]. No absolute recommendations can be formulated on the basis of studies. It would seem wise, however, to avoid streptomycin in the elderly patient, where toxicity of a vestibular nature would be a serious problem. Clearly, the aminoglycosides should be avoided in such patients when vancomycin is the agent used.

"Penicillin-resistant" streptococcal endocarditis due to *S. faecalis* should be treated with both an aminoglycoside and a penicillin or vancomycin. The problem of high-level streptomycin resistance of *S. faecalis* is discussed in Chapter 3. Several groups still recommend penicillin for 6 weeks, plus streptomycin at an initial dose of 1 g twice daily for 2 weeks, followed by 4 weeks of 0.5 g twice daily [94]. Unpublished data from the Mayo Clinic (W. Wilson, personal communication) have shown excellent results in 68 patients treated for 30 days, so the extra 12 days may not be necessary. Other physicians have utilized a program of ampicillin plus streptomycin, but the program that is being used increasingly is that of penicillin plus gentamicin. Some concern over this program has been voiced by the Mayo Clinic group; they have reported a greater incidence of nephrotoxicity than they had encountered with the penicillin-streptomycin program [112]. Whether it is always necessary to use an aminoglycoside when vancomycin is used has not been established. A number of prominent infectious disease experts do not use amino-glycosides when they use vancomycin to treat enterococcal endocarditis (personal communications).

S. aureus endocarditis has been treated with an antistaphylococcal penicillin and an aminoglycoside on the basis of in vitro and animal model synergy experiments. But in recent randomized clinical trials of intravenous drug treatment of *S. aureus* endocarditis, users showed no benefit from the two-drug program over a single penicillin agent, nafcillin [94]. A retrospective study by Watanakunakorn and Baird [108] also failed to show a benefit of combined therapy of aminoglycoside (gentamicin) and penicillin in staphylococcal endocarditis. Indeed, in the population in which mortality from *S. aureus* endocarditis is greatest, those over 50 years [31], the risk from use of an aminoglycoside is greatest. Thus, we recommend use of rifampin with the penicillin for the elderly with *S. aureus* endocarditis.

S. epidermidis endocarditis occurs primarily on prosthetic heart valves. No study has demonstrated a benefit in using an aminoglycoside with the penicillin. It is our custom to utilize rifampin with the penicillin or vancomycin. The National Endocarditis study group is investigating the value of gentamicin 1 mg/kg every 8 hr for the first 7-10 days in the treatment of prosthetic valve endocarditis.

Other forms of endocarditis due to gram-negative bacilli such as *P. aeruginosa* and *S. marcescens* should be treated with aminoglycosides. Combination of a β-lactam such as carbenicillin or ticarcillin with tobramycin would be the recommended therapy for *P. aeruginosa* endocarditis. A cephalosporin and an aminoglycoside should be used for *Klebsiella* endocarditis, and *Serratia* should be treated with any combination of synergistic agents. However, it is clear that endocarditis due to gram-negative bacilli on the left side of the heart will not respond to antibiotics alone and always requires surgery to be successful.

VIII. Infections of the Nervous System

The role of aminoglycosides in the treatment of neonatal and childhood meningitis is discussed in Chapter 23. Gram-negative bacillary meningitis is an uncommon event, and results of therapy are not very satisfactory. Approximately 25% of patients who develop gram-negative rod meningitis do so in association with head trauma, 50% occur after neurosurgical procedures, and the remainder are of various causes. *Klebsiella* seems to be the most common organism, followed by *E. coli* and *Pseudomonas,* but other bacteria, such as various *Proteus,* also occur.

The problem that one faces with this form of meningitis is that the optimal therapy is unknown. Chloramphenicol has not proved as useful in this type of meningitis as it has with the more common forms. Furthermore, this type of meningitis often is a ventriculitis as well. Current recommendations for the intralumbar or intraventricular administration of aminoglycoside vary, but 30 μg/ml of cerebrospinal fluid (CSF) of gentamicin or tobramycin and 100 μg/ml of CSF of amikacin would be injected every 24 hr [51,113, 115]. In the moribund patient it may be necessary to place an Ommaya reservoir and administer the aminoglycoside immediately. It is wise to measure aminoglycosides in the ventricles to make certain that distribution is adequate. Regardless of the agent used, it is necessary to use systemic therapy as well. The intraventricular or intrathecal route should be continued for at least 7-10 days after the CSF culture is sterile. Meningeal irritation does occur when the lumbar route is used, but rarely is it a major problem.

IX. Brain Abscess and Parameningeal Foci of Infection

Aminoglycosides have not been used to any extent in the therapy of brain abscess or parameningeal infections because the infecting bacteria usually are anaerobes or staphylococci. If an infection were truly caused by an aerobic organism, it would seem that instillation of the aminoglycoside would be necessary but that therapy should be surgical.

X. Skin and Soft Tissue Infections

Cellulitis due to gram-negative species can occur after trauma or in the immunosuppressed host. Organisms such as *Aeromonas, Acinetobacter,* and *Pseudomonas* can require use of gentamicin, tobramycin, or amikacin. There are too few studies to indicate a best agent. It is important, however, not to utilize aminoglycosides topically in this situation since this will cause an increased bacterial resistance to the agent used.

Chronic superficial skin ulcers or those that occur in the diabetic often will have facultative bacteria such as *Proteus* and *E. coli* and aerobic species such as *Pseudomonas* cultured from the surface. In recent years it has been realized that the anaerobic species are more important and that unless therapy is directed agaisnt these organisms, cure is unlikely [30].

XI. Burns

Aminoglycosides have played an important role in the therapy of burn wound sepsis which is due to *Pseudomonas* and to various other members of the *Enterobacteriaceae.* Stone [96] utilized gentamicin, as did a number of other investigators. However, earlier use of topical gentamicin, although extremely effective for the individual patient, caused a major epidemic of infection due to gentamicin-resistant *Pseudomonas.* Tobramycin has also been effective treatment for burn wound sepsis [90], and amikacin has been effective in *Serratia* and *Providencia* burn sepsis due to gentamicin-resistant organisms. Extreme attention to the dose of aminoglycoside used in treating the burn patient is necessary since the patients lose the drug through the denuded skin and may have subinhibitory levels (see Chapter 5). Blood levels must be monitored [79].

XII. Gastrointestinal Infections

Although aminoglycosides have been utilized to treat a number of diarrheal diseases, the knowledge of the organism involved as well as the understanding of the pathogenic mechanisms has caused us to reevaluate the role of these agents. At present there is no evidence that aminoglycosides have any role in treatment of infectious diarrhea due to any of the common, important pathogens. The one exception to this would be a possible role of a paromycin in treating intestinal amoebiasis. The widespread indiscriminate use of oral aminoglycosides in diarrhea is to be deplored since it has resulted in the selection of plasmid-bearing aminoglycoside-inactivating bacteria in the community at large.

XIII. Bone and Joint Infections

Although *S. aureus* is the most commonly encountered pathogen in bone and joint infections if all ages are lumped together, gram-negative species have been noted with increasing frequency in recent years [34].

There is no evidence that aminoglycosides add anything to the therapy of *Haemophilus* joint infections. In contrast, the agents are needed in gram-negative enteric rod arthritis. Gentamicin, tobramycin [6], and amikacin [95] have all been utilized to treat infectious arthritis due to *E. coli, Klebsiella, Serratia,* and *Pseudomonas.* Both of the latter organisms are very common in the narcotic addict. Unfortunately, in all the series the failure rate is high and the rate of serious, chronic damage to the joint is great [35,95] regardless of the antibiotic used. It is clear that the joint should be aspirated repeatedly to be certain that it is being cleared of the organisms. Local instillation is rarely necessary but should be employed if the infecting organism has an inhibitory concentration that cannot be achieved by use of the parenteral route alone.

Most of the infections of bone that require aminoglycoside therapy are of a chronic nature and "cures" are not frequent. Indeed, the study by Schurman and Wheeler [95] showed a failure rate of 42% with amikacin. The same rate of failure has been found when any of the aminoglycosides are used to treat chronic osteomyelitis [80]. Good results have been reported for sisomicin therapy, but the total number of patients treated is small [67,81]. In contrast, in acute cases due to *Serratia* or *Pseudomonas* that occur in the addict population, results are much better. The duration of therapy and whether the maximal "septic doses" should be continued for a full 4 or 6 weeks is unclear. Good results depend to a great extent upon the removal of sufficient bone to produce clean tissue and clear all the dead bone present.

The role of perfusion-irrigation with aminoglycosides in the treatment of osteomyelitis is a controversial one since there are no controlled data to support the efficacy of the technique, and there is a hazard that the use of this technique will introduce infection with organisms resistant to the irrigating agent. Thus, we do not recommend the irrigation of bone infections with aminoglycosides.

XIV. Gynecologic Infections

Selection of antimicrobial agents for use in pelvic infections should be based on the fact that common facultative species such as *E. coli* and anaerobic species such as *B. fragilis* are those most often isolated. It is infrequent to find species such as *P. aeruginosa, S. marcescens,* or the indole-positive *Proteus.* The use of an aminoglycoside plus clindamycin has been an effective program. Kanamycin in particular had been evaluated by Ledger and colleagues [61] and has proved to be a successful form of therapy. The other aminoglycosides—gentamicin, tobramycin, and amikacin—have also been combined with antianaerobic agents such as clindamycin, chloramphenicol, carbenicillin, or metronidazole with approximately equivalent efficacy [32,60,98]. Thus, no one agent would seem best in this situation, and the choice of agent should depend upon the antimicrobial susceptibility pattern of the gynecologic service and the considerations of potential toxicity. Operative intervention has been necessary in a larger number of these infections, and it is conceivable that agents such as cefoxitin or the third-generation cephalosporins—moxalactam—would be preferable to aminoglycosides in the treatment of pelvic infections.

XV. Sexually Transmitted Diseases

Aminoglycosides have been used to treat gonorrhoea. Kanamycin was advocated as an alternative therapy in the 1960s [11] and later gentamicin was also used [29]. None of the strict aminoglycosides should be considered as a first-line agent in gonorrhoea, but the aminocyclitol spectinomycin is an excellent agent, particularly for penicillinase-producing *Neisseria gonorrhoeae.* The aminoglycosides are not adequate therapy for syphilis and have not been useful in nongonococcal urethritis due to *Chlamydia trachomatis* or *Ureaplasma urealycitum.*

XVI. Eye Infections

Aminoglycosides have proved to be extremely useful in the therapy of selected infections of the eye. Bacterial conjunctivitis is most often due to gram-positive species or to *Haemophilus* or *Neisseria,* so that agents other than the aminoglycosides would be utilized. But bacterial keratitis most often is due to *P. aeruginosa* or to one of the *Proteus* or *Klebsiella* species, and recently even *Serratia* has been implicated as a cause of keratitis. Topical therapy with a solution of gentamicin has been helpful [50]. Parenteral use of the aminoglycoside is not necessary for keratitis.

Bacterial endophthalmitis is a serious infection that develops suddenly and progresses rapidly to destroy the eye. *S. aureus* is the most common organisms, but *P. aeruginosa* closely follows and other gram-negative rods, such as *Proteus* and *E. coli,* do occur. Gentamicin administered parenterally and also by topical drops and subconjunctival injection has been the most utilized therapy [5,71,90]. Doses of 20–40 mg have been employed. More recently, intravitreal injection of gentamicin has been performed [5,91]. It is too early to tell if the use of intravitreal injection of aminoglycoside will improve the survival of infected eyes.

XVII. Chemoprophylaxis

Various investigators have utilized aminoglycosides as chemoprophylactic agents to prevent infection at the time of surgery. Cohn [18] advocated rinsing the abdomen with kanamycin at the time of surgery. This resulted in postoperative respiratory problems due to neuromuscular blockade and has not been proved effective. Burton et al. [13] recommended gentamicin as prophylaxis in colonic and rectal surgery, but this is not adequately established. However, excellent prophylaxis of infection at the time of colon surgery has followed the use of neomycin and erythromycin administered three times in the 24 hr just before surgery [17]. Neomycin by itself is not effective, but combined with erythromycin it has reduced postoperative wound infections from 35% to 3%.

In general, the aminoglycosides are not suitable chemoprophylactic agents for respiratory, abdominal, or gynecologic infections, and agents of the β-lactam class, which have a lower toxicity, have proved to be more useful.

XVIII. Conclusions

Aminoglycosides are useful antimicrobial agents. They have been of major benefit in the therapy of serious infections, particularly those that have been hospital-acquired and due to multiresistant bacteria. Nonetheless, it is essential that the agents be used only to treat infections in which they have had a clearly demonstrated beneficial effect and in which the toxicities attendant on their use have been fully considered. Careful attention to the proper dosage programs is necessary to achieve clinical and bacteriologic success and to avoid the renal and ototoxicity discussed in other chapters of this book.

References

1. Altucci, P., Abbate, G.F., Gattoni, A., and Leonossa, V.: Clinical evaluation of tobramycin in urinary tract infections. *J. Infect. Dis.* 134(Suppl.):139–141, 1976.
2. Anderson, E.T., Young, L.S., and Hewitt, W.L.: Simultaneous antibiotic levels in "breakthrough" gram-negative bacteremia. *Am. J. Med.* 61:493– , 1976.
3. Anderson, E.T., Young, L.S., and Hewitt, W.L.: Antimicrobial synergism in the therapy of gram-negative rod bacteremia. *Chemotherapy* 24:45–54, 1978.
4. Bartlett, J.G., Miso, P.V.W., and Gorbach, S.L.: Emperic treatment with clindamycin and gentamicin of suspected sepsis due to anaerobic and aerobic bacteria. *J. Infect. Dis.* 135(Suppl.):80–85, 1977.
5. Baum, J.L.: Antibiotic administration in the treatment of bacterial endophthalmitis. 1. Periocular injections. *Surv. Ophthalmol.* 21:332– , 1977.
6. Bendush, C.L., and Weber, R.: Tobramycin sulfate: a summary of worldwide experience from clinical trials. *J. Infect. Dis.* 134(Suppl.):219–234, 1976.
7. Bennett, A.H.: Evaluation of tobramycin in severe urinary tract infection. *J. Infect. Dis.* 134(Suppl.):156–157, 1976.
8. Blair, D.C., Fekety, F.R.J., Bruce, B., Silva, J., and Archer, G.: Therapy of *Pseudomonas aeruginosa* infections with tobramycin. *Antimicrob. Agents Chemother.* 8:22–29, 1975.
9. Bodey, G.P., Middleman, E., Umsawadi, T., and Rodriguez, V.: Intravenous gentamicin therapy for infections in patients with cancer. *J. Infect. Dis.* 124(Suppl.):174–179, 1971.
10. Boxerbaum, B., Pittman, S., Doershuk, C.F., Stern, R.C., and Matthews, L.W.: Use of gentamicin in children with cystic fibrosis. *J. Infect. Dis.* 134(Suppl.):293–295, 1976.
11. Bunn, P.A.: Kanamycin. *Med. Clin. N. Am.* 54:1245–1256, 1970.
12. Bunn, P.A., and Lunn, J.S.: Status of kanamycin therapy of infections. *Ann. N.Y. Acad. Sci.* 132:819– , 1966.
13. Burton, R.C., Hughes, E.S.R., and Cutherbertson, A.M.: Prophylactic use of gentamicin in colonic and rectal surgery. *Med. J. Aust.* 2:597– , 1975.
14. Carmalt, E.D., Cortex, L.M., and Rosenblatt, J.E.: Clinical experience with tobramycin in the treatment of infections due to gram-negative bacilli. *Am. J. Med. Sci.* 271:285–295, 1976.
15. Chandler, R.: Malignant external otitis. Further considerations. *Ann. Otol. (St. Louis)* 86:417–428, 1977.
16. Chisholm, G.D.: The use of gentamicin in urinary tract infections with special reference to drug levels in complicated (urological) infections. *Postgrad. Med. J.* 50(Suppl.):23–26, 1974.
17. Clarke, J.S., Condon, R.E., Bartlett, J.G., Gorbach, S.L., Nichols, R.L., and Ochi, S.:

Preoperative oral antibiotics reduce septic complications of colon operations. *Ann. Surg.* 186:251–259, 1977.

18. Cohn, I., Jr.: Kanamycin as an intestinal antiseptic aid in the treatment of peritonitis: Resume of clinical experience. *Ann. N.Y. Acad. Sci.* 132:860–869, 1966.

19. Cox, C.E.: Gentamicin, a new aminoglycoside antibiotic: Clinical and laboratory study in urinary tract infection. *J. Infect. Dis.* 119:486–491, 1969.

20. Cox, C.E.: Amikacin therapy of urinary tract infections. *J. Infect. Dis.* 134(Suppl.): 362–373, 1976.

21. Cox, C.E., and Garvey, F.K.: Efficacy of gentamicin in treatment of chronic urinary tract infections. *Antimicrob. Agents Chemother.* pp. 169–171, 1967.

22. Cox, C.E., and Harrison, L.H.: Comparison of gentamicin and polymyxin B-kanamycin in therapy of bacteremia due to gram-negative bacilli. *J. Infect. Dis.* 124(Suppl.): 156–163, 1971.

23. Craven, P.C., Jorgensen, J.II., Kasper, R.L., and Drutz, D.J.: Amikacin therapy of patients with multiply antibiotic resistant *Serratia marcescens* infections. *Am. J. Med.* 62(Suppl.):66–74, 1977.

24. Crozier, D.N., and Khan, S.R.: Tobramycin in treatment of infections due to *Pseudomonas aeruginosa* in patients with cystic fibrosis. *J. Infect. Dis.* 134(Suppl.):187–191, 1976.

25. Durach, D.T., Pelletier, L.L., and Petersdorf, R.G.: Chemotherapy of experimental streptococcal endocarditis. II. Synergism between penicillin and streptomycin against penicillin-sensitive streptococci. *J. Clin. Invest.* 53:829–833, 1974.

26. Duwoos, H., Carpentier, P., Nouveau, J., Guyonnaud, C.D., and Abderhaiden, C.: Clinical and bacteriological evaluation of sisomicin in 16 cases of severe bronchopulmonary infection. *Infection* 4(Suppl.):475–482, 1976.

27. Eller, J.J., Klinger, J.D., Hilton, C.B., and Bass, J.A.: Amikacin treatment of *Pseudomonas* pneumonia in cystic fibrosis patients with advanced pulmonary involvement. *Am. J. Med.* 62(Suppl.):179–185, 1977.

28. Fass, R.J.: Treatment of mixed bacterial infections with clindamycin and gentamicin. *J. Infect. Dis.* 135(Suppl.):74–79, 1977.

29. Felarea, A.B., Laqui, E.M., and Ibarra, L.M.: Gentamicin in gonococcal urethritis in Filipino males. *J. Infect. Dis.* 124(Suppl.):287–292, 1971.

30. Finegold, S.M.: Infections of skin, soft tissue and muscle. In *Anaerobic Bacteria in Human Diseases.* New York, Academic Press, 1977, pp. 386–390.

31. Garvey, G.J., and Neu, H.C.: Infective endocarditis: an evolving disease. *Medicine* 57:105–127, 1978.

32. Giamarellou, H., Kanellakopoulou, K., Pragastis, D., Tagaris, N., and Daikos, G.K.: Treatment with metronidazole of 48 patients with serious anaerobic infections. *Antimicrob. Chemother.* 3:347–353, 1977.

33. Gilbert, D.N., Eubanks, N., and Jackson, J.: Comparison of amikacin and gentamicin in the treatment of urinary tract infections. *Am. J. Med.* 62:121–126, 1977.

34. Goldenberg, D.L., Brandt, K.D., and Cathcart, E.S.: Acute arthritis caused by gram-negative bacilli. A clinical characterization. *Medicine* 53:197–208, 1974.

35. Goldenberg, D.L., and Cohen, A.S.: Acute infectious arthritis. A review of patients with nongonococcal joint infections. *Am. J. Med.* 60:369–377, 1976.

36. Gooding, P.G., Berman, E., Lane, A.Z., and Agre, K.: A review of results of clinical trials with amikacin. *J. Infect. Dis.* 134(Suppl.):441–445, 1976.

37. Grunewaldt, G., Arcieri, G., and Gionti, A.: Zusammenfassung der Ergenbnisse der internationalen klinschen Sisomicin-Prufung. *Infection* 4(Suppl.):505–513, 1976.

38. Haas, H., Morris, J.F., and Samson, S.: Bacterial flora of the respiratory tract in chronic bronchitis: comparison of transtracheal fiberbronchoscopic and oropharyngeal sampling methods. *Am. Rev. Resp. Dis.* 116:41–47, 1977.

39. Hamory, B.H., Sande, M.A., and Sydnor, A., Jr.: Etiology and antimicrobial therapy of acute maxillary sinusitis. *J. Infect. Dis.* 139:197-202, 1979.

40. Hawley, H.B., Lewis, R.M., Swartz, D.R., and Gump, D.W.: Tobramycin therapy of pulmonary infections in patients with cystic fibrosis. *Curr. Ther. Res.* 16:414-423, 1974.

41. Helm, E., Shah, P.M., and Stille, W.: Clinical experience with tobramycin. In *Tobramycin:* Selected Proceedings from the 8th International Congress of Chemotherapy, Athens, September 8-15, 1973. Amsterdam, Excerpta Medica, 1974, pp. 55-58.

42. Hodges, G.R., and Saslaw, S.: Gentamicin in the treatment of respiratory tract infection. *Ohio State J. Med.* 68:475-479, 1972.

43. Huang, N.N., Laraya-Cuasay, L.R., Yasmin, N., Keith, H.H., Burden, M., and Cundy, K.R.: Efficacy of sisomicin in patients with cystic fibrosis. *Infection* 4(Suppl.): 465- , 1976.

44. Huang, N.N., Laraya-Cuasay, L.R., Yasmin, N., Keith, H.H., Burden, M., Cundy, K.R.: Clinical experience with amikacin in patients with cystic fibrosis. *Am. J. Med.* 62:186-195, 1977.

45. Inanini, P.B., Claffey, T., and Quintiliani, R.: Bacteremic *Pseudomonas* pneumonia. *JAMA* 230:558- , 1974.

46. Ishiyama, S., Nakayama, I., Iwamoto, H., Iwai, S., Murata, I., and Ohashi, M.: Clinical use of tobramycin in patients with surgical infections due to gram-negative bacilli. *J. Infect. Dis.* 134(Suppl.):178-181, 1976.

47. Jackson, G.G., and Riff, L.J.: *Pseudomonas* bacteremia pharmacologic and other bases for failure of treatment with gentamicin. *J. Infect. Dis.* 124(Suppl.):185-191, 1971.

48. Jaffe, G., Ravneby, W., Meyers, B.R., and Hirschman, S.Z.: Clinical study of the use of new aminoglycoside tobramycin for therapy of infections due to gram-negative bacteria. *Antimicrob. Agents Chemother.* 5:75-81, 1974.

49. Jahre, J., Fu, K.P., and Neu, H.C.: Clinical evaluation of netilmicin therapy in serious infections. *Am. J. Med.* 66:67-73, 1979.

50. Jones, D.B.: A plan for antimicrobial therapy in bacterial keratitis. *Trans. Am. Acad. Ophthalmol. Otolaryngol.* 79:95-99, 1975.

51. Kaiser, A.B., and McGee, Z.A.: Aminoglycoside therapy of gram-negative bacillary meningitis. *N. Engl. J. Med.* 293:1215-1220, 1975.

52. Karchmer, A.W., Moellering, R.C., Jr., Maki, D.G., and Swartz, M.N.: Single-antibiotic therapy for streptococcal endocarditis. *JAMA* 241:1801-1806, 1979.

53. Klastersky, J., Hengens, C., and Debusscher, L.: Emperic therapy for cancer patients: comparative study of ticarcillin-tobramycin, ticarcillin-cephalothin and cephalothin-tobramycin. *Antimicrob. Agents Chemother.* 7:640-645, 1975.

54. Klastersky, J., Hensgens, C., Gerard, M., and Daneau, D.: Comparison of sisomicin and gentamicin in bacteriuric patients with underlying diseases of the urinary tract. *Antimicrob. Agents Chemother.* 7:742-747, 1975.

55. Klastersky, J., Hensgens, C., Nebri, A., and Daneau, D.: Comparative clinical study of tobramycin and gentamicin. *Antimicrob. Agents Chemother.* 5:133-138, 1974.

56. Klastersky, J., Meunier-Carpentier, F., Coppens-Kahan, L., Daneau, D., and Prevost, J.M.: Clinical and bacteriological evaluation of netilmicin in gram-negative infections. *Antimicrob. Agents Chemother.* 12:503-509, 1977.

57. Kreger, B.E., Craven, D.E., and McCabe, W.R.: Gram-negative bacteremia: IV. Re-evaluation of clinical features and treatment in 612 patients. *Am. J. Med.* 68:344-355, 1980.

58. Labovitz, E., Levison, M.E., and Kaye, D.: Single-dose daily gentamicin therapy in urinary tract infection. *Antimicrob. Agents Chemother.* 6:465- , 1974.

59. Lau, W.K., Young, L.S., Black, R.E., Winston, D.J., Linne, S.R., Weinstein, R.J.,

and Hewitt, W.L.: Comparative efficacy and toxicity of amikacin/carbenicillin versus gentamicin/carbenicillin in leukopenic patients. *Am. J. Med.* 62:212–219, 1977.

60. Ledger, W.J., Gee, C.L., Lewis, W.P., and Bobitt, J.R.: Comparison of clindamycin and chloramphenicol in treatment of serious infections of the female genital tract. *J. Infect. Dis.* 135(Suppl.):30–39, 1977.

61. Ledger, W.J., Kriewall, T., Sweet, R., FeKety, F.R., Jr.: A comparison of penicillin-kanamycin and clindamycin-kanamycin in the treatment of severe obstetric-gynecologic infections. *Obstet. Gynecol.* 43:490–497, 1974.

62. Lewis, R.P., Meyer, R.D., and Finegold, S.M.: Amikacin therapy of patients with gentamicin-resistant gram-negative bacillary infection. *Am. J. Med.* 62(Suppl.): 142–150, 1977.

63. Lindberg, A.A., Bucht, H., and Kallings, O.: Treatment of chronic urinary tract infections with gentamicin. *Gentamicin:* Proceedings of the 1st International Symposium, Paris, 1967, p. 75.

64. Louria, D.B., Young, L., Armstrong, D., and Smith, J.K.: Gentamicin in the treatment of pulmonary infections. *J. Infect. Dis.* 119:483–485, 1969.

65. Madsen, P.O., Kjaer, T.B., and Mosegaard, A.: Comparison of tobramycin and gentamicin in the treatment of complicated urinary tract infections. *J. Infect. Dis.* 134(Suppl.):150–152, 1976.

66. Madsen, P.O., Kjaer, T.B., and Mosegaard, A.: Treatment of complicated urinary tract infections with sisomicin. A comparison of once daily and twice daily dosage forms. *Infection* 4(Suppl.):498–500, 1976.

67. Maki, D.G., Graig, W.A., and Agger, W.A.: A comparative clinical trial of sisomicin and gentamicin in major gram-negative infections. *Infection* 7(Suppl.):298–300, 1979.

68. Makris, A.T., Asper, R.F., Schanbacher, K., and Schwartz, A.R.: Treatment of gram-negative bacteremia with sisomicin. *Curr. Chemother.* 1:925–928, 1977.

69. Malacoff, R.F., Frank, E., and Andriole, V.T.: Streptococcal endocarditis (non-enterococcal non-group A) single vs. combination therapy. *JAMA* 241:1807–1810, 1979.

70. Martin, C.M., Cuomo, A.J., Geraghty, M.J., Zager, J.R., and Mandes, T.C.: Gram-negative rod bacteremia. *J. Infect. Dis.* 119:506–517, 1969.

71. Mathalone, B.: Gentamicin in eye infections. *Postgrad. Med. J.* 50(Suppl.):38–40, 1974.

72. Mathias, R.G., Ronald, A.R., and Gurwith, M.J.: Clinical evaluation of amikacin in treatment of infections due to gram-negative aerobic bacilli. *J. Infect. Dis.* 134 (Suppl.):394–   , 1976.

73. McCrae, W.M., Raeburn, J.A., and Hanson, E.J.: Tobramycin therapy of infections due to *Pseudomonas aeruginosa* in patients with cystic fibrosis: effect of dosage and concentration of antibiotic in sputum. *J. Infect. Dis.* 134(Suppl.):191–193, 1976.

74. McHenry, M.C., Alfi, R.J., Deodhar, S.D., Braun, W.E., and Popowniak, K.L.: Hospital-acquired pneumonia. *Med. Clin. North Am.* 58:565–580, 1974.

75. McHenry, M.C., Gaven, T.L., VanOmmen, R.A., and Hawk, W.A.: Therapy with gentamicin for bacteremic infections—results with 53 patients. *J. Infect. Dis.* 124 (Suppl.):164–173, 1971.

76. Meyer, R.D., Lewis, R.P., and Finegold, S.M.: Amikacin therapy for gram-negative septicemia. *Am. J. Med.* 62:127–132, 1977.

77. Minuth, J.N., Musher, D.M., and Thorsteinsson, S.B.: Inhibition of the antibacterial activity of gentamicin by urine. *J. Infect. Dis.* 133:14–21, 1976.

78. Murdock, J., Geddes, A.M., and Synne, J.: Studies with kanamycin sulfate. *Lancet* 1:457–460, 1962.

79. Neu, H.C.: The pharmacology of newer aminoglycosides with a consideration of the application to clinical situations. *Med. G. Aust.* 2(Suppl.):13–18, 1977.

80. Neu, H.C.: Tobramycin, an overview. *J. Infect. Dis.* 134(Suppl.):1–19, 1976.

81. Neu, H.C., and Parry, M.F.: Therapy of serious gram-negative infections with a new aminoglycoside. *Infection* 7(Suppl.):287–288, 1979.

82. Noone, P., Parsons, T.M.C., Pattison, J.R., Slack, R.C.B., Garfield-Davies, D., and Hughes, K.: Experience in monitoring gentamicin therapy during treatment of serious gram-negative sepsis. *Br. Med. J.* 1:477–481, 1974.

83. Noone, P., Perera, M.R., and Amirak, I.D.: Sisomicin in serious gram-negative sepsis. *Curr. Chemother. Infect. Dis.* 2:1051–1053, 1980.

84. Panwalker, A.P., Malow, J.B., Zimelo, V.M., Jackson, G.G.: Netilmicin: clinical efficacy, tolerance and toxicity. *Antimicrob. Agents Chemother.* 13:170–176, 1978.

85. Parry, M.F., and Neu, H.C.: Tobramycin and ticarcillin therapy for exacerbations of pulmonary disease in patients with cystic fibrosis. *J. Infect. Dis.* 134(Suppl.): 194–197, 1976.

86. Parry, M.F., and Neu, H.C.: A comparative study of ticarcillin plus tobramycin versus carbenicillin plus gentamicin for the treatment of serious infections due to gram-negative bacilli. *Am. J. Med.* 64:961–966, 1978.

87. Parry, M.F., Neu, H.C., Merlino, M., Gaerland, P.F., Ores, C.N., and Denning, C.R.: Treatment of pulmonary infections in patients with cystic fibrosis: a comparative study of ticarcillin and gentamicin. *J. Pediatr.* 90:144– , 1977.

88. Petersdorf, R.G., and Featherstone, H.: New antimicrobial drugs and their value in the treatment of respiratory infections. *Am. Rev. Resp. Dis.* 117:1–3, 1978.

89. Petersdorf, R.G., and Turck, M.: Kanamycin in urinary tract infections. *Ann. N.Y. Acad. Sci.* 132:834– , 1966.

90. Peyman, G.A., and Herbst, R.: Bacterial enophthalmitis: treatment with intraocular injection of gentamicin and dexamethasone. *Arch. Ophthalmol.* 91:416–418, 1974.

91. Peyman, G.A., Vastine, D.W., and Couch, E.R.: Clinical use of intravitreal antibiotics to treat bacterial enophthalmitis. *Trans. Am. Acad. Ophthalmol. Otolaryngol.* 78:862– , 1974.

92. Pines, A., Raafat, H., and Pulcinki, K.: Gentamicin and colistin in chronic purulent bronchial infections. *Br. Med. J.* 2:543–545, 1967.

93. Ronald, A.R., Boutros, P., and Mourtada, H.: Bacteriuria localization and response to single-dose therapy in women. *JAMA* 235:1854–1857, 1976.

94. Sande, M.A., and Scheld, W.M.: Combination antibiotic therapy of bacterial endocarditis. *Ann. Intern. Med.* 92:390–395, 1980.

95. Schurman. D.J., and Wheeler, R.: Bone and joint gram-negative infection and amikacin treatment. *Am. J. Med.* 62:160–164, 1977.

96. Stone, H.H.: The diagnosis and treatment of *Pseudomonas* sepsis in major burns. *J. Infect. Dis.* 119:504–505, 1969.

97. Sweedler, D.R., Gravenkenper, C.F., Bulger, R.J., Brodie, J.L., and Kirby, W.M.M.: Laboratory and clinical studies with gentamicin. *Antimicrob. Agents Chemother.* 1963:157–160, 1964.

98. Swensen, R.M., and Lorber, B.: Clindamycin and carbenicillin in treatment of patients with intraabdominal and female genital tract infections. *J. Infect. Dis.* 135(Suppl.):40–45, 1977.

99. Tager, I., and Speizer, F.E.: Role of infection in chronic bronchitis. *N. Engl. J. Med.* 292:563–571, 1975.

100. Tally, F.P.: Review of 152 patients with bacteremias treated with amikacin. *Am. J. Med.* 62:137–141, 1977.

101. Tally, F.P., and Gorbach, S.L.: Amikacin therapy of gram-negative bacteremia. *Am. J. Med.* 62:133–136, 1977.

102. Thompson, I.: Sisomicin, efficacy and tolerance of various dosages in urinary tract infections. *Infection* 4(Suppl.):501–504, 1976.

103. Trenholme, G.M., McKellar, P.P., Rivera, N., and Levin, S.: Amikacin in the treatment of gram-negative pneumonia. *Am. J. Med.* 62(Suppl.):155–159, 1977.

104. Trestman, I., Parsons, J., Santoro, J., Goodhardt, G., and Kaye, D.: Pharmacology and efficacy of netilmicin. *Antimicrob. Agents Chemother.* 13:832–836, 1978.

105. Ueda, Y., Saito, A., Matsumoto, F., Omori, M., Shiba, K., Yamaji, T., and Ihara, H.: Clinical studies on diebakacin. *Curr. Chemother.* 2:931–933, 1978.

106. Waldvogel, F.A., Frochau, M., Vrancheva, S., and Lehmann, H.: Treatment of *Pseudomonas aeruginosa* infections with tobramycin: clinical, pharmacological and bacteriological results. In *Tobramycin:* Selected Proceedings of the 8th International Congress of Chemotherapy, Athens, September 8–15, 1973. Amsterdam, Excerpta Medica, 1974, pp. 110–116.

107. Walker, B.D., and Gentry, L.O.: A randomized comparative study of tobramycin and gentamicin in treatment of acute urinary tract infections. *J. Infect. Dis.* 134 (Suppl.):146–149, 1976.

108. Watanakunatorn, C., and Baird, I.M.: Prognostic factors in *Staphylococcus aureus* endocarditis and results of therapy with a penicillin and gentamicin. *Am. J. Med. Sci.* 273:133–139, 1977.

109. Westernfelder, S.R., Welling, P.G., and Madsen, P.O.: Efficacy and pharmacokinetics of tobramycin in patients with chronic urinary tract infection and various degrees of renal impairments. *Infection* 2:76–79, 1974.

110. Wieser, O., and Gillissen, J.: Sisomicin bei schweren Pulmonalen und bronchaien bakteriell bedingten Erkrankungun. *Infection* 4(Suppl.):472–475, 1976.

111. Wilkinson, A.E., Roce, J.W., and Curtis, F.R.: Kanamycin in the treatment of gonorrhoea in males. *Postgrad. Med. J.* May(Suppl.):65, 1967.

112. Williams, T.W., Jr.: Meningitis: special techniques in treatment. *Mod. Treat.* 7:606–618, 1970.

113. Wilson, W.R., Wilkowskie, C.J., Thompson, R.L., and Geraci, J.E.: Treatment of streptomycin-resistant enterococcal infective endocarditis. Abstr. 1063, 11th International Congress of Chemotherapy, Boston, October 1979.

114. Wirt, T.C., McGee, Z.A., Oldfield, E.H., and Meacham, W.F.: Intraventricular administration of amikacin for complicated gram-negative meningitis and ventriculitis. *J. Neurosurg.* 50:95–99, 1979.

115. Wolfe, J.C., and Johnson, W.D.: Penicillin-sensitive streptococcal endocarditis. In vitro and clinical observations on penicillin-streptomycin therapy. *Ann. Intern. Med.* 81:178–181, 1974.

116. VanderStraetem, M., Pauwels, R., Dijck, J., VanNimmen, L., Khuyskems, P., and Soep, H.: Amikacin in the treatment of gram-negative bronchopulmonary infections. *J. Infect. Dis.* 134(Suppl.):391–393, 1976.

117. Zaky, D.A., Bentley, D.W., and Lowy, K.: Malignant external otitis: a severe form of otitis in diabetic patients. *Am. J. Med.* 61:298–302, 1976.

INDEX

629